SIXTH EDITION

LABORATORY MANUAL FOR NON-MAJORS BIOLOGY

James W. Perry
University of Wisconsin–Fox Valley, Emeritus

David Morton
Frostburg State University, Emeritus

Joy B. Perry
University of Wisconsin–Fox Valley

Contributions to Animal Diversity
by Ronald E. Barry, Jr.

And Evolutionary Agents
by James H. Howard

BROOKS/COLE
CENGAGE Learning·

Australia • Brazil • Japan • Korea • Mexico • Singapore • Spain • United Kingdom • United States

BROOKS/COLE
CENGAGE Learning·

Laboratory Manual for Non-Majors Biology, Sixth Edition
James W. Perry, David Morton, and
Joy B. Perry

Publisher: Yolanda Cossio

Senior Acquisitions Editor: Peggy Williams

Developmental Editor: Suzannah Alexander

Editorial Assistant: Sean Cronin

Assistant Editor: Shannon Holt

Associate Media Editor: Lauren Oliveira

Senior Marketing Manager: Tom Ziolkowski

Marketing Coordinator: Jing Hu

Art and Cover Direction, Production
Management, and Composition:
  PreMediaGlobal

Manufacturing Planner: Karen Hunt

Rights Acquisitions Specialist: Don Schlotman

Cover Image: © Tyler Fox/Shutterstock;
  © T Sebastian Kaulitzki/Shutterstock;
  © Patrizia Tilly/Shutterstock; © Christopher
  Meade/Shutterstock; © Sebastian Duda/
  Shutterstock; © Mirek Srb/Shutterstock

© 2013, 2007 Brooks/Cole Cengage Learning

ALL RIGHTS RESERVED. No part of this work covered by the copyright herein may be reproduced, transmitted, stored or used in any form or by any means graphic, electronic, or mechanical, including but not limited to photocopying, recording, scanning, digitizing, taping, Web distribution, information networks, or information storage and retrieval systems, except as permitted under Section 107 or 108 of the 1976 United States Copyright Act, without the prior written permission of the publisher.

For product information and technology assistance, contact us at
Cengage Learning Customer & Sales Support, 1-800-354-9706
For permission to use material from this text or product,
submit all requests online at **www.cengage.com/permissions**
Further permissions questions can be emailed to
permissionrequest@cengage.com

Library of Congress Control Number: 2012931720

ISBN-13: 978-0-8400-5380-0

ISBN-10: 0-8400-5380-0

Brooks/Cole
20 Davis Drive
Belmont, CA 94002-3098
USA

Cengage Learning is a leading provider of customized learning solutions with office locations around the globe, including Singapore, the United Kingdom, Australia, Mexico, Brazil, and Japan. Locate your local office at **www.cengage.com/global**

Cengage Learning products are represented in Canada by Nelson Education, Ltd.

To learn more about Brooks/Cole, visit **www.cengage.com/brookscole**

Purchase any of our products at your local college store or at our preferred online store **www.cengagebrain.com**

Printed in the United States of America
2 3 4 5 6 7 18 17 16 15 14

Contents

Preface

Greetings from the authors! We're happy that you are examining the results of our efforts to assist you and your students. We believe you'll find the information below a valuable introduction to this laboratory manual.

Audience

This manual is designed for students at the college entry level and assumes that the student has no previous college biology or chemistry. The lab manual's terminology and classification conform to that of Starr/Taggart/Evers/Starr's *Biology: The Unity and Diversity of Life,* Starr/Evers/Starr's *Biology: Concepts and Applications,* and Starr/Evers/Starr's *Biology Today and Tomorrow,* although you'll find that the exercises support virtually any biology text used in an introductory course.

Features of This Edition

Updated, Relevant Exercises

As always, we strive to provide students with exciting, relevant activities, observations, and experiments that allow them to explore some of the rapidly developing areas of biological knowledge and how those relate to their daily lives.

For example, Exercise 7 (Diffusion, Osmosis, and the Functional Significance of Biological Membranes) includes a new activity relating membrane function to the consequences and treatment of cholera. Similarly, Exercise 10 (Respiration: Energy Conversion) includes an activity in which cellular respiration rates are measured as levels of students' physical performance change. Also new in Exercise 10 is an experiment investigating the potential for different plant carbohydrate sources to be used in the production of ethanol.

Updated Taxonomy

An attempt to provide accurate systematic information is like trying to hit a moving target. As new information floods in, especially molecular information, our best understanding of taxonomic and systematic relationships sometimes seems to change daily. The taxonomy in our exercises is completely updated and reflects the most widely accepted information at the time of publication.

Streamlined

Exercises have been updated throughout to streamline procedures and to enhance student comprehension. We have simplified procedures to reduce student frustration and time needed for setup.

Inquiry Experiments and Observations and the Methods of Science

We realize that the best way to learn science is by doing science. Thus, in this edition we further emphasize hypothesis testing and scientific thinking, following the cognitive techniques whose foundations are laid in the first exercise, "The Scientific Method."

Throughout the manual, we place more emphasis on testing predictions generated from hypotheses. We strongly believe that all biology students—both nonscience majors and science majors—benefit by repeating the logical thought processes of science.

For example, in Exercises 8 (Enzymes: Catalysts of Life) and 9 (Photosynthesis: Capture of Light Energy), students learn a simple semiquantitative technique and then employ that technique in an experiment of their own design to investigate factors that affect those processes.

What's Important to Us

In preparing this lab manual, we paid particular attention to pedagogy, clarity of procedures and terminology, illustrations, and practicality of materials used.

Pedagogy

The exercises are written so the conscientious student should be able to accomplish the objectives of each exercise with minimal input from an instructor. As suggested by the publisher, the procedure sections of the exercises are more detailed and step-by-step than in other manuals. Instructions follow a natural progression of thought so the instructor need not conduct every step.

We attempted to make each portion of the exercise part of a continuous flow of thought. Thus, we do not wait until the post-lab questions to ask students to record conclusions when it is more appropriate to do so within the body of the procedure. Answers to in-lab questions are to be found in the print and online *Instructor's Manual*.

Terms required to accomplish objectives are **boldface**. Scientific names and precautionary statements, or those needing emphasis, are *italic*. Potential problems are set aside in their own caution boxes.

The use of scientific names is deemphasized when it is not relevant to understanding the subject. However, these names generally do appear in parentheses because the labels on many prepared microscope slides bear only the scientific name. This is also true for many specimens.

Format

Each exercise includes:

1. *Objectives*: a list of desired outcomes.

2. *Introduction*: to stimulate student interest, indicate relevance, and provide background.

3. *Materials*: a list for each portion of the exercise so a student can quickly gather the necessary supplies. Materials are listed "Per student," "Per pair," "Per group," and "Per lab room."

4. *Procedures*: including precautionary statements and cautions, illustrations of apparatuses, figures to be labeled, drawings to be made, tables to record data, graphs to draw, and questions that lead to conclusions. The procedures are listed in easy-to-follow numbered steps.

5. *Pre-Lab Questions*: ten multiple-choice questions that the student should be able to answer after reading the exercise, but prior to entering the laboratory.

6. *Post-Lab Questions*: questions that draw on knowledge gained from doing the exercise and that the student should be able to answer after finishing the exercise. These post-lab questions facilitate **recall** (preparing students for lab practical assessments), **understanding**, and **application**.

Practical Post-Lab Questions

Virtually all courses use laboratory practical examinations. We explain to our students the difference between lecture-type questions, which they need to read and provide an answer based on the written word, and practical-type questions, for which a response depends on observation.

We believe the post-lab questions should draw on the knowledge gained by observation. Consequently, we've incorporated as many illustrations as possible into the post-lab questions. These illustrations typically are similar, but not identical, to those in the procedures. Thus, they assess the student's ability to use knowledge gained during the exercise in a new situation.

Post-lab questions are identified by exercise section. This allows the instructor to easily assign or use only those questions relevant to portions of exercises performed by students. The questions also have been revised and are directly tied to the learning outcomes expected from each exercise.

Flexible Quiz Options

Each exercise has a set of pre-lab questions. We have found through nearly 90 combined years of experience that students left to their own initiative typically come to laboratory unprepared to do the exercise. Few read the exercise beforehand. One solution to this problem is to incorporate some sort of graded pre-exercise activity. At the same time, we recognize that grading a large number of lab papers each week can put an unreasonable burden on instructors. Consequently, we decided on a multiple-choice format, which is easy to grade but still accomplishes the pedagogical goal.

In our own courses, we have students take a pre-lab quiz consisting of the questions in their lab manual in scrambled order to discourage memorization. These scrambled quizzes are reproduced in our *Instructor's Manual*. Our quiz takes about 10 minutes of lab time and counts as a portion of the lab grade, thus rewarding students for preparation.

Other instructors have told us they use the pre-lab questions to assess learning after the exercise has been completed. We encourage you to be creative with the manual; do what you like best.

Lab Length and Exercise Options

We realize there is wide variation in the amount of time each instructor devotes to laboratory activities. To provide maximum flexibility for the instructor, the procedure portions of the exercises are divided by major headings, with the approximate time it takes to perform each portion indicated. Once the introduction has been studied, portions of the procedures can be deleted or conducted as demonstrations without sacrificing the pedagogy of the exercise as a whole.

It's our experience that if the lecture section covers the topic prior to the lab, students find the exercise much more relevant and understandable. We endeavor to include lab material in the classroom, and thus little time is spent on a lecture-style introduction in the lab itself. Therefore, we have two to three full hours for real scientific investigation and need to delete very little material to complete most exercises in the time allotted.

Illustrations

Perhaps our illustrations are more noticeable than anything else as you thumb through the manual. We continued our incorporation of a generous number of high-quality color illustrations, including everything a student needs visually to accomplish the objectives of each exercise. While there is no need for students to purchase supplemental publications, students may find the *Photo Atlas for Biology*, ISBN 0-534-23556-5, to be a useful reference for this and other biology courses.

Most illustrations of microscopic specimens are labeled to provide orientation and clarity. A few are unlabeled but are provided with leaders for students to attach labels. In other cases, more can be gained by requiring the student to do simple drawings. Space is included in the manual for these, with boxes for drawings of macroscopic specimens and circles for microscopic specimens.

We include a number of relevant figures from Starr/Taggart/Evers/Starr's *Biology: The Unity and Diversity of Life*, Starr/Evers/Starr's *Biology: Concepts and Applications,* and Starr/Evers/Starr's *Biology Today and Tomorrow*. These illustrations further illuminate connections between lab exercise material and biological concepts.

Materials

Most of the equipment and supplies used in the exercises are readily available from biological and laboratory supply houses. Many others can be collected from nature or purchased in local supermarkets and discount or office supply stores. (Our department budgets are not large!) We've attempted to keep instrumentation as simple and inexpensive as possible.

Anyone who lives in a temperate climate knows it may be necessary to adjust the sequence of the laboratory exercises to accommodate seasonal availability of certain materials. However, we provided alternatives, including the use of preserved specimens wherever possible, to avoid this problem.

Instructor's Manual

There is no need to worry about "Where can I get that?" or "How do I prepare this?" Our *Instructor's Manual* includes:

- Material and equipment lists for each exercise
- Procedures to prepare reagents, materials, and equipment
- Scheduling information for materials needing advance preparation
- Approximate quantities of materials needed
- Answers to in-text questions
- Answers for pre-lab questions
- Answers for post-lab questions
- Tear-out sheets of pre-lab questions, in scrambled order from those in the lab manual, for those who wish to duplicate them for quizzes
- Vendors and ordering information for supplies

And in the End...

We would like to express our special thanks to Sean Cronin, Shannon Holt, Suzannah Alexander, Peggy Williams, and other individuals at Cengage Learning | Brooks/Cole involved in this project, and to Pradhiba Kannaiyan and Sara Golden of PreMediaGlobal. We would especially like to thank the following reviewers for their helpful insights and suggestions. Their comments allowed us to correct some errors and improve clarity, and so they have improved the lab manual in this and previous editions: Adebiyi Banjoko, Maricopa Community College; Evelyn Bruce, University of North Alabama; Juville Dario-Becker, Central Virginia Community College; Richard D. Gardner, Southern Virginia University; J. Robert Hatherill, Del Mar College; Rachel M. P. Hopp, Houston Baptist University; Sherry Krayesky, University of Louisiana, Lafayette; Kamau W. Mbuthia, Bowling Green State University; Snehlata Pandey, Hampton University; Marie Panec, Moorpark College; Tom Patterson, South Texas College; Jean Rothe, Quinnipiac University; Jennifer Siemantel, Cedar Valley College; Steve Threlkeld, Calhoun State Community College; Liza Vela, University of Texas–Pan American; Mary E. White, Southeastern Louisiana University; Martin Zahn, Thomas Nelson Community College.

There are very few things in life that are perfect. We don't suppose that this lab manual is one of them. We hope your students will enjoy the exercises. We know they will learn from them. Perhaps you and they will find places where rephrasing will make the activity better. Please contact us with your opinions and any ideas you wish to share; encourage your students to do likewise.

James W. Perry
james.perry@uwc.edu

David Morton
dmorton@frostburg.edu

Joy B. Perry
joy.perry@uwc.edu

To the Student

Welcome! You are about to embark on a journey through the cosmos of life. You will learn things about yourself and your surroundings that will broaden and enrich your life. You will have the opportunity to marvel at the microscopic world, to be fascinated by the cellular events occurring in your body at this very moment, and to gain an appreciation for the environment, including the marvelous diversity of the plant and animal world.

We offer a number of suggestions to make your college experience in biology a pleasant one. We have taken the first step toward that goal; we have written a laboratory guide that is user friendly. You will be able to hear the authors speaking as though we were there to share your experience. The authors share a personal belief that the more comfortable we make you feel, the more likely you will share our enthusiasm for biology. It would be naive for us to suppose that each of you will be biology majors at graduation. But one thing we all must realize is that we are citizens of "spaceship Earth." The fate of our spaceship is largely in your hands because you are the decision makers of the future. As has been so aptly stated, "We inherited the Earth from our parents and grandparents, but we are only the caretakers for our children and grandchildren."

As caretakers, we need to be informed about the world around us. That's why we enroll in colleges and universities with the hope of gaining a broad education. In doing so, we establish a basis on which to make educated decisions about the future of the planet. Each exercise in this manual contains a lesson in life that is of a more global nature than just the surroundings of your biology laboratory.

To enhance your biology education, take the initiative to give yourself the best possible advantage. Don't miss class. Read your text assignment routinely. And read the laboratory exercise before you come to the lab.

Each exercise in the manual is organized in the same way:

1. *Objectives* tell what you should learn from the exercise. If you wish to know what will be on the exam, consult the objectives for each exercise.

2. The *Introduction* provides background information for the exercise and is intended to stimulate your interest.

3. The *Materials* list for each portion of the exercise allows you to determine at a glance whether you have all the necessary supplies needed to do the activity.

4. The *Procedure* for each section, in easy-to-follow step-by-step fashion, describes the activity. Within the procedure, spaces are provided to make required drawings. Questions are posed with spaces for answers, asking you to draw conclusions about an activity you are engaged in. You'll find a lot of illustrations, most of which are labeled and others that are not but have leaders for you to attach labels. The terms to be used as labels are found in the procedure and in a list accompanying the illustration. We believe it best for you to sometimes make a simple drawing, and have inserted boxes or circles for your sketches. Where appropriate, tables and graphs are present to record your data.

5. *Pre-Lab Questions* can be answered easily by simply reading the exercise. They're meant to "set the stage" for the lab period by emphasizing some of the more salient points.

6. *Post-Lab Questions* are intended to be answered after the laboratory is completed. Some are straightforward interpretations of what you have done, while others require additional thought and perhaps some research in your textbook. In fact, some have no "right" or 'wrong" answer at all!

It is our experience that students are much too reluctant to ask questions for fear of appearing stupid. Remember, there is no such thing as a stupid question. Speak up! Think of yourselves as "basic learners" and your instructors as "advanced learners." Interact and ask questions so that you and your instructors can further your/their respective educations.

Laboratory Supplies and Procedures

Materials and Supplies Kept in the Lab at All Times

The following materials will always be available in the lab room. Familiarize yourself with their location prior to beginning the exercises.

- compound light microscopes
- dissection microscopes
- glass microscope slides
- coverslips

- lens paper
- tissue wipes
- plastic 15-cm rulers
- dissecting needles
- razor blades
- glassware of various types and sizes
- assorted glassware-cleaning brushes
- detergent for washing glassware
- distilled water
- hand soap
- paper towels
- safety equipment (see separate list)

Laboratory Safety

None of the exercises in this manual are inherently dangerous. Some of the chemicals are corrosive (causing burns to the skin) and others are poisonous if ingested or inhaled in large amounts. Contact with your eyes by otherwise innocuous substances may result in permanent eye injury. *Remember, once your sight is lost, it's probably lost forever.*

Locate the safety items described below and then study the list of basic safety rules.

1. *Shower*

    Do not be put off by the lack of a drain; this isolates any spill to the room in which it occurs.

2. *Eyewash bottle or eye bath*

    If any substance is splashed in your eyes, wash them thoroughly.

3. *Fire extinguisher*

    Read the directions for use of the fire extinguisher.

4. *Fire blanket*

    If someone's clothing catches on fire, wrap the blanket around the individual and roll the person on the floor to smother the flames.

5. *First-aid kit*

    Minor injuries such as small cuts can be treated effectively in the lab. Open the first-aid kit to determine its contents.

6. *Safety goggles*

    Eye protection should be worn during some exercises as directed in your lab manual or by your instructor.

7. *Laboratory gloves*

    Hand protection should be worn as directed in your lab manual or by your instructor.

Safety Rules

1. Do not eat, drink, or smoke in the laboratory.

2. Wash your hands with soap and warm water before leaving the laboratory.

3. When heating a test tube, point the mouth of the tube away from yourself and other people.

4. Always wear shoes in the laboratory.

5. Keep extra books and clothing in designated places so your work area is as uncluttered as possible.

6. If you have long hair, tie it back when in the laboratory.

7. Read labels carefully before removing substances from a container. *Never return a substance to a container unless directed to do so by your instructor.*

8. Discard used chemicals and materials into appropriately labeled containers. Certain chemicals should not be washed down the sink; these will be indicated by your instructor.

CAUTION

Report all accidents and spills to your instructor immediately!

Instructions for Washing Laboratory Glassware

1. Place contents to be discarded in the proper waste container as described in the exercise or as directed by your instructor.

2. Rinse the glassware with tap water.

3. Add a small amount of glassware-cleaning detergent.

4. Scrub the glassware with an appropriately sized brush.

5. Rinse the glassware with tap water until detergent disappears.

6. Rinse the glassware three times with distilled water (dH$_2$O). When glassware is clean, dH$_2$O sheets off rather than remaining on the surface in droplets.

7. Allow the glassware to dry in an inverted position on a drying rack (if available).

The Scientific Method

After completing this exercise, you will be able to

1. define *scientific method, mechanist, vitalist, cause and effect, induction, deduction, experimental group, control group, independent variable, dependent variable, controlled variables, correlation, theory, principle, bioassay;*

2. explain the nature of scientific knowledge;

3. describe the basic steps of the scientific method;

4. state the purpose of an experiment;

5. explain the difference between cause and effect and correlation;

6. describe the design of a typical research article in biology;

7. do a bioassay as an example of an experiment.

Introduction

To appreciate biology or, for that matter, any body of scientific knowledge, you need to understand how the scientific method is used to gather that knowledge. We use the scientific method to test the predictions of possible answers to questions about nature in ways that we can duplicate or verify. Answers supported by test results are added to the body of scientific knowledge and contribute to the concepts presented in your textbook and other science books. Although these concepts are as up to date as possible, they are always open to further questions and modifications.

One of the roots of the scientific method can be found in ancient Greek philosophy. The natural philosophy of Aristotle and his colleagues was mechanistic rather than vitalistic. A mechanist believes that only natural forces govern living things, along with the rest of the universe. A vitalist believes that the universe is at least partially governed by supernatural powers. Mechanists look for interrelationships between the structures and functions of living things and the processes that shape them. Their explanations of nature deal in **cause and effect**—the idea that one thing is the result of another thing. (For example, fertilization of an egg initiates the developmental process that forms an adult.) In contrast, vitalists often use purposeful explanations of natural events. (The fertilized egg strives to develop into an adult.) Although statements that ascribe purpose to things often feel comfortable to the writer, try to avoid them when writing lab reports and scientific papers.

Aristotle and his colleagues developed three rules to examine the laws of nature: first, carefully observe some aspect of nature; second, examine these observations as to their similarities and differences; and third, produce a principle or generalization about the aspect of nature being studied. An example is the principle that all mammals nourish their young with milk, which made it difficult to accept two egg-laying mammals found in Australia.

The major defect of natural philosophy was that it accepted the idea of absolute truth. This belief suppressed the testing of principles after they had been formulated. Another example is Aristotle's belief in spontaneous generation, the principle that some life can arise from nonliving things (e.g., maggots from spoiled meat). This belief survived more than 2000 years of controversy before being discredited by Louis Pasteur in 1860. Rejection of the idea of absolute truth coupled with the testing of principles either by experimentation or by further pertinent observation is the essence of the modern scientific method.

1.1 Modern Scientific Method *(About 70 min.)*

Although there is not one universal scientific method, Figure 1-1 illustrates the general process.

MATERIALS

Per lab room:

- blindfold
- plastic beakers with an inside diameter of about 8 cm stuffed with cotton wool
- four or five liquid crystal thermometers

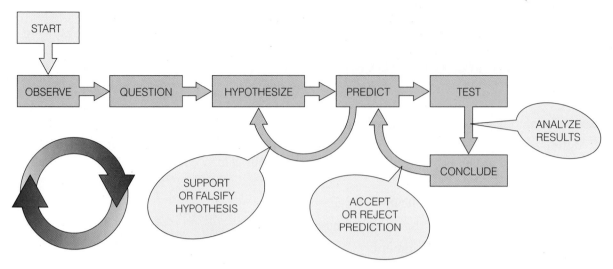

Figure 1-1 The scientific method. Support of the hypothesis usually necessitates further observations, adjustment to the question, and modification of the hypothesis. Once started, the scientific method cycles over and over again, each turn further refining the hypothesis.

PROCEDURE

A. Observation. As with natural philosophy, *we start the scientific method with careful observation*. An investigator may make observations from nature or from the words of other investigators, which are published in books or research articles in scientific journals and are available in the storehouses of human knowledge (e.g., libraries and the Internet). One subject we all have some knowledge of is the human body. The first four rows of Table 1-1 list some observations about the human body. The fifth row is blank so that you can fill in the steps of the scientific method for either the bioassay described at the end of the exercise, another observation about the human body, or anything else you and your instructor wish to investigate.

B. Question. In the second step of the scientific method, *we ask a question* about these observations. The quality of this question will depend on how carefully the observations were made and analyzed. Table 1-1 includes questions raised by the listed observations.

C. Hypothesis. Now *we construct a hypothesis*—that is, we derive by inductive reasoning a possible answer to the question. Induction is a logical process by which all known observations are combined and considered before producing a possible answer. Table 1-1 includes examples of hypotheses.

D. Prediction. Next *we formulate a prediction*—we assume the hypothesis is correct and predict the result of a test that reveals some aspect of it. This is deductive or "if-then" reasoning. Deduction is a logical process by which a prediction is produced from a possible answer to the question asked. Table 1-1 lists a prediction for each hypothesis.

E. Experiment or Pertinent Observations. Now *we perform an experiment or make pertinent observations* to test the prediction.

1. Along with the other members of your lab group, choose one prediction from Table 1-1. Coordinate your choice with the other lab groups so that each group tests a different prediction.

2. In an experiment of classical design, the individuals or items under study are divided into two groups: an **experimental group** that is treated with (or possesses) the independent variable and a **control group** that is not (or does not). Sometimes there is more than one experimental group. Sometimes subjects participate in both groups, experimental and control, and are tested both with and without the treatment.

   In any test there are three kinds of variables. The **independent variable** is the treatment or condition under study. The **dependent variable** is the event or condition that is measured or observed when the results are gathered. The **controlled variables** are all other factors, which the investigator attempts to keep the same for all groups under study.

3. Here are some hints about each of the predictions listed in Table 1-1.
   (a) To test the prediction in row I of Table 1-1, follow these directions:
       (i) With arms dangling at their sides, identify in as many group members as possible a vein segment either on the back of the hands or on the forearms (Figure 1-2).
       (ii) Demonstrate that blood flows in vein segments: Note their collapse due to gravity speeding up blood flow when each subject's arm is raised above the level of the heart.
       (iii) Lower the arm to a position below the heart and stop blood flow through the vein segment by permanently pressing a finger on the swelling farthest from the heart.

TABLE 1-1 Some Observations about the Human Body

Observation	Question	Hypothesis	Prediction
I. Veins containing blood are seen under the skin. Swellings present along the vein are often located where veins join together.[a]	What is the function of the swellings?	Swellings contain one-way valves that allow blood to flow only toward the heart.	If these valves are present, then blood flows only from vein segments farther from the heart to the next segments nearer the heart and never in the opposite direction.
II. People have two ears.	What is the advantage of having two ears?	Two ears allow us to locate the sources of sounds.	If the hypothesis is correct, then blocking hearing in one ear will impair our ability to determine a sound's source.[b]
III. People can hold their breath for only a short period of time.	What factor forces a person to take a breath?	The buildup of carbon dioxide derived from the body's metabolic activity stimulates us to take a breath.	If the hypothesis is correct, then people will hold their breath a shorter time just after exercise compared with when they are at rest.
IV. Normal body temperature is 98.6°F.	Is all of the body at the same 98.6°F temperature?	The skin, or at least some portion of it, is not 98.6°F.	If the hypothesis is correct, then a liquid crystal thermometer will record different temperatures on the forehead, back of the neck, and forearm.
V. _____ _____ _____ _____	_____ _____ _____ _____	_____ _____ _____ _____	_____ _____ _____ _____

© Cengage Learning 2013

[a]The portion of the vein between swellings is called a segment and is illustrated in Figure 1-2.
[b]This is especially the case for high-pitched sounds because higher frequencies travel less easily directly through tissues and bones.

(iv) Use a second finger to squeeze the blood out of the vein segment past the next swelling toward the heart.

(v) Remove the second finger and note whether blood flows back into the vein segment.

(vi) Remove the first finger and note whether blood flows back into the vein segment.

(vii) Repeat v–vi, only reverse the order of the swellings—first stop blood flow through the swelling nearest the heart, then squeeze the blood out of the vein segment in a direction away from the heart past the next swelling.

(b) To test the prediction in row II of Table 1-1, you will need a blindfold and a beaker stuffed with cotton wool to block hearing in one ear. Quantify how well each blindfolded group member can point out the source of a high-pitched sound by estimating the angle (0–180°) between the line from source to subject and the line along which the subject points to identify the source.

(c) To test the prediction in row III, each group member in turn will need to exercise in a safe way. Either run in place or *follow instructions given to you by your instructor*.

(d) To test the prediction in row IV, you will need a liquid crystal thermometer to measure skin temperature in three places for each group member.

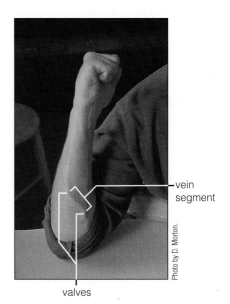

vein segment

valves

Photo by D. Morton.

Figure 1-2   Veins under the skin.

4. In your own words describe your procedure to test your group's prediction and identify its variables below.

_____

_____

Independent variable:

_____

_____

Dependent variable:

_____

_____

Controlled variables:

_____

_____

5. Show the above section to your instructor and get approval to perform your test.

F. Test. After your instructor's approval, perform your procedure and record your results in Table 1-2. You will need to label the column headings for your particular test. You may not need all of the rows and columns to accommodate your data. Describe or present your results in one of the bar charts, etc. in Figure 1-3.

G. Conclusion

1. In the last step of one cycle of the scientific method, *we make a conclusion.* Use the results of the experiment or pertinent observations to evaluate your hypothesis. If your prediction does not occur, it is rejected and your hypothesis or some aspect of it is falsified. If your prediction does occur, you may conditionally accept your prediction, and your hypothesis is supported. However, you can never completely accept or reject any hypothesis; all you can do is state a probability that one is correct or incorrect. To quantify this probability, scientists use a branch of mathematics called *statistical analysis.*

2. Even if the prediction is rejected, this does not necessarily mean that the treatment caused the result. A coincidence or the effect of some unforeseen and thus uncontrolled variable could be causing the result. For this reason, the results of experiments and observations must be *repeatable* by the original investigator and others.

3. Even if the results are repeatable, this does not necessarily mean that the treatment caused the result. *Cause and effect*, especially in biology, is rarely proven in experiments. We can, however, say that the treatment and result are correlated. A correlation is a relationship between the independent and the dependent variables.

   The following example should illustrate these concepts. Severe narrowing of a coronary artery branch reduces blood flow to the heart muscle downstream. This region of heart muscle gets insufficient oxygen and cannot contract and may die, resulting in a heart attack. The initial cause is the narrowing of the artery and the final effect is the heart attack. Perhaps the heart attack victim smoked cigarettes. Smoking cigarettes is one of several factors that make a person more likely to have a heart attack. This is based on a correlation between smoking and heart attacks in the general population but we cannot say for sure that the smoking caused the heart attack.

TABLE 1-2	Generalized Data Sheet
Subject	
1	
2	
3	
4	
5	
Average	

© Cengage Learning 2013

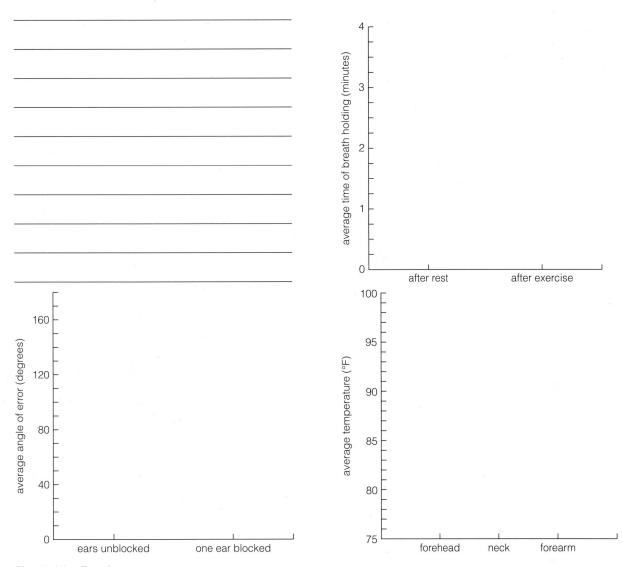

Figure 1-3 Results.

4. Write a likely conclusion for your experiment or pertinent observation. Statistics are not required, but if you know how, apply the correct statistical test before writing your conclusion.

_____

_____

H. Theories and Principles. When exhaustive experiments and observations consistently support an important hypothesis, it is accepted as a theory. A theory that stands the test of time may be elevated to the status of a principle (Figure 1-4). Theories and principles are always considered when new hypotheses are formulated. However, like hypotheses, theories and principles can be modified or even discarded in the light of new knowledge. Biology, like life itself, is not static but is constantly changing.

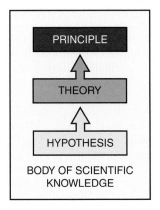

Figure 1-4 Theories and Principles. When exhaustive experiments and observations consistently support an important hypothesis, it is accepted as a **theory.** A theory that stands the test of time may be raised to the status of a **principle.** Theories and principles are always considered when new hypotheses are formulated. However, like hypotheses, theories and principles can be modified or even discarded in the light of new knowledge. Biology, like life itself, is not static but is constantly changing.

The account of one or several related cycles of the scientific method is usually initially reported in depth in a research article published in a scientific journal. The goal of the scientific community is to be cooperative as well as competitive. Writing research articles allows scientists to share knowledge. Scientists provide enough information so that other scientists can repeat the experiments or pertinent observations they describe. The journal *Science* along with several others presents its research articles in narrative form, and many of the details of the scientific method are understood rather than stated. However, adherence to the modern scientific method is expected, and the scientific community understands that it is as important to expose mistakes as it is to praise new knowledge.

MATERIALS

Per student:

- a typical research article in biology
- a research article from *Science*

PROCEDURE

1. Check the design of a typical research article in biology and list the titles of its various sections.

   (a) Example: Abstract (summary of paper)

   (b) _____

   (c) _____

   (d) _____

   (e) _____

   (f) _____

2. Read the article and then fill in the blanks in the following statements or answer the questions.

   (a) Any changes in the dependent variable are described in the _____ section.

   (b) Which section contains the details necessary to repeat this experiment or observation?

   _____

   (c) Which section contains the questions being asked, the predictions, or the hypotheses?

   _____

   (d) The _____ section contains the conclusions.

3. Look at an article from the journal *Science.* What steps of the scientific method are included in the narrative?

   _____

   _____

You no doubt know through various media sources of the myriad chemicals that pervade modern life. These substances can be beneficial or toxic to plants or animals. It's obvious that drinking gasoline is bad for you, but this is an extreme example. The broad questions are: what effects do new chemicals have on living organisms, and at what exposure level do they exert these effects? To answer these questions, we perform experiments known as bioassays. Bioassays are used by the pharmaceutical industry to test new drugs. Agricultural firms determine the effectiveness of new fertilizers and herbicides with bioassays. Some industries measure the effects their waste discharges have on aquatic organisms with bioassays. A **bioassay** establishes the quantity of a substance that results in a defined *effective dose* (ED)—that is, the dose that produces a particular effect. In the case of toxic substances, the standard measurement of toxicity is known as an LD_{50} (the lethal dose causing the death of 50% of the organisms exposed to the substance).

In this experiment you will test the hypothesis that many everyday substances affect germination of the seeds of Wisconsin Fast Plants™, members of the mustard family. Predictions are that if this is so, then the timing of seed germination will change, and if germination occurs, the extent of germination as measured by the length of roots and stems will change. If possible, you will use your results to estimate ED or LD_{50} for the substance tested.

MATERIALS

Per student:

- fine-pointed, water-resistant marker ("Sharpie" or similar)
- 35-mm film canister with lid prepunched with four holes or microcentrifuge tube holder (rack)
- four microcentrifuge tubes with caps
- four paper-towel wicks
- disposable pipet and bulb
- forceps
- eight RCBR seeds (Wisconsin Fast Plants™)
- 15-cm plastic ruler

Per student group (6):

- solution of test substance (paint thinner, household cleaner, perfume, coffee, vinegar, and so on)
- distilled water (dH$_2$O) in dropping bottle

CAUTION

Test substances may be harmful to you. Wear lab gloves and, even if you wear contact lenses, wear safety goggles. Eyeglasses should suffice when goggles do not fit over them.

PROCEDURE

1. Put on a pair of lab gloves and protective goggles.
2. Using a fine-pointed, water-resistant marker, label the caps and sides of the four microcentrifuge tubes 1.0, 0.1, 0.01, and C (for "Control").
3. Place 10 drops of the test solution into the tube labeled 1.0, which now contains 100% of the test solution. A test solution can be anything you and your instructor decide to test. Write the name of your test substance.
4. Make *serial dilutions* of the solution, producing concentrations of 10% and 1% of the test substance (Figure 1-5). Start by adding nine drops of dH$_2$O to the tubes labeled 0.1 and 0.01.
5. Remove a small quantity of the solution from tube 1.0 with the disposable pipet and place one drop into tube 0.1. Mix the contents thoroughly by flicking the sides of the bottom of the tube with the index finger of your dominant hand while holding the sides of the top of the tube between the index finger and thumb of your other hand. Return any solution remaining in the pipet to tube 1.0. Tube 0.1 now contains a concentration 10% of that in tube 1.0.
6. Now, from tube 0.1, remove a small quantity of the solution with the disposable pipet and place one drop in tube 0.01. Mix the contents thoroughly and return any solution left in the pipet to tube 0.1. Tube 0.01 now contains a 1% concentration of the original.
7. From the dropping bottle, place 10 drops of dH$_2$O into tube C. This is the control for the experiment; it contains none of the test solution.
8. Insert a paper-towel wick into each microcentrifuge tube (Figure 1-6). With your forceps, place two RCBR seeds near the top of each wick. Close the caps of the tubes. Do not allow any of the wick to protrude outside the cap.
9. Carefully insert the microcentrifuge assay tubes through the holes in the film canister lid (Figure 1-7). Set the experiment aside in the location indicated by your instructor.
10. At this point, write a prediction for your experiment in Table 1-3.

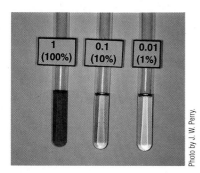

Figure 1-5 Serial dilution of test substance.

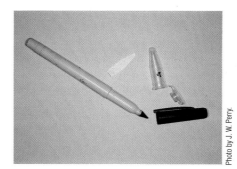

Figure 1-6 One of the four assay tubes.

Figure 1-7 Experimental apparatus.

TABLE 1-3	Effect of _____ on Germination of RCBR™ Seeds						

Prediction:

		My Experiment				Class Averages	
		Roots (mm)		Shoots (mm)			
Tube	Germination	S1	S2	S1	S2	Roots	Shoots
C (dH₂O)							
1.0							
0.1							
0.01							

© Cengage Learning 2013

11. State your experimental variables.

    Independent variable:

    _____

    _____

    Dependent variable:

    _____

    _____

    Controlled variables:

    _____

    _____

12. After 24 hours or more, examine your experiment to see how many and which seeds germinated. Record your results in Table 1-3. Record seeds as "germinated" (G) or "did not germinate" (DNG), and for germinated seeds indicate the extent of germination by measuring and recording the length in millimeters (mm) of the roots and shoots of seed 1 (S1) and seed 2 (S2) for each tube.

13. You have studied the effect of your substance on only a small number of seeds. To make your analysis more reliable, you should analyze a larger number of seeds. In an effort to increase reliability, pool your data with those students who tested the same substance. Calculate the averages and record them in Table 1-3.

14. If germination occurred in more than one group, graph the results in Figure 1-8, plotting the growth of roots and shoots at each concentration.

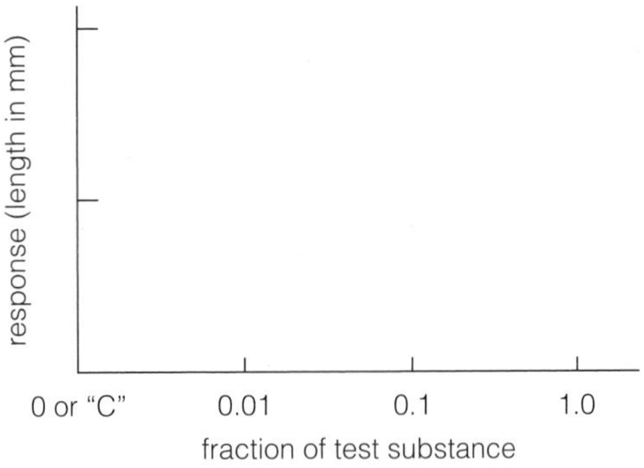

Figure 1-8 Effect of _____ on germination of RCBR seeds (O = roots and ● = shoots).

15. If germination occurred in the lower concentrations but not the higher ones, what effective dose prevents germination? Alternatively, could you have defined any other effective doses?

    _____

    _____

16. Considering your germination results, estimate or make a statement about the LD_{50} for your test substance. If negative effects on germination were observed, describe how you could modify the procedure to better determine the LD_{50}.

    _____

    _____

17. Make a conclusion regarding the effect of the test substance on germination of RCBR seeds.

_____

_____

18. Does your conclusion allow you to accept or reject your prediction? Is the hypothesis supported or falsified? Explain your answer.

_____

_____

_____ 1. The natural philosophy of Aristotle and his colleagues was
  (a) mechanistic.
  (b) vitalistic.
  (c) a belief in absolute truth.
  (d) a and c

_____ 2. A person who believes that the universe is at least partially controlled by supernatural powers can best be described as a(n)
  (a) teleologist.
  (b) vitalist.
  (c) empiricist.
  (d) mechanist.

_____ 3. The first step of the scientific method is to
  (a) ask a question.
  (b) construct a hypothesis.
  (c) observe carefully.
  (d) formulate a prediction.

_____ 4. Which series of letters lists the first four steps of the scientific method (see question 3) in the correct order?
  (a) a, b, c, d
  (b) a, b, d, c
  (c) c, a, b, d
  (d) d, c, a, b

_____ 5. In an experiment the subjects or items being investigated are divided into the experimental group and
  (a) the nonexperimental group.
  (b) the control group.
  (c) the statistics group.
  (d) none of these choices

_____ 6. The variables that investigators try to keep the same for both the experimental and the control groups are
  (a) independent.
  (b) controlled.
  (c) dependent.
  (d) a and c

_____ 7. Variables that are always different between the experimental and the control groups are
  (a) independent.
  (b) controlled.
  (c) dependent.
  (d) a and c

_____ 8. The results of an experiment
  (a) don't have to be repeatable.
  (b) should be repeatable by the investigator.
  (c) should be repeatable by other investigators.
  (d) must be both b and c.

_____ 9. The detailed report of an experiment is usually published in a
  (a) newspaper.
  (b) book.
  (c) scientific journal.
  (d) magazine.

_____ 10. Bioassays can be used to
  (a) test new drugs.
  (b) determine the effectiveness of new fertilizers and herbicides.
  (c) measure the effects their waste discharges have on aquatic organisms.
  (d) do all of these choices.

EXERCISE **1**

The Scientific Method

Post-Lab Questions

Introduction

1. How does the modern scientific method differ from the natural philosophy of the ancient Greeks?

1.1 Modern Scientific Method

2. List the six steps of one full cycle of the scientific method.
   (a) _____
   (b) _____
   (c) _____
   (d) _____
   (e) _____
   (f) _____

3. What is tested by an experiment?

4. Within the framework of an experiment, describe the
   (a) independent variable:

   (b) dependent variable:

   (c) controlled variables:

5. Is the statement, "In most biology experiments, the relationship between the independent and the dependent variable can best be described as cause and effect," true or false? Explain your answer.

6. Is a scientific principle taken as absolutely true? Explain your answer.

1.2 Research Article

7. What is the function of research articles in scientific journals?

1.3 Bioassay

8. Define the design and structure of a bioassay.

Food for Thought

9. Describe how you have applied or could apply the scientific method to an everyday problem.

10. Do you think the differences between religious and scientific knowledge make it difficult to debate points of perceived conflict between them? Explain your answer.

Measurement

OBJECTIVES

After completing this exercise, you will be able to

1. define *length, volume, meniscus, mass, density, temperature, thermometer;*
2. recognize graduated cylinders, beakers, Erlenmeyer flasks, different types of pipets, and a triple beam balance;
3. measure and estimate length, volume, and mass in metric units;
4. explain the concept of temperature;
5. measure and estimate temperature in degrees Celsius;
6. explain the advantages of the metric system of measurement.

Introduction

One requirement of the scientific method is that results be repeatable. As numerical results are more precise than purely written descriptions, scientific observations are usually made as measurements. Of course, sometimes a written description without numbers is the most appropriate way to describe a result.

2.1   Metric System *(About 60 min.)*

Metric system measurements are easy to convert from one unit to another (e.g., centimeters to meters to kilometers for length) and from one related unit to another (e.g., length to area to volume). The metric system is preferred by most of the people and countries in the world, especially by scientists. Governments have largely switched to it, at least in globally traded and shared manufacturing items (e.g., all cars made in the United States have metric parts and the U.S. Department of Defense adopted the metric system in 1957).

The metric reference units are the **meter** for length, the **liter** for volume, the **gram** for mass, and the **degree Celsius** for temperature. Regardless of the type of measurement, the same prefixes are used to designate the relationship of a unit to the reference unit. Table 2-1 lists the prefixes we will use in this and subsequent exercises.

TABLE 2-1	Prefixes for Metric System Units
Prefix of Unit (Symbol)	**Part of Reference Unit**
nano (n)	$1/1,000,000,000 = 0.000000001 = 10^{-9}$
micro (μ)	$1/1,000,000 = 0.000001 = 10^{-6}$
milli (m)	$1/1000 = 0.001 = 10^{-3}$
centi (c)	$1/100 = 0.01 = 10^{-2}$
kilo (k)	$1000 = 10^{3}$

© Cengage Learning 2013

As you can see, the metric system is a decimal system of measurement. Metric units are 10, 100, 1000, and sometimes 1,000,000 or more times larger or smaller than the reference unit. Thus, it's relatively easy to convert from one measurement to another either by multiplying or dividing by 10 or a multiple of 10:

In this section, we examine the metric system and compare it to the American Standard system of measurement (feet, quarts, pounds, and so on).

MATERIALS

Per student pair:

- 30-cm ruler with metric and American (English) Standard units on opposite edges
- nonmercury thermometer(s) with Celsius (°C) and Fahrenheit (°F) scales (about −20°–110°C)
- 250-mL beaker made of heat-proof glass
- hot plate
- 250-mL Erlenmeyer flask
- three boiling chips
- three graduated cylinders: 10-mL, 25-mL, 100-mL
- thermometer holder

Per student group:

- a triple beam balance

Per lab room:

- source of distilled water (dH$_2$O)
- metric bathroom scale
- source of ice
- 1-quart jar or bottle marked with a fill line
- one-piece plastic dropping pipet (not graduated) or Pasteur pipet and bulb
- graduated pipet and safety bulb or filling device
- 1-pound brick of coffee
- ceramic coffee mug
- 1-gallon milk bottle
- metric tape measure
- 1-L measuring cup

PROCEDURE

A. Length (15 min.)

Length is the measurement of a real or imaginary line extending from one point to another. The standard unit is the meter, and the most commonly used related units of length are:

1000 millimeters (mm) = 1 meter (m)
100 centimeters (cm) = 1 m
1000 m = 1 kilometer (km)

For orientation purposes, the yolk of a chicken egg is about 3 cm in diameter. Since the differences between these metric units are based on 10 or multiples of 10, it's fairly easy to convert a measurement in one unit to another.

1. For example, if you want to convert 1.7 km to centimeters, your first step is to determine how many centimeters there are in 1 km. Remember, like units may be cancelled.

$$\frac{100 \text{ cm}}{1 \text{ m}} \times \frac{1000 \text{ m}}{1 \text{ km}} = \frac{100,000 \text{ cm}}{1 \text{ km}}$$

The second step is to multiply the number by this fraction.

$$\frac{1.7 \text{ km}}{1} \times \frac{100,000 \text{ cm}}{1 \text{ km}} = 170,000 \text{ cm}$$

The last calculation can also be done quickly by shifting the decimal point five places to the right.

1.700000 km = 170, 000.0 cm

Alternatively, to convert 1.7 km to centimeters, we add exponents.

$$\frac{1.7 \text{ km}}{1} \times \frac{10^2 \text{ cm}}{\text{m}} \times \frac{10^3 \text{ m}}{\text{km}} = 1.7 \times 10^5 \text{ cm} = 170,000 \text{ cm}$$

Using the method most comfortable for you, calculate how many millimeters there are in 4.8 m.

_____ mm

Now let's convert 17 mm to meters.

step 1 $\dfrac{1\ m}{100\ cm} \times \dfrac{1\ cm}{10\ mm} = \dfrac{1m}{1000\ mm}$

step 2 $\dfrac{17\ mm}{1} \times \dfrac{1\ m}{1000\ mm} = 0.017\ m$

Alternatively, you can shift the decimal point three places to the left,

0017.0 mm = 0.017 m

or add exponents:

$\dfrac{1.7\ mm}{1} \times \dfrac{10^{-2}\ m}{cm} \times \dfrac{10^{-1}\ cm}{mm} = 1.7 \times 10^{-3}\ m = 0.017\ m$

Calculate how many kilometers there are in 16 cm.

_____ km

2. Precisely measure the length of this page in centimeters to the nearest 10th of a centimeter with the metric edge of a ruler. Note that nine lines divide the space between each centimeter into 10 millimeters.

The page is _____ cm long.

Calculate the length of this page in millimeters, meters, and kilometers.

_____ mm _____ m _____ km

   Now repeat the above measurement using the American Standard edge of the ruler. Measure the length of this page in inches to the nearest eighth of an inch.

_____ in

Convert your answer to feet and yards.

_____ ft _____ yd

Explain why it is much easier to convert units of length in the metric system than in the American Standard system.

_____

_____

_____

B. Volume (20 min.)

Volume is the space an object occupies. The standard unit of volume is the liter (L), and the most commonly used subunit, the milliliter (mL). There are 1000 mL in 1 liter. A chicken egg has a volume of about 60 mL.

   The volume of a box is the height multiplied by the width multiplied by the depth. The amount of water contained in a cube with sides 1 cm long is 1 cubic centimeter (cc), which for all practical purposes equals 1 mL (Figure 2-1).

1. How many milliliters are there in 1.7 L?

   _____ mL

   How many liters are there in 1.7 mL?

   _____ L

2. Use Figure 2-2 to recognize **graduated cylinders, beakers, Erlenmeyer flasks,** and the different types of **pipets**. Some of these objects are

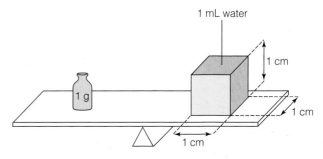

Figure 2-1 Relationship among the units of length, volume, and mass in the metric system.

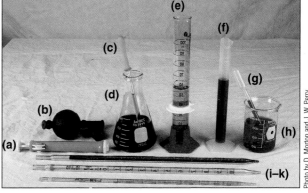

Figure 2-2 Apparatuses commonly used to measure volume: (**a**) pipet filling device, (**b**) pipet safety bulb, (**c**) Pasteur pipet, (**d**) Erlenmeyer flask, (**e**) glass graduated cylinder, (**f**) plastic graduated cylinder, (**g**) plastic dropping pipet, (**h**) beaker, (**i–k**) graduated pipets.

Photo by D. Morton and J. W. Perry.

made of glass; some are plastic. Some will be calibrated in milliliters and liters; others will not be.

3. Pour some water into a 100-mL graduated cylinder and observe the boundary between fluid and air, the **meniscus**. Surface tension makes the meniscus curved, not flat. The high surface tension of water is due to its cohesive and adhesive or "sticky" properties. Draw the meniscus in the plain cylinder outlined in Figure 2-3. The correct reading of the volume is at the *lowest* point of the meniscus.

4. Using the 100-mL graduated cylinder, pour water into a 1-quart jar or bottle. About how many milliliters of water are needed to fill the vessel up to the line?

    _____ mL

5. Pipets are used to transfer small volumes from one vessel to another. Some pipets are not graduated (for example, Pasteur pipets and most one-piece plastic dropping pipets); others are graduated.
    (a) Fill a 250-mL Erlenmeyer flask with distilled water.
    (b) Use a plastic dropping pipet or Pasteur pipet with a bulb to withdraw some water.
    (c) Count the number of drops needed to fill a 10-mL graduated cylinder to the 1-mL mark. Record this number in Table 2-2.
    (d) Repeat steps b and c two more times and calculate the average for your results in Table 2-2.
    (e) Explain why the average of three trials is more accurate than if you only do the procedure once.

    _____

    _____

    _____

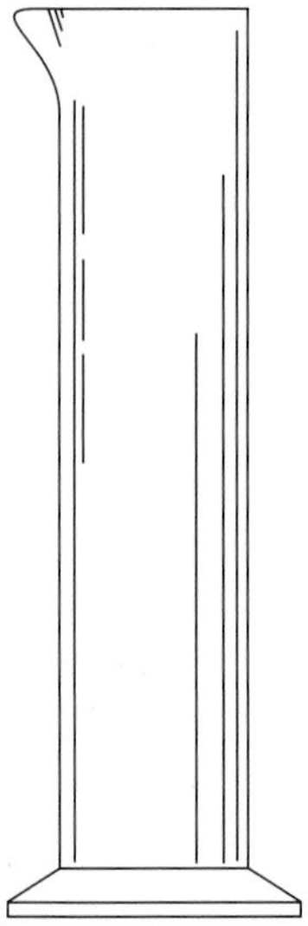

Figure 2-3
Draw a meniscus in this plain cylinder.

TABLE 2-2	Estimate of the Number of Drops in 1 mL
Trial	**Drops/mL**
1	
2	
3	
Total	
Average	

© Cengage Learning 2013

C. Mass (25 min.)

Mass is the quantity of matter in a given object. The standard unit is the **kilogram** (kg), and other commonly used units are the milligram (mg) and gram (g). There are 1,000,000 mg in 1 kg and 1000 g in 1 kg. A chicken egg has a mass of about 60 g. Note that the following discussion avoids the term *weight*. This is because weight depends on the gravitational field in which the matter is located. For example, you weigh less on the Moon but your mass remains the same as it is on Earth. Although it is technically incorrect, mass and weight are often used interchangeably.

1. How many milligrams are there in 1 g?

    _____ mg

    Convert 1.7 g to milligrams and kilograms.

    _____ mg _____ kg

2. A 1-cc cube, if filled with 1 mL of water, has a mass of 1 g (Figure 2-1). The mass of other materials depends on their **density** (water is defined as having a density of 1). The density of any substance is its mass divided by its volume.

    Approximately how many liters are present in 1 cubic meter (m^3) of water? As each of the sides of 1 m^3 are 100 cm in length, it's easy to calculate the number of cubic centimeters (that is, 100 cm × 100 cm × 100 cm = 1,000,000 cc). Now just change cubic centimeters to milliliters and convert 1,000,000 mL to liters.

    _____ L

What is its mass in kilograms?

_____ kg

In your "mind's eye," contemplate calculating how many pounds there are in a cubic yard of water. It's easier to convert between different units of the metric system than between those of the American Standard system.

3. Determine the mass of an unknown volume of water.* Mass may be measured with a **triple beam balance,** which gets its name from its three beams (Figure 2-4). A movable mass hangs from each beam.

   (a) Slide all of the movable masses to zero. Note that the middle and back masses each click into the leftmost notch.

   (b) Clear the pan of all objects and make sure it is clean.

   (c) The balance marks should line up, indicating that the beam is level and that the pan is empty. If the balance marks don't line up, rotate the zero adjust knob until they do.

   (d) Place a 250-mL beaker on the pan. The right side of the beam should rise. Slide the mass on the middle beam until it clicks into the notch at the 100-g mark. If the right end of the beam tilts down below the stationary balance mark, you have added too much mass. Move the mass back a notch. If the right end remains tilted up, additional mass is needed. Keep adding 100-g increments until the beam tilts down; then move the mass back one notch. Repeat this procedure on the back beam, adding 10 g at a time until the beam tilts down, and then backing up one notch. Next, slide the front movable mass until the balance marks line up.

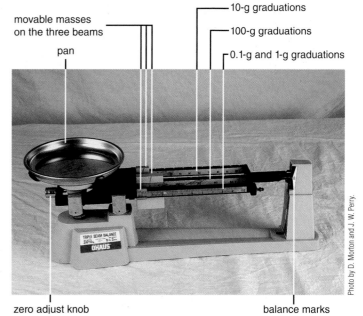

Figure 2-4   A triple beam balance.

   (e) The sum of the masses indicated on the three beams gives the mass of the beaker. Nine unnumbered lines divide the space between the numbered gram markings on the front beam into 10 sections, each representing 0.1 g. Record the mass of the beaker to the nearest 10th of a gram in Table 2-3.

   (f) Add an unknown amount of water and repeat the above procedure. Record the mass of the beaker and water in Table 2-3.

   (g) Calculate the mass of the water alone by subtracting the mass of the beaker from that of the combined beaker and water. Record it in Table 2-3. Do not dispose of the water yet.

   (h) Now measure the volume of the water in milliliters with a graduated cylinder. What is the volume? _____ mL

TABLE 2-3	Weighing an Unknown Quantity of Water with a Triple Beam Balance
Objects	**Masses (g)**
Beaker and water	
Beaker	
Water	

4. Using the triple beam balance, determine the mass of (that is, weigh) a pound of coffee in grams.
   _____ g

---

*Modified from C. M. Wynn and G. A. Joppich, *Laboratory Experiments for Chemistry: A Basic Introduction*, 3rd ed. Wadsworth, 1984.

D. Estimating Length, Volume, and Mass (10 min.)

Now that you have experience using metric units, let's try estimating the measurements of some everyday items. Also, you may consult the metric/American Standard conversion table in Appendix 1 at the end of the lab manual.

1. Estimate the length of your index finger in centimeters. _____ cm
2. Estimate your lab partner's height in meters. _____ m
3. How many milliliters will it take to fill a ceramic coffee mug? _____ mL
4. How many liters will it take to fill a 1-gallon milk bottle? _____ L
5. Estimate the weight of some small personal item (for example, loose change) in grams. _____ g
6. Estimate your weight or your lab partner's weight in kilograms. _____ kg
7. Transfer your estimates to Table 2-4.
8. Then check your results using either a ruler, metric tape measure, 100-mL graduated cylinder, 1-L measuring cup, triple beam balance, or metric bathroom scale, recording your measurements in Table 2-4. Complete Table 2-4 by calculating the difference between each estimate and measurement.

TABLE 2-4	Differences Between Estimates and Measurements		
Number	**Estimate**	**Measurement**	**Estimate – Measurement**
1	cm	cm	cm
2	m	m	m
3	mL	mL	mL
4	L	L	L
5	g	g	g
6	kg	kg	kg

© Cengage Learning 2013

E. Temperature (About 20 min.)

The degree of hot or cold of an object is termed **temperature**. More specifically, it is the average kinetic energy of molecules. Heat always flows from high to low temperatures. This is why hot objects left at room temperature always cool to the surrounding or ambient temperature, while cold objects warm up. Consequently, to keep a heater hot and the inside of a refrigerator cold requires energy. **Thermometers** are instruments used to measure temperature.

1. Using a thermometer with both *Celsius* (°C) and *Fahrenheit* (°F) scales, measure room temperature and the temperature of cold and hot running tap water. Record these temperatures in Table 2-5.
2. Fill a 250-mL beaker with ice about three-fourths full and add cold tap water to just below the ice. Wait for 3 minutes, measure the temperature, and record it in Table 2-5. Remove the thermometer and discard the ice water into the sink.
3. Fill the beaker with warm tap water to about three-fourths full and add three boiling chips. Use a thermometer holder to clip the thermometer onto the rim of the beaker so that the bulb of the thermometer is halfway into the water. Boil the water in the beaker by placing it on a hot plate. After the water boils, record its temperature in Table 2-5. Turn off the hot plate and let the water and beaker cool to below 50°C before pouring the water into the sink.

> **CAUTION**
>
> Your instructor will give you specific instructions on how to set up the equipment in your lab for boiling water.

4. To convert Celsius degrees to Fahrenheit degrees, multiply by $\frac{9}{5}$ and add 32. Is 4°C the temperature of a hot or cool day? _____ What temperature is this in degrees Fahrenheit? _____ °F
5. To convert Fahrenheit degrees to Celsius degrees, subtract 32 and multiply by $\frac{5}{9}$. What is body temperature, 98.6°F, in degrees Celsius? _____ °C
6. In summary, the formulas for these temperature conversions are:
   $$°F = \frac{9}{5} °C + 32$$
   $$°C = \frac{5}{9} (°F - 32)$$

TABLE 2-5	Comparison of Celsius and Fahrenheit Temperatures	
Location	°C	°F
Room		
Cold running tap water		
Hot running tap water		
Ice water		
Boiling water		

© Cengage Learning 2013

2.2 Micromeasurement (About 30 min.)

More and more procedures used in molecular biology and related areas require the measurement of very small masses and volumes. The following procedure introduces you to use of the electronic balance and micropipets or micropipetters. We'll do this as an experiment in part by answering the question: in the hands of students, how exact are instruments that deliver small volumes in doing what they say they do? Our hypothesis is that under these conditions these instruments perform within the manufacturer's stated specifications for the volumes they deliver.

MATERIALS

Per student group (table):

- standardized electronic balance capable of weighing 0.001 g and with at least a readability of 0.001 g, a repeatability (standard deviation) of 0.001 g, and a linearity of 0.002 g
- micropipets or micropipetters with matching tips (capable of delivering 10-μL, 50-μL, and 100-μL volumes)
- nonabsorbent weighing dish
- small test tube in a beaker or rack
- scientific calculator with instructions

Per lab room:

- source of distilled water (dH_2O)

PROCEDURE

1. Remember that in Section 2.1 we established a relationship between the volume and mass of water. If the mass of 1 μL of water is 1 g (Figure 2-1), how much does 1 μL weigh?

   _____ g or _____ mg

CAUTIONS ON THE USE OF *MICROPIPETTERS*

- Never adjust the micropipetter above or below the stated range or volume limit of measurement.
- Never use the micropipetter without a tip in place, as this will damage the internal piston mechanism.
- Never invert the micropipetter with a filled tip, as fluid could run up into its chamber.
- Never allow the plunger to snap back from the depressed position, as this could damage the piston. After drawing up or dispensing the fluid, release the plunger slowly.
- Never immerse the barrel of the micropipetter in fluid.

The electronic balances have been standardized prior to lab. If the micropipets or micropipetters in the hands of students perform within manufacturer's specifications, then we can predict what 100 μL, 50 μL, and 10 μL delivered from one of these instruments will weigh. In Table 2-6, write a prediction for this experiment.

2. Fill your test tube with distilled water.
3. Turn on the electronic balance and wait a moment (usually about 5 seconds) until it is stable.
4. Place a weighing dish on the pan.
5. Press the zero or tare key of the balance. Note that the digital display reads all zeros.
6. Use a 100-μL micropipet or a micropipetter adjusted to 100 μL and deliver this volume into the weighing dish. Wait a second for the balance to stabilize, then read and record the mass in grams or milligrams in

the correct column of Table 2-6. If you are using a toploading balance, the last digit may continue to change. In this case, use the middle number of the range of numbers displayed.

7. Tare or zero your balance and repeat step 6 for a total of 10 trials.
8. Tare or zero your balance and repeat steps 6 and 7 for the 50- and 10-μL volumes.
9. Use a scientific calculator to determine the average or mean and standard deviation for each column.
10. Plot your points on the graph (Figure 2-5a) and the means ± standard deviations on the bar chart (Figure 2-5b).

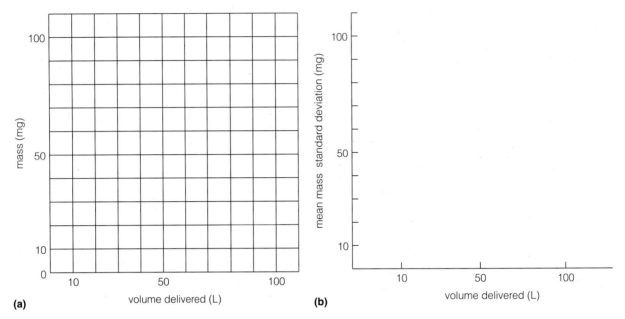

(a)

(b)

Figure 2-5 Micromeasurement results.

11. Write your conclusion in Table 2-6.

TABLE 2-6	Comparison of Volumes of Water with Their Masses		
Trial	100 μL	50 μL	10 μL
1			
2			
3			
4			
5			
6			
7			
8			
9			
10			
Mean			
Standard deviation			
Conclusion:			

© Cengage Learning 2013

_____ 1. The metric system is the measurement system of choice for
(a) scientists.
(b) most countries in the world.
(c) most peoples of the world.
(d) all of the above

_____ 2. A kilowatt, a unit of electrical power, is
(a) 10 watts.
(b) 100 watts.
(c) 1000 watts.
(d) 1,000,000 watts.

_____ 3. A millicurie, a unit of radioactivity, is
(a) a tenth of a curie.
(b) a hundredth of a curie.
(c) a thousandth of a curie.
(d) a millionth of a curie.

_____ 4. Length is the measurement of
(a) a line, extending from one point to another.
(b) the space an object occupies.
(c) the quantity of matter present in an object.
(d) the degree of hot or cold of an object.

_____ 5. Volume is the measurement of
(a) a line, extending from one point to another.
(b) the space an object occupies.
(c) the quantity of matter present in an object.
(d) the degree of hot or cold of an object.

_____ 6. Mass is the measurement of
(a) a line, extending from one point to another.
(b) the space an object occupies.
(c) the quantity of matter present in an object.
(d) the degree of hot or cold of an object.

_____ 7. If 1 cc of a substance has a mass of 1.5 g, its density is
(a) 0.67.
(b) 1.0.
(c) 1.5.
(d) 2.5.

_____ 8. If your mass is 70 kg on Earth, how much is your mass on the Moon?
(a) less than 70 kg
(b) more than 70 kg
(c) 70 kg
(d) none of the above

_____ 9. Above zero degrees, the actual number of degrees Celsius for any given temperature is _____ the degrees Fahrenheit.
(a) higher than
(b) lower than
(c) the same as
(d) a or b

_____ 10. A thermometer measures
(a) the degree of hot or cold.
(b) temperature.
(c) a and b
(d) none of the above

EXERCISE **2**

Measurement

Post-Lab Questions

Introduction

1. What is the importance of measurement to science?

2.1 Metric System

2. Convert 1.24 m to millimeters, centimeters, and kilometers.

   _____ mm

   _____ cm

   _____ km

3. Observe the following carefully and read the volume.

   _____ mL

Photo by D. Morton.

4. Construct a conversion table for mass. Construct it so that if you wish to convert a measurement from one unit to another, you multiply it by the number at the intersection of the original unit and the new unit.

Original Unit	New Unit		
	mg	g	kg
mg	1		
g		1	
kg			1

5. How is it possible for objects of the same volume to have different masses?

6. Today, many packaged items have the volume or weight listed in both American Standard and metric units. Before your next lab period, find and list 10 such items.

Item	American Standard	Metric

7. Each °F is _____ (larger, smaller) than each °C.

2.2 Micromeasurement

8. Describe what it means to tare or zero an electronic balance. What purpose does this function serve?

Food for Thought

9. How are length, area, and volume related in terms of the three dimensions of space?

10. If you were to choose between the metric and American Standard systems of measurement for future generations, which one would you choose? Set aside your familiarity with the American Standard system and consider their ease of use and degree of standardization with the rest of the world.

Microscopy

OBJECTIVES

After completing this exercise, you will be able to

1. define *magnification, resolving power, contrast, field of view, parfocal, parcentral, depth of field, working distance;*

2. describe how to care for a compound light microscope;

3. recognize and give the function of the parts of a compound light microscope;

4. accurately align a compound light microscope;

5. correctly use a compound light microscope;

6. make a wet mount;

7. correctly use a dissecting microscope;

8. describe the usefulness of the phase-contrast, transmission electron, and scanning electron microscopes;

9. use your skills to enjoy a fascinating world unavailable to the unaided eye.

Introduction

Light microscopes contain transparent glass lenses, which focus light rays emanating from a specimen to produce an image of a specimen on the retina, the light-sensitive layer of the eye. Table 3-1 presents the three most important properties of lenses and their images.

TABLE 3-1	**Important Lens and Image Properties**
Property	**Definition**
Magnification	The amount that the image of an object is enlarged (e.g., 100×)
Resolving power	The extent to which object detail in an image is preserved during the magnifying process
Contrast	The degree to which image details stand out against their background

© Cengage Learning 2013

3.1   Compound Light Microscope *(About 80 min.)*

In everyday life, the eye lens projects onto the retina a focused image of an object held no closer than about 10 cm. At this distance, details separated by 0.1 mm are visible. Most cells and related structures are smaller than this, and a light microscope is needed to see them. A microscope placed between the eye and a specimen (usually a section or thin object[s] mounted on a glass slide) acts to bring the specimen very close to the eye so you can focus on its details. Ultimately, the greater the proportion of the retina covered by the final image of the specimen, the greater its magnification. It does this by producing a series of magnified images.

Magnification without enough resolving power is referred to as empty, and with a light microscope, the maximum useful magnification is about 1000 times the diameter of the specimen (1000×). Above this value, additional details are missing. Furthermore, adequate contrast is needed to see the details preserved in an image. Dyes are usually added to sections of biological specimens to increase contrast.

Similar to automobiles, there are many models of compound light microscopes, and these instruments have numerous accessories that may or may not be present. Typical examples are shown in Figure 3-1, and one is diagrammed in Figure 3-2. If your microscope differs significantly from Figure 3-2, your instructor will give you an unlabeled diagram. If the instructor assigns you a specific microscope for your lab work, record its identification code in the second column of the first row in Table 3-2.

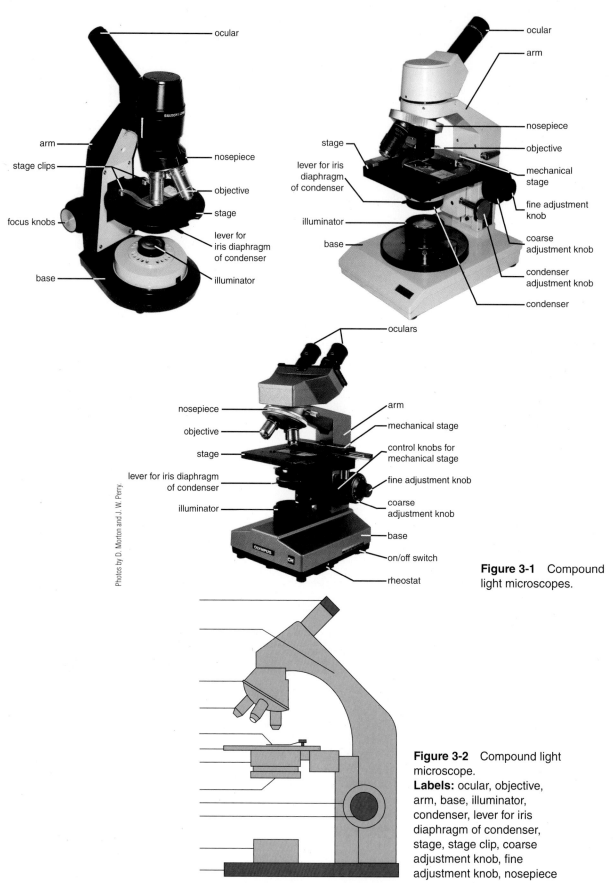

Figure 3-1 Compound light microscopes.

Photos by D. Morton and J. W. Perry.

Figure 3-2 Compound light microscope.
Labels: ocular, objective, arm, base, illuminator, condenser, lever for iris diaphragm of condenser, stage, stage clip, coarse adjustment knob, fine adjustment knob, nosepiece

TABLE 3-2	Characteristics of My Microscope	
Characteristic	**Description**	**Function**
Code		identification of my microscope
Light source		
Condenser		
Stage		
Focusing knobs		
Objectives		
Ocular(s)		

© Cengage Learning 2013

MATERIALS

Per student:

- compound light microscope
- lens paper
- lint-free cloth (optional)
- unlabeled diagram of the compound light micro-scope model used in your course (optional)
- prepared slide with a whole mount of stained diatoms
- Wright-stained smear of mammalian blood
- prepared slide with the mounted letter "e"
- index card
- prepared slide with crossed colored threads coded for thread order
- prepared slide with a section of the mammalian kidney
- prepared slide with unstained fibers

Per student group (4):

- bottle of lens-cleaning solution (optional)
- dropper bottle of immersion oil (optional)

Per lab room:

- labeled chart of a compound light microscope

PROCEDURE

A. Care of a Compound Light Microscope

1. To carry a microscope to and from your lab bench, grasp the arm (Figure 3-1) with your dominant hand and support the **base** (Figure 3-1) with the other hand, always keeping the microscope upright. If the arm is wide, there should be a carrying indent or something similar to help you grasp it. *Do not try to carry any-thing else at the same time.* Label the arm and the base on Figure 3-2 or on the diagram given to you by your instructor.

> **CAUTION**
>
> Never wipe a glass lens with anything other than lens paper.

2. Remove the dust cover and clean the exposed parts of the optical system. Blow off any loose dust that may be on the ocular and then gently brush off any remaining dust with a piece of lens paper.

   If the lens of the ocular is still dirty, breathe on it and gently polish it with a rotary motion using a fresh piece of lens paper. If the ocular lens is still dirty, and *with your instructor's approval*, clean it with a piece of lens paper moistened with lens-cleaning solution.

3. Always remember that your microscope is a precision instrument. Never force any of its moving parts.

4. It is just as difficult to see clearly through a dirty slide as through a dirty microscope. Clean dirty slides with a lint-free cloth or with lens paper before using.

5. At the end of an exercise, make sure the last slide has been removed from the stage and *rotate the nosepiece so that the low-power objective is in the light path.* If your instrument focuses by moving the body tube, turn the coarse adjustment so that it is racked all the way down. If your microscope has an electric cord, neatly fold it up on itself and tie it with a plastic strap or rubber band. Otherwise, wind the cord around the base of the arm of the microscope.

6. Replace the dust cover before returning your microscope to the cabinet.

B. Parts of the Compound Light Microscope

Now that you know how to care for your microscope, remove the instrument assigned to you from the cabinet and place it on your lab bench. Use Figure 3-1 and the chart on the wall of your lab room to identify the various parts of your microscope. Read each step below and manipulate the parts *only where indicated*. Before you start, make sure the low-power objective is in the light path.

1. *Light source.* A compound microscope uses transmitted light to illuminate a transparent specimen usually mounted on a glass slide. Newer microscopes have a built-in **illuminator** (Figure 3-1). Locate the illuminator; the *off/on switch*; and perhaps also a rheostat, which is used to vary the intensity of the light. On some models, the switch and the rheostat are combined. Turn on the light source and look through the ocular. If the illuminator has a rheostat, adjust the intensity so the light is not too bright.

   (a) Label the illuminator on Figure 3-2 or on your instructor's diagram.

   (b) Describe the light source in Table 3-2, and then state its function.

2. *Condenser.* For maximum resolving power, a **condenser**—with a *condenser lens* and *iris diaphragm*—focuses the light source on the specimen so that each of its points is evenly illuminated. **The lever for the iris diaphragm** of the condenser is used to open and close the condenser. Instead of a lever, there may be a rotating ring.

   Establish whether there is a condenser adjustment knob (Figure 3-1) to set the height of the condenser. *Do not turn the knob; you will learn how to use it later.*

   There may be a *filter holder* under the condenser with a blue or frosted glass disk. Many microscope manufacturers believe that blue light is more pleasing to the eye because when used with an incandescent bulb, it produces a color balance similar to daylight conditions. Also, theoretically at least, blue light gives better resolving power because of its shorter wavelength. The frosted glass disk scatters light and can be useful in producing even illumination at low magnifications.

   (a) Label the condenser and lever for the iris diaphragm of the condenser on Figure 3-2 or on your instructor's diagram.

   (b) Describe the condenser and its related parts in Table 3-2 and then state its function.

3. *Stage.* Either a pair of *stage clips* or a *mechanical stage* holds a specimen mounted on a glass slide in place suspended over a central hole (Figure 3-1).

   (a) If your microscope has a mechanical stage, skip to part b. If your microscope has stage clips, place a prepared slide of stained diatoms under their free ends. Never remove the stage clips because they make it easier to move a slide in small increments. Skip b and do part c next.

   (b) Position a prepared slide of stained diatoms on the stage by releasing the tension on the spring-loaded movable arm of the mechanical stage (Figure 3-3a). There are two knobs to the right or left of the stage: one to move the specimen forward and backward and the other to move it

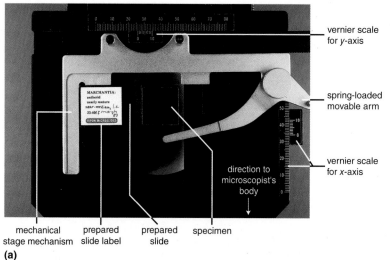

mechanical stage mechanism   prepared slide label   prepared slide   specimen

vernier scale for *y*-axis

spring-loaded movable arm

direction to microscopist's body

vernier scale for *x*-axis

(a)

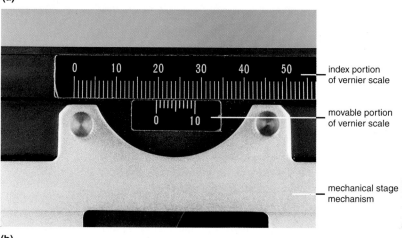

index portion of vernier scale

movable portion of vernier scale

mechanical stage mechanism

(b)

Figure 3-3 (a) Mechanical stage. (b) Vernier scale on mechanical stage of compound light microscope. The correct reading is 19.6 mm.

Photos by J. W. Perry.

laterally (Figure 3-1). Label the stage and stage clips (or mechanical stage) on Figure 3-2 or on your instructor's diagram.

(c) On most mechanical stages, each direction has a vernier scale so you can easily locate interesting fields again and again. A vernier scale consists of two scales running side by side, a long one in millimeters and a short one that is 9 mm in length and divided into 10 equal subdivisions. To take a reading, note the whole number on the long scale coinciding with or just below the zero line of the short scale. If the whole number of the long scale and the zero of the short scale coincide, the first place after the decimal point is zero. Otherwise, the first place after the decimal point is the value of the line on the short scale that coincides (or nearly coincides) with one of the next nine lines after the whole number on the long scale. For example, the correct reading of the vernier scale in Figure 3-3b is 19.6 mm.

(d) Describe the stage and its related parts in Table 3-2 and then state its function.

4. *Focusing knobs* (Figure 3-1). The **coarse-focus adjustment knob** is for use with the lower-power objectives, and the **fine-focus adjustment knob** is for critical focusing, especially with the higher power objectives. On most modern microscopes, you move the stage of the instrument up and down to focus the specimen. Modern microscopes usually have a *preset focus lock*, which stops the stage at a particular height. After setting this lock, you can lower the stage with the coarse focus knob to facilitate changing of the specimen and then raise it to focusing height without fear of colliding the specimen against the objective. There may also be a *focus tension adjustment knob*, usually located inside of the left-hand coarse focus knob.

(a) Turn the coarse focus knob. Do you turn the knob toward you or away from you to bring the slide and objective closer together?

_____

(b) Label the coarse and fine-focus adjustment knobs on Figure 3-2 or on your instructor's diagram.

(c) Describe the focusing knobs in Table 3-2 and then state their function. As part of your description note if a preset focus lock or focus adjustment knob is present.

5. *Objectives*. The compound light microscope has at least two magnifying lenses, the objective and the ocular (Figure 3-1). The objective scans the specimen. Most microscopes have several objectives mounted on a revolving **nosepiece**. The magnifying power of each objective is labeled on its side. Usually included are these objectives: a 4× *low-power* or scanning, a 10× *medium-power* (Figure 3-4a), an about 40× *high-dry*, and perhaps an about 100× *oil-immersion objective* (Figure 3-4b). The other number often labeled on the side of nosepiece objectives is the **numerical aperture** (NA). The larger the NA is, the greater the resolving power and useful magnification.

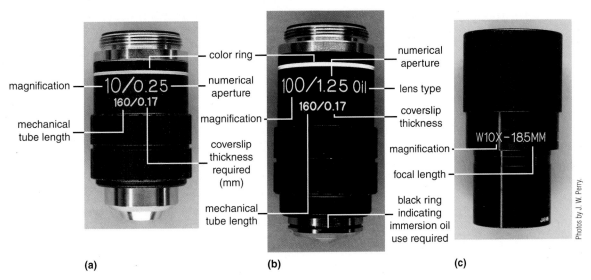

Figure 3-4 (a) 10× objective removed from microscope. (b) 100× oil-immersion objective removed from microscope. (c) Ocular removed from microscope.

In a properly aligned microscope, the objectives are **parfocal**. That is, when an objective has been focused, you can rotate to another one, and the image will remain in coarse focus, requiring only slight movement of the fine-focus knob. Objectives are also **parcentral**, meaning that the center of the field of view remains about the same for each objective. The **field of view** is the circle of light you see when looking into the microscope.

Often, objectives have different lengths; the lower-power objectives are shorter than the higher-power ones. That is, the **working distance** of objectives decreases with magnification. Working distance is the space between the objective lens and the slide. Therefore, the higher the power of the objective in use, the closer the objective is to the slide—and the more careful you must be.

(a) Record the magnifying power and NA of the objectives on your microscope in Table 3-3. If your instrument does not have a particular objective, indicate that it is not present (NP).

TABLE 3-3	Objectives Present on My Compound Light Microscope		
Objective	Objective Magnifying Power (ObMP×)	Total Magnifying Power (ObMP × OcMP = ___ ×)	Numerical Aperture (NA)
Low-power			
Medium-power			
High-dry			
Oil-immersion			

© Cengage Learning 2013

(b) Label the nosepiece and objective on Figure 3-2 or on your instructor's diagram.
(c) Describe the objectives in Table 3-2 and then state their function.

6. *Ocular.* The magnifying lens you look into is called an **ocular** (Figure 3-4c). Oculars are generally 10×.
(a) Because each objective has a different magnifying power, the total magnification is calculated by multiplying the magnifying power of the ocular by that of the objective in use. What is the ocular magnification power (OcMP) of the ocular(s) on your microscope?
_____ ×
Calculate the total magnification for each of your microscope's ocular and objective combinations and then record them in Table 3-3.
(b) Label the ocular on Figure 3-2 or on your instructor's diagram.
(c) Describe the ocular in Table 3-2 and then state its function.
(d) Your microscope will have one or two oculars mounted *on a monocular or binocular head*, respectively. There may be a pointer mounted in an ocular so that you can easily show a specimen detail to your instructor or another student. For a monocular microscope, it's best to use your dominant eye to look down the ocular, keeping your other eye open. Is your microscope monocular or binocular?
_____
(e) If your microscope is monocular, determine your dominant eye.
   ■ Look at a small object on the far wall of your room with both eyes open.
   ■ Form your thumb and index finger of one hand into a circle and place this circle in your line of sight at arm's length so it surrounds the object.
   ■ Close your right eye. If the object shifts out of the circle to your left, your right eye is probably dominant. If the object remains in the circle, your left eye is probably dominant.
   ■ This time, close your left eye and go through the process again. If the object shifts to the right, your left eye is dominant. If the object remains within the circle, your right eye is dominant. The more pronounced the shift, the greater the dominance. If there is no shift, neither eye is dominant.

C. Aligning a Compound Light Microscope with In-Base Illumination and a Condenser with an Iris Diaphragm

Aligning your microscope properly will not only help you see specimen detail clearly but will also protect your eyes from strain.

1. Rotate the nosepiece until the medium-power objective is in the light path. Open the iris diaphragm.
2. If it is not already there, place the prepared slide of stained diatoms on the stage; center and carefully focus on it. *Skip steps 3 and 4 if your microscope is monocular. Skip step 5 if your microscope does not have a control to adjust the height of the condenser.*
3. If your microscope is binocular, adjust the interpupillary distance. Hold a different ocular tube with each hand and while looking at the specimen, pull the tubes apart or push them together until you see one field of view. After making this adjustment, read and record the number off the scale.
My interpupillary distance is _____.
From now on, you can set the interpupillary distance at this number.

4. Now compensate for any difference in diopter between the lenses of each eye.
   (a) *If there is one diopter adjustment ring around the left ocular tube,* cover your left eye with an index card and focus your microscope using the fine-focus knob. Now uncover the left eye and cover your right one. Use the diopter adjustment ring to bring the specimen into focus.
   (b) *If both ocular tubes have a diopter adjustment ring,* set the left one to the same number as the interpupillary distance, cover your right eye with an index card, and focus on the specimen. Then uncover the right eye and cover your left one. Use the diopter adjustment ring on the right ocular tube to bring the specimen into focus.
5. Place a sharp point (pencil, dissecting needle, or some similar object) on top of the illuminator and bring the silhouette into sharp focus by adjusting the height of the condenser.

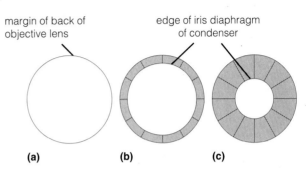

   (a) *If the ocular on your microscope is removable (and with the permission of your instructor),* carefully slide it out and put the ocular open end down on a piece of lens paper in a safe place. Then, while looking down the ocular tube, adjust the iris diaphragm until the edge of the aperture lies just inside the margin of the back lens element of the objective (Figure 3-5). When done, replace the ocular.
   (b) *If the ocular cannot be removed,* close the condenser diaphragm and then open it until there is no further increase in brightness. Now close it again, stopping when you see the brightness begin to diminish.
   (c) Models that have rings that open and close the iris diaphragm of the condenser usually have numbers associated with them that

Figure 3-5 Correct setting for condenser iris diaphragm. Drawing (**b**) is correct. In drawing (**a**), you cannot see the edge of the iris diaphragm. In (**c**), the diaphragm has been closed too much.

   match the magnification of the objective being used. Every time you switch objectives, you need to reset the ring to the correct position.
6. If your microscope has a rheostat, adjust the illumination to a level that lets you see specimen detail and that is comfortable for your eyes. To maintain the same illumination at higher magnifications, you will have to increase its intensity.
7. For best results, repeat steps 5 and 6 every time you use a different objective.

D. Using Different Magnifications

It is safest to observe a specimen on a slide first with low power and then, step by step, with higher power objectives. This way you avoid colliding the objective and slide together.

Because the magnification is in diameters, the area of the field of view decreases dramatically with increasing magnification (Figure 3-6). It follows that it is easier to use a lower power objective to locate a specific specimen detail. This is why the low-power objective is sometimes called the *scanning objective.* Also, if you lose a specimen detail at higher magnification, it is always easier to find it again if you switch to a lower power objective.

Now follow these steps to use each objective.

1. Rotate the low-power objective into the light path.
2. If it is not already there, place a prepared slide of stained diatoms on the stage, securing it with either the stage clips or the movable arm of the mechanical stage.
3. Look through the ocular. Bring the diatoms into focus using the *coarse focus knob.* Adjust the illumination as described in steps 5 and 6 of Section C. At this magnification, the diatoms appear small. Center a diatom by moving the slide.
4. Rotate the nosepiece so the medium-power objective is in the light path. Adjust the illumination. Focus the diatom.
5. Rotate the nosepiece so the high-dry objective is in the light path. Adjust the illumination. Focus the diatom using the *fine-focus adjustment knob.*

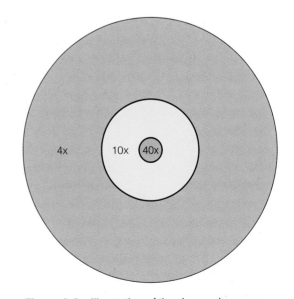

Figure 3-6 Illustration of the decreasing area of the field of view when a 4×, 10×, and 40× objective is used with a 10× ocular. The actual area of each circle has been enlarged 7.5×.

6. If your microscope has an oil-immersion objective and *with the permission of your instructor*, replace the prepared slide of diatoms with a smear of mammalian blood and repeat steps 1–5, otherwise skip steps 6–11 and just remove the prepared diatom slide from the stage. Center and focus on a prominent blue nucleus of a white blood cell. Most of the cells are red blood cells stained a light pink.

7. Rotate the nosepiece so that the light path is midway between the high-dry and oil-immersion objectives.

8. Place a small drop of immersion-oil on to the cover slip, using the circle of light above the specimen as a guide. Rotate the nosepiece so that the oil-immersion objective is in the oil.

9. Adjust the illumination and focus the white blood cell nucleus with the fine-focus knob.

10. When you are done examining the white blood cell nucleus, rotate the oil-immersion objective out of the light path, carefully wiping the oil from the oil-immersion objective with lens paper.

11. Remove the slide from the stage and wipe the cover slip with lens paper.

E. Orientation of the Image Compared with the Specimen

1. If you have not already done so, remove the prepared slide of diatoms and replace it with a prepared slide with the letter *e*. With the medium-power objective in the light path, position the slide with the specimen (the letter *e*) right side up on the stage. Center the *e* in the field of view and carefully bring it into focus.

2. In Figure 3-7, draw the image of the *e* as you see it through the ocular. Record the total magnification used in the line at the end of the legend.
   Is the image right side up or upside down compared with the specimen?

   _____

   Compared with the specimen, is the image backward as well as upside down?
   (yes or no) _____
   In summary, the image is *inverted* with respect to the specimen.

3. Move the specimen to the right while watching it through the microscope. In which direction does the image move?

   _____

4. Move the specimen away from you. In which direction does the image move?

   _____

5. Remove the slide and put it away.

Figure 3-7 Drawing of the letter *e* as seen through the ocular (_____ ×).

F. Depth of Field

The **depth of field** is the distance through which you can move the specimen and still have it remain in focus. Remember, the working distance—the space between the objective lens and the coverslip—decreases with increasing magnifying power. Therefore, the higher the power of the objective in use, the closer the objective is to the slide—and the more careful you must be.

1. Obtain a prepared slide of three crossed colored threads. *This exercise requires care because you are probably not yet adept at focusing on a specimen.* When you have the threads in focus (using first the low-power objective and then the medium-power objective), you need only use the fine-focus knob to focus with the high-dry objective. After switching to the high-dry objective, try rotating the fine-focus knob ½ turn away from you and then a full turn toward you. If you have not found the plane of focus, next try 1½ turns away from you and 2 full turns toward you and so on. If you work deliberately, you will find the plane of focus and will not crack the coverslip.
   (a) How many threads are in focus using the
       low-power objective? _____
       medium-power objective? _____
       high-dry objective? _____
   (b) With which objective is it easiest to focus a specimen? _____
   (c) At which magnification is it most difficult to focus a specimen? _____

2. Specimens have depth. Continue using the prepared slide of three crossed colored threads.
   (a) Use the high-dry objective to determine the order of the three threads mounted on the slide and record the results in Table 3-4. (Each slide label has a code on it. When you believe you have discovered the correct order, check with your instructor to find out if you are correct.)
   (b) Focusing carefully with the fine-focus knob, move from the bottom to the upper thread. Did you move the knob away from or toward you? _____

© Cengage Learning 2013

TABLE 3-4 Order of Threads

Location	Color
Closest to slide	
Middle	
Closest to coverslip	

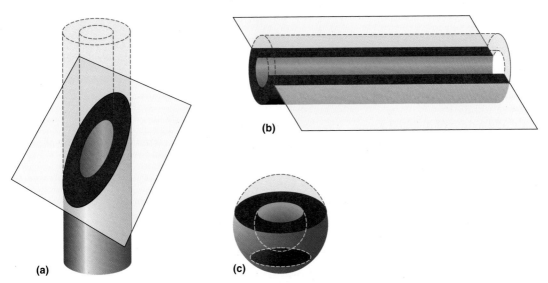

Figure 3-8 Difficulties encountered in interpreting the three-dimensional shape of objects from sections. Compare the transverse (cross) section of the cylinder with the distorted oblique section (**a**) and the longitudinal section (**b**). A similar shape to the transverse section of a cylinder results when a hollow ball is sectioned (**c**). A section through the wall of the hollow ball results in a solid shape.

3. Remove and put away the slide.
4. Viewing sections of three-dimensional structures makes interpretation of the original shape quite difficult.
    (a) Examine a slide of the cortex of the mammalian kidney with your compound microscope (Figure 3-8). Hypothesize as to the three-dimensional shape of the structures labeled *renal corpuscles* and *nephron tubules.*
        The shape of renal corpuscles is _____.
        The shape of nephron tubules is _____.
    (b) Check the textbook to see if your hypotheses are acceptable.

G. Using the Iris Diaphragm to Improve Contrast

1. Place a specimen of unstained fibers on the stage. Locate and focus on these fibers using the medium-power objective. Make sure the condenser and iris diaphragm are correctly set.
2. Close the iris diaphragm.
    Does this procedure increase or decrease contrast? _____
    Although this procedure is useful when viewing specimens with low contrast, it should be used only as a last resort because resolving power is also decreased.
3. Remove and put away the slide.

H. Units of Measurement

The basic metric unit of length at the light-microscopic level is the micrometer (μm).

1000 μm = 1 mm
1 μm = 0.001 mm

How many nanometers are there in 1 mm? _____ nm
How many millimeters are there in 1 nm? _____ mm

In the mid-seventeenth century, Robert Hooke used a microscope to discover tiny, empty compartments in thin shavings of cork. He named them cells. Repeating this historic observation is a good way to learn how to prepare a wet mount.

MATERIALS

Per student:

- compound microscope, lens paper, a bottle of lens-cleaning solution (optional), a lint-free cloth (optional)
- cork
- razor blade

- glass microscope slide
- glass coverslip
- dissecting needle

Per student group (4):

- dropper bottle of distilled water (dH₂O)

PROCEDURE

1. Carefully use a razor blade to cut a number of *very thin shavings* from a cork stopper. Place them on a glass microscope slide.
2. Gently add a drop of distilled water.
3. Place one end of a glass coverslip to the right or left of the specimen so that the rest of the slip is held at a 45-degree angle over the specimen (Figure 3-9a).

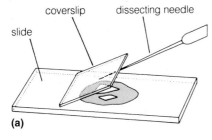
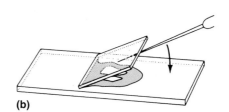

Figure 3-9   How to make a wet mount.

4. Slowly lower the coverslip with a dissecting needle so as not to trap air bubbles (Figure 3-9b).
5. Observe the wet mount, first at low magnification and then with higher power. Air may be trapped either in the cork or as free bubbles (Figure 3-10). Trapped air will appear dark and refractive around its edges. This effect is caused by sharply bending rays of light. Draw what you see in Figure 3-11. Note the total magnification used to make the drawing.
6. Clean and replace the slide and coverslip as indicated by your instructor.

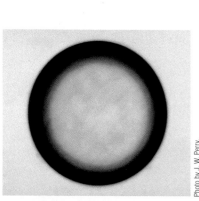

Figure 3-10   Free air bubble (250×).

Figure 3-11   Drawing of the microscopic structure in a cork shaving (_____ ×).

Examining the microscopic world is both challenging and fun. Most of the macroscopic world has been explored, but the microscopic world is barely touched. Yet the microbes in it are essential to our very existence. So be an explorer and see what you can discover!

MATERIALS

Per student:

- compound microscope, lens paper, a bottle of lens-cleaning solution (optional), a lint-free cloth (optional)
- glass microscope slide
- glass coverslip

Per student group (4):

- pond water or some other mixed culture in a dropper bottle
- dropper bottle of Protoslo®

Per lab room:

- reference books for the identification of microorganisms

PROCEDURE

1. Obtain a drop of pond water or other mixed culture from the bottom of the bottle.
2. Add a drop of Protoslo®. This methyl cellulose solution slows down any swimming microorganisms.
3. Make a wet mount (Figure 3-9).
4. Observe the wet mount with your compound microscope. Start at the upper left corner of the coverslip and scan the wet mount with the low-power objective. When you find something interesting, focus on it and switch to the medium-power objective and then, if necessary, the high-dry objective.
5. Draw what you find on Figure 3-12 and note the total magnification.
6. Attempt to identify what you found using the resource books provided by your instructor. If successful, write its name under your drawing.
7. Clean and replace the slide and coverslip as indicated by your instructor.
8. Put away your compound microscope as described on page 27.

Figure 3-12 Drawing of microorganisms (_____ ×).

| **3.4** | **Dissecting Microscope (About 10 min.)** |

Dissecting microscopes (Figure 3-13) have a large working distance between the specimen and the objective lens. They are especially useful in viewing larger specimens (including thicker slide-mounted specimens) and in manipulating the specimen (e.g., when dissection of a small structure or organism is required).

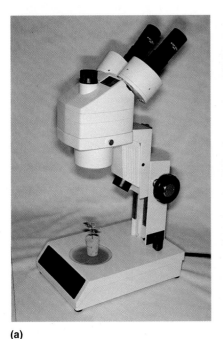

(a)

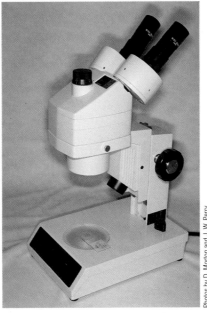

(b)

Figure 3-13 Dissecting microscope (**a**) reflected light, (**b**) transmitted light.

Photos by D. Morton and J. W. Perry.

The large working distance also allows for illumination of the specimen from above (reflected light) as well as from below (transmitted light). Light reflected from the specimen shows surface features better than transmitted light.

MATERIALS

Per student group:

- dissecting microscope
- specimens appropriate for viewing with the dissecting microscope (e.g., a prepared slide with a whole mount of a small organism, bread mold, an insect mounted on a pin stuck in a cork, a small flower)

PROCEDURE

1. Under a dissecting microscope, view one or more of the specimens provided by your instructor. What is the magnification range of this microscope?

   _____ × to _____ ×

2. Is the image of the specimen inverted as in the compound light microscope? (yes or no) _____
3. Describe the type of illumination used by your dissecting microscope. Is there a choice?

   _____

   _____

3.5	Other Microscopes *(About 10 min.)*

In future exercises, you will examine pictures taken with other types of microscopes. Some will be of living cells taken with a phase-contrast microscope (Figure 3-14a) or similar instrument, including those using the Nomarski process (Figure 3-14b). Others will be of very thin-sectioned, heavy metal–stained specimens taken with a transmission electron microscope or TEM (Figure 3-14c). Still others will be of precious metal–coated surfaces produced by signals from the scanning electron microscope or SEM (Figure 3-14d). Table 3-5 summarizes the technology and use of these microscopes.

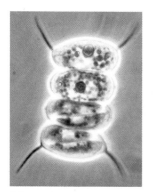  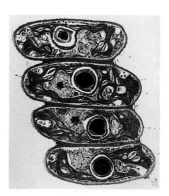 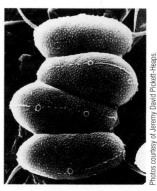

(a) phase contrast   **(b)** Nomarski process   **(c)** transmission electron   **(d)** scanning electron

Photos courtesy of Jeremy David Pickett-Heaps.

Figure 3-14   How different types of microscopes reveal detail in cells of the green alga *Scenedesmus*.

TABLE 3-5	Other Microscopes	
Microscope	**Technology**	**Use**
Phase-contrast and Nomarski process	Converts phase differences in light to differences in contrast	Observation of low-contrast specimens (often living)
Transmission electron	Increases resolving power by using electrons in a vacuum and magnetic lenses instead of light and glass lenses, respectively	Preservation of greater specimen detail allows for magnifications up to 1,000,000× or more (usually dead materials)
Scanning electron	Forms a TV-like picture from a secondary electron signal, which is emitted from surface points excited by a thin beam of electrons drawn across	Investigation of the fine structure of surfaces (usually dead specimens); the surface in a raster pattern

© Cengage Learning 2013

MATERIALS

Per student group:

- photographs of TEM micrographs (negatives) or digital images
- photographs of SEM negatives or digital images

PROCEDURE

1. Examine some photographs of TEM micrographs (negatives) or digital images. The darker areas are more electron dense in the specimen than the lighter areas.
2. Now look at some photographs of SEM negatives or digital images. The lighter areas correspond to the emission of greater numbers of secondary electrons from that part of the specimen; the darker areas emit fewer secondary electrons.
3. What type of microscope (compound light, dissecting, phase-contrast, TEM, or SEM) would you use to examine the specimens listed in Table 3-6?

TABLE 3-6	**Microscope Use**	
Specimen	**Microscope**
Living surface of the finger	
Dye-stained slide of a section of the finger	
Gold-coated bacteria on a single cell of the finger	
Unstained section of a biopsy from the finger	
Heavy metal–stained, very thin section of the finger	

© Cengage Learning 2013

_____ 1. Magnification
    (a) is the amount that an object's image is enlarged.
    (b) is the extent to which detail in an image is preserved during the magnifying process.
    (c) is the degree to which image details stand out against their background.
    (d) focuses light rays emanating from an object to produce an image.

_____ 2. Resolving power
    (a) is the amount that an object's image is enlarged.
    (b) is the extent to which detail in an image is preserved during the magnifying process.
    (c) is the degree to which image details stand out against their background.
    (d) focuses light rays emanating from an object to produce an image.

_____ 3. A lens
    (a) is the amount that an object's image is enlarged.
    (b) is the extent to which detail in an image is preserved during the magnifying process.
    (c) is the degree to which image details stand out against their background.
    (d) focuses light rays emanating from an object to produce an image.

_____ 4. Contrast
    (a) is the amount that an object's image is enlarged.
    (b) is the extent to which detail in an image is preserved during the magnifying process.
    (c) is the degree to which image details stand out against their background.
    (d) focuses light rays emanating from an object to produce an image.

_____ 5. The maximum useful magnification for a light microscope is about
    (a) $100\times$.
    (b) $1000\times$.
    (c) $10,000\times$.
    (d) $100,000\times$.

_____ 6. The two image-forming lenses of a compound light microscope are
    (a) the condenser and objective.
    (b) the condenser and ocular.
    (c) the objective and ocular.
    (d) none of these choices

_____ 7. Dyes are usually added to sections of biological specimens to increase
    (a) resolving power.
    (b) magnification.
    (c) contrast.
    (d) all of the above

_____ 8. If the magnification of the two image-forming lenses are both $10\times$, the total magnification of the image will be
    (a) $1\times$.
    (b) $10\times$.
    (c) $100\times$.
    (d) $1000\times$.

_____ 9. The distance through which a microscopic specimen can be moved and still have it remain in focus is called the
    (a) field of view.
    (b) working distance.
    (c) depth of field.
    (d) magnification.

_____ 10. Electron microscopes differ from light microscopes in that
    (a) electrons are used instead of light.
    (b) magnetic lenses replace glass lenses.
    (c) the electron path has to be maintained in a high vacuum.
    (d) a, b, and c are all true.

EXERCISE **3**

Microscopy

3.1 Compound Light Microscope

1. What is the function of the following parts of a compound light microscope?
   (a) condenser lens

   (b) iris diaphragm

   (c) objective

   (d) ocular

2. In order, list the lenses in the light path between a specimen viewed with the compound light microscope and its image on the retina of the eye.

3. What happens to contrast and resolving power when the aperture of the condenser (i.e., the size of the hole through which light passes before it reaches the specimen) of a compound light microscope is decreased?

4. What happens to the field of view in a compound light microscope when the total magnification is increased?

5. Describe the importance of the following concepts to microscopy.
   (a) magnification

   (b) resolving power

   (c) contrast

6. Which photomicrograph of unstained cotton fibers was taken with the iris diaphragm closed? ___

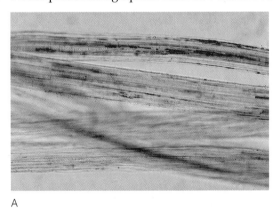
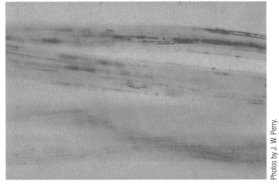

A   B

7. Describe how you would care for and put away your compound light microscope at the end of the lab.

3.2 How to Make a Wet Mount

8. Describe how to make a wet mount.

3.5 Other Microscopes

9. A camera mounted on a _____ microscope took this photo of a cut piece of cork.

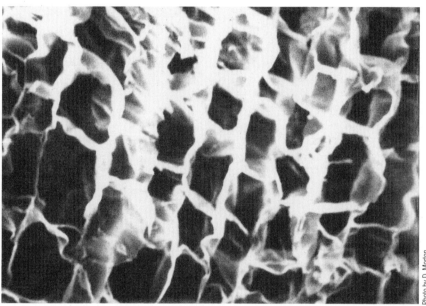

(515×)

Food for Thought

10. Why were humans unaware of microorganisms for most of their history?

Homeostasis

After completing this exercise, you will be able to

1. define *homeostasis, intracellular fluid, extracellular fluid, plasma, interstitial fluid, homeostatic mechanisms, pH scale, neutral solution, alkaline solution, acidic solution, buffer, systolic blood pressure, diastolic blood pressure, sounds of Korotkoff;*

2. explain the logarithmic nature of the pH scale;

3. describe the role of sensory receptors, integrating centers, and effectors in homeostatic mechanisms using the regulation of body temperature and body fluid volume as examples;

4. use a sphygmomanometer and stethoscope to estimate blood pressure;

5. compare negative and positive feedback and feedforward homeostatic mechanisms.

Introduction

The world outside ourselves (our external environment) is constantly changing. For example, consider the seasonal and daily range of air temperature. However, the body must maintain an internal environment that is very constant, not only in its temperature but in all of its other functional and structural characteristics (variables) as well. Maintenance of a stable internal environment is called **homeostasis.**

The internal environment is mostly fluid. An alien in a science fiction story once described us as "dirty bags of water." Body fluid plays a central role in homeostasis. Its major compartments are summarized in Table 4-1. Body fluid is divided into two major compartments: (1) fluid inside of cells, or **intracellular fluid**, and (2) body fluid external to cells, or the **extracellular fluid**. The extracellular fluid is further subdivided into the fluid portion of blood or **plasma** and into the body fluid outside of blood vessels or **interstitial fluid.**

TABLE 4-1	**Fluid Compartments**		
Body Fluid	**Definition**	**Function**	**Average Volume in the Human Body (L)**
Intracellular	All of the fluid in cells	Contains dissolved substances	29
Interstitial	All of the fluid between the cells and the blood vessels	Diffusion of dissolved substances between cells and blood	12
Plasma	All of the fluid portion of the blood	Transport of dissolved substances throughout body	3

© Cengage Learning 2013

4.1 Homeostatic Mechanisms *(About 20 min.)*

Homeostatic mechanisms function throughout the body to keep the chemistry of the body fluids constant and to maintain the various structures of the body.

MATERIALS

Per student:

- text

PROCEDURE

A. Vital Functions of Organ Systems

Each organ system has functions vital to homeostasis. Use your textbook to complete Table 4-2.

TABLE 4-2	Organ Systems of Mammals
Systems	**Vital Functions**
Integumentary	
Nervous	
Endocrine	
Skeletal	
Muscular	
Circulatory	
Lymphatic	
Respiratory	
Digestive	
Urinary	
Reproductive	

© Cengage Learning 2013

B. Regulation of Body Temperature

All homeostatic mechanisms involve feedback loops (Figure 4-1). A typical feedback loop consists of three steps. First, a **sensory receptor** receives a stimulus—a rise or fall in body temperature, for example. In the case of temperature regulation, temperature-sensitive receptors are located in the skin and in the brain. Second, the sensor relays the information that a change has occurred to an **integrating center**. This control center for temperature regulation is also located in the brain. An integrating center compares the intensity of a stimulus with a preset value. Third, the integrating center activates an **effector**, which produces a response that reverses the direction of the original change back toward the preset value. Effectors are usually muscles or glands. Normally, over time, variables such as body temperature are maintained by homeostatic mechanisms within a range of values. The midpoint of the range is the preset point (Figure 4-2).

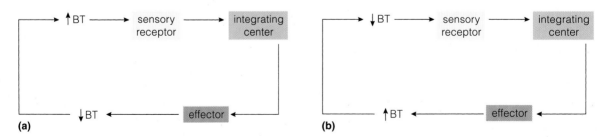

Figure 4-1   Homeostatic mechanisms activated by (**a**) a rise in body temperature (BT) and (**b**) a fall in body temperature.

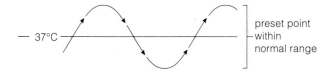

preset point
within
normal range

Figure 4-2 Body temperature trace (*line with arrowheads*) over several seconds showing that it stays within its normal range due to the action of homeostatic mechanisms.

1. From your own experience, identify one effector that is activated when body temperature gets too hot—and its response to cool the body down. Also identify the organ system to which the effector belongs.

   Effector

   _____

   Response

   _____

   Organ system

   _____

2. From your own experience, describe one effector that is activated when temperature gets too cold—and its response to warm up the body. Also identify the organ system to which the effector belongs.

   Effector

   _____

   Response

   _____

   Organ system

   _____

4.2   Regulation of pH (*About 35 min.*)

The **pH scale** (Figure 4-3) measures the concentration of chemically active hydrogen ions (H^+) in solutions. It is logarithmic in nature, and each full unit is 10 times more or less than the next full unit. A **neutral solution** has a pH of 7 and depending on the temperature has or almost has an equal number of H^+ and OH^- (hydroxyl) ions. Pure water and intracellular fluid are examples of neutral solutions. Extracellular fluid is a little more alkaline at a pH of 7.3 to 7.5. An **alkaline solution** (basic solution) has a pH higher than 7, which means a lower concentration of H^+. An **acidic solution** has a pH lower than 7 and a higher concentration of H^+.

The regulation of the pH of our bodies involves homeostatic mechanisms at two levels—molecular and organ system (respiratory and excretory). Molecules of some of the compounds dissolved in our body fluids minimize pH changes because of the addition of acids (H^+ donors) or bases (H^+ acceptors). A compound that does this is called a **buffer.**

MATERIALS

Per student group (4):

- plastic tray
- five 100-mL beakers labeled, respectively, W, M, OJ, D, and V
- 1-mL pipet and safety bulb or filling device
- glass stirring rod
- wide-range pH test paper and dispenser with scale
- pencil

Per lab room:

- 500-mL container of nonphosphate detergent solution (¼ teaspoon/500 mL dH_2O)
- cartons of milk
- orange juice
- 500-mL bottle of white vinegar

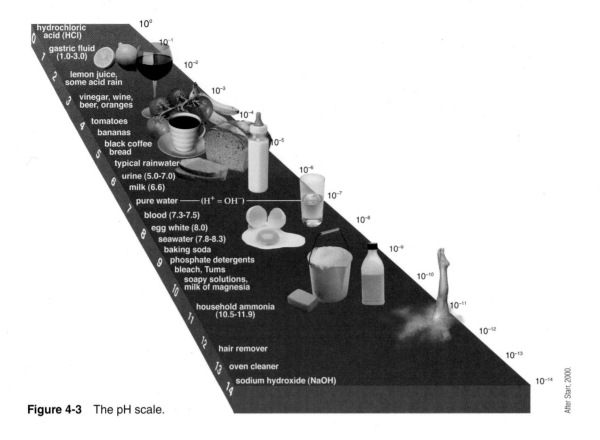

Figure 4-3 The pH scale.

PROCEDURE

1. Pour 20 mL of water, milk, orange juice, detergent solution, and vinegar separately into their own 100-mL beakers, labeled W, M, OJ, D, and V, respectively.
2. Tear off 11 short pieces (about 1 cm in length) of pH indicator paper and label them with a pencil W, W+V, W+2V, W+3V, M, M+V, M+2V, M+3V, OJ, D, and V. Place them on the plastic tray.
3. Dip a clean glass stirring rod into the water and transfer a drop of water to the pH paper labeled W.
4. Wait for the paper to absorb the drop. Find the closest match between the color of the paper and the colored scale of the pH paper dispenser and read the corresponding pH value. Record this value in the second row (Water) of the second column (pH) of Table 4-3.
5. Wash and dry the glass stirring rod.
6. Repeat steps 4 and 5 for the remaining fluids using the appropriately labeled piece of pH paper (M, OJ, D, or V) and record the pH values you read in the third to sixth rows of the second column of Table 4-3.
7. Pipet 1 mL of vinegar into the beaker containing water and stir it with the glass rod to make the W+V solution.

TABLE 4-3	**pH Values of Common Fluids**			
		pH after Addition of:		
Fluid	**pH**	**1 mL of Vinegar**	**2 mL of Vinegar**	**3 mL of Vinegar**
Water				
Milk				
Orange juice				
Detergent				
Vinegar				

© Cengage Learning 2013

8. As before, measure the pH and record it in the second row of the third column of Table 4-3.
9. Pipet another 1 mL of vinegar into the W+V solution and stir with the glass rod to make the W+2V solution.
10. Measure the pH and record it in the second row of the fourth column of Table 4-3.
11. Pipet one last 1 mL of vinegar into the W+2V solution and stir with the glass rod to make the W+3V solution.
12. Measure the pH and record it in the second row of the fifth column of Table 4-3.
13. Wash and dry the glass stirring rod.
14. Repeat steps 7 to 12 for the beaker containing milk (M+V, M+2V, and M+3V) except record the data in the appropriate columns of the third row. The milk will curdle as it sours.
15. Plot the four data points for W, W+V, W+2V, and W+3V and then plot them for milk (M, M+V, M+2V, M+3V) in Figure 4-4.
16. Which fluid, water or milk, is more likely to contain buffers? _____
17. Explain any differences that you observe between the slopes of the two lines in Figure 4-4.

_____

_____

_____

_____

_____

_____

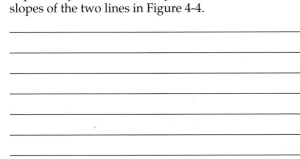

Figure 4-4 Changes in pH due to the addition of vinegar to water and to milk.

4.3 Simulation of Homeostatic Mechanisms Regulating Body Fluid Volume (About 35 min.)

In this section, you will be part of model homeostatic feedback loops that regulate the volume of fluid in the body. Specifically, you will be the sensor, integrating center, and controller of the effectors. Your instructor has prepared a vertically stacked double-buret system, as illustrated in Figure 4-5, for each group of students. The lower buret has been half-filled with alkaline buffer solution, containing the pH indicator phenol red. Imagine that this solution is blood.

MATERIALS

Per student group (4):
- vertically stacked double-buret system
- 1-L Erlenmeyer flask of water
- empty 1-L Erlenmeyer flask

Per lab room:
- 1-L container of red alkaline buffer solution (pH 10) containing phenol red
- 1-L container of colorless solution acid buffer (pH 6)

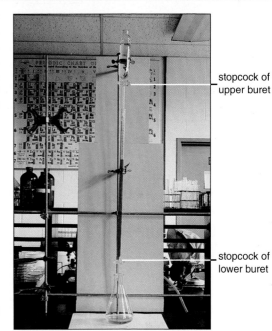

Figure 4-5 Double-buret apparatus used to simulate homeostasis.

PROCEDURE

1. Fill about three-quarters of the reservoir of the upper buret with water. Imagine that this is drinking water, which, after you drink it, adds to the body fluid volume.
2. Pour 100 mL of acid buffer solution into a 1-L Erlenmeyer flask and place it under the lower buret. Imagine its contents are urine.
3. In the body, your kidneys function to filter the blood to make urine. This is a continuous process that results in the obligatory loss of fluid from the body. To simulate this fluid loss, turn the stopcock of the lower buret so that it slowly drips into the flask at about 24 drops per minute.
4. Adjust the stopcock of the upper buret to keep the fluid (blood) level in the lower buret constant. When the dripping of both burets is stable, read and record on the line below the number on the side of the lower buret that corresponds to the fluid level. _____ This number represents the normal value for body fluid volume with water intake equal to that lost in urine.
5. On a hot day, water is lost from the body during temperature regulation and thus is not available for urine formation. Your kidneys produce a smaller amount of more concentrated urine. To simulate this situation, reduce the drop rate from the upper buret to 12 drops per minute. Readjust the drop rate of the lower buret to hold body fluid volume constant, albeit at a lower level.
6. Figuratively, take a drink. That is, open the stopcock on the upper buret to return the fluid level in the lower buret to the number that corresponds to the normal body fluid volume. When this level is reached, reduce the drip rate to maintain this level.
7. When you drink excess fluid, the kidneys increase urine production to maintain the correct amount of fluid in the body. To simulate this situation, increase the drip rate from the upper buret. Adjust urine production to return to and maintain normal body fluid volume.
8. When alcoholic beverages are consumed, regulation of kidney function is disturbed. Urine production is higher than it should be until the alcohol is metabolized. Simulate this by slightly increasing the drip rate of the lower buret for 30 seconds. Read and record on the line below the new number for body fluid volume. _____ The morning after excessive alcohol consumption, a person feels dehydrated and thirsty.
9. In carrying out this activity, you used different body parts to simulate the homeostatic feedback loop that maintains water balance. In Table 4-4, match the body part used with the correct listed part of the homeostatic feedback loop. Use each answer only once.

TABLE 4-4	Parts of Homeostatic Feedback Loops
Body Part	**Homeostatic Feedback Loop**
___ Hands	a. Integration center
___ Eyes	b. Effector
___ Brain	c. Sensor

© Cengage Learning 2013

10. When red fluid drops from the lower buret into the flask, it turns yellow, simulating the color of urine. On the lines below, write a hypothesis as to the mechanism that accomplishes this color change.

_____

_____

_____

11. Dispose of solutions as directed by your instructor.

4.4   Blood Pressure *(About 35 min.)*

The force that causes blood to flow through the heart and vessels of the circulatory system is blood pressure. By means of fast-circulating blood, the effects of homeostatic mechanisms are spread first to the interstitial fluid and then to the intracellular fluid throughout the body. Various effectors regulate the blood pressure so that the right amount of blood is delivered to the body organs. For example, blood pressure in the arteries (relatively large vessels delivering blood to capillary beds in organs) is partly controlled by the number of times the heart contracts in a minute (heart rate).

The most common test done when you visit your physician is the measurement of blood pressure in the artery of the upper arm usually with the body at rest in the sitting position. This blood pressure is reported as **systolic blood pressure** (blood pressure when the heart contracts) over diastolic blood pressure (blood pressure when the heart relaxes). The **diastolic blood pressure** is caused by the elastic rebound of the walls of the largest arteries.

The most common unit of blood pressure is the number of millimeters in a column of mercury (mm Hg) needed to produce a force equal to a particular blood pressure. Normal blood pressure at rest for an 18-year-old young woman is 116/72 mm Hg. Normal blood pressure is slightly higher in males and increases with age for both genders. Many homeostatic mechanisms function to maintain normal blood pressure. Values that are significantly different from the average blood pressure for a person's age, height, and weight or from his or her normal blood pressure may indicate a medical problem that needs immediate attention.

The instrument used to measure blood pressure is the **sphygmomanometer** (Figure 4-6). Blood pressure measurements by a sphygmomanometer are approximations of true blood pressure. Tissue composition of the arm, the size of the sphygmomanometer cuff, and operator experience are three factors that can cause inaccurate measurements.

The specific steps taken to measure blood pressure depend on the type of sphygmomanometer used. The following procedure lists the steps for the measurement of blood pressure. Make sure that you read and understand the complete instructions before you begin.

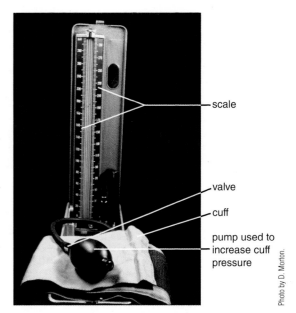

Figure 4-6 Sphygmomanometer with a mercury column to illustrate the basis of blood pressure units (i.e., mm Hg). You will be using a much safer aneroid instrument, but the principle of how it works is the same.

MATERIALS

Per student pair:

- aneroid sphygmomanometer
- stethoscope

> **CAUTION**
>
> Do not leave a fully inflated cuff on a subject's arm for more than a few seconds.

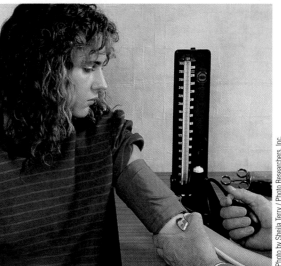

Figure 4-7 Correct use of a stethoscope while taking blood pressure.

PROCEDURE

1. Have your lab partner sit with one arm resting on the lab bench.
2. Wrap the cuff around that arm and position it 2 to 3 cm above the elbow.
3. Feel for the pulse of the brachial artery in the hollow between the muscles of the upper arm, just above the crease of the elbow, and place the membrane end of a stethoscope over this location (Figure 4-7).
4. The sounds you hear as the pressure rises above the diastolic level are called the Korotkoff sounds and are caused by the turbulence created as blood squeezes through the artery. With the valve closed, pump up the cuff pressure to 20 mm Hg above the point where these sounds cease.
5. While watching the scale, open the valve to allow the blood pressure to decrease at a rate of about 2 mm Hg per second.

6. Listen carefully for the muffled return of the Korotkoff sounds. The reading at the instant these sounds are first heard is the systolic blood pressure, and the reading at the instant they disappear is the diastolic blood pressure.
7. Measure and record your lab partner's blood pressure: _____ mm Hg.
8. Have your lab partner measure and record your blood pressure: _____ mm Hg.

| 4.5 | Negative Feedback, Positive Feedback, and Feedforward *(About 10 min.)* |

All of the examples given so far are negative feedback homeostatic mechanisms in that the response reverses the direction of the change. That is, if a change in the value of the variable occurs in either a positive or negative direction away from the preset value, then the value of the variable is turned around back toward the preset point. Buffers help to maintain the correct pH by releasing H^+ when the pH increases and consuming H^+ when the pH decreases. A decrease in body fluid volume is compensated for by producing more concentrated urine and taking a drink. A decrease in blood pressure in an artery delivering blood to an organ or an increase in an organ's activity that requires greater blood flow is compensated partly by an increase in blood pressure elsewhere in the circulatory system.

In a few situations, positive feedback contributes to homeostasis. Positive feedback occurs when the response intensifies the change from a variable's preset point. One example is the increasing contractions of the uterus just before childbirth. All cases of positive feedback are smaller parts of larger negative feedback homeostatic mechanisms. In the case of childbirth, the mother's body is returned to the "not pregnant" condition.

Some other homeostatic mechanisms are feedforward rather than feedback, meaning the response occurs before the change. An example is the small increase in stress hormones that occurs before taking a test. (i.e., a little anxiety is good for your performance, but, of course, too much is detrimental).

TABLE 4-5	Review of Types of Homeostatic Mechanisms	
Homeostatic Mechanism	**Homeostatic Feedback Loop**	**Explanation**
_____ Negative feedback	a. Blood clotting (liquid to solid state)	
_____ Positive feedback	b. Visualization of performance just before an athletic event	
_____ Feedforward	c. Removal of calcium from bones because of low levels in the blood	

© Cengage Learning 2013

_____ **1.** The portion of the body fluid contained in cells is
   (a) interstitial fluid.
   (b) plasma.
   (c) intracellular fluid.
   (d) extracellular fluid.

_____ **2.** All the body fluid located outside of cells is
   (a) interstitial fluid.
   (b) plasma.
   (c) intracellular fluid.
   (d) extracellular fluid.

_____ **3.** Extracellular fluid includes
   (a) interstitial fluid.
   (b) plasma.
   (c) intracellular fluid.
   (d) both a and b

_____ **4.** In the case of temperature regulation, an example of an effector is
   (a) a receptor in the brain.
   (b) a center in the brain.
   (c) a shivering skeletal muscle.
   (d) none of the above

_____ **5.** In the case of temperature regulation, integration is accomplished by
   (a) receptors in the brain.
   (b) centers in the brain.
   (c) muscles and glands.
   (d) none of the above

_____ **6.** A neutral solution has a pH of
   (a) 3.
   (b) 5.
   (c) 7.
   (d) 9.

_____ **7.** The instrument used to measure blood pressure is called
   (a) a sphygmomanometer.
   (b) a hemocytometer.
   (c) a hematocrit centrifuge.
   (d) none of the above

_____ **8.** When measuring blood pressure, the Korotkoff sounds are heard
   (a) below the diastolic pressure.
   (b) above the systolic pressure.
   (c) between the diastolic and systolic pressures.
   (d) at none of the above times.

_____ **9.** Normal blood pressure for an 8-year-old boy is 100/67 mm Hg. The systolic blood pressure is
   (a) 33 mm Hg.
   (b) 100 mm Hg.
   (c) 67 mm Hg.
   (d) 167 mm Hg.

_____ **10.** In positive feedback homeostatic mechanisms,
   (a) the response reverses the change.
   (b) the response intensifies the change.
   (c) the response occurs before the change.
   (d) none of the above occurs.

EXERCISE **4**

Homeostasis

Post-Lab Questions

Introduction

1. Define homeostasis.

2. List and describe the various body fluid components.

4.1 Homeostatic Mechanisms

3. Which organ systems are involved with the acquisition of oxygen and its delivery to the cells of the body?

4.2 Regulation of pH

4. Explain the pH scale. How many hydrogen ions are present in a solution at pH 5 compared with those in a solution at pH 6?

5. Describe how neutral, alkaline, and acidic solutions differ from each other.

4.3 Simulation of Homeostatic Mechanisms Regulating Body Fluid Volume

6. List the components of a homeostatic feedback loop. Use the regulation of fluid volume to illustrate your answer.

4.4 Blood Pressure

7. A normal blood pressure for a man in his late forties is 130/82 mm Hg. Explain the two numbers and what they mean.

4.5 Negative Feedback, Positive Feedback, and Feedforward

8. Describe how negative feedback, positive feedback, and feedforward are similar and dissimilar when it comes to homeostatic mechanisms.

Food for Thought

9. The lowest body pH is found in the fluid of the working stomach. How does a pH of about 3 in this particular location benefit the body as a whole?

10. What is hypertension? List some possible causes of this condition.

Macromolecules and You: Food and Diet Analysis

After completing this exercise, you will be able to

1. define *macromolecule, vitamin, mineral, carbohydrate, monosaccharide, disaccharide, polysaccharide, lipid, protein, amino acid, calorie;*

2. describe the basic structures of carbohydrates, lipids, triglycerides, and proteins;

3. identify positive and negative tests for carbohydrates, lipids, and proteins;

4. list the roles that carbohydrates, lipids, proteins, minerals, and vitamins each play in the body's construction and metabolism;

5. test food substances to determine the presence of some biologically important molecules;

6. identify common dietary sources of nutrients;

7. compare your intake of nutrients and calories with the components and relative proportions of a recommended diet.

Introduction

Why do you eat? Because food tastes good? Because you get hungry? Like all animals, our bodies are programmed to ensure that we provide it with an adequate supply of food. These food items provide us with the energy stored in the chemical bonds of the food molecules, plus the raw materials from which cellular and tissue components are built. Hunger evolved to stimulate us to seek foods that provide enough nutrients for survival and health.

Food nutrients include minerals and vitamins, plus the larger biological (organic, i.e., carbon-based) molecules known as carbohydrates, lipids, and proteins. As it happens, these last nutrients are three of the four major groups of **macromolecules**, large organic molecules of which all cells are made. The fourth group is the nucleic acids that store and control the genetic instructions within a cell. These crucial molecules are not usually utilized by the body as nutrients, and so will be studied in another exercise.

Vitamins are necessary organic molecules that our bodies cannot construct internally; we require vitamins from our diets, though in relatively small amounts. **Minerals** are required inorganic (noncarbon-containing) nutrients such as calcium and potassium. Good health, then, depends upon eating and drinking the proper balance and quantities of all these nutrients.

In this exercise, you will establish tests to identify the presence of carbohydrates, lipids, and proteins, and then determine the presence or absence of these molecules in some common foods. You will also analyze your own diet and compare it to a diet recommended for maintaining good health.

5.1 Identification of Large Biological Molecules

You will learn some simple tests for carbohydrates, lipids, and proteins in a variety of substances, including food products. *Most of the reagents used are not harmful; however, observe all precautions listed and perform the experiments only in the proper location as identified by your instructor. Clean out test tubes between tests in the designated location.*

MATERIALS

Per student group (4):

- china marker
- ten test tubes in test tube rack
- dropper bottles (*or* bottles and plastic pipets) of:
    distilled water (dH₂O)
    onion juice
    potato juice
    hamburger juice
    cream
    colorless nondiet soft drink
    colorless diet soft drink
    glucose solution
    fructose solution
    lemon juice
    starch solution

- vegetable oil
- egg albumin
- benedict's reagent
- biuret reagent
- lugol's solution
- sudan IV dye
- piece of uncoated paper (grocery bag)
- test tube clamp
- hot plate *or* ring stand with wire gauze support and Bunsen burner
- 250- *or* 400-mL beaker with boiling beads or stones
- vortex mixer (optional)

A. Carbohydrates

Carbohydrates are composed of carbon, hydrogen, and oxygen. A **carbohydrate** is a simple sugar or a larger molecule composed of multiple sugar units. Those composed of a single sugar molecule are called **monosaccharides.** Two examples are fructose, sometimes called fruit sugar, and glucose (Figure 5-1), a sugar that commonly provides the most immediate source of energy to cells. Monosaccharides are easily used within cells as energy sources.

Two monosaccharides can be bonded together to form a **disaccharide**. Examples of disaccharides are sucrose (common table sugar), maltose (found in many seeds), and lactose (milk sugar). If many monosaccharides are bonded together, the resulting long carbohydrate molecule is called a **polysaccharide** (Figure 5-2).

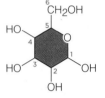

Figure 5-1 Structure of glucose, a monosaccharide.

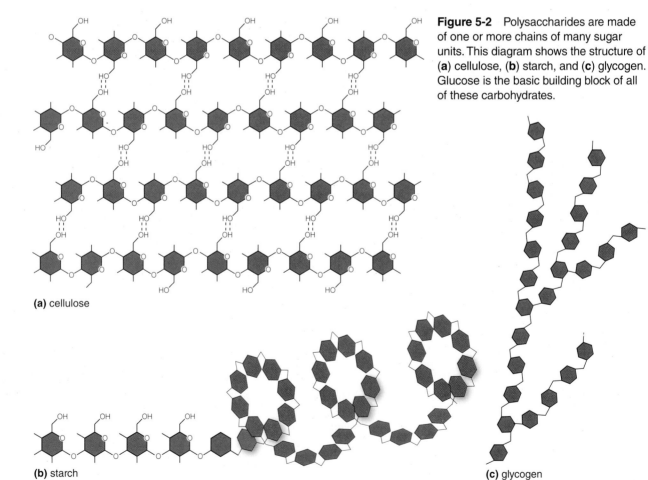

Figure 5-2 Polysaccharides are made of one or more chains of many sugar units. This diagram shows the structure of (**a**) cellulose, (**b**) starch, and (**c**) glycogen. Glucose is the basic building block of all of these carbohydrates.

(a) cellulose

(b) starch

(c) glycogen

Animals, including humans, store glucose energy in the form of glycogen, a highly branched polysaccharide chain of glucose molecules. Starch, a key energy-storing polysaccharide in plants, is an important food molecule for humans. Plant cells are surrounded by a tough cell wall mostly made of another chain of glucose molecules, cellulose. Our digestive system does not break cellulose apart very well. Plant materials thus provide bulky cellulose fiber, which is necessary for good digestive system health.

Carbohydrate digestion occurs in humans as digestive enzymes produced in the salivary glands, pancreas, and lining of the small intestine that break polysaccharides and disaccharides into monosaccharides such as glucose. These simple sugars are small enough to be absorbed into intestinal cells and then carried throughout the bloodstream.

Many monosaccharide molecules react upon heating with a copper-containing compound called Benedict's reagent, changing the reagent color from blue to orange or red. A disaccharide may or may not react with Benedict's solution, depending upon how the bonding of the component monosaccharides took place.

Polysaccharides do not react with Benedict's reagent. Other tests are available for some polysaccharides, the most common of which is Lugol's test for starch. In this test, a dilute solution of potassium iodide (I_2KI) reacts with the starch molecules to form a dark blue (nearly black) colored product.

A.1. Test for Sugars Using Benedict's Solution (About 20 min.)

PROCEDURE

1. Half-fill the beaker with tap water and apply heat with a hot plate or burner to bring the water to a gentle boil.
2. Using the china marker, number the test tubes 1 to 10.
3. Add 10 drops of each test solution into the numbered test tubes as listed in Table 5-1.

TABLE 5-1	Benedict's Test for Simple Sugars			
Tube No.	Contents	Initial Color	Color after Heating	Sugar Content (0, +, ++, +++)
1	Distilled water			
2	Onion juice			
3	Potato juice			
4	Cream			
5	Colorless nondiet soft drink			
6	Colorless diet soft drink			
7	Glucose			
8	Fructose			
9	Lemon juice			
10	Starch			

© Cengage Learning 2013

4. Add 10 drops of Benedict's solution to each test tube, and then agitate the mixture by shaking the tubes from side to side or with a vortex mixer, if available. Record the color of the mixture in Table 5-1 in the column "Initial Color."
5. Heat the tubes in the boiling water bath for about 3 minutes. Remove the tubes with the test tube clamp and place in rack. Record any color changes that have taken place in the column "Color after Heating."

   A cloudy precipitate will form that varies from green or yellow (+) to orange or red (++) to brown (+++) in color, indicating increasing concentrations of sugars. Record your conclusions regarding the presence or absence and concentration of simple sugars.

What is the purpose of the tube containing distilled water?

_____

_____

Rank the test substances in apparent order of sugar concentration, from none (0) to most (+++):

_____

_____

_____

_____

A.2. Test for Starch Using Lugol's Solution *(About 15 min.)*

PROCEDURE

1. Using the china marker, number the test tubes 1 to 8.
2. Add 10 drops of the correct test solutions into the test tubes as described in Table 5-2.

TABLE 5-2	Lugol's Test for Starch			
Tube No.	Contents	Initial Color	Color after Adding Lugol's Solution	Starch Present? (Yes or No)
1	Distilled water			
2	Onion juice			
3	Potato juice			
4	Cream			
5	Colorless nondiet soft drink			
6	Glucose			
7	Lemon juice			
8	Starch			

© Cengage Learning 2013

3. Record the color of each solution in the column "Initial Color" in Table 5-2.
4. Add 3 drops of Lugol's solution to each test tube, and then agitate the mixture by shaking the tubes from side to side or with a vortex mixer, if available. Record the color of the mixture in the column "Color after Adding Lugol's Solution." Also record your conclusions regarding the presence or absence of starch in each test solution.

What is the purpose of the tube with distilled water?

_____

What is the purpose of the tube with starch solution?

_____

B. Lipids

Lipids are oily or greasy compounds that don't dissolve in water but do dissolve in organic solvents like ether or chloroform. Lipids provide long-term energy storage in cells, among other functions, and are very diverse. Substances we think of as fats and oils are examples of lipids.

Those lipids having fatty acids—hydrocarbon chains with an acid [—COOH] group at one end—combine with glycerol to form glycerides, a rich source of stored energy. **Triglycerides** are lipids with three fatty acids attached to a glycerol molecule (Figure 5-3).

Lipids provide long-term energy storage in cells and are very diverse. Lipid digestion occurs primarily in the small intestine where bile produced by the liver breaks lipid globules into smaller droplets, and then pancreatic enzymes break large lipid molecules into smaller components for absorption. The lipid components are then transported throughout the body in lymph, the fluid that bathes the tissues.

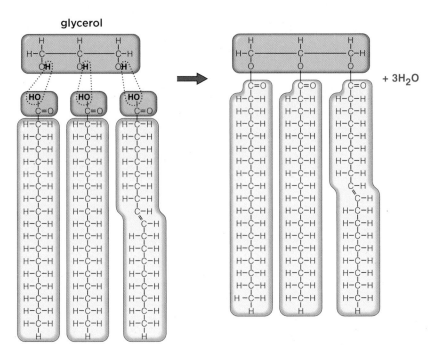

glycerol

three fatty acid tails   triglyceride

+ 3H$_2$O

Figure 5-3 Condensation reaction of three fatty acids and a glycerol molecule to form a triglyceride.

Fats are triglycerides that tend to be solid at room temperature due to the saturation of their fatty acids with single covalent bonds. Some common examples of fats include the animal fats in butter and lard.

Oils are triglycerides that tend to be liquid at room temperature. Like fats, they are composed of fatty acids and glycerol, but their fatty acids have one or more double covalent bonds, making them unsaturated.

One of the simplest tests for lipids is to determine whether or not they leave a grease spot on a piece of uncoated paper, such as a grocery sack. Another employs Sudan IV, a chemical commonly used to identify fats and oils in microscopic preparations. Sudan IV is a reddish liquid that dissolves in lipids and can be used to indicate their presence in a test tube.

B.1. Uncoated Paper Test (About 15 min.)

PROCEDURE

1. Place a drop of each test substance listed in Step 2 on a *labeled* spot on the grocery bag piece. Set aside to dry for 10 minutes.
2. After 10 minutes, describe the appearance of each spot on the paper.

   Distilled water _____

   Onion juice _____

   Potato juice _____

   Hamburger juice _____

   Cream _____

   Vegetable oil _____

   Which substances appear to contain lipids?

   _____

   _____

   What was the purpose of testing vegetable oil?

   _____

B.2. Sudan IV Test (About 15 min.)

PROCEDURE

1. With a china marker, number the test tubes 1 to 6.
2. Add 20 drops of dH$_2$O to each test tube and then 20 drops of each substance indicated in Table 5-3 to the appropriate test tube.

3. Add 9 drops of Sudan IV to each tube, agitate by shaking the tube side to side, and then add 10 more drops of dH₂O to each tube.
4. Record your results in Table 5-3.

TABLE 5-3	Sudan IV Test for Lipids		
Tube No.	Contents	Observations after Addition of Sudan IV	Conclusions
1	Distilled water		
2	Onion juice		
3	Potato juice		
4	Hamburger juice		
5	Cream		
6	Vegetable oil		

© Cengage Learning 2013

Which test, the uncoated paper test or the Sudan IV test, do you think is most sensitive to small quantities of lipids? Why?

_____

_____

What are some limitations of these tests?

_____

_____

C. Proteins

Proteins are a diverse group of biological molecules with a wide range of functions in an organism. Many are structural components of muscle, bone, hair, and nails, among other tissues. Others are enzymes that speed up cellular reactions that would otherwise take years to occur. Movement of structures within cells (such as during cell division) and of sperm cells also depends on proteins. There are still further functions for specific proteins. For example, egg albumin (egg white) protects and nourishes a developing animal embryo.

All proteins are complex chains of **amino acids,** whose general structure is illustrated in Figure 5-4. Humans cannot synthesize about half of the amino acids; these *essential amino acids* must be obtained through our foods.

Though there are only about 20 amino acids, innumerable kinds of proteins result from different amino acid sequences. A protein begins to form when two or more amino acids are linked together by peptide bonds (bonds formed by condensation reactions between the amino group of one amino acid and the acid group of another). Multiple amino acids joined together form a polypeptide chain (Figure 5-5). Then the chain folds and twists as links form between adjacent parts of the chain.

Protein digestion begins in the stomach lining and continues in the small intestine, where various enzymes break protein molecules first into protein fragments, and then into amino acids, which are absorbed across the small intestine wall. The amino acids are then transported throughout the body in the blood.

Several tests are used for proteins, including Biuret reagent, which indicates the presence of peptide bonds. The greater the number of peptide bonds, the more intense the bluish color reaction with Biuret.

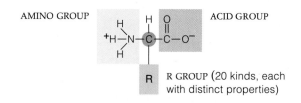

Figure 5-4 General structure of an amino acid. The R-group consists of one or more atoms and is unique for each of the 20 amino acids.

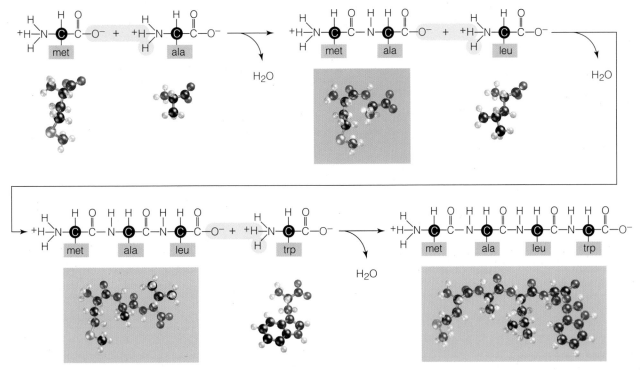

Figure 5-5 A protein is built as peptide bonds form between amino acids.

C.1. Test for Proteins with Biuret Reagent *(About 15 min.)*

PROCEDURE

1. Using the china marker, number the test tubes 1 to 6.
2. Add 20 drops of the correct test solutions into the test tubes as described in Table 5-4.
3. Add 10 drops of Biuret reagent to each test tube and agitate the mixture by shaking the tubes from side to side or with a vortex mixer, if available.
4. Wait 2 minutes and then record the color of the mixture in Table 5-4, in the column "Color Reaction." Also enter your conclusion about the presence of protein in each substance.

TABLE 5-4	Biuret Test for Protein		
Tube No.	**Contents**	**Color Reaction**	**Conclusions**
1	Distilled water		
2	Onion juice		
3	Potato juice		
4	Hamburger juice		
5	Cream		
6	Egg albumin		

© Cengage Learning 2013

D. Testing Unknown Food Substances

Your instructor will provide you with several samples of unidentified food substances. Alternatively, you may be instructed to bring your own food samples in order to attempt to determine some of their macromolecular components.

You may need to grind small quantities of solid substances with a mortar and pestle and then dilute the resulting slurry with water in order to perform some of the tests. Filter the slurry to obtain a relatively clear liquid for your tests. Use the additional equipment that may be supplied by your instructor to accomplish the preparations.

OPTIONAL MATERIALS

Per student group (4):

- mortar and pestle
- funnel and filter paper *or* coffee filters
- dropping pipets

PROCEDURE

1. Following the previous procedures, perform tests for carbohydrates (Benedict's and Lugol's tests), lipids (uncoated paper and Sudan IV tests), and proteins (Biuret reagent) on each unknown item.
2. Record your test results (+ or – for each test) in Table 5-5.

TABLE 5-5	Results of Testing Composition of Unknown Food Materials				
Unknown	Benedict's Test for Sugars	Lugol's Test for Starch	Brown Paper Test for Lipids	Sudan IV Test for Lipids	Biuret Test for Protein
A					
B					
C					
D					
E					

© Cengage Learning 2013

3. What kind(s) of biological molecule(s) are found in significant concentrations in each food? Did you find any of the test food substances that appeared to be relatively simple, providing only one of the biological molecules? Did you find any food substances that appeared to be complex, providing multiple kinds of biological molecules? Describe.

_____

_____

_____

_____

_____

_____

_____

_____

Note: After completing all experiments, take your dirty glassware to the sink and wash it following the directions given in *"Instructions for Washing Laboratory Glassware,"* page x. Invert the test tubes in the test tube rack so they drain. Tidy up your work area, making certain all equipment used in this exercise is there for the next class.

5.2   Food Guides and Diet Analysis

Food must be digested before most nutrients are available within the body. In the digestive system, macromolecules are digested to smaller components (proteins broken down to amino acids, polysaccharides to simple sugars, for example). The resulting smaller molecules are absorbed across the intestinal wall, transported throughout the body, and used to build and power cells. Figure 5-6 summarizes the major digestive pathways for the macromolecule food components we consume. We truly *are* what we eat.

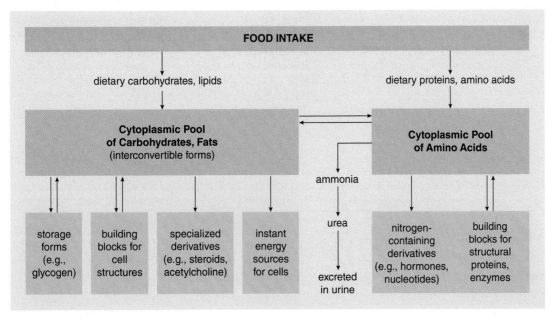

Figure 5-6 Summary of major metabolic pathways. Cells continually synthesize and tear down carbohydrates, lipids, and proteins.

The U.S. Department of Agriculture and health authorities issue dietary guidelines for average people of good health. Several nutrition principles underlie the recommendations: moderation in portion sizes; balance among food groups to ensure that all key nutrients are included but none are in excess; and emphasis on daily exercise for a multitude of health benefits.

The major food groups are vegetables and fruits; grains; dairy and soy; meat, beans, and alternatives; and vegetable oils. There also may be an allotment of "discretionary calories" for each person, calories in excess of basic energy needs. Each type of food provides specific nutrients.

Vegetables and **fruits** are especially important sources of the vitamins and minerals needed for a healthy diet, as well as complex carbohydrates and **phytochemicals**, a catchall term for a diverse group of molecules from plants and whose benefits we are just beginning to understand. Diets that include a colorful variety of vegetables and fruits may help prevent illnesses, including heart disease, stroke, and certain types of cancer.

Grains are high in the polysaccharides (starches and cellulose) that provide ready energy sources plus dietary fiber. Health authorities recommend that you "make half your grains whole," meaning that *at least* half your grain intake should be whole grains, which supply important vitamins and minerals, plus the fiber needed for a healthy digestive system. Whole grains include oatmeal, brown rice, popcorn, and whole grain or whole wheat flour.

Low-fat **dairy** and **soy** products are important sources of calcium and provide significant amounts of proteins. Protein-rich, vitamin-D-fortified dairy and alternative soy foods, along with weight-bearing exercise, build and maintain strong bones.

Smaller quantities are recommended of the "**meat, bean, and alternatives**" group, which includes poultry, fish, red meats, eggs, beans, and nuts. These protein-rich foods supply essential amino acids, vitamins, and minerals, and also help maintain the "full" feeling between meals.

Recommended **vegetable oils** include monounsaturated vegetable oils (olive oil and canola oil, for example) and polyunsaturated oils (corn oil, soybean oil). These oils are considered healthy in moderation, unlike the solid saturated fats (sour cream, butter, palm oil) and especially the harmful trans fats, which can be identified on food labels as "hydrogenated" or "partially hydrogenated" oils.

A **calorie** is a unit of energy content. Specific calorie guidelines vary depending upon your age, gender, and activity level, with more calories of food energy needed to power your cells as your activity level increases. Use Table 5-6 to determine your recommended average daily calorie allowance.

Consult Table 5-7 for guidelines for consumption of each food group for persons of moderate activity level and your age level and gender.

A. Food Diary (About 20 min. daily for two days)

You'll be on the path toward enjoying a healthy diet that provides balanced nutrients and calories if you follow the guidelines and eat a variety of foods in moderate portions. In this activity you will analyze *your* diet, so that you can see in which areas you make good nutrition choices, and which areas need improvement.

TABLE 5-6 Approximate Daily Calorie Requirements

	Sedentary (less than 30 min. moderate activity daily)	Moderate (30 to 60 min. moderate activity daily)	Active (more than 60 min. moderate activity daily)
FEMALES			
Ages 16–18	1750	2100	2400
Ages 19–30	1900	2100	2350
Ages 31–50	1800	2000	2250
MALES			
Ages 16–18	2400	2800	3200
Ages 19–30	2500	2700	3000
Ages 31–50	2350	2600	2900

© Cengage Learning 2013

TABLE 5-7 Recommended Daily Food Group Serving Intake for Moderate Activity Levels

Age in Years	Teens 14–18		Adults 19–50	
Gender	Females	Males	Females	Males
Vegetables and fruits	7	8	7–8	8–10
Grains	6	7	6–7	8
Dairy and soy	5–4	5–4	2	2
Meat and alternatives	2	3	2	3
Oils and fats	2 to 3 tablespoons, or 24 grams, of unsaturated or monounsaturated fats for all uses			

© Cengage Learning 2013

PROCEDURE (To be completed before class)

For two typical days during the week preceding this activity, keep a complete diary of *every* food and drink item you consume other than water, unsweetened coffee or tea, or other zero-calorie beverages. In Table 5-9, record what you eat and drink, and how much (the portion size: ounces, cups, teaspoons, and so on). Enter these data in the first two columns.

It's often difficult to determine portion sizes. To do this accurately, refer to the information on food product labels regarding serving size. You also may want to use a measuring cup for this activity. For unlabeled foods, use the portion guidelines in Table 5-8.

You should record items immediately after eating whenever possible. Be as honest, specific, and descriptive as you can. The more specific you can be, the more accurate your diet analysis will be. For example, include brand names, the method of preparation (baked, fried, canned, frozen), type of food (whole, 2%, 1%, or skim milk, for example). Record the calorie content of each item whenever possible. Do this for *each* meal and snack.

Eat your usual diet. Don't change your eating habits for this exercise.

TABLE 5-8 Visualizing Portion Sizes

Portion	Approximate Size of Item
1 teaspoon (tsp)	Thumb tip to base of nail
1 tablespoon (Tbsp)	Whole thumb
1 ounce (oz)	Two dice; one golf ball; one slice of processed cheese
3 ounces (oz) (one serving meat)	Palm of hand (thickness and size); one deck of playing cards; checkbook
½ cup	Volume within cupped hand; one tennis ball; one lightbulb
1 cup	Volume of a woman's fist; one baseball (not the larger softball)
1 oz grains	½ cup oatmeal or rice; 1 cup cereal flakes; ½ English muffin; 1 slice of commercial bread loaf

© Cengage Learning 2013

© Cengage Learning 2013

TABLE 5-9 Two-Day Food Diary

| Food or Beverage Item | Total Portion Size | Food Group Servings in Each Food Item | | | | | | Calories |
		Vegetables	Fruits	Grains	Dairy and Soy	Meat, Beans, and Alternatives	Fats and Oils (g)	
Example: McDonalds Double Cheeseburger	1	0.2	0	2.4	0.5	1.9	24.4	459
2-day total:								
2-day average:								

B. In-Class or At-Home Food Diary Analysis (About 1 hour)

MATERIALS

Per student group (4):

■ food diary data
■ diet analysis books and/or computers with Internet access

PROCEDURE

1. Complete the food diary table (Table 5-9) by using online diet analysis calculators identified by your instructor and/or reference books. Determine how each food item is allocated as servings of the various food groups: grains, vegetables, fruits, dairy and soy, meat and beans, and vegetable oils. Alternatively, if using online resources such as USDA SuperTracker, record summary serving data for an entire day's food intake.
2. Determine and record the calorie content of each item.
3. **Average** the servings for each food group and calorie content over the two days of data collection, and use the averaged data for the rest of this activity.
4. On Figure 5-7a, create a bar graph to represent the daily food group serving and discretionary calorie *recommendations* for your gender and activity level. Record discretionary calories, those that exceed your recommended calorie intake, in 100-calorie "servings".
5. Now construct a bar graph of your *actual* daily food group serving and discretionary calorie consumption on Figure 5-7b, using **average** 2-day data from your food diary.

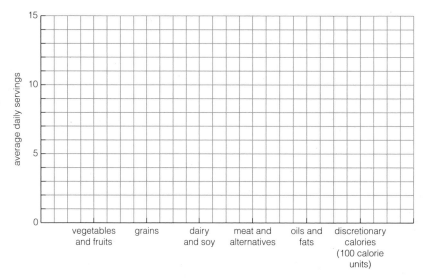

Figure 5-7a (above)   My *recommended* food group serving bar chart.

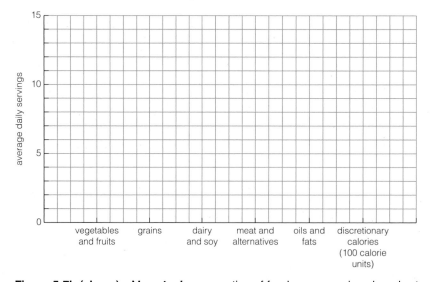

Figure 5-7b (above)   My *actual* consumption of food group servings bar chart.

6. Answer the following questions *after* you have completed your 2-day food diary and bar graphs:

(a) What is your typical level of physical activity (sedentary, moderate, active)? _____

(b) What is your approximate daily calorie requirement? _____

(c) How does your caloric intake compare with the recommendations for your gender and activity level? If your caloric intake is not in balance, suggest *at least two* strategies to correct the imbalance.

_____

_____

_____

_____

(d) Which, if any, food groups are you eating too much of? What major kinds of nutrients are you thus overconsuming?

_____

_____

(e) Which, if any, food groups are you not eating enough of? What kinds of major nutrients are thus underrepresented in your diet?

_____

_____

_____

_____

(f) What proportion of your grain servings included whole grains? _____
How many *different kinds* of vegetables and fruits did you consume? _____

(g) Choose one area in which your diet differs from recommendations. Search the Internet for two sites that describe consequences of that dietary choice. List the two sites and briefly summarize their contents. (If your diet fully conforms to recommendations, choose a dietary issue of interest instead and research that topic.)

http:// _____

_____

_____

_____

Summary: _____

_____

_____

_____

http:// _____

_____

_____

_____

Summary: _____

_____

_____

_____

(h) Given all the above information, describe two *reasonable* changes you could make to improve your diet and your health. (If your diet is currently healthy, describe your two greatest challenges in maintaining that healthy diet, plus a successful strategy for overcoming each challenge.)

_____

_____

_____

_____

_____

_____ 1. A carbohydrate consists of
   (a) amino acid units.
   (b) one or more sugar units.
   (c) lipid droplets.
   (d) glycerol.

_____ 2. A protein is made up of
   (a) amino acid units.
   (b) one or more sugar units.
   (c) lipid droplets.
   (d) Biuret solution.

_____ 3. Benedict's solution is commonly used to test for the presence of
   (a) proteins.
   (b) certain carbohydrates.
   (c) nucleic acids.
   (d) lipids.

_____ 4. Glycogen is
   (a) a polysaccharide.
   (b) a storage carbohydrate.
   (c) found in human tissues.
   (d) all of the above

_____ 5. To test for the presence of starch, one would use
   (a) Benedict's solution.
   (b) uncoated paper.
   (c) Sudan IV.
   (d) Lugol's solution.

_____ 6. Rich sources of stored energy that are dissolvable in organic solvents are
   (a) carbohydrates.
   (b) proteins.
   (c) glucose.
   (d) lipids.

_____ 7. Rubbing a substance on uncoated paper should reveal if it contains
   (a) lipid.
   (b) carbohydrate.
   (c) protein.
   (d) sugar.

_____ 8. Proteins consist of
   (a) monosaccharides linked in chains.
   (b) amino acid units.
   (c) polysaccharide units.
   (d) condensed fatty acids.

_____ 9. Biuret reagent will indicate the presence of
   (a) peptide bonds.
   (b) proteins.
   (c) amino acid units linked together.
   (d) all of the above

_____ 10. The largest number of food servings in your daily diet should be from
   (a) meats and beans.
   (b) dairy products.
   (c) vegetables and fruits.
   (d) grains.

EXERCISE **5**

Macromolecules and You: Food and Diet Analysis

Post-Lab Questions

A.1. Test for Sugars Using Benedict's Solution

1. Let's suppose you are teaching science in a part of the world without easy access to a doctor and you're worried that you may have developed diabetes. (Diabetics are unable to regulate blood glucose levels, and glucose accumulates in urine.) How could you perform a test to gain an indication of whether or not you have diabetes?

2. The test tubes in the photograph contain Benedict's solution and two unknown substances that have been heated. What do the results indicate for each substance?

3. How could you verify that a soft-drink container contains diet soft drink rather than soft drink sweetened with fructose?

A.2. Test for Starch Using Lugol's Solution

4. Observe the photomicrograph accompanying this question. This thin section of a potato tuber has been stained with Lugol's iodine solution. When you eat french fries, the potato material is broken down in your small intestine into what kind of small subunits?

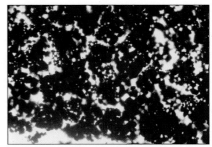

B.2. Sudan IV Test

5. The test tube in the photograph contains water at the bottom and another substance that has been stained with Sudan IV at the top. What is the macromolecular composition of this stained substance?

J. W. Perry

6. You are given a sample of an unknown food. Describe how you would test it for the presence of lipids.

7. You wish to test the same unknown food for the presence of sugars. Describe how you would do so.

Food for Thought

8. What is the purpose of the distilled water sample in each of the chemical tests in this exercise?

9. Many health food stores carry enzyme preparations that are intended to be ingested orally (by mouth) to supplement existing enzymes in various organs like the liver, heart, and muscle. Use your knowledge of the digestion process to explain why these preparations are unlikely to be effective as advertised.

10. A young child grows rapidly, with high levels of cell division and high energy requirements. If you were planning the child's diet, which **food groups** would you emphasize, and why? Which **food groups** would you deemphasize, and why?

Structure and Function of Living Cells

After completing this exercise, you will be able to

1. define *cell, cell theory, plasma membrane, DNA, cytoplasm, prokaryotic, eukaryotic, nucleus, organelle, ribosome, cyanobacteria, cytoplasmic streaming, sol, gel, envelope, mitochondrion, endoplasmic reticulum (rough and smooth), Golgi body, chloroplasts, cell wall, central vacuole;*

2. list the structural features shared by all cells;

3. describe the similarities and differences between prokaryotic and eukaryotic cells;

4. identify the cell parts described in this exercise;

5. state the function(s) for each cell part described in this exercise;

6. describe and identify distinguishing structures between plant and animal cells;

7. identify the structures presented in boldface in the procedure sections.

Introduction

Structurally and functionally, all life has one common feature: all living organisms are composed of **cells**. The development of this concept began with Robert Hooke's seventeenth-century observation that slices of cork were made up of small units. He called these units "cells" because their structure reminded him of the small cubicles that monks lived in. Over the next 100 years, the **cell theory** emerged. This theory has three principles: (1) all organisms are composed of one or more cells; (2) the cell is the basic living unit of organization; and (3) all cells arise from preexisting cells.

Although cells vary in organization, size, and function, all share three structural features: (1) all possess a **plasma membrane** defining the boundary of the living material; (2) all contain a region of **DNA** (deoxyribonucleic acid), which stores genetic information; and (3) all contain **cytoplasm**, which is everything inside the plasma membrane that is not part of the DNA region.

With respect to internal organization, there are two basic types of cells, **prokaryotic** and **eukaryotic**. Study Table 6-1, comparing the more important differences between prokaryotic and eukaryotic cells. The Greek word *karyon* means "kernel," referring to the nucleus. Thus, *prokaryotic* means "before a nucleus," while *eukaryotic* indicates the presence of a "true nucleus." Prokaryotic cells typical of bacteria, cyanobacteria, and archaea are believed to be similar to the first cells, which arose on Earth 3.5 billion years ago. Eukaryotic cells, such as those that comprise the bodies of protists, fungi, plants, and animals, including humans, probably evolved from prokaryotes.

This exercise will familiarize you with the basics of cell structure and function of prokaryotes (prokaryotic cells) and eukaryotes (eukaryotic cells).

6.1 Prokaryotic Cells *(About 20 min.)*

MATERIALS

Per student:

- dissecting needle
- compound microscope
- microscope slide
- coverslip

Per student pair:

- distilled water (dH$_2$O) in dropping bottle

Per student group (table):

- culture of a cyanobacterium (either *Anabaena* or *Oscillatoria*)

Per lab room:

- three bacterium-containing nutrient agar plates (demonstration)
- three demonstration slides of bacteria (coccus, bacillus, spirillum)

	Cell Type	
TABLE 6-1 Comparison of Prokaryotic and Eukaryotic Cells		
Characteristic	**Prokaryotic**	**Eukaryotic**
Genetic material	Located within cytoplasm, not bounded by a special membrane Consists of a single circular molecule of DNA	Located in **nucleus**, a double membrane-bounded compartment within the cytoplasm Multiple molecules of DNA combined with protein Organized into chromosomes
Cytoplasmic structures	Small ribosomes Photosynthetic membranes arising from the plasma membrane (occur in some types only)	Large ribosomes **Organelles**, multiple kinds of membrane-bounded compartments specialized to perform specific functions
Kingdoms represented	Bacteria Archaea	Protista Fungi Plantae Animalia

© Cengage Learning 2013

PROCEDURE

1. Observe the culture plate with bacteria growing on the surface of a nutrient medium. Can you see the individual cells with your naked eye?

_____

2. Observe the microscopic preparations of bacteria on *demonstration* next to the culture plate. The three slides are made from the three basic shapes of bacteria. Which objective lenses are being used to view the bacteria?

_____

Can you discern any detail within the cytoplasm?

_____

   In the space provided in Figure 6-1, sketch what you see through the microscope. Record the magnification you are using in the blank provided in the figure caption.

3. Study Figure 6-2, a three-dimensional drawing of a bacterial cell. Now examine the electron micrograph of the bacterium *Escherichia coli* (Figure 6-3). Locate the **cell wall**, a structure chemically distinct from the wall of plant cells but serving the same primary function to contain and protect the cell's contents.

4. Find the **plasma membrane**, which is lying flat against the internal surface of the cell wall and is difficult to distinguish.

5. Look for two components of the **cytoplasm: ribosomes**, electron-dense particles (they appear black) that give the cytoplasm its granular appearance, and a relatively electron-transparent region (appears light) called the **nucleoid** which contains fine threads of the long DNA molecule.

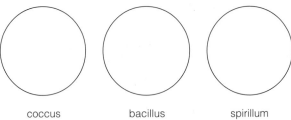

coccus   bacillus   spirillum

Figure 6-1 Drawings of several bacterial cells (_____ ×).

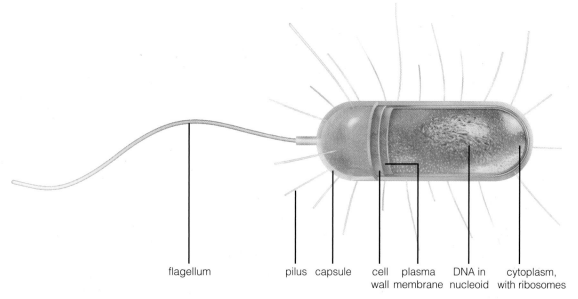

flagellum   pilus   capsule   cell   plasma   DNA in   cytoplasm,
                                   wall   membrane   nucleoid   with ribosomes

Figure 6-2   Three-dimensional drawing of a bacterial cell as seen with the electron microscope.

Another type of prokaryotic cell is exemplified by cyanobacteria, such as *Oscillatoria* and *Anabaena*. Cyanobacteria (sometimes called blue-green algae) are commonly found in water and damp soils. They obtain their nutrition by converting the sun's energy into chemical energy through photosynthesis.

6. With a dissecting needle, remove a few filaments from the cyanobacterial culture, placing them in a drop of water on a clean microscope slide.

7. Place a coverslip over the material and examine it first with the low-power objective and then using the high-dry objective (or oil-immersion objective, if your microscope is so equipped).

8. In the space provided in Figure 6-4, sketch the cells you see at high power. Estimate the size of a *single* cyanobacterial cell and record the magnification you used to make your drawing.

9. Now examine the electron micrograph of *Anabaena* (Figure 6-5), which identifies the **cell wall, cytoplasm,** and **ribosomes**. These cyanobacteria also possess membranes that function in photosynthesis. Identify the **photosynthetic membranes**, which look like tiny threads within the cytoplasm. Because the electron micrograph of *Anabaena* is of relatively low magnification, the plasma membrane is not obvious, but if you could see it, it would be found just under the cell wall.

10. Look at the captions for Figures 6-3 and 6-5. Judging by the magnification of each electron micrograph, which cell is larger, the bacterium *E. coli* or the cyanobacterium *Anabaena*?

_____

_____

_____

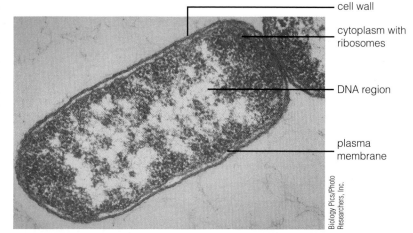

cell wall

cytoplasm with ribosomes

DNA region

plasma membrane

Biology Pics/Photo Researchers, Inc.

Figure 6-3   Electron micrograph of the common intestinal bacterium *Escherichia coli* (28,300×; color added).

Figure 6-4   Drawing of several cells of a prokaryotic cyanobacterium (_____×).

MATERIALS

Per student:

- textbook
- toothpick
- microscope slide
- coverslip
- culture of *Physarum polycephalum*
- compound microscope
- forceps
- dissecting needle

Per student pair:

- methylene blue in dropping bottle
- distilled water (dH₂O) in dropping bottle

Per student group (table):

- *Elodea* in water-containing culture dish
- onion bulb
- tissue paper
- container of ice *or* refrigerator
- Celsius thermometer
- timer or watch with second hand

Per lab room:

- model of animal cell
- model of plant cell

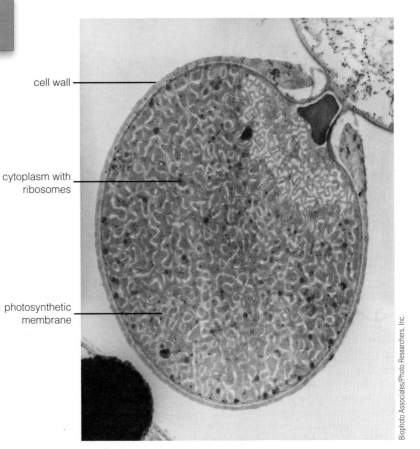

cell wall

cytoplasm with ribosomes

photosynthetic membrane

Biophoto Associates/Photo Researchers, Inc.

Figure 6-5   Electron micrograph of *Anabaena* (11,600×).

A. Cells of a Slime Mold, the Protist, *Physarum polycephalum*

Physarum polycephalum, a member of the Kingdom Protista, is a unicellular organism. As such, it contains all the metabolic machinery for independent existence. Furthermore, its single cell can grow to relatively gigantic size, allowing easy observation of its cytoplasm and cellular organelles.

   Physarum absorbs its organic food substances by engulfing particles and digesting them within organelles called *food vacuoles*. Your culture will be growing across the surface of agar gel in a petri dish. The organism grows branches out across the substrate and then comes together on the surface of a food item. The "food" provided to your culture is bacterial cells naturally occurring on the surfaces of oatmeal flakes.

PROCEDURE

1. Place a plain microscope slide on the stage of your compound microscope. This will serve as a platform on which you can place a culture dish.
2. Obtain a petri dish culture of *Physarum*, remove the lid, and place it on the platform. Observe initially with the low-power objective and then with the medium-power objective. If you choose to view the culture with the high-dry objective, place a coverslip over part of the organism before rotating the objective into place. (This prevents the agar from getting on the lens.)

Physarum is *multinucleate*, meaning that more than one nucleus occurs within the cytoplasm. Unfortunately, the nuclei are tiny; you will not be able to distinguish them from other organelles in the cytoplasm.

3. Locate the **plasma membrane**, which is the outer boundary of the cytoplasm. Again, the resolving power of your microscope is not sufficient to allow you to actually view the membrane.
4. Watch the cytoplasm of the organism move. This intracellular motion is called **cytoplasmic streaming**. Contractile proteins called *microfilaments* are believed to be responsible for cytoplasmic streaming. The cytoplasmic motion carries nutrients, proteins, organelles, and other cytoplasmic components throughout the cell.

5. Note that the outer portion of the cytoplasm appears solid; this is the **gel** state of the cytoplasm. Notice that the granules closer to the interior are in motion within a fluid; this interior portion of the cytoplasm is in the **sol** state. Movement of the organism occurs as the sol-state cytoplasm at the advancing tip pushes against the plasma membrane, causing the region to swell outward. The sol-state cytoplasm flows into the region, converting to the gel state along the margins.

6. In Figure 6-6, sketch a portion of *Physarum* that you have been observing and label it.

B. Experiment: Temperature Effects on Cytoplasmic Streaming (About 25 min.)

Temperature affects many cellular and organismal processes. For example, reptiles and insects are ectotherms (animals that gain heat from the environment), unlike humans, whose body heat comes primarily from cellular metabolism. You may have observed that in nature, these animals are relatively sluggish during cold weather. Is the same true for other organisms, such as the slime mold?

Figure 6-6 Drawing of a portion of *Physarum* (_____×).

This simple experiment addresses the hypothesis that *cold slows cytoplasmic streaming in P. polycephalum*. Before starting this experiment, you may wish to review the discussion in Exercise 1, "The Scientific Method."

PROCEDURE

1. Place the *Physarum* culture on the stage of your compound microscope as described in Section 6.2.A.
2. Time the duration of cytoplasmic streaming in one direction and then in the other direction. Do this for five cycles of back-and-forth motion. Calculate the average duration of flow in either direction. Record the temperature and your observations in Table 6-2.

TABLE 6-2	Effect of Temperature on Cytoplasmic Streaming	
Prediction of observations:		
Temperature (°C)	**Time/Cycle Number**	**Observations and Duration of Directional Flow (sec)**
	1	
	2	
	3	
	4	
	5	
	Average	
	1	
	2	
	3	
	4	
	5	
	Average	
Conclusions:		

© Cengage Learning 2013

3. Remove your culture from the microscope's stage, replace the cover, and place it and the thermometer in a refrigerator or atop ice for 15 minutes.
4. While you are waiting, in Table 6-2, write a prediction for the effect on the duration of cytoplasmic streaming resulting from reducing the temperature of the culture of *P. polycephalum*.
5. After 15 minutes have elapsed, remove the culture from the cold treatment, record the temperature of the experimental treatment, and repeat the observations in step 2.
6. Record your observations and make a conclusion in Table 6-2, accepting or rejecting the hypothesis.

Why was it a good idea to time cytoplasmic streaming for more than a single direction cycle?

_____

How could this experiment have been improved so that it allows a more reliable conclusion?

_____

A logical question to ask at this time is *why* temperature has the effect you observed. If you perform Exercise 8, "Enzymes: Catalysts of Life," you may be able to make an educated guess (a hypothesis).

C. Human Cheek Cells Observed with the Light Microscope

PROCEDURE

1. Using the broad end of a clean toothpick, gently scrape the inside of your cheek. Stir the scrapings into a drop of distilled water on a clean microscope slide and add a coverslip. Dispose of used toothpicks in the jar containing alcohol.
2. Because the cells are almost transparent, decrease the amount of light entering the objective lens to increase the contrast. (See Exercise 3, page 33.) Find the cells using the low-power objective of your microscope; then switch to the high-dry objective for detailed study.
3. Find the **nucleus,** a centrally located spherical body within the **cytoplasm** of each cell.
4. Now stain your cheek cells with a dilute solution of methylene blue, a dye that stains the nucleus darker than the surrounding cytoplasm and further increases the contrast of the transparent cells. To stain your slide, follow the directions illustrated in Figure 6-7. Without removing the coverslip, add a drop of the stain to one edge of the coverslip. Then draw the stain under the coverslip by touching a piece of tissue paper to the *opposite* side of the coverslip. Search your slide for cheek cells near the central edge of the area colored by the dye.
5. In Figure 6-8, sketch the cheek cells, labeling the **cytoplasm, nucleus,** and the location of the **plasma membrane**. (A light microscope cannot resolve the plasma membrane, but the boundary between the cytoplasm and the external medium indicates its location.) Many of the cells will be folded or wrinkled due to their thin, flexible nature.

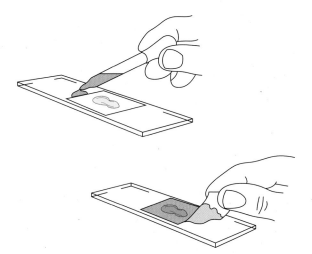

Figure 6-7   Method for staining specimen under coverslip on microscope slide.

Figure 6-8   Drawing of human cheek cells (_____×).
Labels: cytoplasm, nucleus, plasma membrane

D. Animal Cells as Observed with the Electron Microscope

Studies with the electron microscope have yielded a wealth of information on the structure of eukaryotic cells. Many structures too small to be seen with the light microscope have been identified. These include many **organelles**, structures in the cytoplasm that have been separated ("compartmentalized") by enclosure in membranes. Examples of organelles are the nucleus, mitochondria, endoplasmic reticulum, and Golgi bodies. Although the cells in each of the eukaryotic kingdoms have some unique peculiarities, electron microscopy has revealed that all eukaryotic cells are fundamentally similar.

PROCEDURE

1. Study Figure 6-9, a three-dimensional drawing of an animal cell.
2. With the aid of Figure 6-9, identify the parts on the model of the animal cell that is on *demonstration*.
3. Figure 6-10 is an electron micrograph (EM) of an animal cell (kingdom Animalia). Study the electron micrograph and, with the aid of Figure 6-9 and any electron micrographs in your textbook, label each structure listed.
4. Pay particular attention to the membranes surrounding the nucleus and mitochondria. Note that these two are each bounded by *two* membranes, which are commonly referred to collectively as an **envelope**.
5. Using your textbook as a reference, list the function for the following cellular components:

   (a) plasma membrane _____

   _____

   (b) cytoplasm _____

   _____

   (c) nucleus (the plural is *nuclei*) _____

   _____

   (d) nuclear envelope _____

   _____

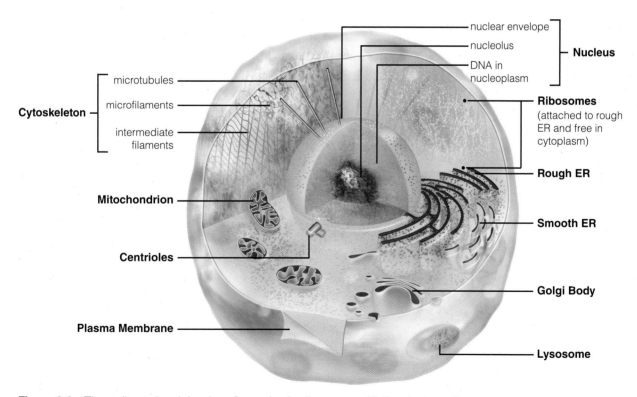

Figure 6-9 Three-dimensional drawing of an animal cell as seen with the electron microscope.

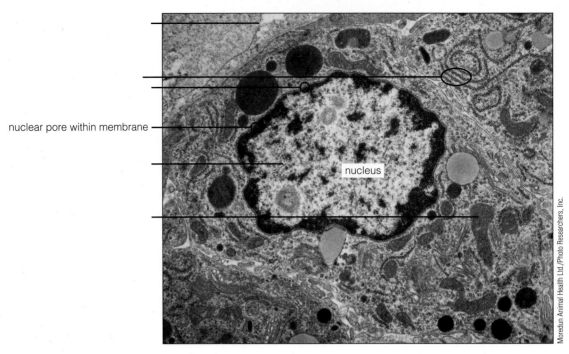

nuclear pore within membrane

nucleus

Moredun Animal Health Ltd./Photo Researchers, Inc.

Figure 6-10 Electron micrograph of an animal cell (1600×).
Labels: plasma membrane, nuclear envelope, chromatin, rough ER, mitochondrion

(e) nuclear pores _____

_____

(f) chromatin _____

_____

(g) nucleolus (the plural is *nucleoli*) _____

_____

(h) rough endoplasmic reticulum (RER) _____

_____

(i) smooth endoplasmic reticulum (SER) _____

_____

(j) Golgi body _____

_____

(k) mitochondrion (the plural is *mitochondria*) _____

_____

E. Plant Cells Seen with the Light Microscope

E.1. Elodea *leaf cells*

Young leaves at the growing tip of the aquatic pondweed, *Elodea*, are particularly well suited for studying cell structure because these leaves are only a few cell layers thick. Alternatively, your instructor may provide you with similarly thin leaves from a different plant.

PROCEDURE

1. With a forceps, remove a single young leaf, mount it on a slide in a drop of distilled water, and cover with a coverslip.
2. Examine the leaf first with the low-power objective. Then concentrate your study on several cells using the high-dry objective. Refer to Figure 6-11.
3. Observe the abundance of green bodies in the cytoplasm. These are the **chloroplasts**, organelles that function in photosynthesis and that are typical of green plants.
4. Locate the numerous dark lines running parallel to the long axis of the leaf. These are the air-containing *intercellular spaces*.
5. Find the **cell wall**, a structure distinguishing plant from animal cells, visible as a clear area surrounding the cytoplasm.
6. After the cells have warmed a bit, notice the **cytoplasmic streaming** taking place. Movement of the chloroplasts along the cell wall is the most obvious visual evidence of cytoplasmic streaming. Microfilaments (much too small to be seen with your light microscope) are responsible for this intracellular motion.
7. Be aware that you are looking at a three-dimensional object. In the middle portion of the cell is the large, clear **central vacuole**, which can take up from 50% to 90% of the cell interior. Because the vacuole in *Elodea* is transparent, it cannot be seen directly with the light microscope; it simply appears as a large empty space.

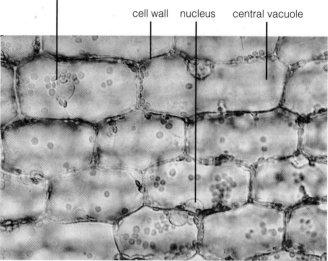

chloroplasts (surrounding a nucleus)

cell wall   nucleus   central vacuole

Figure 6-11   *Elodea* leaf cells (400×).

Photo by J. W. Perry.

8. The chloroplasts occur in the cytoplasm surrounding the vacuole, so they will appear to be in different locations, depending on where you focus in the cell. Focus in the upper or lower surface and observe that the chloroplasts appear to be scattered throughout the cell.
9. Now focus in the center of the cell (by raising or lowering the objective with the fine focus knob), and note that the chloroplasts lie in a thin layer of cytoplasm along the wall. The central vacuole usually pushes the rest of the cytoplasm toward the outer edges of the cell.
10. Locate the **nucleus** within the cytoplasm. It will appear as a clear or slightly amber body that is larger than the chloroplasts. You may need to examine several cells to find a clearly defined nucleus. Often, the clearest clue to the location of a nucleus is a traffic jam of chloroplasts that have clumped against it.
11. Describe the three-dimensional shape of the *Elodea* leaf cell.

_____

_____

12. What are the shapes of the chloroplasts and of the nucleus? _____

_____

13. Now add a drop of methylene blue stain to make the cell wall more obvious. Add the stain as shown previously in Figure 6-7.
14. Look for the very, very tiny **mitochondria**. (If you have an oil-immersion lens on your microscope, you should use that lens.) They will appear like numerous tiny granules within the cytoplasm.
15. How does the size of the mitochondria compare to that of the chloroplasts?

_____

_____

E.2. Onion scale cells

PROCEDURE

1. Make a wet mount of a leaf of an onion bulb, using the technique described in Figure 6-12. The *inner* face of the leaf is easiest to remove, as shown in Figure 6-12d.
2. Observe your preparation with your microscope, focusing first with the low-power objective. Continue your study, switching to the medium-power and finally the high-dry objective. Refer to Figure 6-13.
3. Identify the **cell wall** and **cytoplasm**.
4. Find the **nucleus**, a prominent sphere within the cytoplasm.
5. Examine the nucleus more carefully at high magnification. Within it, find one or more nucleoli (the singular is *nucleolus*). Nucleoli are rich in a nucleic acid known as RNA (ribonucleic acid), while the nucleus as a whole is largely DNA (deoxyribonucleic acid), the genetic material.
6. You may see numerous *oil droplets* within the cytoplasm, visible in the form of granule-like bodies. These oil droplets are a form of stored food material. You may be surprised to learn that onion "rings" are actually leaves! Which cellular components present in *Elodea* leaf cells are absent in onion leaf cells?

_____

7. If you are using the pigmented tissue from a red onion, you should see a purple pigment located in the vacuole. In this case, the cell wall appears as a bright line.
8. In Figure 6-14, sketch and label several cells from onion bulb leaves.

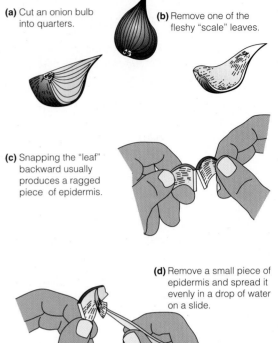

(a) Cut an onion bulb into quarters.

(b) Remove one of the fleshy "scale" leaves.

(c) Snapping the "leaf" backward usually produces a ragged piece of epidermis.

(d) Remove a small piece of epidermis and spread it evenly in a drop of water on a slide.

(e) Gently lower a coverslip to prevent trapping air bubbles. Examine with your microscope. Add more water to the edge of the coverslip with an eye dropper if the slide begins to dry.

From Peter Abramoff and Robert G. Thomson, Laboratory Outlines in Biology III. Copyright © 1962, 1963 Peter Abramoff and Robert G. Thomson. Copyright © 1964, 1966, 1972, 1982 W. H. Freeman and Company. Used by permission.

Figure 6-12 Method for obtaining onion bulb leaf cell specimens.

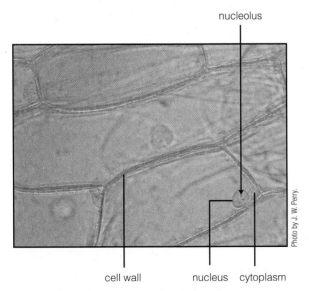

Photo by J. W. Perry.

Figure 6-13 Onion bulb leaf cells (67×).

Figure 6-14 Drawing of onion leaf cells (_____×). **Labels:** cell wall, cytoplasm, nucleus

F. Plant Cells as Seen with the Electron Microscope

The electron microscope has made obvious some of the unique features of plant cells.

PROCEDURE

1. Study Figure 6-15, a three-dimensional drawing of a typical plant cell.
2. With the aid of Figure 6-15, identify the structures present on the model of a plant cell that is on *demonstration*.
3. Now examine Figure 6-16, a transmission electron micrograph from a corn leaf. Label all of the structures listed. *Caution: Many plant cells do not have a large central vacuole. This is one of them.* Notice that the chloroplast has an envelope, just as do the nucleus and mitochondria.
4. With the help of Figure 6-15 and any transmission electron micrographs and text in your textbook and/or websites, list the function of the following structures:

    (a) cell wall _____
    _____

    (b) chloroplast _____
    _____

    (c) vacuole _____
    _____

    (d) vacuolar membrane _____
    _____

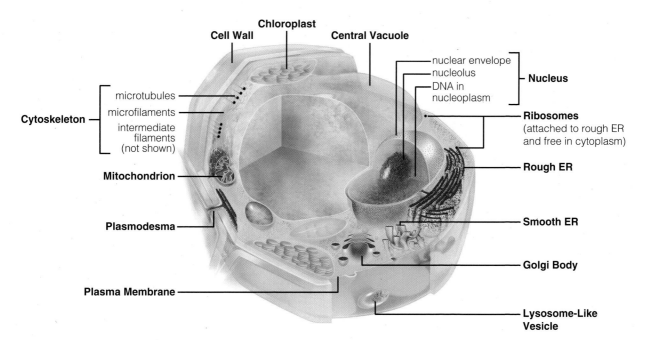

Figure 6-15   Three-dimensional drawing of a plant cell as seen with the electron microscope.

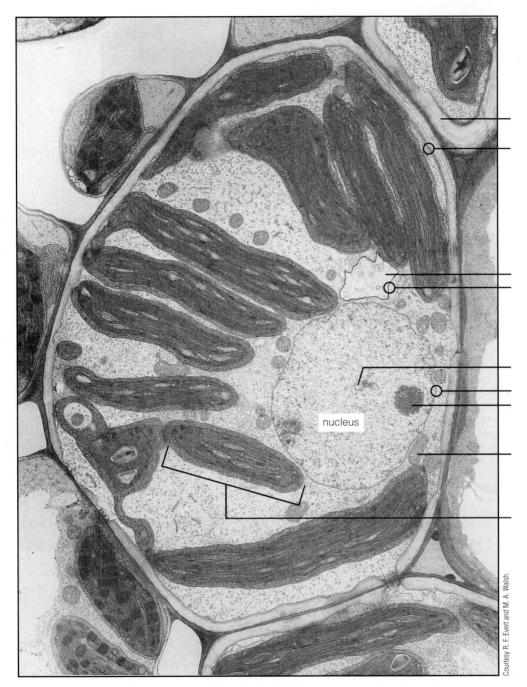

nucleus

Courtesy R. F. Evert and M. A. Walsh.

Figure 6-16 Electron micrograph of a corn leaf cell (2700×).
Labels: cell wall, chloroplast, vacuole, vacuolar membrane, plasma membrane, nuclear envelope, chromatin, nucleolus, mitochondrion

_____ 1. The person who first used the term *cell* was
(a) Darwin.
(b) Leeuwenhoek.
(c) Hooke.
(d) Watson.

_____ 2. All cells contain
(a) a nucleus, plasma membrane, and cytoplasm.
(b) a cell wall, nucleus, and cytoplasm.
(c) DNA, plasma membrane, and cytoplasm.
(d) mitochondria, plasma membrane, and cytoplasm.

_____ 3. Prokaryotic cells *lack*
(a) DNA.
(b) a true nucleus.
(c) a cell wall.
(d) none of the above

_____ 4. The word *eukaryotic* refers specifically to a cell containing
(a) photosynthetic membranes.
(b) a true nucleus.
(c) a cell wall.
(d) none of the above

_____ 5. A bacterium is an example of a
(a) prokaryotic cell.
(b) eukaryotic cell.
(c) plant cell.
(d) all of the above

_____ 6. Methylene blue
(a) is used to kill cells that are moving too quickly to observe.
(b) detoxifies cells so they are safe to handle.
(c) is used as food for *Physarum* cells.
(d) is a biological stain used to increase contrast of transparent cells.

_____ 7. Components typical of plant cells but **not** of animal cells are
(a) nuclei.
(b) cell walls.
(c) mitochondria.
(d) ribosomes.

_____ 8. A central vacuole
(a) is found only in plant cells.
(b) may take up between 50% and 90% of the cell's interior.
(c) regulates water balance.
(d) all of the above

_____ 9. The intercellular spaces between plant cells
(a) contain air.
(b) are responsible for cytoplasmic streaming.
(c) produce energy for the cell.
(d) contain chloroplasts.

_____ 10. An envelope
(a) surrounds the nucleus.
(b) surrounds mitochondria.
(c) consists of two membranes.
(d) all of the above

Name _____ Section Number _____

Structure and Function of Living Cells

Post-Lab Questions

6.1 Prokaryotic Cells

1. Did all living cells that you saw in lab contain mitochondria? If not, explain.

2. Below is a high-magnification photomicrograph of an organism similar to one you observed in this exercise. Each rectangular "box" is a single cell. What normally visible organelle is absent from each cell that makes it "prokaryotic?"

(750×)

6.2 Eukaryotic Cells

3. Is it possible for a cell to contain more than one nucleus? Explain.

4. When students are asked to write an essay explaining how to distinguish between an animal cell and a plant cell, they typically answer that plant cells contain chloroplasts and animal cells do not. If you were the professor reading that essay answer, what sort of grade would you assign, and why?

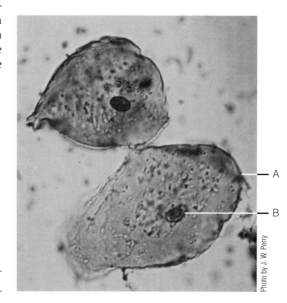

5. Identify the indicated structures.

   A. _____

   B. _____

6. Look at the photomicrograph to the right, which was taken with a technique that gives a three-dimensional impression. Identify the structures labeled A, B, and C.

   A. _____

   B. _____

   C. _____

7. Describe one function for each structure you identified in (6).

   A. _____

   B. _____

   C. _____

8. In the electron micrograph below, identify structures labeled A, B, and C.

   A. _____

   B. _____

   C. _____

C   B   A   (750×)

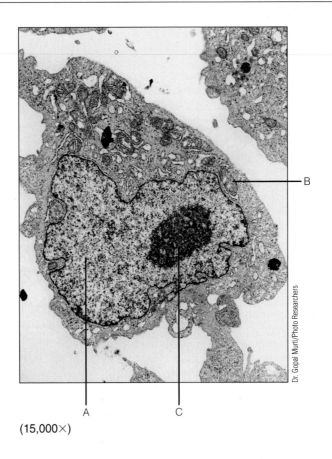

A   C

(15,000×)

Food for Thought

9. One botulinum toxin produced by the bacterium *Clostridium botulinum* prevents microfilaments from forming in affected cells. If a *Physarum* cell were exposed to that chemical, would you expect to see cytoplasmic streaming? Explain.

10. Mitochondrial diseases are common, affecting up to 4000 children per year in the United States. Mild forms cause "exercise intolerance." Search two Internet sites for information on "exercise intolerance," and learn about the connections between the role(s) of defective mitochondria and disease symptoms. List the two sites below and briefly summarize the relevant information from each.

http:// _____

_____

_____

_____

_____

_____

http:// _____

_____

_____

_____

_____

_____

Diffusion, Osmosis, and the Functional Significance of Biological Membranes

OBJECTIVES

After completing this exercise, you will be able to

1. define *solvent, solute, solution, selectively permeable, diffusion, osmosis, concentration gradient, equilibrium, hypertonic, isotonic, hypotonic, plasmolysis, turgor pressure, cholera, oral rehydration therapy;*

2. describe the structure of cellular membranes;

3. distinguish between diffusion and osmosis;

4. determine the effects of solute size and concentration on diffusion and selective permeability;

5. describe the effects of hypertonic, isotonic, and hypotonic solutions on red blood cells and *Elodea* leaf cells;

6. describe how transport across membranes is affected by cholera, causing dehydration, as well as how to reverse the effects.

Introduction

Living cells are made up of 75% to 85% water. Virtually all substances entering and leaving cells are dissolved in water, making it the **solvent** most important for life processes. The substances dissolved in water are called **solutes** and include such molecules as salts and sugars. The combination of a solvent and dissolved solute is a **solution**. The cytoplasm of living cells is a solution of many solutes, including ions, sugars, and proteins.

All cells are surrounded by a cell membrane made of a phospholipid bilayer with different kinds of embedded and surface proteins (Figure 7-1). Membranes are boundaries that solutes must cross to reach the cellular site where they are used in the processes of life. These membranes regulate the passage of substances into and out of the cell. They are **selectively permeable**, allowing some substances to move across easily while completely or partially excluding others.

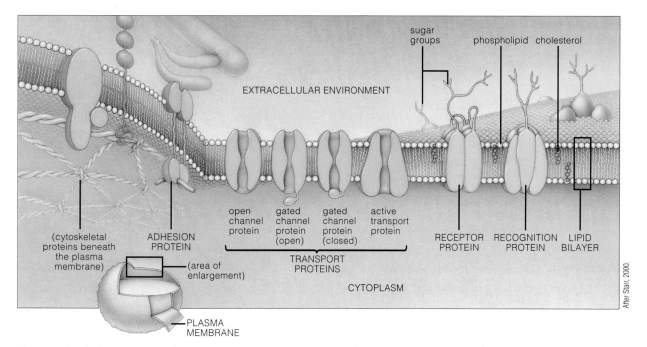

Figure 7-1 Artist's drawing of a cutaway view of part of a plasma membrane.

The simplest way in which solutes enter a cell is **diffusion**, the movement of solute molecules from a region of high concentration to one of lower concentration. Diffusion occurs spontaneously without the expenditure of cellular energy. Oxygen, carbon dioxide, small nonpolar molecules, and water freely cross biological membranes. Channel and transporter proteins allow the passage of some types of polar, water-soluble molecules and ions across biological membranes. Large molecules, such as proteins and polysaccharides, cannot pass membranes; they must be digested and hydrolyzed into smaller component subunits.

When they are inside the cell, solutes move through the cytoplasm by diffusion, sometimes assisted by cytoplasmic streaming (review Exercise 6).

Water (the solvent) also moves across the membrane. **Osmosis** is the movement of water molecules across selectively permeable membranes. Think of osmosis as a special form of diffusion, with net movement of water molecules from a region of higher water concentration (and thus lower solute concentration) to one of lower water concentration (and thus higher solute concentration).

The difference in concentration of like molecules between two regions is called a **concentration gradient**. Diffusion and osmosis take place down concentration gradients, with net movements of molecules from regions of higher concentration to regions of lower concentration. Over time, the concentration of solvent and solute molecules becomes equally distributed, and the gradient ceases to exist. At this point, the system is said to be at **equilibrium**.

Molecules are always in motion, even at equilibrium. Thus, solvent and solute molecules continue to move because of random collisions. However, at equilibrium, there is no net change in their concentrations.

This exercise introduces you to principles of diffusion and osmosis and their application to states of disease and health.

7.1 Experiment: Effect of Solute Concentration on Osmosis (About 20 min. for setup)

In this experiment, you will learn about osmosis using dialysis membrane, a selectively permeable cellulose sheet that permits the passage of water and other small molecules but obstructs passage of larger molecules. If you examined the membrane with a scanning electron microscope, you would see that it is porous; it thus prevents molecules larger than the pores from passing through the membrane. Specifically, you will be attempting to answer the question, *"Does the concentration of a solute affect the rate of osmosis?"* One reasonable hypothesis is that *solute concentration has no effect on the rate of osmosis."* You will test this hypothesis.

MATERIALS

Per student group (4):

- four 15-cm lengths of dialysis tubing, soaking in dH_2O
- eight 10-cm pieces of string or waxed dental floss
- ring stand and funnel apparatus (Figure 7-2)
- 25-mL graduated cylinder
- four small string tags
- china marker
- four 400-mL beakers

Per student group (table):

- dishpan half-filled with dH_2O
- paper toweling
- balance

Per lab room:

- source of dH_2O (at each sink)
- 15% and 30% sucrose solutions
- scissors (at each sink)

PROCEDURE

Work in groups of four for this experiment.

1. Obtain four sections of dialysis tubing, each 15 cm long, that have been presoaked in dH_2O. Recall that the dialysis tubing is permeable to water molecules but not to sucrose.
2. Fold over one end of each tube and tie it tightly with string or dental floss.
3. Attach a string tag to the tied end of each bag and number them 1–4.

4. Slip the open end of the bag over the stem of a funnel (Figure 7-2). Use a graduated cylinder to measure volume (be sure to rinse it well if it has been previously used to measure sucrose) and fill the bags as follows:

Bag 1. 10 mL of dH_2O Bag 3. 10 mL of 30% sucrose
Bag 2. 10 mL of 15% sucrose   Bag 4. 10 mL of dH_2O

5. As each bag is filled, force out excess air by squeezing the bottom end of the tube.
6. Fold the end of the bag and tie it securely with another piece of string or dental floss.
7. Rinse each filled bag in the dishpan containing dH_2O; gently blot off the excess water with paper toweling.
8. Weigh each bag to the nearest 0.5 g.
9. Record the weights in the column marked "0 min." in Table 7-1.
10. Number four 400-mL beakers with a china marker.
11. Add 200 mL of dH_2O to beakers 1–3.
12. Add 200 mL of 30% sucrose solution to beaker 4.
13. Place bags 1–3 in the correspondingly numbered beakers.
14. Place bag 4 in the beaker containing 30% sucrose.
15. After 15 minutes, remove each bag from its beaker, blot off the excess fluid, and weigh each bag.

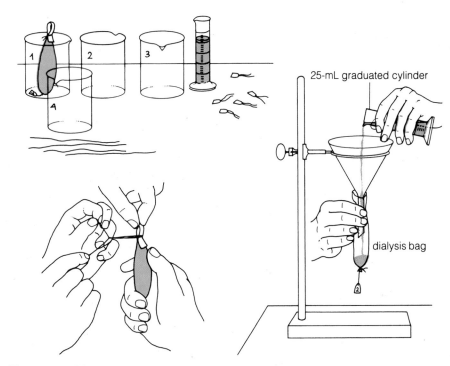

Figure 7-2   Method for filling dialysis bags.

16. Record the weight of each bag in Table 7-1.
17. Return the bags to their respective beakers immediately after weighing.
18. Repeat steps 15–17 at 30, 45, and 60 minutes from time zero.

At the end of the experiment, take the bags to the sink, cut them open, pour the contents down the drain, and discard the bags in the wastebasket. Pour the contents of the beakers down the drain and wash them according to the instructions given on page x.

TABLE 7-1	Change in Weight as a Consequence of Osmosis							

Prediction:

			Bag Weight (g)					
Bag	Bag Contents	Beaker Contents	0 min.	15 min.	30 min.	45 min.	60 min.	Weight Change (g)
1	dH_2O	dH_2O						
2	15% sucrose	dH_2O						
3	30% sucrose	dH_2O						
4	dH_2O	30% sucrose						

Conclusions:

© Cengage Learning 2013

Make a qualitative (descriptive) statement about what you have observed. _____

_____

Was the direction of net movement of water in bags 2–4 into or out of each bag? _____

_____

Which bag gained the most weight? Why? _____

_____

| 7.2 | Experiment: Effect of Solute Size on Diffusion and Selective Permeability of Membranes (*About 15 min. for setup*) |

This experiment will demonstrate the effect of solute molecule sizes on their ability to pass across a selectively permeable membrane. Given what you know about dialysis membrane and diffusion, write a reasonable prediction about the relationship between solute size and membrane permeability, entering it in Table 7-2.

MATERIALS

Per student group (4):
- bottle of dilute iodine solution (I_2KI)
- bottle of 10% soluble starch solution
- two 15-cm lengths of dialysis tubing, soaking in dH_2O
- four 10-cm pieces of string or waxed dental floss
- dishpan half-filled with dH_2O
- two 400-mL graduated beakers
- ring stand and funnel apparatus (Figure 7-2)
- china marker
- 25-mL graduated cylinder
- two plastic dropping pipets
- spot plate or plastic petri dish half
- scissors

PROCEDURE
Work in groups of four.

1. Label one beaker "#1, Starch Bag in Iodine" and the second beaker "#2, Iodine Bag in Starch."
2. Obtain a 15-cm section of dialysis tubing that has been soaked in dH_2O.
3. Fold over one end of the tubing and tie it securely with string or dental floss to form a leakproof bag (Figure 7-2).
4. Slip the open end of the bag over the stem of a funnel.
5. Measure 15 mL of 1% soluble starch and pour it through the funnel to fill the bag.
6. Remove the bag from the funnel; fold and tie the open end of the bag.
7. Rinse the tied bag in a dishpan partially filled with dH_2O; place the bag into Beaker 1.
8. Make and fill a second bag with the iodine solution. Place that dialysis tubing bag into Beaker 2.
9. Pour 200 mL of iodine solution into Beaker 1.
10. Pour 200 mL of starch solution into Beaker 2.
11. Record the time: _____
12. Now, and every 15 minutes over the next hour, record in Table 7-2 your observations of the appearance of the contents inside each dialysis bag and beaker.
13. In the meantime, perform a visual test for the presence of starch by adding 1 drop of iodine solution to several drops of starch solution in the spot plate or petri dish half. Describe the result of this test. How can you identify the presence of starch?

_____

_____

TABLE 7-2 Observations of Contents of Dialysis Bags and Beakers

Prediction:

	At Start of Experiment	After 15 Min.	After 30 Min.	After 45 Min.	After 60 Min.
Beaker 1 dialysis bag contents					
Beaker 1 contents					
Beaker 2 dialysis bag contents					
Beaker 2 contents					

Conclusions:

© Cengage Learning 2013

To which substance(s) was the dialysis tubing permeable?

_____

_____

I_2KI has a molecular weight of approximately 420, and that of starch, a large and variable polysaccharide, ranges from 20,000 to 225,000. What physical property of the dialysis tubing might explain its differential permeability to those two substances?

_____

_____

_____

14. Puncture the dialysis bags and discard the contents of the bags and beakers down the sink drain. Wash the glassware by using the technique described on page x.
15. Discard the dialysis tubing in the wastebasket.

7.3 Osmosis and Red Blood Cells

Recall that the composition of a solution can be described in terms of the proportions of the total made up of solute and solvent. For example, a 5% glucose solution is made of 5% glucose molecules and 95% water molecules.
  Fill in the following:
  A 10% salt (NaCl) solution is composed of _____ % NaCl and _____ % H_2O.
  A 0.9% NaCl solution is composed of _____ % NaCl and _____ % H_2O.
  Tonicity describes one solution's solute concentration *compared with that of another solution*. The solution containing the lower concentration of solute molecules than the other is **hypotonic** *relative to the second solution*. Solutions containing equal concentrations of solute are **isotonic** to each other, and one containing a greater concentration of solute relative to another is **hypertonic** to that second solution.
  Fill in the following:
  A 10% salt solution is _____ relative to a 0.9% salt solution.
  A 0.9% salt solution is _____ relative to a 5% salt solution.
  Osmosis occurs when different concentrations of water are separated by a selectively permeable membrane. Water molecules flow spontaneously (no cellular energy is needed) from the area of higher water concentration to the area of lower water concentration, with a net flow continuing across the membrane until water concentrations are equal on both sides.
  Circle the best answers: This means that water exhibits net flow by osmosis from a (hypertonic, hypotonic, *or* isotonic) solution to a (hypertonic, hypotonic, or isotonic) solution.

An animal cell increases in size as water enters the cell. However, because the plasma membrane is relatively fragile, it may rupture when too much water enters the cell. This is because of excessive pressure pushing against the membrane. A blood cell that has burst is said to have been **hemolyzed**. Conversely, if water moves out of the cell, it becomes *plasmolyzed* (the cell undergoes the process of **plasmolysis**), shrinking in size. In the case of red blood cells, plasmolysis is given a special term, *crenation;* the blood cell is said to be *crenate.*

In this exercise, you will view red blood cells to see effects of osmosis in animal cells.

MATERIALS

Per student:
- compound microscope

Per student group (4):
- three clean screw-cap test tubes
- test tube rack
- metric ruler
- china marker
- bottle of 0.9% sodium chloride (NaCl)
- bottle of 10% NaCl
- bottle of dH$_2$O
- three disposable plastic pipets
- three clean microscope slides
- three coverslips

Per student group (table):
- bottle of sheep blood (in an ice bath)

Per lab room:
- source of dH$_2$O

PROCEDURE

Work in groups of four for this experiment but do the microscopic observations individually.

1. Observe the scanning electron micrographs in Figure 7-3.

Figure 7-3a illustrates the normal appearance of red blood cells. They are biconcave disks; that is, they are circular in outline with a depression in the center of both surfaces. Cells in an isotonic solution will appear like these blood cells.

Figure 7-3b shows cells that have been plasmolyzed.

Figure 7-3c represents a cell that has taken in water but has not yet burst. (Burst red blood cells are said to be *hemolyzed*, and of course they are not visible.) Note the swollen, spherical appearance of the cell.

2. Obtain three clean screw-cap test tubes.
3. Lay test tubes 1 and 2 against a metric ruler and mark lines indicating 5 cm *from the bottom of each tube.*
4. Fill each tube as follows:
   Tube 1: 5 cm of 0.9% sodium chloride (NaCl)
            5 drops of sheep blood
   Tube 2: 5 cm of 10% NaCl
            5 drops of sheep blood
5. Lay test tube 3 against a metric ruler and mark lines indicating 0.5 cm and 5 cm *from the bottom of the tube.*

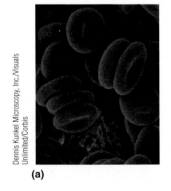
(a)

(b)

(c)

Figure 7-3 Scanning electron micrographs of red blood cells: (**a**) in isotonic solution, (**b**) plasmolyzed, in hypertonic solution, and (**c**) just after placing in hypotonic solution, before hemolysis.

6. Fill tube 3 to the 0.5-cm mark with 0.9% NaCl and to the 5-cm mark with dH_2O. Then add 5 drops of sheep blood. Enter the contents of each tube in the appropriate column of Table 7-3.

7. Replace the caps and mix the contents of each tube by inverting them several times (Figure 7-4a).

8. Hold each tube flat against the printed page of your lab manual (Figure 7-4b). *Only if the blood cells are hemolyzed should you be able to read the print.*

9. In Table 7-3, record your observations in the column "Print Visible?"

10. Number three clean microscope slides.

11. With three *separate* disposable pipets, remove a small amount of blood from each of the three tubes. Place 1 drop of blood from tube 1 on slide 1, 1 drop from tube 2 on slide 2, and 1 drop from tube 3 on slide 3.

12. Cover each drop of blood with a coverslip.

13. Observe the three slides with your compound microscope, focusing first with the medium-power objective and finally with the high-dry objective. (Hemolyzed cells are virtually unrecognizable; all that remains are membranous "ghosts," which are difficult to see with the microscope. To see red blood cells placed in a hypotonic solution, you can make a wet mount slide of the preparation immediately after mixing.)

14. In Figure 7-5, sketch the cells from each tube. Label the sketches, indicating whether the cells are normal, plasmolyzed (crenate), or hemolyzed.

15. Record the microscopic appearance in Table 7-3.

16. Record in Table 7-3 the tonicity of the sodium chloride solutions you added to the test tubes, relative to the cytoplasm of the red blood cells.

If a hypotonic and a hypertonic solution are separated by a selectively permeable membrane, in which direction will there be net movement of water molecules due to osmosis? _____

_____

_____

(a)

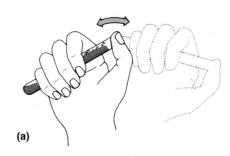

(b)

After Abramoff and Thomson, 1982.

Figure 7-4 Method for studying effects of different solute concentrations on red blood cells.

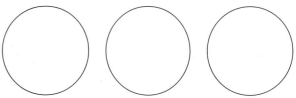

Figure 7-5 Microscopic appearance of red blood cells in different solute concentrations (_____×).
Labels: normal, plasmolyzed (crenate), hemolyzed

If in this same situation, the solute molecules are able to pass through the membrane, in which direction will there be net movement of solute molecules due to diffusion? _____

_____

_____

After completing all experiments, take your dirty glassware to the sink and wash it as directed on page x. Invert the test tubes in the test tube rack so they drain. Reorganize your work area, making certain all materials used in this exercise are present for the next class.

TABLE 7-3	Effects of Salt Solutions on Red Blood Cells			
Tube	Contents	Print Visible?	Microscopic Appearance of Cells	Tonicity of External Solution[a]
1				
2				
3				

© Cengage Learning 2013

[a]With respect to that inside the red blood cell at the start of the experiment.

Plant cells are surrounded by a rigid cell wall, composed primarily of the glucose polymer cellulose. Recall from Exercise 6 that many plant cells have a large central vacuole surrounded by the vacuolar membrane. The vacuolar membrane is selectively permeable. Normally, the solute concentration within the cell's central vacuole is greater than that of the external environment. Consequently, water moves into the cell, creating turgor pressure, which presses the cytoplasm against the cell wall. Such cells are said to be turgid. Many nonwoody plants (like beans and peas) rely on turgor pressure to maintain their rigidity and erect stance.

In this experiment, you will discover the effect of external solute concentration on the structure of plant cells.

MATERIALS

Per student:

- forceps
- two microscope slides
- two coverslips
- compound microscope

Per student group (table):

- *Elodea* in tap water
- two dropping bottles of dH_2O
- two dropping bottles of 20% sodium chloride (NaCl)

PROCEDURE

1. With forceps, remove two young leaves from the tip of an *Elodea* plant.
2. Mount one leaf in a drop of distilled water on a microscope slide and the other in 20% NaCl solution on a second microscope slide.
3. Place coverslips over both leaves.
4. Observe the leaf in distilled water with the compound microscope. Focus first with the medium-power objective and then switch to the high-dry objective.
5. Label the photomicrograph of turgid cells (Figure 7-6).
6. Now observe the leaf mounted in 20% NaCl solution. After several minutes, the cell will have lost water, causing it to become plasmolyzed. (This process is called plasmolysis.) Label the plasmolyzed cells shown in Figure 7-7.

Figure 7-6 Turgid *Elodea* cells (400×).
Labels: cell wall, chloroplasts in cytoplasm, central vacuole

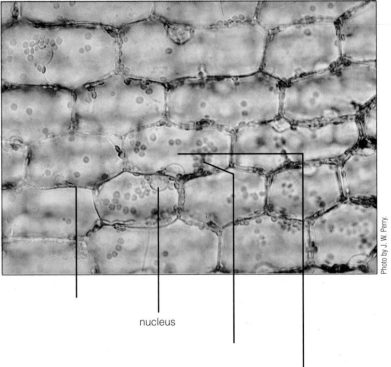

nucleus

Photo by J. W. Perry.

Were the contents of the vacuole in the Elodea leaf in distilled water hypotonic, isotonic, or hypertonic compared to the dH_2O? _____

Was the 20% NaCl solution hypertonic, isotonic, or hypotonic relative to the cytoplasm? _____

If a hypotonic and a hypertonic solution are separated by a selectively permeable membrane, in which direction will the water move? _____

Name two selectively permeable membranes that are present within the *Elodea* cells and that were involved in the plasmolysis process.

1. _____

2. _____

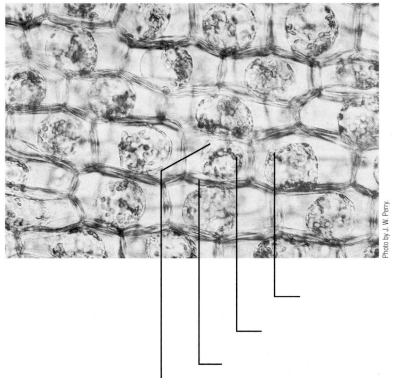

Figure 7-7 Plasmolyzed *Elodea* cells (400×).
Labels: cell wall, chloroplasts in cytoplasm, plasma membrane, space between cell wall and plasma membrane

Photo by J. W. Perry.

7.5   When Osmosis Malfunctions: Understanding the Disease, Cholera

Osmosis and diffusion are homeostatic mechanisms that maintain the water and solute contents of cells and the entire body. If plasma membranes are defective or damaged because of genetic mutations or toxins produced by bacteria or viruses, the consequences can be serious.

Cholera is a disease caused by infection of the small intestine with the bacterium *Vibrio cholerae* (Figure 7-8), often caused by contaminated drinking water. An infected person experiences severe fluid loss through diarrhea. The dehydration proceeds rapidly and can be fatal if not treated, especially in infants and children.

The cholera bacterium produces a toxin that affects plasma membrane proteins in the cells lining the interior of the small intestine. The toxin attaches to specific cell surface receptors and transfers a portion of the toxin molecule into the cell, which responds by continuously secreting chloride (Cl^-) ions into the intestinal lumen (interior). The resulting osmosis causes water to move from tissues into the lumen, producing large volumes of watery diarrhea.

We can visualize this process using a dialysis membrane bag as a stand-in for the small intestine. The contents of the small intestine are normally a watery soup of molecules represented by the liquid that fills the dialysis bag.

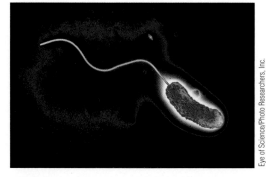

Eye of Science/Photo Researchers, Inc.

Figure 7-8 False color micrograph of *Vibrio cholera*, the bacterium that causes cholera.

MATERIALS

Per student group (4):

- bottle of 80% sucrose
- two 15-cm lengths of dialysis tubing, soaking in dH_2O
- four 10-cm pieces of string or waxed dental floss
- two short lengths of plastic tube
- two large conical tubes standing in rack or two larger beakers
- ring stand and funnel apparatus (Figure 7-2)
- china marker
- 25-mL graduated cylinder
- source of dH_2O

PROCEDURE

1. Label one large tube or beaker "#1 Normal" and the second tube or beaker "#2 Cholera Affected."
2. Obtain a 15-cm section of dialysis membrane that has been soaked in dH_2O.
3. Fold over one end of the membrane and tie it securely with string or dental floss to form a leakproof bag (Figure 7-2).
4. Slip the open end of the bag over the stem of a funnel.
5. Measure 15 mL of dH_2O and pour through funnel to nearly fill the bag.
6. Remove the bag from the funnel. Press out excess air, insert piece of plastic tubing into the opening, and use dental floss or string to tie the open end of the bag tight around the tubing. The plastic tubing end inside the bag should be immersed in the liquid (Figure 7-9).
7. Place the bag into labeled Tube or Beaker 1.
8. Make and fill a second bag with the sucrose solution. Ensure the end of the tubing is immersed into the sucrose before tying the bag closed. Place the dialysis tubing bag into labeled Tube or Beaker 2.
9. Pour dH_2O into each tube or beaker so that the dialysis bag is covered and the water line is just above the dental floss tie.

   Fill in the following using the terms hypertonic, hypotonic, or isotonic:

   In set #1 the contents of the dialysis bag are _____ relative to the liquid outside the bag.

   In set #2 the contents of the dialysis bag are _____ relative to the liquid outside the bag.

   Describe your expectations and explanation regarding net flow of water in set #1, "Normal": _____
   _____
   _____

   Describe your expectations and explanation regarding net flow of water in set #2, "Cholera Affected":
   _____
   _____
   _____

10. Observe both tubes every 5 minutes over the next half hour, taking note of apparent changes in bag volume as well as whether or not any liquid is seen overflowing through the tubing. At the end of the time, record your observations and explanations for each set:

    Set #1, "Normal":
    _____
    _____
    _____

Figure 7-9 Dialysis bag and tube apparatus for visualizing effects of cholera.

    Set #2, "Cholera Affected":
    _____
    _____
    _____

    How do your observations relate to the symptoms of cholera?
    _____
    _____
    _____

Although the *Vibrio* bacterium is sensitive to antibiotics, dehydration is too rapid and severe to wait for the time needed for antibiotics to take effect. **Oral rehydration therapy** is a simple, quick, and effective treatment. A common recipe for oral rehydration solution (ORS) is 1 level tsp of salt (NaCl) plus 8 level tsp of sugar (sucrose) dissolved in 1 L of clean drinking water. This inexpensive solution is drunk by the infected person, with the goals of replacing lost fluid and maintaining the internal ion concentrations of body cells.

ORS is isotonic to body fluids. Sodium ions are actively moved (a transport process against a concentration gradient and requiring expenditure of cellular energy) into the intestine cells, replacing ions lost during diarrhea. Maximum water and ion absorption is achieved with the mixture of salt and sugar, though, through a membrane transport mechanism known as "cotransport."

Recall that table sugar, sucrose, is a disaccharide made of fructose and glucose subunits. Sucrose is hydrolyzed into fructose and glucose in the upper small intestine; glucose moves through channel proteins in the plasma membrane of some of the small intestine lining cells. This glucose transport only occurs in the presence of sodium ions, though, with Na^+ ions entering cells from the intestine lumen at the same time as glucose. The ORS composition thus ensures maximal restoration of cytoplasmic tonicity. Sports drinks, such as Gatorade™, are not adequate substitutes for ORS because they contain too high sugar concentration and too little salt to be effective.

_____ 1. If one were to identify the most important compound for sustenance of life, it would probably be
  (a) salt.
  (b) sugar.
  (c) water.
  (d) I_2KI.

_____ 2. A solvent is
  (a) the substance in which solutes are dissolved.
  (b) a salt or sugar.
  (c) one component of a biological membrane.
  (d) selectively permeable.

_____ 3. Diffusion
  (a) is a process requiring cellular energy.
  (b) is the movement of molecules from a region of higher concentration to one of lower concentration.
  (c) occurs only across selectively permeable membranes.
  (d) none of the above

_____ 4. Cellular membranes
  (a) consist of a phospholipid bilayer containing embedded proteins.
  (b) control the movement of substances into and out of cells.
  (c) are selectively permeable.
  (d) all of the above

_____ 5. An example of a solute is
  (a) NaCl.
  (b) water.
  (c) sucrose.
  (d) both a and c

_____ 6. Dialysis membrane
  (a) is selectively permeable.
  (b) is used in these experiments to simulate cellular membranes.
  (c) has pores that allow passage to specific-sized molecules.
  (d) all of the above

_____ 7. Specifically, osmosis
  (a) requires the expenditure of cellular energy.
  (b) is diffusion of water from lower to higher concentration.
  (c) is diffusion of water across a selectively permeable membrane.
  (d) none of the above

_____ 8. When the cytoplasm of a red blood cell has lost water to its surroundings, the cell is said to be
  (a) isotonic.
  (b) burst.
  (c) hemolyzed.
  (d) crenate.

_____ 9. When the cytoplasm of a plant cell is pressed against the cell wall, the cell is said to be
  (a) turgid.
  (b) plasmolyzed.
  (c) hemolyzed.
  (d) crenate.

_____ 10. Cholera is a disease whose main symptom is
  (a) increase in water uptake by intestine cells.
  (b) rapid dehydration.
  (c) uptake of glucose by intestine cells.
  (d) none of the above

EXERCISE **7**

Diffusion, Osmosis, and the Functional Significance of Biological Membranes

Post-Lab Questions

7.1 Experiment: Effect of Solute Concentration on Osmosis

1. If a 10% sugar solution is separated from a 20% sugar solution by a selectively permeable membrane, in which direction will there be a net movement of water? Explain.

2. Based on your observations in this exercise, do you think that dialysis membrane is permeable to sucrose? Explain.

3. How does osmosis differ from diffusion?

7.2 Experiment: Effect of Solute Size on Diffusion and Selective Permeability of Membranes

4. Based on your observations in this exercise, would you expect dialysis membrane to be permeable to egg albumen, a protein with molecular weight of approximately 36,800? Why or why not?

7.3 Osmosis and Red Blood Cells

5. Explain, using terminology of tonicity and membrane function, why disaster would result if a patient were given intravenous dH$_2$O rather than 0.9% "physiological saline" solution.

7.4 Plasmolysis in Plant Cells

6. This drawing represents what happens inside a plant cell after it is placed in a solution.

   (a) What *process* is taking place as the cell progresses through the stages shown in the direction of the arrows? _____

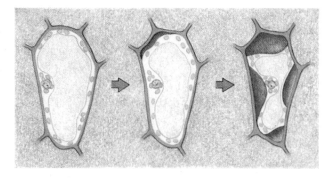

(b) Is the solution in which the cells have been placed hypotonic, isotonic, or hypertonic relative to the cytoplasm? How do you know?

(c) What happens at the cellular level when a wilted plant is watered and begins to recover from the wilt?

7.5 When Osmosis Malfunctions: Understanding the Disease, Cholera

7. A human lost at sea without fresh drinking water is effectively lost in an osmotic desert. Why is drinking salt water harmful?

8. Explain why a person whose intestinal lining cells continuously "leak" chloride ions into the intestinal lumen experiences dehydration.

9. Oral rehydration therapy is ancient. In 1000 BCE, an Indian physician recommended that his patients with diarrhea drink large amounts of water containing dissolved rock salt and molasses. Describe in your own words how oral rehydration solution acts to prevent death from cholera.

Food for Thought

10. Cystic fibrosis is a common genetic disease, affecting about one of every 3200 Caucasian births in the United States. This disease is recognizable in the ancient Northern European folk saying, "Woe to that child which when kissed on the forehead tastes salty. He is bewitched and soon must die." Search two Internet sites for information on "cystic fibrosis and osmosis" or "cystic fibrosis and membrane transport." List the two sites below and briefly summarize what you learn from each of them regarding the connections between this disease and malfunctioning plasma membranes.

http:// _____

_____

_____

_____

_____

_____

http:// _____

_____

_____

_____

_____

_____

Enzymes: Catalysts of Life

OBJECTIVES

After completing this exercise, you will be able to

1. define *enzyme, catalyst, activation energy, substrate, enzyme–substrate complex, product, active site, enzyme specificity, denaturation;*

2. explain how an enzyme operates;

3. identify the substrate and products of the reaction catalyzed by the enzyme catalase;

4. design and carry out experiments to test hypotheses about factors that affect enzyme action;

5. describe how enzyme activity is generally affected by changes in enzyme concentration, pH, and temperature.

Introduction

Life would be impossible without enzymes. **Enzymes** are proteins that function as biological **catalysts**. A catalyst is a substance that lowers the amount of energy necessary for a chemical reaction to occur. You might think of this so-called **activation energy** as a mountain to be climbed. Enzymes decrease the height of the mountain, in effect turning it into a molehill. (Figure 8-1.) By lowering the activation energy, an enzyme speeds up the rate at which a reaction occurs.

Enzymes catalyze condensation and hydrolysis reactions in cells, rearrange chemical bonds, and perform many other tasks essential for cellular metabolism and maintenance of the body's homeostasis, or stable internal balance.

In an enzyme-catalyzed reaction, the reactant (the substance being acted upon) is called the **substrate**. Substrate molecules combine with enzyme molecules to form a temporary **enzyme–substrate complex**. Products are then formed, and the enzyme molecule is released unchanged. The enzyme molecule is not used up in the process and is capable of catalyzing the same reaction again and again. This can be summarized as follows:

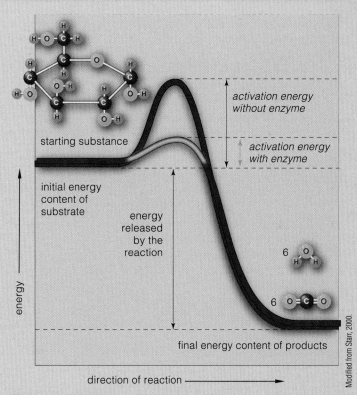

Figure 8-1 Enzymes and activation energy.

Substrate + Enzyme → Enzyme–substrate complex → Products + Enzyme

An enzyme is similar to a key in that it has a specifically shaped "active site" into which only specific substrates fit. After it is bound in the active site, the enzyme–substrate complex changes shape slightly, allowing the chemical reaction to proceed more quickly. The products of the chemical reaction are released, and the enzyme returns to its original shape.

Before we proceed, let's visualize this process (Figure 8-2). Think of an enzyme as the key that unlocks a padlock. Imagine that you have many locks, and all of them use the same key. You would only need to reuse your single key to unlock them all.

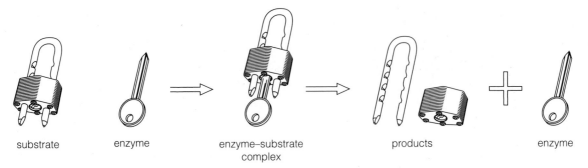

substrate   enzyme   enzyme–substrate complex   products   enzyme

Figure 8-2   Analogy of enzyme action.

Changing the shape of the key just a tiny bit may still allow the key to function in the lock, but you may have to fumble with the key a while to get the lock open. Changing the key shape still further may prevent the key from opening the lock at all.

Most enzymes are specific and catalyze reactions with only one or a very few kinds of substrates and products. This **enzyme specificity** lies in the shape of the protein molecule; anything that changes the shape of an enzyme alters its ability to function. Enzymes function best within a certain range of temperature, pH, or salinity (saltiness), for example. Although thousands of kinds of enzymes are present in living cells, we'll examine the activity of one, catalase.

Catalase is a widespread enzyme found in nearly all aerobic cells (plants, animals, and microbes). It protects cells from the toxic effects of hydrogen peroxide (H_2O_2) generated as a by-product of cell metabolism. It does this by catalyzing the decomposition of the substrate hydrogen peroxide into oxygen gas and water products as follows:

$$2H_2O_2 \rightarrow 2H_2O + O_2 \text{ (gas)}$$

Catalase activity can be visualized in Figure 8-3.

In this exercise, you will examine the activity of catalase, and design and carry out an experiment to examine the effect of one or more factors on the reaction. You will use liver homogenate (a slurry of liver and water blended together) as a source of catalase. The oxygen produced by the breakdown of hydrogen peroxide will be measured as an indication of catalase activity.

MATERIALS

Per student:

- safety goggles

Per student group (4):

- beaker of liver extract containing catalase
- wash bottle of dH_2O
- bottle or stoppered flask of fresh hydrogen peroxide, H_2O_2
- reaction chamber
- 25-mL graduated cylinder
- two 10-mL graduated cylinders
- deep square or rectangular pan
- filter paper
- hole punch
- forceps
- stopwatch
- source of distilled water
- buffers pH 4, 7, and 10 or others as provided by your instructor
- ice
- insulated container
- water bath at 37°C
- thermometer

Work in groups of four for all sections in this exercise.

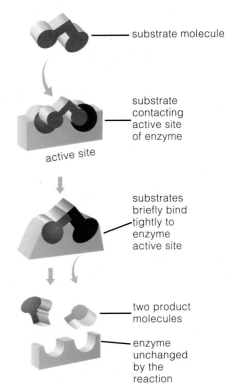

substrate molecule

substrate contacting active site of enzyme

active site

substrates briefly bind tightly to enzyme active site

two product molecules

enzyme unchanged by the reaction

Figure 8-3   Induced-fit model of enzyme–substrate interactions. The hydrogen peroxide substrate physically fits into the active site of catalase. The substrate briefly binds to the active site, and an enzyme shape change allows water and oxygen gas to be formed. These products are released, and the enzyme resumes its original shape.

> **CAUTION**
>
> Hydrogen peroxide is a reactive material. Wear goggles and avoid eye and skin contact.

PROCEDURE

1. Punch several filter paper discs with the hole punch. Do not touch the discs with your fingers because skin oils will interfere with the process; handle the discs only with forceps.
2. Fill the pan with water. Lay the 25-mL graduated cylinder on its side in the pan so that it fills with water. Tilt the mouth of the cylinder upward slightly to remove all trapped air bubbles. Keep the mouth of the cylinder underwater while you stand the cylinder upside down in the pan.
3. Pour 10 mL of hydrogen peroxide into the reaction chamber. Take note if the hydrogen peroxide runs down a side wall of the chamber.
4. Hold a filter paper disc with the forceps and hold it in the liver solution for several seconds. Drain the excess solution from the disc by blotting it against the side of the container.

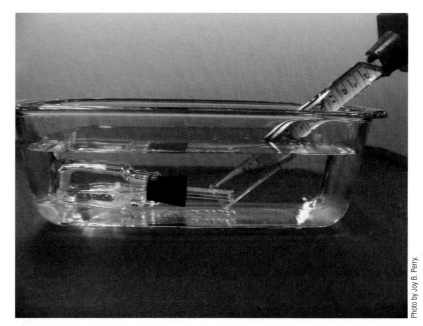

Figure 8-4 Equipment to measure catalase activity. Note bubbles in reaction chamber and accumulating oxygen bubble at top of graduated cylinder.

5. Place the disc high on a side wall of the reaction chamber; do not allow the disc to touch the hydrogen peroxide or an area that it flowed over.
6. Insert the stopper in the chamber and close it tightly.
7. Place the reaction chamber on its side in the pan of water, *making sure that the filter paper disc is on the top side.*
8. Tip the graduated cylinder into position so that its mouth lies directly over the tip of the pipet protruding from the reaction chamber. This will allow any oxygen released to rise into the graduated cylinder. Prop or hold the cylinder in position for the duration of data collection (see Figure 8-4).
9. Rotate the reaction chamber 180 degrees so the paper disc comes in contact with the hydrogen peroxide solution.
10. Allow the reaction to proceed for 10 minutes and then tip the graduated cylinder upside down again with the mouth still underwater so that you can measure the accumulated oxygen.

   Describe the appearance of the filter paper disc and surrounding solution as the reaction begins and explain this observation in terms of the action of catalase.

   _____

   _____

   _____

   How much oxygen was generated during this 10-minute experiment? _____

   _____

11. Discard the reaction chamber contents into the sink or other area designated by your instructor. Rinse the reaction chamber with water and drain it upside down before proceeding with further investigations.

In the remainder of this exercise, you will work with your team to determine what, if any, changes in catalase activity result from changing various conditions in the reaction chamber. Recall from Exercise 1, The Scientific Method, that you construct a *hypothesis* (tentative, possible general explanation) to answer a *question*. This hypothesis is often written in the form of an "if-then" statement as a *prediction* of results from an experimental test.

Your experiment will allow you to make observations and collect data to test the prediction. You can then compare your predicted results with what you've actually observed and then make a conclusion about whether the hypothesis was supported or falsified.

In an experiment, the factor that you change or manipulate is the *independent variable*. The factor that you measure as a result of the experiment is the *dependent variable*. In every experiment, you will also have a number of *controlled variables*, which are all of the other conditions and factors that are kept the same for all treatment groups. In this way, your experiment will determine if the dependent variable changes based on your manipulations of the independent variable.

8.2.A. Experiment: Effect of Enzyme Concentration on Reaction Rate (About 30 min.)

Is the rate of oxygen production affected by concentration of catalase? In this activity, you will use the method from Section 8.1 to design an experiment to test the hypothesis that *reaction rate increases with increasing enzyme concentration*. In this experimental system, placing multiple soaked filter paper discs in the reaction chamber effectively multiplies the enzyme concentration correspondingly.

PREDICTION: If enzyme concentration is (changed how?) _____, then the rate of oxygen production will (change how?) _____

EXPERIMENTAL DESIGN:

The independent variable is _____

What levels of the independent variable will you test? _____

The dependent variable is _____

Controlled variables include _____

_____

Record your data in Table 8-1.

What sources of error or uncertainty were there in your experiment?

_____

_____

How might another experiment to test this hypothesis be improved?

_____

_____

TABLE 8-1	Effect of Enzyme Concentration on Reaction Rate
Prediction:	
Enzyme concentration	Oxygen production at _____ minutes (mL)
1× enzyme concentration	
___× enzyme concentration	
___× enzyme concentration	
Conclusion:	

© Cengage Learning 2013

8.2.B. Experiment: Effect of pH on Catalase Activity
(About 30 min.)

pH is a measure of the hydrogen ion concentration in a cell on a scale ranging from pH of 1 (extreme acidity) to 7 (neutral) to 14 (extreme base). A change of 1 pH unit reflects a 10-fold change in hydrogen ions and acidity (see Figure 8-5). pH affects the three-dimensional shape of enzymes, thus regulating their function. Most enzymes operate best when the pH of the solution is near neutrality (pH 7). Others, however, have pH optima in the acidic or basic range, corresponding to the environment in which they normally function. When an enzyme's structure is changed sufficiently to destroy its function, the enzyme is said to be **denatured** (has undergone **denaturation**). Most enzymatically controlled reactions have an *optimum* pH and temperature—that is, one pH and temperature at which activity is maximized.

How is catalase activity affected by pH changes? In this activity, you will use the method from Section 8.1 and the pH buffers present in your lab to design an experiment to test the hypothesis that *enzyme activity is greatest in a neutral pH range.*

The pH of hydrogen peroxide solution can be changed by adding 8 mL of H_2O_2 to the reaction chamber and then adding 4 mL of the desired buffer solution and swirling to mix well.

PREDICTION: If pH is (changed how?) _____, then the rate of oxygen production will (change how?)

_____

EXPERIMENTAL DESIGN:

The independent variable is _____

What levels of the independent variable will you test?

_____

The dependent variable is _____

Controlled variables include _____

_____

Record your data in Table 8-2.

What sources of error or uncertainty were there in your experiment?

_____

_____

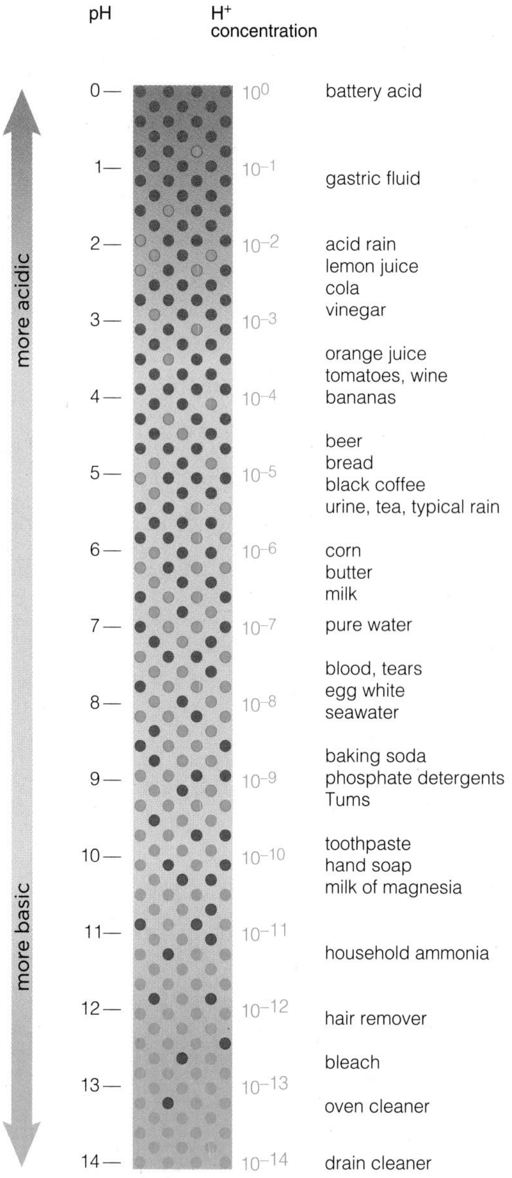

How might another experiment to test this hypothesis be improved?

_____

_____

Figure 8-5   The pH scale.

pH	H⁺ concentration	
0	10^0	battery acid
1	10^{-1}	gastric fluid
2	10^{-2}	acid rain / lemon juice / cola
3	10^{-3}	vinegar
		orange juice
4	10^{-4}	tomatoes, wine / bananas
		beer
5	10^{-5}	bread / black coffee / urine, tea, typical rain
6	10^{-6}	corn / butter / milk
7	10^{-7}	pure water
8	10^{-8}	blood, tears / egg white / seawater
9	10^{-9}	baking soda / phosphate detergents / Tums
10	10^{-10}	toothpaste / hand soap / milk of magnesia
11	10^{-11}	household ammonia
12	10^{-12}	hair remover
13	10^{-13}	bleach / oven cleaner
14	10^{-14}	drain cleaner

more acidic / more basic

TABLE 8-2	**Effect of pH on Catalase Activity**
Prediction:	
pH	Oxygen production at _____ minutes (mL)
4	
7	
10	
Conclusion:	

© Cengage Learning 2013

The rate at which chemical reactions take place is largely determined by the temperature of the environment. *Generally, for every 10°C increase in temperature, the reaction rate doubles.* Within a rather narrow range, this is true for enzymatic reactions as well. However, because enzymes are proteins, excessive temperature alters their structure, interferes with activity, and may denature them if too hot.

In this activity, you will use the methods in Section 8.1 to design an experiment to determine the *effect of different temperatures on the activity of catalase.* This may help you determine the best (optimum) temperature for the reaction. Ice is available to create low temperature conditions surrounding the reaction chamber, and a heated waterbath will increase the water temperature to approximately body temperature, 37°C.

PREDICTION: If temperature is (changed how?) _____, then the rate of oxygen production will (change how?) _____

EXPERIMENTAL DESIGN:

The independent variable is _____

What levels of the independent variable will you test? _____

The dependent variable is _____

Controlled variables include _____

_____

Record your data in Table 8-3.

What sources of error or uncertainty were there in your experiment? _____

_____

_____

How might another experiment to test this hypothesis be improved? _____

_____

_____

TABLE 8-3	Effect of Temperature on Enzyme Activity

Prediction:

Temperature (°C)	Oxygen production at _____ minutes (mL)

Conclusion:

© Cengage Learning 2013

Note: **After completing all experiments, take your dirty glassware to the sink and wash it following the directions on page x. Invert the glassware so it drains. Tidy up your work area, making certain all materials used in this exercise are there for the next class.**

_____ 1. Enzymes are
   (a) biological catalysts.
   (b) agents that speed up cellular reactions.
   (c) proteins.
   (d) all of the above

_____ 2. Enzymes function by
   (a) being consumed (used up) in the reaction.
   (b) lowering the activation energy of a reaction.
   (c) combining with otherwise toxic substances in the cell.
   (d) adding heat to the cell to speed up the reaction.

_____ 3. The substance that an enzyme combines with is
   (a) another enzyme.
   (b) a product.
   (c) an active site.
   (d) the substrate.

_____ 4. _Enzyme specificity_ refers to the
   (a) specific name of each kind of enzyme.
   (b) fact that enzymes catalyze one particular substrate or a small number of structurally similar substrates.
   (c) effect of temperature on enzyme activity.
   (d) effect of pH on enzyme activity.

_____ 5. For every 10°C increase in temperature, the rate of most chemical reactions will
   (a) double.
   (b) triple.
   (c) increase by 100 times.
   (d) stop.

_____ 6. When an enzyme becomes denatured, it
   (a) increases in effectiveness.
   (b) loses its requirement for a substrate.
   (c) forms an enzyme–substrate complex.
   (d) loses its ability to function.

_____ 7. An enzyme may lose its ability to function because of
   (a) excessively high temperatures.
   (b) a change in its three-dimensional structure.
   (c) a large change in the pH of the environment.
   (d) all of the above

_____ 8. pH is a measure of
   (a) an enzyme's effectiveness.
   (b) enzyme concentration.
   (c) hydrogen ion concentration.
   (d) none of the above

_____ 9. Catalase
   (a) is an enzyme found in most cells, including liver cells.
   (b) catalyzes the production of hydrogen peroxide.
   (c) has as its substrate water and oxygen gas.
   (d) is a substance that encourages the growth of microorganisms.

_____ 10. The oxygen production measured in the experiments of this exercise
   (a) is a consequence of the breakdown of hydrogen peroxide.
   (b) is an index of enzyme activity.
   (c) may depend on pH, temperature, or enzyme concentration.
   (d) all of the above

EXERCISE **8**

Enzymes: Catalysts of Life

Post-Lab Questions

Introduction

1. Explain the difference between *substrate* and *active site*.

8.2.A. Experiment: Effect of Enzyme Concentration on Reaction Rate

2. Given your observations in this activity, predict the effect on the catalase reaction rate if the hydrogen peroxide solution were diluted by half. Explain.

8.2.B. Experiment: Effect of pH on Catalase Activity

3. Explain what happens to the catalase molecule and activity when the pH is on either side of the optimum level.

4. (a) While the pH of saliva ranges from 6.5 to 11.5, the pH in the stomach is about pH 2. What would you expect the optimum pH to be for the enzymes secreted into your stomach that digest proteins into amino acids?

   (b) What would you expect the optimum pH to be for the salivary enzyme amylase, which digests starches into sugars?

8.2.C. Experiment: Effect of Temperature on Enzyme Activity

5. Eggs can contain bacteria such as *Salmonella*. Considering what you've learned in this exercise, explain why cooking eggs makes them safe to eat.

6. Why do you think high fevers alter cellular functions?

7. Some surgical procedures involve lowering a patient's body temperature during periods when blood flow must be restricted. What effect might this have on enzyme-controlled cellular metabolism?

Food for Thought

8. Is it necessary for a cell to produce one enzyme molecule for every substrate molecule that needs to be catalyzed? Why or why not?

9. What happened to the composition of the solution inside the reaction chamber as a catalase reaction proceeded? Why was it a good practice to start with fresh hydrogen peroxide for each experimental trial?

10. Hydrogen peroxide solution is often used to sterilize cuts and scrapes. Why does hydrogen peroxide bubble when it is poured on an open wound but *not* when it is poured on intact skin?

Photosynthesis: Capture of Light Energy

OBJECTIVES

After completing this exercise, you will be able to

1. define *photosynthesis, autotroph, heterotroph, light-dependent reactions, light-independent reactions, thin layer chromatography, chlorophyll* a, *chlorophyll* b, *carotenes, xanthophylls, chloroplast, stroma, thylakoid disks, grana, starch grains;*

2. use a pH indicator to make visible the uptake of carbon dioxide in photosynthesis;

3. use a leaf disk assay to quantify oxygen production during photosynthesis;

4. design and perform a leaf disk assay experiment to measure photosynthetic response to change in an environmental factor;

5. separate and identify the pigments in chloroplast extract using thin layer chromatography;

6. identify the structures composing the chloroplast and indicate the function of each structure in photosynthesis.

Introduction

Photosynthesis is the process by which light energy converts inorganic compounds to organic substances with the subsequent release of elemental oxygen. It may very well be the most important biological event sustaining life. Without it, most living things would starve, and atmospheric oxygen would become depleted to a level incapable of supporting animal life. Ultimately, the source of light energy is the sun, although on a small scale we can substitute artificial light.

Nutritionally, two types of organisms exist in our world, autotrophs and heterotrophs. **Autotrophs** (*auto* means self, *troph* means feeding) synthesize organic molecules (carbohydrates) from inorganic carbon dioxide. The vast majority of autotrophs are the photosynthetic organisms that you're familiar with—plants, as well as some protistans and bacteria. These organisms use light energy to produce carbohydrates. (A few bacteria produce their organic carbon compounds chemosynthetically, that is, using chemical energy.)

By contrast, **heterotrophs** must rely directly or indirectly on autotrophs for their nutritional carbon and metabolic energy. Heterotrophs include animals, fungi, many protistans, and most bacteria.

In both autotrophs and heterotrophs, carbohydrates originally produced by photosynthesis are broken down by *cellular respiration* (Exercise 10), releasing the energy captured from the sun for metabolic needs.

The photosynthetic reaction is often conveniently summarized by the equation:

$$12H_2O + 6CO_2 \xrightarrow{\text{light energy}} 6O_2 + C_6H_{12}O_6 + 6H_2O$$

water carbon   oxygen glucose water
      dioxide

The **light-dependent reactions** capture light energy and store it temporarily in the bonds of ATP and NADPH. These molecules then provide the energy required for the **light-independent reactions (Calvin-Benson cycle)**, in which environmental carbon dioxide is "fixed" (incorporated into a biological molecule) and used to build carbohydrates.

Although glucose is often produced during photosynthesis, it is usually converted to another transport or storage compound unless it is to be used immediately for carbohydrate metabolism. In plants and many protists, the most common storage carbohydrate is *starch*, a compound made up of numerous glucose units linked together. Most plants transport carbohydrate as sucrose.

This investigation will acquaint you with several aspects of the process of photosynthesis and provide you with practice in thinking scientifically.

9.1 Uptake of Carbon Dioxide during Photosynthesis
(About 15 min. to set up)

Phenol red is a pH indicator substance that is yellow in an acidic solution (pH < 6.8) but is red in a basic solution above pH 8. In this experiment you'll use phenol red to detect the uptake of CO_2 by a submerged aquatic plant, *Elodea*.

To detect the carbon dioxide uptake, you will put the *Elodea* stems into a solution that you've made slightly acidic with your breath. Carbon dioxide in your breath will dissolve in water to form carbonic acid, which lowers the pH of the solution.

As the plant fixes CO_2 and removes it from the water, the pH of the solution rises. As the pH rises above 6.8, the solution will become orange, becoming red if pH exceeds 8. This provides visible evidence of the uptake of CO_2.

MATERIALS

Per student group (4):
- two 50-mL graduated cylinders
- 250-mL beaker
- dilute solution of phenol red
- straw
- *Elodea*
- razor blade
- ruler
- light source

PROCEDURE

1. Pour about 100 mL of phenol red solution into a beaker.
2. Slowly blow into the beaker with a straw. Stop blowing into the beaker as soon as the color changes to yellow. (Continuing to acidify the solution with your breath will create excess carbonic acid that will lengthen this experiment.)
3. Cut two stems of *Elodea* long enough to reach to the 50-mL mark of a cylinder.
4. Add **both** stems to **one** of the cylinders.
5. Pour the acidified phenol red solution to just below the brim of **each** cylinder.
6. Place both tubes in bright light, but not so close that the solution will become warm.
7. Observe every 10 minutes.

What is the purpose of the tube without *Elodea*? _____

_____

What happened over time to the color of the indicator in the tube with *Elodea*? _____

_____

What happened over time to the color of the indicator in the tube without *Elodea*? _____

_____

Explain these results, including discussion of the reactants and/or products of photosynthesis. _____

_____

_____

_____

9.2 Release of Oxygen: Floating Leaf Disk Assay for Quantitative Investigation of Photosynthesis *(About 100 min. total)*

You will learn a procedure called a "floating leaf disk assay" to illustrate oxygen production in photosynthesis today, and then employ that technique in an experiment of your own design. First you will compare the rate of photosynthesis in leaves with a carbon dioxide source present vs. the rate in leaves without carbon dioxide.

MATERIALS

Per student group (4):

- 0.2% sodium bicarbonate solution in flask or bottle (produces dissolved carbon dioxide in solution)
- dilute liquid soap in dropper bottle
- 100-mL beakers
- two plastic 10-cc syringes
- leaf material
- hole punch
- Sharpie or other fine-tipped marker

Per lab room:

- clock
- light source
- source of distilled water

9.2.A. Release of Oxygen in Photosynthesis (About 20 min. to set up)

PROCEDURE

(a) Pour about 50 mL of bicarbonate solution into one labeled beaker, and add 1 drop of dilute liquid soap to it. The soap will wet the hydrophobic leaf surfaces, allowing the solution to be drawn into the leaves. Avoid suds. If your solution generates suds as you use it, then dilute it with more bicarbonate solution.

(b) Pour about 50 mL of distilled water into the second labeled beaker, and add a drop of diluted soap to it.

(c) Cut 20 or more uniform leaf disks from the plant material provided with the hole punch. Avoid major veins and try to make the cuts as clean and sharp as possible.

(d) Infiltrate the leaf disks with solution as follows:

1. Label one syringe "with CO_2" and the other "dH_2O." Remove the plunger from each syringe, and place 10 leaf disks into each, tapping them down to the bottom. Replace the plunger, being careful not to crush the leaf disks. Push on the plunger until only a small volume of air and leaf disk material remain in the barrel (<10% of total volume).

2. Pull 10 mL of sodium bicarbonate solution into the "with CO_2" syringe. Tap the syringe to suspend the leaf disks in the solution.

3. Hold a finger over the syringe opening, draw back on the plunger to create a vacuum. Hold this vacuum for about 5 seconds. While holding the vacuum, swirl the leaf disks to suspend them in the solution. Release the vacuum by quickly releasing the plunger.

   The bicarbonate solution should infiltrate the air spaces in the leaf disks, causing the disks to sink. You will probably have to repeat this procedure several times to get all the disks to sink. If you still have difficulty getting the disks to sink, you may need a bit more soap in the solution. Alternatively, if you pull a vacuum on the disks too often, you can injure the tissues and ruin the discs for further use.

4. Repeat procedure with dH_2O-soap solution in second syringe.

(e) Remove the plunger from the syringe barrel and release the infiltrated "with CO_2" disks into the bicarbonate solution beaker. Do not pour the disks through the air or create air bubbles in the solution. Repeat with the water-infiltrated disks and the dH_2O beaker. Adjust total volume in each beaker to 100 mL.

(f) Place beakers under the light source and note time. Every 10 minutes, record the number of floating disks in each beaker. Very gently swirl or tap syringes to dislodge any that are stuck against the sides, if necessary. Continue until all of the disks are floating or an hour has elapsed. The point at which 80% of the leaf disks are floating is the point of reference for this procedure. If leaf discs don't float within an hour, assume that they will not. Record data in Table 9-1.

How many minutes elapsed before 80% of the leaf disks were floating in the sodium bicarbonate solution?

_____

For the distilled water solution? _____

Explain your outcome in terms of the reactants and/or products of photosynthesis. _____

_____

_____

This simple activity was designed to illustrate the leaf disk bioassay method and to demonstrate a product of photosynthesis. Ten leaf disks per treatment provided a modest sample size and more reliable results than if you had only included one or two disks per syringe. An experiment should also include replication, using multiple "subjects" per trial. In this activity, replication could have been provided by setting up multiple syringes with both control (dH_2O) and treatment (sodium bicarbonate) conditions.

© Cengage Learning 2013

TABLE 9-1	Leaf Disk Flotation	
Time (min.)	Number of Disks Floating (bicarbonate solution)	Number of Disks Floating (distilled water)
0		
10		
20		
30		
40		
50		
60		

9.2.B. Investigation of Factors that Affect Rate of Photosynthesis (About 90 min.)

Using the floating leaf disk assay technique, design and carry out an experiment to test the effect of a chosen variable on the rate of photosynthesis. Work with a partner or team as assigned by your instructor and choose which variable to investigate (other than the differences between using a carbon dioxide source vs. distilled water as just completed). Table 9-2 lists possible variables that provide a starting point for developing questions to investigate the light-dependent reactions of photosynthesis. Think about how each of these variables might affect the rate of photosynthesis. You may think of additional variables to test. Consult with your instructor about materials and conditions available. (Note that disks from hairy leaves are difficult to infiltrate.)

What question do you wish to answer about a factor that may affect photosynthesis rate? _____

_____

What independent variable will you manipulate in order to answer your question? What changes will you make in that variable? _____

What is the dependent variable? _____
What factors will you control? _____

_____

How will you provide adequate sample size and replication? _____

_____

Title and label Table 9-3 on the next page. Conduct your experiment and record data.

TABLE 9-2	Variables that Might Affect the Rate of Photosynthesis	
Plant or Leaf Variables	**Environmental Variables**	**Method Variables**
■ Leaf color (chlorophyll presence/absence or presence of other pigments)	■ Light color (wavelength)	■ Size of leaf disk
■ Leaf age	■ Light intensity (brightness)	■ Methods of cutting disks
■ Leaf health	■ Temperature	
■ Leaf position (grown in full sun vs. grown in shade)	■ Solution pH	

© Cengage Learning 2013

TABLE 9-3	Effect of ▭▭▭▭▭▭▭▭▭▭▭▭ on Leaf Disk Flotation

Prediction:

© Cengage Learning 2013

Conclusion:

Describe several sources of error *and* several ideas for improvement of your experiment. _____
_____
_____
_____

9.3 Separation of Photosynthetic Pigments by Thin Layer Chromatography *(25 min.)*

Chromatography separates complex mixtures based on mixture components' solubility in different kinds of solvents. This process can be used to help separate and identify some of the pigments used in photosynthesis. You will utilize **thin layer chromatography** (TLC) by spotting a leaf extract onto a TLC strip (a thin layer of absorbent material bound to a plastic sheet) and then allowing a solvent mix to be drawn up the TLC strip. Different pigments vary in their ability to dissolve in the solvent mix and will travel up the chromatography strip at different rates. The result will be a series of pigment stripes separating out along the length of the strip.

CAUTION

The solvents used in this exercise are flammable and toxic. Wear safety glasses and gloves, and work in a fume hood if possible to minimize your exposure.

MATERIALS

Per student pair:

- TLC strip
- forceps
- pencil
- ruler
- screw-cap bottle or tube
- mortar and pestle
- hair dryer
- dropping pipets
- watch glass
- toothpicks
- bottle of acetone
- dropping pipet
- Sharpie or other fine marking pen
- chromatography solvent

Per lab room:

- colored pencils

Photosynthesis: Capture of Light Energy　**117**

PROCEDURE

(a) Obtain a TLC strip approximately 1.25 cm × 10 cm, or sized to fit into the test tubes provided. The oils from your fingers will interfere with the chromatography process, so handle the strip by holding it on the edges of the top corners only, preferably with forceps.

(b) With a *pencil* and ruler, gently mark a line across the TLC strip about 1 cm above bottom on the powdered side. This will mark the location for loading pigment extract onto the strip.

(c) Tear a small portion of a leaf (about 1 cm square) from the plant material provided into small pieces and drop them into the mortar. Add a couple milliliters of acetone and grind into a paste. Use another pipet to transfer the dark green liquid in the mortar to the watch glass. Repeat if needed until you have transferred about 20 drops of pigment extract. Work quickly to prevent formation of pigment breakdown products.

(d) Use a hair dryer to evaporate the solvent from the extract in the watch glass; take care not to blow the extract from the glass!

(e) When the extract is completely dry, add 3–4 drops of acetone and mix into the dried extract with a toothpick.

(f) Use a micropipet to transfer small drops of extract to the pencil line on the TLC strip. Try to create a spot no more than 2 mm wide. Dry the spot thoroughly, and reload. Repeat several times until the small pigment spot is very dark green.

(g) Use forceps to slide the strip into the chromatography chamber. Mark the tube just below the level of the pigment spot. Remove the strip.

(h) Add chromatography solvent up to the mark on the chamber. Put the TLC strip back into the test tube, making sure to not touch the surface of the strip, or to allow the strip to touch the sides of the tube. Seal the tube and watch the chromatogram develop as the solvent is drawn up the surface.

(i) Note that the pigments move up the strip more slowly than the solvent. When the solvent front has nearly reached the top of the strip, remove the chromatogram and air-dry it. Discard the excess solvent into the designated waste bottles.

(j) Look carefully at the chromatogram. The pigments will fade within a few hours. Ordinarily, the pigments can be observed in the following order from the top: orange-yellow **carotenes**, blue-green **chlorophyll** *a*, yellow-green **chlorophyll** *b*, orange-yellow **xanthophylls**. You may also observe grayish phaeophytin, a breakdown product, below the carotenes.

(k) Use colored pencils to sketch the results in Figure 9-1, showing the relative position of the pigment colors along the TLC strip. Label the pigment bands.

If you cannot see all those pigments, it may be because you had insufficient pigment extract loaded onto the strip. Try again. You may preserve your chromatogram briefly by enclosing it in the pages of a book or by keeping it in a similar dark place. Light causes the chromatogram to quickly fade.

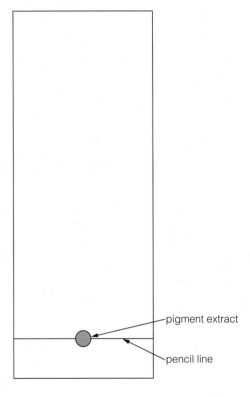

Figure 9-1 Chloroplast pigment thin layer chromatogram.
Labels: chlorophyll *a*, chlorophyll *b*, carotenes, xanthophylls

---

9.4	Structure of the Chloroplast *(15 min.)*

Work alone.

The chloroplast is the organelle concerned with photosynthesis. Study Figure 9-2, which includes an artist's conception of the three-dimensional structure of a chloroplast.

Like the mitochondrion and the nucleus, the chloroplast is surrounded by two membranes. Within the **stroma** (semifluid matrix), identify the **thylakoid disks** stacked into **grana** (a single stack is a **granum**). The chloroplast pigment molecules are located on the surface of the thylakoid disks. Hydrogen ion buildup occurs within the interior of the disks. As these ions are expelled back into the stroma, ATP is formed. Within the stroma, the ATP is used to generate organic compounds. These compounds are converted to carbohydrates, lipids, and amino acids from carbon dioxide, water, and other raw materials.

Now examine Figure 9-3, a high-magnification electron micrograph of a chloroplast.

If the plant is killed and fixed for electron microscopy after being exposed to strong light, the chloroplasts will contain **starch grains**. Note the large starch grain present in this chloroplast. (Starch grains appear as ellipsoidal white structures in electron micrographs.) With the aid of Figure 9-2, label the electron micrograph.

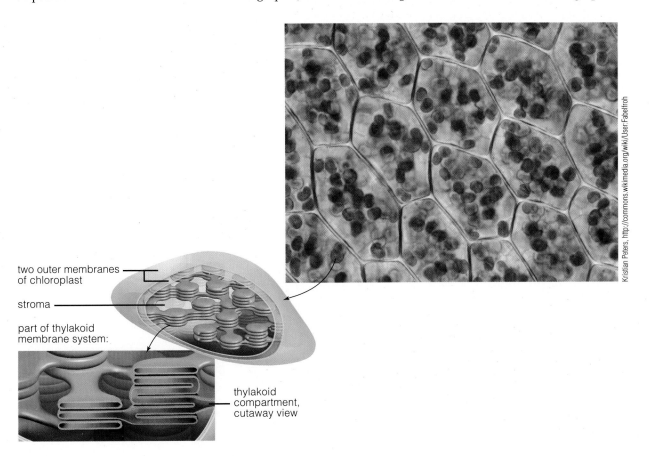

two outer membranes of chloroplast

stroma

part of thylakoid membrane system:

thylakoid compartment, cutaway view

Kristian Peters, http://commons.wikimedia.org/wiki/User:Fabelfroh

Figure 9-2   The chloroplasts, sites of photosynthesis in the cells of plants.

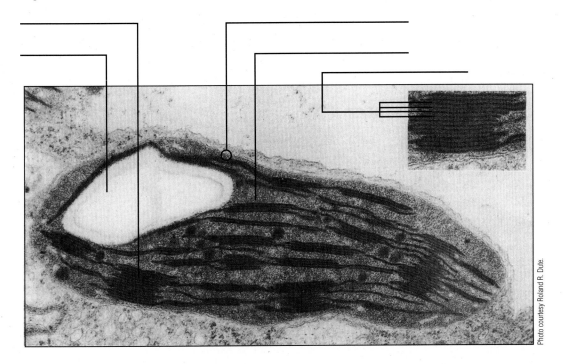

Photo courtesy Roland R. Dute.

Figure 9-3   Electron micrograph of chloroplast (10,000×). Inset: a single granum (20,000×).
Labels: chloroplast membrane, thylakoid disks, stroma, starch

_____ 1. Autotrophs include which of the following?
(a) fungi
(b) plants
(c) animals
(d) all of the above

_____ 2. The raw materials used for photosynthesis include
(a) O_2.
(b) $C_6H_{12}O_6$.
(c) CO_2 and H_2O.
(d) CH_2Ot.

_____ 3. Products and by-products of photosynthesis do NOT include
(a) O_2.
(b) $C_6H_{12}O_6$.
(c) CO_2.
(d) H_2O.

_____ 4. Most plants store carbohydrates in the form of
(a) glucose.
(b) sucrose.
(c) starch.
(d) water.

_____ 5. A thin layer chromatogram is useful for
(a) measuring the amount of photosynthesis.
(b) determining the amount of gas produced during photosynthesis.
(c) separating pigments based on their physical characteristics.
(d) determining the presence of sugars in a leaf.

_____ 6. Which of the following pigments would you find in a typical leaf?
(a) chlorophyll, xanthophyll, phycobilins
(b) chlorophyll a, chlorophyll b, carotenoids
(c) phycocyanin, xanthophyll, fucoxanthin
(d) carotenoids, chlorophylls, phycoerythrin

_____ 7. Organisms capable of producing their own food are known as
(a) autotrophs.
(b) heterotrophs.
(c) omnivores.
(d) herbivores.

_____ 8. Grana are
(a) the same as starch grains.
(b) the site of ATP production within chloroplasts.
(c) part of the outer chloroplast membrane.
(d) contained within mitochondria and nuclei.

_____ 9. The ultimate source of energy trapped during photosynthesis is
(a) CO_2.
(b) H_2O.
(c) O_2.
(d) sunlight.

_____ 10. When carbon dioxide is absorbed into water, the solution becomes more
(a) acidic.
(b) basic.
(c) warm.
(d) red.

EXERCISE **9**

Photosynthesis: Capture of Light Energy

Post-Lab Questions

Introduction

1. Defend the statement "Photosynthesis is the most important biological process on the planet Earth."

9.1 Uptake of Carbon Dioxide during Photosynthesis

2. Explain why the yellow solution in the graduated cylinder became a reddish color after it was incubated in the light with stems of *Elodea*.

3. Predict what would happen if you used the pH indicator thymophthalein instead of phenol red in the above setup. Thymophthalein solution appears colorless below pH 9.3, but blue at higher pH.

9.2 Release of Oxygen: Floating Leaf Disk Assay for Quantitative Investigation of Photosynthesis

4. Why was it important to infiltrate the leaf disks with liquid before performing this assay?

5. The floating leaf disk assay measures oxygen production as an indication of photosynthetic activity. What other substances could be measured as alternate ways of measuring photosynthesis?

9.3 Separation of Photosynthetic Pigments by Thin Layer Chromatography

6. How is the color of each of the major photosynthetic pigments (chlorophyll *a*, chlorophyll *b*, carotenes, and xanthophylls) related to the light wavelengths absorbed by that pigment?

9.4 Structure of the Chloroplast

7. As you saw while performing thin layer chromatography, photosynthetic pigments are mostly hydrophobic substances soluble in organic solvents but not in water-based solutions. How does this feature explain the location of these pigments within the chloroplast?

8. Examine this electron micrograph of a chloroplast.

   (a) Identify the stack of membranes labeled A.

   (b) Identify the region labeled B.

   (c) Would the production of organic compounds during the light-independent reactions occur in region B *or* on the membranes labeled A?

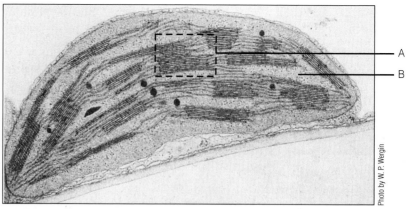

(17,257×)

Photo by W. P. Wergin

   (d) Would you expect the plant in which this structure was found to have been illuminated with strong light immediately before it was prepared for electron microscopy? Why or why not?

Food for Thought

9. Numerous hypotheses have been proposed for the mass extinction of about 60% of all species, including nearly all the dinosaurs, about 65 million years ago. Evidence has been found of the impact of a large meteor at about the time of this mass extinction. The amount of dust and debris put into the atmosphere upon impact, as well as atmospheric heating, would have been enormous. Using your knowledge of photosynthesis, speculate about why the dinosaurs subsequently became extinct.

10. Explain and justify this statement: "Without autotrophic organisms, heterotrophic life would cease to exist."

Respiration: Energy Conversion

OBJECTIVES

After completing this exercise, you will be able to

1. define *metabolism, respiration, aerobic respiration, alcoholic fermentation, efficiency, ATP, pH indicator, muscle fiber, ethanol, mitochondrion;*

2. give the overall balanced equations for aerobic respiration and alcoholic fermentation;

3. distinguish among the inputs, products, and efficiency of aerobic respiration and those of fermentation;

4. explain how carbon dioxide input acidifies water and pH indicators can be used to visualize that change;

5. explain how increasing intensity of physical activity affects carbon dioxide production due to aerobic respiration;

6. describe the efficiency of various carbohydrate sources, including corn and sugarcane, for ethanol production through alcoholic fermentation;

7. identify the structures and list the functions of each part of a mitochondrion.

Introduction

All living organisms have a constant energy requirement. Because of this they have mechanisms to gather, store, and use energy. Collectively, these mechanisms are called **metabolism**.

In Exercise 9, we investigated the metabolic pathways by which green plants capture light energy and use it to make carbohydrates such as glucose. Carbohydrates are temporary energy stores. The process by which energy stored in carbohydrates is released to the cell is **respiration**. (Don't confuse this term with the *respiratory cycle* of land vertebrates, during which air is inhaled into lungs and then exhaled.)

Both autotrophs and heterotrophs undergo respiration. Photoautotrophs such as plants use the carbohydrates they have produced by photosynthesis to build new biomass and maintain cellular processes. Heterotrophic organisms may obtain materials for respiration in two ways: by digesting plant biomass or by digesting the tissues of animals that have previously digested plants.

Several different forms of respiration have evolved. The particular respiration pathway used depends on the specific organism and/or environmental conditions. In this exercise, we will investigate two different pathways: (1) **aerobic respiration,** an oxygen-dependent pathway common in most organisms; and (2) **alcoholic fermentation,** an ethanol-producing process occurring in some yeasts (single-celled fungi).

Perhaps the most important aspect to remember about these two processes is that aerobic respiration is by far the most energy-efficient. **Efficiency** refers to the amount of energy captured in the form of ATP relative to the amount available within the bonds of the carbohydrate. **ATP, adenosine triphosphate,** is the so-called universal energy currency of the cell. Energy contained within the bonds of carbohydrates is transferred to ATP during respiration. This stored energy can be released later to power a wide variety of cellular reactions.

For aerobic respiration, the general equation is:

$$C_6H_{12}O_6 + 6O_2 \xrightarrow{\text{enzymes}} 6CO_2 + 6H_2O + 36ATP^*$$
glucose   oxygen   carbon dioxide water chemical energy

If glucose is completely broken down to CO_2 and H_2O, about 686,000 calories of energy are released. Each ATP molecule produced represents about 7500 calories of usable energy. The 36 ATP represent 270,000 calories of energy (36 × 7500 calories). Thus, aerobic respiration is about 39% efficient [(270,000/686,000) × 100%].

By contrast, fermentation yields only 2 ATP for each molecule of glucose entering the process. Thus, fermentation is only about 2% efficient [(2 × 7500/686,000) × 100%]. Obviously, breaking down carbohydrates by aerobic respiration gives a bigger energy payback than through fermentation.

---

*Depending on the tissue, as many as 38 ATP may be produced.

During the process of aerobic respiration, relatively high-energy carbohydrates are broken down in stepwise fashion, ultimately producing the low-energy products of carbon dioxide and water and transferring released energy into ATP. But what is the role of oxygen?

During aerobic respiration, the carbohydrate undergoes a series of oxidation–reduction reactions. Whenever one substance is oxidized (loses electrons), another must be reduced (accept, or gain, those electrons). The final electron acceptor in aerobic respiration is oxygen. Tagging along with the electrons as they pass through the electron transport process are protons (H^+). When the electrons and protons are captured by oxygen, water (H_2O) is formed:

$$2H^+ + 2e^- + \tfrac{1}{2}O_2 \longrightarrow H_2O$$

In the following experiment, we examine aerobic respiration first in germinating seeds and then in humans exercising at different levels.

10.1.A. Experiment: Carbon Dioxide Production (About 25 min. for setup, 1½ hr to complete)

Seeds contain stored food material, usually in the form of some type of carbohydrate. When a seed germinates, the carbohydrate is broken down by aerobic respiration, liberating the energy (ATP) required for each embryo to grow into a seedling.

Two days ago, one set of dry corn seeds was soaked in water to start the germination process. Another set was not soaked and remains dormant. A third set began to germinate but was then boiled to kill them. In this experiment, you will compare carbon dioxide production among germinating corn seeds, ungerminated (dry) corn seeds, and germinating boiled seeds.

This experiment investigates the hypothesis that *germinating seeds produce carbon dioxide from aerobic respiration.*

MATERIALS

Per student group (4):
- three respiration bottle apparatuses (Figure 10-1)
- Sharpie or china marker
- phenol red solution
- 600-mL beaker
- dropper bottle of tetrazolium chloride solution
- petri dish
- single-edged razor blade
- plastic gloves (optional)
- forceps
- magnifying lens *or* dissecting microscope

Per lab room:
- germinating corn seeds
- germinated, boiled seeds
- ungerminated (dry) corn seeds

PROCEDURE
Work in groups.

1. Obtain three respiration bottle setups (Figure 10-1). With a marker, label one "Germ" for germinating corn seeds, a second "Ungerm" for ungerminated seeds, and the third "Boiled" for germinated, boiled seeds.
2. From the class supply, obtain and put enough germinating corn seeds into the "Germ"-labeled respiration bottle to fill it approximately halfway. Fill the "Ungerm" bottle half full with ungerminated (dry) corn seeds. Fill the "Boiled" bottle half full with boiled seeds.
3. Fit the rubber stoppers with attached glass tubes into the respiration bottles. Add enough water to each test tube to cover the end of the glass tubing that comes out of the respiration bottle. (This keeps gases from escaping from the respiration bottle.)
4. Insert a rubber stopper into each thistle tube.
5. Set the bottles aside for the next 1¼ hours and do the other experiments in this exercise.
6. Make a prediction about carbon dioxide production in each bottle and record in Table 10-1.

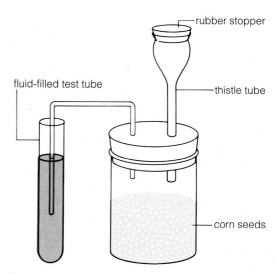

Figure 10-1 Respiration bottle apparatus.

Now start other activities while you allow this one to proceed.

7. After 1¼ hours, pour the water in each test tube into the sink and replace it with an equal volume of dilute phenol red solution. Phenol red solution, which should appear pinkish in the stock bottle, will be used to test for the presence of carbon dioxide (CO_2) within the respiration bottles. If CO_2 is bubbled through water, carbonic acid (H_2CO_3) forms:

$$CO_2 + H_2O \longrightarrow H_2CO_3$$

Phenol red is a **pH indicator**. When the phenol red solution is basic (pH > 8), it is pink; when it is acidic (pH < 6.8), the solution is yellow; and when neutral, it appears orange. The phenol red solution in the stock bottle is _____ (color); therefore, the stock solution is _____ (acidic/basic/neutral).

8. Put several hundred milliliters of tap water in the beaker.
9. Remove the stopper in the top of the thistle tube and *slowly* pour water from the beaker into each thistle tube. The water will force out gases present in the bottles. If CO_2 is present, the phenol red will become yellow.
10. Record your observations in Table 10-1.

TABLE 10-1	CO_2 Production by Corn Seeds	
Predictions		
Germinating corn:		
Nongerminating corn:		
Boiled corn:		
Corn Seeds	**Indicator Color (Phenol Red)**	**Conclusion (CO_2 Present or Absent)**
Germinating		
Ungerminated		
Boiled		
Conclusions:		

© Cengage Learning 2013

Seeds consist of a protective outer seed coat surrounding an embryo plant and stored food. In a corn grain, the embryo is located just under the seed coat near the pointed end. Most of the interior of the seed is filled with a starchy tissue called endosperm; the endosperm provides a rich source of stored energy for the embryo during germination and its early growth.

Tetrazolium chloride is used by the agricultural seed industry to check the viability of seeds before planting. Tetrazolium is colorless when oxidized but becomes reddish when reduced; the color change happens when a seed's electron transport system of aerobic cellular respiration is working and so indicates a living seed. If the seed is dead, the seeds will remain unstained. We will use tetrazolium chloride to visualize where cellular respiration is proceeding in the germinating corn seeds.

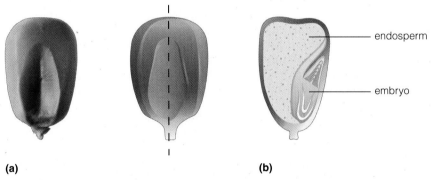

(a) (b)

AGStockUSA/Alamy

Figure 10-2 Corn grain (**a**) whole, and (**b**) internal structure.

11. Remove a grain of germinating corn. Place the kernel flat on a hard surface oriented as in Figure 10-2a and cut in half vertically as indicated by the dashed line.
12. Put several drops of tetrazolium chloride in the petri dish, and use forceps to place the seed halves with the cut side down in the drops. Cover the dishes and allow the seeds to soak for about 15 minutes.
13. Turn the seed halves over and examine them, using a magnifying lens or dissecting microscope. Compare with Figure 10-2b to determine which portions of the seeds have the highest rates of aerobic respiration.

    What part(s) of the seed is(are) experiencing the highest rate of respiration?

CAUTION

Tetrazolium chloride is a poison; wear gloves and avoid contact with your skin. Wash immediately and thoroughly if you touch the solution.

10.1.B. Effect of Physical Activity on Cellular Respiration (About 30 min.)

You use large skeletal muscles in your legs, as well as other muscles in your body, when you walk or run. The muscles attach to bones and actively contract, then relax, in order to move the joint (see Exercise 39.) Skeletal muscles are each composed of many long individual muscle cells (fibers). A **muscle fiber** contracts when protein filaments inside slide past each other, shortening the length of the cell. An entire muscle contracts when its component fibers contract in unison.

The sliding action of the filaments is powered by ATP, so muscle fibers have many mitochondria to churn out ATP fuel via cellular respiration. The cells first utilize any glucose circulating in your blood, then break down glycogen, a storage polysaccharide, to produce more glucose. If the physical activity lasts more than about 10 minutes, fats will begin to be broken down to supply energy from that stored in fatty acids.

In this activity you'll see how physical activity affects carbon dioxide production as an indicator of aerobic cellular respiration. You will determine relative carbon dioxide production while one member of your group is at rest, then after a short period of mild exercise, and then again after a short period of moderate activity. What is your prediction regarding a human body's production of carbon dioxide from these three levels of activity? Will more *or* less carbon dioxide be produced as exercise intensity increases? Record in Table 10-2.

We will utilize another pH indicator in this activity. Bromthymol blue (BTB) is blue-colored above pH 7.6, yellow below pH 6.0, and green between pH 6 and 7.6. Recall from Part 10.1.A. that carbon dioxide acidifies water (lowers the pH). We will use the time required for BTB to change from blue to green or yellow as an indication of relative aerobic respiration activity.

MATERIALS

Per student group (4):

- three small beakers
- 10-mL graduated cylinder
- straws
- bromthymol blue (BTB) solution in flask
- large beaker for wastes
- Sharpie or other marker
- stopwatch or clock for timing

PROCEDURE

1. Designate at least one exerciser, a data recorder, an exercise timer, and a setup manager from your team.
2. Label the three small beakers "Trial 1," "Trial 2," and "Trial 3."
3. Measure and pour 10 mL of BTB solution in each beaker.
4. At time 0, the exerciser **slowly** blows through a straw into the bottom of Trial 1 beaker.
5. Repeat the resting control test *twice more*, using the corresponding beakers. Try to be consistent in using approximately the same color endpoint for all timing.
6. Discard used BTB solutions in the waste beaker and replace with fresh BTB.
7. Repeat steps 4 to 6 with the exerciser first walking quickly for one minute around the classroom or in a nearby corridor (mild exercise), and then running in place or up and down a nearby staircase for one minute (moderate activity).
8. Calculate the average of the three trials of each exercise level and record in Table 10-2.

> **CAUTION**
>
> Do not inhale! When the solution changes color to green or yellow, record the elapsed time (in seconds) in Table 10-2.

TABLE 10-2	Effect of Activity Level on Carbon Dioxide Production from Aerobic Respiration				
	Time (seconds) to Color Change of Bromthymol Blue Solution				
Exercise Level	Trial 1	Trial 2	Trial 3	Average	Ratio Difference
Resting					
Walking					
Running					

© Cengage Learning 2013

9. Calculate the ratio between each of the two activity levels and the resting control average (activity level average divided by resting average). Record in Table 10-2.
10. What can you conclude about the effect of activity intensity on aerobic cellular respiration? Record in Table 10-2.

CAUTION

Do not do this activity if you have a medical condition that interferes with your ability to exercise. If you feel weak or dizzy while exercising, stop immediately and sit down.

10.2 Experiment: Alcoholic Fermentation and Ethanol Production
(About 20 min. to set up, 1½ hr to complete)

Alcoholic fermentation is an anaerobic (nonoxygen requiring) process occurring in some yeast and a few bacterial cells. Fermentation provides sufficient energy for those organisms to survive despite relatively low ATP yield; most of the energy originally contained in glucose remains in the ethanol product. The chemical equation for this process is:

$$C_6H_{12}O_6 \longrightarrow 2C_2H_5OH + CO_2 + 2ATP$$

glucose   ethanol   carbon   energy
dioxide

Alcoholic fermentation by yeast is the basis for the baking, wine-making, and brewing industries. Additionally, fermentation is used industrially to produce ethanol for transportation fuel. Gasoline engines can be modified to operate on blends of gasoline with up to 85% ethanol added, suggesting the possibility of a biofueled transportation future.

The United States and Brazil are the world's leading producers of ethanol. In the United States, corn grain is the major feedstock for the fermentation process. The endosperm of the corn grain (see Figure 10-2) is made mostly of starch. This starch is enzymatically broken down into glucose subunits; yeasts then ferment the glucose into ethanol. In contrast, Brazil produces ethanol from yeast fermentation of sugars extracted directly from sugarcane stalks (sugarcane is a major crop in wet, tropical Brazil.) Other plant materials are also being investigated and used for commercial ethanol production. Especially promising is "cellulosic" ethanol, produced by digesting the cellulose plant cell walls abundant in nonfood materials such as straw and cornstalks into glucose for fermentation. But are all carbohydrate sources equally efficient in yielding bioethanol? This activity will offer insight into some answers to that question.

MATERIALS

Per student group (4):
- Sharpie or other marker
- five 50-mL beakers
- 500-mL flask
- 25-mL graduated cylinder
- 200-mL graduated cylinder
- five fermentation tubes *or* alternative provided by your instructor
- balance and weighing paper
- dry rapid-rise yeast
- NaCl
- spoons *or* metal spatulas
- 15-cm metric ruler

Per lab room:
- glucose
- ground corn grain
- pulverized sugarcane stem
- at least two additional ground carbohydrate materials (All Bran cereal, straw, etc.)
- warm distilled water (about 38°–43°C)
- source of distilled water

PROCEDURE
Work in groups.

1. Prepare the active yeast culture as follows:
   (a) Weigh 0.2 g of NaCl and add to 500-mL flask.
   (b) Weigh 7.0 g of dry rapid-rise yeast and add to flask.
   (c) Measure 200 mL of warm water and add to flask. Swirl to mix.

2. Using a marker, number five 50-mL beakers. Also number the five fermentation tubes.
3. Weigh 2 g of each of the following and place into the corresponding beaker: #2, glucose; #3, ground corn; #4, pulverized sugarcane stem; #5, carbohydrate source of your choice. (Beaker #1 has no carbohydrate added.)
4. With a 25-mL graduated cylinder, measure out and pour 20 mL[†] of yeast solution into each of the five beakers. Swirl to mix.
5. When each is thoroughly mixed, pour the contents into five correspondingly numbered fermentation tubes (Figure 10-3). Cover the opening of the fermentation tube with your thumb and invert each fermentation tube so that the "tail" portion is filled with the solution. This is "time zero" for this experiment.

   What is the purpose of Tube 1 (yeast culture only)? _____

   What is the purpose of Tube 2 (yeast culture plus glucose)? _____

   What is the independent variable in this experiment? _____

   What is the dependent variable in this experiment? _____
6. What gas will be produced by the process of fermentation? _____ Note that we will be measuring the relative amounts of this gas in each tube as an indirect measure of ethanol production. Write a prediction about relative amounts of gas production in each tube in Table 10-3.
7. At intervals of 10 minutes over the next hour, measure the height of the gas bubble (if any) that forms in the closed end of the tube. Record data in Table 10-3.
8. Graph data from each fermentation tube in Figure 10-4. Be careful to clearly distinguish between treatments.
9. What conclusions can you draw regarding the efficiency of ethanol production with each carbohydrate source in Beakers 3–5 relative to that from glucose in Beaker 2? Record in Table 10-3.

Figure 10-3 Fermentation tube.

[†]The amount of fluid needed to fill the fermentation tube depends on its size. Your instructor may indicate the required volume.

TABLE 10-3	Production of Gas by Yeasts with Various Carbohydrate Sources

Predictions

Tube 1:

Tube 2:

Tube 3:

Tube 4:

Tube 5:

Tube	Solution	Distance from Tip of Tube to Fluid Level (mm)					
		10 min.	20 min.	30 min.	40 min.	50 min.	60 min.
1	Yeast culture						
2	Yeast culture plus glucose						
3	Yeast culture plus ground corn						
4	Yeast culture plus pulverized sugarcane						
5	Yeast culture plus _____						

Conclusions:

© Cengage Learning 2013

Did your results conform to your predictions? If not, speculate on reasons why this might be so.

_____

_____

_____

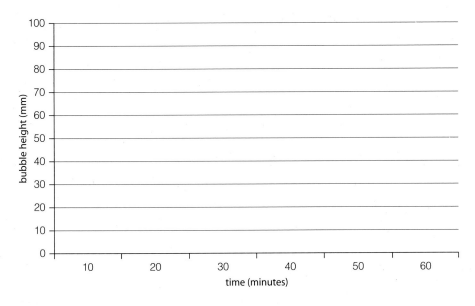

Figure 10-4 Gas production (mm) from fermentation of various carbohydrate sources.

Pool your results with those of other groups in your class. Which carbohydrate source(s) produced the greatest volume of gas, and by extension, the most ethanol? _____

Which industrial raw material, corn or sugarcane, seems to be the more efficient for ethanol production?

_____

10.3   Ultrastructure of the Mitochondrion

1. Study Figure 10-5b, which shows the three-dimensional structure of a mitochondrion, the respiratory organelle of all living eukaryotic cells. The mitochondrion has frequently been referred to as the "powerhouse of the cell," because most of the cell's chemical energy (ATP) is produced here.

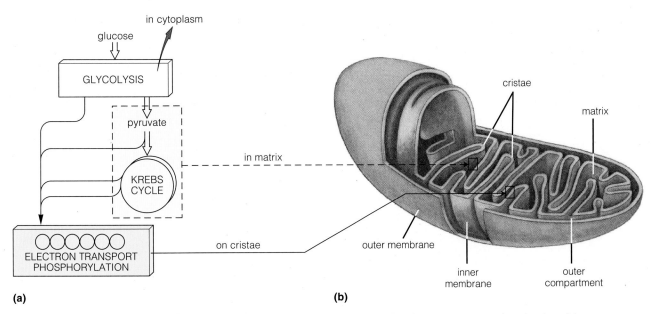

(a)   (b)

Figure 10-5   (**a**) The pathways in aerobic respiration. (**b**) The membranes and compartments of a mitochondrion.

2. Now observe Figure 10-6, a high-magnification electron micrograph of a mitochondrion. Identify and label *the outer membrane* separating the organelle from the cytoplasm.

3. Note the presence of an inner membrane, folded into fingerlike projections. Each projection is called a *crista* (the plural is *cristae*). The folding of the inner membrane greatly increases the surface area on which many of the chemical reactions of aerobic respiration take place. Label a crista.

4. Identify and label the *outer compartment*, the space between the inner and outer membranes. The outer compartment serves as a reservoir for hydrogen ions.

5. Finally, identify and label the *inner compartment* (filled with the *matrix*), the interior of the mitochondrion.

6. Study Figure 10-5a, a diagram of the pathways in aerobic respiration and where in the cell they take place.

Now that you know the structure of the mitochondrion, you can visualize the events that take place to produce the chemical energy needed for life. Glycolysis, the first step in *all* respiration pathways, takes place in the cytoplasm. Pyruvate, a carbohydrate, and energy carriers formed during glycolysis enter the mitochondrion during aerobic respiration, moving through both the outer and inner membranes to the matrix within the inner compartment.

Within the inner compartment, the pyruvate is broken down in the Krebs cycle, forming more energy-carrier molecules as well as CO_2. A small amount of ATP is also produced during these reactions.

Electron transport molecules are embedded on the inner membrane, and ATP production occurs as hydrogen ions cross from the outer compartment to the inner compartment. Water is also formed. This accomplishes the third portion of aerobic respiration, electron transfer phosphorylation.

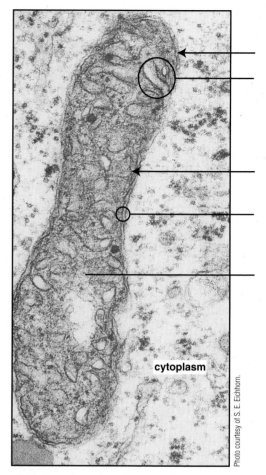

cytoplasm

Photo courtesy of S. E. Eichhorn.

Figure 10-6 Transmission electron micrograph of a mitochondrion (18,600×).
Labels: outer membrane, inner compartment, crista, outer compartment, matrix

Note: After completing all labs, take your dirty glassware to the sink and wash it following the directions given in *"Instructions for Washing Laboratory Glassware,"* page x. Invert the test tubes in the test tube racks so that they drain. Tidy up your work area, making certain all equipment used in this exercise is there for the next class.

_____ 1. The energy-releasing process that occurs in most organisms when oxygen is present is
(a) alcoholic fermentation.
(b) aerobic fermentation.
(c) aerobic respiration.
(d) adenosine triphosphate.

_____ 2. The "universal energy currency" of the cell is
(a) O_2.
(b) $C_6H_{12}O_6$.
(c) ATP.
(d) H_2O.

_____ 3. Products of aerobic respiration include
(a) glucose.
(b) oxygen.
(c) carbon dioxide.
(d) starch.

_____ 4. "Efficiency" of a respiration pathway refers to the
(a) number of steps in the pathway.
(b) amount of CO_2 produced relative to the amount of carbohydrate entering the pathway.
(c) amount of H_2O produced relative to the amount of carbohydrate entering the pathway.
(d) amount of ATP energy produced relative to the energy content of the carbohydrate entering the pathway.

_____ 5. Phenol red is used in Part 10.1.A. to visualize
(a) O_2 production from aerobic respiration.
(b) CO_2 production from aerobic respiration.
(c) sugar production from aerobic respiration.
(d) O_2 consumption in aerobic respiration.

_____ 6. Tetrazolium chloride is used by the seed industry to indicate
(a) whether a seed is undergoing photosynthesis.
(b) if seeds are undergoing aerobic respiration.
(c) if seeds are alive.
(d) both b and c

_____ 7. Which of these is a direct source of energy for muscle contraction?
(a) ATP
(b) glycogen
(c) glucose
(d) fatty acids

_____ 8. Bromthymol blue (BTB) is _____-colored below pH 6.0 and _____-colored above pH 7.6.
(a) blue, yellow
(b) yellow, green
(c) red, blue
(d) yellow, blue

_____ 9. Yeast cells undergoing alcoholic fermentation produce
(a) glucose.
(b) ethanol.
(c) H_2O.
(d) all of the above

_____ 10. Alcoholic fermentation is the basis for
(a) bread baking.
(b) beer brewing.
(c) ethanol fuel production.
(d) all of the above

EXERCISE **10**

Respiration: Energy Conversion

Post-Lab Questions

10.1 Aerobic Respiration

10.1.A. Experiment: Carbon Dioxide Production

1. Explain the role of the following components in the experiment on carbon dioxide production:
    germinating corn seeds _____
    ungerminated (dry) corn seeds _____
    germinating, boiled corn seeds _____
    phenol red solution _____

2. Cyanide affects the electron transport system of mitochondria by binding to components of the system and inhibiting the transfer of electrons to oxygen. If seeds were treated with cyanide, what results would they show in the tetrazolium test? Explain.

10.1.B. Effect of Physical Activity on Cellular Respiration

3. Describe at least two reasons why the time it takes your breath to produce a color change in BTB after running 1 minute might be different from the time recorded for another person?

4. As physical activity ramps up, a person typically begins to breathe more rapidly, to sweat, and to flush red. Describe how each of these physical responses of the body is connected to the reactants, products, and byproducts of the process of aerobic respiration.

10.2 Experiment: Alcoholic Fermentation and Ethanol Production

5. Sucrose (table sugar) is a disaccharide composed of glucose and fructose. Glycogen is a polysaccharide composed of many glucose subunits. Which of the following fermentation tubes would you expect to produce the greatest gas volume over a 1-hour period? Why?

    Tube 1: glucose plus yeast

    Tube 2: sucrose plus yeast

    Tube 3: glycogen plus yeast

6. Bread is made by mixing flour, water, sugar, and yeast to form a dense dough. Why does the dough rise? What gas is responsible for the holes in bread?

10.3 Ultrastructure of the Mitochondrion

7. Examine the artificially-colored electron micrograph of the mitochondrion on the right.

   (a) What portions of aerobic respiration occur in region A?

   _____

   (b) What substance is produced as hydrogen ions cross from the space between the inner and outer membranes?

   _____

   (c) What portion(s) of cellular respiration takes(s) place in the cytoplasm *outside* of this organelle?

   _____

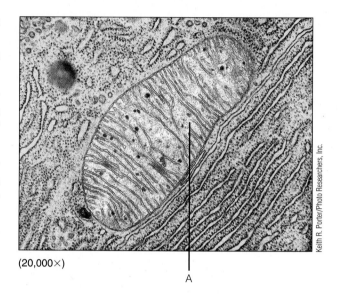

(20,000×)

A

Keith R. Porter/Photo Researchers, Inc.

Food for Thought

8. Compare aerobic respiration and fermentation in terms of

   (a) efficiency of obtaining energy from glucose

   (b) end products

9. The first law of thermodynamics seems to conflict with what we know about ourselves. For example, after strenuous exercise we run out of "energy." We must eat to replenish our energy stores. Where has that energy gone? What form has it taken?

10. Government subsidy programs for corn ethanol stimulated great production increases in the U.S. in the early years of the twenty-first century. The programs were controversial due to disagreements over a number of issues, including the efficiency of energy production from ethanol relative to that required to grow corn crop and transport and manufacture it, as well as the economic effects caused by diverting a food crop to energy production (more than half of U.S. corn grain is fed to livestock, and many ingredients of processed foods are derived from corn). Search two Internet sites for information on benefits and drawbacks of producing ethanol from corn in the United States. List the two sites below and briefly summarize what you learn from each.

   http:// _____

   _____

   _____

   _____

   http:// _____

   _____

   _____

   _____

Mitosis and Cytokinesis: Nuclear and Cytoplasmic Division

OBJECTIVES

After completing this exercise, you will be able to

1. define *fertilization, zygote, DNA, chromosome, fission, mitosis, cytokinesis, cell cycle, interphase, chromatin, sister chromatid, centromere, duplicated chromosome, HeLa cells, meristem, nuclear envelope, prophase, metaphase, anaphase, telophase, spindle fibers, spindle, cell plate, daughter cells, centriole, furrowing;*

2. identify the events of the cell cycle;

3. distinguish between mitosis and cytokinesis as they take place in animal and plant cells;

4. identify the structures involved in nuclear and cell division (those in boldface) and describe the role each plays;

5. demonstrate and/or describe chromosomal numbers and movements during the cell cycle.

Introduction

"All cells arise from preexisting cells." This is one part of the cell theory. It's easy to understand this concept if you think of a single-celled *Amoeba* or bacterium in which each cell divides to give rise to two entirely new individuals. It is fascinating that each of us began life as *one* single cell and developed into an astonishingly complex animal, a human, with trillions of cells. Our first cell has *all* the hereditary information we'll ever get.

In higher plants and animals, **fertilization**, the fusion of egg and sperm nuclei, produces a single-celled **zygote**. The zygote divides into two cells, those two into four, and so on to produce a multicellular organism. During cell division, each new cell receives a complete set of hereditary information and an assortment of cytoplasmic components.

Recall from Exercise 6 that there are two basic cell types, prokaryotic and eukaryotic. The genetic material of both consists of **DNA (deoxyribonucleic acid)**. In prokaryotes, the DNA molecule is organized into a single circular chromosome. Prior to cell division, the **chromosome** duplicates. Then the cell undergoes **fission**, the splitting of a preexisting cell into two, with each new cell receiving a full complement of the genetic material.

In eukaryotes, the process of cell division is more complex, primarily because of the much more complex nature of the hereditary material. Here the chromosomes consist of DNA and proteins complexed together within the nucleus. Cell division is preceded by duplication of the chromosomes and usually involves two processes: **mitosis** (nuclear division) and **cytokinesis** (cytoplasmic division). Whereas mitosis results in the production of two nuclei, both containing identical chromosomes, cytokinesis ensures that each new cell contains all the metabolic machinery necessary for sustenance of life.

In this exercise, we consider only nuclear and cell division as it occurs in eukaryotic cells.

Dividing cells pass through a regular sequence of events called the **cell cycle** (Figure 11-1). Notice that the majority of the time is spent in **interphase** and that actual nuclear division—mitosis—is but a brief portion of the cycle.

Interphase is comprised of three parts (Figure 11-1): the G1 period, during which cytoplasmic growth takes place; the S period, when the DNA is duplicated; and the G2 period, when structures directly involved in mitosis are synthesized.

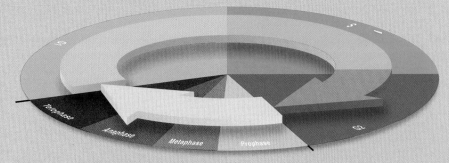

Figure 11-1 The eukaryotic cell cycle. The length of the intervals differs among cells. G1, S, and G2 are parts of interphase.

Unfortunately, because of the apparent relative inactivity that early microscopists observed, interphase was given the misnomer "resting stage." In fact, we know now that interphase is anything but a resting period. The cell is producing new DNA, assembling proteins from amino acids, and synthesizing or breaking down carbohydrates and many other compounds. In short, interphase is a very busy time in the life of a cell.

11.1 Chromosomal Structure *(About 25 min.)*

During much of a cell's life, each DNA–protein complex, the *nucleoprotein,* is extended as a thin strand within the nucleus. In this form, it is called **chromatin**. Prior to the onset of nuclear division, the genetic material duplicates itself. As nuclear division begins, the chromatin condenses. The two identical condensed nucleoproteins are called **sister chromatids** and are attached at the **centromere**. The centromere often gives the appearance of dividing each chromatid into two "arms." Collectively, the two attached sister chromatids are referred to as a **duplicated chromosome**. Before looking at actual chromosomes, let's examine the background of the materials we'll use.

In 1951, Henrietta Lacks died from cervical cancer. Prior to her death, some of the cancerous (tumor) cells were removed from her body and grown in culture. The cells were allowed to divide repeatedly in this artificial culture environment and resulted in the formation of what is known as a *cell line*. These **HeLa cells** (the abbreviation coming from *Henrietta Lacks*) live on today and are used for research on cell and tumor growth. They're also used to show chromosomal structure in the biology laboratory.

Normal human body cells contain 23 pairs of chromosomes (46 total chromosomes). Human cells of tumor origin (such as the HeLa cells) often produce greater chromosome numbers when grown in culture. Cells containing up to 200 chromosomes have been observed, although 50–70 per cell are most frequently found. In this section, you will observe the chromosomes of the descendant tumor cells of Henrietta Lacks.

Note: There is no danger from this activity. The tumor-producing properties are nontransmissible, and the cells have been killed and preserved.

MATERIALS

Per student:

- clean microscope slide prechilled in cold 40% methanol
- coverslip
- disposable pipet
- paper toweling
- compound microscope (with oil-immersion objective preferable)

Per student pair:

- metric ruler
- tube of HeLa cells
- dropper bottle of dH_2O

Per lab room:

- Coplin staining jars containing stains 1 and 2
- 1-L beaker containing dH_2O

PROCEDURE

Refer to Figure 11-2.

1. Place a paper towel on your work surface.
2. Remove a clean glass microscope slide from the cold methanol and lean it against a surface at a 45° angle, with one short edge resting on the paper towel.
3. Most of the cells are at the bottom of the culture tube. Resuspend the cells by inserting your pipet and gently squeezing the bulb. This expels air from the pipet, which disperses the cells. Remove a small cell sample with the pipet. Holding the pipet about 18–36 cm above the slide, allow 8–10 drops of the cell suspension to "splat" onto the upper edge of the wet slide and to tumble down the slide.

Note: The slide must still be wet when "splatting" takes place.

4. Allow the cells to *air-dry completely.*
      You will now stain the slide three times with both stains, for *1 second each time.* The time in the stain is critical. Count "one thousand one"; this will be 1 second.

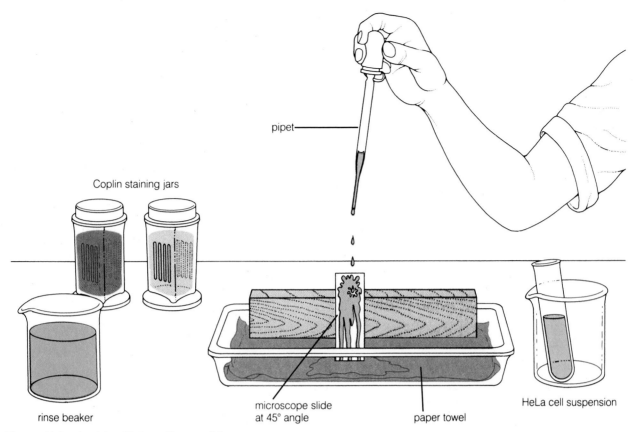

pipet

Coplin staining jars

rinse beaker

microscope slide
at 45° angle

paper towel

HeLa cell suspension

Figure 11-2 Applying HeLa cells to a slide.

5. Go to the location of the staining solutions and dip the slide in stain 1 for 1 second. Withdraw the slide and then dip twice more. Drain the slide of the excess stain, blotting the bottom edge of the slide on the paper toweling before proceeding.

6. Immediately dip the slide in stain 2 for 1 second. (Repeat twice more.) Drain and blot the excess stain as before.

7. Rinse the slide in dH_2O by swishing it gently back and forth in the beaker.

8. Allow the slide to *air-dry completely*.

9. Add a drop of water onto the preparation and make a traditional wet mount slide.

10. Place the slide on the microscope stage and observe your chromosome spread, focusing first with the medium-power objective and then with the high-dry objective. Locate cells that appear to have burst and have the chromosomes spread out (see Figure 11-3.) The number of good spreads will be low; so careful observation of many cells is necessary.

Note: If your microscope has an oil-immersion objective, proceed to step 11. If not, skip to step 12.

11. Once a good spread has been located, rotate the high-dry objective out of the light path and place a drop of immersion oil on that spot. Rotate the oil-immersion objective into the light path. Be very careful to focus only with the fine-focus of the microscope.

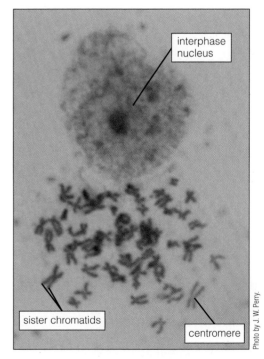

interphase nucleus

sister chromatids

centromere

Photo by J. W. Perry.

Figure 11-3 Stained chromosome spread of HeLa cells. Sister chromatids connected at their centromeres are clearly visible (w.m., 1000×).

12. Observe the structure of the chromosomes, identifying sister chromatids and centromeres. Draw several of the chromosomes in Figure 11-4, labeling these parts.
13. Select 10 cells in which the chromosomes are visible and count the number of chromosomes you observe in each cell, recording your data in Table 11-1. When you have counted the chromosomes in 10 individual cells, calculate the average number of chromosomes per cell.

　　Now examine Figure 11-5, an electron micrograph of a human chromosome. Label the two chromatids, the centromere, and the duplicated chromosome.

Note: Use illustrations in your textbook to aid you in the following study.

Figure 11-4　Drawing of human chromosomes from HeLa cells (＿＿×).
Labels: chromosome, sister chromatid, centromere

TABLE 11-1	Chromosome Numbers in HeLa Cells											
Cell		1	2	3	4	5	6	7	8	9	10	Average
Number of chromosomes												

© Cengage Learning 2013

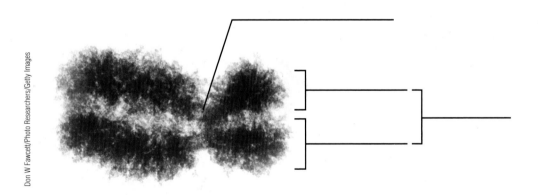

Don W Fawcett/Photo Researchers/Getty Images

Figure 11-5　Electron micrograph of a chromosome (26,000×).
Labels: chromatid (2), centromere, duplicated chromosome

11.2　The Cell Cycle in Plant Cells: Broad Bean Root Tip Squash
(About 45 min.)

Nuclear and cell divisions in plants are, for the most part, localized in specialized regions called meristems. **Meristems** are regions of active growth. A meristem contains cells that have the capability to divide repeatedly.

　Plants have two types of meristems: apical and lateral. Apical meristems are found at the tips of plant organs (shoots and roots) and increase length. Lateral meristems, located beneath the bark of woody plants, increase girth. In this section, you will examine structures related to the cell cycle in the apical meristem of broad bean roots. The tips of actively growing roots may have been prestained for you, or your instructor may provide you with instructions for staining them yourself.

MATERIALS

Per student:

- clean microscope slides
- forceps
- pencil with eraser
- compound microscope
- razor blade or scalpel (optional)

Per student group:

- beaker with stained root tips of broad bean
- dropper bottle of 45% acetic acid
- Kimwipes™ or other tissues

PROCEDURE

1. Obtain a single stained root of broad bean and place it on a clean slide. With a second slide or razor blade, cut off all but the stained tip (approximately 1–3 mm). Discard the remainder and keep the stained tip.
2. Add a drop of 45% acetic acid onto the stained root tip and add a coverslip. Try to prevent air bubbles in your preparation. Allow the root tip to remain in the acetic acid for one or two minutes.
3. The root tip preparation process softened the tissue so that it can be spread out into a single layer of cells. Tap the root tip area with a pencil eraser until the spot is much paler in color and has spread to 8 to 10 mm in diameter. Try not to move the coverslip; if it moves, it will cause some of the cells to roll and fold.

> **CAUTION**
>
> 45% acetic acid can cause burns. Rinse your hands well with water if you get acid on them.

4. With a folded Kimwipe or tissue on top of the coverslip, use the ball of your thumb or the heel of your hand to press down firmly, absorbing excess acetic acid into the tissue and flattening the root tip.
5. Observe the squash preparation with low magnification. If the cells are in a single layer with little overlapping, continue. If most of the cells are in two or more layers, continue tapping with the pencil until the squash is light pink and well spread.
6. Look for cuboidal cells that are actively dividing under low, and then medium power. Use high power to scrutinize cells closely. Locate and study cells in all the phases of the cell cycle, such as in Figure 11-6. Keep in mind that the cell cycle is a continuous cycle; we separate events into different stages as a convenience to aid in their study, but it is often difficult to say definitively when one phase begins and another ends.

A. Interphase and Mitosis

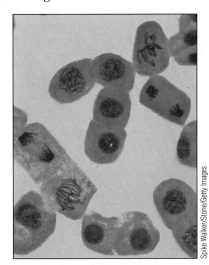

Figure 11-6 Broad bean root tip squash.

1. **Interphase**. Use the medium-power objective to scan the squash. Note that most of the nuclei are in interphase.

    Switch to the high-dry objective, focusing on a single interphase cell. Note the distinct **nucleus**, with one or more **nucleoli**, and the **chromatin** dispersed within the bounds of the **nuclear envelope**. Label these features in cell 1 of Figure 11-7.

2. **Mitosis**

    (a) *Prophase*. During **prophase** the chromatin folds and condenses; the duplicated chromosomes become visible as threadlike structures. At the same time, microtubules outside the nucleus are beginning to assemble into **spindle fibers.** Collectively, the spindle fibers make up the **spindle,** a three-dimensional structure widest in the middle and tapering to a point at the two *poles* (opposite ends of the cell). *You will not see the spindle during prophase.*

    Find a nucleus in prophase. Draw and label a prophase nucleus in cell 2 of Figure 11-7.

    The transition from prophase to metaphase is marked by the fragmentation and disappearance of the nuclear envelope. At about the same time, the nucleoli also disappear.

    (b) *Metaphase*. When the nuclear envelope is no longer distinct, the cell is in metaphase. Identify a metaphase cell by locating a cell with the duplicated chromosomes, each consisting of two **sister chromatids,** lined up midway between the two poles. This imaginary midline is called the **spindle equator**. (You may not be able to distinguish the chromatids.) The spindle has moved into the space the nucleus once occupied. The microtubules have become attached to the chromosomes at the *kinetochores,* groups of proteins that form the outer faces of the centromeres (most chromosomes of broad bean have the centromere near one end). Find a cell in metaphase. Label cell 3 of Figure 11-7.

(c) *Anaphase*. During **anaphase**, sister chromatids of each chromosome separate, each chromatid moving toward an opposite pole. Find an early anaphase cell, recognizable by the slightly separated chromatids. Notice that the chromatids begin separating at the centromere. The last point of contact before separation is complete is at the ends of the "arms" of each chromatid. Although incompletely understood, the mechanism of chromatid separation is based on action of the spindle-fiber microtubules. Once separated, each chromatid is referred to as an individual daughter chromosome. Note that now the chromosome consists of a *single* chromatid. Find a late anaphase cell and draw it in cell 4 of Figure 11-7.

(d) *Telophase*. When the daughter chromosomes arrive at opposite poles, the cell is in **telophase**. The spindle disorganizes. The chromosomes expand again into chromatin form, and a nuclear envelope re-forms around each newly formed daughter nucleus. Find a telophase cell and label individual chromosomes, nuclei, and nuclear envelopes on cell 5 of Figure 11-7.

B. Cytokinesis in Broad Bean Cells

Cytokinesis, division of the cytoplasm, usually follows mitosis. In fact, it often overlaps with telophase. Find a cell undergoing cytokinesis in the broad bean root tip. In plants, cytokinesis takes place by **cell plate formation** (Figure 11-8). During this process, Golgi body–derived vesicles filled with cellulose and other cell wall materials migrate to the spindle equator, where they fuse. Their contents contribute to the formation of a new cell wall, and their membranes make up the new plasma membranes. In most plants, cell plate formation starts in the *middle* of the cell.

1. Examine Figure 11-8, an electron micrograph showing cell plate formation. Note the microtubules that are part of the spindle apparatus.
2. Find a cell undergoing cytokinesis on the prepared slide of broad bean root tips. With your light microscope, the developing **cell plate** appears as a line running horizontally between the two newly formed nuclei. Return to cell 5 of Figure 11-7 and label the developing cell plate.

   Recently divided cells are often easy to distinguish by their square, boxy appearance. Find two recently divided **daughter cells**; then draw and label their contents in cell 6 of Figure 11-7. Include cytoplasm, nuclei, nucleoli, nuclear envelopes, and chromatin.

   Following cytokinesis, the cell undergoes a period of growth and enlargement, during which time it is in interphase. Interphase may be followed by another mitosis and cytokinesis, or in some cells interphase may persist for the rest of a cell's life.

C. Duration of Phases of the Cell Cycle

A broad bean root tip preparation can be thought of as a snapshot capturing cells in various phases of the cell cycle at a particular moment in time. The frequency of occurrence of a cell cycle phase is directly proportional to the length of a phase. You can therefore estimate the amount of time each phase takes by tallying the proportions of cells in each phase.

The length of the cell cycle for cells in actively dividing broad bean root tips is approximately 24 hours, with mitosis lasting for about 90 minutes.

1. Examine a region of the broad bean root tip squash slide where cells in mitosis are abundant. Count the number of cells in each of the stages of mitosis plus interphase in one field of view. Repeat this procedure for other fields of view until you count 100 cells. Record your data in Table 11-2.

TABLE 11-2	Determining Duration of Cell Cycle Phases		
Phase	**Number Observed**	**% of Total**	**Duration (hrs)**
Interphase			
Prophase			
Metaphase			
Anaphase			
Telophase			
Total			

© Cengage Learning 2013

2. Now calculate the time spent in each stage based on a 24-hour cell cycle by dividing the number of cells in each stage by the total number of cells counted to determine percent of total cells in each stage.
3. Multiply the fraction obtained in step 2 by 24 to determine duration.

Although your calculations are only a rough approximation of the time spent in each stage, they do illustrate the differences in duration of each stage in the cell cycle.

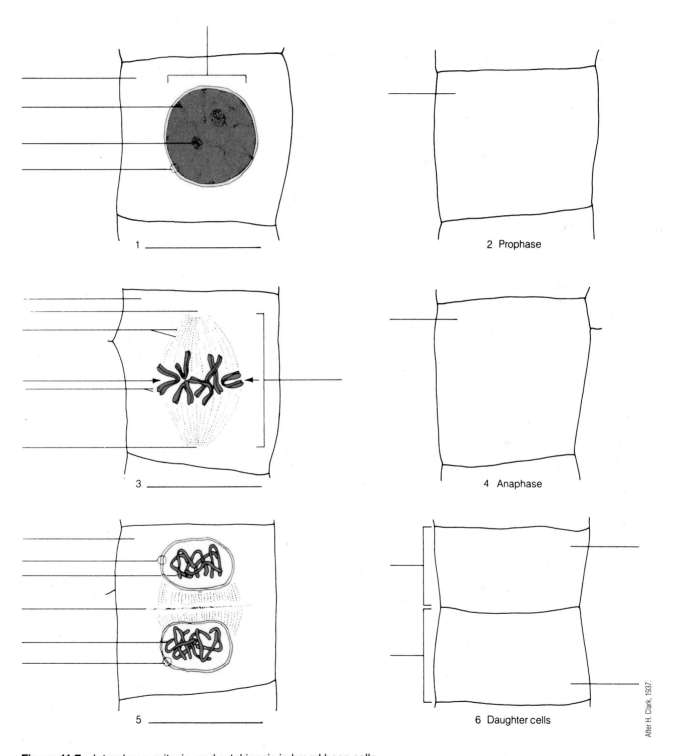

1 _____

2 Prophase

3 _____

4 Anaphase

5 _____

6 Daughter cells

After H. Clark, 1937.

Figure 11-7 Interphase, mitosis, and cytokinesis in broad bean cells.
Labels: interphase, cytoplasm, nucleus, nucleolus, chromatin, nuclear envelope, metaphase, spindle fibers, spindle, pole, spindle equator (between arrows), sister chromatids, telophase and cell plate formation, chromosome, cell plate, daughter cell
(*NOTE:* Some terms are used more than once.)

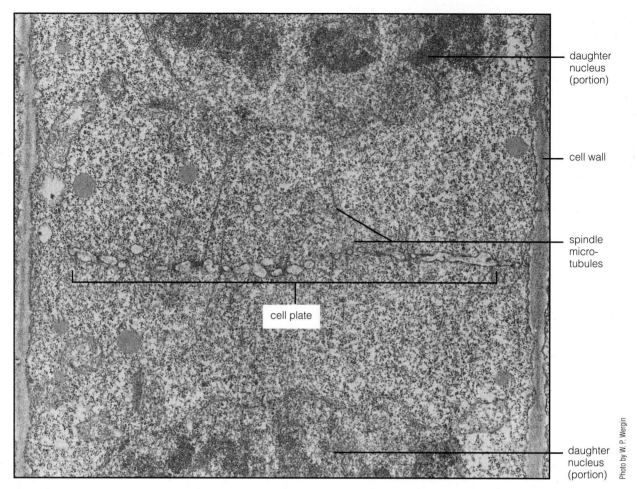

Figure 11-8 Transmission electron micrograph of cytokinesis by cell plate formation in a plant cell (2000×).

11.3 The Cell Cycle in Animal Cells: Whitefish Blastula *(About 20 min.)*

Fertilization of an ovum by a sperm produces a zygote. In animal cells, the zygote undergoes a special type of cell division, *cleavage*, in which no increase in cytoplasm occurs between divisions. A ball of cells called a *blastula* is produced by cleavage. Within the blastula, repeated nuclear and cytoplasmic divisions take place; consequently, the whitefish blastula is an excellent example in which to observe the cell cycle of an animal.

Note a key difference between plants and animals: whereas plants have meristems where divisions continually take place, animals do not have specialized regions to which mitosis and cytokinesis are limited. Indeed, divisions occur continually throughout many tissues of an animal's body, replacing worn-out or damaged cells.

With several important exceptions, mitosis in animals is remarkably like that in plants. These exceptions will be pointed out as we go through the cell cycle in the whitefish blastula.

MATERIALS

Per student:

- prepared slide of whitefish blastula mitosis
- compound microscope

PROCEDURE

Obtain a slide labeled "whitefish blastula." Scan it with the low-power objective and then at medium power. This slide has numerous thin sections of a blastula. Select one section (Figure 11-9) and then switch to the high-dry objective for detailed observation.

As you examine the slides, draw the cells to show the correct sequence of events in the cell cycle of whitefish blastula.

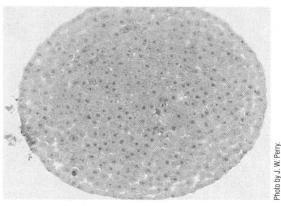

Figure 11-9 Section of a whitefish blastula (75×).

1. **Interphase.** Locate a cell in **interphase**. As you observed in the broad bean root tip, note the presence of the nucleus and chromatin within it. Note also the absence of a cell wall.

   Draw an interphase cell above the word "Interphase" in Figure 11-10 and label the cytoplasm, nucleus, and plasma membrane.

2. **Mitosis**

   (a) *Prophase.* The first obvious difference between mitosis in plants and animals is found in **prophase**.

   Unlike the broad bean cells, those of whitefish contain **centrioles** (Figure 11-11). As seen with the electron microscope, centrioles are barrel-shaped structures consisting of nine radially arranged triplets of microtubules.

   One pair of centrioles was present in the cytoplasm in the G1 stage of interphase. These centrioles duplicated during the S stage of interphase. Subsequently, one new and one old centriole migrated to each pole.

   Although the centrioles are too small to be resolved with your light microscope, you can see a starburst pattern of spindle fibers that appear to radiate from the centrioles. Other microtubules extend between the centrioles, forming the **spindle** (Figure 11-12). The chromosomes become visible as the chromatin condenses.

   Find a prophase cell, identifying the spindle and starburst cluster of fibers about the centriole.

   Draw the prophase cell in the proper location on Figure 11-10. Label the spindle, chromosomes, cytoplasm, and the position of the plasma membrane.

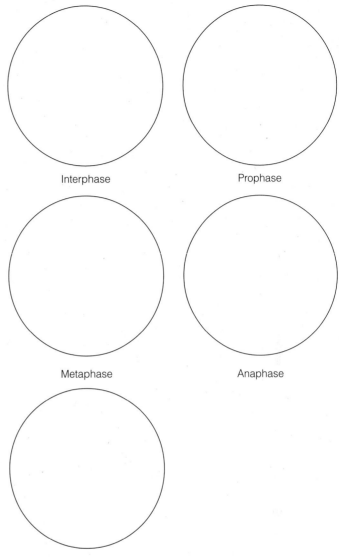

Interphase   Prophase

Metaphase   Anaphase

Telophase and cytokinesis

Figure 11-10   Drawings of cell cycle stages in whitefish blastula. **Labels:** cytoplasm, nucleus, plasma membrane, spindle, chromosomes, spindle equator, sister chromatids, daughter nuclei, chromatin, furrow. (*NOTE:* Some terms are used more than once.)

   (b) *Metaphase.* As was the case in plant cells, during **metaphase** the spindle fiber microtubules become attached to the *kinetechore* of each centromere region, and the duplicated chromosomes (each consisting of two **sister chromatids**) line up on the **spindle equator**. Locate a metaphase cell.

   Draw the metaphase cell in the proper location on Figure 11-10. Label the chromosomes on the spindle equator, spindle, and plasma membrane.

   (c) *Anaphase.* Again similar to that observed in plant cells, **anaphase** begins with the separation of sister chromatids into individual (daughter) chromosomes. Observe a blastula cell in anaphase. Draw the anaphase cell in the proper location on Figure 11-10. Label the separating sister chromatids, spindle, cytoplasm, and plasma membrane.

   (d) *Telophase.* **Telophase** is characterized by the arrival of the individual (daughter) chromosomes at the poles. A nuclear envelope forms around each daughter nucleus. Find a telophase cell.

   Is the spindle still visible? _____

   Is there any evidence of a nuclear envelope forming around the chromosomes?_____

   Draw the telophase cell in Figure 11-10. Label daughter nuclei, chromatin, cytoplasm, and plasma membrane.

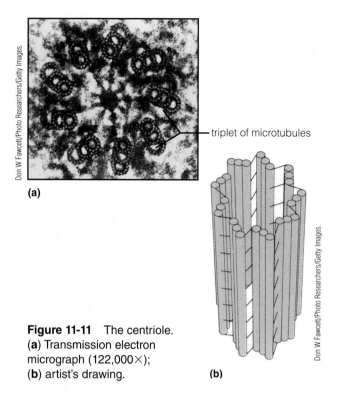

triplet of microtubules

Figure 11-11 The centriole.
(**a**) Transmission electron
micrograph (122,000×);
(**b**) artist's drawing.

(**a**)

(**b**)

Don W Fawcett/Photo Researchers/Getty Images.

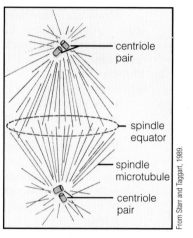

centriole
pair

spindle
equator

spindle
microtubule

centriole
pair

From Starr and Taggart, 1989.

Figure 11-12 Spindle
apparatus in animal cell.

B. Cytokinesis in Animal Cells

A second major distinction between cell division in plants and animals occurs during cytoplasmic division. Cell plates are absent in animal cells. Instead, cytokinesis takes place by **furrowing**.

To visualize how furrowing takes place, imagine wrapping a string around a balloon and slowly tightening the string until the balloon has been pinched in two. In life, the animal cell is pinched in two, forming two discrete cytoplasmic entities, each with a single nucleus. Figure 11-13 illustrates the cleavage furrow in an animal cell.

Find a cell in the blastula undergoing cytokinesis. The telophase cell that you drew in Figure 11-10 may also show an early stage of cytokinesis. Label the cleavage furrow if it does.

 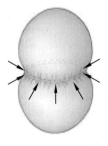 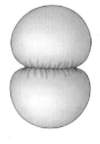  

(**a**) Mitosis is completed, and the bipolar spindle is starting to disassemble.

(**b**) At the former spindle equator, a ring of actin filaments attached to the plasma membrane contracts.

(**c**) The diameter of the contractile ring continues to shrink and pull the cell surface inward.

(**d**) The contractile mechanism continues to operate until the cytoplasm is partitioned.

(**e**)

D. M. Phillips/Visuals Unlimited

Figure 11-13 (**a–d**) Cytoplasmic division of an animal cell. (**e**) Scanning electron micrograph of the cleavage furrow.

Understanding chromosome movements is crucial to understanding mitosis. You can simulate mitosis with a variety of materials. This is a simple activity, but a valuable one. It will be especially helpful when comparing the events of mitosis with those of meiosis in the next exercise.

MATERIALS

Per student pair:
- 44 pop beads each of two colors
- eight magnetic centromeres

PROCEDURE

1. Build the components for two pairs of chromosomes by assembling strings of pop beads as follows:
   (a) Assemble two strands of pop beads with eight pop beads of one color on each arm, with a magnetic centromere connecting the two arms.
   (b) Repeat step a, but use pop beads of the second color.
   (c) Assemble two more strands of pop beads, but with three pop beads of one color on each arm.
   (d) Repeat step c, using pop beads of the second color.

   You should have four long strings, two of each color, and four short strings, with two of each color. Each pop bead string should have a magnetic centromere at its midpoint. Note that pop bead strings can attach to each other at the magnetic centromere. Each pop bead string represents a single molecule of DNA plus proteins.

2. Place **one** of each kind of strand in the center of your workspace, which represents the nucleus. You have created a nucleus with four "chromosomes," two long and two short.

3. Manipulate these model chromosomes through the phases of the cell cycle, beginning in the G1 phase of interphase and proceeding through the rest of interphase, mitosis, and cytokinesis.

4. Check with your instructor to be sure that you are modeling the process correctly.

_____ 1. Reproduction in prokaryotes occurs primarily through the process known as
(a) mitosis.
(b) cytokinesis.
(c) furrowing.
(d) fission.

_____ 2. The genetic material (DNA) of eukaryotes is organized into
(a) centrioles.
(b) spindles.
(c) chromosomes.
(d) microtubules.

_____ 3. The process of cytoplasmic division is known as
(a) meiosis.
(b) cytokinesis.
(c) mitosis.
(d) fission.

_____ 4. The product of chromosome duplication is
(a) two chromatids.
(b) two nuclei.
(c) two daughter cells.
(d) two spindles.

_____ 5. The correct sequence of stages in _mitosis_ is
(a) interphase, prophase, metaphase, anaphase, telophase.
(b) prophase, metaphase, anaphase, telophase.
(c) metaphase, anaphase, prophase, telophase.
(d) prophase, telophase, anaphase, interphase.

_____ 6. During prophase, duplicated chromosomes
(a) consist of chromatids.
(b) contain centromeres.
(c) consist of nucleoproteins.
(d) contain all of the above.

_____ 7. During the S period of interphase,
(a) cell growth takes place.
(b) nothing occurs because this is a resting period.
(c) chromosomes divide.
(d) synthesis (or replication) of the nucleoproteins takes place.

_____ 8. Chromatids separate during
(a) prophase.
(b) telophase.
(c) cytokinesis.
(d) anaphase.

_____ 9. Cell plate formation
(a) occurs in plant cells but not in animal cells.
(b) usually begins during telophase.
(c) is a result of fusion of Golgi vesicles.
(d) is all of the above.

_____ 10. Centrioles and a starburst cluster of spindle fibers would be found in
(a) both plant and animal cells.
(b) only plant cells.
(c) only animal cells.
(d) none of the above

Name _____ Section Number _____

EXERCISE 11

Mitosis and Cytokinesis: Nuclear and Cytoplasmic Division

Post-Lab Questions

Introduction

1. Distinguish among interphase, mitosis, and cytokinesis.

11.1 Chromosomal Structure

2. Distinguish between the structure of a duplicated chromosome before mitosis and the chromosome produced by separation of two chromatids during mitosis.

11.2 The Cell Cycle in Plant Cells: Broad Bean Root Tip Squash

3. If the chromosome number of a typical broad bean root tip cell is 12 before mitosis, what is the chromosome number of each newly formed nucleus after mitosis has taken place?

4. In plants, what name is given to a region where mitosis occurs most frequently?

5. The plant cells in the following photomicrographs have been stained to show microtubules comprising the spindle apparatus. Identify the stage of mitosis in each and label the region indicated on **(b)**.

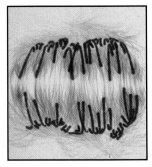

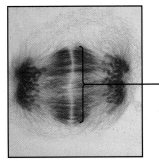

— Region? _____

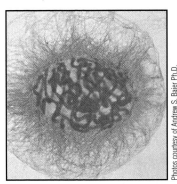

Photos courtesy of Andrew S. Bajer Ph.D.

(a) Stage? _____   **(b)** Stage? _____   **(c)** Stage? _____

11.3 The Cell Cycle in Animal Cells: Whitefish Blastula

6. Name two features of animal cell mitosis and cytokinesis you can use to distinguish these processes from those occurring in plant cells.

   (a)

   (b)

Food for Thought

7. Observe photomicrographs **(a)** and **(b)** below. Note the double nature of the blue "threads" in **(a)**. Each individual component of the doublet is called a(an) _____.
   Is **(b)** an image of a plant or an animal cell? How do you know?

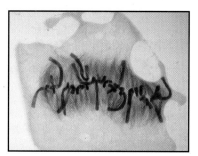

(a) Name of structure? _____

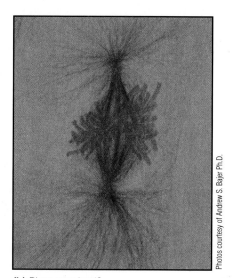

Photos courtesy of Andrew S. Bajer Ph.D.

(b) Plant or animal? _____
How do you know? _____

8. Why do you suppose cytokinesis generally occurs in the cell's midplane?

9. Why must the DNA be duplicated during the S phase of the cell cycle, prior to mitosis?

10. Tumor cells arise from normal body cells after a series of cellular "mistakes" disrupts the controls over, and processes of, mitosis and cytokinesis. Describe two errors that might happen during the processes of mitosis and/or cytokinesis and that would produce a daughter cell with "extra" chromosomes.

Meiosis: Basis of Sexual Reproduction

After completing this exercise, you will be able to

1. define *meiosis, homologue (homologous chromosome), gene, diploid, haploid, gamete, ovum, sperm, fertilization, zygote, spore, allele, gene pair, locus, synapsis, genotype, crossing over, nonlinked, linked, nondisjunction;*

2. describe differences and similarities between meiosis and mitosis;

3. describe the basic similarities and differences between the life cycles of higher plants and higher animals;

4. describe the process of meiosis, and recognize events that occur during each stage;

5. discuss the significance of crossing over, segregation, and independent assortment;

6. describe the process of nondisjunction and how it produces chromosome number abnormalities in resulting gametes and zygotes.

Introduction

Mitosis alone is not adequate for every nuclear division job required to complete a life cycle. Sexual reproduction requires **meiosis**, which, like mitosis, is a process of nuclear division. During mitosis, the number of chromosomes in the daughter nuclei remains the same as that in the parental nucleus. In meiosis, however, the genetic complement is halved, resulting in daughter nuclei containing only one-half the number of chromosomes as the parental nucleus. Thus, while mitosis is sometimes referred to as an *equational division*, meiosis is often called *reduction division*. Moreover, while mitosis is completed after a single nuclear division, two divisions, called meiosis I and meiosis II, occur during meiosis. Table 12-1 summarizes differences between mitosis and meiosis.

TABLE 12-1 Comparison of Mitosis and Meiosis	
Mitosis	**Meiosis**
Equational division: amount of genetic material remains constant.	Reduction division: amount of genetic material is halved.
Completed in one division cycle.	Requires two division cycles for completion.
Produces two genetically identical nuclei.	Produces two to four genetically different nuclei.
Generally produces cells not directly involved in sexual reproduction.	Ultimately produces cells used for sexual reproduction.

© Cengage Learning 2013

In the body cells of most eukaryotes, chromosomes exist in pairs called **homologues** (homologous chromosomes); that is, there are two chromosomes that are physically similar and contain the same **genes**, which are sections of DNA that are units of inheritance.

When both homologues are in the *same* nucleus, the nucleus is **diploid** (2n); when only one of the homologues is present, the nucleus is **haploid** (n). A parental nucleus normally contains the diploid (2n) chromosome number before meiosis; all four daughter nuclei contain the haploid (n) number at the completion of meiosis.

The reduction in chromosome number is the basis for sexual reproduction. In animals, the cells containing the daughter nuclei produced by meiosis are called **gametes**: **ova** (singular is *ovum*) if the parent is female, **sperm** cells if male. As you probably know, gametes are produced in the gonads—ovaries and testes, respectively. In fact, this is the *only* place where meiosis occurs in higher animals. Figure 12-1 shows where meiosis occurs in humans, and Figure 12-2 shows where and when meiosis occurs during the life cycle of a higher animal.

Note when meiosis occurs—during gamete production. During **fertilization** (the fusion of a sperm nucleus with an ovum nucleus), the diploid chromosome number is restored as the two haploid gamete nuclei fuse to form the **zygote**, the first cell of the new diploid generation.

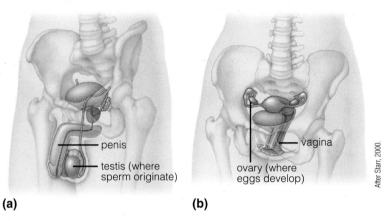

(a)   **(b)**

Figure 12-1 Gamete-producing structures in humans: (**a**) human male, (**b**) human female.

What about plants? Do plants have sex? Indeed they do. However, the plant life cycle is a bit more complex than that of animals. Plants of a single species have two completely different body forms. The primary function of one is the production of gametes. This plant is called a *gametophyte* ("gamete-producing plant") and all of its cells are haploid. Because the entire plant is haploid, gametes are produced in specialized organs by mitosis. The other body form, a *sporophyte*, is diploid. This diploid sporophyte has specialized organs in which meiosis occurs, producing haploid spores (hence the name *sporophyte*, "spore-producing plant"). When spores germinate and produce more cells by mitosis, they grow into haploid gametophytes, completing the life cycle.

Figure 12-3 shows the structures in a typical flower that produce sperm and eggs.

Examine Figure 12-4, which shows the gametophyte and sporophyte of a fern plant. Remember, the gametophyte and sporophyte are different, free-living stages of the *same* species of fern. Now look at Figure 12-5, which diagrams a typical plant life cycle. Again, note the consequence of meiosis. In plants, it results in the production of **spores**, not gametes.

You should understand an important concept from these diagrams: *Meiosis always halves the chromosome number. The diploid chromosome number is eventually restored when two haploid nuclei fuse during fertilization.*

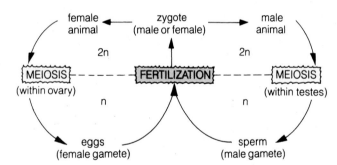

Figure 12-2 Life cycle of higher animals.

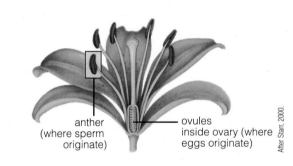

Figure 12-3 Gamete-producing structures in a flowering plant.

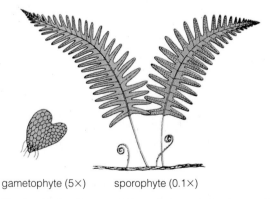

gametophyte (5×)   sporophyte (0.1×)

Figure 12-4 Gametophyte and sporophyte phases of the same fern species. Note size differences.

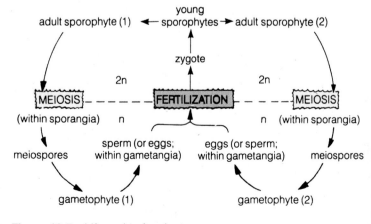

Figure 12-5 Life cycle of a plant.

Understanding meiosis is an absolute necessity for understanding the patterns of inheritance in Mendelian genetics, the subject of Exercise 13. Gregor Mendel, an Austrian monk, spent years deciphering the complexity of simple genetics. Although he knew nothing of genes and chromosomes, he noted certain patterns of inheritance and formulated three principles, now known as Mendel's principles of recombination, segregation, and independent assortment. The following activities will demonstrate the events of meiosis and the genetic basis for Mendel's principles.

12.1 Demonstrations of Meiosis Using Pop Beads *(About 60 min.)*

MATERIALS

Per student pair:

- 44 pop beads each of two colors (red and yellow, for example), as used in Exercise 11
- eight magnetic centromeres
- marking pens
- eight pieces of string, each 40 cm long
- meiotic diagram cards similar to those used here
- colored pencils

Per student group (table):

- bottle of 95% ethanol to remove marking ink
- tissues

PROCEDURE

Work in pairs.

Within the nucleus of an organism, each chromosome bears the genes, the units of inheritance. Genes may exist in two or more alternative forms called **alleles**. Each homologue bears *genes* for the same traits; these are the **gene pairs**. However, the homologues may or may not have the same *alleles.*

An example will help here: suppose the trait in question is flower color and that a flower has only two possible colors, red or white (Figure 12-6a, b). The gene, designated with a letter "R," is coding (providing the information) for flower color. There are two homologues in the same nucleus, so each bears the gene for flower color. *But,* on one homologue, the *allele* might code for red flowers ("*R*"), while the allele on the other homologue might code for white flowers ("*r*," Figure 12-6c). There are two other possibilities. The alleles on *both* homologues might be coding for red flowers (Figure 12-6d), or they *both* might be coding for white flowers (Figure 12-6e). Note that these three possibilities are mutually exclusive in a nucleus.

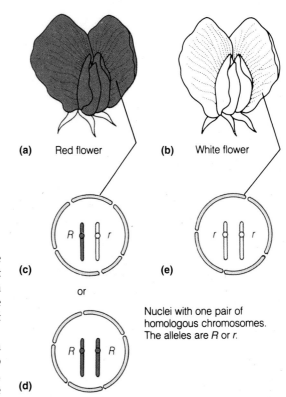

Figure 12-6 Chromosomal control of flower color. **(c–e)** show a nucleus with one pair of homologous chromosomes. The alleles for flower color are *R* or *r*.

1. Build the components for two pairs of homologous chromosomes by assembling strings of pop beads as follows:
   (a) Assemble two strands of pop beads with eight pop beads of one color on each arm, with a magnetic centromere connecting the two arms.
   (b) Repeat step a, but use pop beads of the second color.
   (c) Assemble two more strands of pop beads, but with three pop beads of one color on each arm.
   (d) Repeat step c, using pop beads of the second color.
   You should have four long strings, two of each color, and two short strings, also two of each color. Each pop-bead string should have a magnetic centromere at its midpoint, by which pop-bead strings can attach to each other. Each pop-bead string represents a single molecule of DNA plus proteins, with each bead representing a gene.
2. Place **one** of each kind of strand in the center of your workspace, which represents the interphase nucleus of a cell that will undergo meiosis. You have created a nucleus with four "chromosomes," two long and two short. The long strands represent one homologous pair, and the short strands represent a second homologous pair of chromosomes.

We start by assuming that these chromosomes represent the diploid condition. The two colors represent the origin of the chromosomes: one homologue (color: _____) came from the male parent, and the other homologue (color: _____) came from the female parent.

3. The four single-stranded chromosomes represent four unduplicated chromosomes. Now simulate DNA duplication during the S-phase of interphase (Figure 11-1), when each DNA molecule and its associated proteins are copied exactly. The two copies, called sister chromatids, remain attached to each other at their centromeres (Figure 12-7). During chromosome replication, the genes also duplicate. Thus, alleles on sister chromatids are identical.

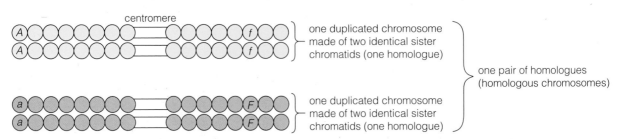

Figure 12-7   One pair of homologous chromosomes.

How many sister chromatids are there in a duplicated chromosome? _____

How many chromosomes are represented by four sister chromatids? _____ By eight? _____

What is the diploid number of the starting (parental) nucleus you've created? (*Hint:* Count the number of homologues to obtain the diploid number.) _____

4. As mentioned previously, genes may exist in two or more alternative forms, called alleles. The location of an allele on a chromosome is its **locus** (plural: *loci*). Using the marking pen, mark two loci on each long chromatid with letters to indicate alleles for a common trait. Suppose the long pair of homologous chromosomes codes for two traits, skin pigmentation and the presence of attached earlobes in humans. We'll let the capital letter *A* represent the allele for normal pigmentation and a lowercase *a* the allele for albinism (the absence of skin pigmentation); *F* will represent free earlobes and *f* attached earlobes. A suggested marking sequence is illustrated in Figure 12-7.

5. Let's assign a gene to our second homologous pair of chromosomes, the short pair. We'll suppose this gene codes for the production of an enzyme necessary for metabolism. On one homologue (consisting of two chromatids) mark the letter *E*, representing the allele causing enzyme production. On the other homologue, *e* represents the allele that interferes with normal enzyme production.

6. Obtain a meiotic diagram card like the one in Figure 12-8. Manipulate your model chromosomes through the stages of meiosis described below, moving the chromosomes to the correct diagram circles (representing nuclei) as you go along. Reference to Figure 12-8 will be made at the proper steps. *DO NOT* draw on the meiotic diagram cards.

A. Meiosis without Crossing Over

Although crossing over is a nearly universal event during meiosis, we will first work with a simplified model to illustrate chromosomal movements and separations during meiosis. Refer to Figure 12-9 as you manipulate your model.

1. **Late interphase.** During interphase, the nuclear envelope is intact and the chromosomes are randomly distributed throughout the nucleoplasm (semifluid substance within the nucleus). All duplicated chromosomes (eight chromatids) should be in the parental nucleus, indicating that DNA duplication has taken place. The sister chromatids of each homologue should be attached by their magnetic centromeres, but the four homologues should be separate. Your model nucleus contains a diploid number 2n = 4.

   The pop-bead chromosomes should appear during interphase in the parental nucleus as shown in Figure 12-8. Be sure to mark the location of the alleles. Use different pencil or pen colors to differentiate the homologues on your drawings.

2. **Meiosis I.** During meiosis I, homologues are separated from each other into different nuclei. Daughter nuclei created are thus haploid.

   (a) *Prophase I.* During the first prophase, the parental nucleus contains four duplicated homologous chromosomes, each comprised of two sister chromatids joined at their centromeres. The chromatin condenses to form discrete, visible chromosomes. The homologues pair with each other. This pairing is called **synapsis**. Slide the two homologues together.

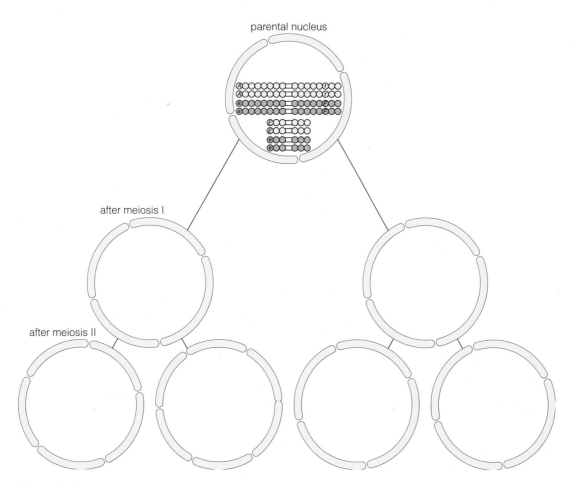

parental nucleus

after meiosis I

after meiosis II

Figure 12-8   Meiosis without crossing over.

Twist the chromatids about one another to simulate synapsis.

The nuclear envelope disorganizes at the end of prophase I.

(b)   *Metaphase I.* Homologous chromosomes now move toward the spindle equator, the centromeres of each homologue coming to lie *on either side of the equator.* Spindle fibers, consisting of aggregations of microtubules, attach to the centromeres. One homologue attaches to microtubules extending from one pole, and the other homologue attaches to microtubules extending from the opposite spindle pole.

To simulate the spindle fibers, attach one piece of string to each centromere. Then lay the free ends of strings from two homologues toward one spindle pole and the ends of the other homologues toward the opposite pole.

(c)   *Anaphase I.* During anaphase I, the homologous chromosomes separate, one homologue moving toward one pole, the other toward the opposite pole. The movement of the chromosomes is apparently the result of shortening of some spindle fibers and lengthening of others. Each homologue is still in the duplicated form, consisting of two sister chromatids.

Pull the two strings of one homologous pair toward its spindle pole and the other toward the opposite spindle pole, separating the homologues from one another. Repeat with the second pair of homologues.

(d)   *Telophase I.* Continue pulling the string spindle fibers until each homologue is now at its respective pole. The first meiotic division is now complete. You should have two nuclei, each containing two chromosomes (one long and one short) consisting of two sister chromatids.

Draw your pop-bead chromosomes as they appear after meiosis I on the two nuclei labeled "after meiosis I" of Figure 12-8. Depending on the organism involved, an interphase (interkinesis) and cytokinesis may precede the second meiotic division, *or* each nucleus may enter directly into meiosis II. The chromosomes decondense into chromatin form.

It is important to note here that DNA synthesis *does not* occur following telophase I (between meiosis I and meiosis II).

Before meiosis II, the spindle is rearranged into two spindles, one for each nucleus.

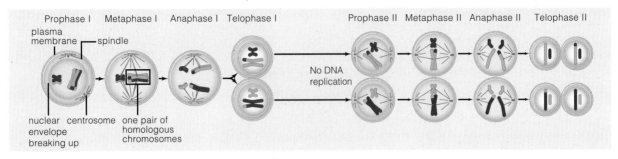

Figure 12-9 Meiosis in a generalized diploid germ cell. Two pairs of chromosomes are shown. Maternal chromosomes are shaded *pink*. Paternal chromosomes are shaded *blue*.

3. **Meiosis II.** During meiosis II, sister chromatids are separated into different daughter nuclei. The result is four haploid nuclei.

   (a) *Prophase II.* At the beginning of the second meiotic division, the sister chromatids are still attached by their centromeres. During prophase II, the nuclear envelope disorganizes, and the chromatin recondenses.

   (b) *Metaphase II.* Within each nucleus, the duplicated chromosomes align with the equator, the centromeres lying *on the equator.* Spindle fiber microtubules attach the centromeres of each chromatid to opposite spindle poles.

   Your string spindle fibers should be positioned so that the two spindle fiber strings from sister chromatids lie toward opposite poles. Note that each nucleus contains only *two* duplicated chromosomes (one long and one short) consisting of *two* sister chromatids each.

   (c) *Anaphase II.* The sister chromatids separate, moving to opposite poles. Pull on the string until the two sister chromatids separate. After the sister chromatids separate, each is an individual (not duplicated) daughter chromosome.

   (d) *Telophase II.* Continue pulling on the string spindle fibers until the two daughter chromosomes are at opposite poles. The nuclear envelope reforms around each chromosome and the chromosomes decondense back into chromatin form. Four daughter nuclei now exist. Note that each nucleus contains two individual unduplicated chromosomes (each formerly a chromatid) originally present within the parental nucleus. These nuclei and the cells they're in generally undergo a differentiation and maturation process to become gametes (in animals) or spores (in plants).

   Draw your pop-bead chromosomes as they appear after meiosis II in the "gamete nuclei" of Figure 12-8. Your diagram should indicate the genetic (chromatid) complement *before* meiosis and *after* each meiotic division, *not* the stages of each division.

   Remember that meiosis takes place in both male and female organisms. (See Figure 12-2.)

   If the parental nucleus was from a male, what is the gamete called? _____

   If female? _____

   Is the parental nucleus diploid or haploid? _____

   Are the nuclei produced after the *first* meiotic division diploid or haploid? _____

   Are the nuclei of the gametes diploid or haploid? _____

   What is the **genotype** of each gamete nucleus after meiosis II? (The genotype is the genetic composition of an organism, or the alleles present. Another way to ask this question is, What alleles are present in each gamete nucleus? Write these in the format: *AFE, afe,* and so on.)

_____

If you answered the preceding questions correctly, you might logically ask, "If the chromosome number of the gametes is the same as that produced after the first meiotic division, why bother to have two separate divisions? After all, the genes present are the same in both gametes and first-division nuclei."

There are two answers to this apparent paradox. The first, and perhaps the most obvious, is that the second meiotic division ensures that a *single* chromatid (nonduplicated chromosome) is contained within each gamete. After gametes fuse, producing a zygote, the genetic material duplicates prior to the zygote's undergoing mitosis. If gametes contained two chromatids, the zygote would have four, and duplication prior to zygote division would produce eight, twice as many as the organism should have. If DNA duplication within the zygote were not necessary for the onset of mitosis, this problem would not exist. Alas, DNA synthesis apparently is a necessity to initiate mitosis.

You can discover the second answer for yourself by continuing with the exercise, for although you have simulated meiosis, you have done so without showing what happens in *real* life. That's the next step.

B. Meiosis with Crossing Over

A very important event that results in a reshuffling of alleles on the chromatids occurs during prophase I. Recall that synapsis results in pairing of the homologues. During synapsis, the chromatids break, and portions of chromatids bearing genes for the same characteristic (but perhaps *different* alleles) are exchanged between *nonsister* chromatids. This event is called **crossing over**, and it results in recombination (shuffling) of the alleles on a chromatid.

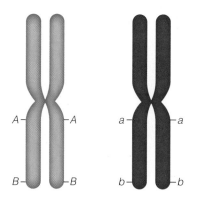

1. Look again at Figure 12-7. Distinguish between sister and nonsister chromatids. Now look at Figure 12-10, which demonstrates crossing over in one pair of homologues.
2. Return your chromosome models to the nucleus format with two pairs of homologues entering prophase I.
3. To simulate crossing over, break four beads from the arms of two nonsister chromatids in the long homologue pair, exchanging bead color between the two arms. During actual crossing over, the chromosomes may break anywhere within the arms.

    Crossing over is virtually a universal event in meiosis. Each pair of homologues may cross over in several places simultaneously during prophase I.

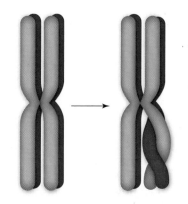

4. Manipulate your model chromosomes through meiosis I and II again and watch what happens to the distribution of the alleles as a consequence of the crossing over. Fill in Figure 12-11 as you did before, but this time show the effects of crossing over. Again, use different colors in your sketches.

    What are the genotypes of the gamete nuclei? _____
    Is the distribution of alleles present in the gamete nuclei after crossing over the same as that which was present without crossing over?

    _____

    Is the distribution of alleles present in the gamete nuclei after crossing over the same as that in the nuclei after the first meiotic division?

    _____

    Crossing over provides for genetic recombination, resulting in increased variety. How many different genetic *types* of daughter chromosomes are present in the gamete nuclei without crossing over (Figure 12-8)? _____
    How many different types are present with crossing over (Figure 12-11)? _____
    We think you would agree that a greater number of *types* of daughter chromosomes indicates greater *variety*.
    Recall that the parental nucleus contained a pair of homologues, each homologue consisting of two sister chromatids. Because sister chromatids are identical in all respects, they have the same alleles of a gene (see Figure 12-7). As your models showed, the alleles on nonsister chromatids may not (or may) be identical; they bear the same genes but may have different alleles, different forms of some genes.
    What is the difference between a gene and an allele?

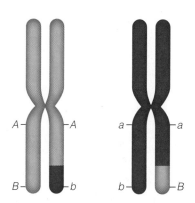

Figure 12-10 Crossing over in one pair of homologues. *Pink* signifies a maternal chromosome, and *blue*, its paternal homologue.

Let's look at the single set of alleles on your model chromosomes that are those for pigmentation, *A* and *a*. Both alleles were present in the parental nucleus. How many are present in the gametes? _____
This illustrates Mendel's first principle, segregation. Segregation means that during gamete formation, pairs of alleles are separated (segregated) from each other and end up in different gametes.

C. Demonstrating Independent Assortment

Manipulate your model chromosomes again through meiosis with crossing over (Figure 12-11), searching for different possibilities in chromosome distribution that would make the gametes genetically different.
    Does the distribution of the alleles for enzyme production to different gametes on the second set of homologues have any bearing on the distribution of the alleles on the first set (alleles for skin pigmentation and earlobe condition)? _____
    This distribution demonstrates the principle of independent assortment, which states that segregation of alleles into gametes is independent of the segregation of alleles for other traits, *as long as the genes are on*

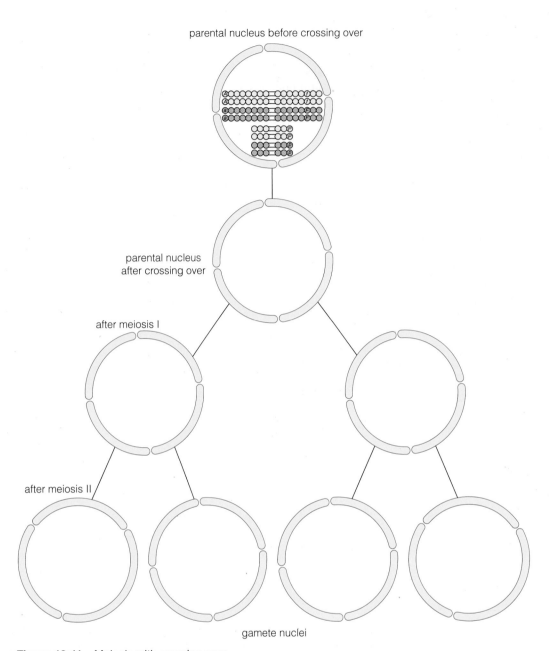

parental nucleus before crossing over

parental nucleus
after crossing over

after meiosis I

after meiosis II

gamete nuclei

Figure 12-11 Meiosis with crossing over.

different sets of homologous chromosomes. Genes that are on *different* (nonhomologous) chromosomes are said to be **nonlinked**. By contrast, genes for different traits that are on the *same* chromosome are **linked**.

Because the genes for enzyme production and those for skin pigmentation and earlobe attachment are on different homologous chromosomes, these genes are _____, while the genes for skin pigmentation and earlobe attachment are _____ because they are on the same chromosome.

In reality, most organisms have many more than two sets of chromosomes. Humans have 23 pairs (2n = 46), while some plants literally have hundreds!

A thorough understanding of meiosis is necessary to understand genetics. With this foundation, you'll find that problems involving Mendelian genetics are easy and fun to do. Without an understanding of meiosis, Mendelian genetics will be hopelessly confusing.

D. Nondisjunction and the Production of Gametes with Abnormal Chromosome Number

Errors in the process of meiosis can occur in many ways. Perhaps the best understood error process is that of **nondisjunction,** when one or more pairs of chromosomes fail to separate in anaphase. The result is gamete nuclei with too few or too many chromosomes.

1. Begin to manipulate your model chromosomes to show meiosis without crossing over (Section 12.1.A). In modeling events at metaphase I, however, arrange the spindle fiber threads for the long pair of homologues so that they all extend to the same pole.
2. Model anaphase I, pulling the chromosomes toward their respective poles. Nondisjunction occurs in the long pair of homologues, with both duplicated chromosomes being pulled to the same pole. See Figure 12-12.

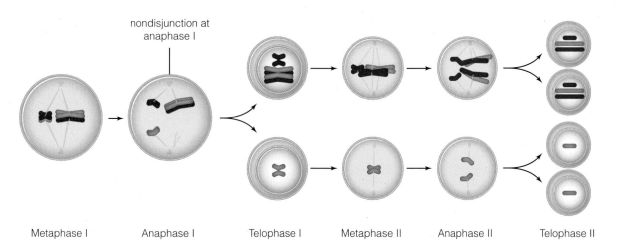

| Metaphase I | Anaphase I | Telophase I | Metaphase II | Anaphase II | Telophase II |

Figure 12-12 An example of nondisjunction. Of the two pairs of homologous chromosomes shown, one pair fails to separate at anaphase I of meiosis. The chromosome number is altered in the resulting gametes.

3. Continue to manipulate the model chromosomes through the remainder of the meiotic process.
    How many chromosomes are found in gamete nuclei? _____
    How does this compare to the chromosome number in normal gametes? _____
    Recall that each chromosome bears a unique set of genes and speculate about the effect of nondisjunction on the resulting zygotes formed from fertilization with such a gamete.

    _____
    _____
    _____

    One of the most common human genetic disorders arises from nondisjunction during gamete (usually ovum) formation. Down syndrome results from nondisjunction in one of the 23 pairs of human chromosomes, chromosome 21. An individual with Down syndrome has three copies of chromosome 21 instead of the normal two copies. While symptoms of this genetic disorder vary greatly, most individuals show moderate to severe mental impairment and a host of associated physical disorders. Relatively few other human genetic disorders arise from nondisjunction, probably because the consequences of abnormal chromosome number are often lethal.

Note: Remove marking ink from pop beads with 95% ethanol and tissues.

_____ 1. In meiosis, the number of chromosomes _____, while in mitosis, it _____.
(a) is halved/is doubled
(b) is halved/remains the same
(c) is doubled/is halved
(d) remains the same/is halved

_____ 2. The term "2n" means
(a) the diploid chromosome number is present.
(b) the haploid chromosome number is present.
(c) chromosomes within a single nucleus exist in homologous pairs.
(d) both a and c

_____ 3. In higher animals, including humans, meiosis results in the production of
(a) egg cells (ova).
(b) gametes.
(c) sperm cells.
(d) all of the above

_____ 4. Recombination of alleles on nonsister chromatids occurs during
(a) anaphase I.
(b) meiosis II.
(c) telophase II.
(d) crossing over.

_____ 5. Alternative forms of genes are called
(a) homologues.
(b) locus.
(c) loci.
(d) alleles.

_____ 6. If both homologous chromosomes of each pair exist in the same nucleus, that nucleus is
(a) diploid.
(b) unable to undergo meiosis.
(c) haploid.
(d) none of the above

_____ 7. DNA duplication occurs during
(a) interphase.
(b) prophase I.
(c) prophase II.
(d) interkinesis.

_____ 8. Nondisjunction
(a) results in gametes with abnormal chromosome numbers.
(b) occurs at anaphase.
(c) results when homologues fail to separate properly in meiosis.
(d) is all of the above.

_____ 9. The daughter nuclei produced by meiosis are
(a) genetically identical.
(b) diploid.
(c) haploid.
(d) both a and b

_____ 10. Meiosis differs from mitosis in that meiosis
(a) requires two cycles of division for completion.
(b) produces spores in plants but gametes in animals.
(c) is only found in animals and not in plants.
(d) both a and b

EXERCISE 1 2

Meiosis: Basis of Sexual Reproduction

Post-Lab Questions

Introduction

1. If a cell of an organism has 46 chromosomes before meiosis, how many chromosomes will exist in each nucleus after meiosis?

2. What basic difference exists between the life cycles of higher plants and higher animals?

3. In animals, meiosis results directly in gamete production, while in plants spores are produced. How are the gametes produced in the life cycle of a plant?

4. How would you argue that meiosis is the basis for sexual reproduction in plants, even though the *direct* result is a spore rather than a gamete?

12.1 Demonstrations of Meiosis Using Pop Beads

5. Suppose one sister chromatid of a chromosome has the allele *H*. What allele will the other sister chromatid have? (Assume crossing over has not taken place.) _____

6. Suppose that two alleles on one homologous chromosome are *A* and *B*, and the other homologous chromosome's alleles are *a* and *b*.
   (a) How many different genetic types of gametes would be produced *without* crossing over? _____
   (b) What are the genotypes of the gametes? _____
   (c) If crossing over were to occur, how many different genetic types of gametes could occur? _____
   (d) List them. _____

7. Assume that you have built a homologous pair of *duplicated* chromosomes, one chromosome red and the other yellow. Describe or draw the appearance of two nonsister chromatids after crossing over.

8. Examine the meiotic diagram at right. Describe in detail what's wrong with it.

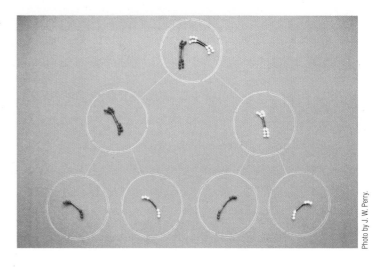

Photo by J. W. Perry.

Food for Thought

9. From a genetic viewpoint, of what significance is fertilization?

10. One of your friends has several brothers and sisters, each quite different in appearance, interests, and abilities, even though all have the same parents. Explain in detail how these siblings reflect crossing over and independent assortment.

Heredity

OBJECTIVES

After completing this exercise, you will be able to

1. define *true-breeding, hybrid, monohybrid cross, law of segregation, diploid, haploid, genotype, phenotype, dominant, recessive, complete dominance, homozygous, heterozygous, probability, chi-square test, dihybrid cross;*

2. solve monohybrid and dihybrid cross problems;

3. use sampling to determine phenotypic ratios of a visible trait in the gametophytes of an F_1 C-fern hybrid;

4. observe sperm release and fertilization events that lead to an F_2 C-fern sporophyte generation;

5. form hypotheses about genotypic and phenotypic ratios in the F_2 C-fern sporophyte generation;

6. use a chi-square test to determine whether observed results are consistent with expected results;

7. determine your phenotype and provide your probable genotype for some common traits.

Introduction

In 1866, an Austrian monk, Gregor Mendel, presented the results of painstaking experiments on the inheritance of the garden pea, but the scientific community ignored them, possibly because they didn't understand their significance. Now, more than a century later, Mendel's work seems elementary to modern-day geneticists, but its importance cannot be overstated. The principles generated by Mendel's pioneering experimentation are the foundation for the genetic counseling so important today to families with genetically based health disorders. They are also the framework for the modern research that is making inroads into treating diseases previously believed to be incurable. In this era of genetic engineering—the incorporation of foreign DNA into chromosomes of other species—it's easy to lose sight of the concepts underlying the processes that make it all possible. These experiments and genetics problems should give you a good basic understanding of these processes.

13.1 Monohybrid Crosses

Garden peas have both male and female parts in the same flower and are able to self-fertilize. For his experiments, Mendel chose parental plants that were **true-breeding,** meaning that all self-fertilized offspring displayed the same form of a trait as their parent. For example, if a true-breeding purple-flowered plant self-fertilizes, all of its offspring will have purple flowers.

When parents that are true-breeding for *different* forms of a trait are crossed—for example, purple flowers and white flowers—the offspring are called **hybrids.** When only one trait is being studied, the cross is a **monohybrid cross.** We'll look first at monohybrid problems and crosses.

A. Monohybrid Problems with Complete Dominance (About 20 min.)

MATERIALS
- simulated chromosomes, consisting of pop beads with magnetic centromeres, and meiotic diagram cards (Page 151, Exercise 12)
- bottle of 70% ethanol

1. Most organisms are diploid; that is, they contain homologous chromosomes with genes for the same traits. The location of a gene on a chromosome is its locus (plural: loci). Two genes at homologous loci are called a gene pair. Chromosomes have numerous genes, as shown in Figure 13-1. Genes exist in different forms, called *alleles*. Let's consider one gene pair at the *F* locus. There are three possibilities for the allelic makeup at the *F* locus.

   Both alleles are *FF*:

   Both alleles are *ff*:

   One allele is *F*, and the other is *f*:

   Gametes, on the other hand, are **haploid**; they contain only one of the two homologues and thus only one of the two alleles for a specific trait. According to Mendel's first law of inheritance, the **law of segregation**, each organism contains two alleles for each trait, and the alleles segregate (separate) during the formation of gametes during meiosis. Each gamete then contains only one allele of the pair.

   The **genotype** of an organism represents its genetic constitution—that is, the alleles present, either for each locus, or taken cumulatively as the genotype of the entire organism.

   For each of these diploid genotypes, indicate all possible genotypes of the gametes that can be produced by the organism:

Diploid Genotype	Potential Gamete Genotype(s)
FF	_____
ff	_____
Ff	_____ , _____

   In order to review the process that gives rise to the gamete genotypes, manipulate the pop-bead models that you used in Exercise 12. Using a marking pen, label one bead of each chromosome and go through the meiotic divisions that give rise to the gametes. *It is imperative that you understand meiosis before you attempt to do genetics problems.*

2. During fertilization, two haploid gamete nuclei fuse, and the diploid condition is restored. Give the diploid genotype produced by fusion of the following gamete genotypes.

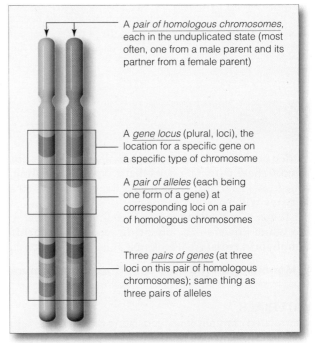

Figure 13-1   A few genetic terms illustrated.

A *pair of homologous chromosomes*, each in the unduplicated state (most often, one from a male parent and its partner from a female parent)

A *gene locus* (plural, loci), the location for a specific gene on a specific type of chromosome

A *pair of alleles* (each being one form of a gene) at corresponding loci on a pair of homologous chromosomes

Three *pairs of genes* (at three loci on this pair of homologous chromosomes); same thing as three pairs of alleles

After Starr, 2000.

Gamete Genotype	×	Gamete Genotype	⟶	Diploid Genotype
F		F		_____
F		f		_____
F		f		_____

3. Now let's attach some meaning to genotypes. As you see, the genotype is the actual genetic makeup of the organism. The **phenotype** is the outward expression of the genotype—that is, what the organism looks like because of its genotype, as well as its physiological traits and behavior. (Although phenotype is determined primarily by genotype, in many instances environmental factors can modify phenotype.)

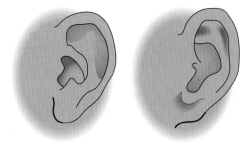

Human earlobes are either attached or free (Figure 13-2). This trait is determined by a single gene consisting of two alleles, *F* and *f*. An individual whose genotype is *FF* or *Ff* has free earlobes. This is the **dominant** condition. Note that the presence of one *or* two *F* alleles results in the dominant phenotype, free earlobes. The allele *F* is said to be dominant over its allelic partner, *f*. The **recessive** phenotype, attached earlobes, occurs only when the genotype is *ff*. In the case of **complete dominance,** the dominant allele completely masks the expression or effect of the recessive allele.

Figure 13-2 Free (left) and attached (right) earlobes in humans.

When both alleles in a nucleus are identical, the nucleus is **homozygous**. Those with both dominant alleles are homozygous dominant.

When both recessives are present in the same nucleus, the individual is said to be *homozygous recessive* for the trait.

When both the dominant and recessive alleles are present in a single nucleus, the individual is **heterozygous** for that trait.

A man has the genotype *FF*. What is the genotype of his gamete (sperm) nuclei? _____

A woman has attached earlobes. What is her genotype? _____

What allele(s) does(do) her gametes (ova) carry? _____

These two individuals produce a child. Show the genotype of the child by diagramming the cross:

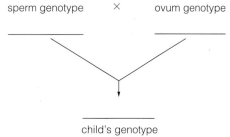

sperm genotype   ×   ovum genotype

child's genotype

What is the phenotype of the child? (That is, does this child have attached or free earlobes?) _____

4. In garden peas, purple flowers are dominant over white flowers. Let *A* represent the allele for purple flowers, *a* the allele for white flowers.

(a) What is the phenotype (color) of the flowers with the following genotypes?

Genotype	Phenotype
AA	
aa	
Aa	

Note: Always distinguish clearly between upper- and lowercase letters.

A white-flowered garden pea is crossed with a homozygous dominant purple-flowered plant.

(b) Name the genotype(s) of the gametes of the white-flowered plant. _____

(c) Name the genotype(s) of the gametes of the purple-flowered plant. _____

(d) Name the genotype(s) of the plants produced by the cross. _____

(e) Name the phenotype(s) of the plants produced by the cross. _____

(f) The Punnett square is a convenient way to perform the mechanics of a cross. The circles along the top and side of the Punnett square represent the possible gamete nuclei. Insert the proper letters indicating the genotypes of the possible gamete nuclei for the white/purple flowered cross in the circles, then fill in the following Punnett square for all the possible genetic outcomes represented by each combination of gametes.

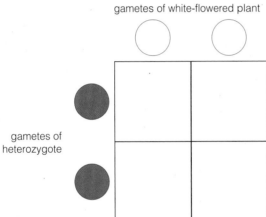

(g) A heterozygous plant is crossed with a white-flowered plant. Fill in the Punnett square below, then give the genotypes and phenotypes of all the possible genetic outcomes.

gametes of white-flowered plant

gametes of
heterozygote

Possible genotypes: _____

Possible phenotypes: _____

Draw a line connecting each possible genotype listed above to its respective phenotype.

(h) It's unlikely that every cross between two pea plants will produce four seeds that will in turn grow into four offspring plants every time. Rather, one of the most useful facets of problems such as these is that they allow you to *predict* the chances of a particular genetic outcome occurring. Genetics is really a matter of **probability,** the likelihood of the occurrence of a particular outcome.

To take a simple example, consider that the probability of coming up with heads in a single toss of a coin is one chance in two, or ½. Now let's apply this idea to the probability that offspring will have a certain genotype. Look at your Punnett square in part (g). The probability of having a genotype is the sum of all occurrences of that genotype. For example, the genotype *Aa* occurs in two of the four boxes. The probability that the genotype *Aa* will be produced from that particular cross is thus ²⁄₄, or 50%.

(i) What is the probability of an individual from part (g) having the genotype *aa*? _____

B. An Observable Monohybrid Cross (About 15 min.)

MATERIALS

Per student group (table):

- genetic corn ears illustrating monohybrid cross
- hand lens or magnifying lens (optional)

Examine the monohybrid genetic corn demonstration. This illustrates a monohybrid cross between plants producing purple kernels and ones producing yellow kernels. By convention, P stands for the parental generation. The offspring are called the *first filial generation*, abbreviated F_1. If these F_1 offspring are crossed, their offspring are the *second filial generation*, designated F_2, and are shown as follows:

PROCEDURE

1. Note that all the first-generation kernels (F_1) in the genetic corn demonstration are purple, while the second-generation ear (F_2) has both purple and yellow kernels. Count the purple kernels and then the yellow ones in the F_2 ear. _____ purple; _____ yellow. When reduced to the lowest common denominator, is this ratio closest to 1:1, 2:1, 3:1, or 4:1? _____ This is called the *phenotypic ratio*.

2. A corncob with kernels represents the products of multiple instances of sexual reproduction. Each kernel represents a single instance; fertilization of one egg by one sperm produced *each* kernel. Thus each kernel represents a different cross.

P × P   and F_1 × F_1

F_1 F_2

Let the letter *P* represent the gene for kernel color.

(a) What genotypes produce a purple phenotype? _____
(b) Which allele is dominant? _____
(c) What is the genotype of the yellow kernels on the F_2 ear? _____
(d) You are given an ear with purple kernels. How can you determine its genotype with a single cross? Explain.

_____

_____

_____

C. Experiment: Monohybrid Heredity in a Fern

A significant limitation of carrying out genetics experiments in the biology lab is that most take several months or years to collect relevant data. Even though Mendel could raise two generations of peas in a growing season, the experiments conducted in his garden plot often lasted several years. Two growing seasons were required to produce the corn monohybrid cross studied above. Fortunately, we can now look at inheritance in organisms with a much shorter life cycle. We'll use C-ferns to investigate a monohybrid cross.

Like all ferns, C-ferns have two independent life cycle phases: a structurally simple, haploid gametophyte and a more complex diploid sporophyte. (See Figures 12-4 and 12-5, to review the generalized fern life cycle.) A mature C-fern plant produces haploid spores via the process of meiosis. The spores germinate under suitable environmental conditions, and begin to divide mitotically to produce the gametophyte phase (Figure 13-3). This haploid phase develops very rapidly, with gametophytes maturing within 2 weeks.

At maturity, the gametophyte consists of a small (2 mm), simple, photosynthetic flattened structure with sex organs that produce *by mitosis* eggs in structures called archegonia and/or sperm in structures called antheridia. In the presence of water, flagellated sperm are discharged. The sperm are attracted to substances produced by the archegonia and swim toward the egg. Eventually one sperm fertilizes the egg, producing the first cell of the next diploid sporophyte generation, the zygote.

The photosynthetic sporophyte also develops rapidly, with roots and leaves visible within 1–2 weeks. The C-fern sporophytes reach heights of 10–40+ cm. Spores are produced by meiosis in structures on the leaves, completing the life cycle.

C.1. Week 1—Observation of F_1 Hybrid Gametophytes (About 30 min.)

MATERIALS

Per student:

- 2-week-old F_1 C-fern gametophyte culture in petri dish
- dissecting microscope
- sterile dH$_2$O

- sterile pipet
- marking pen
- calculator (optional)

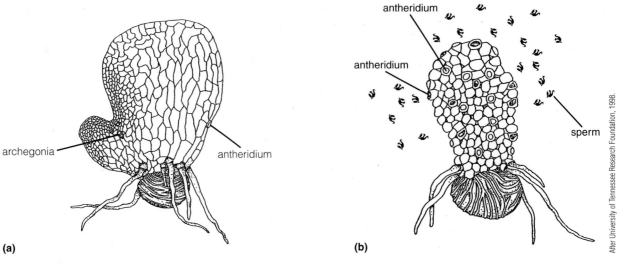

Figure 13-3 Mature C-fern gametophytes and gametes. (**a**) Hermaphroditic (bisexual) gametophytes produce both eggs and sperm and are somewhat heart-shaped. (**b**) Male gametophytes produce only sperm and appear tongue-shaped.

After University of Tennessee Research Foundation, 1998.

Prior to this class, petri dishes with nutrient medium were inoculated with spores from an F_1 hybrid C-fern sporophyte plant. The spores germinated and have grown into mature haploid gametophytes.

PROCEDURE

1. Observe the F_1 cultures under the dissecting microscope with the highest magnification possible and transmitted light (light from below.) You can prevent your cultures from drying out from the heat of the microscope by leaving the lid on the culture as much as possible and by turning the light off or removing the culture to the lab bench when it is not being observed.
2. While you are observing the culture, tilt the lid up and use a sterile pipet to add 1–2 mL sterile distilled water. Lower the lid and tilt the plate back and forth to cover all of the gametophytes with water. Observe the release of swimming sperm from antheridia and their attempts to find and fertilize mature eggs within archegonia; the sperm often spin in place briefly once they're released from an antheridium before zooming off and homing in on chemical signals produced by a mature archegonium.

    Do all the gametophytes have the same phenotype? Describe any differences you observe.

    _____

    _____

    For this experiment, we will focus only on the larger, heart-shaped hermaphroditic gametophytes. Which of the phenotypes would you designate as a mutant? Why?

    _____

3. Take a random sample of the hermaphroditic F_1 gametophyte population by counting up to 50 individuals and tallying their phenotypes. First, think of the following questions about collecting such data:
    (**a**) Why is it important to take a random sample from the cultures?

    _____

    (**b**) What is a suitable method of collecting data that would ensure a random sample?

    _____

4. Record your data in Table 13-1 and in the location designated by your instructor for the class data. Leave the lid on during this procedure if possible. If it becomes fogged, quickly exchange the fogged lid for a clean lid from an unused dish. After scoring, replace the old lid over the culture.
5. When you've finished your observations, remove any excess water from the culture by lifting the lid slightly and pouring off the excess. Place the culture in the location designated by your instructor. Be sure the petri dish lid is in place.
    (**a**) Are the F_1 plants in the culture dish haploid or diploid?

    _____

    (**b**) What products will result from the fertilization events in the culture?

    _____

TABLE 13-1 **Gametophyte Phenotypes**

Description of Phenotypes	Number of Gametophytes	Class Total

© Cengage Learning 2013

Will the products of fertilization be haploid or diploid?

(c)   If an F_1 sporophyte is heterozygous for a single mutant trait, use your own gene and allele designations to list the genotypes that will be present in the spores produced by that sporophyte.

What genotypes will be present in the F_3 gametophytes that grow from those spores?

(d)   What is the *expected* ratio of genotypes in the F_3 gametophytes?

(e)   What is the approximate *actual* phenotypic ratio of the gametophytes you counted?

(f)   Can you determine the dominance relationships from the data in Table 13-1? Explain and describe.

Biologists have assigned the designation *CP* to the single gene responsible for the two different phenotypes seen in this experiment. The dominant allele is thus designated *CP*, and the recessive allele *cp*.

(g)   Predict the genetic outcome of the fertilizations taking place in your culture by formulating a hypothesis to explain the inheritance of the trait. Indicate expected ratios of both the gametophyte and the F_2 generations.

C.2. Data Analysis (About 15 min.)

Gametes from true-breeding gametophyte parents (the P generation) combine to produce hybrid F_1 sporophyte fern plants. Meiosis within the F_1 plants produces spores. Each gametophyte you observe resulted from mitotic divisions of one of those spores. Their gametes will combine to produce the F_2 generation sporophytes.

In the space below, diagram the crosses involved in the F_1 and F_2 generations, indicating which generations and organisms are haploid, and which are diploid. (Refer to the crosses diagrammed in Part 13.1.B. as an example.)

You can now test your hypothesis concerning the method of trait inheritance to determine whether the data you collected support or do not support your model. Geneticists typically use the chi-square (χ^2) statistical test to determine whether experimentally obtained data are a satisfactory approximation of the expected data. In short, this test expresses the difference between expected (hypothetical) and observed (collected) numbers as a single value, χ^2. If the difference between observed and expected results is large, a large χ^2 results, while a small difference results in a small χ^2. Chi-square values are calculated according to the formula

$$\chi^2 = \sum \frac{(O-E)^2}{E}$$

where O = *observed* number of individuals
E = *expected* number of individuals
Σ = the sum of all values of $(O-E)^2/E$ for the various categories of phenotypes

Let's use this formula in our garden-pea monohybrid cross, question 4 from pages 163–164. Suppose 81 flowers are counted in a cross. Our hypothesis (expectation) is that three-fourths of them will be purple:

$$\frac{3}{4} \times 81 = 60.75$$

Similarly, we expect one-fourth to be white:

$$\frac{1}{4} \times 81 = 20.25$$

Suppose we actually count 64 purple flowers and 17 white flowers. Examine Table 13-2, noting how these values are used.

$$\chi^2 = \sum (0.174 + 0.522) = 0.696$$

TABLE 13-2	Calculations of Chi Square for Garden-Pea Monohybrid Cross					
Phenotype	Genotype	O	E	$(O-E)$	$(O-E)^2$	$(O-E)^2/E$
Purple	$P\_$	64	60.75	3.25	10.56	0.174
White	pp	17	20.25	−3.25	10.56	0.522
Total		81	81	0		0.696

© Cengage Learning 2013

Now, how do we interpret the χ^2 value we found? Suppose the expected and observed values were identical. Then $\chi^2 = 0$. You might guess that a number very close to zero indicates close agreement between observed and expected and a large χ^2 value suggests that "something unusual" is taking place. The problem is that chance alone almost always causes small deviations between observed and expected results, *even when the hypothesis being tested is correct.*

When does the χ^2 value indicate that chance alone cannot explain the deviation? Geneticists generally agree on a probability value of 1 in 20 (or 5% = .05) as the lowest acceptable value derived from the χ^2 test. This number indicates that if the experiment is repeated many times, the deviations expected due to chance alone will be as large as or larger than those observed only about 5% or less of the time. Probabilities equal to or greater than .05 are considered to support the hypothesis, while probabilities lower than .05 do **not** support the hypothesis. Here we must consult a table of χ^2 values to make our decision (Table 13-3).

In our example, the χ^2 value is 0.696. Since this is a monohybrid problem with only two categories of possible outcomes (purple or white flowers), the number of degrees of freedom (n in the left-hand column of Table 13-3) is 1. Read across the table until you come to .05 and find the χ^2 value 3.84. Because 0.696, our calculated χ^2 value, is less than 3.841, it is likely that the variation in the observed and expected is the result of chance, and that our hypothesized outcome is correct. A value *greater than* 3.841, however, would indicate that chance alone cannot explain the deviation between observed and expected, and we would reject our hypothesis.

The term *degrees of freedom* requires further explanation. The number of degrees of freedom is always 1 *less* than the number of categories of possible outcomes. Thus, if you are dealing with a dihybrid problem with a ratio of 9:3:3:1 (four possible phenotypes), $n = 3$.

TABLE 13-3	Distribution of χ^2			
	Probability of Obtaining a χ^2 Value as Large or Larger			
Degrees of Freedom, n	.10	.05	.01	.001
1	2.71	3.84	6.63	10.83
2	4.61	5.99	9.21	13.82
3	6.25	7.82	11.35	16.27
4	7.78	9.49	13.28	18.47

© Cengage Learning 2013

1. Transfer your individual or group data from the Total columns in Table 13-1 to Table 13-4 and calculate χ^2.

TABLE 13-4	χ^2 Calculation from F_1 Gametophyte Data				
Phenotype	Observed (O)	Expected (E)[a]	(O – E)	(O – E)^2	(O – E)^2/E
Totals					$\chi^2=$

© Cengage Learning 2013

[a]This number should be based on the hypothesis you developed in week 1 observations pages 166–167.

2. Use Table 13-3 to determine the probability of obtaining this χ^2 value for the gametophyte data in Table 13-4. How many degrees of freedom are there? _____

Is your hypothesis supported or not supported? _____ If not, what might be changed in your hypothesis or in the experimental design?

_____
_____

C.3. Week 3—Observation of F_2 Sporophytes (About 45 min.)

MATERIALS

Per student:

- 4-week-old F_2 C-fern sporophyte culture in petri dish
- dissecting microscope
- dissecting needle or toothpicks
- calculator (optional)

PROCEDURE

1. Examine your cultures with the dissecting microscope. Mutant and wild-type phenotypes are best observed using reflected light from the top or the side. Carefully observe the oldest leaves. Can you see mutant and wild-type phenotypes? _____
   Are the young sporophytes haploid or diploid? How do you know?

_____

2. Sketch what you are observing, and label it with the following terms: gametophyte, sporophyte leaf, sporophyte root.
3. Take a random sample of the sporophyte population in a dish by counting up to 50 individuals and identifying their phenotype. You can remove the lid from the culture to do this. It may be easier to score the phenotype after gently and randomly pulling up individual sporophytes with a dissecting needle or toothpick and laying them out in a row on empty areas of the culture plate. Observe the largest leaf on each sporophyte and examine the differences carefully before recording data in Table 13-5 and in the location designated for class data.
4. Following scoring of phenotypes, place the lid back on the plate and return the culture to the designated location, or take it home so that you can observe it over the next several weeks to determine whether the phenotype of older sporophytes is apparent without use of a microscope.

TABLE 13-5	F_2 Sporophyte Phenotypes	
Description of Phenotypes	**Number of Sporophytes**	**Class Total**

5. Restate your hypothesis regarding the inheritance of the mutant and wild-type alleles, and your prediction of the genetic outcome in the F_2 sporophytes.

_____

_____

_____

6. Transfer your individual or class total data to Table 13-6 to calculate χ^2.

TABLE 13-6	χ^2 Calculation from F_2 Sporophyte Data				
Phenotype	**Observed (O)**	**Expected (E)**[a]	**(O – E)**	**(O – E)^2**	**(O – E)^2/E**
Totals					$\chi^2 =$

_____

[a]This number should be based on the hypothesis you developed in this week's observations.

7. Use Table 13-3 to determine the probability of obtaining this χ^2 value for the sporophyte data in Table 13-6.
   How many degrees of freedom are there? _____
   What is the approximate probability? _____
   Is your hypothesis supported or not supported? _____
   Which allele is the dominant allele? _____
   Which is recessive? _____
   If gametophytes had not expressed the phenotype, would you be able to form a hypothesis from observations of the gametophyte generation? _____ Why or why not? _____

   _____

   _____

13.2 Dihybrid Inheritance

All the problems and experiments so far have involved the inheritance of only one trait; that is, they are monohybrid problems. Now we'll examine cases in which two traits are involved: **dihybrid problems.**

Note: We will assume that the genes for these traits are carried on different (nonhomologous) chromosomes.

A. Dihybrid Problems (About 15 min.)

1. Let's consider these two traits:
   - In humans, a pigment in the front part of the eye masks a blue layer at the back of the iris. The dominant allele P causes production of this pigment. Those who are homozygous recessive (*pp*) lack the pigment,

and the back of the iris shows through, resulting in blue eyes. (Other genes determine the color of the pigment, but in this problem we'll consider only the presence or absence of *any* pigment at the front of the eye.)

- Dimpled chins (D = allele for dimpling) are dominant over undimpled chins (d = allele for lack of dimple).

  (a) List all possible genotypes for an individual with pigmented iris and dimpled chin. _____

  (b) List the possible genotypes for an individual with pigmented iris but lacking a dimpled chin. _____

  _____

  (c) List the possible genotypes of a blue-eyed, dimple-chinned individual. _____

  (d) List the possible genotypes of a blue-eyed individual lacking a dimpled chin. _____

2. An individual is heterozygous for both traits (eye pigmentation and chin form).

   (a) What is the genotype of such an individual? _____

   (b) What are the possible genotypes of that individual's gametes? _____

   If determining the answer for question 2 was difficult, recall from Exercise 12 that the principle of independent assortment states that genes on different (nonhomologous) chromosomes are separated out independently of one another during meiosis. That is, the occurrence of an allele for eye pigmentation in a gamete has *no bearing* on which allele for chin form will occur in that same gamete.

   There is a useful method for determining possible gamete genotypes produced during meiosis from a given parental genotype. Using the genotype *PpDd* as an example, follow the four arrows below to determine the four possible gamete genotypes:

   (c) Two individuals heterozygous for both eye pigmentation and chin form have children. What are the possible genotypes of those F_1 offspring?

   _____

   gametes of one parent

   You can set up a Punnett square to do dihybrid problems just as you did with monohybrid problems. However, depending on the parental genotypes, the square may have as many as 16 boxes, rather than just 4. Insert the possible genotypes of the gametes from one parent in the top circles and the gamete genotypes of the other parent in the circles to the left of the box.

   gametes of other parent

   (d) Possible genotypes of children produced by two parents heterozygous for both eye pigmentation and chin form:

   _____

   What is the ratio of the genotypes? _____

   What is the phenotypic ratio? _____

   (e) Recalling the discussion of probability in Section 13.1, state the probability of a child from part **(d)** having the following genotypes.

   ppDD _____
   PpDd _____
   PPDd _____

   To extend the probability discussion, let's reconsider flipping a coin by asking the question, What is the probability of flipping heads twice in a row? The chance of flipping heads the first time is $\frac{1}{2}$. The same is true for the second flip. The chance (probability) that we'll flip heads twice in a row is $\frac{1}{2} \times \frac{1}{2} = \frac{1}{4}$. The probability that we could flip heads three times in a row is $\frac{1}{2} \times \frac{1}{2} \times \frac{1}{2} = \frac{1}{8}$.

   (f) State the probability that three children born to the parents in part **(d)** will have the genotype *ppdd*.

   _____

   What is the probability that three children born to these parents will have dimpled chins and pigmented eyes? _____

   (g) What is the genotype of the F_1 generation when the father is homozygous for both pigmented eyes and dimpled chin, but the mother has blue eyes and no dimple? _____

   What is the phenotype of this individual? _____

B. An Observable Dihybrid Cross (About 20 min.)

MATERIALS

Per student group (table):

- genetic corn ears illustrating a dihybrid cross

PROCEDURE

1. Examine the demonstration of dihybrid inheritance in corn. Notice that not only are the kernels two different colors (one trait), but they are also differently shaped (second trait). Kernels with starchy endosperm (the carbohydrate-storing tissue) are smooth, while those with sweet (sugary) endosperm are shriveled. Notice that all *four* possible phenotypic combinations of color and shape are present in the F_2 generation.

   The P gene is involved in pigment production, with two alleles P and p. The S gene determines carbohydrate (sugar) storage, with two alleles S and s.

   Which genotypes of the parents produced the F_2 generation kernels? _____

2. Set up a Punnett square of this dihybrid cross:

   What is the predicted phenotypic ratio? _____

3. Count the number of kernels of each possible phenotype and record in Table 13-7. To increase your sample size, count three ears.

   Which traits seem dominant?

   _____

   Which traits seem recessive?

   _____

4. Calculate the actual phenotypic ratio you observed:

   _____

   Do your observed results differ from the expected results? _____

gametes of one parent

gametes of other parent

TABLE 13-7	Phenotypes in Dihybrid Corn Cross			
	Number of Kernels with Phenotypes			
Ear	Yellow Smooth	Yellow Shriveled	Purple Smooth	Purple Shriveled
1				
2				
3				
Totals				

© Cengage Learning 2013

5. Use the chi-square test to determine if the deviation from the expected results can be accounted for by chance alone.

   Chi-square test results: _____

   _____

13.3 Some Readily Observable Human Traits (About 15 min.)

In the preceding pages, we examined several human traits that are fairly simple and that follow the Mendelian pattern of inheritance. Most of our traits are much more complex, involving many genes or interactions between genes. For example, hair color is determined by at least four genes, each one coding for the production of melanin, a brown pigment. Because the effect of these genes is cumulative, hair color can range from blond (little melanin) to very dark brown (much melanin).

Clearly, human traits are of great interest to us. Table 13-8 lists a number of traits that seem to exhibit Mendelian inheritance. For each trait, work with a lab partner to determine your phenotype, then record in Table 13-8. List your possible genotype(s) for each trait. When convenient, examine your parents' phenotypes and attempt to determine your actual genotype.

| | | | Mom's | | Dad's | | |
TABLE 13-8 Summary of My Mendelian Traits							
Trait	My Phenotype	My Possible Genotypes	Phenotype	Possible Genotype	Phenotype	Possible Genotype	My Possible/ Probable Genotype
Mid-digital hair							
Tongue rolling							
Widow's peak							
Earlobe attachment							
Hitchhiker's thumb							
Relative finger length							

© Cengage Learning 2013

1. *Mid-digital hair* (Figure 13-4a). Examine the joint of your fingers for the presence of hair, the dominant condition (*MM, Mm*). Complete absence of hair is due to the homozygous-recessive condition (*mm*). You may need a hand lens to determine your phenotype. Even the slightest amount of hair indicates the dominant phenotype.

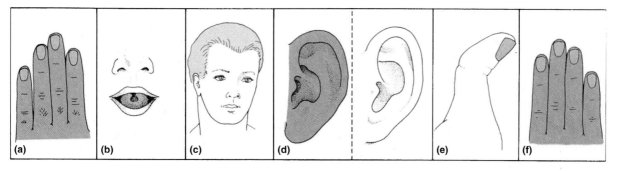

Figure 13-4 Some readily observable human Mendelian traits.

2. *Tongue rolling* (Figure 13-4b). The ability to roll one's tongue is due to a dominant allele, *T*. The homozygous-recessive condition, *tt*, results in inability to roll the tongue (Figure 13-5).
3. *Widow's peak* (Figure 13-4c). Widow's peak describes a distinct downward point in the frontal hairline and is due to the dominant allele, *W*. The recessive allele, *w*, results in a continuous hairline. (Omit study of this trait if baldness is affecting the hairline.)
4. *Earlobe attachment* (Figure 13-4d). Most individuals have free (unattached) earlobes (*FF, Ff*). Homozygous recessives (*ff*) have earlobes attached directly to the head.
5. *Hitchhiker's thumb* (Figure 13-4e). Although considerable variation exists in this trait, we'll consider those

Figure 13-5 A family with a tongue-rolling father and tongue-rolling-challenged mother. What are the genotype and phenotype of their son?

Graham Dunn/Alamy

individuals who *cannot* extend their thumbs backward to approximately 45° to be carrying the dominant allele, *H*. Homozygous-recessive persons (*hh*) can bend their thumbs at least 45°, if not farther.

6. *Relative finger length* (Figure 13-4f). An interesting sex-influenced (*not* sex-linked) trait relates to the relative lengths of the index and ring finger. In males, the allele for a short index finger (*S*) is dominant. In females, it's recessive. In rare cases, each hand is different. If one or both index fingers are greater than or equal to the length of the ring finger, the recessive genotype is present in males, and the dominant one present in females.

   Genetics Problems. One of the best ways to solidify your understanding of different patterns of inheritance is to work genetics problems. Your instructor may assign you portions of Appendix 2, which is a collection of these useful and interesting problems.

_____ **1.** In a monohybrid cross,
  (a) only one trait is being considered.
  (b) the parents are always dominant.
  (c) the parents are always heterozygous.
  (d) no hybrid is produced.

_____ **2.** The genetic makeup of an organism is its
  (a) phenotype.
  (b) genotype.
  (c) locus.
  (d) gamete.

_____ **3.** An allele whose expression is completely masked by the expression or effect of its allelic partner is
  (a) homologous.
  (b) homozygous.
  (c) dominant.
  (d) recessive.

_____ **4.** The physical appearance and physiology of an organism, resulting from interactions of its genetic makeup and its environment, is its
  (a) phenotype.
  (b) hybrid vigor.
  (c) dominance.
  (d) genotype.

_____ **5.** When both dominant and recessive alleles are present within a single nucleus, the organism is _____ for the trait.
  (a) diploid
  (b) haploid
  (c) homozygous
  (d) heterozygous

_____ **6.** A Punnett square is used to determine
  (a) probable gamete genotypes.
  (b) possible parental phenotypes.
  (c) possible parental genotypes.
  (d) possible genetic outcomes of a cross.

_____ **7.** The gametophyte of a fern is
  (a) haploid.
  (b) photoautotrophic.
  (c) a structure that produces eggs and/or sperm.
  (d) all of the above

_____ **8.** A chi-square test is used to
  (a) determine if experimental data adequately matches what was expected.
  (b) analyze a Punnett square.
  (c) determine parental genotypes producing a given offspring genotype.
  (d) determine if a trait is dominant or recessive.

_____ **9.** Possible gamete genotypes produced by an individual of genotype $PpDd$ are
  (a) Pp and Dd.
  (b) all $PpDd$.
  (c) PD and pd.
  (d) PD, Pd, pD, and pd.

_____ **10.** If you can roll your tongue,
  (a) you have at least one copy of the dominant allele T.
  (b) you have two copies of the recessive allele t.
  (c) you must be male.
  (d) you are haploid.

EXERCISE **13**

Heredity

Post-Lab Questions

13.1 Monohybrid Crosses

1. Explain how Mendel's law of segregation applies to the distribution of alleles in gametes.

2. Assume that production of hairs on a plant's leaves is controlled by a single gene with two alleles, H (dominant) and h (recessive). Hairy leaves are dominant to smooth (nonhairy) leaves.
   (a) Name the genotype(s) of a smooth-leaved plant. _____
   (b) Name the genotype(s) of a hairy-leaved plant. _____
   (c) What are the possible genotypes of gametes produced by the smooth-leaved plant? _____
   (d) What are the possible genotypes of gametes produced by the hairy-leaved plant? _____

3. *Non*-true-breeding hairy-leaved plants are crossed with smooth-leaved plants.
   (a) What genotypic and phenotypic ratios would you expect for the potential offspring? _____
   (b) Suppose you perform such a cross, collect data, and do a chi-square test to aid in data analysis. How many degrees of freedom would there be? _____
   (c) Suppose your chi-square value is very large (>25). What does this indicate about your experiment and/or hypothesis?

4. What genotypic ratio would you expect in the gametophyte generation of C-ferns produced by F_1 spores if two traits on separate chromosomes were being followed?

5. Were dominant and recessive traits observed equally in both gametophytes and sporophytes of C-ferns? How did you determine which character was dominant and which was recessive?

13.2 Dihybrid Inheritance

6. Suppose you have two traits controlled by genes on separate chromosomes. If sexual reproduction occurs between two heterozygous parents, what is the genotypic ratio of all possible gametes?

Food for Thought

7. Explain the purpose and uses of the chi-square test.

8. Suppose students in previous semesters had removed some of the corn kernels from the genetic corn ears before you counted them. What effect would this have on your results?

9. Assume that one allele is completely dominant over the other for the following questions.
   (a) Two individuals heterozygous for a *single* trait have children. What is the expected phenotypic ratio of the possible offspring? _____
   (b) Two individuals heterozygous for *two* traits have children. What would be the expected phenotypic ratio of the possible offspring? _____
   (c) Crossing two individuals heterozygous for two traits results in the same phenotypic ratio as for a single trait. Are the genes for these two traits on separate chromosomes or on the same chromosome? Explain your answer. (Remember that the gene for each trait is located at a locus, a physical region on the chromosome.)

10. How does probability differ from actuality?

Nucleic Acids: Blueprints for Life

OBJECTIVES

After completing this exercise, you will be able to

1. define *DNA, RNA, purine, pyrimidine, principle of base pairing, replication, transcription, translation, codon, anticodon, peptide bond, gene;*

2. identify the components of deoxyribonucleotides and ribonucleotides;

3. distinguish between DNA and RNA according to their structure and function;

4. describe DNA replication, transcription, and translation;

5. give the base sequence of DNA or RNA when presented with the complementary strand;

6. identify a codon and anticodon on RNA models and describe the location and function of each;

7. give the base sequence of an anticodon when presented with that of a codon, and vice versa;

8. describe what is meant by the *one-gene, one-polypeptide hypothesis;*

9. describe the process of DNA recombination by bacterial conjugation;

10. explain the difference between DNA recombination by bacterial conjugation and the technique by which eukaryotic gene products are produced by bacteria.

Introduction

By 1900, Gregor Mendel had demonstrated patterns of inheritance, based solely on careful experimentation and observation. Mendel had no clear idea how the traits he observed were passed from generation to generation, although the seeds of that knowledge had been sown as early as 1869, when the physician-chemist Friedrich Miescher isolated the chemical substance of the nucleus. Miescher found the substance to be an acid with a large phosphorus content and named it "nuclein." Subsequently, nuclein was identified as **DNA**, short for **deoxyribonucleic acid**. Some 75 years would pass before the significance of DNA would be revealed.

Few would argue that the demonstration of DNA as the genetic material and the subsequent determination of its molecular structure are among the most significant discoveries of the twentieth century. Since the early 1950s, when James Watson and Francis Crick built on discoveries of others before them to construct their first model of DNA, tremendous advances in molecular biology have occurred, many of them based on the structure of DNA. Today we speak of gene therapy and genetic engineering in household conversations. In the minds of some, these topics raise hopes for curing or preventing many of the diseases plaguing humanity. For others, thoughts turn to "playing with nature," undoing the deeds of God, or creating monstrosities that will cause great harm.

This exercise will familiarize you with the basic structure of nucleic acids and their role in the cell. Understanding the function of nucleic acids—both DNA and **RNA (ribonucleic acid)**—is central to understanding life itself. We hope you will gain an understanding that will allow you to form educated opinions concerning what science should do with its biotechnology.

14.1	Isolation and Identification of Nucleic Acids *(About 30 min.)*

In this section, you'll isolate and identify a nucleic acid component of strawberries (*Fragaria ananassa*). Like all organisms, strawberries are composed of cells containing genetic material. Commercial garden strawberries are especially interesting because they are 8N, that is, each nucleus has eight copies of each of its seven different chromosomes.

MATERIALS

Per student:

- frozen strawberry, thawed
- Ziploc bag
- funnel with cheesecloth square
- wooden skewer or wire loop
- 9 mL of ice cold 95% ethanol in test tube
- two test tubes
- test tube rack
- agar gel plate with methylene blue stain
- dropping pipettes

Per student group (4):

- detergent (10% Woolite) and salt (1% NaCl) solution in flask fitted with a 5-mL pipet and Pi-pump
- 10-mL graduated cylinder
- TBE (Tris/borate/EDTA) buffer solution
- fine-pointed marker
- DNA standard solution in dropper bottle
- 1% albumin solution in dropper bottle
- paper towels

Per lab room:

- source of dH_2O
- white light transilluminator or other bottom source of white light

PROCEDURE

A. Isolation of Nucleic Acids

1. Thoroughly mash up a thawed, frozen strawberry by kneading the berry in a Ziploc bag with 5 mL of the detergent and salt solution. Continue kneading for at least 2 minutes. This action exposes the strawberry fruit cells to the detergent and salt solution.
   Review from Exercise 7: What are the major molecular components of cell membranes? _____ _____
   The detergent-salt solution is useful in this procedure because detergents break down lipids, and salt causes proteins that would otherwise obscure the nucleic acids to precipitate out of solution.
2. Filter the liquefied berry through cheesecloth into a test tube.
3. Slowly pour the ice cold ethanol down the side of the test tube so the ethanol forms a clear layer on top of the watery strawberry filtrate. *Be careful not to mix the water and alcohol layers.*
4. Allow the test tube to sit for several minutes. You should observe a frothy material begin to form at the interface between the two layers.

Nucleic acids precipitate at the boundary between alcohol and water. The nucleic acids you have extracted are not pure; they contain cellular debris as well as adhering proteins. You will test your extracted precipitate to identify its major component.

B. Identification—Test for DNA

1. Use a 10-mL graduated cylinder to measure 3 mL of TBE buffer solution into a clean test tube.
2. Slowly twirl the skewer or loop to wrap up the precipitated material and withdraw the skewer from the test tube.
3. Place the slimy material from the skewer into the TBE buffer. Swirl the skewer in the buffer solution to thoroughly dissolve the viscous material in the buffer.
4. Obtain an agar plate with methylene blue stain. (Agar is an inert gel-like substance; methylene blue stain binds specifically with DNA, causing a visible purplish color to develop.) Turn the unopened plate over and use the marking pen to draw four small, widely separated circles on the *underside* of the plate. Label the circles U, D, A, and C. These will be visible when looking at the plate from above to mark the locations where different test substances will be applied.

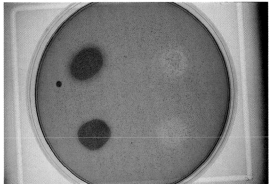

Figure 14-1 Agarose-methylene blue plate showing positive (left) and negative (right) tests for DNA.

5. Open the plate, and with a dropping pipette, apply a drop of the precipitated strawberry material dissolved in TBE buffer solution onto the area marked U, for "unknown."
6. Similarly, apply a drop of DNA standard solution onto the area marked D.
7. Apply a drop of 1% albumin protein solution onto the area marked A.
8. Apply a control drop of TBE buffer onto the area marked C (for control).
9. Set the plate aside for 20 to 30 minutes to allow time for color development.
10. View the petri plate on the white light transilluminator or other source of bottom light. A purple color indicates the presence of DNA (Figure 14-1). Record your results in Table 14-1.

TABLE 14-1	**Identification of Contents Extracted from Strawberry**	
Droplet Code	**Droplet Contents**	**Color**
U	Unknown strawberry precipitate in TBE buffer	
D	DNA standard	
A	1% albumin protein	
C	TBE buffer	

© Cengage Learning 2013

Name the substance(s) present in the material you isolated from the strawberry.

_____

What is the purpose of the DNA standard droplet?

_____

What is the purpose of the TBE buffer droplet?

_____

14.2 Modeling the Structure and Function of Nucleic Acids and Their Products *(About 90 min.)*

MATERIALS

Per student pair or group:
- DNA puzzle kit

Per lab room:
- DNA model

PROCEDURE

Work in pairs or groups.

Note: **Clear your work surface of everything except your lab manual and the DNA puzzle kit.**

In this section, you will model the three processes by which DNA carries out its vital functions: *replication, transcription,* and *translation.* But before you study these three *per se,* formulate an idea of the structure of DNA itself.

A. Nucleic Acid Structure

1. Obtain a DNA puzzle kit. It should contain the following parts, which you can identify from the figure below:

   - 18 deoxyribose sugars
   - nine ribose sugars
   - 18 phosphate groups
   - four adenine bases
   - six guanine bases
   - six cytosine bases
   - four thymine bases
   - two uracil bases
   - three transfer RNA (tRNA)
   - three amino acids
   - three activating units
   - ribosome template sheet

2. Group the components into separate stacks. Select a single deoxyribose sugar, an adenine base (labeled A), and a phosphate, fitting them together as shown in Figure 14-2. This is a single nucleotide (specifically a *deoxy*ribonucleotide), a unit consisting of a sugar (deoxyribose), a phosphate group, and a nitrogen-containing base (adenine).

   Let's examine each component of the nucleotide.

   Deoxyribose (Figure 14-3) is a sugar compound containing five carbon atoms. Four of the five are joined by covalent bonds into a ring. Each carbon is given a number, indicating its position in the ring. (These numbers are read "1-prime, 2-prime," and so on. "Prime" is used to distinguish the carbon atoms from the position of atoms that are sometimes numbered in the nitrogen-containing bases.) This structure is usually drawn in a simplified manner, without actually showing the carbon atoms within the ring (Figure 14-4).

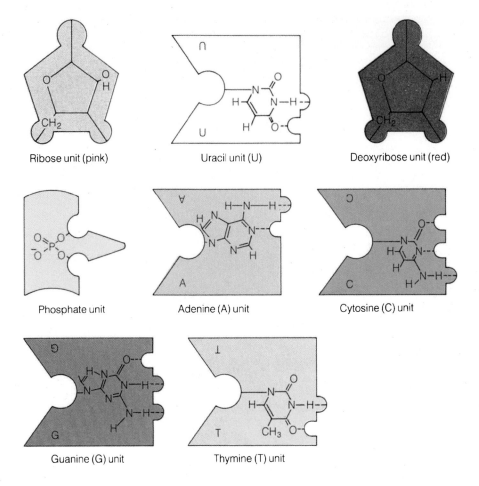

Ribose unit (pink)

Uracil unit (U)

Deoxyribose unit (red)

Phosphate unit

Adenine (A) unit

Cytosine (C) unit

Guanine (G) unit

Thymine (T) unit

There are four kinds of nitrogen-containing bases in DNA. Two are **purines** and are double-ring structures. Specifically, the two purines are *adenine* and *guanine* (abbreviated A and G, respectively; Figure 14-5).

The other two nitrogen-containing bases are **pyrimidines,** specifically *cytosine* and *thymine* (abbreviated C and T, respectively). Pyrimidines are single-ring compounds, as shown in Figure 14-6.

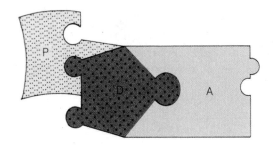

Figure 14-2 One deoxyribonucleotide.

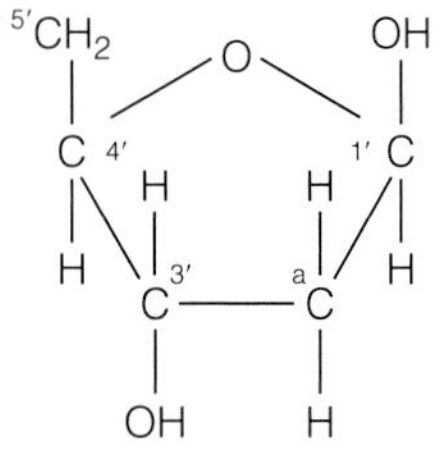

Figure 14-3 Deoxyribose.

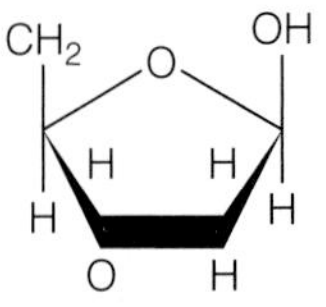

Figure 14-4 Simplified representation of deoxyribose.

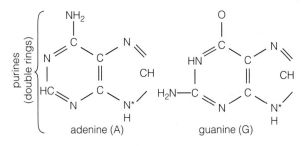

Figure 14-5 Double-ringed purines found in DNA.

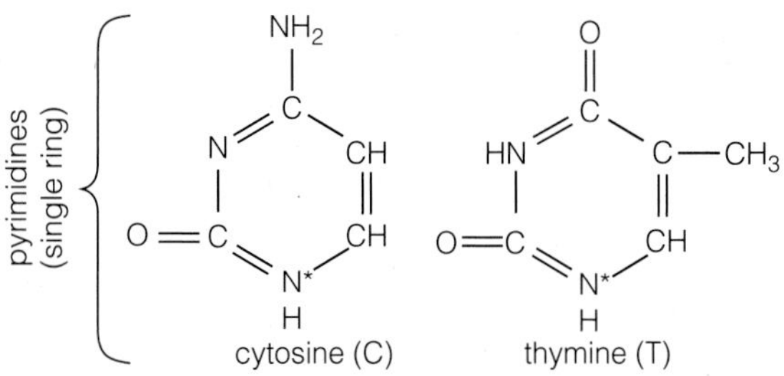

Figure 14-6 Pyrimidines found in DNA.

The symbol * indicates where a bond forms between each nitrogen-containing base and the 1′ carbon atom of the sugar ring structure. Although deoxyribose and the nitrogen-containing bases are *organic* compounds (that is, they have a carbon backbone), the phosphate group is an *inorganic* compound (without a carbon backbone) with the structural formula shown in Figure 14-7.

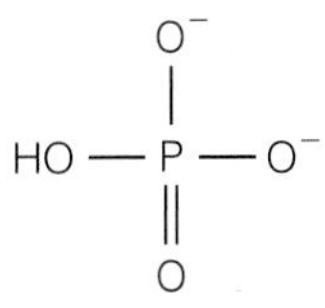

The phosphate end of the deoxyribonucleotide is referred to as the 5′ end, because the phosphate group bonds to the 5′ carbon atom.

There are four kinds of deoxyribonucleotides, each differing only in the type of base it possesses. Construct the other three kinds of deoxyribonucleotides, then draw them in Figures 14-8b–d. Rather than drawing the somewhat complex shape of the model, in this and other drawings, just give the correct position and letters. Use D for deoxyribose, P for a phosphate group, and A, C, G, and T for the different bases (as shown in Figure 14-8a).

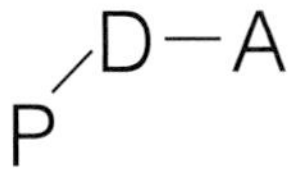

Note the small notches and projections in the nitrogen-containing bases. Will the notches of adenine and thymine fit together? _____

Will guanine and cytosine? _____

Will adenine and cytosine? _____

Will thymine and guanine? _____

The notches and projections represent bonding sites. Make a prediction about which bases will bond with one another.

Will a purine base bond with another purine? _____

Will a purine base bond with both types of pyrimidines?

Assemble the three additional deoxyribonucleotides, linking them with the adenine-containing unit, to form a nucleotide strand of DNA. Note that the sugar backbone is bonded together by phosphate groups. Your strand should appear like that shown in Figure 14-9.

3. Now assemble a second four-nucleotide strand, similar to that of Figure 14-9. However, this time make the base sequence T-A-C-G, from bottom to top. DNA molecules consist of *two* strands of nucleotides, each strand the *complement* of the other.

4. Assemble the two strands by attaching (bonding) the nitrogen bases of complementary strands. Note that the adenine of one nucleotide always pairs with the thymine of its complement; similarly, guanine always pairs with cytosine. This phenomenon is called the **principle of base pairing**. On Figure 14-10, attach letters to the model pieces indicating the composition of your double-stranded DNA model.
   What do you notice about the *direction* in which each strand is running? (That is, are both 5′ carbons at the same end of the strands?)

(Does the second strand of your drawing show this? It should.)

In life, the purines and pyrimidines are joined together by hydrogen bonds. Note again that the sugar backbone is linked by phosphate groups. Your model illustrates only a very small portion of a DNA molecule. The entire molecule may be tens of thousands of nucleotides in length!

5. Slide your DNA segment aside for the moment.
6. Examine the three-dimensional model of DNA on display in the laboratory (Figure 14-11). Notice that the two strands of DNA are twisted into a spiral-staircaselike pattern. This is why DNA is known as a *double helix*. Identify the deoxyribose sugar, nitrogen-containing bases, hydrogen bonds linking the bases, and the phosphate groups.

Figure 14-7 Phosphate group found in nucleic acids.

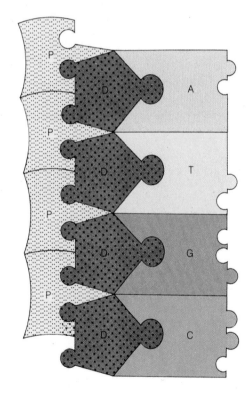

Deoxyribonucleotide containing adenine
(a)

Deoxyribonucleotide containing guanine
(b)

Deoxyribonucleotide containing cytosine
(c)

Deoxyribonucleotide containing thymine
(d)

Figure 14-8 Drawings of deoxyribonucleotides containing guanine, cytosine, and thymine.

Figure 14-9 Four-nucleotide strand of DNA.

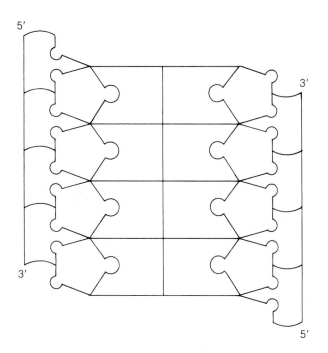

Figure 14-10 Drawing of a double strand of DNA.
Labels: A, T, G, C, D, P (all used more than once)

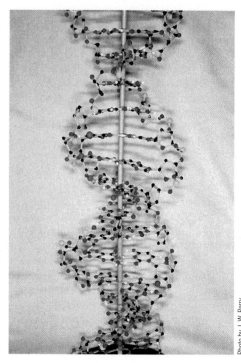

Figure 14-11 Three-dimensional model of DNA.

The second type of nucleic acid is RNA, short for ribonucleic acid. There are three important differences between DNA and RNA:
(a) RNA is a *single strand* of nucleotides.
(b) The sugar of RNA is **ribose**.
(c) RNA lacks the nucleotide that contains thymine. Instead, it has one containing the pyrimidine uracil (U) (see Figure 14-12).

Compare the structural formulas of ribose (Figure 14-13) and deoxyribose (Figure 14-4). How do they differ? _____

Why is the sugar of DNA called *deoxy*ribose?

---

7. From the remaining pieces of your model kit, select four ribose sugars, an adenine, uracil, guanine, and cytosine, and four phosphate groups. Assemble the four **ribonucleotides** and draw each in Figure 14-14. (Use the convention illustrated in Figure 14-8 rather than drawing the actual shapes.)
Disassemble the RNA models after completing your drawing.

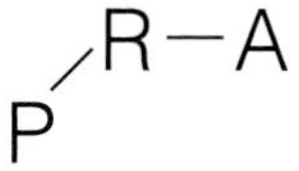

$$P \overset{R-A}{\diagup}$$

Ribonucleotide containing adenine
(a)

Ribonucleotide containing guanine
(b)

Ribonucleotide containing cytosine
(c)

Ribonucleotide containing uracil
(d)

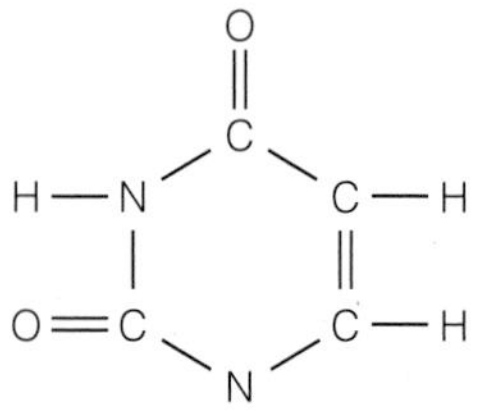

Figure 14-12 The pyrimidine uracil.

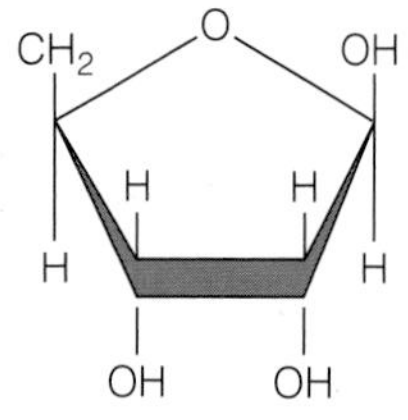

Figure 14-13 Ribose.

Figure 14-14 Drawings of four possible ribonucleotides.

184 EXERCISE 14

B. Modeling DNA Replication

DNA **replication** takes place during the S-stage of interphase of the cell cycle (see Exercise 11). Recall that the DNA is aggregated into chromosomes. Before mitosis, the chromosomes duplicate themselves so that the daughter nuclei formed by mitosis will have the same number of chromosomes (and hence the same amount of DNA) as did the parent cell.

Replication begins when hydrogen bonds between nitrogen bases break and the two DNA strands "unzip." Free nucleotides within the nucleus bond to the exposed bases, thus creating *two* new strands of DNA (as described below). The process of replication is controlled by enzymes called **DNA polymerases**.

1. Construct eight more deoxyribonucleotides (two of each kind) but don't link them into strands.
2. Now return to the double-stranded DNA segment you constructed earlier. Separate the two strands, imagining the zipperlike fashion in which this occurs within the nucleus.
3. Link the free deoxyribonucleotides to each of the "old" strands. When you are finished, you should have two double-stranded segments.

Note that one strand of each is the parental ("old") strand and the other is newly synthesized from free nucleotides. This illustrates the *semiconservative* nature of DNA replication. Each of the parent strands remains intact—it is *conserved*—and a new complementary strand is formed on it. Two "half-old, half-new" DNA molecules result.

4. Draw the two replicated DNA molecules in Figure 14-15, labeling the old and new strands. (Once again, use the convention shown in Figure 14-8.)

Figure 14-15 Drawing of two replicated DNA segments, illustrating their semiconservative nature.

C. Transcription: DNA to RNA

DNA is an "information molecule" residing *within* the nucleus. The information it provides is for assembling proteins *outside* the nucleus, within the cytoplasm. The information does not go directly from the DNA to the cytoplasm. Instead, RNA serves as an intermediary, carrying the information from DNA to the cytoplasm.

Synthesis of RNA takes place within the nucleus by **transcription**. During transcription, the DNA double helix unwinds and unzips, and a single strand of RNA, designated **messenger RNA (mRNA)**, is assembled using the nucleotide sequence of *one* of the DNA strands as a pattern (template). Let's see how this happens.

1. Disassemble the replicated DNA strands into their component deoxyribonucleotides.
2. Construct a new DNA strand consisting of nine deoxyribonucleotides. With the purines and pyrimidines pointing away from you, lay the strand out horizontally in the following base sequence: T-G-C-A-C-C-T-G-C
3. Now assemble RNA ribonucleotides complementary to the exposed nitrogen bases of the DNA strand. Don't forget to substitute the pyrimidine uracil for thymine.

What is the sequence from left to right of nitrogen bases on the mRNA strand?

_____

After the mRNA is synthesized within the nucleus, the hydrogen bonds between the nitrogen bases of the deoxyribonucleotides and ribonucleotides break.

4. Separate your mRNA strand from the DNA strand. (You can disassemble the deoxyribonucleotides now.) At this point, the mRNA moves out of the nucleus and into the cytoplasm.

By what avenue do you suppose the mRNA exits the nucleus? (*Hint:* Reexamine the structure of the nuclear membrane, as described in Exercise 6.)

_____

To *transcribe* means to "make a copy of." Is transcription of RNA from DNA the formation of an *exact* copy? Explain. _____

_____

_____

You will use this strand of mRNA in the next section. Keep it close at hand.

D. Translation—RNA to Polypeptides

Once in the cytoplasm, mRNA strands attach to *ribosomes,* on which translation occurs. To *translate* means to change from one language to another. In the biological sense, **translation** is the conversion of the linear message encoded on mRNA to a linear strand of amino acids to form a polypeptide. (A *peptide* is two or more amino acids linked by a peptide bond.)

Translation is accomplished by the interaction of mRNA, ribosomes, and **transfer RNA (tRNA)**, another type of RNA. The tRNA molecule is formed into a four-cornered loop. You can think of tRNA as a baggage-carrying molecule. Within the cytoplasm, tRNA attaches to specific free amino acids. This occurs with the aid of activating enzymes, represented in your model kit by the pieces labeled "glycine activating" or "alanine activating." The amino acid–carrying tRNA then positions itself on ribosomes where the amino acids become linked together to form polypeptides.

1. Obtain three tRNA pieces, three amino acid units, and three activating units.
2. Join the amino acids first to the activating units and then to the tRNA. Will a particular tRNA bond with *any* amino acid, or is each tRNA specific? Explain. _____
3. Now let's do some translating. In the space below, list the sequence of bases on the *messenger* RNA strand, starting at the left. (left, 3′ end)_____(right, 5′ end)

Translation occurs when a *three*-base sequence on mRNA is "read" by tRNA. This three-base sequence on mRNA is called a **codon**. Think of a codon as a three-letter word, read right (5′) end to left (3′) end. What is the order of the rightmost (first) mRNA codon? (Remember to list the letters in the *reverse* order of that in the mRNA sequence.)
   The first codon on the mRNA model is (5′ end)_____(3′ end)
4. Slide the mRNA strand onto the ribosome template sheet, with the first codon at the 5′ end.
5. Find the tRNA–amino acid complex that complements (will fit with) the first codon. The complementary three-base sequence on the tRNA is the **anticodon**. Binding between codons and anticodons begins at the P site of the 40s subunit (the smaller subunit) of the ribosome. The tRNA–amino acid complex with the correct anticodon positions itself on the P site.
6. Move the tRNA–amino acid complex onto the P site on the ribosome template sheet and fit the codon and anticodon together. In the boxes below, indicate the codon, anticodon, and the specific amino acid attached to the tRNA.

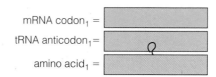

7. Now identify the second mRNA codon and fill in the boxes.

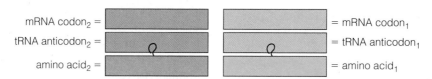

8. The second tRNA–amino acid complex moves onto the A site of the 40s subunit. Position this complex on the A site. An enzyme now catalyzes a condensation reaction, forming a **peptide bond** and linking the two amino acids into a dipeptide. (Water, HOH, is released by this condensation reaction.)
9. Separate amino acid₁ from its tRNA and link it to amino acid₂. (In reality, separation occurs somewhat later, but the puzzle doesn't allow this to be shown accurately; see below for correct timing.) Fill in the boxes.

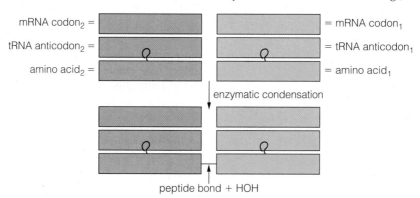

One tRNA–amino acid complex remains. It must occupy the A site of the ribosome in order to bind with its codon. Consequently, the dipeptide must move to the right.

10. Slide the mRNA to the right (so that $tRNA_2$ is on the P site) and fit the third mRNA codon and tRNA anti-codon to form a peptide bond, creating a model of a tripeptide. At about the same time that the second peptide bond is forming, the first tRNA is released from both the mRNA and the first amino acid. Eventually, it will pick up another specific amino acid.

What amino acid will $tRNA_1$ pick up? _____ Fill in the boxes below.

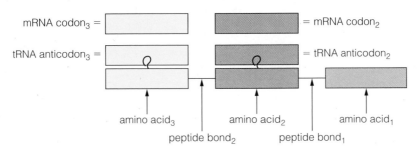

Record the tripeptide that you have just modeled. _____

You have created a short polypeptide. Polypeptides may be thousands of amino acids in length. As you see, the amino acid sequence is ultimately determined by DNA, because it was the original source of information.

Finally, let's briefly turn our attention to the concept of a gene. A **gene** is a unit of inheritance. Our current understanding of a gene is that a gene codes for one polypeptide. This is appropriately called the **one-gene, one-polypeptide hypothesis**. Given this concept, do you think a gene consists of one, several, or many deoxyribonucleotides?

A gene probably consists of _____ deoxyribonucleotides.

Note: **Please disassemble your models and return them to the proper location.**

_____ 1. The individuals responsible for constructing the first model of DNA structure were
   (a) Wallace and Watson.
   (b) Lamarck and Darwin.
   (c) Mendel and Meischer.
   (d) Crick and Watson.

_____ 2. Deoxyribose is
   (a) a five-carbon sugar.
   (b) present in RNA.
   (c) a nitrogen-containing base.
   (d) one type of purine.

_____ 3. A nucleotide may consist of
   (a) deoxyribose or ribose.
   (b) purines or pyrimidines.
   (c) phosphate groups.
   (d) all of the above

_____ 4. Which of the following is consistent with the principle of base pairing?
   (a) purine–purine
   (b) pyrimidine–pyrimidine
   (c) adenine–thymine
   (d) guanine–thymine

_____ 5. Nitrogen-containing bases between two complementary DNA strands are joined by
   (a) polar covalent bonds.
   (b) hydrogen bonds.
   (c) phosphate groups.
   (d) deoxyribose sugars.

_____ 6. The difference between deoxyribose and ribose is that ribose
   (a) is a six-carbon sugar and deoxyribose has only five carbons.
   (b) bonds only to thymine, not uracil.
   (c) has one more oxygen atom than deoxyribose has.
   (d) is all of the above.

_____ 7. Replication of DNA
   (a) takes place during interphase.
   (b) results in two double helices from one.
   (c) is semiconservative.
   (d) is all of the above.

_____ 8. Transcription of DNA
   (a) results in formation of a complementary strand of RNA.
   (b) produces two new strands of DNA.
   (c) occurs on the surface of the ribosome.
   (d) is semiconservative.

_____ 9. An anticodon
   (a) is a three-base sequence of nucleotides on tRNA.
   (b) is produced by translation of RNA.
   (c) has the same base sequence as does the codon.
   (d) is the same as a gene.

_____ 10. Each amino acid is specified by a set of _____ base(s) in mRNA.
   (a) one
   (b) two
   (c) three
   (d) four

EXERCISE **1 4**

Nucleic Acids: Blueprints for Life

Post-Lab Questions

14.1 Isolation and Identification of Nucleic Acids

1. The photo below shows the result of mixing part of a banana with detergent and salt, and then layering ice-cold ethanol on top of the mixture. What is the composition of the material containing all the bubbles?

14.2 Modeling the Structure and Function of Nucleic Acids and Their Products

2. The following diagram represents some of the puzzle pieces used in this section.
   (a) Assembled in this form, do they represent an amino acid, a base, a portion of messenger RNA, or a deoxyribonucleotide?

   (b) Explain your answer.

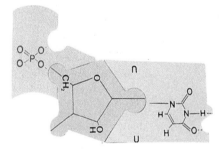

3. Why is DNA often called a *double helix?*

4. State the following ratios.
   (a) Guanine to cytosine in a double-stranded DNA molecule: _____
   (b) Adenine to thymine: _____

5. Define the following terms.
   (a) replication

   (b) transcription

   (c) translation

   (d) codon

   (e) anticodon

6. What does it mean to say that DNA replication is *semiconservative*?

7. **(a)** If the base sequence on one DNA strand is ATGGCCTAG, what is the sequence on the other strand of the helix?
   (b) If the original strand serves as the template for transcription, what is the sequence on the newly formed RNA strand?
8. **(a)** What amino acid would be produced if *transcription* takes place from a nucleotide with the three-base sequence ATA? _____
   (b) A genetic mistake takes place during *replication* and the new DNA strand has the sequence ATG. What is the three-base sequence on an RNA strand transcribed from this series of nucleotides?

   _____

   (c) Which amino acid results from this codon?

Food for Thought

9. What kinds of molecules are present in strawberry cells in addition to the DNA you might have isolated in Section 14.1?

10. Mutations permanently change the nucleotide sequence of DNA. Most that occur within genes are harmful. Two kinds of "frameshift mutations" are implicated in the devastating disease, cystic fibrosis. Search two Internet sites for information on "frameshift mutation" and "cystic fibrosis." List the two sites below and briefly summarize what you learn from them regarding the nature of the mutations and their effects on proteins.

   http:// _____

   _____

   _____

   http:// _____

   _____

   _____

Genetic Engineering: Bacterial Transformation

OBJECTIVES

After completing this exercise, you will be able to

1. define *bacterial transformation, plasmids, vectors, recombinant plasmids, competence, genomic libraries, gene therapy, genetic engineering*;

2. explain how restriction enzymes and DNA ligase are used to insert a foreign gene into a plasmid;

3. diagram a typical plasmid used by scientists to transform bacteria;

4. state what can be harvested from cultures of transformed bacteria;

5. describe how a plasmid gene that confers antibiotic resistance is used to identify transformed bacteria;

6. discuss how a restriction site in the *lacZ* gene can be used to identify bacteria transformed with a plasmid that contains an inserted foreign gene;

7. explain how bacterial transformation and similar procedures are used in health, agriculture, and other industries.

Introduction

Bacterial transformation occurs when bacterial cells pick up **plasmids**—tiny circles of double-stranded DNA—that carry additional genes. They occur naturally and are the smallest gene-carrying vehicles (**vectors**) that can enter, replicate, and express themselves within bacteria. Scientists use manipulated plasmids to introduce foreign genes into bacteria. They cut the plasmids open with **restriction enzymes**, allow foreign DNA fragments to fill the breach, and seal the cut ends with another enzyme, **DNA ligase** (Figure 15-1).

Plasmids containing new DNA, or **recombinant plasmids**, and similar vectors are used in biotechnology to carry foreign genes into bacteria where (1) replication of the plasmid makes many copies of the foreign gene or

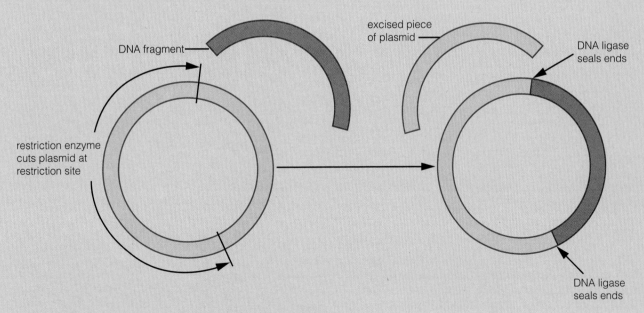

Figure 15-1 The insertion of a DNA fragment into a plasmid.

(2) expression of the duplicated genes produces a high concentration of gene product. In either case, the plasmids or the gene products can be harvested later.

A typical plasmid ring used for making additional recombinant plasmids contains a gene that confers antibiotic resistance and one or more **restriction sites** where one or more foreign genes can be inserted into the plasmid (Figure 15-2).

When recombinant plasmids are made, plasmids and DNA fragments are mixed in a solution that contains all the substances needed to open the plasmid and splice in a DNA fragment. However, not all plasmids are successfully cut and spliced back together with a DNA fragment. Similarly, when the recombinant plasmids are mixed with bacteria, typically only 1 in 1000 bacteria is transformed—that is, gets a plasmid. Thus as part of any procedure involving transformation, there must be ways to identify whether it has taken place.

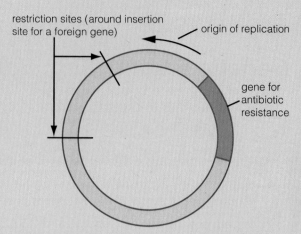

Figure 15-2 Typical plasmid prior to foreign gene insertion.

15.1 Experiment: Bacterial Transformation (2 hours over parts of 2–3 days)

The principle of bacterial transformation is based on the observation that cells in a transient state of competence can be penetrated with plasmid DNA (Figure 15-3). Exposing bacteria to an ice-cold solution of calcium chloride and then briefly heating them up produces competence. Cells transformed with plasmids that confer antibiotic resistance are easily selected because they produce bacterial colonies when grown on solid media containing this antibiotic (Figure 15-4).

The plasmid DNA stock provided in this experiment, wild-type pUC19, gives blue colonies of transformed cells. This plasmid contains a gene for ampicillin resistance and the *lacZ* gene, which activates β-galactosidase. This enzyme cleaves a substrate (X-gal) also present in the media. It is the products of this reaction that in the presence of an inducer (IPTG) give rise to the blue color of the bacterial colonies.

The hypothesis for this experiment is that under the conditions described above bacteria will be transformed when exposed to a plasmid that contains genes for antibiotic resistance and β-galactosidase activation. A question to keep in mind is: how can the active *lacZ* gene be used to identify bacteria with a plasmid that contains an inserted foreign gene?

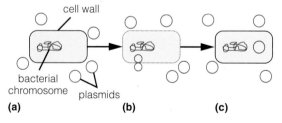

Figure 15-3 Transformation of a bacterial cell with plasmid DNA. Exposing a cell to an ice-cold solution of calcium chloride and then briefly heating makes the cell wall porous (**a, b**). If plasmid DNA is present, it can now enter the cell (**b**). The plasmid DNA can add genes to the now transformed cell (**c**).

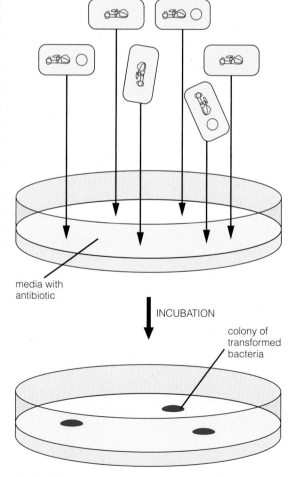

Figure 15-4 Identification of transformed bacteria.

MATERIALS

Note: Kits for similar experiments can be purchased from various vendors. Though generally more expensive in cost per student, the ordering process and preparation time is more efficient.

Per lab room:

- 50-mL polypropylene tube containing bacterial culture (*E. coli*) labeled BaCu in an ice bucket
- wall clock
- sterile transfer hood (optional)
- 37°C incubator (optional)

Per student group (4):

- 3–4 empty microfuge tubes labeled BC, C1, and C2 (optional)
- microfuge tube containing ~10 μL of wild-type pUC19 plasmid (labeled BP)
- microfuge tube containing 1 mL transformation solution labeled TS
- microfuge rack
- ice bucket and ice

- disposal beaker
- microcentrifuge
- 6–10 transfer pipets
- 3–4 plastic micropipets or micropipetter and 3–4 tips to transfer 200 μL
- micropipetter and tips to transfer 10 μL
- water bath set at 42°C or other source of this temperature water
- three bacterial plates containing X-gal/IPTG LB agar containing antibiotic
- bacterial plate containing X-gal/IPTG LB agar (optional)
- 3–4 inoculating loops or inoculating loop and burner/starter

PROCEDURE

1. Locate the labeled tubes listed in Table 15-1.
2. Use a transfer pipet to transfer 1.5 mL of the bacterial culture into the empty microfuge tubes labeled BC and C1 and C2 (optional). Store the used transfer pipet in the disposal beaker.
3. Pellet the bacterial cells by spinning the tubes in a microcentrifuge for approximately 15 seconds.
4. With the used transfer pipet, gently remove the supernatant (the fluid above the pellet) from the microfuge tubes and discard it into the disposal beaker along with the used transfer pipet. Immediately place the tubes in a bucket of ice.
5. Use a clean plastic micropipet or a micropipetter with a clean tip to add 200 μL of ice-cold transformation solution buffer (TS buffer) to each microfuge tube. If you use a micropipet, your instructor will tell you how many drops approximate 200 μL. Keep the tubes as much as possible in the ice while gently resuspending the bacterial pellets by mixing the solutions up and down with the pipet tip. Discard the micropipet into the disposal beaker, or, if you used a micropipetter, discard the tip.
6. Wait 20 minutes for the bacterial cells to become competent.

TABLE 15-1	**Tubes and Solutions Needed for Transformation Procedure**			
Container	**Location**	**Contents**	**Label**	**Comment**
50-mL polypropylene tube	Ice bucket on common bench	Bacterial culture	BaCu	Contains incubated cells and media
Microfuge tube	Microfuge rack on student station	Empty	BC	For production of competent cells for transformation with blue plasmid
Microfuge tube	Microfuge rack on student station	Empty	C1	For production of control competent cells not exposed to blue plasmid
Microfuge tube (optional)	Microfuge rack on student station	Empty	C2	For production of control competent cells not exposed to plasmid or antibiotic
Microfuge tube	Ice bucket on student station	Blue plasmid in sterile H$_2$O	BP	
Microfuge tube	Ice bucket on student station	Transformation solution buffer	TS	

© Cengage Learning 2013

Note: If a C2 tube is included, a sterile transfer hood will be needed.

7. **(a)** Use a clean transfer pipet to transfer the contents of the BC microfuge tube (200 μL of competent cells) to the tube labeled BP, which contains 10 μL of blue plasmid DNA solution. Dispose of the transfer pipet and BC tube. Keep the BP tube on ice for approximately 30 minutes.

   (b) Alternatively, use a micropipetter with a clean tip to transfer 10 μL of the solution containing blue plasmid DNA (tube labeled BP) to the tube labeled BC, which contains 200 μL of competent cells. Dispose of the used tip and BP tube. Keep the BC tube on ice for approximately 30 minutes.

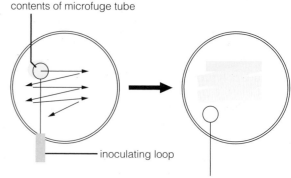

Figure 15-5 How to streak a bacterial plate.

8. Heat-shock the bacterial cells by incubating the microfuge tubes at 42°C for 1 minute. After heat-shock, put the tubes back into ice for about 5 minutes. During this process, the temporary porous cell walls of the bacteria permit the uptake of plasmid DNA into the cells.

9. Using a clean transfer pipet for each microfuge tube, transfer its contents to the appropriate bacterial plate as described in Table 15-2. For each plate, use a clean inoculating loop to streak the contents gently over the media, being careful not to dig into the media (Figure 15-5). *Your instructor may demonstrate how to sterilize a metal inoculating loop in a burner flame.* Let the streaked solution dry by keeping the plates open (without their covers) for 15 minutes. If a C2 plate is included, dry open plates in a sterile transfer hood.

TABLE 15-2	Transfer Contents of Microfuge Tubes onto Bacterial Plates with the Following Media	
Tube Code	**Plate Code**	**Media**
BC or BP	B	X-gal/IPTG LB agar containing antibiotic
C1	C1	X-gal/IPTG LB agar containing antibiotic
C2 (optional)	C2	X-gal/IPTG LB agar

© Cengage Learning 2013

10. Replace the covers. Incubate the bacterial plates upside down (media side up) in a 37°C incubator for 15 hours or at room temperature for 1–2 days for colonies to appear.

11. Write some predictions about the presence or absence of colonies and if present their color after incubation.

12. Make arrangements with your instructor to check the bacterial plates after the appropriate period of incubation has passed. Record the date and time. _____

13. After you check your results, fill in Table 15-3.

TABLE 15-3	Results of Transformation Experiment		
Predictions:			
Plate	**Colonies (Yes or No)**	**Color of Colonies**	**Other Comments**
B			
C1			
C2 (optional)			
Conclusions:			

© Cengage Learning 2013

14. How do your results compare with those of the rest of the class?

15. What can you conclude from the results of the C1 control plate?

16. *Optional.* What can you conclude from the results of the C2 control plate?

17. Look over these results and draw some conclusions about whether your predictions should be accepted or rejected. Is your hypothesis supported or falsified?

18. If you were to use the plasmid DNA from the blue colonies to insert a foreign gene and the insertion site was within the *lacZ* gene (Figure 15-6), how could you detect bacteria with a plasmid that contains the inserted foreign gene?

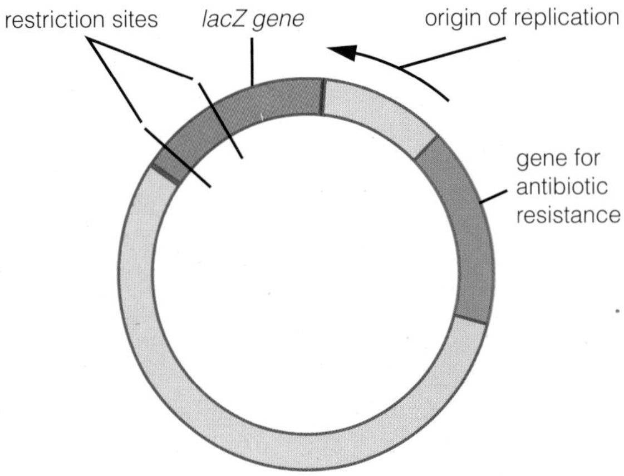

Figure 15-6 Plasmid with both antibiotic resistance and *lacZ* genes with restriction sites surrounding an insertion site.

15.2 Bacterial Transformation on the Internet *(30 min. or more)*

Bacterial transformation plays a central role in many areas of biotechnology, including basic research, health, agriculture, and industry. One use for plasmids is to store cloned DNA fragments in **genomic libraries**. Like words and sentences in library books, such plasmids contain DNA sequences and genes. Now that the human genome is known, other vectors such as viruses may be commonly used to transform human cells to counter diseases caused by defective genes (**gene therapy**). Biotechnology industry harvests proteins produced by transformed bacteria cultured in large steady-state incubators for use in many fields. **Genetic engineering** or tinkering with the genes of naturally and artificially selected plants has already produced enhanced cereals, fruits, and vegetables, which among other things offer better nutrition, longer storage lives, and resistance to disease and insects. As you probably know, all of this is, and will continue to be, increasingly controversial.

MATERIALS

Per student:

- Internet access

PROCEDURE

1. Visit the National Center for Biotechnology Information. The current address is http://www.ncbi.nlm.nih.gov/.

2. Search across databases by making sure that plasmid is in the box to the right and then click on the GO button. What general types of databases are available to search?

_____

_____

_____

Scientists from all over the world use these databases to report, share, and gather genetic information.
3. Go to several of the databases and open some of the files. Don't try to understand anything in particular; scan through the material instead. You should be able to find uses and gene maps of plasmids.
4. Now search for the plasmid pUC19, which is often used to store DNA fragments for clonal and subclonal surveys.
5. Visit the Human Genome Project. The current address is
   http://www.ornl.gov/sci/techresources/Human_Genome/home.shtml.
What was the Human Genome Project?

_____

_____

_____

The decoding of our DNA sequence by government and private laboratories was one of the most important achievements in human history. Use this site to better understand its significance, including any ethical issues.
6. Another great site is managed by the Cold Spring Harbor Laboratory:
   http://www.cshl.org/.
Be sure to search for plasmids and visit the DNA Learning Center. What is Barbara McClintock famous for in the field of plant genetics?

_____

_____

_____

7. Since the preceding experiment involves bacterial growth, use your favorite search engine to find a video or cell cam of bacterial growth. The following site typically has excellent examples:
   http://www.cellsalive.com/.
8. Use your favorite search engine to find five sites that have something to do with plasmids or bacterial transformation. List them below with a brief summary of their contents.

   http://_____

   http://_____

   http://_____

   http://_____

   http://_____

Pre-Lab Questions

_____ 1. Transformed cells contain
   (a) transformants.
   (b) new genes.
   (c) bacteria.
   (d) none of these choices

_____ 2. A tiny loop of double-stranded DNA describes a
   (a) bacterium.
   (b) DNA ligase.
   (c) plasmid.
   (d) antibiotic.

_____ 3. Enzymes used to insert a foreign gene into a plasmid include
   (a) DNA ligase.
   (b) vectors.
   (c) restriction enzymes.
   (d) both a and c

_____ 4. A molecule or anything else that transports new genes into a cell is called a
   (a) vector.
   (b) plasmid.
   (c) restriction enzyme.
   (d) bacterium.

_____ 5. In this treatment of bacteria with cold transformation solution followed by heat-shock, the procedure
   (a) makes the cell wall porous.
   (b) allows for plasmids to enter the cells.
   (c) creates competent cells.
   (d) does all of these choices.

_____ 6. The presence of an antibiotic in the media where the bacteria are grown enables the selection of colonies grown from
   (a) transformed cells.
   (b) nontransformed cells.
   (c) cells that have the gene for resistance to the antibiotic.
   (d) both a and c

_____ 7. In the preceding experiment, bacteria that form blue colonies on X-gal/IPTG LB agar containing antibiotic also contain
   (a) a gene for antibiotic resistance.
   (b) an active _lacZ_ gene.
   (c) an inactive _lacZ_ gene.
   (d) both a and b

_____ 8. The wild-type pUC19 plasmid (blue plasmid) contains genes for
   (a) antibiotic resistance.
   (b) activation of β-galactosidase production.
   (c) both a and b
   (d) none of these choices

_____ 9. In genomic libraries,
   (a) naturally and artificially selected plants are genetically engineered.
   (b) DNA fragments are stored in plasmids.
   (c) a virus is used to transform human cells.
   (d) none of the above are done.

_____ 10. Characteristics of genetically engineered plants and plant products include
   (a) better nutrition.
   (b) longer storage lives.
   (c) resistance to disease and insects.
   (d) all of the above

EXERCISE 15

Genetic Engineering: Bacterial Transformation

Post-Lab Questions

Introduction

1. Define bacterial transformation.

2. What does the term *vector* mean in the field of biotechnology?

3. Draw and label a typical plasmid used by scientists to transform bacteria.

4. How are restriction and DNA ligase enzymes used to insert a foreign gene into a plasmid?

5. When is a plasmid considered a recombinant plasmid?

6. What two general types of products can be harvested from cultures of transformed bacteria?

15.1 Experiment: Bacterial Transformation

7. In the process of bacterial transformation, how are the plasmid-containing bacteria identified?

8. How are transformed bacteria that contain a plasmid with an inserted foreign gene identified?

15.2 Bacterial Transformation on the Internet

9. Name the two elements needed for procedures similar to bacterial transformation to be useful in human gene therapy.

Food for Thought

10. When this exercise was first written, the media campaign in support of golden rice was being launched. Search the Internet for the roots of the controversy over this genetically enhanced plant. Briefly describe the benefits of, and the concerns over, its introduction to the rice-growing areas of the world.

Evolutionary Agents

OBJECTIVES

After completing this exercise, you will be able to

1. define *evolutionary agent, natural selection, fitness, directional selection, stabilizing selection, disruptive selection, gene flow, divergence, speciation, mutation, genetic drift, bottleneck effect, founder effect*;

2. determine the allele frequencies for a gene in a model population;

3. calculate expected ratios of phenotypes based on Hardy–Weinberg proportions;

4. describe the effects of nonrandom mating, natural selection, migration, genetic drift, and mutation on a model population;

5. describe the effects of different selection pressures on identical model populations;

6. identify the level at which selection operates in a population;

7. describe the impact of the founder effect on the genetic structure of populations.

Introduction

Heredity itself cannot cause changes in the frequencies of alternate forms of the same gene (alleles). If certain conditions are met, then the proportions of genotypes that make up a population of organisms should remain constant generation after generation according to the equation that describes the Hardy–Weinberg equilibrium. For two alleles, the Hardy–Weinberg equilibrium is:

$$p^2 + 2pq + q^2 = 1.0$$

If p is the frequency of one allele, and q is the frequency of the other allele, then

$$p + q = 1.0$$

For example, if two alleles for coat color exist in a population of mice and the allele for white coats is present 70% of the time, then the alternate allele (black) must be present 30% of the time.

The Hardy–Weinberg equation describes the proportions of phenotypes present in succeeding generations, as long as conditions don't change. In our example, since $p = .7$, we would expect 49% (p^2) of the mice in our population to be homozygous for white coats. Thus, 42% ($2pq$) would have one of each allele and would appear gray if neither allele is dominant (that is, both alleles would have equal expression in the phenotype). What percentage of our population is homozygous for black coats? _____%

In nature, however, the frequencies of genes in populations change over time. Natural populations never meet all of the conditions assumed for Hardy–Weinberg equilibrium. *Evolution is a process resulting in changes in the genetic makeup of populations through time*; therefore, factors that disrupt Hardy–Weinberg equilibrium are referred to as **evolutionary agents**. This exercise demonstrates the effect of these agents on the genetic structure of a simplified model population.

16.1 Natural Selection *(75 min.)*

The populations you will work with are composed of colored beads. White beads in our model represent individuals that are homozygous for the white allele ($C^W C^W$). Red beads are individuals homozygous for the red allele ($C^R C^R$), and pink beads are heterozygotes ($C^W C^R$). These beads exist in "ponds"—plastic dishpans filled with smaller beads. Counts of white, pink, and red individuals in the pond are made after straining the beads through a sieve. The smaller beads pass through the mesh, which retains the larger beads. Figure 16-1 shows the initial experimental setup.

When the individuals are recovered, the frequencies of the color alleles are determined using the Hardy–Weinberg equation. The alleles in our population are codominant. Thus, each white bead contains two white alleles; each pink bead, one white and one red allele; and each red bead, two red alleles. The total number of color alleles in a population of 40 individuals is 80. If such a population contains 10 white beads, 20 pink beads, and 10 red beads, the frequency of the white allele is

$$p = \frac{(2 \times 10) + 20}{80} = .5$$

Because $p + q = 1.0$, the frequency of the red allele (q) must also be .5 if there are only two color alleles in this population.

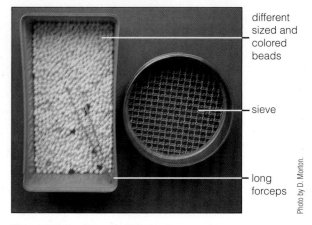

different sized and colored beads

sieve

long forceps

Figure 16-1 Experimental setup used to demonstrate natural selection.

MATERIALS

Per student group (4):

■ plastic dishpan (12″ × 7″ × 2″)
■ 50 large (10-mm diameter) white beads
■ 50 large red beads
■ 50 large pink beads
■ 4000 small (8-mm diameter) white beads
■ 4000 small red beads (optional)

■ ruler with a centimeter scale
■ pair of long forceps
■ coarse sieve (9.5 mm)
■ scientific calculator

Per lab room:

■ clock with a second hand

PROCEDURE

A. Experiment: Natural Selection Acting Alone

Natural selection disturbs the Hardy–Weinberg equilibrium by discriminating between individuals with respect to their ability to produce young. Those individuals that survive to reproduce perpetuate more of their genes in the population. These individuals exhibit greater **fitness** than do those who leave no offspring or fewer offspring.

This experiment addresses the hypothesis that *individuals are more likely to survive and reproduce when their coloration makes it easier to hide from predators in the environment.*

1. Work in groups of four with each group member assuming one of the following roles: Predator, Data Recorder/Timer, Calculator, or Caretaker. Predators search for prey. Data Recorders/Timers record numerical results and time events. Calculators use a scientific calculator to crunch numbers as needed. Caretakers look after and manipulate the experimental setup.
2. Create a white pond by filling a dishpan with small white beads to a depth of about 5 cm and establish an initial population by mixing into the pond 10 large white beads, 10 large red beads, and 20 large pink beads. These numbers have been recorded for you in Table 16-1. The Predator will prey on the large beads, removing as many as possible in a limited amount of time. The survivors will then reproduce the next generation and predation will begin again. This cycle will be repeated several times. Make a prediction as to the changing frequency in the population of the red allele over time and write it in the Prediction (Selection Alone) row of Table 16-3.
3. Search the pond for prey (large beads) and, using a pair of long forceps, remove as many of them as possible in 30 seconds.
4. Strain the pond with the sieve, and count the number of large white, pink, and red beads. Record the totals in the After row under Population/Initial in Table 16-1.
5. Calculate the frequencies of the white (p) and red (q) alleles remaining in the population after selection and record them in First generation after selection of Table 16-2. For example, if 6 white, 8 pink, and 8 red beads remain, the frequency of the white allele is

$$p = \frac{(2 \times 6) + 8}{44} = .45$$

TABLE 16-1	Large Bead Counts Before and After Four Rounds of Simulated Predation			
Population	White Beads	Pink Beads	Red Beads	Total Beads
Initial				
Before	10	20	10	40
After	_____	_____	_____	_____
Second Generation				
Before	_____	_____	_____	50
After	_____	_____	_____	_____
Third Generation				
Before	_____	_____	_____	50
After	_____	_____	_____	_____
Fourth Generation				
Before	_____	_____	_____	50
After	_____	_____	_____	_____

© Cengage Learning 2013

TABLE 16-2	Allele and Genotype Frequencies Due to Selection by Simulated Predation				
Population	p	q	p^2	$2pq$	q^2
Initial	.5	.5	.25	.5	.25
First generation after selection					
Second generation after selection					
Third generation after selection					
Fourth generation after selection					

© Cengage Learning 2013

6. Using the new values for allele frequencies, calculate genotype frequencies for homozygous white (p^2), heterozygous pink ($2pq$), and homozygous red (q^2) individuals, and record them in Table 16-2. For example, if p now equals .45, the frequency of homozygous white individuals is

$$p^2 + (.45)^2 = .20$$

Assuming that 50 individuals comprise the next and succeeding generations (maximum number of individuals the pond can sustain), calculate the number of white, pink, and red individuals needed to create the population of a new pond and record these numbers in the Before row under Second Generation in Table 16-1. Here and in future calculations to generate new numbers of individuals, round up or down to the nearest whole number. For example, if $p^2 = .20$, the number of white beads needed is

$$p^2 \times 50 = .20 \times 50 = 10 \text{ white beads}$$

Using these numbers, construct a new pond.

7. Repeat steps 2–6 for three more rounds, filling in the remaining rows in Tables 16-1 and 16-2. When you are finished, copy the frequency of the red allele from Table 16-1 to Table 16-3 (Selection Alone column) and plot this data in Figure 16-2.

TABLE 16-3	Frequency of Red Allele (q) Due to Selection and Migration	

Prediction (Selection Alone):

Prediction (Selection and Migration):

Generation	Selection Alone	Selection and Migration
1		
2		
3		
4		

Conclusion (Selection Alone):

Conclusion (Selection and Migration):

© Cengage Learning 2013

8. Write your conclusion as to your prediction in the Conclusion (Selection Alone) row of Table 16-3.
9. If you had started with a pond filled with small red beads as a background, how would the frequency of the red allele change?

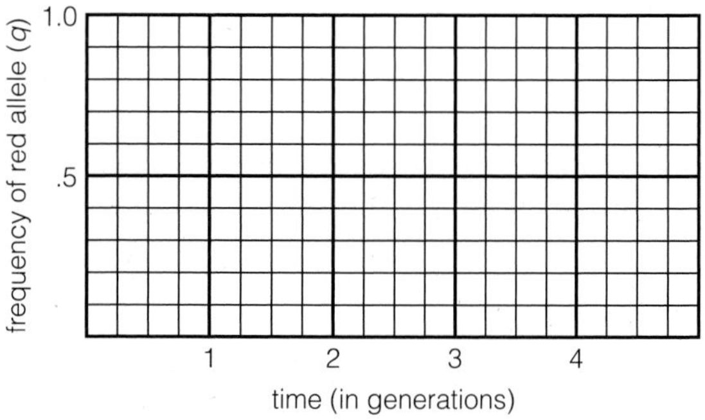

_____

_____

_____

Figure 16-2 Effects of predation on allele frequencies.

10. Selection that favors one extreme phenotype over the other and causes allele frequencies to change in a predictable direction is known as **directional selection**. When selection favors an intermediate phenotype rather than one at the extremes, it's known as **stabilizing selection**. Selection that operates against the intermediate phenotype and favors the extreme ones is called **disruptive selection**. Which kind of selection is illustrated by simulated predation of white, pink, and red beads in a white pond? Explain why you made this choice.

_____

_____

_____

It is important to realize that selection operates on the entire phenotype so that the overall fitness of an organism is based on the result of interactions of thousands of genes.

11. If two identical populations (such as the mix of beads described in step 2) inhabited different environments (such as red and white ponds), how would the frequency of the color genes in each pond compare after a large number of generations?

_____

_____

_____

As two populations become genetically different through time **(divergence)**, individuals from these populations can lose the ability to interbreed. If this happens, two species can form from one ancestral species. This process is called **speciation**.

B. Experiment: Effect of Gene Flow on Natural Selection

The frequencies of alleles in a population also change if new organisms immigrate and interbreed, or when old breeding members emigrate. **Gene flow** due to *migration* may be a powerful force in evolution. This activity demonstrates its effect.

1. Establish an initial population as in the previous section, step 2.
2. Begin selection as before, except add five new red beads to each generation before the new allele frequencies are determined. These beads represent migrants from a population where the red allele confers greater fitness. This experiment addresses the hypothesis that *gene flow resulting from the migration of a significant number of individuals into a population undergoing predation affects the change in allele frequencies expected from selection alone.* Write your prediction as to how the change in frequency of the red allele will be affected in the Conclusion (Selection and Migration) row of Table 16-3.
3. For each generation, record the frequencies of the red allele obtained with both selection and migration in Table 16-3.
4. How does migration influence the effectiveness of selection in this example?

_____

_____

_____

5. Write your conclusion as to your prediction in the Conclusion (Selection and Migration) row of Table 16-3.
6. How would migration have influenced the change in gene frequencies if white instead of red individuals had entered the population?

_____

_____

_____

Gene flow keeps local populations of the same species from becoming more and more different from each other. Things that serve as barriers to gene flow can accelerate the production of new species. Migration can also introduce new genes into a population and produce new genetic combinations. Imagine the result of a black allele being introduced into our model population and the new heterozygotes (perhaps gray and dark red) it would produce.

Note: If time is short, any or all of the remaining sections may be done as thought experiments, i.e. doing it in your head rather than actually setting it up.

16.2 Mutation *(About 15 min.)*

Another way new genetic information enters a population is through **mutation**. This usually represents an actual change in the information encoded by the DNA of cells involved with the reproduction of an organism. As such, most mutations are harmful and will be eliminated by natural selection. Nevertheless, mutations do provide the raw material for evolution.

MATERIALS

Per student group (4):

- small bowl
- 10 large white beads
- 10 large red beads
- 20 large pink beads
- one large gray bead
- scientific calculator

PROCEDURE

1. Establish an initial population by placing 10 large white beads, 10 large red beads, and 20 large pink beads in a small bowl (without the small beads).
2. For the sake of expediency, establish a new generation by one group member picking, without looking, 20 large beads from the bowl. Replace one white bead with a gray bead. This represents a mutation in a gamete that one parent contributed to this generation.
3. Calculate the allele frequencies of the new generation, including the frequency of the new color allele (r). Record them in Table 16-4.

TABLE 16-4	Change in Allele Frequencies Due to Mutation		
Population	p	q	r
Initial	.5	.5	
New generation with mutation			

© Cengage Learning 2013

4. Three alleles are present ($p + q + r = 1.0$), so the Hardy–Weinberg equation is expanded to $p^2 + 2pq + q^2 + 2pr + 2qr + r^2 = 1.0$, and in addition to white, pink, and red phenotypes, we now have gray, dark red, and potentially, in subsequent generations, black. If the next generation contains 50 individuals, how many offspring of each phenotype would you expect? Use Table 16-5 to calculate these numbers.

TABLE 16-5	Numbers of Each Phenotype Two Generations after a Single Mutation			
Color	Genotype	Frequency	$\times$ 50	Number of Individuals
White	p^2			
Pink	$2pq$			
Red	q^2			
Gray	$2pr$			
Dark red	$2qr$			
Black	r^2			

© Cengage Learning 2013

Imagine a population made up of individuals in these proportions. What effect will natural selection have on these phenotypes in a white pond?

_____

_____

_____

How could conditions change to favor the selection of the rare black allele?

_____

_____

_____

16.3 Genetic Drift *(About 30 min.)*

Chance is also a factor that results in shifts in gene frequencies over several generations (**genetic drift**). This is primarily due to the random aspects of reproduction and fertilization. Genetic drift is often a problem for small populations in that they can lose much of their genetic variability. In very small populations, chance can eliminate an allele from a population, such that p becomes 0 and the other allele becomes fixed ($q = 1.0$). This loss of genetic variation due to a small population size is known as the **bottleneck effect**.

MATERIALS

Per student group (4):
- small bowl
- 10 large white beads
- 10 large red beads
- 20 large pink beads
- scientific calculator

PROCEDURE

1. As in the previous section, place 10 large white beads, 10 large red beads, and 20 large pink beads in a small bowl. Listed in Table 16-6 are the expected allele frequencies for color in this population given all individuals participate in reproduction.

TABLE 16-6 Allele Frequencies Produced by Genetic Drift

		Actual Frequency in	
	Expected Frequency	Small Cluster	Large Cluster
p	.5		
q	.5		

© Cengage Learning 2013

2. Have a group member choose, without looking, 10 reproductively lucky individuals from the bowl.
3. In the second column of Table 16-6, record the allele frequencies present in this cluster.
4. Now replace the 10 beads you removed in step 2. Select beads at random again, but this time select 30 beads representing a larger cluster of reproductively lucky individuals.
5. Calculate the allele frequencies for this larger cluster and record them in the third column of Table 16-6.
6. Compare the allele frequencies in the three columns of Table 16-6. Sometimes chance determines whose gametes contribute to the next generation. What effect does the size of the number of individuals participating in reproduction have on gene flow to the next generation?

_____

_____

_____

7. Another way in which chance affects allele frequencies in a population is when migrants from old populations establish new populations. To model this effect, choose at random six individuals from an initial population of red, pink, and white beads to represent the migrants.
8. Move these individuals to a new unoccupied pond. (It is not necessary to actually set up a new pond for this demonstration. Use your imagination.)
9. Now calculate the allele frequencies in the new pond and record them in Table 16-7. How do they compare with the frequencies that characterized the pond from which these migrants came?

_____

_____

_____

The genetic makeup in future generations in the new population will more closely resemble the six migrants than the population from which the migrants came. This situation is known as the **founder effect**. The founder effect may not be an entirely random process because organisms that migrate from a population may be genetically different from the rest of the population to begin with. For example, if wing length in a population of insects is variable, we might expect insects with longer wings to be better at founding new populations because they can be carried farther by winds.

TABLE 16-7 Allele Frequencies in a Founder Population

	p	q
Initial Population	.5	.5
Founder Population		

© Cengage Learning 2013

Hardy–Weinberg equilibrium is also disturbed if individuals in a population don't choose mates randomly. Some members of a population may show a strong preference for mates with similar genetic makeups. This activity models this effect.

MATERIALS

Per student group (4):

- small bowl
- large white beads
- large red beads
- 20 large pink beads
- scientific calculator

PROCEDURE

1. Establish an initial population as in Section 16.3.
2. Assume that individuals will mate only with individuals of the same color. Arbitrarily assign sex to every bead so there are equal numbers of males and females in each color group.
3. If each pair of beads produces four offspring, record the number individuals with the same phenotype present in the next generation in Table 16-8. Remember that the pink pairs will produce one red, one white, and two pink individuals on average.

TABLE 16-8	Phenotype Changes Due to Nonrandom Mating Color	
	Number in	
Color	Initial Generation	Next Generation
White	10	
Pink	20	
Red	10	

© Cengage Learning 2013

4. Calculate the genotype frequencies in this generation, record them in Table 16-9, and compare these with the frequencies in the initial generation.

TABLE 16-9	Genotype Frequency Changes Due to Nonrandom Mating	
Genotype Frequency	Initial Generation	Next Generation
p^2		
$2pq$		
q^2		

© Cengage Learning 2013

5. What happens to the frequency of the heterozygote genotype in subsequent generations?

_____

_____

_____

_____ 1. If all conditions of Hardy–Weinberg equilibrium are met,
 (a) allele frequencies move closer to .5 each generation.
 (b) allele frequencies change in the direction predicted by natural selection.
 (c) allele frequencies stay the same.
 (d) all allele frequencies increase.

_____ 2. If a population is in Hardy–Weinberg equilibrium and $p = .6$,
 (a) $q = .5$.
 (b) $q = .4$.
 (c) $q = .3$.
 (d) $q = .16$.

_____ 3. Natural selection operates directly on
 (a) the genotype.
 (b) individual alleles.
 (c) the phenotype.
 (d) color only.

_____ 4. The process that discriminates between phenotypes with respect to their ability to produce offspring is known as
 (a) natural selection.
 (b) gene flow.
 (c) genetic drift.
 (d) migration.

_____ 5. Two populations that have no gene flow between them are likely to
 (a) become more different with time.
 (b) become more alike with time.
 (c) become more alike if the directional selection pressures are different.
 (d) stay the same unless mutations occur.

_____ 6. A process that results in individuals of two populations losing the ability to interbreed is referred to as
 (a) stabilizing selection.
 (b) fusion.
 (c) speciation.
 (d) differential migration.

_____ 7. Two ways in which new alleles can become incorporated in a population are
 (a) mutation and genetic drift.
 (b) selection and genetic drift.
 (c) selection and mutation.
 (d) mutation and gene flow.

_____ 8. If a new allele appears in a population, the Hardy–Weinberg formula
 (a) cannot be used because no equilibrium exists.
 (b) can be used but only for two alleles at a time.
 (c) can be used by lumping all but two phenotypes in one class.
 (d) can be expanded by adding more terms.

_____ 9. A shift from expected allele frequencies, resulting from chance, is known as
 (a) natural selection.
 (b) genetic drift.
 (c) mutation.
 (d) gene flow.

_____ 10. Genetic drift is a process that has a greater effect on populations that
 (a) are large.
 (b) are small.
 (c) are not affected by mutation.
 (d) do not go through bottlenecks.

EXERCISE **16**

Evolutionary Agents

16.1 Natural Selection

1. What effect does increasing gene flow between two populations have on their genetic makeup?

2. How can selection cause two populations to become different with time?

3. Describe how the effects of directional selection can be offset by gene flow.

16.2 Mutation

4. In addition to mutation, what other mechanism allows for new genetic information to be introduced into a population? Explain your answer.

5. What is the fate of most new mutations?

6. If a population has three codominant color alleles, how many phenotypes are possible?

16.4 Nonrandom Mating

7. What effects can nonrandom mating exert on a population?

Food for Thought

8. What two evolutionary agents are most responsible for decreases in genetic variation in a population?

9. If a population has three color alleles and one is dominant over the other two, how many phenotypes are possible?

10. In humans, birth weight is an example of a characteristic affected by stabilizing selection. What does this mean to the long-term average birth weight of human babies? How might the increasing number of Caesarean sections be affecting this characteristic?

Evidences of Evolution

OBJECTIVES

After completing this exercise, you will be able to

1. define *evolution, natural selection, population, species, fitness, fossil, hominid, Australopithecus, Homo, Homo sapiens;*

2. explain how natural selection operates to alter the genetic makeup of a population over time;

3. describe the general sequence of evolution of life forms over geologic time;

4. recognize primitive and advanced characteristics of skull structure of human ancestors and relatives.

Introduction

The process by which the incredible diversity of mammals living and extinct species came to exist is called *evolution*, the focus of this exercise. **Evolution**, the process that results in changes in the genetic makeup of populations of organisms through time, is the unifying framework for the whole of biology. Less than 200 years ago, it seemed obvious to most people that living organisms had not changed over time—that oak trees looked like oak trees and humans looked like humans, year after year, generation after generation, without change. As scientists studied the natural world more closely, however, evidence of change and the relatedness of all living organisms emerged from geology and the fossil record, as well as from comparative morphology, developmental patterns, and biochemistry.

Charles Darwin (and, independently, Alfred Russel Wallace) postulated the major mechanism of evolution to be **natural selection**, the difference in survival and reproduction that occurs among individuals of a population that differ in one or more alleles. (Review Exercise 12 for the definition of *allele*.) A **population** is a group of individuals of the same species occupying a given area. A **species** is one or more populations that closely resemble each other, interbreed under natural conditions, and produce fertile offspring.

As we understand natural selection, genetic modifications that place an organism at a disadvantage in its environment will be "selected against," that is, those members of a population will die off, reproduce less, or both. Other individuals may have allele combinations that are advantageous; these are "selected for" and tend to live longer, have more offspring, or both. They have greater **fitness**. Their offspring will receive those advantageous alleles from their parents. As physical, chemical, or biological aspects of the populations' environment change, those modifications that are best adapted to that environment tend to be found more often within the population. The population evolves as some traits become more common and others decrease or disappear over time.

Many people think of evolution in historic terms—as something that produced the dinosaurs but that no longer operates in today's world. Remember, though, that *the process of genetic change in populations over time continues today* and is a dominant force shaping the living organisms of our planet.

The scientific evidence for evolution is overwhelming, although scientists continue to debate the exact mechanisms by which natural selection and other evolutionary agents change allele frequencies. In this exercise, you will track the effects of natural selection in a simulated population, consider the time over which evolution has occurred, and examine the fossil record for evidence of large-scale trends and change among our human ancestors and relatives.

17.1 How Natural Selection Works *(About 1 hour)*

In this activity, you will simulate interacting populations of predators and prey. You will study how predators exert selective pressure on prey populations and vice versa and consider how the environment affects the interaction. The simulated prey are populations of different dry beans, and the simulated predator populations are familiar utensils.

This experiment tests two hypotheses: (1) that *prey populations of different forms vary in fitness as a consequence of predation* and (2) *predator populations of different forms vary in fitness as a consequence of prey characteristics.* Each student group will act as one predator species and forage for prey over a short time period. Analysis of changes in prey population numbers will allow us to draw conclusions about the validity of these hypotheses.

MATERIALS

Per student group (4):

- predator utensils of one form (knife, fork, spoon, forceps, or chopsticks)
- collection of dry beans (50 each of five types)
- four cups to hold prey
- calculator

Per lab room:

- patterned fabric or carpet pieces
- reservoirs of additional beans of each type
- stopwatch or timer

PROCEDURE

1. Obtain a fabric or carpet piece to serve as the habitat. Scatter the beans randomly over the habitat. What prey species (bean type) do you expect to be most able to survive predation? Why?

   _____

   _____

   Which prey species do you expect to be least successful? Why?

   _____

   _____

   What predator species (utensil type) do you expect to be most successful in capturing prey? Why?

   _____

   _____

   Which predator do you expect to be least successful? Why?

   _____

   _____

2. At the signal from your instructor, students (predators) will collect prey for 30 seconds. Prey may be collected only with your predator utensil and only one at a time. Place each prey bean into your cup and continue foraging until the signal to stop.

3. Record the number of each kind of prey in your cup after this first round foraging session:

   Black bean prey _____

   Garbanzo prey _____

   Lentil prey _____

   Lima bean prey _____

   Navy bean prey _____

   In Table 17-1, in the "1st round" column, record the total number of each prey species taken by your whole group. Also record your group's results on the class master data sheets.

4. Calculate the number of each prey species remaining. These surviving prey reproduce so that the total number of each species in the habitat doubles. For example, if 10 garbanzo prey were captured from the habitat, the remaining 40 will reproduce and add 40 new garbanzo prey for a total of 80 in the habitat. Repeat this process for each species.

5. Repeat foraging and reproduction of the survivors for two more generations, entering group data in Table 17-1 and on class master data sheets. Double the number of surviving prey of each species in the habitat after each round of predation.

   My round two prey capture:

   Black bean prey _____

   Garbanzo prey _____

   Lentil prey _____

   Lima bean prey _____

   Navy bean prey _____

   My round three prey capture:

   Black bean prey _____

   Garbanzo prey _____

   Lentil prey _____

   Lima bean prey _____

   Navy bean prey _____

TABLE 17-1	Group Totals: Prey Population Changes Caused by Predation by _____ (Predator Species)										
	Black		Garbanzo		Lentil		Lima		Navy		
Prey Species	Number Captured	Number Remaining	Number Captured	Number Remaining	Number Captured	Number Remaining	Number Captured	Number Remaining	Number Captured	Number Remaining	Total Prey Captured
1st round											
2nd round											
3rd round											
Total											

© Cengage Learning 2013

6. Graph the total population size over time for each prey species in Figure 17-1. Be sure to distinguish among the five species by using different symbols or colors for the trend lines for each.

Do your group's results support the hypothesis that *prey populations of different forms vary in fitness as a consequence of predation*? Write a conclusion accepting or rejecting the hypothesis and explaining your reasoning.

_____

_____

Which prey species was most vulnerable to your team's predator species?

_____

Which prey species was least vulnerable to your predator?

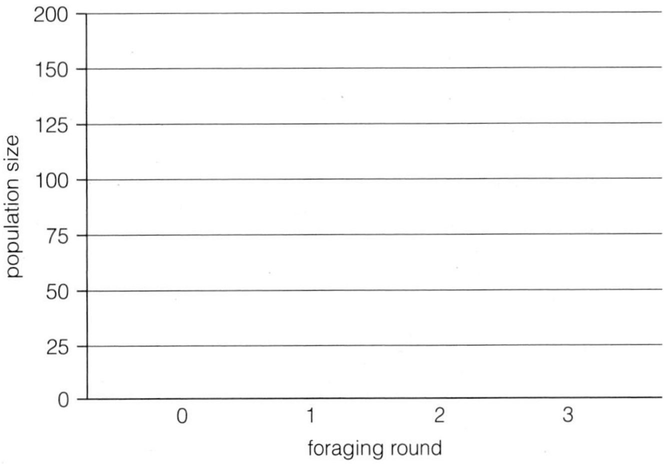

Figure 17-1 Prey population sizes after predator foraging rounds.

How well do these results agree with your expectations?

_____

Study the class total results. Did any prey species become extinct? _____ If so, speculate why.

_____

_____

Which prey species appears to be most fit in all environments with all predator species? What trait(s) may contribute to that fitness?

_____

_____

Which predator species appears to be most fit? Which predator species appears to be least fit? Explain.

_____

_____

What effect did the habitat characteristics have on your group's results? If the habitat had been uniformly white, speculate on the effect of this environmental change on the outcome for predator and prey species.

_____

_____

_____

If you failed to detect any differences in fitness among prey species, speculate on the reason(s) for this failure. Does this failure necessarily mean that natural selection would not occur in these populations?

_____

_____

Explain the results of this activity in terms of the effects of natural selection on the genetic makeup of the five prey populations.

_____

_____

_____

17.2 Geologic Time *(About 30 min.)*

Earth is an ancient planet that formed from a cloud of dust and gas approximately 4.6 billion years ago. Life formed relatively quickly on the young planet, and the first cells emerged in the seas by 3.8 billion years ago. Initially, the only living organisms on Earth were prokaryotic bacteria and bacteria-like cells. Eventually, though, more complex, eukaryotic organisms evolved. The seas were colonized by single-celled organisms first followed by multicelled and colonial plants and animals (Figure 17-2). Only much later did life move onto the land.

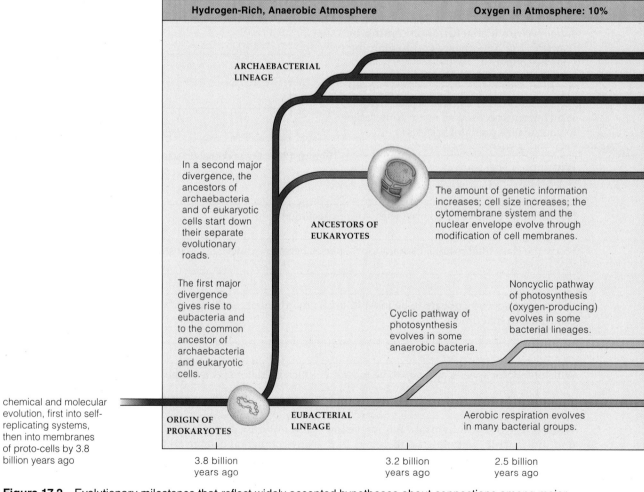

Figure 17-2 Evolutionary milestones that reflect widely accepted hypotheses about connections among major groups of organisms. (Timeline not shown to scale.)

Geologic time, "deep time," is difficult for humans to grasp given our recent evolution and 75-year life spans. In this section, you will construct a geologic timeline to help you gain perspective on the almost incomprehensible sweep of time over which evolution has operated.

MATERIALS

Per student pair:

- one 4.6-m rope or string
- meter stick or metric ruler (or both)
- masking tape
- calculator

PROCEDURE

1. Obtain a 4.6 m length of rope or string. The string represents the entire length of time (4.6 billion years) since the Earth was formed. On this timeline, you will attach masking tape labels to mark the points when the events below occurred. Measure from the starting point with the meter stick or metric ruler.
2. Locate and mark with tape on the timeline the events listed in Table 17-2 (bya = billion years ago, mya = million years ago, ya = years ago).

   Over what *proportion* of the Earth's history were there only single-celled living organisms? _____

   Over what *proportion* of the Earth's history have multicelled organisms existed? _____

   Over what *proportion* of the Earth's history have mammals been a dominant part of the fauna?

   _____

   Over what *proportion* of the Earth's history have modern humans existed? _____

   _____

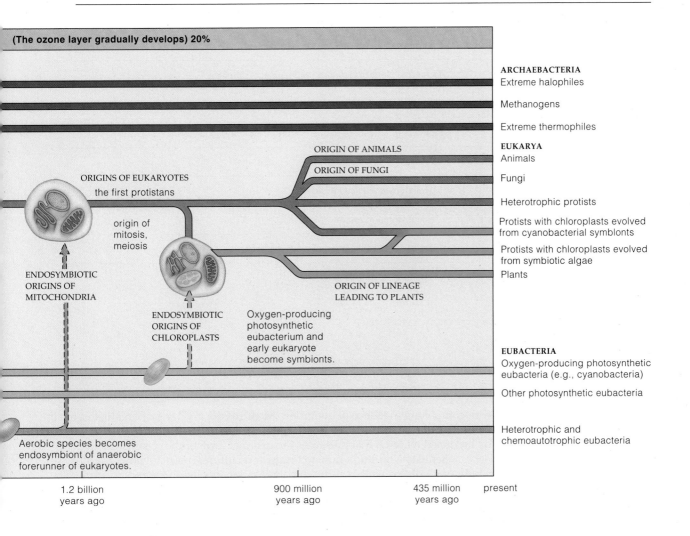

TABLE 17-2   Events in Evolution of Life on Earth

Time Frame	Event
4.6 bya*	Formation of Earth
3.8 bya	First living prokaryotic cells
2.5 bya	Oxygen-releasing photosynthetic pathway
1.2 bya	Origin of eukaryotic cells
580 mya	Origin of multicellular marine plants and animals
500 mya	First vertebrate animals (marine fishes)
435 mya	First vascular land plants and land animals
345 mya	Great coal forests; amphibians and insects undergo great diversification
240 mya	Mass extinction of nearly all living organisms
225 mya	Origin of mammals, dinosaurs; seed plants dominate
135 mya	Dinosaurs reach peak; flowering plants arise
65 mya	Mass extinction of dinosaurs and most marine organisms
63 mya	Flowering plants, mammals, birds, insects dominate
6 mya	First hominid (upright walking) human ancestors
2 mya	Humans (members of genus *Homo*) using stone tools, forming social life
200,000 ya	Evolution of *Homo sapiens* (modern humans)
6400 ya	Egyptian pyramids constructed
2000 ya	Peak of Roman Empire
A.D. 1776	Signing of the Declaration of Independence
A.D. 1969	Humans step on Earth's moon

© Cengage Learning 2013

*bya = billion years ago, mya = million years ago, ya = years ago

17.3   The Fossil Record and Human Evolution (About 30 min.)

The fossil record provides us with compelling evidence of evolution. Most **fossils** are parts of organisms, such as shells, teeth, and bones of animals and stems and seeds of plants. These parts are replaced by minerals to form stone or are surrounded by hardened material that preserves the external form of the organism.

Anthropologists and *archeologists*, scientists who study human origin and cultures, piece together the story of human evolution through the study of fossils, as well as through comparative biochemistry and anatomy. The fossil record of our human lineage is fragmentary and generates much discussion and differing interpretations. However, scientists generally agree that **hominids,** all species on the evolutionary branch leading to modern humans, arose in Africa by about 7 million years ago from the same genetic line that also produced great apes (gorillas, orangutans, and chimpanzees).

Many anatomic changes occurred in the course of evolution from hominid ancestor to modern humans. Arms became shorter, feet flattened and then developed arches, and the big toe moved in line with the other toes. The legs moved more directly under the pelvis. These features allowed upright posture and bipedalism (walking on two legs), and they also allowed the use of the hands for tasks other than locomotion.

Still other anatomic changes were associated with the head. In this section, you will observe some of the evolutionary trends associated with skull structure by studying images or reproductions of the skulls of several ancestors and relatives of modern humans.

MATERIALS

Per laboratory room:

- images or sets of skull reproductions and replicas, including *Australopithecus afarensis*, *Homo habilis*, *Homo erectus*, *Homo sapiens*, and chimpanzee (*Pan troglodytes*)
- collection of fossil plants and animals (optional)

PROCEDURE

1. If available, examine the fossil collection. Note which fossils lived in the seas and which lived on land.
   (a) Do any of the fossils resemble organisms now living? If so, describe the similarities between fossil and current forms.

   _____

   _____

   _____

   (b) Are there any fossils in the collection that are unlike any organisms currently living? If so, describe the features that appear to be most unlike today's forms.

   _____

   _____

2. Examine the hominid skulls in the laboratory or images available online as instructed. (Be aware that physical skull specimens are not actual fossils, but are plaster or plastic restoration designed partly from fossilized remains and partly from reconstruction. The *Homo sapiens* and chimpanzee skulls are replicas of modern specimens.)
3. Identify the scientific name of each organism and their approximate date of origin. As you can see, some of these hominids appear to have lived during the same approximate time span.
4. Study the skull specimens and Figure 17-3. For each skull, rate the characteristic listed in Table 17-3 (use a scale of (– – –) for the most primitive and (+ + +) for the most advanced). Record your observations of the changes that occurred over time in Table 17-4.

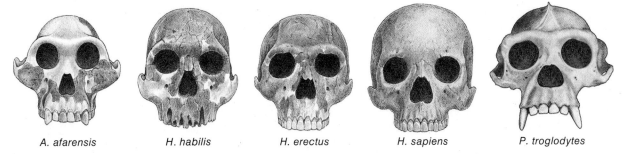

| A. afarensis | H. habilis | H. erectus | H. sapiens | P. troglodytes |

Figure 17-3 Comparison of skull shapes of hominid relatives and ancestors.

TABLE 17-3	**Primitive and Advanced Characteristics of Primate Skulls**	
Characteristic	**Primitive Form**	**Advanced Form**
Teeth	Many large, specialized	Fewer, smaller, less specialized
Jaw	Jaw and muzzle long	Jaw and muzzle short
Cranium (braincase)	Cranium small, forehead receding	Cranium large, prominent vertical forehead
Muscle attachments	Prominent eyebrow ridges and cranial keel (ridge)	Much reduced eyebrow ridges, no keel
Nose	Nose not protruding from face	Prominent nose with distinct bridge

© Cengage Learning 2013

Genus and species	Teeth	Jaw	Structure	Cranium	Muscle Attachments	Nose
Australopithecus afarensis						
Homo habilis						
Homo erectus						
Homo sapiens						
Pan troglodytes						

TABLE 17-4 — Skull Characteristics of Human Ancestors and Relatives, from (– – –) = Most Primitive to (+ + +) = Most Advanced

© Cengage Learning 2013

5. The first true hominids found in the fossil record have been assigned to the genus **Australopithecus** (from the Greek *australis* for "southern" and *pithecus* for "ape" because the first specimens were found in southern Africa). Some of the oldest known fossils of *Australopithecus* are of the species *A. afarensis*. The most famous skeleton has been named Lucy. Lucy and others of her species lived approximately 3.6 to 2.9 million years ago.

6. Now turn your attention to the skulls of the *Homo* lineage. There is debate about the lineage of our own genus, *Homo* ("Man"). *Homo habilis*, the early toolmaker, lived in eastern and southern Africa from about 1.9 to 1.6 million years ago. *Homo erectus* ("upright man") also arose in Africa. However, it was *Homo erectus* whose populations left Africa in waves between 2 million and 500,000 years ago. Some still lived in southeast Asia until at least 53,000 years ago. Our own species, *Homo sapiens* ("wise man"), seems to have evolved from *Homo erectus* ancestors by 100,000 years ago.

(a) What changes from primitive to advanced characteristics do you see in the evolution from *A. afarensis* to *Homo habilis*?

_____

_____

_____

(b) What is the trend in cranial capacity between the ancestral *A. afarensis* and *Homo erectus*?

_____

_____

_____

(c) Between *Homo erectus* and *Homo sapiens*?

_____

_____

_____

7. Finally, study the skulls of the modern human form (*Homo sapiens*) and our nearest living relative, the chimpanzee (*Pan troglodytes*). Fossil evidence, biochemistry, and genetic analyses indicate that chimpanzees and humans are the most closely related of all living primates. In fact, comparisons of amino acid sequences of proteins and DNA sequences show that chimpanzees and humans share about 99% of their genes! Other evidence shows that the separation from the hominid line of the lineage that led to chimpanzees and other great apes occurred about 6 million years ago.

(a) How does the chimpanzee skull compare with the skull of *Australopithecus afarensis*? What features are most alike? Which are most different?

_____

_____

(b) How does the chimpanzee skull compare with the human skull with respect to primitive and advanced anatomical features?

_____

_____

_____

_____ **1.** Evolution is
(a) the survival of populations.
(b) the process that results in changes in the genetic makeup of a population over time.
(c) a group of individuals of the same species occupying a given area.
(d) change in an individual's genetic makeup over its lifetime.

_____ **2.** The major mechanism of evolution is
(a) species development.
(b) predation.
(c) fitness.
(d) natural selection.

_____ **3.** An organism whose genetic makeup allows it to produce more offspring than another of its species is said to have greater
(a) fitness.
(b) evolution.
(c) foraging.
(d) selection.

_____ **4.** The Earth formed approximately
(a) 4600 years ago.
(b) 1 million years ago.
(c) 1 billion years ago.
(d) 4.6 billion years ago.

_____ **5.** The first living organisms appeared on Earth approximately
(a) 4.6 billion years ago.
(b) 3.8 billion years ago.
(c) 1 billion years ago.
(d) 6400 years ago.

_____ **6.** Modern humans evolved about
(a) 200,000 years ago.
(b) 2,000,000 years ago.
(c) 2,000,000,000 years ago.
(d) 4,600,000,000 years ago.

_____ **7.** Fossils
(a) are remains of organisms.
(b) are formed when organic materials are replaced with minerals.
(c) provide evidence of evolution.
(d) are all of the above.

_____ **8.** The first true hominids in the fossil record are
(a) _Australopithecus._
(b) _Homo habilis._
(c) the chimpanzee.
(d) _Homo sapiens._

_____ **9.** Hominids with evolutionarily advanced skull characteristics would have
(a) many large, specialized teeth.
(b) a large cranium with prominent vertical forehead.
(c) prominent eyebrow ridges.
(d) a long jaw and muzzle.

_____ **10.** The human ancestor whose populations dispersed from Africa to other parts of the world was
(a) _Homo sapiens._
(b) _Australopithecus afarensis._
(c) _Pan troglodytes._
(d) _Homo erectus._

EXERCISE **17**

Evidences of Evolution

Post-Lab Questions

Introduction

1. Define the following terms:
   (a) Evolution

   (b) Natural selection

17.1 How Natural Selection Works

2. Describe how natural selection might operate to change the genetic makeup of a population of rabbits if a new, faster predator was introduced into their habitat.

17.2 Geologic Time

3. Arrange the following events in their order of occurrence:
   (a) Photosynthesis *or* eukaryotic cells?
   (b) Vertebrate animals *or* flowering plants?
   (c) Extinction of dinosaurs *or* origin of hominids?

17.3 The Fossil Record and Human Evolution

4. What primitive characteristics are visible in the skull pictured here?

5. Describe anatomical changes in the skull that occurred in human evolution between an *Australopithecus afarensis*–like ancestor and *Homo sapiens*.

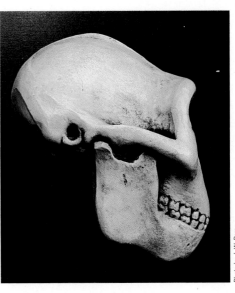

Photo by J. W. Perry.

6. Compare the slope of the forehead of chimpanzees with that of modern humans.

7. Compare the teeth of *Australopithecus afarensis* and *Homo sapiens*.
   (a) Describe similarities and differences between the two species.

   (b) Write a hypothesis about dietary differences between the two species based on their teeth.

Food for Thought

8. Why is the development of bipedalism considered to be a major advancement in human evolution?

9. Do you believe humans are still evolving? If not, why not? If yes, explain in what ways humans may be evolving.

10. Humans have selectively bred many radically different domestic animals (e.g., St. Bernard and Chihuahua dog breeds). Does this activity result in evolution? Why or why not?

Taxonomy: Classifying and Naming Organisms

OBJECTIVES

After completing this exercise, you will be able to

1. define *common name, scientific name, binomial, genus, specific epithet, species, taxonomy, phylogenetic system, dichotomous key, herbarium;*

2. distinguish common names from scientific names;

3. explain why scientific names are preferred over common names in biology;

4. identify the genus and specific epithet in a scientific binomial;

5. write out scientific binomials in the form appropriate to the Linnean system;

6. construct a dichotomous key;

7. explain the usefulness of an herbarium;

8. use a dichotomous key to identify plants, animals, or other organisms as provided by your instructor.

Introduction

We are all great classifiers. Every day, we consciously or unconsciously classify and categorize the objects around us. We recognize an organism as a cat or a dog, a pine tree or an oak tree. But there are numerous kinds of oaks, so we refine our classification, giving the trees distinguishing names such as "red oak," "white oak," or "bur oak." These are examples of **common names**, names with which you are probably most familiar.

Scientists are continually exchanging information about living organisms. But not all scientists speak the same language. The common name "white oak," familiar to an American, is probably not familiar to a Spanish biologist, even though the tree we know as white oak may exist in Spain as well as in our own backyard. Moreover, even within our own language, the same organism can have several common names. For example, within North America a gopher is also called a ground squirrel, a pocket mole, and a groundhog. On the other hand, the same common name may describe many different organisms; there are more than 300 different trees called "mahogany"! To circumvent the problems associated with common names, biologists use **scientific names** that are unique to each kind of organism and that are used throughout the world.

A scientific name is two-parted, a **binomial**. The first word of the binomial designates the group to which the organism belongs; this is the **genus** name (the plural of genus is *genera*). All oak trees belong to the genus *Quercus*, a word derived from Latin. Each kind of organism within a genus is given a **specific epithet**. Thus, the scientific name for white oak is *Quercus alba* (specific epithet is *alba*), while that of bur oak is *Quercus macrocarpa* (specific epithet is *macrocarpa*).

Notice that the genus name is always capitalized; the specific epithet usually is not capitalized (although it can be if it is the proper name of a person or place). The binomial is written in *italics* (since these are Latin names); if italics are not available, the genus name and specific epithet are underlined.

You will hear discussion of "species" of organisms. For example, on a field trip, you may be asked "What species is this tree?" Assuming you are looking at a white oak, your reply would be *"Quercus alba."* The scientific name of the **species** includes *both* the genus name and specific epithet.

If a species is named more than once within a document, it is accepted convention to write out the full genus name and specific epithet the first time and to abbreviate the genus name every time thereafter. For example, if white oak is being described, the first use is written *Quercus alba*, and each subsequent naming appears as *Q. alba*.

Similarly, when a number of species, all of the same genus, are being listed, the accepted convention is to write both the genus name and specific epithet for the first species and to abbreviate the genus name for each species listed thereafter. Thus, it is acceptable to list the scientific names for white oak and bur oak as *Quercus alba* and *Q. macrocarpa*, respectively.

Taxonomy is the science of classification (categorizing) and nomenclature (naming). Biologists prefer a system that indicates the evolutionary relationships among organisms. To this end, classification became a **phylogenetic system**; that is, one indicating the current understanding of evolutionary ancestry among organisms.

Current taxonomic thought separates all living organisms into six kingdoms:

1. Kingdom Bacteria (prokaryotic cells of great diversity, and that include pathogens)
2. Kingdom Archaea (prokaryotic organisms that are evolutionarily between eukaryotes and bacteria)
3. Kingdom Protista (as collection of lineages, some single-celled and some multi-celled; includes euglenoids, ciliates, dinoflagellates, amoebas, and many groups of "algae")
4. Kingdom Fungi (fungi)
5. Kingdom Plantae (plants)
6. Kingdom Animalia (animals)

Let's consider the scientific system of classification, using ourselves as examples. All members of our species belong to Kingdom Animalia (animals):

- Phylum Chordata (animals with a notochord)
- Class Mammalia (animals with mammary glands and skin with hair)
- Order Primates (mammals that walk upright on two legs)
- Family Hominidae (human forms, existing and extinct)
- Genus *Homo* (mankind)
- Specific epithet *sapiens* (wise)
- Species: *Homo sapiens*

The more closely related evolutionarily two organisms are, the more categories they share. You and I are different individuals of the same species. We share the same genus and specific epithet, *Homo* and *sapiens*. A creature believed to be a close, extinct ancestor walked the Earth 1.5 million years ago. That creature shared our genus name but had a different specific epithet, *erectus*. Thus, *Homo sapiens* and *H. erectus* are *different* species.

Like all science, taxonomy is subject to change as new information becomes available. Modifications are made to reflect revised interpretations.

18.1   Constructing a Dichotomous Key *(About 45 min.)*

To classify organisms, you must first identify them. A *taxonomic key* helps to identify an object or organism unknown to you but that someone else has described. The user chooses between alternative characteristics of the unknown object and, by making the correct choices, arrives at the name of the object.

Keys that are based on successive choices between two alternatives are known as **dichotomous keys** (*dichotomous* means "to fork into two equal parts"). When using a key, always read both choices even though the first appears to describe the subject. Don't guess at measurements; use a ruler. Since living organisms vary in their characteristics, don't base your conclusion on a single specimen if more are available.

MATERIALS

Per lab room:

- several meter sticks or metric height charts taped to a wall

PROCEDURE

1. Suppose the geometric shapes below have unfamiliar names. Look at the dichotomous key following the figures. Notice there is a 1a and a 1b. Start with 1a. If the description in 1a fits the figure you are observing better than description 1b, then proceed to the choices listed under 2, as shown at the end of line 1a. If 1a does *not* describe the figure in question, 1b does. Looking at the end of line 1b, you see that the figure would be called an Elcric.
2. Use the key provided to determine the hypothetical name for each object. Write the name beneath the object and then check with your instructor to see if you have made the correct choices.

    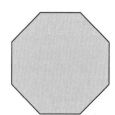

Key		
1a.	Figure with distinct corners	2
1b.	Figure without distinct corners	Elcric
2a.	Figure with 3 sides	3
2b.	Figure with 4 or more sides	4
3a.	All sides of equal length	Legnairt
3b.	Only 2 sides equal	Legnairtosi
4a.	Figure with only right angles	Eraqus
4b.	Figure with other than right angles	Nogatco

3. Now you will construct a dichotomous key, using your classmates as subjects. The class should divide up into groups of eight (or as evenly as the class size will allow). Working with the individuals in your group, fill in Table 18-1, measuring height with a metric ruler or the scale attached to the wall.

4. To see how you might plan a dichotomous key, examine the following branch diagram below. If there are both men and women in a group, the most obvious first split is male/female (although other possibilities for the split could be chosen as well). Follow the course of splits for two of the men in the group.

   Note that each choice has *only* two alternatives. Thus, we split into "under 1.75 m" and "1.75 m or taller." Likewise, our next split is into "blue eyes" and "nonblue eyes" rather than all the possibilities.

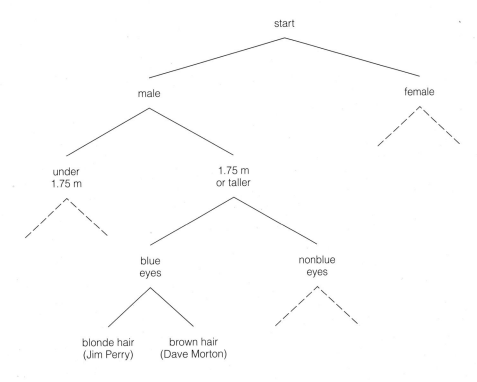

5. On a separate sheet of paper, construct a branch diagram for your group using the characteristics in Table 18-1 and then condense it into the dichotomous key that follows. When you have finished, exchange your key with that of an individual in another group. Key out the individuals in the other group without speaking until you believe you know the name of the individual you are examining. Ask that individual if you are correct. If not, go back to find out where you made a mistake, or possibly where the key was misleading. (Depending on how you construct your key, you may need more or fewer lines than have been provided.)

TABLE 18-1 Characteristics of Students

Student (name)	Sex (m/f)	Height (m)	Eye Color	Hair Color	Shoe Size
1.					
2.					
3.					
4.					
5.					
6.					
7.					
8.					

© Cengage Learning 2013

Key to Students in Group _____

1a.

1b.

2a.

2b.

3a.

3b.

4a.

4b.

5a.

5b.

6a.

6b.

7a.

7b.

8a.

8b.

18.2   Using a Taxonomic Key

A. Some Microscopic Members of the Freshwater Environment (About 30 min.)

Suppose you want to identify the specimens in some pond water. The easiest way is to key them out with a dichotomous key, now that you know how to use one. In this section, you will do just that.

MATERIALS

Per student:

- compound microscope
- microscope slide
- coverslips
- dissecting needle

Per student group (table):

- cultures of freshwater organisms
- one disposable plastic pipet per culture
- methylcellulose in dropping bottle

PROCEDURE

1. Obtain a clean glass microscope slide and clean coverslip.
2. Using a disposable plastic pipet or dissecting needle, withdraw a small amount of the culture provided.
3. Place *one* drop of the culture on the center of the slide.
4. Gently lower the coverslip onto the liquid.
5. Using your compound light microscope, observe your wet mount. Focus first with the low-power objective and then with the medium or high-dry objective, depending on the size of the organism in the field of view.
6. Concentrate your observation on a single specimen, keying out the specimen using the Key to Selected Freshwater Inhabitants that follows.
7. In the space provided, write the scientific name of each organism you identify. After each identification, have your instructor verify your conclusion.
8. Clean and reuse your slide after each identification.

Key to Selected Freshwater Inhabitants

1a.	Filamentous organism consisting of green, chloroplast-bearing threads	2
1b.	Organism consisting of a single cell or nonfilamentous colony	4
2a.	Filament branched, each cell mostly filled with green chloroplast	*Cladophora*
2b.	Filament unbranched	3
3a.	Each cell of filament containing 1 or 2 spiral-shaped green chloroplasts	*Spirogyra*
3b.	Each cell of filament containing 2 star-shaped green chloroplasts	*Zygnema*
4a.	Organism consisting of a single cell	5
4b.	Organism composed of many cells aggregated into a colony	6
5a.	Motile, teardrop-shaped or spherical organism	*Chlamydomonas*
5b.	Nonmotile, elongate cell on either end; clear, granule-containing regions at ends	*Closterium*
6a.	Colony a hollow round ball of more than 500 cells; new colonies may be present inside larger colony	*Volvox*
6b.	Colony consisting of less than 50 cells	7
7a.	Organism composed of a number of tooth-shaped cells	*Pediastrum*
7b.	Colony a loose square or rectangle of 4–32 spherical cells	*Gonium*

Organism 1 is _____

Organism 2 is _____

Organism 3 is _____

Organism 4 is _____

Organism 5 is _____

Organism 6 is _____

Organism 7 is _____

Organism 8 is _____

B. Common Trees and Shrubs (About 1 hour)

Suppose you want to identify the trees growing on your campus or in your yard at home. Without having an expert present, you can now do that, because you know how to use a taxonomic key. But how can you be certain that you have keyed your specimen correctly?

Typically, scientists compare their tentative identifications against *reference specimens*—that is, preserved organisms that have been identified by an expert *taxonomist* (a person who names and classifies organisms). If you are identifying fishes or birds, the reference specimen might be a bottled or mounted specimen with the name on it. In the case of plants, reference specimens most frequently take the form of *herbarium mounts* (Figure 18-1) of the plants. An **herbarium** (plural, *herbaria*) is a repository, a museum of sorts, of preserved plants. The taxonomist flattens freshly collected specimens in a plant press. They are then dried and mounted on sheets of paper.

Herbarium labels are affixed to the sheets, indicating the scientific name of the plant, the person who collected it, the location and date of collection, and often pertinent information about the habitat in which the plant was found.

It is likely that your school has an herbarium. If so, your instructor may show you the collection. To some, this endeavor may seem boring, but herbaria serve a critical function. The appearance or disappearance of plants from the landscape often gives a very good indication of environmental change. An herbarium records the diversity of plants in the area, at any point in history since the start of the collection.

MATERIALS

Per student group (table):

- set of eight tree twigs with leaves (fresh or herbarium specimens) *or*
- trees and shrubs in leafy condition (for an outdoor lab)

PROCEDURE

Use the appropriate following key to identify the tree and shrub specimens that have been provided in the lab or that you find on your campus. Refer to the *Glossary to Tree Key* (page 236) and Figures 18-2 through 18-9 (pages 230–232) when you encounter an unfamiliar term. When you have finished keying a specimen, confirm your identification by checking the herbarium mounts or asking your instructor.

Note: Some descriptions within the key have more characteristics than your specimen will exhibit. For example, the key may describe a fruit type when the specimen doesn't have a fruit on it. However, other specimen characteristics are described, and these should allow you to identify the specimen in most cases.

Note: The keys provided are for *selected* trees of your area. In nature, you will find many more genera than can be identified by these keys.

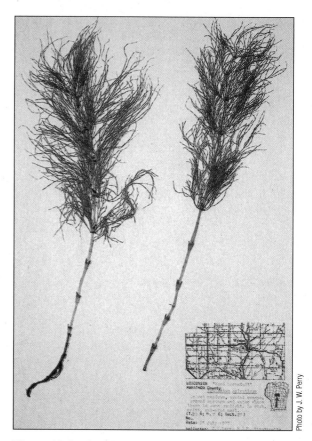

Figure 18-1 A typical herbarium mount.

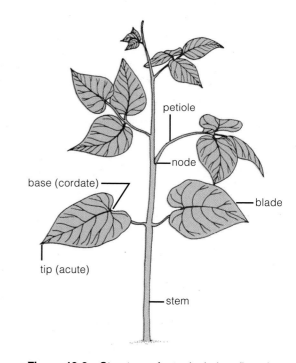

Figure 18-2 Structure of a typical plant (bean).

Common names within parentheses follow the scientific name. A metric ruler is provided below for use where measurements are required.

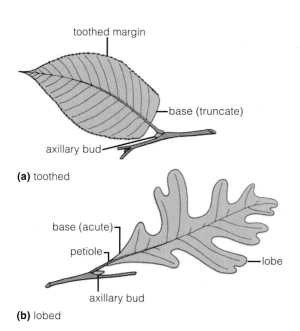

(a) toothed

(b) lobed

Figure 18-3 Simple leaves.

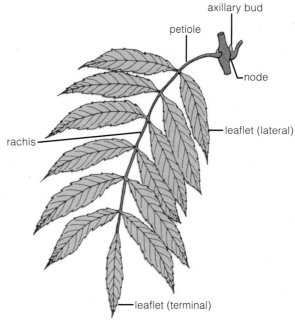

Figure 18-4 Pinnately compound leaf.

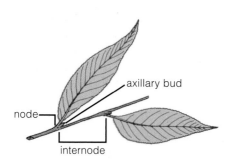

Figure 18-5 Simple leaves—alternating.

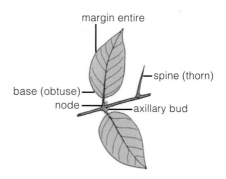

Figure 18-6 Simple leaves—opposite.

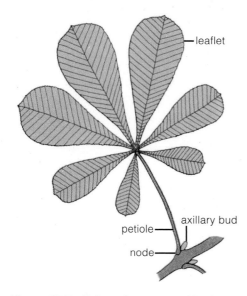

Figure 18-7 Palmately compound leaf.

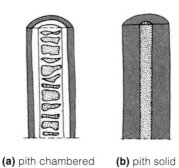

(a) pith chambered   (b) pith solid

Figure 18-8 Pith types.

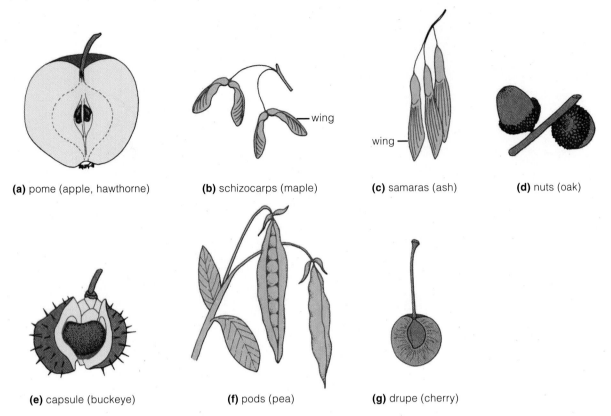

(a) pome (apple, hawthorne) **(b)** schizocarps (maple) **(c)** samaras (ash) **(d)** nuts (oak)

(e) capsule (buckeye) **(f)** pods (pea) **(g)** drupe (cherry)

Figure 18-9 Fruit types.

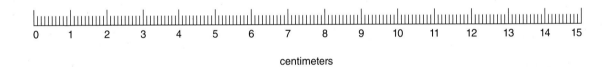

centimeters

Key to Some Common Genera of Trees of the Midwestern and Eastern United States and Canada		
1a.	Leaves broad and flat; plants producing flowers and fruits (angiosperms)	2
1b.	Leaves needlelike or scalelike; plants producing cones, but no flowers or fruits (gymnosperms)	22
2a.	Leaves compound	3
2b.	Leaves simple	9
3a.	Leaves alternate	4
3b.	Leaves opposite	7
4a.	Leaflets short and stubby, less than twice as long as broad; branches armed with spines or thorns; fruit a beanlike pod	5
4b.	Leaflets long and narrow, more than twice as long as broad; trunk and branches unarmed; fruit a nut	6
5a.	Leaflet margin without teeth; terminal leaflet present; small deciduous spines at leaf base	*Robinia* (black locust)
5b.	Leaflet margin with fine teeth; terminal leaflet absent; large permanent thorns on trunk and branches	*Gleditsia* (honey locust)

6a.	Leaflets usually numbering less than 11; pith of twigs solid	*Carya* (hickory)
6b.	Leaflets numbering 1 or more; pith of twigs divided into chambers	*Juglans* (walnut, butternut)
7a.	Leaflets pinnately arranged; fruit a light-winged samara	8
7b.	Leaflets palmately arranged; fruit a heavy leathery spherical capsule	*Aesculus* (buckeye)
8a.	Leaflets mostly 3–5; fruit a schizocarp with curved wings	*Acer* (box elder)
8b.	Leaflets mostly more than 5; samaras borne singly, with straight wings	*Fraxinus* (ash)
9a.	Leaves alternate	10
9b.	Leaves opposite	21
10a.	Leaves very narrow, at least 3 times as long as broad; axillary buds flattened against stem	*Salix* (willow)
10b.	Leaves broader, less than 3 times as long as broad	11
11a.	Leaf margin without small, regular teeth	12
11b.	Leaf margin with small, regular teeth	13
12a.	Fruit a pod with downy seeds; leaf blade obtuse at base; petioles flattened, or if rounded, bark smooth	*Populus* (poplar, popple, aspen)
12b.	Fruit an acorn; leaf blade acute at base; petioles rounded; bark rough	*Quercus* (oaks)
13a.	Leaves (at least some of them) with lobes or other indentations in addition to small, regular teeth	14
13b.	Leaves without lobes or other indentations except for small, regular teeth	16
14a.	Lobes asymmetrical, leaves often mitten-shaped	*Morus* (mulberry)
14b.	Lobes or other indentations fairly symmetrical	15
15a.	Branches thorny (armed); fruit a small applelike pome	*Crataegus* (hawthorne)
15b.	Branches unarmed	17
16a.	Bark smooth and waxy, often separating into thin layers; leaf base symmetrical	*Betula* (birch)
16b.	Bark rough and furrowed; leaf base asymmetrical	*Ulmus* (elm)
17a.	Leaf base asymmetrical, strongly heart-shaped, at least on one side	*Tilia* (basswood or linden)
17b.	Leaf base acute, truncate, or slightly cordate	18
18a.	Leaf base asymmetrical; bark on older stems (trunk) often warty	*Celtis* (hackberry)
18b.	Leaf base symmetrical	19
19a.	Leaf blade usually about twice as long as broad, generally acute at base; fruit fleshy	20
19b.	Leaf not much longer than broad, generally truncate at base; fruit a dry pod	*Populus* (poplar, popple, aspen)
20a.	Leaf tapering to a pointed tip, glandular at base	*Prunus* (cherry)
20b.	Leaf spoon-shaped with a rounded tip, no glands at base	*Crataegus* (hawthorne)
21a.	Leaf margins with lobes and points; fruit a schizocarp	*Acer* (maple)
21b.	Leaf margins without lobes or points; fruit a long capsule	*Catalpa* (catalpa)

22a.	Leaves needlelike, with 2 or more needles in a cluster	23
22b.	Leaves needlelike or scalelike, occurring singly	24
23a.	Leaves more than 5 in a cluster, soft, deciduous, borne at the ends of conspicuous stubby branches	*Larix* (larch, tamarack)
23b.	Leaves 2–5 in a cluster	*Pinus* (pines)
24a.	Leaves soft, not sharp to the touch	25
24b.	Leaves stiff or sharp and often unpleasant to touch	27
25a.	Leaves about 0.2 cm and scalelike, overlapping	*Thuja* (white cedar, arbor vitae)
25b.	Leaves needlelike, appear to form two ranks on twig	26
26a.	Leaves with distinct petioles, 0.8–1.5 cm long; twigs rough; female cones drooping from branches	*Tsuga* (hemlock)
26b.	Leaves without distinct petioles, 1–3 cm long; twigs smooth; female cones erect on branches	*Abies* (firs)
27a.	Leaves appear triangular-shaped, about 0.5 cm, and tightly pressed to twig; cone blue, berrylike	(juniper, Eastern red cedar)
27b.	Leaves elongated and needlelike	28
28a.	Tree; leaves 4-sided, protrude stiffly from twig; female cones droop from branch	*Picea* (spruces)
28b.	Shrub; leaves flattened, pressed close to twig at base; seed partially covered by a fleshy coat, usually red	*Taxus* (yew)

Key to Some Common Genera of Trees of the Pacific Region of the United States and Canada

1a.	Leaves broad and flat; plants producing flowers and fruits (angiosperms)	2
1b.	Leaves needlelike or scalelike; plants producing cones, but no flowers or fruits (gymnosperms)	15
2a.	Leaves compound	3
2b.	Leaves simple	6
3a.	Leaves pinnately arranged	4
3b.	Leaves palmately arranged	5
4a.	Leaflets numbering 7; fruit a samara	*Fraxinus* (ash)
4b.	Leaflets 15–17; fruit a nut	*Juglans* (walnut)
5a.	Leaflets numbering 3, lobed; fruit a schizocarp	*Acer* (box elder)
5b.	Leaflets numbering more than 3; fruit a smooth or spiny capsule	*Aesculus* (buckeye)
6a.	Three or more equal-sized veins branching from leaf base	7
6b.	Leaf with single large central vein with other main veins branching from the central vein	9
7a.	Leaves opposite; fruit a schizocarp	*Acer* (maple)
7b.	Leaves alternate	8

8a.	Leaves nearly round in outline; fruit a pod	*Cercis* (redbud)
8b.	Leaves deeply lobed, very hairy beneath; fruit consisting of an aggregation of many 1-seeded nutlets surrounded by long hairs	*Platanus* (sycamore)

9a.	Leaf lobed; fruit a nut	*Quercus* (oak)
9b.	Leaf not lobed	10

10a.	Leaves opposite; fruit a drupe; halves of leaf remain attached by "threads" after blade has been creased and broken	*Cornus* (dogwood)
10b.	Leaves alternate	11

11a.	Upon crushing, blade gives off strong, penetrating odor	*Eucalyptus* (eucalyptus)
11b.	Blade not strongly odiferous upon crushing	12

12a.	Branch bark smooth, conspicuously red-brown; fruit a red or orange berry	*Arbutus* (madrone)
12b.	Branch rough, not colored red-brown	13

13a.	Undersurface of leaves golden-yellow; fruit a spiny, husked nut	*Catanopsis* (golden chinquapin)
13b.	Leaves green beneath	14

14a.	Petiole hairy; leathery blade with a stubby spine at end of each main vein; fruit a nut	*Lithocarpus* (tanoak)
14b.	Petiole and leaf lacking numerous hairs, leaves long and narrow, more than twice as long as wide	*Salix* (willow)

15a.	Leaves needlelike	16
15b.	Leaves scalelike	23

16a.	Leaves needlelike, with 2 or more needles in a cluster	17
16b.	Leaves needlelike, occurring singly	18

17a.	Needles 2–5 in a cluster	*Pinus* (pine)
17b.	Needles 6 or more per cluster, soft, deciduous, borne at the ends of stubby, conspicuous branches	*Larix* (larch)

18a.	Round scars on twigs where old needles have fallen off; twigs smooth; needles soft to the grasp; cones pointing upward with reference to stem	*Abies* (fir)
18b.	Twigs rough, with old needle petioles remaining	19

19a.	Needles angled, stiff, sharp, pointed, unpleasant to grasp; cones hanging	*Picea* (spruce)
19b.	Needles soft, not sharp when grasped	20

20a.	Needles round in cross section, can be rolled easily between thumb and index finger; needles less than 1.3 cm long; cones small, less than 1.5 cm	*Tsuga* (hemlock)
20b.	Needles too flat to be rolled easily	21

21a.	Tips of needles blunt or rounded, undersurface with 2 white bands; cones with long, conspicuous, 3-lobed bracts	*Pseudotsuga* (Douglas fir)
21b.	Tips of needles pointed	22

22a.	Tops of needles grooved; woody seed cones broadly oblong in outline	*Sequoia* (redwood)
22b.	Tops of needles with ridges; lacking in cones, instead having a red, fleshy, cuplike seed covering	*Taxus* (yew)

| 23a. | Twig ends appear as if jointed | *Calocedrus* (incense cedar) |
| 23b. | Tips of branches flattened, not jointed in appearance | 24 |

| 24a. | Leaves glossy and fragrant | *Thuja* (Western red cedar) |
| 24b. | Leaves awl-shaped, arranged spirally on twig | *Sequoiadendron* (giant sequoia) |

Glossary to Tree Key

- *Acorn*—The fruit of an oak, consisting of a nut and its basally attached cup (Figure 18-9d)
- *Acute*—Sharp-pointed (Figure 18-2)
- *Alternate*—Describing the arrangement of leaves or other structures that occur singly at successive nodes or levels; not opposite or whorled (Figure 18-5)
- *Angiosperm*—A flowering seed plant (e.g., bean plant, maple tree, grass)
- *Armed*—Possessing thorns or spines
- *Asymmetrical*—Not symmetrical
- *Axil*—The upper angle between a branch or leaf and the stem from which it grows
- *Axillary bud*—A bud occurring in the axil of a leaf (Figures 18-3 through 18-7)
- *Basal*—At the base
- *Blade*—The expanded, more or less flat portion of a leaf (Figure 18-2)
- *Bract*—A much reduced leaf
- *Capsule*—A dry fruit that splits open at maturity (e.g., buckeye; Figure 18-9e)
- *Compound leaf*—Blade composed of two or more separate parts (leaflets) (Figures 18-4, 18-7)
- *Cordate*—Heart-shaped (Figure 18-2)
- *Deciduous*—Falling off at the end of a functional period (such as a growing season)
- *Drupe*—Fleshy fruit containing a single hard stone that encloses the seed (e.g., cherry, peach, or dogwood; Figure 18-9g)
- *Fruit*—A ripened ovary, in some cases with associated floral parts (Figure 18-9a–g)
- *Glandular*—Bearing secretory structures (glands)
- *Gymnosperm*—Seed plant lacking flowers and fruits (e.g., pine tree)
- *Lateral*—On or at the side (Figure 18-4)
- *Leaflet*—One of the divisions of the blade of a compound leaf (Figures 18-4, 18-7)
- *Lobed*—Separated by indentations (sinuses) into segments (lobes) larger than teeth (Figure 18-3b)
- *Node*—Region on a stem where leaves or branches arise (Figures 18-2 through 18-7)
- *Nut*—A hard, one-seeded fruit that does not split open at maturity (e.g., acorn; Figure 18-9d)
- *Obtuse*—Blunt (Figure 18-6)
- *Opposite*—Describing the arrangement of leaves of other structures that occur two at a node, each separated from the other by half the circumference of the axis (Figure 18-6)
- *Palmately compound*—With leaflets all arising at apex of petiole (Figure 18-7)
- *Petiole*—Stalk of a leaf (Figures 18-2, 18-3, 18-4, 18-7)
- *Pinnately compound*—A leaf constructed somewhat like a feather, with the leaflets arranged on both sides of the rachis (Figure 18-4)
- *Pith*—Internally, the centermost region of a stem (Figure 18-8a, b)
- *Pod*—A dehiscent, dry fruit; a rather general term sometimes used when no other more specific term is applicable (Figure 18-9f)
- *Pome*—Fleshy fruit containing several seeds (e.g., apple or pear; Figure 18-9a)
- *Rachis*—Central axis of a pinnately compound leaf (Figure 18-4)
- *Samara*—Winged, one-seeded, dry fruit (e.g., ash fruits; Figure 18-9c)
- *Schizocarp*—Dry fruit that splits at maturity into two one-seeded halves (Figure 18-9b)
- *Simple leaf*—One with a single blade, not divided into leaflets (Figures 18-3, 18-5, 18-6)
- *Spine*—Strong, stiff, sharp-pointed outgrowth on a stem or other organ (Figure 18-6)
- *Symmetrical*—Capable of being divided longitudinally into similar halves
- *Terminal*—Last in a series (Figure 18-4)
- *Thorn*—Sharp, woody, spinelike outgrowth from the wood of a stem; usually a reduced, modified branch
- *Tooth*—Small, sharp-pointed marginal lobe of a leaf (Figure 18-3a)
- *Truncate*—Cut off squarely at end (Figure 18-3a)
- *Unarmed*—Without thorns or spines
- *Whorl*—A group of three or more leaves or other structures at a node

Each year millions of "evergreen" trees become the center of attraction in human dwellings during the Christmas season. The process of selecting the all-important tree is the same whether you reside in the city where you buy your tree from a commercial grower, or whether you cut one off your "back forty." You ponder and evaluate each specimen until, with the utmost confidence, you bring home that perfect tree. Now that you have it, just what kind of tree stands in your home, looking somewhat like a cross between Old Glory and the Sistine Chapel? This key contains most of the trees that are used as Christmas trees; other gymnosperm trees are included, too. The common "Christmas trees" have an asterisk after their scientific name. Note that this key, unlike those in the preceding sections, indicates actual species designations.

1a.	Tree fragrant, boughs having supported (on clear moonlit nights) masses of glistening snow on their green needles; tree a product of nature	2
1b.	Tree not really a tree but rather a product of a cold and insensitive society; tree never giving life and never having life	17
2a.	Leaves persistent and green throughout the winter, needlelike, awl-shaped, or scalelike	3
2b.	Leaves deciduous; for this reason not a desirable Christmas tree	4
3a.	Leaves in clusters of 2–5, their bases within a sheath	5
3b.	Leaves borne singly, not in clusters	9
4a.	Cones 1.25–1.8 cm long, 12–15 scales making up cone	*Larix laricina* (tamarack)
4b.	Cones 1.8–3.5 cm long, 40–50 scales comprising cone	*Larix decidua* (larch)
5a.	Leaves 5 in a cluster, cones 10–25 cm long	*Pinus strobus* (white pine)
5b.	Leaves 2 in a cluster, cones less than 10 cm long	6
6a.	Leaves 2.5–7.5 cm long	7
6b.	Leaves 7.5–15 cm long	8
7a.	Leaves with a bluish cast; cones with a stout stalk, pointing away from the tip of the branch; bark orange in the upper part of the tree	*Pinus sylvestris** (scotch pine)
7b.	Leaves 1.25–3.75 cm long; cones stalkless, pointing forward toward the tip of branch	*Pinus banksiana* (jack pine)
8a.	Leaves slender, shiny; bark of trunk red-brown; cones 5–7.5 cm long; scales of cones without any spine at tip	*Pinus resinosa** (red pine)
8b.	Leaves thickened, dull; bark of trunk gray to nearly black; cones 5 to 7.5 cm long; scales of cone armed with short spine at tip	*Pinus nigra** (Austrian pine)
9a.	Leaves scalelike or awl-shaped	10
9b.	Leaves needlelike	11
10a.	Twigs flattened, leaves all of one kind, scalelike, extending down the twig below the point of attachment	*Thuja occidentalis* (white cedar, arbor vitae)
10b.	Twigs more or less circular in cross section; leaves of two kinds, either scalelike or awl-shaped, often both on same branch, not extending down the twig; coneless but may have a blue berrylike structure	*Juniperus virginiana* (red cedar)
11a.	Leaves with petioles	12
11b.	Leaves lacking petioles, leaf tip notched, needles longer than 1.25 cm	*Abies balsamea** (balsam fir)

12a.	Leaves angular, 4-sided in cross section, harsh to the touch; petiole adheres to twig	13
12b.	Leaves flattened	15
13a.	Leaves 3–10 mm long, blunt-pointed; twigs rusty and hairy	*Picea mariana** (black spruce)
13b.	Leaves 2 cm long, sharp-pointed; twigs smooth	14
14a.	Cones 2.5–5 cm long; leaves ill-scented when bruised or broken; smaller branches mostly horizontal	*Picea glauca** (white spruce)
14b.	Cones 7.5–15 cm long; scales comprising cone with finely toothed markings; leaves not ill-scented when bruised or broken; smaller branches drooping	*Picea abies** (Norway spruce)
15a.	Leaves pointed, over 1.25 cm long; red fleshy, berrylike structures present	16
15b.	Leaves rounded at tip, less than 1.25 cm long, with two white lines on underside	*Tsuga canadensis* (hemlock)
16a.	Leaves 2–2.5 cm long, dull dark green on top, with two broad yellow bands on undersurface; petiole yellowish	*Taxus cuspidata* (Japanese yew)
16b.	Leaves 1.25–2 cm long, without yellow bands on underside	*Taxus canadensis* (American yew)
17a.	Tree a glittering mass of structural aluminum, sometimes illuminated by multicolored floodlights	*Aluminous ersatzenbaum** (aluminum substitute)
17b.	Tree green, made from petroleum products; used year after year; exactly like all others of its manufacture	*Plasticus perfectus** (plastic substitute)

_____ **1.** The name "human" is an example of a
    (a) common name.
    (b) scientific name.
    (c) binomial.
    (d) polynomial.

_____ **2.** Current scientific thought places organisms in one of ___ kingdoms.
    (a) two
    (b) four
    (c) five
    (d) six

_____ **3.** The scientific name for the ruffed grouse is _Bonasa umbellus. Bonasa_ is
    (a) the family name.
    (b) the genus.
    (c) the specific epithet.
    (d) all of the above

_____ **4.** A binomial is always a
    (a) genus.
    (b) specific epithet.
    (c) scientific name.
    (d) two-part name.

_____ **5.** The science of classifying and naming organisms is known as
    (a) taxonomy.
    (b) phylogeny.
    (c) morphology.
    (d) physiology.

_____ **6.** Which scientific name for the wolf is presented correctly?
    (a) Canis lupus
    (b) canis lupus
    (c) _Canis lupus_
    (d) Canis Lupus

_____ **7.** A road that dichotomizes is
    (a) an intersection of two crossroads.
    (b) a road that forks into two roads.
    (c) a road that has numerous entrances and exits.
    (d) a road that leads nowhere.

_____ **8.** Most scientific names are derived from
    (a) English.
    (b) Latin.
    (c) Italian.
    (d) French.

_____ **9.** One problem with using common names is that
    (a) many organisms may have the same common name.
    (b) many common names may exist for the same organism.
    (c) the common name may not be familiar to an individual not speaking the language of the common name.
    (d) all of the above are true.

_____ **10.** Phylogeny is the apparent
    (a) name of an organism.
    (b) ancestry of an organism.
    (c) nomenclature.
    (d) dichotomy of a system of classification.

EXERCISE **18**

Taxonomy: Classifying and Naming Organisms

Post-Lab Questions

Introduction

1. If you were to use a binomial system to identify the members of your family or the family of a friend, (mother, father, sisters, brothers), how would you write their names so that your system would most closely approximate that used to designate species?

2. Describe several advantages of using scientific names instead of common names.

3. Based on the following classification scheme, which two organisms are most closely phylogenetically related? Why?

	Organism 1	Organism 2	Organism 3	Organism 4
Kingdom	Animalia	Animalia	Animalia	Animalia
Phylum	Arthopoda	Arthropoda	Arthropoda	Arthropoda
Class	Insecta	Insecta	Insecta	Insecta
Order	Coleoptera	Coleoptera	Coleoptera	Coleoptera
Genus	*Caulophilus*	*Sitophilus*	*Latheticus*	*Sitophilus*
Specific epithet	*oryzae*	*oryzae*	*oryzae*	*zeamaize*
Common name	Broadnosed grain weevil	Rice weevil	Longheaded flour beetle	Maize weevil

18.2 Using a Taxonomic Key

Consider the drawing of plants A and B in answering questions 4–6.

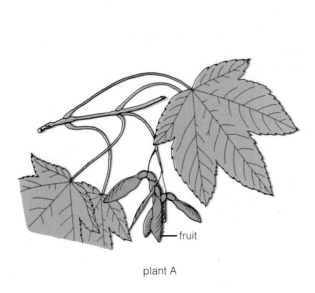

fruit

plant A

cone

plant B

4. Using the taxonomic key in the exercise, identify the two plants as either angiosperms or gymnosperms.

   Plant A is a (an) _____.

   Plant B is a (an) _____.

5. To what genus does plant A belong? What is its common name?

   genus: _____

   common name: _____

6. To what genus does plant B belong? What is its common name?

   genus: _____

   common name: _____

Consider the drawing of plants C and D in answering questions 7–9.

plant C

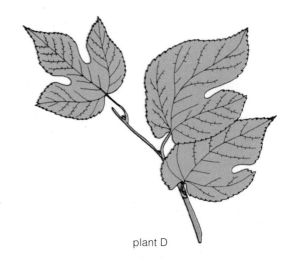

plant D

7. As completely as possible, describe the leaf of plant C.

8. To what genus does plant C belong? What is its common name?

    genus: _____

    common name: _____

9. Using the taxonomic key to freshwater organisms in the exercise, identify the genus of the organism below.

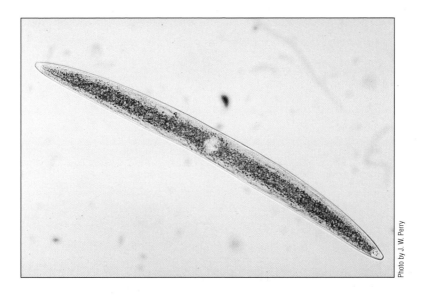

Photo by J. W. Perry

    genus: _____

Food for Thought

10. If you owned a large, varied music collection, how might you devise a key to keep track of all your different kinds of music?

Bacteria and Protists I

OBJECTIVES

After completing this exercise, you will be able to

1. define *prokaryotic, eukaryotic, pathogen, decomposer, coccus, bacillus, spirillum, Gram stain, antibiotic, symbiosis, parasitism, commensalism, mutualism, nitrogen fixation, monoecious, dioecious, obligate mutualism, gametangia, vector, agar, fucoxanthin, algin, kelp;*

2. describe characteristics distinguishing bacteria from protists;

3. identify and classify the organisms studied in this exercise;

4. identify structures (those in **boldface** in the procedure sections) in the organisms studied;

5. distinguish Gram-positive and Gram-negative bacteria, indicating their susceptibility to certain antibiotics;

6. suggest measures that might be used to control malaria.

Introduction

The bacteria (Domain Bacteria, Kingdom Bacteria) and archaea (Domain Archaea, Kingdom Archaea) and protists (Domain Eukarya, Kingdom Protista) are among the simplest of living organisms. These have unicellular organisms within them, but that's where the similarity ends.

Bacteria and archaea are **prokaryotic** organisms, meaning that their DNA is free in the cytoplasm, unbounded by a membrane. They lack organelles (cytoplasmic structures surrounded by membranes). By contrast, the protists are **eukaryotic** organisms: the genetic material contained within the nucleus and many of their cellular components are compartmentalized into membrane-bound organelles.

Both bacteria and archaea, and unicellular protists, are at the base of the food chain. From an ecological standpoint, these simple organisms are among the most important organisms on our planet. Ecologically, they're much more important than we are.

19.1   **Domain Bacteria, Kingdom Bacteria *(About 40 min.)***

The Kingdom Bacteria consists of bacteria and cyanobacteria. Most bacteria are heterotrophic, dependent upon an outside source for nutrition, while cyanobacteria are autotrophic (photosynthetic), able to produce their own carbohydrates.

Some heterotrophic bacteria are **pathogens**, causing plant and animal diseases, but most are **decomposers**, breaking down and recycling the waste products of life. Others are nitrogen fixers, capturing the gaseous nitrogen in the atmosphere and making it available to plants via a symbiotic association with their roots.

A. Bacteria: Are Bacteria Present in the Lab? (About 15 min.)

MATERIALS

Per student group (4):

■ dH$_2$O in dropping bottle
■ four nutrient agar culture plates
■ two bottles labeled A and B (A contains tap water, B contains 70% ethyl alcohol)
■ sterile cotton swabs
■ transparent adhesive tape
■ china marker
■ paper towels

PROCEDURE

Work in groups of four.

1. Obtain four petri dishes containing sterile nutrient agar. Using a china marker, label one dish "Dish 1: Control." Label the others "Dish 2: Dry Swab," "Dish 3: Treatment A," and "Dish 4: Treatment B." Also include the names of your group members.
2. Run a sterile cotton swab over a surface within the lab. Some examples of things you might wish to sample include the surface of your lab bench, the floor, and the sink. Be creative!
3. Lift the lid on Dish 2 as little as is necessary to run the swab over the surface of the agar.

   Note: **Be careful that you don't break the agar surface.**

4. Tape the lid securely to the bottom half of the dish.
5. Soak one paper towel with liquid A and a second paper towel with liquid B.
6. Wipe down one-half of the surface you just sampled with liquid A (tap water), the other half with liquid B (70% ethyl alcohol). After the areas have dried, using dishes 3 and 4, repeat the procedures described in step 2. Place the cultures in a desk drawer to incubate until the next lab period. At that time, examine your culture for bacterial colonies, noting the color and texture of the bacterial growth.
7. Make a prediction of what you will find in the culture plates after the next lab period.

   _____

   _____

8. Describe what you see.

   Source of sample: _____

   Dish 1: Control. _____

   _____

   _____

   Dish 2: Dry Swab. _____

   _____

   Dish 3: Treatment A. _____

   _____

   _____

   Dish 4: Treatment B. _____

   > **CAUTION**
   >
   > Leave the lid on as you examine the cultures to prevent the spread of any potentially pathogenic (disease-causing) organisms. While the probability is small that pathogens are present, you should always err on the side of caution.

9. Make a conclusion about the usefulness of tap water and 70% ethyl alcohol as disinfectants.

   _____

   _____

B. Bacteria: Bacterial Shape and Sensitivity to Antibiotics (About 10 min.)

MATERIALS

Per student:

■ bacteria type slide
■ compound microscope

Per lab room:

■ Gram-stained bacteria (three demonstration slides)

Work alone for this and the rest of these activities.

Bacteria exist in three shapes: **coccus** (plural, *cocci;* spherical), **bacillus** (plural, *bacilli;* rods), and **spirillum** (plural, *spirilla;* spirals).

PROCEDURE

1. Study a bacteria type slide illustrating these three shapes. You'll need to use the highest magnification available on your compound microscope. In Figure 19-1, draw the bacteria you are observing.

In addition to being differentiated by shape, bacteria can be separated according to how they react to a staining procedure called **Gram stain,** named in honor of a nineteenth-century microbiologist, Hans Gram. *Gram-positive* bacteria are purple after being stained by the Gram stain procedure, while *Gram-negative* bacteria appear pink. The Gram stain reaction is important to bacteriologists because it is one of the first steps in identifying an unknown bacterium. Furthermore, the Gram stain reaction indicates a bacterium's susceptibility or resistance to certain **antibiotics**, substances that inhibit the growth of bacteria.

2. Examine the *demonstration slides* illustrating Gram-stained bacteria. Gram-positive bacteria are susceptible to penicillin, while Gram-negative bacteria are not. In Table 19-1, list the species of bacteria that you have examined and their staining characteristics.

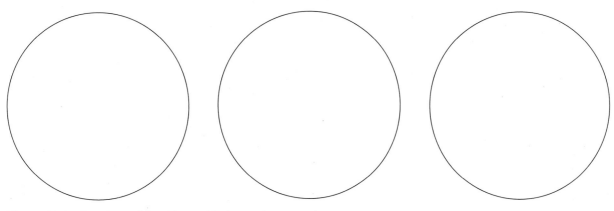

Figure 19-1   Drawings of three bacterial shapes (_____×).

TABLE 19-1	Gram Stain Reaction of Various Bacteria	
Bacterial Species		**Gram Reactions (+ or −)**

© Cengage Learning 2013

C. Cyanobacteria (Blue-Green Algae) (About 15 min.)

The cyanobacteria (sometimes called blue-green algae) are distinguished from the heterotrophic bacteria by being photosynthetic (photoautotrophic).

C.1. Oscillatoria

MATERIALS

Per student:
- microscope slide
- coverslip
- dissecting needle
- compound microscope

Per lab room:
- *Oscillatoria*—living culture; disposable pipet

PROCEDURE

1. Examine the culture provided of *Oscillatoria*. Describe what the culture looks like, including the texture and the color.

_____

_____

The color of the culture is a consequence of the photosynthetic accessory pigments that for the most part mask the chlorophyll.

(If you have immediate access to a greenhouse, you might walk through it now to see if you can find any *Oscillatoria* growing on the flowerpots or even the floor.)

2. Using a dissecting needle, scrape some of the culture from the surface and prepare a wet mount slide.
3. Examine your preparation with your compound microscope, starting with the medium-power objective and finally with the highest magnification available (oil immersion, if possible). Note that the individual cells are joined and so form the filament.

Do all the *Oscillatoria* cells look alike, or is there differentiation of certain cells within the filament?

_____

_____

4. Draw a portion of the filament in Figure 19-2.

C.2. Anabaena

Some cyanobacteria live as *symbionts* within other organisms. Literally, **symbiosis** means "living together." There are three types of symbiosis. In a parasitic symbiosis (**parasitism**), one organism lives at the expense of the other; that is, the parasite benefits while the host is harmed. A commensalistic symbiosis (**commensalism**) occurs when effects are positive for one species and neutral for the other. In a mutualistic symbiosis (**mutualism**) both organisms benefit from living together. In this section, you will examine one such symbiotic relationship.

Figure 19-2   Drawing of *Oscillatoria* (_____×).

MATERIALS

Per student:
- microscope slide
- coverslip
- dissecting needle
- compound microscope

Per lab room:
- *Azolla*—living plants

PROCEDURE

1. Place a leaf of the tiny water fern *Azolla* on a clean glass slide. Use a dissecting needle to crush the leaf into very small pieces. Now add a drop of water and a coverslip.
2. Scan your preparation with the medium-power objective of your compound microscope, looking for long chains (composed of numerous beadlike cells) of the filamentous cyanobacterium *Anabaena*. Switch to higher magnification when you find *Anabaena*.
3. Within the filament, locate the **heterocysts,** cells that are a bit larger than the other cells.
4. Draw what you see in Figure 19-3 and label it.
5. Examine the very high magnification electron micrograph of *Anabaena* in Figure 19-4. Identify the heterocyst with its distinctive *polar nodules* at either end of the cell. In the other cells, note the numerous wavy **thylakoids,** membranes on and in which the photosynthetic pigments are found. Identify the large electron-dense carbohydrate **storage granules** within the cytoplasm and the cell wall.

Is a nucleus present within the cells of *Anabaena?* Explain.

_____

Figure 19-3   Drawing of *Anabaena* that was within *Azolla* leaves (_____×).
Label: heterocyst

Based upon this electron micrograph, would you hypothesize that the heterocyst is photosynthetic?

_____

6. Now look again at your microscope slide of *Anabaena*. Find the polar nodules, appearing as bright spots within the heterocysts.

Heterocysts convert nitrogen in the air (or water) to a form that the cyanobacterium can use for cellular metabolism. This process is called **nitrogen fixation**. Nitrogen-containing compounds accumulate in the water fern, some of it being used for its own metabolic needs. There is also an advantage to the *Anabaena*: it not only gets a place to live, but by being within the water fern leaf it is shielded from very high light levels that would otherwise do it damage.

Which type of symbiosis is the association between *Anabaena* and the water fern?

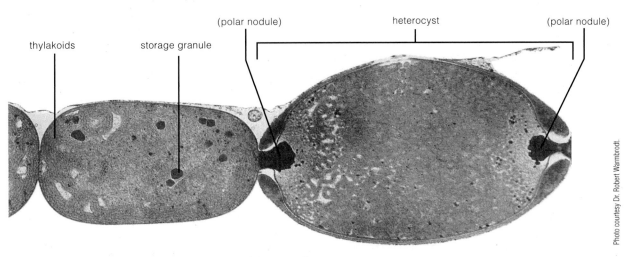

Figure 19-4 Electron micrograph of *Anabaena* (14,220×).

Photo courtesy Dr. Robert Warmbrodt.

19.2   Domain Eukarya, Kingdom Protista *(About 2 hours)*

The protists are a diverse assemblage of organisms, both photoautotrophic and heterotrophic. They are so diverse that different protists have previously been classified as fungi, animals, and plants! The following organisms are examined in this, the first of two protist exercises:

Phylum	Common Name
Parabasalia	Parabasalids
Euglenozoa	Trypanosomes and euglenoids
Alveolata	Ciliates, dinoflagellates, apicomplexans
Stramenophila	Water molds, diatoms, and brown algae

A. Phylum Parabasalia (About 5 min.)

These protistans have one or more flagella to provide motility. The group includes some fairly notorious human parasites that cause disease, including giardiasis (from drinking water contaminated with the causal protozoans) and some sexually transmitted diseases (caused by *Trichomonas vaginalis*).

MATERIALS

Per student:
- prepared slide of *Trichonympha*
- compound microscope

PROCEDURE

Trichonympha

The flagellated protozoan *Trichonympha* (Figure 19-5) inhabits the gut of termites. The association of the termite and *Trichonympha* is an example of **obligate mutualism,** in which neither organism is capable of surviving without the other. Termites lack the enzymes to metabolize cellulose, a major component of wood. Wood particles ingested by termites are engulfed by *Trichonympha*, whose enzymes break the cellulose into soluble carbohydrates that are released for use by the termite.

Study a prepared slide of these organisms. Examine the gut of the termite with the high-dry objective to find *Trichonympha*. Note the large number of **flagella** covering the upper portion of the cell, the more or less centrally located **nucleus** and wood fragments in the cytoplasm.

B. Phylum Euglenozoa (About 25 min.)

B.1. Trypanosoma

African sleeping sickness is caused by *Trypanosoma brucei.*

MATERIALS

Per student:

- prepared slide of *Trypanosoma* in blood smear
- compound microscope

PROCEDURE

Examine a prepared slide of human blood that contains the parasitic flagellate *Trypanosoma* (Figure 19-6), the cause of African sleeping sickness. This flagellate is transmitted from host to host by the bloodsucking tsetse fly. Note the **flagellum** arising from one end of the cell.

B.2. Euglenoids

Euglenoids are motile, unicellular protists. Most species are photosynthetic, but there are also some heterotrophic species.

MATERIALS

Per student:

- microscope slide
- coverslip
- compound microscope

Per group (table):

- methylcellulose in two dropping bottles
- dH₂O in two dropping bottles

Per lab room:

- *Euglena*-living culture; disposable pipet

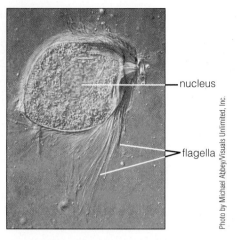

Figure 19-5 *Trichonympha* (900×).

Photo by Michael Abbey/Visuals Unlimited, Inc.

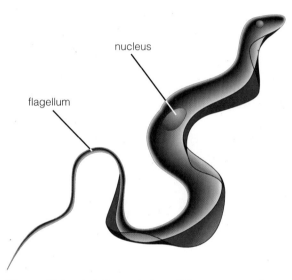

Figure 19-6 Artist's rendering of *Trypanosoma*, the causal agent of African sleeping sickness.

PROCEDURE

1. From the culture provided, prepare a wet mount slide of *Euglena.*
2. Observe with the medium-power objective of your compound microscope. Notice the motion of these green cells as they swim through the medium. If they're swimming too rapidly, prepare another slide, but add a drop of methylcellulose to the cell suspension before adding a coverslip.
3. Switch to the high-dry objective for more detailed observation. Figure 19-7a will serve as a guide in your study. This specimen was photographed by a special technique to give it a three-dimensional appearance.
4. Within the **cytoplasm,** identify the green **chloroplasts** and, if possible, the centrally located **nucleus.**
5. By closing the microscope's diaphragm to increase the contrast, you may be able to locate the **flagellum** at one end of the cell.
6. Search for the orange **eyespot**, a photoreceptive organelle located within the cytoplasm at the base of the flagellum.

In the world of microscopic swimmers, there are two types of flagella. One type, the *whiplash flagellum,* pushes the organism through the medium. An example of this is a human sperm cell. The other type, with which you're probably less familiar, is the *tinsel flagellum,* which pulls the organism through its watery environment. Tinsel flagella have tiny hairlike projections, visible only with the electron microscope.

7. Notice the direction of motion of the *Euglena* cells. Which type of flagellum does *Euglena* have? _____

Besides seeing the swimming motion caused by the flagellum, you may observe a contractionlike motion of the entire cell (*euglenoid movement*). *Euglena* is able to contort like this because it lacks a rigid cell wall. Instead, flexible helical interlocking proteinaceous strips within the cell membrane delimit the cytoplasm. These strips plus the cell membrane form the **pellicle**. Euglenoid movement provides a means of locomotion for mud-dwelling organisms.

8. Examine Figure 19-7b, a scanning electron micrograph showing the *flagellum* and the helical strips of the pellicle.
9. In Figure 19-8, make a series of sketches showing the different shapes *Euglena* takes during euglenoid movement.

C. Phylum Alveolata

C.1. Ciliated Protozoans (About 20 min.)

Ciliates are covered with numerous short locomotory structures called *cilia*. In this section, you will examine one of the largest ciliates, the predatory *Paramecium caudatum*.

Paramecium

MATERIALS

Per student:

- microscope slide
- coverslip
- compound microscope

Per group (table):

- methylcellulose in two dropping bottles
- tissue paper
- acetocarmine stain in two dropping bottles
- box of toothpicks

Per lab room:

- *Paramecium caudatum*—living culture; disposable pipet
- Congo red—yeast mixture; disposable pipet

PROCEDURE

1. From the culture provided, prepare a wet mount of *Paramecium* on a clean microscope slide. Observe their rapid movement with the medium-power objective of your microscope. Look along the edge of the *Paramecium's* body to see evidence of the locomotory structures, the **cilia**. (Increasing the contrast by closing the microscope's iris diaphragm may help.)
2. Examine Figure 19-9, a scanning electron micrograph of *Paramecium*. Note the numerous **cilia** covering the body surface.

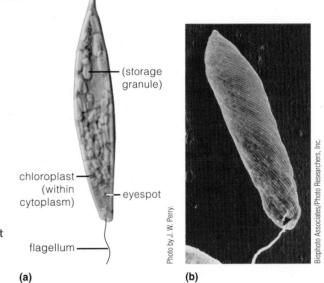

(a)　　　　　(b)

Figure 19-7 *Euglena*. (**a**) Light micrograph (1100×). (**b**) Scanning electron micrograph (1500×).

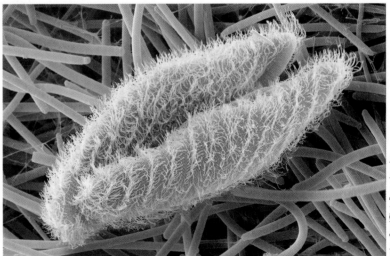

Figure 19-9 Scanning electron micrograph of *Paramecium caudatum* (1170×).

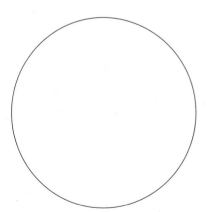

Figure 19-8 Different shapes possible in living *Euglena* exhibiting euglenoid movement.

3. Use Figure 19-10 as an aid as you proceed with your observation of your living specimen.

4. Prepare another wet mount of *Paramecium*, but before you place a coverslip on the slide, add a drop of acetocarmine stain and a drop of methylcellulose to slow the organisms' movement.

5. Find the large, centrally located **macronucleus**. It's likely that the second, a smaller **micronucleus adjacent** to the macronucleus, will be difficult to see. Surprise! This eukaryotic cell is *binucleate*. Each nucleus has its own distinctive functions.

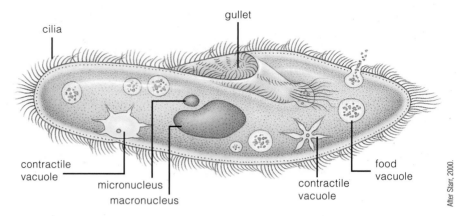

Figure 19-10 Artist's rendering of *Paramecium caudatum* (1700×).

After Starr, 2000.

6. Make a third preparation. First place a drop of methylcellulose on the slide, then a drop of Congo red/yeast mixture. Stir the mixture with a toothpick. Now add a drop of *Paramecium* culture. Place a coverslip on your wet mount and observe with the medium-power objective.

7. Locate the **oral groove,** a depression in the body that leads to the **gullet.**

8. Observe the yeast cells as they are eaten by *Paramecium*. The yeast will be taken into **food vacuoles**, where they are digested. Congo red is a pH indicator. As digestion occurs within the food vacuoles, the indicator will turn blue because of the increased acidity.

9. At either end of the organism, find the **contractile vacuoles**, which regulate water content within the organism. Watch them in operation.

C.2. Dinoflagellates (About 15 min.)

Dinoflagellates spin as they move through the water due to the position of their flagellum. On occasion, populations of certain dinoflagellates may increase dramatically, causing the seas to turn red or brown. These are the **red tides**, which can devastate fish populations because neurotoxins produced by the dinoflagellates poison fish.

MATERIALS

Per student:

■ prepared slide of a dinoflagellate (for example, *Gymnodinium*, *Ceratium*, or *Peridinium*)
■ compound microscope

Per lab room:

■ dinoflagellate—living culture; disposable pipet (optional)

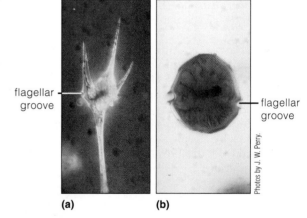

Figure 19-11 Two representative dinoflagellates. (**a**) *Ceratium* (240×). (**b**) *Peridinium* (640×).

Photos by J. W. Perry.

PROCEDURE

1. With the high-dry objective of your compound microscope, examine a prepared slide or living representative.
2. Compare what you are observing with Figures 19-11a and b.
3. Attempt to identify the stiff cellulosic plates encasing the cytoplasm.
4. Locate the two grooves formed by the junction of the plates. This is where the flagella are located. If you are examining living specimens, chloroplasts may be visible beneath the cellulose plates.
5. Examine the scanning electron micrograph of a dinoflagellate in Figure 19-12.

C.3. Apicomplexans (About 20 min.)

All apicomplexans are parasites, infecting a wide range of animals, including humans. *Plasmodium vivax* causes one type of malaria in humans. In this section, you will study its life cycle with demonstration slides. Refer to Figure 19-13 as you proceed.

Figure 19-12 Scanning electron micrograph of a dinoflagellate (1000×).

Dennis Kunkel Microscopy, Inc./Visuals Unlimited, Inc.

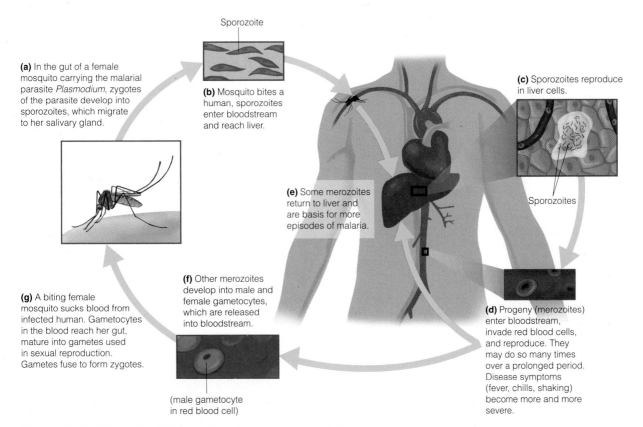

Figure 19-13 Life cycle of *Plasmodium vivax*, causal agent of malaria.

(a) In the gut of a female mosquito carrying the malarial parasite *Plasmodium*, zygotes of the parasite develop into sporozoites, which migrate to her salivary gland.

Sporozoite

(b) Mosquito bites a human, sporozoites enter bloodstream and reach liver.

(c) Sporozoites reproduce in liver cells.

Sporozoites

(e) Some merozoites return to liver and are basis for more episodes of malaria.

(g) A biting female mosquito sucks blood from infected human. Gametocytes in the blood reach her gut, mature into gametes used in sexual reproduction. Gametes fuse to form zygotes.

(f) Other merozoites develop into male and female gametocytes, which are released into bloodstream.

(male gametocyte in red blood cell)

(d) Progeny (merozoites) enter bloodstream, invade red blood cells, and reproduce. They may do so many times over a prolonged period. Disease symptoms (fever, chills, shaking) become more and more severe.

Plasmodium vivax

MATERIALS

Per lab room:

- demonstration slide of *Plasmodium vivax*, sporozoites
- demonstration slide of *P. vivax*, merozoites
- demonstration slide of *P. vivax*, immature gametocytes

PROCEDURE

1. Examine Figure 19-13a. *P. vivax* is transmitted to humans through the bite of an infected female *Anopheles* mosquito. The mosquito serves as a **vector**, a means of transmitting the organism from one host to another. Male mosquitos cannot serve as vectors, because they lack the mouth parts for piercing skin and sucking blood. If the mosquito carries the pathogen, **sporozoites** enter the host's bloodstream with the saliva of the mosquito.
2. Examine the demonstration slide of sporozoites (Figure 19-13b).
3. The sporozoites travel through the bloodstream to the liver, where they penetrate certain cells, grow, and multiply. When released from the liver cells, the parasite is in the form of a **merozoite** and infects the red blood cells. Figures 19-13c and 19-13d show this process.

(There are two intervening stages between sporozoites and merozoites. These are the *trophozoites* and *schizonts*, both developmental stages in red blood cells.)

4. Now examine the demonstration slide illustrating merozoites in red blood cells (Figure 19-13d).
5. Within the red blood cells, merozoites divide, increasing the merozoite population. At intervals of 48 or 72 hours, the infected red blood cells break down, releasing the merozoites. At this time, the infected individual has disease symptoms, including fever, chills, and shaking caused by the release of merozoites and metabolic wastes from the red blood cells. Some of these merozoites return to the liver cells, where they repeat the cycle and are responsible for recurrent episodes of malaria.
6. Merozoites within the bloodstream can develop into **gametocytes**. For development of a gametocyte to be completed, the gametocyte must enter the gut of the mosquito. This occurs when a mosquito feeds upon an infected (diseased) human.
7. Observe the demonstration slide of an **immature gametocyte** in a red blood cell (Figure 19-13f).
8. Within the gut of the mosquito, the gametocyte matures into a gamete (Figure 19-13g). When gametes fuse, they form a zygote that matures into an **oocyst**. Within each oocyst, sporozoites form, completing the life cycle of *Plasmodium vivax*. These sporozoites migrate to the mosquito's salivary glands to be injected into a new host.

D. Phylum Stramenopila

D.1. Oomycotes: Water Molds (About 30 min.)

Oomycotes are commonly referred to as water molds because they are primarily aquatic organisms. While not all members of this division grow in freestanding water, they all rely on the presence of water to spread their asexual spores.

During sexual reproduction, all members produce large nonmotile female gametes, the eggs. These eggs are contained in sex organs (**gametangia**; singular, *gametangium*) called **oogonia.** By contrast, the male gametes are nothing more than *sperm nuclei* contained in **antheridia,** the male gametangia.

MATERIALS

Per student:

- culture of *Saprolegnia* or *Achlya*
- glass microscope slide
- compound microscope

D.1.a. Saprolegnia or Achlya

If you've ever kept goldfish, you may have seen a white, cottony mass growing on the sides of certain fish. This is a parasitic water mold that is easily controlled by the addition of chemicals to the aquarium.

PROCEDURE

1. Examine the water cultures of either *Saprolegnia* or *Achlya.* These fungi are chiefly saprophytes and are growing on a sterilized hemp seed that provides a carbohydrate source (Figure 19-14).
2. Examine its life cycle (Figure 19-15) as you study this organism.
3. Notice the numerous filamentous hyphae that radiate from the hemp seed. A **hypha** (plural, *hyphae*) is the basic unit of the fungal body. Collectively, all the hyphae constitute the **mycelium** (plural, *mycelia*).

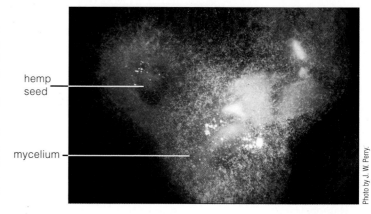

Figure 19-14 Mycelium of water mold growing on hemp seed (3×).

4. Identify the hyphae and mycelium in Figures 19-14 and 19-15a.

The mycelium of the water mold is **multinucleate**, meaning that there are many nuclei in each cell, not just one. Each nucleus is diploid (2n). Moreover, cross walls separating the mycelium into distinct cells are infrequent, forming only when reproductive organs are formed.

5. Place a glass slide on the stage of your compound microscope. This slide will serve as a platform for the culture dish that contains the fungal culture, allowing you to use the mechanical stage of the microscope to move the culture (if the microscope is so equipped).
6. Remove the lid from the culture dish, carefully place the culture dish on the platform, and examine with the low-power objective. Look first at the edge of the mycelium—that is, at the tips of the youngest hyphae.
7. Find the tips that appear denser (darker) than the rest of the hyphae. These dense tips are cells specialized for asexual reproduction. They are the **zoosporangia** (singular, *zoosporangium;* Figures 19-16, 19-15b), which produce biflagellate **zoospores** (Figure 19-15c). Note the cross wall separating the zoosporangium from the rest of the hyphae.
8. Examine Figure 19-15d. Zoospores are released from the zoosporangia, swim about for a period of time, lose their flagella, and encyst, meaning that they form a thick wall around the cytoplasm.
9. Look at Figure 19-15e. When the encysted zoospore germinates, it produces a second type of zoospore. This one also will eventually encyst, as shown in Figure 19-15f.
10. Look at Figure 19-15g. When the second cyst stage germinates, it produces a hypha that proliferates into a new mycelium (Figure 19-15a). As you see, this is asexual reproduction; no sex organs were involved in the formation of zoospores.

Sexual reproduction in the water molds occurs in the older portion of the mycelium. In your culture the older portion is located nearer the hemp seed.

11. Scan the colony to find the spherical female gametangia, the oogonia (singular, **oogonium;** Figures 19-16, 19-15h). Meiosis takes place within the oogonium to form haploid eggs.

12. Switch to the medium-power objective and study a single oogonium in greater detail. Depending on the stage of development, you will find either **eggs** (Figure 19-15i) or **zygotes** (fertilized eggs) (Figures 19-16, 19-15j) within the oogonium.

13. Now you'll need to find the male gametangium, the **antheridium**. It's a short fingerlike hypha that attaches itself to the wall of the oogonium (Figures 19-17, 19-15i), much as you would wrap your finger around a baseball. Find an antheridium.

Because the nuclei of the antheridium have undergone meiosis, each nucleus is haploid.

14. Examine Figure 19-15i, which shows fertilization. Fertilization takes place when tiny fertilization tubes penetrate the wall of the oogonium. Rather than forming special male gametes, the haploid nuclei within the antheridium flow through the fertilization tubes to fuse with the egg nuclei.

15. Search your culture to see if you can locate any thick-walled zygotes within oogonia (Figures 19-18, 19-15j). Following a maturation period that can last several months, the zygote germinates by forming a germ tube (Figure 19-15k) that grows into a new mycelium (Figure 19-15a).

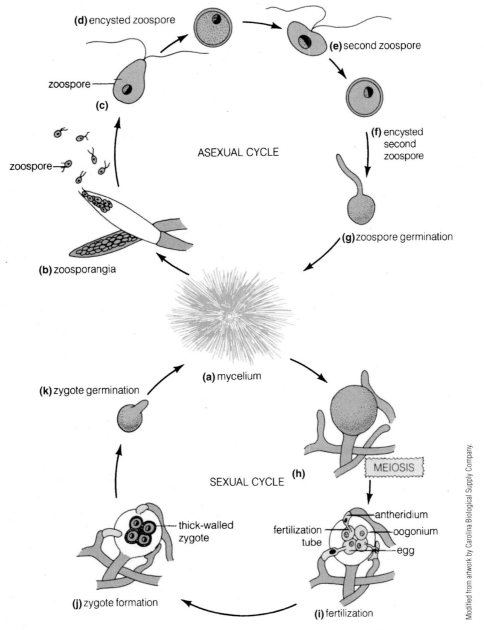

Modified from artwork by Carolina Biological Supply Company.

Figure 19-15 Life cycle of a water mold such as *Saprolegnia* or *Achyla*. (Structures colored green are 2n.)

Notice that both male and female gametangia are produced on the same mycelium. The term describing this condition is **monoecious** (from the Greek words for "one house"). Organisms producing one type of sex organ on one body and the other type of sex organ on another body are **dioecious** (from the Greek words for "two" and "house").

Are humans monoecious or dioecious?

D.1.b. Phytophthora

Some of the most notorious and historically important plant pathogens known to humans are water molds. Included in this group is *Phytophthora infestans*, the fungus that causes the disease known as late blight of potato. This disease spread through the potato fields of Ireland between 1845 and 1847. Most of the potato plants died, and 1 million Irish working-class citizens who had depended on potatoes as their primary food source starved to death. Another 2 million emigrated, many to the United States.

Perhaps no other plant pathogen so poignantly illustrates the importance of environmental factors in causing disease. While *Phytophthora infestans* had been present previous to 1845 in the potato-growing fields of Ireland, it was not until the region experienced several consecutive years of wet and especially cool growing seasons that late blight became a major problem. We can easily observe the importance of these climate-associated factors by studying another *Phytophthora* species, *P. cactorum*.

MATERIALS

Per student:

- culture of *Phytophthora cactorum*
- glass microscope slide
- dissecting needle
- compound microscope

Per lab room:

- refrigerator
  or

Per student group (4):

- ice bath

PROCEDURE

1. Obtain a culture of *P. cactorum* that has been flooded with distilled water. Note that the agar has been removed from the edges of the petri dish and that the mycelium has grown from the agar edge into the water.
2. Place a glass slide on the stage of your compound microscope. This slide will serve as a platform for the culture, allowing you to use the mechanical stage of the microscope to move the culture (if the microscope is so equipped).

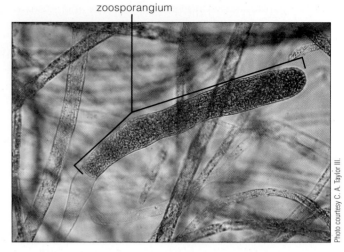

Figure 19-16   Water mold zoosporangium (272×).

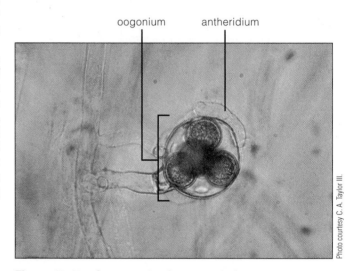

Figure 19-17   Gametangia of water mold (272×).

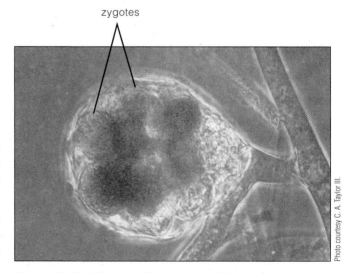

Figure 19-18   Water mold oogonium with zygotes (720×).

3. Remove the lid from the culture, carefully place the culture dish on the platform, and examine the culture with the low-power objective.
4. Search the surface of the mycelium, especially at the edges, until you find the rather *pear-shaped* **zoosporangia**. Switch to the medium-power objective for closer observation and then draw in Figure 19-19 a single zoosporangium.
5. Return your microscope to the low-power objective, remove the culture, replace the cover, and place it in a refrigerator or on ice for 15–30 minutes.
6. After the incubation time, again observe the zoosporangia microscopically. Find one in which the **zoospores** are escaping from the zoosporangium. Each zoospore has the potential to grow into an entirely new mycelium! Draw the zoospores in Figure 19-19.

Like the other water molds, *Phytophthora* reproduces sexually. Your cultures contain the sexual structures as well as asexual zoosporangia.

7. Using a dissecting needle, cut a section about 1 cm square from the agar colony and *invert* it on a glass slide (so that the bottom side of the agar is now uppermost).
8. Place a coverslip on the agar block and observe with your compound microscope, first with the low-power objective, then with the medium-power, and finally with the high-dry objective.
9. Identify the spherical **oogonia** that contain **eggs** or thick-walled **zygotes** (depending on the stage of development). If present, the **antheridia** are club shaped and plastered to the wall of the oogonium.
10. In Figure 19-20, draw an oogonium, eggs (zygotes), and antheridia.

D.2. Chrysophytes: Diatoms (About 15 min.)

Diatoms are called the "organisms that live in glass houses" because their cell walls are composed largely of opaline *silica* (SiO_2 nH_2O). Diatoms are important as primary producers in the food chain of aquatic environments, and their cell walls are used for a wide variety of industrial purposes, ranging from the polishing agent in toothpaste to a reflective roadway paint additive. Massive deposits of cell walls of long-dead diatoms make up diatomaceous earth (Figure 19-21).

MATERIALS

Per student:

- microscope slide
- coverslip
- dissecting needle
- prepared slide of freshwater diatom
- compound microscope

Per group (table):

- diatomaceous earth
- dH₂O in two dropping bottles

Per lab room:

- diatoms—living culture; disposable pipet

PROCEDURE

1. Using a dissecting needle, scrape a small amount of diatomaceous earth onto a microscope slide and prepare a wet mount slide.

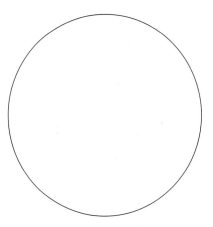

Figure 19-19 Drawing of zoosporangium and zoospores of *Phytophthora cactorum* (_____×). **Labels:** zoosporangium, zoospores

Figure 19-20 Drawing of gametangia of *Phytophthora cactorum* (_____×). **Labels:** oogonium, eggs (zygotes), antheridium

Figure 19-21 Diatomaceous earth quarry near Quincy, Washington.

2. Examine your slide, starting with the low magnification objective, proceeding through the highest magnification available on your compound microscope. Close the iris diaphragm to increase the contrast.

3. You are looking at only the silica opaline shell. The cytoplasm of this organism disappeared hundreds of thousands, if not millions of years ago.

4. Note the tiny holes in the shells.

5. Now prepare a wet mount of living diatoms. Use the high-dry objective to note the golden brown chloroplasts within the cytoplasm and the numerous holes in the cell walls (shells).

6. In Figure 19-22, sketch several of the diatoms you are observing.

7. Now obtain a prepared slide of diatoms. These cells have been "cleaned," making the perforations in the cell wall especially obvious if you close the iris diaphragm on your microscope's condenser to increase the contrast. Study with the high-dry objective.

Figure 19-22 Drawing of diatoms (_____×).

The pattern of the holes in the walls is characteristic of a given species. Before the advent of electronic techniques, microscopists observed diatom walls to assess the quality of microscope lenses. The resolving power (see discussion of resolving power in Exercise 3) could be determined if one knew the diameter of the holes under observation.

D.3. Phaeophytes: Brown Algae (About 10 min.)

The vast majority of the brown algae are found in cold, marine environments. All members are multicellular, and most are macroscopic. Their color is due to the accessory pigment fucoxanthin, which is so abundant that it masks the green chlorophylls. Some species are used as food, while others are harvested for fertilizers. Of primary economic importance is **algin**, a cell wall component of brown algae that makes ice cream smooth, cosmetics soft, and paint uniform in consistency, among other uses.

D.3.a. Kelps—Laminaria and Macrocystis

Kelps are large (up to 100 m long), complex brown algae. They are common along the seashores in cold waters.

MATERIALS

Per lab room:

- demonstration specimen of *Laminaria*
- demonstration specimen of *Macrocystis*

PROCEDURE

Examine specimens of *Laminaria* (Figure 19-23) and *Macrocystis* (Figure 19-24). On each, identify the rootlike **holdfast** that anchors the alga to the substrate; the **stipe**, a stemlike structure; and the leaflike **blades**.

D.3.b. Rockweed—Fucus

Fucus is a common brown alga of the coastal shore, especially abundant attached to rocks where the plants are periodically wetted by splashing waves and the tides.

MATERIALS

Per lab room:

- demonstration specimen of *Fucus*

Figure 19-23 *Laminaria* (0.06×).

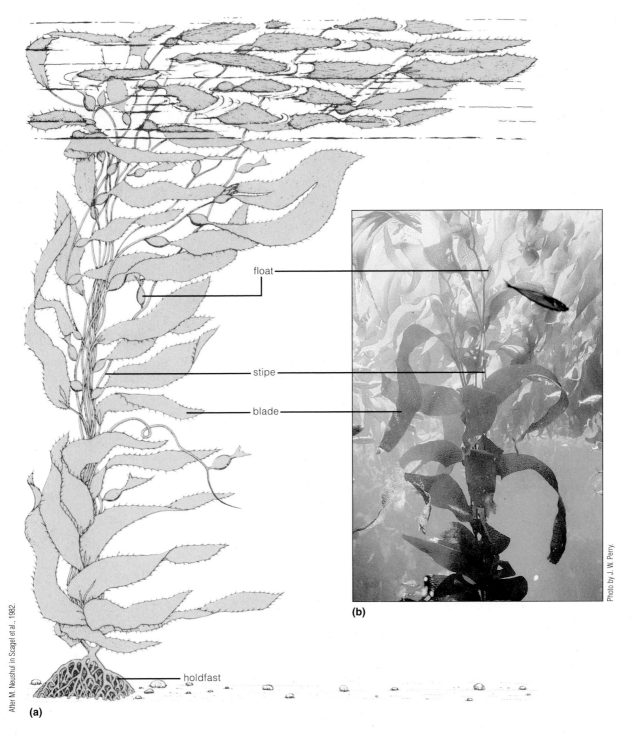

float —

stipe —

blade —

holdfast —

After M. Neushul in Scagel et al., 1982.

Photo by J. W. Perry.

(a)

(b)

Figure 19-24 *Macrocystis.* (**a**) Artist's conception. (**b**) Photograph taken at Monterey Bay Aquarium, Monterey, CA. (Both about 0.01×.)

PROCEDURE

1. Examine demonstration specimens of *Fucus* (Figure 19-25), noting the branching nature of the body.
2. Locate the short **stipe** and **blade**.

3. Examine the swollen and inflated blade tips (Figure 19-26), housing the sex organs of the plant and apparently also serving to keep the plant buoyant at high tide. Notice the numerous tiny dots on the surface of these inflated ends. These are the openings through which motile sperm cells swim to fertilize the enclosed, nonmotile egg cells.

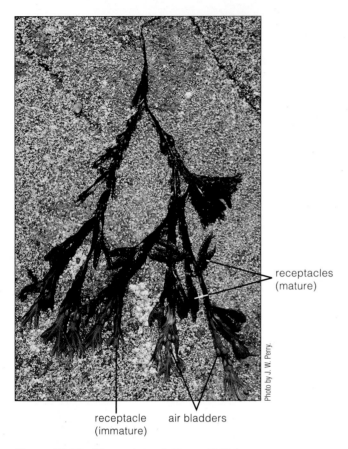

Figure 19-25 The rockweed, *Fucus* (0.6×).

Figure 19-26 *Fucus* colony, showing swollen tips that house sex organs (0.35×).

_____ **1.** Some members of the Domain Bacteria have
 (a) a nucleus.
 (b) membrane-bound organelles.
 (c) chloroplasts.
 (d) photosynthetic ability.

_____ **2.** A pathogen is
 (a) a disease.
 (b) an organism that causes a disease.
 (c) a substance that kills bacteria.
 (d) the same as a heterocyst.

_____ **3.** Gram stain is used to distinguish between different
 (a) bacteria.
 (b) protistans.
 (c) dinoflagellates.
 (d) all of the above

_____ **4.** Organisms capable of nitrogen fixation
 (a) include some bacteria.
 (b) include some cyanobacteria.
 (c) may live as symbionts with other organisms.
 (d) all of the above

_____ **5.** Those organisms that are covered by numerous, tiny locomotory structures belong to the phylum
 (a) Euglenozoa.
 (b) Stramenopila.
 (c) Parabasalia.
 (d) Alveolata.

_____ **6.** Which organisms cause "red tides?"
 (a) bacteria
 (b) cyanobacteria
 (c) euglenoids
 (d) dinoflagellates

_____ **7.** If you found a fish floating on the water covered with a white fuzzy material and tiny dark spheres you would suspect that the fish has been infected by a(an)
 (a) diatom.
 (b) euglenoid.
 (c) water mold.
 (d) cynanobacterium.

_____ **8.** A vector is
 (a) an organism that causes disease.
 (b) a disease.
 (c) a substance that prevents disease.
 (d) an organism that transmits a disease causing organism.

_____ **9.** The organism that causes malaria is
 (a) a pathogen.
 (b) _Plasmodium vivax._
 (c) carried by a mosquito.
 (d) all of the above

_____ **10.** The cell wall component algin is
 (a) found in the brown algae.
 (b) used in the production of ice cream.
 (c) used as a medium on which microorganisms are grown.
 (d) both a and b

EXERCISE **19**

Bacteria and Protists I

Post-Lab Questions

Introduction

1. What major characteristic distinguishes bacteria from protists?

19.1.A. Bacteria

2. What form of bacterium is shown in this photomicrograph?

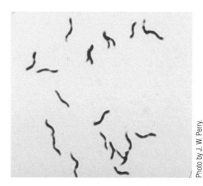

(1100×)

19.1.C. Cyanobacteria

3. Examine the photomicrograph on the right, taken using an oil-immersion objective, the highest practical magnification of a light microscope. (The final magnification is 770×.)
   (a) Based on your observation, identify the kingdom to which the organism belongs.

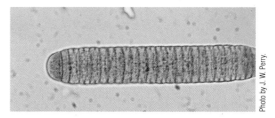

(770×)

   (b) Justify your answer.

4. Observe the photomicrograph on the right of an organism that was found growing symbiotically within the leaves of the water fern *Azolla*.
   (a) What is the common name given to an organism of this type?

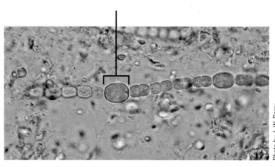

   (b) Give the name *and* function of the cell depicted by the line.

(432×)

19.2.C. Phylum Alveolata

5. The scanning electron micrograph on the right contains two organisms in the phylum. What are the structures at the end of the arrow?

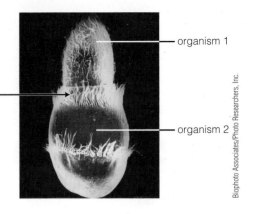

organism 1

organism 2

19.2.D. Phylum Stramenopila

6. You've never seen the protistan whose sexual structures appear here, but you have seen one very similar to it. What are the circular structures in the photo?

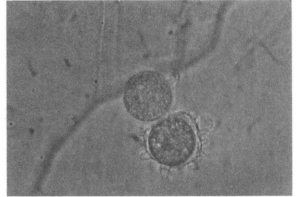

(287×)

7. While the organism that resulted in the Irish potato famine of 1845–1847 had long been present, environmental conditions that occurred during this period resulted in the destructive explosion of disease. Indicate what those environmental conditions were and why they resulted in a major disease outbreak.

8. While walking along the beach at Point Lobos, California, the fellow pictured walks up to you with alga in hand. Figuring you to be a college student who has probably had a good introductory biology course, he asks if you know what it is.
   (a) What is the cell wall component of the organism that has commercial value?

   (b) Name three uses for that cell wall component.

(425×)

Food for Thought

9. If you were to travel to a region where rice is grown in paddies, you would see lots of the water fern *Azolla* growing in the water. A farmer would tell you this is done because *Azolla* is considered a "natural fertilizer." Explain why this is the case.

10. Based on your knowledge of the life history of *Plasmodium vivax*, suggest two methods for controlling malaria. Explain why each method would work.

Protists II

After completing this exercise, you will be able to

1. define *phytoplankton, phyt, phycobilin, agar, gametangium, oogonia, antheridia;*

2. recognize and classify selected members of the phyla represented in this exercise;

3. distinguish among the structures associated with asexual and sexual reproduction described in this exercise;

4. identify the structures of the algae (those in **boldface** in the procedure sections).

Introduction

The organisms in this exercise make an enormous impact on the biosphere, both positively and negatively. Among their greatest contribution is the production of oxygen, because most are the photosynthetic protistans. Many are given the common name "algae."

Pond scum, frog spittle, seaweed, the stuff that clogs your aquarium if it's not cleaned routinely, the debris on an ocean beach after a storm at sea, the nuisance organisms of a lake—these are the images that pop into our minds when we first think about the organisms called algae. But many algae are also **phytoplankton**, the weakly swimming or floating algae, at the base of the aquatic food chain.

Also included are amoebozoans, organisms lacking cell walls in their nonreproductive state that are the shape-shifters of the protistan world.

We'll examine the following organisms in this exercise:

Phylum	Common Name
Rhodophyta	Red algae
Chlorophyta	Green algae
Charophyta	Desmids and stoneworts
Amoebozoa	Amoebas and slime molds

Note the ending *-phyta* for several of the phylum names. This derives from the Greek root word *phyt*, which means "plant." In the not too distant past, these organisms were considered plants because they were photosynthetic. By contrast, the ending *-zoa* means "animal." While the endings remain, today these organisms are considered to be members of the kingdom Protista.

20.1   Phylum Rhodophyta: Red Algae (About 15 min.)

Although commonly called red algae, members of the Rhodophyta vary in color from red to green to purple to greenish-black. The color depends on the quantity of their photosynthetic accessory pigments, the **phycobilins**, which are blue and red. These accessory pigments allow capture of light energy across the entire visible spectrum. This energy is passed on to chlorophyll for photosynthesis. One phycobilin, the red phycoerythrin, allows some red algae to live at great depths where red wavelengths, those of primary importance for green and brown algae, fail to penetrate.

Which wavelengths (colors) would be absorbed by a red pigment?

_____

Most abundant in warm marine waters, red algae are the source of **agar** and carrageenan substances extracted from their cell walls. Agar is the solidifying agent in microbiological media and carrageenan is a thickener used in ice cream and to make beer clear by causing suspended proteins to precipitate out.

MATERIALS

Per lab room:

■ culture dish containing dried agar and petri dish of hydrated agar
■ demonstration slide of *Porphyridium*
■ demonstration specimen and slide of *Porphyra* (nori)

PROCEDURE

1. Observe the dried agar that is on demonstration.
2. Now observe the petri dish containing agar that has been hydrated, heated, and poured into the dish.
3. Examine these cells of the unicellular *Porphyridium* at the demonstration microscope, noting the reddish chloroplast.
4. In Figure 20-1, sketch a cell of *Porphyridium*.
5. Examine a portion of the multicellular membranous *Porphyra* specimen on demonstration. Draw what you see in Figure 20-2.
6. Compare the specimen with the photo of live *Porphyra* (Figure 20-3).
7. In the adjacent demonstration microscope, note the microscopic appearance of *Porphyra* (Figure 20-4). Note that the clear areas are actually the cell walls.

Porphyra is used extensively as a food substance in Asia, where it's commonly sold under the name *nori*. In Japan, nori production is valued at $20 million annually.

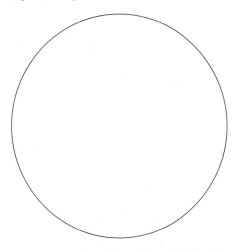

Figure 20-1   Drawing of *Porphyridium* (_____×).

Figure 20-2   Drawing of the macroscopic appearance of *Porphyra* (nori) (_____×).

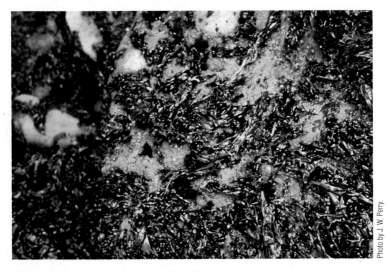

Figure 20-3   Living *Porphyra* (0.5×).

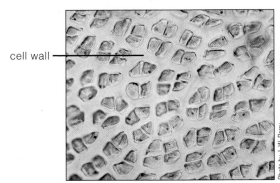

cell wall

Figure 20-4   Microscopic appearance of *Porphyra* (nori) (278×).

The Chlorophyta is a diverse assemblage of green organisms, ranging from motile and nonmotile unicellular forms to colonial, filamentous, membranous, and multinucleate forms. Not only do they include the most species, they are also important phylogenetically because ancestral green algae gave rise to the land plants. As you would expect, the photosynthetic pigments—both primary and accessory—and the stored carbohydrates are identical in the green algae and land plants.

In this section, you will examine the green algae from a morphological standpoint, starting with unicellular forms. For some, you'll also study their modes of reproduction and identify their characteristic sex organs, the gametangia (the singular is *gametangium*).

A. *Chlamydomonas*—A Motile Unicell

MATERIALS

Per student:

- microscope slide
- coverslip
- compound microscope

Per student pair:

- tissue paper
- I₂KI in dropping bottle

Per lab room:

- living culture of *Chlamydomonas*, disposable pipet

PROCEDURE

1. Prepare a wet mount of *Chlamydomonas* cells from the culture provided.
2. First examine with the low-power objective of your compound microscope. Notice the numerous small cells swimming across the field of view.
3. It will be difficult to study the fast-swimming cells, so kill the cells by adding a drop of I₂KI to the edge of the coverslip; draw the I₂KI under the coverslip by touching a folded tissue to the opposite edge. To observe the cells, switch to the highest power.
4. Identify the large, green cup-shaped **chloroplast** filling most of the cytoplasm.
5. *Chlamydomonas* stores its excess photosynthate as *starch grains,* which appear dark blue or black when stained with I₂KI. Locate the starch grains.
6. If the orientation of the cell is just right, you may be able to detect an orange **stigma** (eyespot) that serves as a light receptor.
7. Close the iris diaphragm to increase the contrast and find the two **flagella** at the anterior end of the cell.
8. Examine Figure 20-5, a transmission electron micrograph of *Chlamydomonas*. The electron microscope makes much more obvious the structures you can barely see with your light microscope. Notice the magnification.
9. The green algae have a specialized center for starch synthesis located within their chloroplast, the **pyrenoid.**

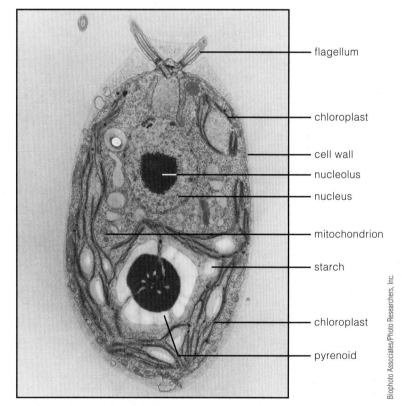

Figure 20-5   Transmission electron micrograph of *Chlamydomonas* (9750×).

Biophoto Associates/Photo Researchers, Inc.

B. *Volvox*—A Motile Colony

MATERIALS

Per student:

- depression slide
- compound microscope

Per lab room:

- living culture of *Volvox*, disposable pipet or prepared slide of *Volvox*

PROCEDURE

1. From the culture provided, place a drop of *Volvox*-containing culture solution on a depression slide. (A prepared slide may be substituted if living specimens are not available.)
2. Observe the large motile **colonies** (Figure 20-6), first with the microscope's low- and medium-power objectives. The colony is a hollow cluster of mostly identical, *Chlamydomonas*-like cells that are held together by a **gelatinous matrix**.
3. Identify the gelatinous matrix, which appears as the transparent region between individual cells.

Each cell possesses flagella. As the flagella beat, the entire colony rolls through the water. (The scientific name, *Volvox*, comes from the Latin word *volvere*, which means "to roll.")

4. Asexual reproduction takes place by **autocolony (daughter colony) formation**. Certain cells within the colony divide and then round up into a sphere called an **autocolony**. Find the autocolonies within your specimen.

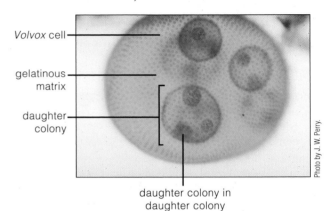

Figure labels: *Volvox* cell, gelatinous matrix, daughter colony, daughter colony in daughter colony

Figure 20-6 *Volvox*, with autocolonies (56×).

C. *Oedogonium*—A Nonmotile Filament

Oedogonium (Figure 20-7) is a filamentous, zoospore-producing green alga that is found commonly attached to sides of aquaria and slow-moving freshwater streams.

MATERIALS

Per student:

- microscope slide
- coverslip
- prepared slide of *Oedogonium*
- compound microscope

Per lab room:

- living culture of *Oedogonium*

PROCEDURE

1. Prepare a wet mount slide from the living culture provided and observe with the medium-power and high-dry objectives of your compound microscope.

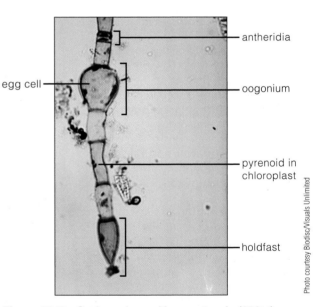

Figure labels: egg cell, antheridia, oogonium, pyrenoid in chloroplast, holdfast

Figure 20-7 *Oedogonium*, with gametangia (250×).

2. Within each cell, locate the single, netlike **chloroplast** with its many **pyrenoids**. The cell that attaches the filament to the substrate is specialized as a **holdfast**. Scan your specimen to determine whether any holdfasts are present at the ends of the filaments. Each cell in the filament contains a single nucleus and a large central vacuole (both of which are difficult to distinguish).

Oedogonium reproduces both asexually and sexually. Asexual reproduction takes place by zoospore production. Sexual reproduction is easy to observe on prepared slides, and illustrates features commonly found in many other algae, land plants, and animals. These features include the presence of a very large nonmotile egg and motile sperm.

3. Obtain a prepared slide of a bisexual species of *Oedogonium*, which bears both male and female gametangia on the same filament. Examine with the medium-power objective of your compound microscope.
4. Find the large, spherical **oogonia**. These are the female sex organs. Within each oogonium locate a single, large **egg cell**, which virtually fills the oogonium.

Now find the male sex organs, the **antheridia**, which appear as short, boxlike cells. Each antheridium produces two *sperm*.

When the egg cell is fertilized by a sperm, a *zygote* is formed. The zygote develops a thick, heavy wall. Some oogonia on your slide may contain zygotes. When the zygote germinates, the diploid nucleus undergoes meiosis, producing motile spores. When the spore settles down, mitosis and cell division occur, producing the haploid filament you have examined.

Eventually, the two nuclei of each gamete fuse to form a **zygote**, which develops a thick wall. This thick-walled zygote serves as an overwintering structure. In the spring, the zygote nucleus undergoes meiosis. Three of the four nuclei die, leaving one functional, haploid nucleus. Germination of this haploid cell results in the formation of a haploid filament.

D. *Ulva*—A Membranous Form

A final representative of the green algae illustrates the fourth morphological form in the group, those having a membranous (tissuelike) body.

MATERIALS

Per lab room:

- demonstration specimen of *Ulva*

PROCEDURE

Examine living or preserved specimens of *Ulva*, commonly known as sea lettuce (Figure 20-8). The broad, leaflike body is called a *thallus*, a general term describing a vegetative body with relatively little cell differentiation. The *Ulva* body is only two cells thick!

Figure 20-8 The sea lettuce, *Ulva* (0.5×).

20.3   Phylum Charophyta: Desmids and Stoneworts

A. *Spirogyra*

A filamentous charophyte common to freshwater ponds is *Spirogyra*. This alga is often called pond scum because it forms a bright green, frothy mass on and just below the surface of the water.

MATERIALS

Per student:

- microscope slide
- coverslip
- dissecting needle
- prepared slide of *Spirogyra*
- compound microscope

Per student pair:

- diluted India ink in dropping bottle

Per lab room:

- living culture of *Spirogyra*

PROCEDURE

1. Observe the *Spirogyra* in the large culture dish. Pick up some of the mass, noting the slimy sensation. This is due to the watery sheath surrounding each filament.
2. Using a dissecting needle, place a few filaments on a slide and add a drop of diluted India ink before adding a coverslip. (Prepared slides may be used if living filaments are not available.)
3. Observe the filaments with the medium-power and high-dry objectives of your compound microscope (Figure 20-9).
4. Identify the **sheath**, which will appear as a bright area off the edge of the cell wall.
5. Note the spiral-shaped **chloroplast** with the numerous **pyrenoids**.
6. Each cell contains a large central **vacuole** and a single **nucleus**. Remember, the chloroplast is located within the cytoplasm, as is

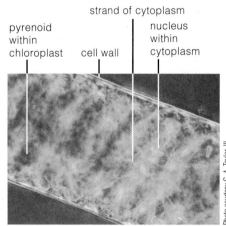

Figure 20-9 *Spirogyra* (550×).

the nucleus. The nucleus, however, is suspended in strands of cytoplasm, much as a spider might be found in the center of a web. Look again at Figure 20-9 and locate the nucleus.

Asexual reproduction occurs when a small portion of the filament simply breaks off and continues to grow. Zoospores are not formed. Sexual reproduction is illustrated in Figure 20-9. Male and gametes, and for that matter sex organs, look identical to one another.

7. Obtain a prepared slide illustrating sexual reproduction in *Spirogyra.* Examine the slide with your medium or high-dry objective.
8. Find two filaments that are joined by cytoplasmic bridges known as **conjugation tubes**. The entire cytoplasmic contents serve as **isogametes** (that is, gametes of similar size) in *Spirogyra,* with one isogamete moving through the conjugation tube into the other cell, where it fuses with the other gamete.
9. Find stages illustrating conjugation as in Figure 20-10.

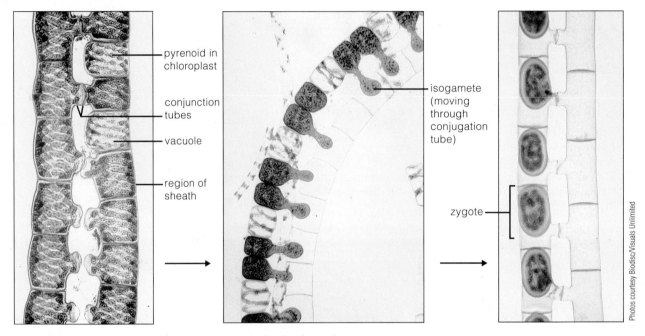

pyrenoid in chloroplast

conjunction tubes

vacuole

region of sheath

isogamete (moving through conjugation tube)

zygote

Photos courtesy Biodisc/Visuals Unlimited

Figure 20-10 *Spirogyra*, stages in sexual reproduction (170×).

B. *Chara* or *Nitella*—Stoneworts

Stoneworts are within a green alga lineage that million of years ago gave rise to the land plants. They have a distinctive body form and sex organs, which are unlike those found in other green algae, and indeed remind one of tiny aquatic plants.

Stoneworts are interesting organisms found "rooted" in brackish and fresh waters, particularly those high in calcium. (Two of your lab manual authors—the Perrys—have ponds that are delightfully full of stoneworts.) They get their common names by virtue of being able to precipitate calcium carbonate over their surfaces, encrusting them and rendering them somewhat stony and brittle. (The suffix -*wort* is from a Greek word meaning "herb.")

MATERIALS

Per student:

- small culture dish
- dissecting microscope

Per lab room:

- living culture of (or preserved) *Chara* or *Nitella*
- demonstration slide of *Chara* with sex organs

PROCEDURE

1. From the classroom culture provided, obtain some of the specimen and place it in a small culture dish partially filled with water.
2. Observe the specimen with a dissecting microscope. Note that the stoneworts resemble what we would think of as a plant. They are divided into "stems" and "branches" (Figure 20-11).
3. Search for flask-shaped and spherical structures along the stem. These are the sex organs. If none is present on the specimen, observe the demonstration slide (Figure 20-12) that has been selected to show these structures.

4. The flask-shaped structures are **oogonia**, each of which contains a single, large egg (Figure 20-13). Notice that the oogonium is covered with cells that twist over the surface of the gametangium. Because of the presence of these cells, the oogonium is considered to be a *multicellular* sex organ. Multicellular sex organs are present in all land plants.

Based on your study of previously examined specimens, would you say the egg is motile or nonmotile?

5. Now find a spherical **antheridium** (Figure 20-13). Like the oogonium, the antheridium is covered by cells and is also considered to be multicellular. Cells within the interior of the antheridium produce numerous flagellated sperm cells.

Figure 20-11 The stonewort, *Chara*, with sex organs (2×).

Fertilization of eggs by sperm produces a *zygote* within the oogonium. The zygote-containing oogonium eventually falls off from the parent plant. The zygote can remain dormant for some time before the nucleus undergoes meiosis in preparation for germination. Apparently, three of the four nuclei produced during meiosis disintegrate.

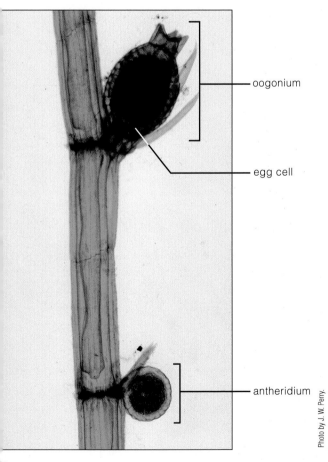

Figure 20-12 *Chara*, showing sex organs (30×).

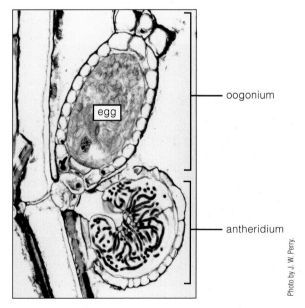

Figure 20-13 Antheridia and oogonia of *Chara* (80×).

The slime molds have both plantlike and animal-like characteristics. They engulf their food and lack a cell wall in their nonreproductive state. However, when they reproduce, they produce spores with a rigid cell wall, similar to plants.

A. *Physarum*: A Plasmodial Slime Mold (About 15 min.)

The vegetative (nonreproductive) body of the plasmodial slime molds consists of a naked multinucleate mass of protoplasm known as a **plasmodium**.

MATERIALS

Per student:
- petri dish culture of slime mold (*Physarum*)
- compound microscope

Per lab room:
- demonstration of slime mold (*Physarum*) sporangia

PROCEDURE

1. Obtain a petri dish culture of *Physarum* (Figure 20-14) and remove the cover. After examining it with your unaided eye, place the culture dish on the stage of your compound microscope and examine it with the low-power objective.
2. Watch the cytoplasm. The motion that you see within the plasmodium is cytoplasmic streaming (Exercise 6). Is the cytoplasmic streaming unidirectional, or does the flow reverse? _____

As you noticed, the *Physarum* culture was stored in the dark. That's because light (along with other factors) stimulates the plasmodium to switch to the reproductive phase.

3. Examine the spore-containing sporangia of *Physarum* that are on demonstration (Figure 20-15). Spores released from the sporangia germinate, producing a new plasmodium.

Because the plasmodium is multinucleate (whereas the spores are uninucleate), what event must occur following spore germination?

_____

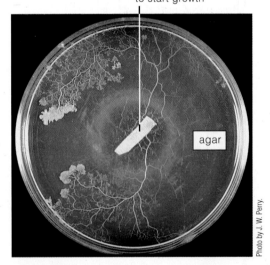

Figure 20-14   Culture dish containing the plasmodial slime mold, *Physarum* (0.7×).

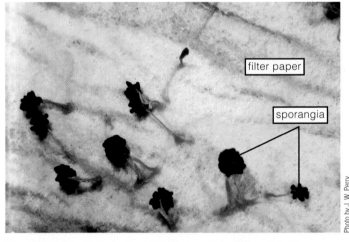

Figure 20-15   Sporangia of the plasmodial slime mold, *Physarum* (1.5×).

B. *Amoeba* (About 20 min.)

Amoebas continually change shape by forming projections called *pseudopodia* (singular, *pseudopodium*, "false foot"). One notorious human pathogen is *Entamoeba histolytica*, the cause of amoebic dysentery.

MATERIALS

Per student:
- depression slide
- coverslip
- dissecting needle
- compound microscope

Per group (table):
- carmine, in two screw-cap bottles
- tissue paper

Per lab room:
- *Amoeba*—living culture on demonstration at dissecting microscope; disposable pipet

PROCEDURE

1. Observe the *Amoeba*-containing culture on the stage of a demonstration dissecting microscope. (One species of amoeba has the scientific name *Amoeba*.) The microscope has been focused on the bottom of the culture dish, where the amoebas are located. Look for gray, irregularly shaped masses moving among the food particles in the culture.
2. Using a clean pipet, remove an amoeba and place it, along with some of the culture medium, in a depression slide.
3. Examine it with your compound microscope using the medium-power objective. You will need to adjust the diaphragm to increase the contrast (see Exercise 3) because *Amoeba* is nearly transparent.
4. Refer to Figure 20-16. Locate the **pseudopodia**. At the periphery of the cell, identify the **ectoplasm,** a thin, clear layer that surrounds the inner, granular **endoplasm**. Watch the organism as it changes shape. Which region of the endoplasm appears to stream, the outer or the inner? _____

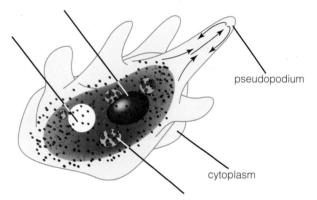

Figure 20-16 Artist's rendering of *Amoeba*.
Labels: ectoplasm, endoplasm, nucleus, contractile vacuole, food vacuole

This region, called the *plasmasol,* consists of a fluid matrix that can undergo phase changes with the semisolid *plasmagel,* the outer layer of the endoplasm. Pseudopodium formation occurs as the plasmasol flows into new environmental frontiers and then changes to plasmagel.

Numerous granules will be found within the endoplasm. Some of these are organelles; others are food granules.

5. Within the endoplasm, try to locate the **nucleus**, a densely granular, spherical structure around which the cytoplasm is streaming.
6. Find the clear, spherical **contractile vacuoles**, which regulate water balance within the cell. Watch for a minute or two to observe the action of contractile vacuoles. Label Figure 20-16.

Amoeba feeds by a process called **phagocytosis**, engulfing its food. Pseudopodia form around food particles, and then the pseudopodia fuse, creating a **food vacuole** within the cytoplasm. Enzymes are then emptied into the food vacuole, where the food particle is digested into a soluble form that can pass through the vacuolar membrane. You can stimulate feeding behavior by drawing carmine under the coverslip.

7. Place a drop of distilled water against one edge of the coverslip.
8. Pick up some carmine crystals by dipping a dissecting needle into the bottle; and deposit them into the water droplet.
9. Draw the suspension beneath the coverslip by holding a piece of absorbent tissue against the coverslip on the side *opposite* the carmine suspension.
10. Observe the *Amoeba* with your microscope again—you may catch it in the act of feeding.

_____ 1. Agar is derived from
   (a) red algae.
   (b) brown algae.
   (c) green algae.
   (d) all of the above

_____ 2. Which is the correct plural form of the word for the organisms studied in this exercise?
   (a) alga
   (b) algae
   (c) algas
   (d) algaes

_____ 3. Phycobilins are
   (a) photosynthetic pigments.
   (b) found in the red algae.
   (c) blue and red pigments.
   (d) all of the above

_____ 4. Specifically, female sex organs are known as
   (a) oogonia.
   (b) gametangia.
   (c) antheridia.
   (d) zygotes.

_____ 5. A reagent that stains the stored food of a green alga black is
   (a) India ink.
   (b) I_2KI.
   (c) methylene blue.
   (d) both a and b

_____ 6. The starch production center within many algal cells is the
   (a) nucleus.
   (b) cytoplasm.
   (c) stipe.
   (d) pyrenoid.

_____ 7. The phylum of organisms _most_ closely linked to the evolution of land plants is the
   (a) Chlorophyta.
   (b) Charophyta.
   (c) Phaeophyta.
   (d) Euglenozoa.

_____ 8. Gametangia are
   (a) chlorophyll-producing structures.
   (b) sex organs.
   (c) a name for colony-forming algae.
   (d) found in slime molds.

_____ 9. Which of the following terms does **not** apply to organisms in the Phylum Amoeboza?
   (a) chlorophyll
   (b) phagocytosis
   (c) pseudopodia
   (d) slime mold

_____ 10. Amoebic dysentery is caused by
   (a) _Ulva._
   (b) _Volvox._
   (c) a desmid.
   (d) _Entamoeba._

Name _____ Section Number _____

EXERCISE **20**

Protists II

Post-Lab Questions

20.1 Phylum Rhodophyta: Red Algae

1. While wading in the warm salt water off the beaches of the Florida Keys on spring break, you stoop down to look at the feathery alga shown here.
   (a) What pigment gives this organism its coloration?

   (b) What commercial products are derived from this phylum?

Photo by J. W. Perry.

20.2 Phylum Chlorophyta: Green Algae

2. What two characteristics found within the green algae are clues that this group of organisms is more closely related to the land plants than red algae?
   (a)

   (b)

Photo courtesy C. A. Taylor, Ill.

20.3 Phylum Charophyta: Desmids and Stoneworts

3. Your class takes a field trip to a freshwater stream, where you collect the organism shown microscopically here.
   (a) What is the genus name of this organism?

   (b) Identify and give the function of the structure within the chloroplast at the end of the leader (line).

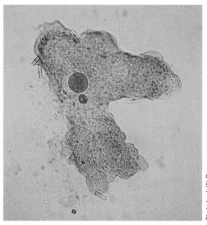

20.4 Phylum Amoebozoa: Slime Molds and Amoebas

4. This photomicrograph was taken from a prepared slide of a stained specimen. You observed unstained living specimens in lab. Describe the mechanism by which it moves from place to place.

Photo by J. W. Perry.

5. What is phagocytosis? What function does it serve?

Food for Thought

6. Some botanists consider the stoneworts to be a link between the higher plants and the algae. As you will learn in future exercises, higher plants, such as the mosses, have both haploid and diploid stages that are *multicellular*.

   (a) Describe the multicellular organism in the charophytes.

   (b) Is this organism haploid or diploid?

   (c) Is the zygote haploid or diploid?

   (d) Is the zygote unicellular or multicellular?

7. (a) What color are the marker lights at the edge of an airport taxiway?

   (b) Are the wavelengths of this color long or short, relative to the other visible wavelengths?

   (c) Which wavelengths penetrate deepest into water, long or short?

   (d) Make a statement regarding why phycobilin pigments are present in deep-growing red algae.

   (e) What benefit is there to the color of airport taxiway lights for a pilot attempting to taxi during foggy weather?

8. Why do you suppose the Swedish automobile manufacturer Volvo chose this company name?

9. List two reasons why algae are useful and important to life.
   (a)

   (b)

10. It's generally accepted that the first life on Earth was unicellular and, while autotrophic, did not have chlorophyll. Later, multicellular, chlorophyll-bearing organisms evolved. Speculate on how amoeba-like organisms that lacked chlorophyll might have obtained the ability to use chlorophyll to capture the energy of the sun.

Bryophytes—Liverworts and Mosses

OBJECTIVES

After completing this exercise, you will be able to

1. define *alternation of generations, isomorphic, heteromorphic, dioecious, antheridium, archegonium, hygroscopic, protonema*;

2. diagram the plant life cycle to show alternation of generations;

3. distinguish between alternation of isomorphic and heteromorphic alternation of generations;

4. list evidence supporting the evolution of land plants from green algae;

5. recognize charophytes, mosses, and liverworts;

6. identify the sporophytes, gametophytes, and associated structures of liverworts and mosses (those in **boldface** in the procedure sections);

7. describe the function of the sporophyte and the gametophyte;

8. postulate why bryophytes live only in environments where free water is often available.

Introduction

This is the first in a series of exercises that addresses true plants—the kingdom Plantae.

Plants evolved from the green algae, specifically now extinct relatives of living green algae called **charophytes**. Evidence for this includes identical food reserves (starch), the same photosynthetic pigments (chlorophylls *a* and *b*, carotenes, and xanthophylls), similarities in structure of their flagella, and the way cell division takes place. Some biologists have gone so far as to suggest that the land plants are nothing more than highly evolved green algae.

One major mystery is the origin of a feature common to *all* true plants (but *not* charophytes), **alternation of generations**. In alternation of generations, two distinct phases exist: a diploid **sporophyte** alternates with a haploid **gametophyte**, as summarized in Figure 21-1.

As animals, we find the concept of alternation of generations difficult to envision. But think of it as the existence of two body forms of the same organism. The primary reproductive function of one body form, the gametophyte, is to produce gametes (eggs and/or sperm) by *mitosis*. The primary reproductive function of the other, the sporophyte, is to produce spores by *meiosis*.

A fundamental distinction exists between the green algae and land plants with respect to alternation of generations. Although some green algae have alternation of generations, the sporophyte and gametophyte look *identical*. To the naked eye, they are indistinguishable. This is called alternation of **isomorphic** generations. (*Iso*- comes from a Greek word meaning "equal"; *morph*- is Greek for "form.")

By contrast, the alternation of generations in land plants is **heteromorphic** (*hetero*- is Greek for "different"). The sporophytes and gametophytes of land plants, including the bryophytes, are distinctly different from one another. The contrast between isomorphic and heteromorphic alternation of generations is modeled in Figure 21-2.

During the course of evolution, two major lines of divergence took place in the plant kingdom. In one line, the plants had as their dominant (most conspicuous) phase the gametophytic generation, meaning that the sporophyte was permanently attached to and dependent on the gametophyte for nutrition, never living an independent existence. Today these plants are represented by the bryophytes, mosses and liverworts. It seems that this line exemplifies *dead-end evolution*— no other group of plants present today arises from it.

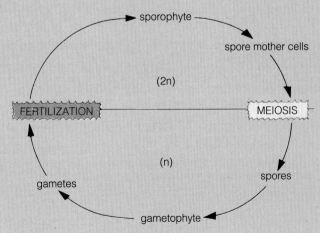

Figure 21-1 Summary of alternation of generations.

In the other line of evolution, the sporophyte led an independent existence, the gametophyte being quite small and inconspicuous. Figure 21-3 summarizes these two evolutionary lines.

As you will see in this exercise, the land plants known as liverworts and mosses are often fairly dissimilar in appearance. They do share similarities, including the following:

1. Both exhibit alternation of heteromorphic generations, in which the gametophyte is the dominant organism. The sporophyte remains attached to the gametophyte, deriving most of its nutrition from the gametophyte.
2. Both are dependent on water for fertilization, since their sperm must swim to a nonmotile egg.
3. Both lack true vascular tissues, xylem and phloem, and hence are relatively small organisms.

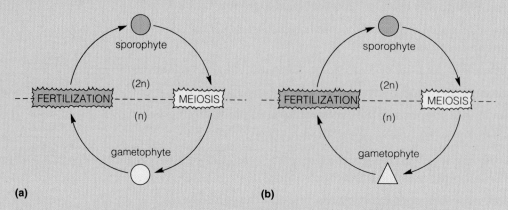

Figure 21-2 Types of alternation of generations. **(a)** Isomorphic. **(b)** Heteromorphic.

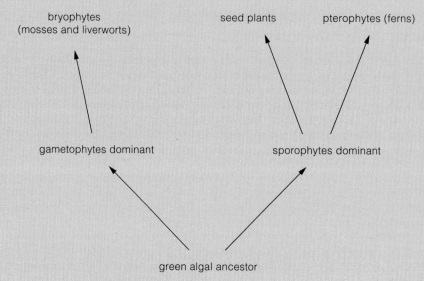

Figure 21-3 Evolution of land plants.

Before beginning study of true plants (kingdom Plantae), let's look at their closest algal relatives. Charophytes (kingdom Protista) called stoneworts are within a lineage that millions of years ago gave rise to the land plants.

Stoneworts are interesting organisms found "rooted" in brackish and fresh waters, particularly those high in calcium. (Two of your lab manual authors—the Perrys—have ponds that are delightfully full of stoneworts.) They get their common names by virtue of being able to precipitate calcium carbonate over their surfaces, encrusting them and rendering them somewhat stony and brittle. (The suffix *-wort* is from a Greek word meaning "herb.")

MATERIALS

Per student:

- small culture dish
- dissecting microscope

Per lab room:

- living culture (or preserved) *Chara* or *Nitella*
- demonstration slide of *Chara* with sex organs

PROCEDURE

1. From the classroom culture provided, obtain some of the specimen and place it in a small culture dish partially filled with water.
2. Observe the specimen with a dissecting microscope. Note that the stoneworts resemble what we would think of as a plant. They are divided into "stems" and "branches" (Figure 21-4).
3. Search for flask-shaped and spherical structures along the stem. These are the sex organs. If none is present on the specimen, observe the demonstration slide (Figure 21-5) that has been selected to show these structures.
4. The flask-shaped structures are **oogonia**, each of which contains a single, large egg (Figure 21-6). Notice that the oogonium is covered with cells that twist over the surface of the gametangium. Because of the presence of these cells, the oogonium is considered to be a *multicellular* sex organ. Multicellular sex organs are present in all land plants.

Figure 21-4   The stonewort, *Chara*, with sex organs (2×).

Based on your study of previously examined specimens, would you say the egg is motile or nonmotile?

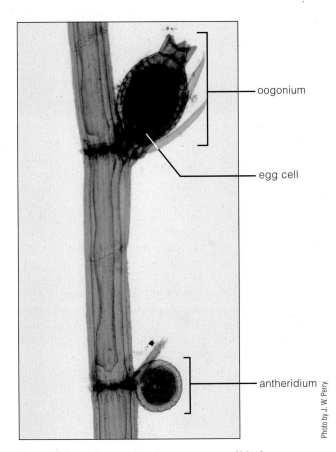

Figure 21-5   *Chara*, showing sex organs (30×).

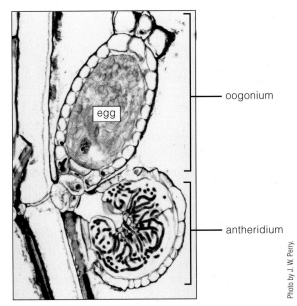

Figure 21-6   Antheridia and oogonia of *Chara* (80×).

Bryophytes—Liverworts and Mosses   **281**

5. Now find a spherical **antheridium** (Figure 21-6). Like the oogonium, the antheridium is covered by cells and is also considered to be multicellular. Cells within the interior of the antheridium produce numerous flagellated sperm cells.

Fertilization of eggs by sperm produces a *zygote* within the oogonium. The zygote-containing oogonium eventually falls off from the parent plant. The zygote can remain dormant for some time before the nucleus undergoes meiosis in preparation for germination. Apparently, three of the four nuclei produced during meiosis disintegrate.

21.2   Phylum Hepatophyta: Liverworts *(About 20 min.)*

Liverworts!? What in the world is a liverwort? Knowing why these plants are called liverworts is a start to knowing them.

The Greek suffix -*wort* means "herb," a nonwoody plant. The "liver" portion of the common name comes about because the plants look a little like a lobed human liver. These are small plants that you have likely overlooked (until now!) because they're inconspicuous and don't look like anything important. You'll find them growing on rocks that are periodically wetted along clean freshwater streams.

MATERIALS

Per student:

- living or preserved *Marchantia* thalli, with gemma cups, antheridiophores, archegoniophores, and mature sporophytes
- dissecting microscope

PROCEDURE

1. Examine the living or preserved plants of the liverwort, *Marchantia*, that are on demonstration. The body of this plant is called a **thallus** (the plural is *thalli*), because it is flattened and has little internal tissue differentiation (Figure 21-7).
2. The thallus is the gametophyte portion of the life cycle.

Is it haploid or diploid? _____

3. Obtain and examine one thallus of the liverwort. Notice that the thallus is lobed. Centuries ago, herbalists believed that plants that looked like portions of the human anatomy could be used to treat ailments of that part of the body. This plant reminded them of the liver, and hence the plant was called a "liverwort."

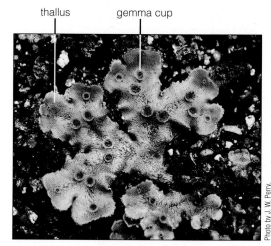

Figure 21-7 Thallus of a liverwort with gemma cups (1×).

4. Now place the thallus on the stage of a dissecting microscope. Looking first at the top surface, find the *pores* that lead to the interior of the plant and serve as avenues of exchange for atmospheric gases (CO_2 and O_2; Figure 21-7).
5. Identify **gemma cups** (Figure 21-7) on the upper surface. Look closely within a gemma cup to find the **gemmae** inside (the singular is *gemma*). Gemmae are produced by *mitotic* divisions of the thallus. They are dislodged by splashing water. If they land on a suitable substrate, they grow into new thalli; hence they are a means of asexual reproduction.
6. Are gemmae haploid or diploid? _____
7. Turn the plant over to find the hairlike **rhizoids** that anchor the organism to the substrate.

As noted above, the thalli are the gametophyte generation. Similar to higher animals (humans, for example), gametes are produced in special organs on the plant. In this liverwort, the organs are contained within elevated branches. The general name for these elevated branches is *gametangiophores*. (The suffix -*phore* is derived from a Greek word meaning "branch.") There are male and female gametangiophores.

8. On a male thallus, find the male **antheridiophores**. They look like umbrellas (Figure 21-8a). The sperm-producing male sex organs called **antheridia** (the singular is *antheridium*) are within the flattened splash platform of the umbrella.
9. On female plants, find the female **archegoniophores** (Figure 21-8b); these look somewhat like an umbrella that has lost its fabric and has only ribs remaining. Underneath these ribs are borne the female gametangia, **archegonia** (singular, *archegonium*).

archegoniophore

antheridiophore

thallus

Figure 21-8 Liverwort thalli with gametangiophores. (**a**) Male thalli with antheridiophores (0.25×). (**b**) Female thalli with archegoniophores (0.5×).

21.3 Experiment: Effect of Photoperiod *(About 10 min. to set up; requires 6 weeks to complete)*

Photoperiod, the length of light and dark periods, exerts a significant effect upon most living things. Flowering in many plants is dependent on photoperiod, some plants being classified as "long day (short night)," others as "short day (long night)," and still others as "day neutral." Flowering is a reproductive event. The reproductive behavior of many animals, such as deer, is also dependent on day length.

This experiment examines the effect of photoperiod on the production of gametangiophores and gemma cups in the liverwort, *Marchantia*. One possible experimental hypothesis is that *reproduction of* Marchantia *is stimulated by long days (short nights)*. You may wish to read the discussion in Exercise 1, "The Scientific Method," before proceeding.

MATERIALS

Per student group:

- vigorously growing colony of the liverwort *Marchantia polymorpha*
- pot label
- china or other marker

Per lab room:

- growth chambers set to long day (16 hr light, 8 hr darkness) and short day (8 hr light, 16 hr darkness) conditions

PROCEDURE

1. In Table 21-1, write a prediction about the effect of day length on *Marchantia* reproduction.
2. Obtain a colony of the liverwort, *Marchantia polymorpha*. On the pot label, write your name, date, and the experimental condition (long day or short day) to which you are going to subject the colony.
3. Place the colony in the growth chamber that is designated as "short day" or "long day," depending on which condition has been assigned to your experimental group. You will be responsible for watering the colonies on a regular basis. Failure to do so will result in unreliable experimental results and/or death of the plants.
4. Observe the colonies weekly over a period of 6 weeks, recording your observations in Table 21-1.
5. After 6 weeks of experimental treatment, compare observations with other experimental groups using the same and different independent variables.
6. List the results and make conclusions, writing them in Table 21-2.

Did the experimental results support your hypothesis? _____

Can you suggest any reasons why it might be beneficial to the liverwort to display the results that the different experimental groups found?

_____

_____

TABLE 21-1   The Effect of Photoperiod on the Reproduction of the Liverwort, *Marchantia polymorpha*

Prediction:

Week	Observations of the Effect of _____ (Long or Short) Day Illumination
1	
2	
3	
4	
5	
6	

© Cengage Learning 2013

TABLE 21-2   Summary of Results: The Effect of Photoperiod on the Reproduction of the Liverwort, *Marchantia polymorpha*

Photoperiod	Results
Long day	
Short day	

Conclusions:

© Cengage Learning 2013

21.4   Phylum Bryophyta: Mosses (*About 60 min.*)

Mosses are plants that you've probably noticed before. They grow on trees, rocks, and soil. Mosses have a wide variety of human uses, the most common ones being as a soil additive to potting soil and as a major component of peat, which is used for fuel in some parts of the world.

MATERIALS

Per student:

- *Polytrichum*, male and female gametophytes, the latter with attached sporophytes
- prepared slide of moss antheridial head, l.s.
- prepared slide of moss archegonial head, l.s.
- prepared slide of moss sporangium (capsule), l.s.
- glass microscope slide
- coverslip

- compound microscope
- dissecting microscope
- dissecting needle

Per student pair:

- dH₂O in dropping bottle

Per student group (table):

- moss protenemata growing on culture medium

PROCEDURE

Refer to Figure 21-15 as you study the life cycle of the mosses.

1. Obtain a living or preserved specimen of a moss that grows in your area. The hairy-cap moss, *Polytrichum*, is a good choice because one of the 10 species in this genus is certain to be found near wherever you live in North America.

Lacking vascular tissues, the mosses do not have true roots, stems, or leaves, although they do have structures that are rootlike, stemlike, and leaflike and perform the same functions as the true organs.

2. Identify the rootlike **rhizoids** at the base of the plant. What function do you think rhizoids perform?

_____

_____

3. Notice that the leaflike organs are arranged more or less radially about the stemlike *axis*. You are examining the **gametophyte generation** of the moss. In terms of the life cycle of the organism, what function does the gametophyte serve?

_____

_____

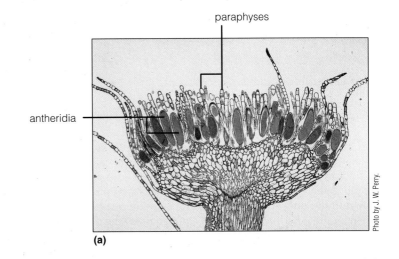

Figure 21-9 Colony of Polytrichum, male gametophytes (0.5×).

Polytrichum is **dioecious,** meaning that there are separate male and female plants. (Some other mosses are monoecious.) The male gametophyte can usually be distinguished by the flattened rosette of leaflike structures at its tip.

4. Examine a male gametophyte (Figures 21-9, 21-15a), noting this feature. Embedded within the rosette are the male sex organs, **antheridia** (singular, *antheridium*).

5. With the low- and medium-power objectives of your compound microscope, examine a prepared slide with the antheridia of a moss (Figure 21-15c). Identify the *antheridia* (Figure 21-10).

6. Study a single antheridium (Figure 21-10), using the high-dry objective. Locate the jacket layer surrounding the sperm-forming tissue that, with maturity, gives rise to numerous biflagellate *sperm.* Scattered among the antheridia, find the numerous sterile *paraphyses.* These do not have a reproductive role (and hence are called sterile) but instead function to hold capillary water, preventing the sex organs from drying out.

7. Examine a female gametophyte of *Polytrichum* (Figure 21-15b). Before the development of the sporophyte, the female gametophyte can usually be distinguished by the absence of the rosette at its tip. Nonetheless, the apex of the female gametophyte contains the female sex organs, **archegonia** (singular, *archegonium*).

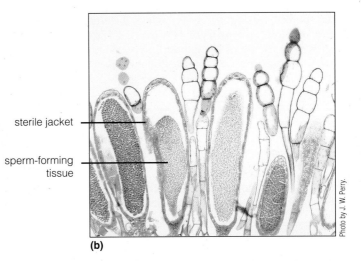

Figure 21-10 Longitudinal section of antheridial head of a moss (a) 8×. (b) 93×.

8. Obtain a prepared slide of the archegonial tip of a moss (Figures 21-11a, 21-15d). Start with the low-power objective of your compound microscope to gain an impression of the overall organization. Find the sterile *paraphyses* and the *archegonia*.

9. Switch to the medium-power objective to study a single archegonium, identifying the long **neck** and the slightly swollen base. Within the base, locate the **egg cell** (Figure 21-15d).

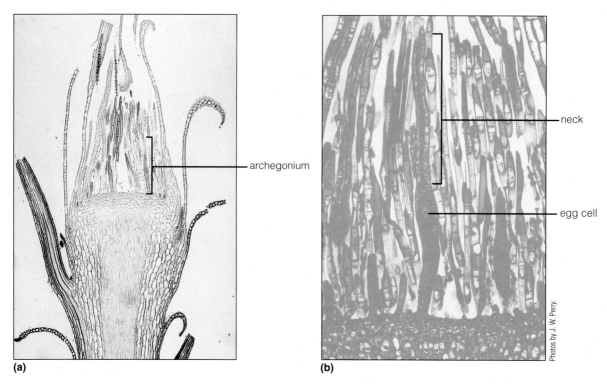

(a) (b)

Figure 21-11 (**a**) Longitudinal section of archegonial head of a moss (22×). (**b**) Detail of archegonium (88×).

Remember, you are looking at a *section* of a three-dimensional object. The archegonium is very much like a long-necked vase, except that it's solid. The egg cell is like a marble suspended in the middle of the base.

The central core of the archegonial neck contains cells that break down when the egg is mature, liberating a fluid that is rich in sucrose and that attracts sperm that are swimming in dew or rainwater. (Sperm are capable of swimming only short distances and so male plants must be close by.) Fertilization of the egg produces the diploid zygote (Figure 21-15e), the first cell of the **sporophyte generation.**

Numerous mitotic divisions produce an **embryo** (embryo sporophyte, Figure 21-15f), which differentiates into the **mature sporophyte** (Figure 21-15g) that protrudes from the tip of the gametophyte. What is the function of the sporophyte (that is, what does it do, or what reproductive structures does it produce)?

_____

_____

10. Now examine a female gametophyte that has an attached *mature sporophyte* (Figures 21-12, 21-15g).

Is the sporophyte green? _____

What can you conclude about the sporophyte's ability to produce at least a portion of its own food?

_____

_____

Figure 21-12 Colony of Polytrichum, female gametophytes with attached sporophytes (0.25×).

11. Grasp the stalk of the sporophyte and detach it from the gametophyte. The base of the stalk absorbs water and nutrients from the gametophyte.
12. At the tip of the sporophyte, locate the **sporangium** covered by a papery hood. The cover is a remnant of the tissue that surrounded the archegonium and is covered with tiny hairs—hence the common name, hairy-cap moss, for *Polytrichum.*
13. Remove the hood to expose the sporangium (Figure 21-15h). Notice the small cap at the top of the sporangium.
14. Remove the cap and observe the interior of the sporangium with a dissecting microscope. Find the *peristomal teeth* that point inward from the margin of the opening.

The peristomal teeth are **hygroscopic**, meaning that they change shape as they absorb water. As the teeth dry, they arch upward, loosening the cap over the spore mass. The teeth may subsequently shrink or swell, thus regulating how readily the spores inside may escape.

15. Study a prepared slide of a longitudinal section of a sporangium (Figure 21-13). At the top, you will find sections of the peristomal teeth. Internally, locate the **sporogenous tissue**, which differentiates into *spores* when mature. The sporogenous tissue of the sporangium is diploid, but the spores are haploid and are the first cells of the gametophyte generation.

What type of nuclear division must take place for the sporogenous tissue to become spores?

<hr>

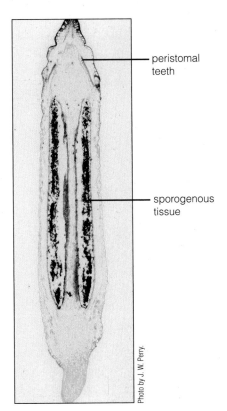

peristomal teeth

sporogenous tissue

Photo by J. W. Perry.

Figure 21-13 Longitudinal section of moss sporangium (12×).

When the spores are shed from the sporangium, they are carried by wind and water currents to new sites. If conditions are favorable, the spore germinates to produce a filamentous protonema (Figures 21-14, 21-15j; plural, *protonemata*) that looks much like a filamentous green alga.

Is the protonema part of the gametophyte or sporophyte generation?

<hr>

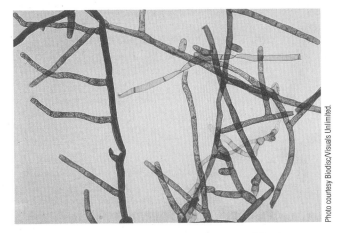

Photo courtesy Biodisc/Visuals Unlimited.

Figure 21-14 Moss protonema, whole mount (70×).

16. Use a dissecting needle to remove a protonema from the culture provided and make a wet mount. Examine it with the medium-power and high-dry objectives of your compound microscope. Note the cellular composition and numerous green *chloroplasts*. If the protonema is of sufficient age, you should find *buds* (Figure 21-15k) that grow into the leafy gametophyte (Figures 21-15a, b, and l).

21.5 Experiment: Effect of Light Quality on Moss Spore Germination and Growth (*About 20 min. to set up; requires 8 weeks to complete*)

What causes plants to grow and reproduce? One significant environmental trigger is light. Both light *quantity* (the length of illumination) and light *quality* (the wavelengths or color of light) trigger growth and reproduction in organisms. In this experiment, you will grow moss protonemata from spores and examine the effect of different wavelengths of light on germination of spores and subsequent growth of the protonema. One hypothesis this simple experiment might address is that *moss spores germinate and protonemata grow best in short (blue) wavelength light.* You may wish to read the discussion in Exercise 1, "The Scientific Method," before proceeding.

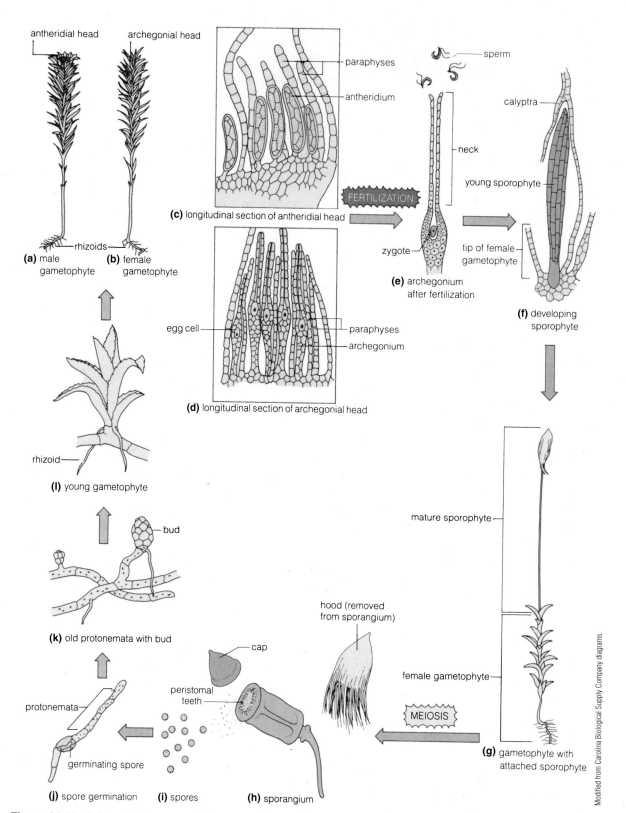

Figure 21-15 Life cycle of representative moss. (Green structures are 2n.)

(a) male gametophyte
antheridial head
rhizoids

(b) female gametophyte
archegonial head
rhizoids

(c) longitudinal section of antheridial head
paraphyses
antheridium

(d) longitudinal section of archegonial head
egg cell
paraphyses
archegonium

(e) archegonium after fertilization
sperm
neck
zygote

FERTILIZATION

(f) developing sporophyte
calyptra
young sporophyte
tip of female gametophyte

(g) gametophyte with attached sporophyte
mature sporophyte
female gametophyte

MEIOSIS

(h) sporangium
hood (removed from sporangium)
cap
peristomal teeth

(i) spores

(j) spore germination
protonemata
germinating spore

(k) old protonemata with bud
bud

(l) young gametophyte
rhizoid

Modified from Carolina Biological Supply Company diagrams.

MATERIALS

Per student group:

- culture plate with moss agar
- water suspension containing moss spores
- china or other marker
- Pi-pump (1 mL or 10 mL)
- 1-mL pipet
- 8-cm strip of parafilm

- compound microscope
- glass microscope slide
- coverslip

Per lab room:

- four incubation chambers (one each for white, blue, green, and red light)

PROCEDURE

1. Obtain the following: one culture plate containing moss agar, china or other marker, a Pi-pump, a 1-mL pipet, and an 8-cm strip of parafilm.
2. Prepare a wet mount of the moss spore suspension provided by your instructor. Examine the spores with all magnifications available with your compound microscope.
3. Draw the spores in Figure 21-16, noting color.
4. Examine Figure 21-17, which shows you how to open the culture plate.
5. Pipet 1 mL of the spore suspension onto the culture plate.
6. Seal the culture plate with the strip of parafilm to prevent drying: remove the paper backing, place one end of the parafilm on the edge of the culture plate, hold the edge down with index finger pressure of the hand that is holding the culture plate, and stretch the parafilm around the circumference of the plate with a rolling motion.
7. With the marker, write your name, the date, and the light regime you will be using on the cover of the culture plate. Place the culture in the incubator with the proper light regime. Your instructor will designate the light regime (white, blue, green, or red) under which you should incubate your culture. This is the independent variable.
8. Formulate a prediction regarding this experiment. Write it in the proper space in Table 21-3.
9. Observe your cultures weekly: remove the lid (if necessary), place the culture on the stage of a dissecting microscope, and observe the spores. Record your observations in Table 21-3. Suggested dependent variables include percent germination, abundance of growth, color of protonemal cells, and formation of buds.
10. After 8 weeks, compare your results with other students using the same and different light regimes.
11. Summarize your observations and make conclusions in Table 21-4.

Was your hypothesis supported by the experimental results? _____

Figure 21-16 Drawing of ungerminated moss spores (_____×).

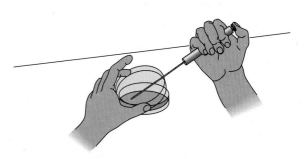

Figure 21-17 Method for inoculating petri plate culture so as to avoid contamination.

TABLE 21-3	Effect of Light Quality on Moss Spore Germination and the Growth of Protonemata

Prediction:

Week	Observations of Dependent Variables (list them here)
1	
2	
3	
4	
5	
6	
7	
8	

© Cengage Learning 2013

TABLE 21-4	Summary of Results on the Effect of Light Quality on the Germination of Moss Spores and Growth of Protonemata

Color of Light	Results
White	
Blue	
Green	
Red	

Conclusions:

© Cengage Learning 2013

_____ 1. Land plants are believed to have evolved from
   (a) mosses.
   (b) ferns.
   (c) charophytes.
   (d) fungi.

_____ 2. In the bryophytes, the sporophyte is
   (a) the dominant generation.
   (b) dependent on the gametophyte generation.
   (c) able to produce all of its own nutritional requirements.
   (d) both a and c

_____ 3. Liverworts and mosses utilize which of the following pigments for photosynthesis?
   (a) chlorophylls _a_ and _b_
   (b) carotenes
   (c) xanthophylls
   (d) all of the above

_____ 4. Which statement _best_ describes the concept of alternation of generations?
   (a) One generation of plants is skipped every other year.
   (b) There are two phases, a sporophyte and a gametophyte.
   (c) The parental generation alternates with a juvenile generation.
   (d) A green sporophyte phase produces food for a nongreen gametophyte.

_____ 5. Alternation of heteromorphic generations
   (a) is found only in the bryophytes.
   (b) is common to all land plants.
   (c) is typical of most green algae.
   (d) occurs in the liverworts but not mosses.

_____ 6. An organ that is hygroscopic
   (a) is sensitive to changes in moisture.
   (b) is exemplified by the peristomal teeth in the sporophyte of mosses.
   (c) may aid in spore dispersal in mosses.
   (d) is all of the above.

_____ 7. Gemmae function for
   (a) sexual reproduction.
   (b) water retention.
   (c) anchorage of a liverwort thallus to the substrate.
   (d) asexual reproduction.

_____ 8. Sperm find their way to the archegonium
   (a) by swimming.
   (b) due to a chemical gradient diffusing from the archegonium.
   (c) as a result of sucrose being released during the breakdown of the neck canal cells of the archegonium.
   (d) by all of the above.

_____ 9. A protonema
   (a) is part of the sporophyte generation of a moss.
   (b) is the product of spore germination of a moss.
   (c) looks very much like a filamentous brown alga.
   (d) produces the sporophyte when a bud grows from it.

_____ 10. The suffix _-phore_ is derived from a Greek word meaning
   (a) branch.
   (b) moss.
   (c) liverwort.
   (d) male.

EXERCISE **21**

Bryophytes—Liverworts and Mosses

Post-Lab Questions

Introduction

1. Explain why water must be present for the bryophytes to complete the sexual portion of their life cycle.

2. Green algae are believed to be the ancestors of the bryophytes. Cite four distinct lines of evidence to support this belief.

   (a)

   (b)

   (c)

   (d)

3. Complete this diagram of a "generic" alternation of generations.

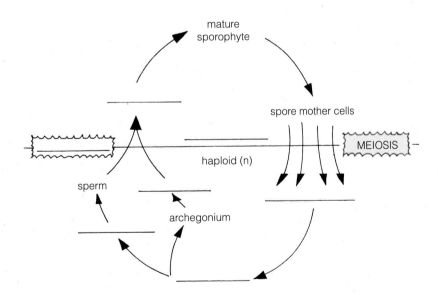

4. Describe in your own words the difference between a sporophyte and a gametophyte.

21.2 Phylum Hepatophyta: Liverworts

5. While the plant shown here is one that you didn't study specifically in lab, you should be able to identify it. Is this the gametophyte or the sporophyte of the plant?

(1×)

6. In what structure does meiosis occur in the bryophytes?

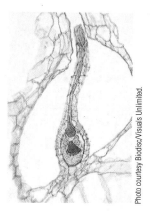

21.4 Phylum Bryophyta: Mosses

7. Identify the type of gametangium shown here.

(150×)

8. Walking along a stream in a damp forest, you see the plants shown at the right. Is this the sporophyte or the gametophyte?

(0.38×)

9. **(a)** What are the golden stalks seen in this photo?

(b) Are they the products of meiosis or fertilization?

(0.25×)

10. Identify the plants in the photo at the right as male or female, gametophyte or sporophyte, moss or liverwort.

(0.25×)

Seedless Vascular Plants: Club Mosses and Ferns

OBJECTIVES

After completing this exercise, you will be able to

1. define *tracheophyte, rhizome, mutualism, sporophyte, sporangium, gametophyte, gametangium, antheridium, archegonium, epiphyte, strobilus, node, internode, frond, sorus, annulus, hygroscopic, chemotaxis*;

2. recognize whisk ferns, club mosses, horsetails, and ferns when you see them, placing them in the proper taxonomic phylum;

3. identify the structures of the fern allies and ferns in **boldface**;

4. describe the life cycle of ferns;

5. explain the mechanism by which spore dispersal occurs from a fern sporangium;

6. describe the significant differences between the life cycles of the bryophytes, and the club mosses and ferns.

Introduction

If you did the exercise on liverworts and mosses, you learned about alternation of generations. (If you did not do this exercise, you now need to read the Introduction for Exercise 21, pages 279–280.) Indeed, all plants have alternation of generations. However, a major distinction exists between the bryophytes (liverworts and mosses) and the fern allies, ferns, gymnosperms, and flowering plants: whereas the dominant and conspicuous portion of the life cycle in the bryophytes is the **gametophyte** (the gamete-producing part of the life cycle), in all other plants that you will examine in this and subsequent exercises, it is the **sporophyte** (that portion of the life cycle producing spores).

A second, and perhaps more important, distinction also exists between bryophytes and the fern allies, ferns, gymnosperms, and flowering plants: the latter group contains *vascular tissue*. Vascular tissues include phloem, the tissue that conducts the products of photosynthesis, and *xylem*, the tissue that conducts water and minerals. As a result of the presence of xylem and phloem, fern allies, ferns, gymnosperms, and flowering plants are sometimes called **tracheophytes**.

Push gently on the region of your throat at the base of your larynx (voicebox or Adam's apple). If you move your fingers up and down, you should be able to feel the cartilage rings in your *trachea* (windpipe). To visualize its structure, imagine that your trachea is a pipe with donuts inside of it (Figure 22-1). This same arrangement exists in some cell types within the xylem of plants. Thus, early botanical microscopists called plants having such an arrangement *tracheophytes*.

In this exercise, you will study the tracheophytes that lack seeds. Two phyla of plants make up the seedless vascular plants:

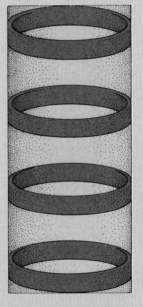

Phylum	Common Names
Lycophyta	Club mosses
Moniliophyta	Whisk ferns, horsetails

Figure 22-1 Three-dimensional representation of the cartilage in your trachea and some xylem cells of vascular plants.

We will study most closely the life cycle of the ferns, since they are common in our environment and have gametophytes and sporophytes that beautifully illustrate the concept of alternation of generations.

Lycopodium and *Selaginella* are the most commonly found genera in the phylum Lycophyta, the club mosses. The common name of this phylum comes from the presence of the so-called **strobilus** (plural, *strobili*), a region of the stem specialized for the production of spores. The strobilus looks like a very small club. Strobili are sometimes called cones.

MATERIALS

Per lab room:

- *Lycopodium*, living, preserved, or herbarium specimens with strobili
- *Lycopodium* gametophyte, preserved (optional)
- *Selaginella*, living, preserved, or herbarium specimens

- *Selaginella lepidophylla*, resurrection plant, two dried specimens
- culture bowl containing water

A. *Lycopodium*

The club moss *Lycopodium* is a small, forest-dwelling plant. These plants are often called ground pines or trailing evergreens. The scientific name comes from the appearance of the rhizome—it looked like a wolf's paw to early botanists. Lycos is Greek for "wolf," and *podium* means "foot."

strobilus ──

Figure 22-2 Sporophyte of *Lycopodium* (0.6×).

Photo by J. W. Perry.

PROCEDURE

1. Observe the living, preserved, or herbarium specimens of *Lycopodium* (Figure 22-2).
2. On your specimen, identify the true **roots**, **stems**, and **leaves**.
3. The adjective *true* is used to indicate that the organs contain vascular tissue—xylem and phloem.
4. Identify the **rhizome** to which the upright stems are connected. In the case of *Lycopodium*, the rhizome may be either beneath or on the soil surface, depending on the species. Notice that leaves cover the rhizome, as are the upright stems.
5. At the tip of an upright stem, find the **strobilus** (Figure 22-2). Look closely at the strobilus. As you see, it too is made up of leaves, but these leaves are much more tightly aggregated than the sterile (nonreproductive) leaves on the rest of the stem.

The leaves of the strobilus produce *spores* within a *sporangium*, the spore container. As was pointed out in the introduction, the plant you are looking at is a sporophyte.

Is the sporophyte haploid (n) or diploid (2n)? _____

Since the spores are haploid, what process must have taken place within the sporangium?

As maturation occurs, the stem regions between the sporangium-bearing leaves elongate slightly, the sporangium opens, and the spores are carried away from the parent plant by wind. If they land in a suitable habitat, the spores germinate to produce small and inconspicuous subterranean gametophytes, which bear sex organs—antheridia and archegonia. For fertilization and the development of a new sporophyte to take place, free water must be available, since the sperm are flagellated structures that must swim to the archegonium.

6. If available, examine a preserved *Lycopodium* gametophyte (Figure 22-3).

B. *Selaginella*

Selaginella is another genus in the phylum Lycophyta. These plants are usually called spike mosses. (This points out a problem with common names—the common name for the phylum is club mosses, but individual genera have their own common names.) Some species are grown ornamentally for use in terraria.

PROCEDURE

1. Examine living representatives of *Selaginella* (Figure 22-4).
2. Look closely at the tips of the branches and identify the rather inconspicuous strobili.
3. Examine the specimens of the resurrection plant (yet another common name!), a species of *Selaginella* sold as a novelty, often in grocery stores. A native of the southwestern United States, this plant grows in environments that are subjected to long periods without moisture. It becomes dormant during these periods.
4. Describe the color and appearance of the dried specimen.

_____

_____

5. Now place a dried specimen in a culture bowl containing water. Observe what happens in the next hour or so, and describe the change in appearance of the plant.

_____

_____

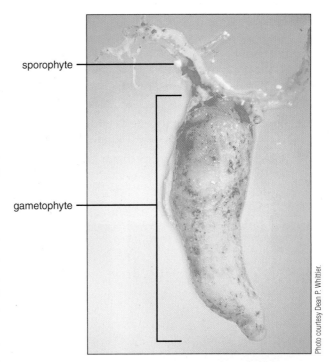

sporophyte

gametophyte

Photo courtesy Dean P. Whittier.

Figure 22-3 Gametophyte with young sporophyte of *Lycopodium* (2×).

(a)

strobili

Photos by J. W. Perry.

(b)

Figure 22-4 (a) Sporophyte of *Selaginella* (0.4×). (b) Higher magnification showing strobili at branch tips (0.9×).

22.2	Phylum Moniliophyta, Subphylum Psilophyta: Whisk Ferns (About 15 min.)

The whisk ferns consist of only two genera of plants, *Psilotum* (the *P* is silent) and *Tmesipteris* (the T is silent). Neither has any economic importance, but they have intrigued botanists for a long time, especially because *Psilotum* resembles the first vascular plants that colonized the Earth. Current evidence indicates that the psilophytes are not directly related to those very early land plants, but instead are ferns. We'll examine only *Psilotum*.

MATERIALS

Per lab room:

- *Psilotum*, living plant
- *Psilotum*, herbarium specimen showing sporangia and rhizome
- *Psilotum* gametophyte
- dissecting microscope

PROCEDURE

1. Observe first the **sporophyte** in a potted *Psilotum* (Figure 22-5). Within the natural landscapes of the United States, this plant grows in parts of Florida and Hawaii.
2. If the pot contains a number of stems, you can see how it got its common name, the whisk fern, because it looks a bit like a whisk broom. Closely examine a single aerial stem.

What color is it? _____

3. From this observation, make a conclusion regarding one function of this stem.

_____

4. Observe the herbarium specimen (mounted plant) of *Psilotum*. Identify the nongreen underground stem, called a **rhizome**.

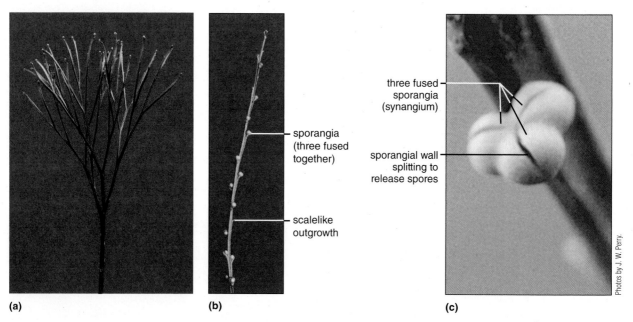

(a)　　　　　　　　　　**(b)**　　　　　　　　　　**(c)**

Figure 22-5 Sporophytes of *Psilotum*. (**a**) Single stem without sporangia. (**b**) Portion of stem with sporangia (0.3×). (**c**) Sporangia (10×).

5. *Psilotum* is unique among vascular plants in that it lacks roots. Absorption of water and minerals takes place through small rhizoids attached to the rhizome. Additionally, a fungus surrounds and penetrates into the outer cell layers of the rhizome. The fungus absorbs water and minerals from the soil and transfers them to the rhizome. This beneficial association is similar to the mycorrhizal association between plant roots and fungi described in the fungi exercise (page 366). A beneficial relationship like that between the fungus and *Psilotum* is called a **mutualistic symbiosis** or, more simply, **mutualism**.
6. On either the herbarium specimen or the living plant, identify the tiny scalelike outgrowths that are found on the aerial stems. Because these lack vascular tissue, they are not considered true leaves. Their size precludes any major role in photosynthesis.
7. The plant you are observing is a sporophyte. Thus, it must produce spores. Find the three-lobed structures on the stem (Figure 22-5b). Each lobe is a single **sporangium** containing spores.
8. If these spores germinate after being shed from the sporangium, they produce a small and infrequently found gametophyte that grows beneath the soil surface. The gametophyte survives beneath the soil thanks to a symbiotic relationship similar to that of the sporophyte's rhizome.
9. If a gametophyte is available, observe it with the aid of a dissecting microscope (Figure 22-16). Note the numerous **rhizoids**. If you look very carefully, you may be able to distinguish **gametangia** (sex organs; singular, *gametangium*).
10. The male sex organs are **antheridia** (Figure 22-6b; singular, *antheridium*); the female sex organs are **archegonia** (singular, *archegonium*). Both antheridia and archegonia are on the same gametophyte.
11. Fertilization of an egg within an archegonium results in the production of a new sporophyte.

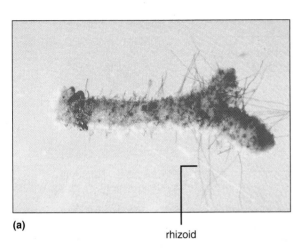

(a)

rhizoid

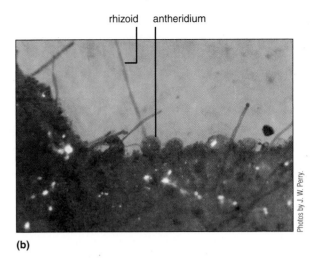

rhizoid   antheridium

(b)

Photos by J. W. Perry.

Figure 22-6 (a) *Psilotum* gametophyte (7.5×). (b) Higher magnification showing antheridia (23×).

22.3 **Phylum Moniliophyta, Subphylum Sphenophyta: Horsetails**
(About 15 min.)

During the age of the dinosaurs, tree-sized representatives of this phylum flourished. But like the dinosaurs, they too became extinct. Only one genus remains. However, the remnants of these plants, along with others that were growing at that time, remain in the form of massive coal deposits. *Speno-*, a Greek prefix meaning *"wedge,"* gives the phylum its name, presumably because the leaves of these plants are wedge shaped.

A single genus, *Equisetum*, is the only living representative of this subphylum. Different species of *Equisetum* are common throughout North America. Many are highly branched, giving the appearance of a horse's tail, and hence the common name. (Equus is Latin for "horse"; *saeta* means "bristle.") Horsetails are very unlike what we often think of as ferns, but molecular evidence confirms their relationship to more typical ferns.

MATERIALS

Per lab room:

- *Equisetum*, living, preserved, or herbarium specimens with strobili
- *Equisetum* gametophytes, living or preserved

PROCEDURE

1. Examine the available specimens of *Equisetum* (Figure 22-7). Depending on the species, it will be more or less branched.
2. Note that the plant is divided into **nodes** (places on the stem where the leaves arise) and **internodes** (regions on the stem between nodes). Identify the small leaves at the node.

If the specimen you are examining is a highly branched species, don't confuse the branches with leaves. The leaves are small, scale-like structures, often brown. (They *do* have vascular tissue, so they are true leaves.) Identify the leaves.

3. Look at the herbarium mount and identify the underground **rhizome** that bears **roots**.
4. Examine both the aerial stem and rhizome closely.

strobilius

leaves
at node

branches

Photo by J. W. Perry.

Figure 22-7 Sporophyte of the horsetail, *Equisetum* (0.2×).

Do both have nodes? _____

Do both have leaves? _____

Which portion of *Equisetum* is primarily concerned with photosynthesis? _____

5. Find the **strobilus** (Figure 22-7). Where on the plant is it located? _____

Based on the knowledge you've acquired in your study of *Lycopodium*, what would you expect to find within the strobilus? _____

When spores fall to the ground, what would you expect them to grow into after germination? _____

6. Now observe the horsetail gametophytes (Figure 22-8).

What color are they? _____

Would you expect them to be found on or below the soil surface? Why?

_____

_____

Figure 22-8 Horsetail gametophyte (5×).

22.4   Phylum Moniliophyta, Subphylum Pterophyta: Ferns *(About 60 min.)*

The variation in form of ferns is enormous, as is their distribution and ecology. We typically think of ferns as inhabitants of moist environments, but some species are also found in very dry locations. Tropical species are often grown as houseplants.

This subphylum gets its name from the Greek word pteri, meaning "fern."

MATERIALS

Per student:

- fern sporophytes, fresh, preserved, or herbarium specimens
- prepared slide of fern rhizome, c.s.
- fern gametophytes, living, preserved, or whole mount prepared slides
- fern gametophyte with young sporophyte, living or preserved
- microscope slide

- compound microscope
- dissecting microscope

Per lab room:

- squares of fern sori, in moist chamber (*Polypodium aureum* recommended)
- demonstration slide of fern archegonium, median l.s.
- other fern sporophytes, as available

PROCEDURE

1. Obtain a fresh or herbarium specimen of a typical fern **sporophyte**. As you examine the structures described in this experiment, refer to Figure 22-16, a diagram of the life cycle of a fern.

The sporophyte of many ferns (Figures 22-9, 22-16a) consists of true roots, stems, and leaves; that is, they contain vascular tissue.

2. Identify the horizontal stem, the **rhizome** (which produces true **roots**), and upright leaves. The leaves of ferns are called **fronds** and are often highly divided.

Ferns, unlike bryophytes, are vascular plants, their sporophytes containing xylem and phloem. Let's examine this vascular tissue.

3. Obtain a prepared slide of a cross section of a fern stem (rhizome). Study it using the low-power objective of your compound microscope. Use Figure 22-10 as a reference.
   (a) Find the epidermis, cortex, and vascular tissue.
   (b) Within the vascular tissue, distinguish between the **phloem** (of which there are outer and inner layers) and the thick-walled **xylem** sandwiched between the phloem layers.
4. Now examine the undersurface of the frond. Locate the dotlike **sori** (singular, *sorus*; Figure 22-9).
5. Each sorus is a cluster of sporangia. Using a dissecting microscope, study an individual sorus. Identify the sporangia (Figures 22-11, 22-16b).

Each sporangium contains spores (Figure 22-16c). Although the sporangium is part of the diploid (sporophytic) generation, spores are the first cells of the haploid (gametophytic) generation.

What process occurred within the sporangium to produce the haploid spores?

_____

Figure 22-9 (**a**) Morphology of a typical fern (0.25×). (**b**) Enlargement showing the sori (0.5×).

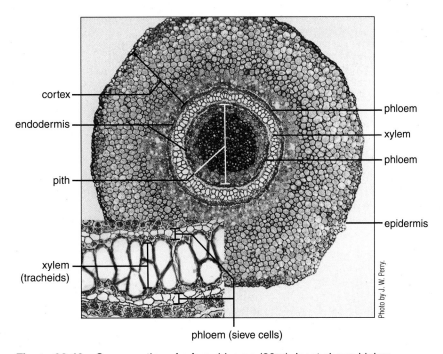

Figure 22-10 Cross section of a fern rhizome (29×). Inset shows higher magnification of vascular tissue (80×).

6. Obtain a single sorus-containing square of the hare's foot fern (*Polypodium aureum*).
    (a) Place it sorus-side up on a glass slide (DON'T ADD A COVERSLIP), and examine it with the low-power objective of your compound microscope.
    (b) Note the row of brown, thick-walled cells running over the top of the sporangium, the **annulus** (Figures 22-11, 22-16b).

The annulus is **hygroscopic**, meaning changes in moisture content within the cells of the annulus cause them to change shape, causing the sporangium to crack open. Watch what happens as the sporangium dries out.

As the water evaporates from the cells of the annulus, a tension develops that pulls the sporangium

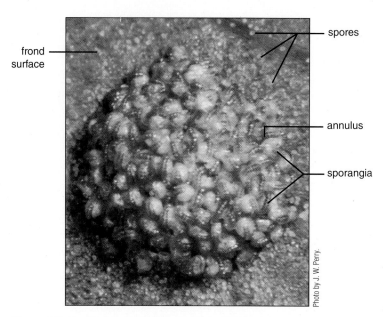

Figure 22-11 Sorus containing sporangia (30×).

apart. Separation of the halves of the sporangium begins at the thin-walled *lip cells*. As the water continues to evaporate, the annulus pulls back the top half of the sporangium, exposing the spores. The sporangium continues to be pulled back until the tension on the water molecules within the annulus exceeds the strength of the hydrogen bonds holding the water molecules together. When this happens, the top half of the sporangium flies forward, throwing the spores out. The fern sporangium is a biological catapult!

If spores land in a suitable environment, one that is generally moist and shaded, they germinate (Figures 22-16d and e), eventually growing into the heart-shaped adult gametophyte (22-16f).

7. Using your dissecting microscope, examine a living, preserved, or prepared slide whole mount of the gametophyte (Figures 22-12, 22-16f).

What color is the gametophyte? _____

What does the color indicate relative to the ability of the gametophyte to produce its own carbohydrates?

_____

(a) Examine the undersurface of the gametophyte. Find the **rhizoids**, which anchor the gametophyte and perhaps absorb water.
(b) Locate the gametangia (sex organs) clustered among the rhizoids (Figure 22-16f). There are two types of gametangia: **antheridia** (Figure 22-16g), which produce the flagellated *sperm*, and **archegonia** (Figures 22-13, 22-16h), which produce *egg cells*.

8. Study the demonstration slide of an archegonium (Figure 22-14). Identify the **egg** within the swollen basal portion of the archegonium. Note that the neck of the archegonium protrudes from the surface of the gametophyte.

The archegonia secrete chemicals that attract the flagellated sperm, which swim in a water film down a canal within the neck of the archegonium. One sperm fuses with the egg to produce the first cell of the sporophyte generation, the *zygote* (Figure 22-16i). With subsequent cell divisions, the zygote develops into an embryo (embryo sporophyte). As the embryo grows, it pushes out of the gametophyte and develops into a young sporophyte (Figure 22-16j).

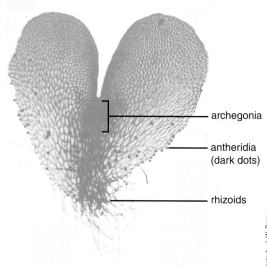

Figure 22-12 Whole mount of fern gametophyte, undersurface (10×).

9. Obtain a specimen of a young sporophyte that is attached to the gametophyte (Figures 22-15, 22-16j). Identify the **gametophyte** and then the *primary leaf* and *primary root* of the young sporophyte. As the sporophyte continues to develop, the gametophyte withers away.

10. Examine any other specimens of ferns that may be on demonstration, noting the incredible diversity in form. Look for sori on each specimen.

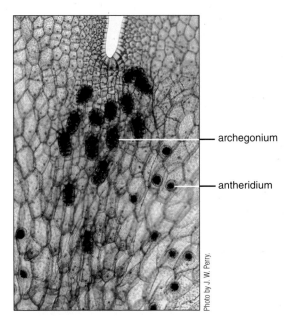

Photo by J. W. Perry.

— archegonium

— antheridium

Figure 22-13 Gametangia (antheridia and archegonia) on undersurface of fern gametophyte (57×).

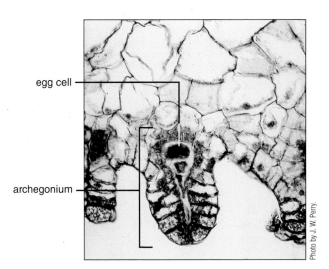

egg cell —

archegonium —

Photo by J. W. Perry.

Figure 22-14 Longitudinal section of a fern archegonium (218×).

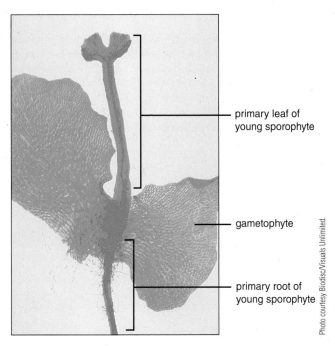

— primary leaf of young sporophyte

— gametophyte

— primary root of young sporophyte

Photo courtesy Biodisc/Visuals Unlimited.

Figure 22-15 Fern gametophyte with attached sporophyte (12×).

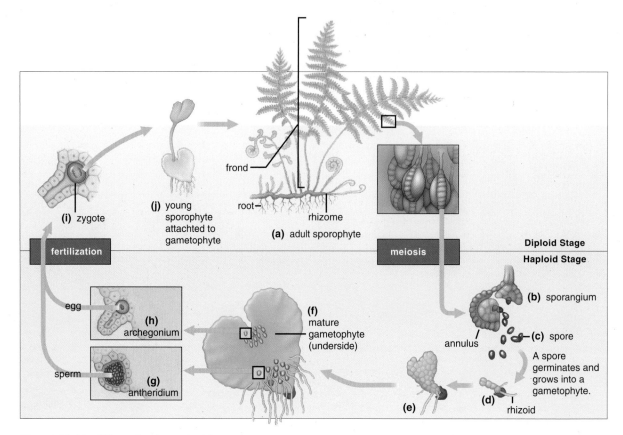

Figure 22-16 Life cycle of a typical fern.

22.5   Experiment: Fern Sperm Chemotaxis *(About 60 min.)*

Chemotaxis is a movement in response to a chemical gradient. This is an important biological process as it allows organisms to find their way or locate things. A dog following a scent is a common example of chemotaxis. How does a sperm cell shed from an antheridium find an archegonium-containing egg cell? The archegonium secretes a sperm-attracting chemical. The farther away the sperm from the archegonium, the more dilute the chemical. The sperm swim up the chemical concentration gradient, right to the egg cell.

In this experiment, you will determine whether certain chemical substances evoke a chemotactic response by fern sperm. This experiment addresses the hypothesis that *some chemicals will attract sperm more strongly and for greater duration than others.*

MATERIALS

Per experimental group (pair of students):

- sharpened toothpicks
- test solutions
- dissecting needle

- depression slide
- dissecting microscope
- petri dish culture containing *C-Fern*™ gametophytes

PROCEDURE

Work in pairs.

1. Obtain six wooden toothpicks whose tips have been sharpened to a fine point.

CAUTION

Do not touch the sharpened end with your fingers!

---

2. To identify the toothpicks, place one to five small dots along the side of the toothpicks near the unsharpened end so that each toothpick has a unique marking (one dot, two dots, and so on). One toothpick remains unmarked.
3. Dip the sharpened end of the toothpick with one dot into test solution 1. Insert it with the sharpened end up in the foam block toothpick holder.
4. Repeat step 3 with the four remaining toothpicks and four other test solutions. Leave the unmarked toothpick dry.

Which toothpick serves as the control? _____

5. Obtain a depression slide and pipet one drop of Sperm Release Buffer (SRB) into the depression.
6. Obtain a 12- to 18-day-old culture of *C-Fern*™ gametophytes. Remove the cover, place it on the stage of a dissecting microscope, and observe the gametophytes using transmitted light. (See Exercise 3, pages 28 and 35, if you need to learn how to choose transmitted light.)
7. Identify the two types of gametophytes present—the smaller thumb-shaped male gametophytes that have many bumps on them and the larger, heart-shaped bisexual gametophytes (Figure 22-17).
8. Using a dissecting needle, pick up 7–10 male gametophytes and place them into the SRB in the depression slide well. They don't need to be completely submerged.

CAUTION

If you wound or damage a gametophyte, discard it.

---

9. Place the petri dish cover on the stage of the dissecting microscope, edges up, and then place the gametophyte-containing depression slide on the cover. (This keeps the gametophytes cooler, particularly important if your dissecting microscope has an incandescent light.) Adjust the magnification so it is at least 12×.
10. Within 5 minutes, you should be able to observe sperm being released from the antheridia.
11. Wait another 5 minutes to ensure that a sufficient number of sperm have been released, and then begin testing for chemotactic response by following steps 12–15.
12. While using 1220× magnification, focus on the TOP surface of the sperm suspension droplet in an area devoid of male gametophytes.

CAUTION

As you make the tests, do not insert the toothpick fully into the drop—just touch the surface briefly (Figure 22-18).

---

13. Remove one test toothpick from the block and, while looking through the oculars of the dissecting microscope, gently and briefly touch the sharpened end of the toothpick to the surface of the drop (Figure 22-18).
14. Observe whether any chemotactic response takes place. Record your observation in Table 22-1.
15. Repeat the procedure with the remaining toothpicks, stirring the sperm suspension (if necessary) with a dissecting needle after each test to redistribute the sperm.
16. Place your data on the lab chalk or marker board. Your instructor will pool the data for all experimental groups.

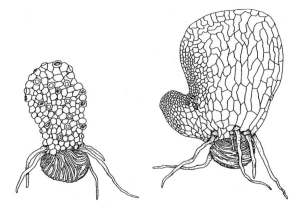

Figure 22-17   Mature male (left) and bisexual (right) *C-Fern*™ gametophytes.

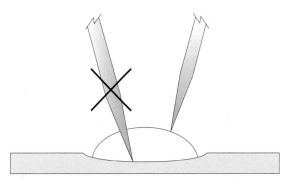

Figure 22-18   Method for applying test substances. Proper technique is important. Gently and briefly touch the end of the toothpick to the surface of the sperm suspension. *Note:* The size of the drop of suspension is exaggerated to show detail.

17. Make a conclusion about the chemotactic properties of the various chemical test substances, and note it in Table 22-1.

Examine the chemical structures of each test solution (Figure 22-19). Relate the biological (chemotactic) response you observed to any of the chemical structural differences of the test substances. _____

_____

_____

_____

What is the advantage to using the pooled data to draw your conclusion? _____

_____

_____

What do you hypothesize the naturally occurring chemical produced by mature archegonia to be? _____

_____

_____

_____

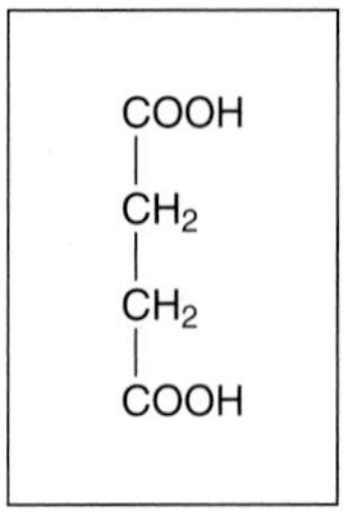

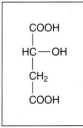

1 succinic acid 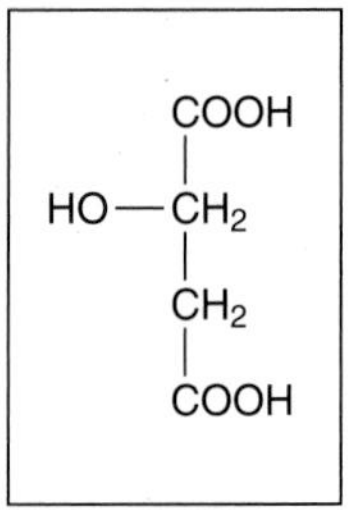   2 L-malic acid   3 D-malic acid

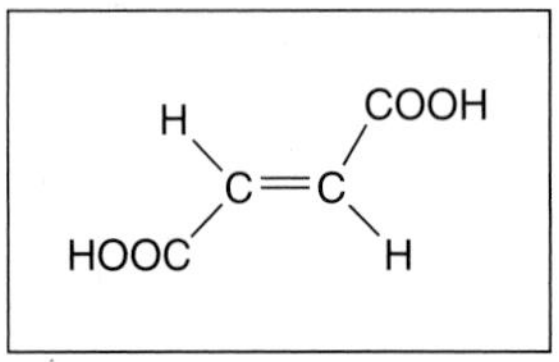

4 fumaric acid 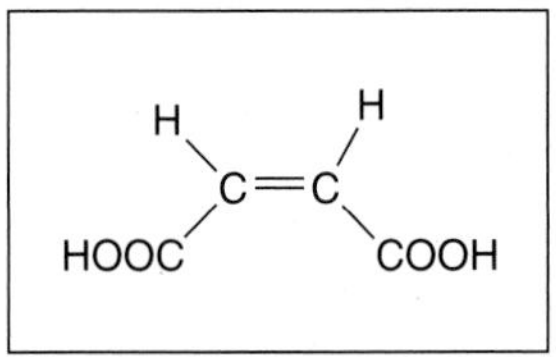   5 maleic acid

Figure 22-19 Chemical structures of test substance.

TABLE 22-1	My Data: Chemotactic Response of *C-Fern*™ Sperm				

Prediction:

	Swarming Response[b]				
	Intensity (low, medium, high)			Duration (short, medium, long)	
Test Substance[a]	My Data	Pooled Data		My Data	Pooled Data
0					
1					
2					
3					
4					
5					

Conclusion:

© Cengage Learning 2013

_____

[a]Identified by number of dots on toothpick.

[b]Quantify the response by using the symbols (0) for no response, (–) if the sperm move away from the test site, and (+, ++, +++) for varying degrees of attraction to the test site.

_____ 1. The sporophyte is the dominant and conspicuous generation in
(a) the fern allies and ferns.
(b) gymnosperms.
(c) flowering plants.
(d) all of the above

_____ 2. A tracheophyte is a plant that has
(a) xylem and phloem.
(b) a windpipe.
(c) a trachea.
(d) the gametophyte as its dominant generation.

_____ 3. *Psilotum* lacks
(a) roots.
(b) a mechanism to take up water and minerals.
(c) vascular tissue.
(d) alternation of generations.

_____ 4. Spore germination followed by cell divisions results in the production of
(a) a sporophyte.
(b) an antheridium.
(c) a zygote.
(d) a gametophyte.

_____ 5. Which phrase *best* describes a plant that is an epiphyte?
(a) a plant that grows upon another plant
(b) a parasite
(c) a plant with the gametophyte generation dominant and conspicuous
(d) a plant that has a mutually beneficial relationship with another plant

_____ 6. Club mosses
(a) are placed in the subphylum Pterophyta.
(b) are so-called because of the social nature of the plants.
(c) do not produce gametophytes.
(d) are so called because most produce strobili.

_____ 7. The resurrection plant
(a) is a species of *Selaginella*.
(b) grows in the desert Southwest of the United States.
(c) is a member of the phylum Lycophyta.
(d) all of the above

_____ 8. Which statement is **not** true?
(a) Nodes are present on horsetails.
(b) The rhizome on horsetails bears roots.
(c) The internode of a horsetail is the region where the leaves are attached.
(d) Horsetails are members of the subphylum Sphenophyta.

_____ 9. In ferns,
(a) xylem and phloem are present in the sporophyte.
(b) the sporophyte is the dominant generation.
(c) the leaf is called a frond.
(d) all of the above are true.

_____ 10. The spores of a fern are
(a) produced by mitosis within the sporangium.
(b) diploid cells.
(c) the first cells of the gametophyte generation.
(d) both a and b

EXERCISE **22**

Seedless Vascular Plants: Club Mosses and Ferns

Post-Lab Questions

Introduction

1. List two features that distinguish the seedless vascular plants from the bryophytes.
   (a)

   (b)

2. After consulting a biological or scientific dictionary, explain the derivation from the Greek of the word *symbiosis*.

22.1 Phylum Lycophyta: Club Mosses

3. Both *Lycopodium* and *Equisetum* have strobili, roots, and rhizomes. What did you learn in this exercise that enables you to distinguish these two plants?

4. Examine the photo at the right.
   (a) Give the genus of this plant.
   _____

   (b) Label the structure indicated.

5. Some species of *Lycopodium* produce gametophytes that grow beneath the surface of the soil, while others grow on the soil surface. Basing your answer on what you've learned from other plants in this exercise, predict how each respective type of *Lycopodium* gametophyte might obtain its nutritional needs.

(0.3×)

22.2 Phylum Moniliophyta, Subphylum Psilophyta: Whisk Ferns

6. Using a biological or scientific dictionary, or a reference in your textbook, determine the meaning of the root word *psilo*, and relate it to the appearance of *Psilotum*.

22.3 Phylum Moniliophyta, Subphylum Sphenophyta: Horsetails

7. Examine the photo on the right. You studied this genus in lab, but this is a different species. This species has two separate stems produced at different times in the growing season. The stem on the left is a reproductive branch, while that on the right is strictly vegetative. Based on the characteristics obvious in this photo, identify the plant, give its scientific name, and then identify the labeled structures.

   Scientific name _____

   A. _____

   B. _____

   C. _____

8. Explain the distinction between a node and an internode.

(0.75×)

22.4 Phylum Moniliophyta, Subphylum Pterophyta: Ferns

9. While rock climbing, you encounter the plant shown in the photo on the right growing out of a crevice.

   (a) Is this the sporophyte or the gametophyte?

   (b) What special name is given to the leaf of this plant?

(0.5×)

10. The environments in which ferns grow range from standing water to very dry areas. Nonetheless, all ferns are dependent upon free water in order to complete their life cycles. Explain why this is the case.

Seed Plants I: Gymnosperms

After completing this exercise, you will be able to

1. define *gymnosperm, pulp, heterosporous, homosporous, pollination, fertilization, dioecious, monoecious;*

2. describe the characteristics that distinguish seed plants from other vascular plants;

3. produce a cycle diagram of heterosporous alternation of generations;

4. list human uses for conifers;

5. recognize the structures in **boldface** and describe the life cycle of a pine;

6. distinguish between a male and a female pine cone;

7. describe the method by which pollination occurs in pines;

8. describe the process of fertilization in pines;

9. recognize members of the four phyla: Coniferophyta, Cycadophyta, Ginkgophyta, and Gnetophyta.

Introduction

The development of the seed was a significant event in the evolution of vascular plants. Seeds have remarkable ability to survive under adverse conditions, a reason for the dominance of seed plants in today's flora.

Let's examine the characteristics of seeds and seed plants.

1. All seed plants produce **pollen grains**. Pollen grains are carriers for sperm, delivering the sperm to the egg-bearing female gametophyte. As a benefit of pollen production, the sperm of seed plants do *not* need free water to swim to the egg. Thus, seed plants can reproduce in harsh climates where nonseed plants are much less successful.

2. Seeds have some type of **stored food** for the embryo to use as it emerges from the seed during germination. (The sole exception is orchid seeds, which rely on symbiotic association with a fungus to obtain nutrients.)

3. All seeds have a **seed coat**, a protective covering enclosing the embryo and its stored food.

A seed coat and stored food are particularly important for survival. An embryo within a seed is protected from an inhospitable environment. Consider a seed produced during a severe drought. Water is necessary for the embryo to grow. If none is available, the seed can remain dormant until growing conditions are favorable. When germination occurs, a ready food source is present to get things underway, providing nutrients until the developing plant can produce its own carbohydrates by photosynthesis.

4. Like the bryophytes, fern allies, and ferns, seed plants exhibit **alternation of generations**.

5. All seed plants are **heterosporous**; that is, they produce *two* types of spores. Bryophytes and most fern allies and ferns are **homosporous**, producing only *one* spore type.

Examine Figure 23-1, a diagram of the heterosporous alternation of generations.

Gymnosperms are one of two groups of seed plants. *Gymnosperm* translates literally as "naked seed," referring to the production of seeds on the *surface* of reproductive structures. This contrasts with the situation in the angiosperms (see Exercise 24), whose seeds are contained within a fruit.

The general assemblage of plants known as gymnosperms contains plants in four separate phyla:

Phylum	Common Names
Coniferophyta	Conifers
Cycadophyta	Cycads
Ginkgophyta	Ginkgos
Gnetophyta	Gnetophytes or vessel-containing gymnosperms

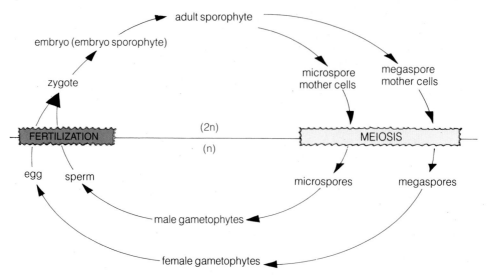

Figure 23-1 Heterosporous alternation of generations.

23.1 Phylum Coniferophyta: Conifers *(About 90 min.)*

By far the most commonly recognized gymnosperms are the conifers. Among the conifers, perhaps the most common is the pine (*Pinus*). Many people think of conifers and pines as one and the same. However, while all members of the genus *Pinus* are conifers, not all conifers are pines. Examples of other conifers include spruce (*Picea*), fir (*Abies*), and even some trees that lose their leaves in the fall such as larch (*Larix*).

The conifers are very important economically, because their wood is used in building construction. Millions of hectares (1 hectare = 2.47 acres) are devoted to growing conifers for this purpose, not to mention the numerous plantations that grow Christmas trees. In many areas conifers are used for **pulp**, the moistened cell wall–derived cellulose used to manufacture paper.

The structures and events associated with reproduction in pine (*Pinus*) will be studied here as a representative conifer.

MATERIALS

Per student:

- cluster of male cones
- prepared slide of male strobilus, l.s., with microspores
- young female cone
- prepared slide of female strobilus, l.s., with megaspore mother cell
- prepared slide of pine seed, l.s.
- compound microscope
- dissecting microscope
- single-edged razor blade

Per student group (table):

- young sporophyte, living or herbarium specimen

Per lab room:

- demonstration slide of female strobilus with archegonium
- demonstration slide of fertilization
- pine seeds, soaking in water
- pine seedlings, 12 weeks old
- pine seedlings, 36 weeks old

PROCEDURE

As you do this activity, examine Figure 23-12, representing the life cycle of a pine tree. To refresh your memory, look at a specimen of a small pine tree. This is the adult **sporophyte** (Figure 23-12a). Identify the stem and leaves. You probably know the main stem of a woody plant as the trunk. The leaves of conifers are often called needles because most are shaped like needles.

A. Male Reproductive Structures and Events

1. Obtain a cluster of **male cones** (Figures 23-2, 23-12b). The function of male cones is to produce **pollen**; consequently, male cones are typically produced at the ends of branches, where the wind currents can catch the pollen as it is being shed.

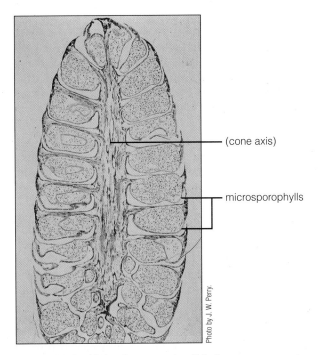

Figure 23-2 Cluster of male
cones shedding pollen (0.75×).

Figure 23-3 Male pine cone, l.s. (12×).

2. Note all the tiny scalelike structures that make up the male cones. These are *microsporophylls*.

Translated literally, a sporophyll would be a "spore-bearing leaf." The prefix *micro-* refers to "small." But the literal interpretation "small, spore-bearing leaf" does not convey the full definition; biologists use *microsporophyll* to mean a leaf that produces *male* spores, called *microspores*. These develop into winged, immature male *gametophytes* called **pollen grains**. (Why they are immature is a logical question. The male gametophyte is not mature until it produces sperm.)

3. Remove a single microsporophyll and examine its lower surface with a dissecting microscope. Identify the two **microsporangia**, also called **pollen sacs** (Figure 23-12c).

4. Study a prepared slide of a longitudinal section of a male cone (also called a male *strobilus*), first with a dissecting microscope to gain an impression of the cone's overall organization and then with the low-power objective of your compound microscope. Identify the *cone axis* bearing numerous **microsporophylls** (Figures 23-3).

5. Switch to the medium-power objective to observe a single microsporophyll more closely. Note that it contains a cavity; this is the **microsporangium** (also called a *pollen sac*), which contains numerous *pollen grains.*

As the male cone grows, several events occur in the microsporophylls that lead to the production of pollen grains. Microspore mother cells within the microsporangia undergo meiosis to form *microspores*. Cell division within the microspore wall and subsequent differentiation result in the formation of the pollen grain.

6. Examine a single **pollen grain** with the high-dry objective (Figures 23-4, 23-12d). Identify the earlike *wings* on either side of the body.

7. Identify the four cells that make up the body: the two most obvious are the **tube cell** and smaller **generative cell**. (The nucleus of the tube cell is almost as large as the entire generative cell.) Note that this male gametophyte is made up of only four cells!

In conifers, **pollination**, the transfer of pollen from the male cone to the female cone, is accomplished by wind and occurs during spring. Pollen grains are caught in a sticky *pollination droplet* produced by the female cone.

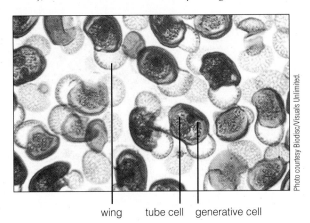

wing   tube cell   generative cell

Figure 23-4 Pine pollen grains (400×).

B. Female Reproductive Structures and Events

Development and maturation of the female cone take 2–3 years; the exact time depends on the species. Female cones are typically produced on higher branches of the tree. Because the individual tree's pollen is generally shed downward, this arrangement favors crossing between *different* individuals.

1. Obtain a young **female cone** (Figures 23-5, 23-12e), and note the arrangement of the cone scales. Unlike the male cone, the female cone is a complex structure, each scale consisting of an **ovuliferous** scale fused atop a *sterile bract*.
2. Remove a single scale-bract complex (Figure 23-12f). On the top surface of the complex find the two **ovules**, the structures that eventually will develop into the **seeds**.
3. Examine a prepared slide of a longitudinal section of a female cone (Figures 23-6, 23-12g) first with the dissecting microscope. Distinguish the smaller *sterile bract* from the *ovuliferous scale*.

Figure 23-5 Female cones of pine (1×).

4. Now examine the slide with the low-power objective of your compound microscope. Look for a section through an ovuliferous scale containing a very large cell; this is the **megaspore mother cell** (Figures 23-7, 23-12g). The tissue surrounding the megaspore mother cell is the **megasporangium**. Protruding inward toward the cone axis are "flaps" of tissue surrounding the megasporangium, the **integument**. Find the integuments and the opening between them, the **micropyle**.

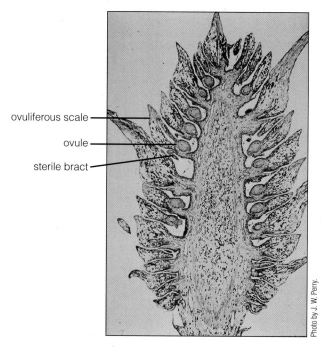

ovuliferous scale

ovule

sterile bract

Figure 23-6 Female pine cone, l.s. (5×).

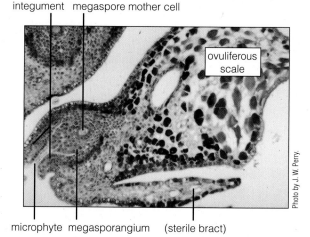

integument   megaspore mother cell

ovuliferous scale

microphyte   megasporangium   (sterile bract)

Figure 23-7 Portion of female cone showing megaspore mother cell (75×).

Think for a moment about the three-dimensional nature of the ovule: it's much like a short vase lying on its side on the ovuliferous scale. The neck of the vase is the integument, the opening of the micropyle. The integument extends around the base of the vase. If you poured liquid rubber inside the base of a vase, suspended a marble in the middle, and allowed the rubber to harden, you'd have a model of the megasporangium and the megaspore mother cell. Figure 23-8 gives you an idea of the three-dimensional structure.

The megaspore mother cell undergoes meiosis to produce four haploid *megaspores* (Figure 23-12h), but only one survives; the other three degenerate. The functional megaspore repeatedly divides mitotically to produce the multicellular **female gametophyte** (Figure 23-12i). At the same time, the female cone is continually increasing in size to accommodate the developing female gametophytes. (Remember, there are numerous ovuliferous scale/sterile bract complexes on each cone.)

The female gametophyte of pine is produced _____ (within *or* outside of) the megasporangium.

5. Examine the demonstration slide of **archegonia** (Figures 23-9, 23-12i) that have developed within the female gametophyte. Identify the single large **egg cell** that fills the entire archegonium. (The nucleus of the egg cell may be visible as well. The other generally spherical structures are protein bodies within the egg.)

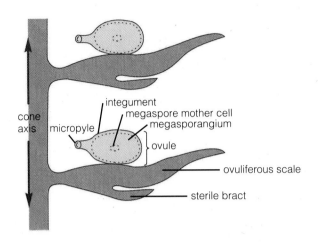

Figure 23-8 Ovule and ovuliferous scale/sterile bract complex.

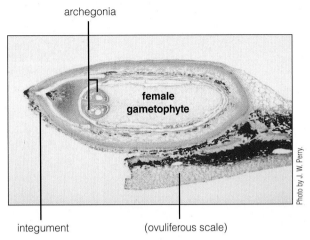

Figure 23-9 Pine ovule with female gametophyte and archegonia, l.s. (12×).

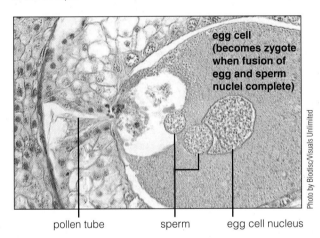

Figure 23-10 Fertilization in pine, l.s. (175×).

Recall that the pollen grains produced within male cones are caught in a sticky pollination droplet produced by the female cone. As the pollination droplet dries, the pollen grain is drawn through the micropyle and into a cavity called the *pollen chamber*.

Fertilization—the fusion of egg and sperm—occurs after the *pollen tube*, an outgrowth of the pollen grain's tube cell, penetrates the megasporangium and enters the archegonium. The generative cell of the pollen grain has divided to produce two sperm, one of which fuses with the egg. (The second sperm nucleus degenerates.)

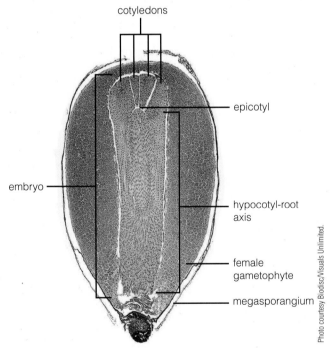

Figure 23-11 Pine seed, l.s. (24×).

6. Examine the demonstration slide of fertilization in *Pinus*. Identify the **zygote**, the product of fusion of egg and sperm (Figures 23-10, 23-12j).

After fertilization, numerous mitotic divisions of the zygote take place producing an **embryo** (*embryo sporophyte*). Fertilization also triggers changes in the integument, causing it to harden and become the seed coat.

7. With the low-power objective of your compound microscope, study a prepared slide of a longitudinal section through a pine seed (Figures 23-11, 23-12k). Starting from the outside, identify the **seed coat**, **megasporangium** (a very thin, papery remnant), the female gametophyte, and **embryo** (embryo sporophyte).

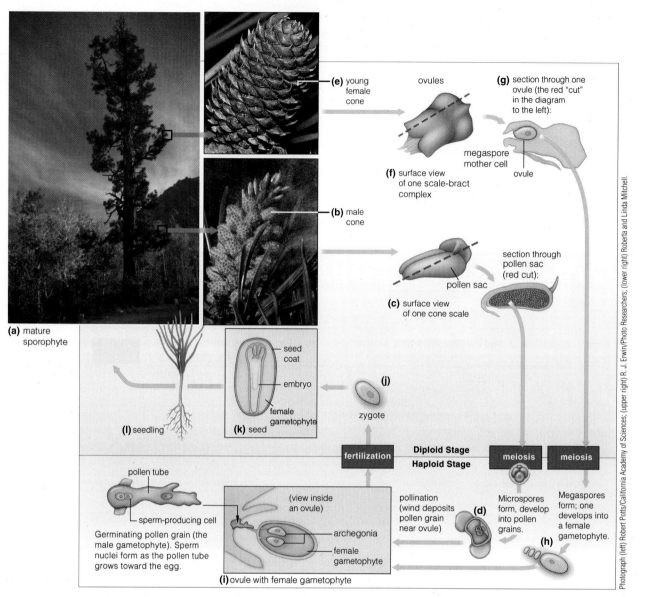

Figure 23-12 Pine life cycle.

8. Within the embryo, identify the **hypocotyl-root axis** and numerous **cotyledons**, in the center of which is the **epicotyl**. (*Hypo-* and *epi-* are derived from Greek, meaning "under" and "over," respectively. Thus, these terms refer to orientation with reference to the cotyledons.) The female gametophyte will serve as a food source for the embryo sporophyte when germination takes place.

9. Obtain a pine seed that has been soaked in water to soften the seed coat. Remove it and make a freehand longitudinal section with a sharp razor blade.

10. Identify the papery remnant of the **megasporangium**, the white **female gametophyte**, and **embryo** (embryo sporophyte).

How many cotyledons are present? _____

11. Examine the culture of pine seeds that were planted in sand 12 weeks ago. Note the germinating seeds (Figure 23-12l).

12. Identify the **hypocotyl-root axis, cotyledons, female gametophyte,** and **seed coat.**

The cotyledons serve two functions. One is to absorb the nutrients stored in the female gametophyte during germination. As the cotyledons are exposed to light, they turn green. What then is the second function of the cotyledons?

_____

_____

_____

13. Finally, examine the 36-week-old sporophyte seedlings. Notice that the cotyledons eventually wither away as the epicotyl produces new leaves.

Phylum Cycadophyta: Cycads *(About 10 min.)*

During the age of the dinosaurs (240 million years ago), cycads were extremely abundant in the flora. In fact, botanists often call this "the age of the cycads."

   Zamia and *Cycas* are two cycads most readily available in North America. In nature, these two species are limited to the subtropical regions.

   They're often planted as ornamentals in Florida, Gulf Coast states, and California.

   All cycads have separate male and female plants, unlike pine.

   The female cones of some other genera become extremely large, weighing as much as 30 kg!

MATERIALS

Per lab room:

■ demonstration specimens of *Zamia* and/or *Cycas*

PROCEDURE

1. Examine the demonstration specimen of *Zamia* and/or *Cycas* (Figure 23-13). Both have the common name *cycad*. Do these plants resemble any of the conifers you know? _____

   _____

2. Notice the leaves of the cycads. They more closely resemble the leaves of the ferns than those of the conifers.

(a)

(b)

Figure 23-13 Two cycad species. (**a**) *Cycas* (0.01×). (**b**) *Zamia* (0.1×).

Phylum Ginkgophyta: *Ginkgo (About 10 min.)*

A single species of this phylum is all that remains of what was once a much more diverse assemblage of plants. *Ginkgo biloba* is sometimes called a living fossil because it has changed little in the last 80 million years. In fact, at one time it was believed to be extinct; the Western world knew it from the fossil record before living trees were discovered in Buddhist temple courtyards of remote China. Today *Ginkgo* is a highly prized ornamental tree that is commonly planted in our urban areas. The tree has a reputation for being resistant to most insect pests and atmospheric pollution. The seeds are ground up and sold as a memory-enhancing dietary supplement, although the supplement's ability to do that is in dispute.

MATERIALS

Per lab room:

■ demonstration specimen of *Ginkgo* (living plant or herbarium specimen)

PROCEDURE

1. Examine the demonstration specimen of *Ginkgo biloba*, the maidenhair tree (Figure 23-14a)
2. Note the fan-shaped leaves (Figure 23-14b).

(a)　　　　　　　　　　　　　　　　　　　　　**(b)**

Figure 23-14　*Ginkgo.* **(a)** Tree. **(b)** Branch with leaves (0.5×).

23.4 Phylum Gnetophyta: Gnetophytes (Vessel-Containing Gymnosperms)
(About 10 min.)

The gnetophytes are a small assemblage of plants that have several characteristics found only in the flowering plants (angiosperms). Their reproductive structures look much more like flowers than cones. "Double fertilization" (pages 330 and 343), a feature thought to be unique to the flowering plants, has been reported in one gnetophyte species. And the water-conducting xylem tissue contains vessels in all species. Consequently, many scientists now believe these plants are very closely related to the flowering plants. Most species are found in desert or arid regions of the world. In the desert Southwest of the United States, *Ephedra* is a common shrub known as Mormon tea, because its stems were once harvested by Mormon settlers in Utah and used to make a tea.

MATERIALS

Per lab room:

■ demonstration specimen of *Ephedra* and/or *Gnetum* (living plant or herbarium specimen)

PROCEDURE

1. Examine the demonstration specimen of *Ephedra* (Figure 23-15).
2. A second representative gnetophyte is the genus *Gnetum* (pronounced "neat-um"). A few North American college and university greenhouses—and some tropical gardens—keep these specimens. *Gnetum* is native to Brazil, tropical west Africa, India, and Southeast Asia. Different species vary in form from vines to trees (Figures 23-16a, b). Examine the living specimen if one is on display, noting particularly the broad, flat leaves (Figure 23-16c).

(a)

(b)

Figure 23-15 *Ephedra.* (**a**) Several plants (1×). (**b**) Close-up of stems (0.5×).

(a)

(b)

(c)

Figure 23-16 *Gnetum.* (**a**) A species that is a vine (0.02×). (**b**) A species that is a tree (0.02×). (**c**) Leaves (0.38×).

_____ **1.** Which statement is *not* true about conifers?
  (a) Conifers are gymnosperms.
  (b) All conifers belong to the genus *Pinus.*
  (c) All conifers have naked seeds.
  (d) Conifers are heterosporous.

_____ **2.** Seed plants
  (a) have alternation of generations.
  (b) are heterosporous.
  (c) develop a seed coat.
  (d) are all of the above.

_____ **3.** A pine tree is
  (a) a sporophyte.
  (b) a gametophyte.
  (c) diploid.
  (d) both a and c

_____ **4.** The male pine cone
  (a) produces pollen.
  (b) contains a female gametophyte.
  (c) bears a megasporangium containing a megaspore mother cell.
  (d) gives rise to a seed.

_____ **5.** The male gametophyte of a pine tree
  (a) is produced within a pollen grain.
  (b) produces sperm.
  (c) is diploid.
  (d) is both a and b.

_____ **6.** Which of these are produced directly by meiosis in pine?
  (a) sperm cells
  (b) pollen grains
  (c) microspores
  (d) microspore mother cells

_____ **7.** An ovule
  (a) is the structure that develops into a seed.
  (b) contains the microsporophyll.
  (c) is produced on the surface of a male cone.
  (d) is all of the above.

_____ **8.** The process by which pollen is transferred to the ovule is called
  (a) transmigration.
  (b) fertilization.
  (c) pollination.
  (d) all of the above

_____ **9.** Which statement is true of the female gametophyte of pine?
  (a) It's a product of repeated cell divisions of the functional megaspore.
  (b) It's haploid.
  (c) It serves as the stored food to be used by the embryo sporophyte upon germination.
  (d) All of the above are true.

_____ **10.** The seed coat of a pine seed
  (a) is derived from the integuments.
  (b) is produced by the micropyle.
  (c) surrounds the male gametophyte.
  (d) is divided into the hypocotyl-root axis and epicotyl.

EXERCISE **23**

Seed Plants I: Gymnosperms

Post-Lab Questions

Introduction

1. What survival advantage does a seed have that has enabled the seed plants to be the most successful of all plants?

2. In the diagram below of a seed, give the ploidy level (n or 2n) of each part listed.

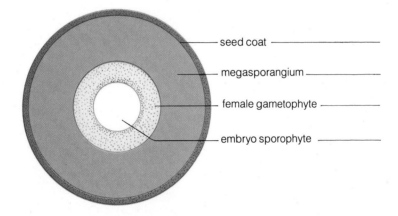

seed coat ——————————————

megasporangium ——————————

female gametophyte ——————————

embryo sporophyte ——————————

3. Distinguish between a homosporous and a heterosporous type of life cycle.

23.1 Phylum Coniferophyta: Conifers

4. List four uses for conifers.

   (a)

   (b)

   (c)

   (d)

5. While snowshoeing through the winter woods, you stop to look at a tree branch pictured at the right. Specifically, what are the brown structures hanging from the branch?

(0.5×)

6. Distinguish between *pollination* and *fertilization*.

7. Are antheridia present in conifers? _____
   Are archegonia present? _____

8. Suppose you saw the seedling pictured at the right while walking in the woods.

   (a) To which gymnosperm phylum does the plant belong?

   (b) Identify structures A and B.

(0.5×)

23.2 Phylum Cycadophyta: Cycads

9. On spring break, you are strolling through a tropical garden in Florida and encounter the plant pictured at the right. Would the xylem of this plant contain tracheids, vessels, or both?

(0.01×)

23.3 Phylum Ginkgophyta: Ginkgo

10. A friend of yours picks up a branch like the one pictured at the right. Knowing that you have taken a biology course and studied plants, she asks you what the plant is. Identify this branch, giving your friend the full scientific name for the plant.

(0.75×)

Seed Plants II: Angiosperms

After completing this exercise, you will be able to

1. define *heterosporous, angiosperm, fruit, pollination, double fertilization, endosperm, seed, germination, annual, biennial, perennial;*

2. describe the significance of the flower, fruit, and seed for the success of the angiosperms;

3. identify the structures of the flower;

4. recognize the structures and events (those in **boldface**) that take place in angiosperm reproduction;

5. describe the origin and function of fruit and seed;

6. identify the characteristics that distinguish angiosperms from gymnosperms.

OBJECTIVES

Introduction

The **angiosperms**, Phylum Anthophyta, are seed plants that produce flowers. *"Antho"* means flower, and *"phyta"* plant. The word *angiosperm* literally means "vessel seed" and refers to the seeds borne within a fruit.

There are more flowering plants in the world today than any other group of plants. Assuming that numbers indicate success, flowering plants are the most successful plants to have evolved thus far.

The most important characteristic that distinguishes the Anthophyta from other seed plants is the presence of flower parts that mature into a **fruit**, a container that protects the seeds and allows them to be dispersed without coming into contact with the rigors of the external environment. In many instances, the fruit also contributes to the dispersal of the seed. For example, some fruits stick to fur (or clothing) of animals and are brushed off some distance from the plant that produced them. Animals eat others; the undigested seeds pass out of the digestive tract and fall into environments often far removed from the seeds' source.

Our lives and diets revolve around flowering plants. Fruits enrich our diet and include such things as apples, oranges, tomatoes, beans, peas, corn, wheat, walnuts, pecans . . . the list goes on and on. Moreover, even when we are not eating fruits, we're eating flowering plant parts. Cauliflower, broccoli, potatoes, celery, and carrots all are parts of flowering plants.

Biologists believe that flower parts originated as leaves modified during the course of evolution to increase the probability for fertilization. For instance, some flower parts are colorful and attract animals that transfer the sperm-producing pollen to the receptive female parts.

As seed plants, and like gymnosperms, the angiosperms are **heterosporous**; that is, they produce *two* spore types. Review the diagram of heterosporous alternation of generations shown in Figure 23-1, page 314. Figure 24-15 depicts the life cycle of a typical flowering plant. Refer to it as you study the specimens in this exercise.

Note: **This exercise provides two alternative paths to accomplish the objectives described above, a traditional approach (24.1–24.2) and an investigative one (24.3). Your instructor will indicate which alternative you will follow.**

24.1 External Structure of the Flower *(About 20 min.)*

The number of different kinds of flowers is so large that it's difficult to pick a single example as representative of the entire phylum. Nonetheless, there is enough similarity among flowers that, once you've learned the structure of one representative, you'll be able to recognize the parts of most.

MATERIALS

Per student:

- flower for dissection (gladiolus or hybrid lily, for example)
- single-edged razor blade
- dissecting microscope

PROCEDURE

1. Obtain a flower provided for dissection.
2. At the base of the flower, locate the swollen stem tip, the **receptacle**, upon which the whorls of floral parts are arranged.
3. Identify the **calyx**, comprising the outermost whorl. Individual components of the calyx are called **sepals.** The sepals are frequently green (although not always). The calyx surrounds the rest of the flower in the bud stage (see Figure 24-15a).
4. Moving inward, locate the next whorl of the flower, the usually colorful **corolla** made up of **petals**.

It is usually the petals that we appreciate for their color. Remember, however, that the evolution of colorful flower parts was associated with the presence of color-visioned *pollinators,* animals that carry pollen from one flower to another. The colorful flowers attract those animals, enhancing the plants' chance of being pollinated, producing seeds, and perpetuating their species.

Both the calyx and corolla are sterile, meaning they do not produce gametes.

5. The next whorl of flower parts consists of the male, pollen-producing parts, the **stamen**s (also called *microsporophylls,* "microspore-bearing leaves"; Figure 24-15b). Examine a single stamen in greater detail. Each stamen consists of a stalklike **filament** and an **anther.** The anther consists of four **microsporangia** (also called *pollen sacs*).
6. Next locate the female portion of the flower, the **pistil** (Figures 24-15a and c). A pistil consists of one or more **carpels**, also called *megasporophylls,* "megaspore-bearing leaves."

If the pistil consists of more than one carpel, they are usually fused, making it difficult to distinguish the individual components. However, you can usually determine the number of carpels by counting the number of lobes of the stigma.

7. How many carpels does your flower contain? _____
8. Identify the different parts of the pistil (Figure 24-15c): at the top, the **stigma**, which serves as a receptive region on which pollen is deposited; a necklike **style**; and a swollen **ovary**. The only members of the plant kingdom to have ovaries are angiosperms!
9. With a sharp razor blade, make a section of the ovary. (Some students should cut the pistil longitudinally; others should cut the ovary crosswise. Then compare the different sections.)
10. Examine the sections with a dissecting microscope, finding the numerous small **ovules** within the ovary. The diagram in Figure 24-15c is oversimplified; many flowers have more than one ovule per ovary.
11. In Figure 24-1, make and label two sketches: one of the cross section (c.s.) and the other of a longitudinal section (l.s.) of the ovary. Notice that the ovules are completely enclosed within the ovary. After fertilization, the ovules will develop into *seeds,* and the ovary will enlarge and mature into the *fruit.*

There are two groups of flowering plants, monocotyledons and dicotyledons. The number of flower parts indicates which group a plant belongs to. Generally, monocots have the flower parts in threes or multiples of three. Dicots have their parts in fours or fives or multiples thereof.

12. Count the number of petals or sepals in the flower you have been examining. Are you studying a monocot or dicot?

_____

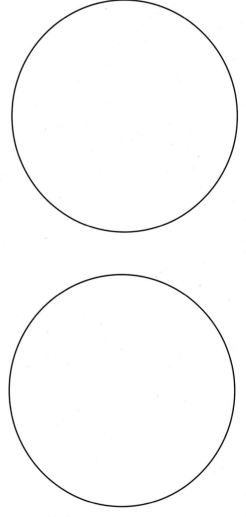

Figure 24-1 Drawings of cross section and longitudinal section of an ovary.
Label: ovules

A. Male Gametophyte (Pollen Grain) Formation in the Microsporangia (About 20 min.)

The male gametophyte in flowering plants is the pollen grain; a microscopic sperm-producing structure.

MATERIALS

Per student:

- prepared slide of young lily anther, c.s.
- prepared slide of mature lily anther (pollen grains), c.s.
- *Impatiens* flowers, with mature pollen
- glass microscope slide

- coverslip
- compound microscope

Per student pair:

- 0.5% sucrose, in dropping bottle

PROCEDURE

1. With the low-power objective of your compound microscope, examine a prepared slide of a cross section of an immature anther (Figures 24-2, 24-15d). Find sections of the four **microsporangia** (also called *pollen sacs*), which appear as four clusters of densely stained cells within the anther.

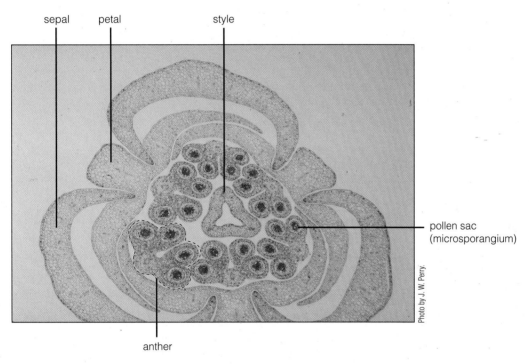

Figure 24-2   Immature anthers within flower bud, c.s. (4×).

2. Study the contents of a single microsporangium. Depending on the stage of development, you will find either diploid *microspore mother cells* (Figure 24-15e) or haploid *microspores* (Figures 24-15f and g).
3. Obtain a prepared slide of a cross section of a mature anther (Figure 24-3). Observe it first with the low-power objective, noting that the walls have split open to allow the pollen grains to be released, as shown in Figure 24-15i.
4. Pollen grains are immature *male gametophytes*, and very small ones at that. Switch to the high-dry objective to study an individual pollen grain more closely (Figure 24-4).
5. The pollen grain consists of only two cells. Identify the large **tube cell** and a smaller, crescent-shaped **generative cell** that floats freely in the cytoplasm of the tube cell (Figure 24-15h).
6. Note the ridged appearance of the outer wall layer of a pollen grain. Within the ridges and valleys of the wall, glycoproteins are present that appear to play a role in recognition between the pollen grain and the stigma.

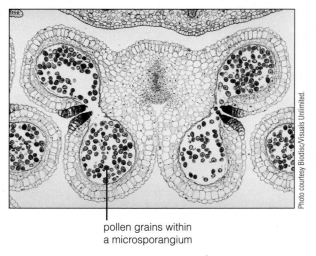

pollen grains within
a microsporangium

Photo courtesy Biodisc/Visuals Unlimited.

Figure 24-3   Mature anther, c.s. (21×).

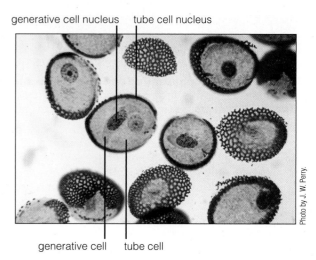

generative cell nucleus   tube cell nucleus

generative cell   tube cell

Photo by J. W. Perry.

Figure 24-4   Pollen grain, c.s. (145×).

Transfer of pollen from the microsporangia to the stigma, called **pollination**, occurs by various means—wind, insects, and birds being the most common carriers of pollen. When a pollen grain lands on the stigma of a compatible flower, it germinates and produces a **pollen tube** that grows down the style (Figure 24-15j). The generative cell flows into the pollen tube, where it divides to form two *sperm* (Figure 24-15k). Because it bears two gametes, the pollen grain is now considered to be a *mature* male gametophyte.

7. Obtain an *Impatiens* flower and tap some pollen onto a clean microslide. Add a drop of 0.5% sucrose, cover with a coverslip, and observe with the medium-power objective of your compound microscope.

8. Look for the *pollen tube* as it grows from the pollen grain. (You may wish to set the slide aside for a bit and reexamine it 15 minutes later to check the progress of the pollen tubes.)

In Figure 24-5, draw a sequence showing the germination of an *Impatiens* pollen grain.

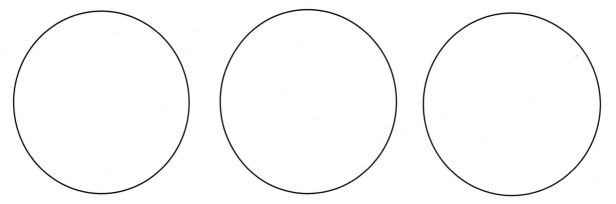

Figure 24-5   Germination of *Impatiens* pollen grain.

B. Female Gametophyte Formation in the Megasporangia (About 20 min.)

The female gametophyte is formed inside each ovule within the ovary of the flower. Like the male gametophyte, the female gametophyte in flowering plants consists of only a few cells.

MATERIALS

Per student:

■ prepared slide of lily ovary, c.s., megaspore mother cell
■ compound microscope

Per lab room:

■ demonstration slide of lily ovary, c.s., seven-celled, eight-nucleate gametophyte
■ demonstration slide of lily ovary, c.s., double fertilization

PROCEDURE

1. With the medium-power objective of your compound microscope, examine a prepared slide of a cross section of an ovary (Figures 24-6, 24-15l).

2. Find the several **ovules** that have been sectioned. One ovule will probably be sectioned in a plane so that the very large, diploid **megaspore mother cell** is obvious (Figures 24-7, 24-15m).

3. The megaspore mother cell is contained within the **megasporangium**, the outer cell layers of which form two flaps of tissue called **integuments**. Identify the structures in boldface print. After fertilization, the integuments develop into the *seed coat*.

4. Identify the **placenta**, the region of attachment of the ovule to the ovary wall.

As the megasporangium develops, the integuments grow, enveloping the megasporangium. However, a tiny circular opening remains. This opening is the **micropyle**. (The micropyle is obvious in Figure 24-15n.) Remember, the micropyle is an opening in the ovule. After pollination, the pollen grain germinates on the surface of the stigma, and the pollen tube grows down the style, through the space surrounding the ovule, through the micropyle, and penetrates the megasporangium (Figure 24-15p).

5. Identify the micropyle present in the ovule you are examining.

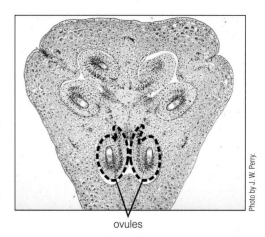

ovules

Figure 24-6   Lily ovary, c.s. (19×).

Considerable variation exists in the next sequence of events; we describe the pattern found in the lily.

The diploid megaspore mother cell undergoes meiosis, producing four haploid *nuclei* (Figure 24-15n). (Note that cytokinesis does *not* follow meiosis, and thus only nuclei—not cells—are formed.) The cell containing the four nuclei (the old megaspore mother cell) is now called the **female gametophyte** (also called the *embryo sac*).

Three of these four nuclei fuse. Thus, the female gametophyte contains one triploid (3n) nucleus and one haploid (n) nucleus (Figure 24-15o). Subsequently, the nuclei undergo two *mitotic* divisions, forming eight nuclei in the female gametophyte. Cell walls form around six of the eight nuclei; the large cell remaining—the **central cell**—contains two nuclei, one of which is triploid, the other haploid. These two nuclei are called the **polar nuclei**. At the micropylar end of the ovule there are three cells, all haploid. One of these is the **egg cell**. Opposite the micropylar end are three 3n cells. This stage of development is often called the *seven-celled, eight-nucleate female gametophyte* (Figures 24-8, 24-15p). Fertilization takes place at this stage.

6. Study the demonstration slide of the seven-celled, eight-nucleate female gametophyte. Identify the **placenta, integuments, micropyle, egg cell, central cell**, and **polar nuclei** (Figures 24-8, 24-15p).

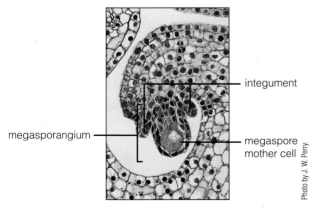

Figure 24-7   Megaspore mother cell within megasporangium (94×).

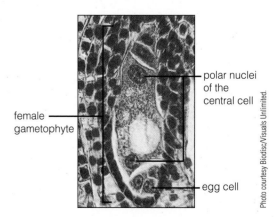

Figure 24-8   Seven-celled, eight-nucleate female gametophyte (170×).

As the pollen tube penetrates the female gametophyte, it discharges the sperm; one of the sperm nuclei fuses with the haploid egg nucleus, forming the *zygote*. Figure 24-15q shows the female gametophyte after fertilization has occurred.

The zygote is a _____ (haploid, diploid) cell.

The other sperm nucleus enters the central cell and fuses with the two polar nuclei, forming the primary endosperm nucleus (Figure 24-15q). Thus, the primary endosperm nucleus is _____ (haploid, diploid, triploid, tetraploid, pentaploid).

The cell containing the primary endosperm nucleus (the old central cell) is now called the **endosperm mother cell**. Traditionally, the process whereby one sperm nucleus fuses with the egg nucleus and the other sperm nucleus fuses with the two polar nuclei has been called **double fertilization**.

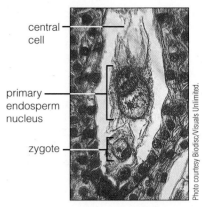

central cell

primary endosperm nucleus

zygote

Photo courtesy Biodisc/Visuals Unlimited.

Figure 24-9 Double fertilization (170×).

7. Observe the demonstration slide of double fertilization (Figure 24-9), and identify the **zygote, primary endosperm nucleus**, and **central cell** of the female gametophyte.

Numerous mitotic and cytoplasmic divisions of the endosperm mother cell form the **endosperm**, a tissue used to nourish the embryo sporophyte as it develops within the seed.

C. Embryogeny (About 10 min.)

The zygote undergoes mitosis and cytokinesis to produce a two-celled **embryo** (also called the *embryo sporophyte*). Numerous subsequent divisions produce an increasingly large and complex embryo. This is the process of "embryogeny," literally embryo genesis.

MATERIALS

Per lab room:

- demonstration slides of *Capsella* embryogeny: globular embryo, emerging cotyledons, torpedo-shaped embryo, mature embryo

PROCEDURE

1. Observe the series of four demonstration slides, which show the stages of embryo development in the female gametophyte of *Capsella* (Figure 24-10).
2. The first slide shows the so-called globular stage (Figure 24-10a), in which a chain of cells (the suspensor) attaches the embryo, a spherical mass of cells, to the wall of the female gametophyte (embryo sac). The very enlarged cell at the base of the suspensor is the *basal cell* and is active in uptake of nutrients that the developing embryo will use. Note the endosperm within the female gametophyte.
3. The second slide shows the heart-shaped stage (Figure 24-10b). Now you can distinguish the emerging **cotyledons** (seed leaves). In many plants, the cotyledons absorb nutrients from the endosperm and thus serve as a food reserve to be used during seed germination.
4. Further development of the embryo has occurred in the third slide, the torpedo stage (Figure 24-10c). Notice that the entire embryo has elongated. Find the *cotyledons*. Between the cotyledons, locate the **epicotyl**, which is the *apical meristem of the shoot*. Beneath the epicotyl and cotyledons find the **hypocotyl-root axis**. At the tip of the hypocotyl-root axis, locate the *apical meristem of the root* and the **root cap** covering it.
5. The final slide (Figure 24-10d) shows a mature embryo, neatly packaged inside the **seed coat**. Identify the seed coat and other regions previously identified in the torpedo stage.

How many cotyledons were there in the slides you examined? _____

Thus, is *Capsella* a monocot or dicot? _____

D. Fruit and Seed (About 20 min.)

Simply stated, a **fruit** is a matured ovary, while a **seed** is a matured ovule. Bear in mind that for each seed produced, a pollen grain had to fertilize the egg cells in the ovules. Fertilization not only causes the integuments of the ovule to develop into a seed coat, it also causes the ovary wall to expand into the fruit.

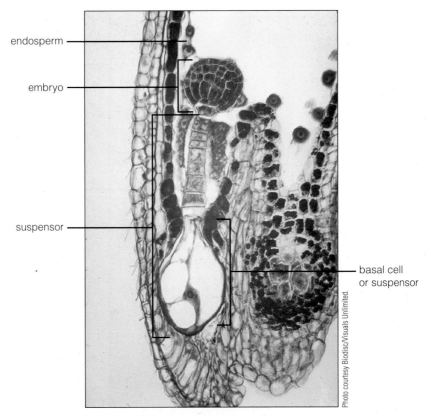

endosperm

embryo

suspensor

basal cell
or suspensor

Photo courtesy Biodisc/Visuals Unlimited.

(a) Globular embryo stage (232×).

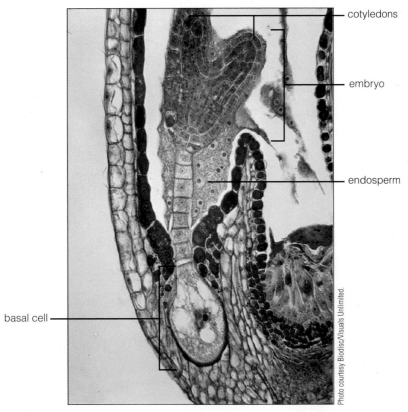

cotyledons

embryo

endosperm

basal cell

Photo courtesy Biodisc/Visuals Unlimited.

(b) Heart-shaped stage (240×).

Figure 24-10 Embryogeny in *Capsella* (shepherd's purse). *Continues on page 332.*

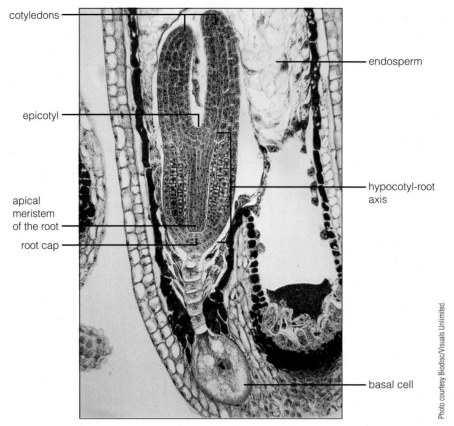

cotyledons

endosperm

epicotyl

hypocotyl-root axis

apical meristem of the root

root cap

basal cell

Photo courtesy Biodisc/Visuals Unlimited.

(c) Torpedo stage (170×).

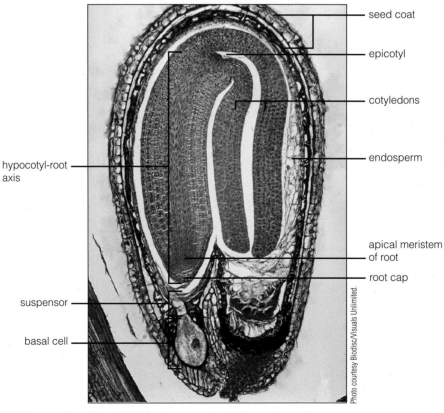

seed coat

epicotyl

cotyledons

endosperm

hypocotyl-root axis

apical meristem of root

root cap

suspensor

basal cell

Photo courtesy Biodisc/Visuals Unlimited.

(d) Mature embryo stage (113×).

Figure 24-10   Embryogeny in *Capsella* (shepherd's purse). *Continued.*

MATERIALS

Per student:

- bean fruits
- soaked bean seeds
- iodine solution (I₂KI), in dropping bottle

Per lab room:

- herbarium specimen of *Capsella*, with fruits
- demonstration slide of *Capsella* fruit, c.s.

PROCEDURE

1. Examine the herbarium specimen and demonstration slide of the fruits of *Capsella*. On the herbarium specimen, identify the **fruits**, which are shaped like the bag that shepherds carried at one time (Figure 24-11; the common name of this plant is "shepherd's purse").

2. Now study the demonstration slide of a cross section through a single fruit (Figure 24-12). Note the numerous **seeds** in various stages of embryo development.

3. Obtain a bean pod and carefully split it open along one seam.

   The pod is a matured ovary and thus is a _____.

   Find the *sepals* at one end of the pod and the shriveled *style* at the opposite end. The "beans" inside are

   _____

4. Note the point of attachment of the bean to the pod. This is the placenta, a term shared by the plant and animal kingdom that describes the nutrient bridge between the "unborn" offspring and parent.

5. In Figure 24-13, draw the split-open bean pod, labeling it with the correct scientific terms. Figure 24-15r shows a section of a typical fruit.

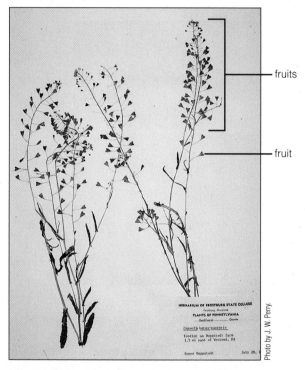

Figure 24-11 Herbarium specimen of *Capsella*, shepherd's purse (0.25×).

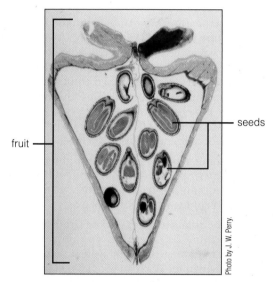

Figure 24-12 Cross section of *Capsella* fruit (8×).

Figure 24-13 Drawing of an open bean pod.
Labels: fruit, seeds, placenta

6. Closely study one of the beans from within the pod or a bean that has been soaked overnight to soften it. Find the scar left where the seed was attached to the fruit wall.

7. Near the scar, look for the tiny opening left in the seed coat.

What is this tiny opening? _____ (*Hint:* The pollen tube grew through it.)

8. Remove the seed coat to expose the two large **cotyledons**. Split the cotyledons apart to find the **epicotyl** and **hypocotyl-root axis**. During maturation of the bean embryo, the cotyledons absorb the endosperm. Thus, bean cotyledons are very fleshy because they store carbohydrates that will be used during seed germination. Add a drop of I₂KI to the cotyledon.

What substance is located in the cotyledon? _____
(*Hint:* Return to Exercise 7 if you've forgotten what is stained by I₂KI.)

E. Seedling (About 20 min.)

When environmental conditions are favorable for growth (adequate moisture, oxygen, and proper temperatures), the seed germinates; that is, the seedling (young sporophyte) begins to grow.

MATERIALS

Per student:
- germinating bean seeds
- bean seedlings

Per table:
- dishpan of water

PROCEDURE

1. Obtain a germinating bean seed from the culture provided (Figure 24-14). Wash the root system in the dishpan provided, *not* in the sink.

2. Identify the **primary root** with the smaller **secondary roots** attached to it. Emerging in the other direction will be the **hypocotyl**, the **cotyledons**, the **epicotyl** (above the cotyledons), and the first **true leaves** above the epicotyl. Identify these structures.

When some seeds (like the pea) germinate, the cotyledons remain *below* ground. Others, like the bean, emerge from the ground because of elongation of the hypocotyl-root axis.

3. Now, obtain a bean seedling from the growing medium. Be careful as you pull it up so as not to damage the root system. Wash the root system in the dishpan provided.

Your seedling should be in a stage of development similar to that shown in Figure 24-15s. Much of the growth that has taken place is the result of cellular elongation of the parts present in the seed.

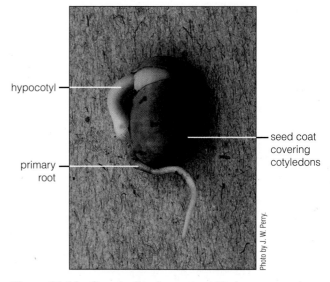

Figure 24-14 Germinating bean seed (2×).

4. On the root system, identify the **primary** and **secondary roots**. As the root system merges into the shoot (aboveground) system, find the **hypocotyl**. (The prefix *hypo-* is derived from Greek, meaning "below" or "underneath.")

5. Next, identify the cotyledons. Notice their shriveled appearance. Now that you know the function of the cotyledons from your previous study, why do you suppose the cotyledons are shriveled?

_____

6. Above the cotyledons, find that portion of the stem called the **epicotyl**.

Knowing what you do about the prefix *hypo-*, speculate on what the prefix *epi-* means.

_____

As noted earlier, the seedling has *true leaves*. Contrast the function of the cotyledons ("seed leaves") with the true leaves. _____

_____

Depending on the stage of development, your seedling may have even more leaves that have been produced by the shoot apex.

The amount of time between seed germination and flowering largely depends on the particular plant. Some plants produce flowers and seeds during their first growing season, completing their life cycle in that growing season. These plants are called **annuals**. Marigolds are an example of an annual. Others, known as **biennials**, grow vegetatively during the first growing season and do not produce flowers and seeds until the second growing season (carrots, for example). Both annuals and biennials die after seeds are produced.

Perennials are plants that live several to many years. The time between seed germination and flowering (seed production) varies, some requiring many years. Moreover, perennials do not usually die after producing seed, but flower and produce seeds many times during their lifetime.

24.3 Experiment: An Investigative Study of the Life Cycle of Flowering Plants

Recent developments with a fast-growing and fast-reproducing plant in the mustard family allow you to study the life cycle of a flowering plant over the course of only 35 days. The plants were bred by a plant scientist at the University of Wisconsin—Madison and are sometimes called Wisconsin Fast Plants™. You will grow these rapid-cycling *Brassica rapa* (abbreviated RCBr for Rapid Cycling *Brassica rapa*) plants from seeds, following the growth cycle and learning about plant structure, the life cycle of a flowering plant, and adaptive mechanisms for pollination.

Because this investigation occurs over several weeks, all Materials and Procedures are listed by the day the activity begins, starting from day 0. Record your measurements and observations in tables at the end of this exercise.

A. Germination

Day 0 (About 30 min.)

MATERIALS

Per student:

- one RCBr seed pod
- metric ruler
- 3 × 5 in. index card
- transparent adhesive tape
- petri dish
- paper toweling or filter paper disk to fit petri dish
- growing quad *or* two 35-mm film canisters
- 12 N-P-K fertilizer pellets (24 if using canister method)
- four paper wicks (growing quad) *or* two cotton string wicks (canister method)
- disposable pipet with bulb
- pot label (two if using canister method)
- marker

- forceps
- model seed
- 500-mL beaker
- two wide-mouth bottles (film canister method)

Per student group (4):

- 2-L soda bottle bottom or tray
- moistened soil mix
- watertight tray
- water mat

Per lab room:

- light bank

PROCEDURE

The life cycle of a seed plant has no beginning or end; it's a true cycle. For convenience, we'll start our examination of this cycle with germination of nature's perfect time capsule, the seed. **Germination** is the process of switching from dormancy to growth.

1. Obtain an RCBr seed pod that has been sandwiched between two strips of transparent adhesive tape. This pod is the **fruit** of the plant. A fruit is a matured ovary. Measure and record the length of the pod in Table 24-1.

2. Harvest the seeds by scrunching up the pod until the seeds come out. Then carefully separate the halves of the pod to expose the seeds.

3. Roll a piece of adhesive tape into a circle with the sticky side out. Attach the tape to an index card and then place the seeds on the tape. (This keeps the small seeds from disappearing.) You may wish to handle the seeds with forceps. Count the number of seeds and record the number in Table 24-1 on page 346.

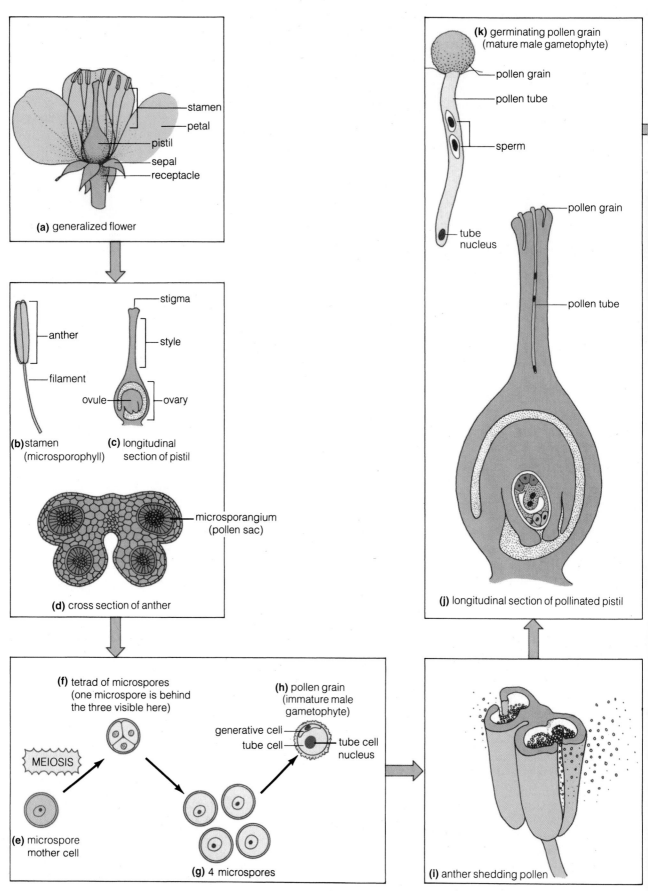

Figure 24-15 Life cycle of an angiosperm. [Color scheme: Haploid (n) structures are yellow, diploid (2n) are green, gold, or red; triploid (3n) are purple.]

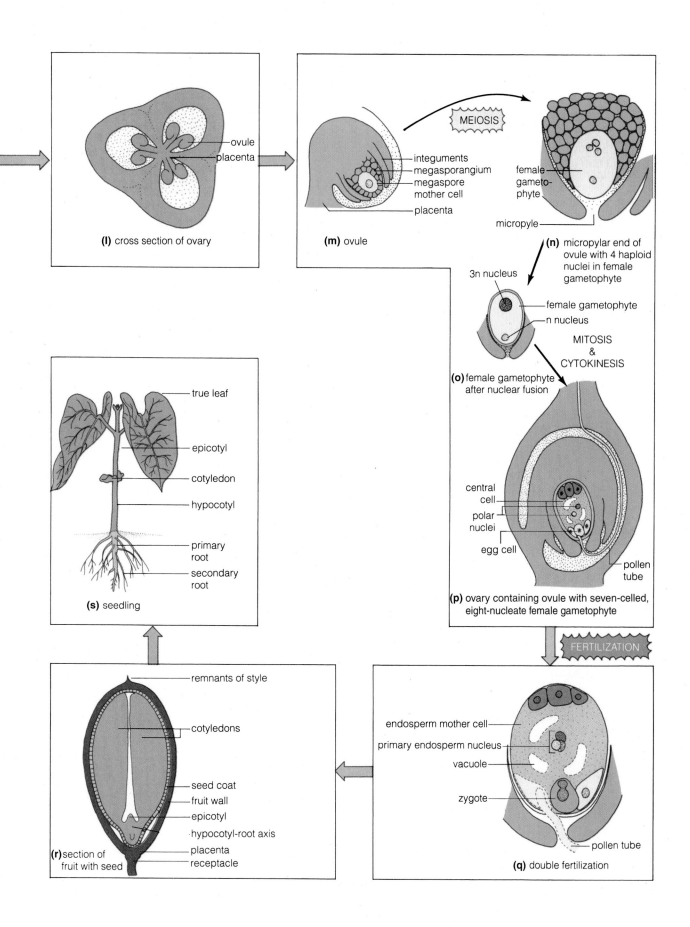

(l) cross section of ovary

(m) ovule

integuments
megasporangium
megaspore mother cell
placenta

ovule
placenta

MEIOSIS

female gametophyte

micropyle

(n) micropylar end of ovule with 4 haploid nuclei in female gametophyte

3n nucleus

female gametophyte
n nucleus

MITOSIS
&
CYTOKINESIS

(o) female gametophyte after nuclear fusion

central cell
polar nuclei
egg cell

pollen tube

(p) ovary containing ovule with seven-celled, eight-nucleate female gametophyte

FERTILIZATION

true leaf

epicotyl

cotyledon

hypocotyl

primary root

secondary root

(s) seedling

remnants of style

cotyledons

seed coat
fruit wall
epicotyl
hypocotyl-root axis
placenta
receptacle

(r) section of fruit with seed

endosperm mother cell
primary endosperm nucleus
vacuole

zygote

pollen tube

(q) double fertilization

4. Obtain a petri dish to use as a germination chamber. Obtain a filter paper disk or cut a disk from paper toweling and insert it into the larger (top) part of the petri dish. With a *pencil* (**DO NOT** use ink), label the bottom of the circle with your name, the date, and current time of day.

5. Moisten the toweling in the petri dish with tap water. Pour out any excess water.

6. Place five seeds on the top half of the toweling and cover with the bottom (smaller half) of the petri dish.

7. Place the germination chamber at a steep angle in shallow water in the base of a 2-L soda bottle or tray so that about 1 cm of the lower part of the towel is below the water's surface. This will allow water to be evenly "wicked" up to keep the seeds uniformly moist.

8. Set your experiment in a warm location (20°–25°C is ideal). In Table 24-1, record the temperature and number of seeds placed in the germination plate.

9. To envision the germination process, obtain a model seed (Figure 24-16). A **seed** consists of a **seed coat, stored food,** and an **embryo**.

10. Germination in the RCBr seed starts with the imbibition of water. Toss your model seed into a beaker of water. Out pops the embryo, consisting of two **cotyledons** and a **hypocotyl-root axis**. (*Hypocotyl* literally means "below the cotyledons.")

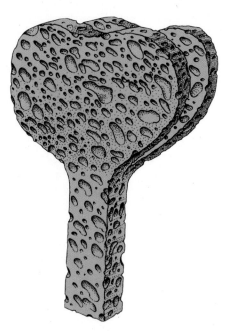

Figure 24-16 Model seed.

You'll now plant seeds for additional studies. If your instructor indicates that you will be using growth quads, proceed to step 11. If you're using the canister method, skip to step 12.

11. *Option A*: Polystyrene Growth Quad Method
    (a) Obtain four diamond-shaped paper towel wicks, growth quad, moistened soil mix, 12 fertilizer pellets, and a pot label.
    (b) Drop one wick into each cell in the quad. The tip of the wick should protrude about 1 cm from the hole in the bottom of the quad.
    (c) Fill each cell *halfway* with soil.
    (d) Add three N-P-K fertilizer pellets to *each* cell.
    (e) Fill each cell to the top with *loose* soil. **DO NOT** compact the soil.
    (f) With your thumb, make a shallow depression in the soil at the top of each cell.
    (g) Drop three seeds into each depression. In Table 24-1, record the total number of seeds planted.
    (h) Sprinkle enough soil to cover the seeds in each cell.
    (i) Water very gently with a pipet until water drips from each wick.
    (j) Write your name and date on a pot label and insert the label against the side of one cell of the quad.
    (k) Place the quad on a wet water mat inside a tray under the light bank. The top of the quad should be approximately 5 cm below the lights.

12. *Option B*: Film Canister Method
    (a) Obtain two cotton string wicks, two film canisters (with holes in bottoms), 24 fertilizer pellets, and two pot labels.
    (b) Using tap water, wet the string wicks thoroughly and insert them through the holes in the canisters' bottoms. The wicks should extend about halfway up the length of the canister.
    (c) Fill each canister *halfway* with soil mix.
    (d) Add 12 N-P-K fertilizer pellets atop the soil of each canister.
    (e) Fill each canister to the top with soil mix. **DO NOT** compact the soil.
    (f) With your thumb, make a shallow depression in the soil of each canister.
    (g) Drop six seeds atop the soil in each canister. Distribute the seeds so they're about equally spaced.
    (h) Sprinkle enough soil to cover the seeds.
    (i) Water the containers very gently with a pipet until water drips from each wick.
    (j) Write your name and date on pot labels and insert them into the soil against the side of each canister.
    (k) Fill two wide-mouth bottles that have the same diameter as the canisters with tap water and place each canister in its own water bottle.
    (l) Place the canisters in their water bottles under the light bank. The top of the canister should be approximately 5 cm below the lights.

Day 1 (About 10 min.)

MATERIALS

Per student:

- dissecting microscope
- tap water
- disposable pipet with bulb
- metric ruler

PROCEDURE

1. Water each quad cell or canister from the top using a pipet.
2. Observe your germination experiment in the petri dish chamber with a dissecting microscope. Note the **radicle** (primary root) and **hypocotyl** extending from a split in the brown **seed coat**. What colors are the radicle and the hypocotyl? _____
3. What does the color indicate about the source of carbohydrates for the germinating seeds?

   _____

4. Count the number of seeds that have germinated, and record it in Table 24-1. Measure the length of each hypocotyl and attached radicle. Record this in Table 24-1. Replace the cover and return your experiment to its place.

Day 2 (About 15 min.)

MATERIALS

Per student:

- germination experiment started on day 0
- disposable pipet with bulb
- tap water
- metric ruler
- forceps
- dissecting microscope
- two additional germination chambers (for gravitropism experiment)
- two RCBr seed pods

PROCEDURE

1. Water each quad cell or canister from the top using a pipet.
2. Observe your germination experiment. Count the number of germinated seeds and calculate the percentage of seeds that germinated. Record both numbers in Table 24-1. With a forceps, remove one of the germinating seeds from the toweling, place it on the stage of a dissecting microscope, and observe the numerous **root hairs** attached to the radicle. Root hairs greatly increase the surface area for absorption of water and minerals. Measure the lengths of the hypocotyls and attached radicles and record their average length in Table 24-1.

In what direction (up, down, or horizontal) are the hypocotyl and radicle growing?

    hypocotyl: _____

    radicle: _____

This directional growth is called **gravitropism**, a plant growth response to gravity.

Day 3 (About 10 min.)

MATERIALS

Per student:

- germination experiment started on day 0
- metric ruler
- disposable pipet with bulb
- tap water

PROCEDURE

1. Water your plants in the quads or canisters from the top using a pipet. Is any part of the plant visible yet? _____ (yes or no) If yes, which part? _____
2. Observe your germination experiment. By this time, the seed coat should be shed, and the **cotyledons**, the so-called seed leaves that were packed inside the seed coat, should be visible. The cotyledons absorb food stored within the seed, making the food available for initial stages of growth before photosynthesis starts to provide carbohydrates necessary for growth.

What color are the cotyledons? _____

What function does this color indicate for the cotyledons? _____

3. Measure the length of the hypocotyls and attached radicles and record their average length in Table 24-1.

B. Growth

Day 4 or 5 (About 10 min.)

MATERIALS

Per student:

- forceps
- metric ruler

- dissecting microscope
- germination experiment started on day 0

PROCEDURE

1. Count the number of seedlings present, and record it in Table 24-1. Use a forceps to thin the plants to one per cell if you are using the quad growing method. If a cell has no plants, transplant one of the extra plants from another cell by inserting the tip of a pencil into the soil to form a hole and then inserting the transplant's root. Gently firm the soil around the transplant. Measure the height of each plant's shoot system—that is, from soil to the apex. Compute the average height and record it in Table 24-1.
2. Observe your germination experiment. Note that the radicle has produced additional roots. These are **lateral roots** (secondary roots). Remove one plant and observe it with the dissecting microscope.

Do lateral roots have root hairs? _____ (yes or no)

At this time, you may discard your germination experiment. Your instructor will indicate what you should do with the materials.

Day 7 (About 10 min.)

MATERIALS

Per student:

- metric ruler

PROCEDURE

1. By now your plant should have its first **true leaves**. Describe how the true leaves differ from the cotyledons.

_____

_____

_____

2. Measure the length of the true leaves, and record their average length in Table 24-1.
3. Measure the height of your plants from soil line to shoot apex, and record the average height in Table 24-1.

C. The Pollinator and Pollination

Pollen grains are actually sperm conveyors that deliver these male gametes to the egg-containing ovules within the ovary. Transfer of pollen from the anther to the stigma, called **pollination,** occurs by various means in different species of flowers, with wind, birds, and insects being the most common carriers of pollen. In the case of *Brassica*, the main pollinator is the common honeybee.

Pollinators and flowers are probably one of the best-known examples of **coevolution**. Coevolution is the process of two organisms evolving together, their structures and behaviors changing to benefit both organisms. Flowering plants and pollinators work together in nearly perfect harmony: the bee receives the food it needs for its respiratory activities from the flower, and the plant receives the male gametes (delivered by the bee) directly to the female structure—it doesn't have to rely on the presence of water, which is necessary in lower plants, algae, and fungi, to get the gametes together.

Flowering plants have considerable advantage over gymnosperms. While gymnosperms also produce pollen grains, the pollinator in these plants is the wind. No organism carries pollen directly to the female structure. Rather, chance and the production of enormous quantities of pollen (which takes energy that might be used otherwise for growth) are integral in assuring reproduction.

In preparation for pollinating your plants' flowers, let's first study the pollinator, the common honeybee.

Day 10 (About 20 min.)

MATERIALS

Per student:

- metric ruler
- three toothpicks
- three dried honeybees
- dissecting microscope

Per student group (4):

- tube of glue
- scissors

PROCEDURE

1. Measure the lengths of the second set of true leaves and plant height, and record their averages in Table 24-1.
2. Obtain a dried honeybee and observe it with your dissecting microscope. Use Figure 24-2 to aid you in this study. Note the featherlike hairs that cover its body. As a bee probes for nectar, it brushes against the anthers and stigma. Pollen is entrapped on these hairs.
3. Closely study the foreleg of the bee, noticing the notch. This is the *antenna cleaner*. Quick movements over the antennae remove pollen from these sensory organs.
4. Examine the mid- and hind legs, identifying their *pollen brushes*, which are used to remove the pollen from the forelegs, thorax, and head.
5. On the hind leg, note the *pollen comb* and *pollen press*. The comb rakes the pollen from the midleg, and the press is a storage area for pollen collected.

Pollen is rich in the vitamins, minerals, fats, and proteins needed for insect growth. When the baskets are filled, the bee returns to the hive to feed the colony. Partially digested pollen is fed to larvae and to young, emerging bees.

As you can imagine, during pollen- and nectar-gathering at numerous flowers, cross-pollination occurs as pollen grains are rubbed off onto stigmas.

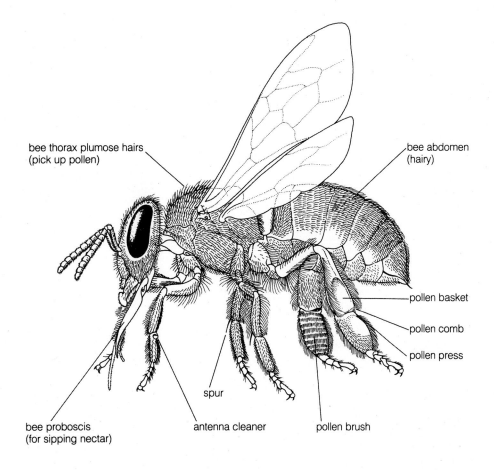

Figure 24-17   Structure of the honeybee.

On day 12, you will start pollinating your flowers. What better means to do it than to use a bee? (In this case, a dead bee that is attached to a stick.) Let's make some beesticks.

6. Add a drop of glue to one end of three round toothpicks. Insert the glue-bearing end of one stick into the top (dorsal) side of the thorax of a bee. Repeat with two more dried bees. Set your beesticks aside to dry.

D. Pollination

It's time to do a pollination study. Some species of flowers are able to **self-pollinate**, meaning that the pollen from one flower of a plant will grow on stigmas of other flowers of that same plant. Hence, the sperm produced by a single plant will fertilize its own flowers. Other species are **self-incompatible**. For fertilization to occur in the latter, pollen must travel from one plant to another plant of the same species. This is **cross-pollination**. In this portion of the exercise, you will determine whether RCBr is self-incompatible or is able to self-pollinate.

Day 12 (About 30 min.)

MATERIALS

Per student:

- beestick prepared on day 10
- index card divider
- metric ruler

- pot label
- forceps
- marker

PROCEDURE

1. Measure the height of each plant, and record the average height in Table 24-1.
2. Separate the cells of your quad or canister with a divider constructed with index cards as shown in Figure 24-18. This keeps the plants from brushing against each other.

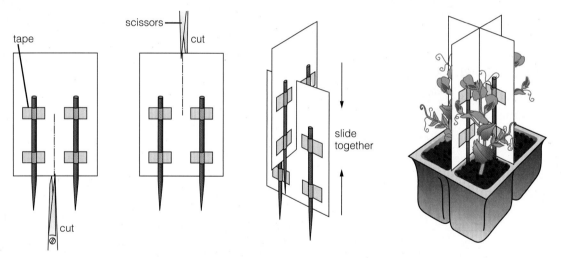

Figure 24-18 Method for separating flowers for pollination study.

3. In one cell of the quad, or in one canister, insert a pot label on which you have written the word ("SELF"), to indicate that this plant (plants if using canister) will be self-pollinated. The other three plants (quad method) or those in the other canister (canister method) will be cross-pollinated.
4. Now use one of your beesticks to pick up pollen from the flowers on the plant in the "SELF" cell. Rub the bee's body against the pollen-containing anthers. Examine your beestick using the dissecting microscope, and note the pollen adhering to the hairs of the bee's body.
5. Brush some of the pollen on the stigmas of the "SELF" plant's flowers. Next, cross-pollinate the other three plants in the quad (or canister) with the *same* beestick.
6. Return your plants to the light-bank growing area and discard the beestick.

> **CAUTION**
>
> It is important that you DO NOT retouch any of the self-pollinated flowers with the beestick once you have touched the flowers of any of the other plants.

Day 14 (About 15 min.)

MATERIALS

Per student:

- metric ruler
- beestick

PROCEDURE

1. Measure the height of each plant, and record their average height in Table 24-1.
2. Pollinate the newly opened flowers using the same techniques as described for day 12.
3. Return your plants to the light-bank growing area.

> **CAUTION**
>
> Use a new beestick and be sure not to transfer pollen from any of the cross-pollinated plants to the self-pollinated plant.

Day 17 (About 15 min.)

MATERIALS

Per student:

- metric ruler
- beestick

PROCEDURE

1. Measure the height of each plant, and record their average height in Table 24-1.
2. Count the number of flowers on each plant (opened and unopened), calculate the average number per plant, and record it in Table 24-1. Pinch off the shoot apex of each plant to remove all unopened buds and pinch off any side shoots the plant has produced.
3. Do a final pollination of your flowers, using a fresh beestick and taking the same precautions as on days 12 and 14.
4. Return your plants to the light-bank growing area.

E. Fertilization and Seed Development

Within 24 hours of pollination, a pollen grain that has landed on a compatible stigma will produce a pollen tube that grows down the style and through the micropyle of the ovule.

Within the pollen tube are two **sperm**, the male gametes. As the pollen tube enters the ovule, the two sperm are discharged. One fertilizes the haploid egg nucleus, forming a diploid **zygote**. The second sperm unites with two other nuclei within the ovule, forming a triploid nucleus known as the **primary endosperm nucleus**.

The fusion of one sperm nucleus to form the zygote and the other to form the primary endosperm nucleus is called **double fertilization**.

1. Reexamine Figure 24-9, which illustrates double fertilization.

What is the fate of cells formed by double fertilization? The zygote undergoes numerous cell divisions during a process called **embryogeny** (the suffix *-geny* is derived from Greek, meaning "production") to form the diploid embryo. The triploid primary endosperm nucleus divides numerous times without cytokinesis. The result is the formation of a multinucleate milky **endosperm** (each nucleus is 3n), which serves as the food source for the embryo as it grows. At the same time, changes are taking place in the cell layers of the ovule—they are becoming hardened and will eventually form the seed coat. Also, the ovary is elongating as it matures into the **fruit**—in this case, a pod.

2. Examine your plants, looking for small, elongated pods.
3. At this time, you should be able to tell whether RCBr can self-pollinate or is self-incompatible, because the fruit—a pod—will develop only if fertilization has taken place. Do the cross-pollinated plants have pods? (yes or no) _____
4. Does the self-pollinated plant have pods? (yes or no) _____
5. Make a conclusion regarding the need for cross-pollination in RCBr. _____

_____

The young embryo goes through a continuous growth process, starting as a ball of cells (the globular stage) and culminating in the formation of a mature embryo. The early stages are somewhat difficult to observe, but the later stages can be seen quite easily with a microscope. By day 24, many of the ovules will be at the globular stage, similar to that seen in Figure 24-10a.

Day 26 (About 30 min.)

MATERIALS

Per student:

- dissecting microscope
- two dissecting needles
- dH$_2$O in dropping bottle

- glass microscope slide
- iodine solution (I$_2$KI) in dropping bottle

PROCEDURE

1. Remove a pod from one of the plants, place it on the stage of your dissecting microscope, and tease it open using dissecting needles. RCBr flowers have pistils consisting of two carpels; each carpel makes up one longitudinal half of the pod, and the "seam" represents where the carpels fuse.
2. Carefully remove the ovules (immature seeds) inside the pod, placing them in a drop of water on a clean microscope slide.
3. Rupture the ovule by mashing it with dissecting needles and place the slide on the stage of a dissecting microscope.
4. Attempt to locate the embryo, which should be at the heart-shaped stage, similar to Figure 24-10b, and which may be turning green. A portion of the endosperm has become cellular and may be green, too, so be careful with your observation. The "lobes" of the heart-shaped embryo are the young cotyledons. Identify these. Find the trailing **suspensor**, a strand of eight cells that serves as a sort of "umbilical cord" to transport nutrients from the endosperm to the embryo.
5. Add a drop of I$_2$KI to stain the endosperm. What color does the endosperm stain? _____
6. What can you conclude about the composition of the endosperm? (*Hint*: If you've forgotten, consult Exercise 7.)

___

Day 29 (About 20 min.)

MATERIALS

Per student:

- dissecting microscope
- two dissecting needles

- dH$_2$O in dropping bottle
- glass microscope slide

PROCEDURE

1. Remove another pod from a plant and examine it as you did on day 26.
2. Attempt to find the torpedo stage of embryo development, similar to that illustrated in Figure 24-10c.
3. Describe the changes that have taken place in the transition from the heart-shaped to the torpedo stage.

___

Day 31 (About 20 min.)

MATERIALS

Per student:

- dissecting microscope
- two dissecting needles

- dH$_2$O in dropping bottle
- glass microscope slide

PROCEDURE

1. Again, remove and dissect a pod and ovules, and look for the embryo whose cotyledons have now curved, forming the so-called walking-stick stage.
2. Identify the **cotyledons** and the **hypocotyl-root axis** that has curved within the ovules' integuments.

Days 35–37 (About 20 min.)

MATERIALS

Per student:

■ dissecting microscope
■ two dissecting needles

PROCEDURE

The pods of your plants should be drying and have much the same appearance as the pod you started with about a month ago.

1. Remove seeds from mature pods and examine the seeds with your dissecting microscope. Note the bumpy **seed coat**, derived from the integument of the ovule. Look for a tiny hole in the seed coat. This is the **micropyle**!

2. Crack the seed coat and remove the embryo, identifying the cotyledons and hypocotyl-root axis. Separate the cotyledons and look between them, identifying the tiny **epicotyl**, which is the *apical meristem of the shoot*.

3. Harvest the remaining pods and count the number of seeds in each pod. Calculate the average number of seeds per pod and number of seeds per plant. Record this information in Table 24-1.

Congratulations! You have completed the life cycle study of a rapid cycling *Brassica rapa*. Virtually all of the 200,000 species of flowering plants have stages identical to that which you have examined over the course of the past 35–40 days. Most just take longer—from a few days to nearly a century longer!

Conclude your study by preparing the required graphs and answering the questions following Table 24-1.

4. In Figure 24-19, graph the increase in average length of the hypocotyl-root axis produced in the germination chamber.

Does length increase in a constant fashion (linearly), or does extension occur more quickly during a particular time period (logarithmically)? _____

Using the data from Table 24-1, graph in Figure 24-20 the rate of growth of your plants' shoot system as a function of time.

How does the number of seeds produced per pod compare with the number in the original pod from which you harvested your seed at the beginning of the study? _____

What is the *reproductive potential* (the number of possible plants coming from one parent plant) of your RCBr? _____

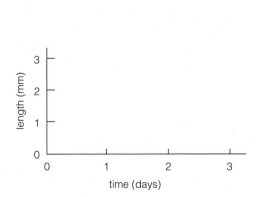

Figure 24-19 Growth of hypocotyl-root axis.

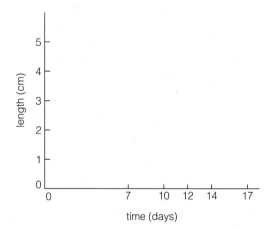

Figure 24-20 Growth of shoot system.

Day	Parameter	Measurement
0	Pod length	mm
	Seed number (single pod)	
	Number of seeds sown in germination chamber	
	Temperature, °C	
	Number of seeds sown in quad (total)	
1	Germinated seeds (germination chamber)	
	Average length of hypocotyl-root axis	mm
2	Germinated seeds (germination chamber)	
	Percentage seeds germinated	
	Average length of hypocotyl-root axis	mm
3	Average length of hypocotyl-root axis	mm
4 or 5	Number of seedlings (quad)	
	Average height of shoot system	mm
7	Average length of first set of true leaves	mm
	Average height of shoot system	mm
10	Average length of second set of true leaves	mm
	Average height of shoot system	mm
12	Average number of leaves per plant	
	Average height of shoot system	mm
14	Average height of shoot system	mm
17	Average height of shoot system	mm
	Average number of flowers per plant	
35–37	Average total number of seeds produced per plant	
	Average number of seeds per pod	

TABLE 24-1 Data for RCBr Life Cycle

© Cengage Learning 2013

_____ 1. Plants that produce flowers are
   (a) members of the Anthophyta.
   (b) angiosperms.
   (c) seed producers.
   (d) all of the above

_____ 2. All of the petals of a flower are collectively called the
   (a) corolla.
   (b) stamens.
   (c) receptacles.
   (d) calyx.

_____ 3. Which group of terms refers to the microsporophyll, the male portion of a flower?
   (a) ovary, stamens, pistil
   (b) stigma, style, ovary
   (c) anther, stamen, filament
   (d) megasporangium, microsporangium, ovule

_____ 4. A carpel is the
   (a) same as a megasporophyll.
   (b) structure producing pollen grains.
   (c) component making up the anther.
   (d) synonym for microsporophyll.

_____ 5. The portion of the flower containing pollen grains is
   (a) the pollen sac.
   (b) the microsporangium.
   (c) the anther.
   (d) all of the above

_____ 6. Which group of terms is in the correct developmental sequence?
   (a) microspore mother cell, meiosis, megaspore, female gametophyte
   (b) microspore mother cell, meiosis, microspore, pollen grain
   (c) megaspore, mitosis, female gametophyte, meiosis, endosperm mother cell
   (d) all of the above

_____ 7. Where does germination of a pollen grain occur in a flowering plant?
   (a) in the anther
   (b) in the micropyle
   (c) on the surface of the corolla
   (d) on the stigma

_____ 8. Double fertilization refers to
   (a) fusion of two sperm nuclei and two egg cells.
   (b) fusion of one sperm nucleus with two polar nuclei and fusion of another with the egg cell nucleus.
   (c) maturation of the ovary into a fruit.
   (d) none of the above

_____ 9. Ovules mature into _____, while ovaries mature into _____.
   (a) seeds/fruits
   (b) stamens/seeds
   (c) seeds/carpels
   (d) fruits/seeds

_____ 10. A bean pod is
   (a) a seed container.
   (b) a fruit.
   (c) a part of the stamen.
   (d) both a and b

Name _____   Section Number _____

EXERCISE **2 4**

Seed Plants II: Angiosperms

Post-Lab Questions

Introduction

1. There are two major groups of seed plants, gymnosperms (Exercise 23) and angiosperms. Compare these two groups of seed plants with respect to the following features:

Feature	Gymnosperms	Angiosperms
(a) type of reproductive structure	_____	_____
(b) source of nutrition for developing embryo	_____	_____
(c) enclosure of mature seed	_____	_____

24.1 External Structure of the Flower

2. What *event*, critical to the production of seeds, is shown at the right?

3. Distinguish between *pollination* and *fertilization*.

(0.84×)

4. Identify the parts of the trumpet creeper flower shown at the right.

   A.

   B.

   C.

   D.

   E.

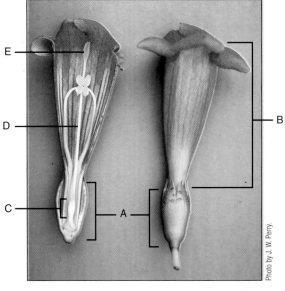

(1×)

5. Examine the photo of the *Trillium* flower pictured at the right.

   (a) Is *Trillium* a monocotyledon or dicotyledon?

   (b) Justify your answer.

(0.75×)

6. Based on your observation of the stigma of the daylily flower in the photo at the right, how many carpels would you expect to comprise the ovary?

(1×)

24.2 The Life Cycle of a Flowering Plant

7. The photo is a cross section of a (an) _____
   _____.

   The numerous circles within the four cavities are
   _____.

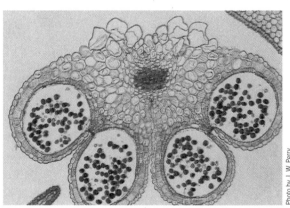

(74×)

8. The photo shows a flower of the pomegranate some time after fertilization. Identify the parts shown.

A. _____

B. _____

C. _____

D. _____

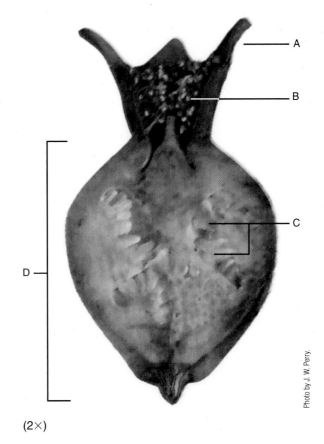

(2×)

Photo by J. W. Perry.

9. Some biologists contend that the term *double fertilization* is a misnomer and that the process should be called *fertilization* and *triple fusion*. Why do they argue that the fusion of the one sperm nucleus and the two polar nuclei is *not* fertilization?

Food for Thought

10. Your roommate says that you need vegetables and asks you to pick up tomatoes at the store. To your roommate's surprise, you say a tomato is not a vegetable, but a fruit. Explain.

Fungi

After completing this exercise, you will be able to

1. define *parasite, saprobe, mutualist, gametangium, hypha, mycelium, multinucleate, sporangium, rhizoid, zygosporangium, ascus, conidium, ascospore, ascocarp, basidium, basidiospore, basidiocarp, lichen, mycorrhiza;*

2. recognize representatives of the major phyla of fungi;

3. distinguish structures that are used to place various representatives of the fungi in their proper phyla;

4. list reasons why fungi are important;

5. distinguish between the structures associated with asexual and sexual reproduction described in this exercise;

6. identify the structures (in **boldface**) of the fungi examined;

7. determine the effect of light on certain species of fungi.

Introduction

As you walk in the woods following a warm fall rain, you are likely to be met by a vast assemblage of colorful fungi. Some grow on dead or diseased trees, some on the surface of the soil, others in pools of water. Some are edible, some deadly poisonous.

Fungi (kingdom Fungi) are *heterotrophic* organisms; that is, they are incapable of producing their own food material. They secrete enzymes from their bodies that digest their food externally. The digested materials are then absorbed into the body.

Depending on the relationship between the fungus and its food source, fungi can be characterized in one of three ways:

1. Parasitic fungi (**parasites**) obtain their nutrients from the organic material of another *living* organism, and in doing so adversely affect the food source, sometimes causing death.
2. Saprotrophic fungi (**saprobes**) grow on nonliving organic (carbon-containing) matter.
3. Mutualistic fungi (**mutualists**) form a partnership *beneficial* to both the fungus and its host.

Fungi, along with the bacteria, are essential components of the ecosystem as decomposers. These organisms recycle the products of life, making the products of death available so that life may continue. Without them we would be hopelessly lost in our own refuse. Fungi and fungal metabolism are responsible for some of the food products that enrich our lives—the mushrooms of the field, the blue cheese of the dairy case, even the citric acid used in making soft drinks.

The kingdom Fungi is divided into four separate phyla, based on structures formed during sexual reproduction (Table 25-1). Certain fungi (the Imperfect Fungi) do not reproduce sexually; hence they are considered an informal "group."

TABLE 25-1	Classification of the Fungi
Phylum	**Common Name**
Chytridiomycota	Chytrids
Zygomycota	Zygosporangium-forming fungi
Ascomycota	Sac fungi
Basidiomycota	Club fungi
"Imperfect Fungi"	Fungi without sexual reproduction

© Cengage Learning 2013

25.1   Phylum Chytridiomycota *(About 5 min.)*

Commonly known as chytrids, these fungi are among the most simple in body form. While most live on dead organic matter, some are parasitic on economically important plants, resulting in damage and/or death of the host plant. Parasitic organisms that cause disease are called **pathogens**.

MATERIALS

Per lab room:

- preserved specimen of potato tuber with black wart disease

PROCEDURE

1. Examine the preserved specimen (if available) of the potato tuber that exhibits the disease known as "black wart." Notice the warty eruptions on the surface of the tuber. The warts are caused by the presence of numerous cells of a chytrid infecting the tuber (Figure 25-1).
2. Now look at Figure 25-2, a photomicrograph of a chytrid similar to that which causes black wart disease.

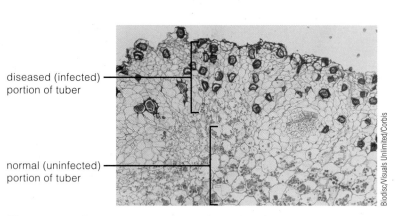

diseased (infected) portion of tuber

normal (uninfected) portion of tuber

Biodisc/Visuals Unlimited/Corbis

Figure 25-1   Section of potato tuber infected by black wart pathogen.

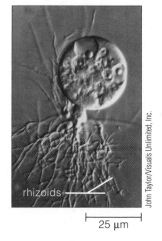

rhizoids

25 μm

John Taylor/Visuals Unlimited, Inc.

Figure 25-2   Chytrid (71×).

25.2   Phylum Zygomycota: Zygosporangium-Forming Fungi *(About 30 min.)*

Commonly called "zygomycetes," all members of this phylum produce a thick-walled zygote called a **zygosporangium**. Most zygomycetes are saprobes. The common black bread mold, *Rhizopus*, is a representative zygomycete. Before the introduction of chemical preservatives into bread, *Rhizopus* was an almost certain invader, especially in high humidity.

MATERIALS

Per student:

- culture of *Rhizopus*
- prepared slide of *Rhizopus*
- dissecting needle
- glass microscope slide
- coverslip
- compound microscope

Per student pair:

- dH$_2$O in dropping bottle

Per lab room:

- demonstration culture of *Rhizopus* zygosporangia, on dissecting microscope

PROCEDURE

Examine Figure 25-7, the life cycle of *Rhizopus*, as you study this organism.

1. Obtain a petri dish culture of *Rhizopus*.
2. Observe the culture, noting that the body of the organism consists of many fine strands. These are called **hyphae** (Figure 25-3). All of the hyphae together are called the **mycelium** Biologists use the term *mycelium* instead of saying "the fungal body."
3. Now note that the culture contains numerous black "dots." These are the **sporangia** (singular, *sporangium*; Figures 25-7a and b). Sporangia contain **spores** (Figures 25-7c–d) by which *Rhizopus* reproduces asexually.
4. Using a dissecting needle, remove a small portion of the mycelium and prepare a wet mount. Examine your preparation with the high-dry objective of your compound microscope.
5. It's likely that when you added the coverslip, you crushed the sporangia, liberating the spores. Are there many or few spores within a single sporangium? _____
6. Identify the **rhizoids** (Figure 25-7a) at the base of a sporangium-bearing hypha. Rhizoids anchor the mycelium to whatever the fungus is growing on.
7. Now observe the demonstration culture of sexual reproduction in *Rhizopus* (Figure 25-4). Two different mycelia must grow in close proximity before sexual reproduction can occur. (The difference in the mycelia is genetic rather than structural. Because they are impossible to distinguish, the mycelia are simply referred to as + and − mating types, as indicated in Figure 25-7.)

Figure 25-3 Bread mold culture (0.5×).

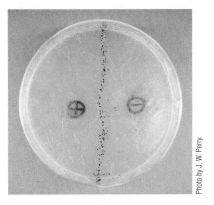

Figure 25-4 Sexual reproduction in the bread mold. Black line consists of zygosporangia (0.5×).

8. Note the black "line" running down the center of the culture plate. These are the numerous **zygosporangia**, the products of sexual reproduction. Here's how the zygosporangia are formed:
   (a) As the hyphae from each mating type grow close together, chemical messengers produced within the hyphae signal them to produce protuberances (Figure 25-7f).
   (b) When the protuberances make contact, gametangia (Figures 25-5, 25-7g) are produced at their tips. Each gametangium contains many haploid nuclei of a single mating type.
   (c) The wall between the two gametangia then dissolves, and the cytoplasms of the gametangia mix.
   (d) Eventually the many haploid nuclei from each gametangium fuse (Figure 25-7h). The resulting cell contains many diploid nuclei resulting from the fusion of gamete nuclei of opposite mating types. Each diploid nucleus is considered a *zygote*. This multinucleate cell is called a zygospore.
   (e) Eventually a thick, bumpy wall forms around this zygospore. Because germination of the zygospore results in the production of a sporangium, the thick-walled zygospore-containing structure is called a **zygosporangium**. Meiosis takes place inside the zygopsporangium so that the spores formed are haploid, some of one mating type, some of the opposite type (Figures 25-6, 25-7i).

9. Obtain a prepared slide of *Rhizopus*. Examine it with the medium- and high-power objective of your compound microscope.
10. Find the stages of sexual reproduction in *Rhizopus*, including gametangia, zygospores, and zygosporangia.

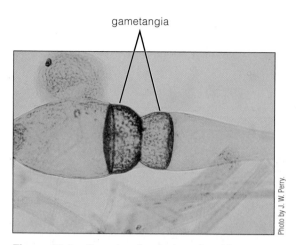

Figure 25-5 Gametangia of a bread mold (230×).

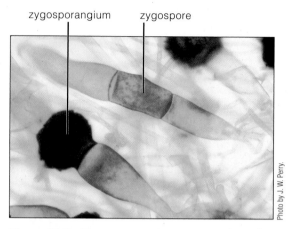

Figure 25-6 Zygospore and zygosporangium of the bread mold (230×).

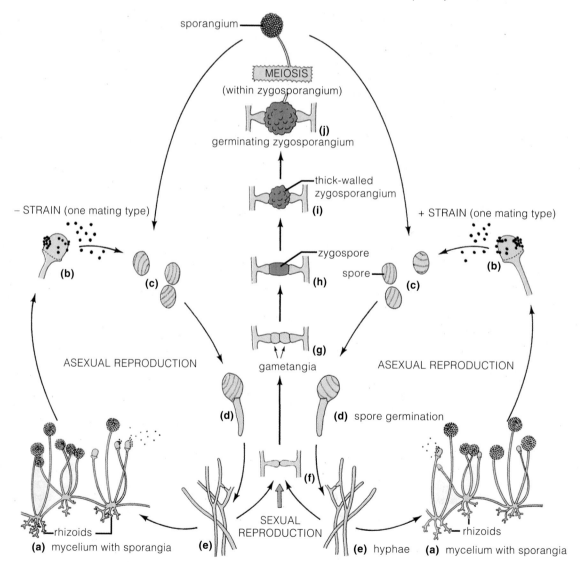

Figure 25-7 Life cycle of *Rhizopus*, black bread mold. (Green structures are 2n.)

Knowing that fungal spores are all around us, why don't we have fungi covering everything we own? Well, given the proper environmental conditions just such a situation can occur. This is particularly true of food products. That's one reason we refrigerate many of our foodstuffs. Producers often incorporate chemicals into some of our foods to retard spoilage, which includes preventing the growth of fungi.

This experiment addresses the hypothesis that *the chemicals placed in bread retard the growth of* Rhizopus. You may wish to review the discussion concerning experimental design in Exercise 1, "The Scientific Method."

MATERIALS

Per student group:

- one slice of bread without preservatives
- one slice of bread with preservatives
- suspension culture of bread mold fungus
- sterile dH$_2$O
- two large culture bowls

- plastic film
- pipet and bulb
- china marker
- metric ruler

PROCEDURE

1. With the china marker, label the side of one culture bowl "NO PRESERVATIVES," the other "WITH PRESERVATIVES." Place the appropriate slice of bread in its respective bowl.
2. Add a small amount of dH$_2$O to the bottom of each culture bowl. This provides humidity during incubation.
3. Add 5 drops of the spore suspension of the bread mold fungus to the center of each slice of bread.
4. Cover the bowls with plastic film and place them in a warm place designated by your instructor.
5. In Table 25-2, list the ingredients of both types of bread.
6. Make a prediction of what you think will be the outcome of the experiment, writing it in Table 25-2.
7. Check your cultures over the next three days, looking for evidence of fungal growth. Measure the diameter of the mycelium present in each culture, using a metric ruler.
8. Record your measurements and conclusion in Table 25-2, accepting or rejecting the hypothesis.

Design an experiment that would allow you to identify the substance that has the effect you observed.

TABLE 25-2	Growth of Bread Mold		
Ingredients			
Bread without preservatives:			
Bread with preservatives:			
Prediction:			
		Mycelial Diameter (cm)	
Culture	Day 1	Day 2	Day 3
With preservatives			
No preservatives			
Conclusion:			

© Cengage Learning 2013

Members of the "ascomycetes" produce spores in a sac, the **ascus** (plural, *asci*), which develops as a result of sexual reproduction. Asexual reproduction takes place when the fungus produces asexual spores called **conidia** (singular, *conidium*). The phylum includes organisms of considerable importance, such as the yeasts crucial to the baking and brewing industries, as well as numerous plant pathogens. A few are highly prized as food, including morels and truffles. Truffles cost in excess of $400 per pound!

MATERIALS

Per student:

- glass microscope slide
- coverslip
- dissecting needle
- prepared slide of *Peziza*
- compound microscope

Per student pair:

- culture of *Eurotium*
- dH₂O in dropping bottle
- large preserved specimen of *Peziza* or another cup fungus

A. *Eurotium:* A Blue Mold

The blue mold *Eurotium* gets its common name from the production of blue-walled asexual conidia.

PROCEDURE

1. Use a dissecting needle to scrape some **conidia** from the agar surface of the culture provided, then prepare a wet mount. (Try to avoid the yellow bodies—more about them in a bit.)
2. Refer to Figure 25-8 as you observe your preparation with the high-dry objective of your compound microscope.
3. Note that the conidia are produced at the end of a specialized hypha that has a swollen tip.

This arrangement has been named *Aspergillus.* This structure, the *conidiophore,* somewhat resembles an aspergillum used in the Roman Catholic Church to sprinkle holy water, from which its name is derived.

(You may be confused about why we italicize *Aspergillus.* The reason is that this is a scientific name. Fungi other than *Eurotium* produce the same type of asexual structure; hence the asexual structure itself is given a scientific name.)

These tiny conidia are carried by air currents to new environments, where they germinate to form new mycelia.

4. Note the yellow bodies on the culture medium. These are the *fruiting bodies,* known as **ascocarps,** which are the products of sexual reproduction (Figure 25-9).
5. With your dissecting needle, remove an ascocarp from the culture and prepare a wet mount.
6. Using your thumb, carefully press down on the coverslip to rupture the ascocarp.

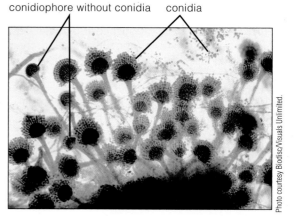

conidiophore without conidia   conidia

Figure 25-8 Conidiophore and conidia of the blue mold, *Eurotium* (138×).

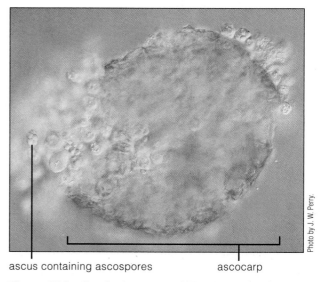

ascus containing ascospores   ascocarp

Figure 25-9 Crushed asocarp of *Eurotium* (287×).

7. Observe your preparation with the medium-power and high-dry objectives of your compound microscope. Identify the **asci**, which contain dark-colored, spherical **ascospores** (Figure 25-9).

The sexual cycle of the sac fungi is somewhat complex and is summarized in Figure 25-10.

The female gametangium, the **ascogonium** (Figure 25-10a; plural, *ascogonia*) is fertilized by male nuclei from antheridia (Figure 25-10a).

The male nuclei (darkened circles in Figure 25-10a) pair with the female nuclei (open circles) but do not fuse immediately.

Papillae grow from the ascogonium, and the paired nuclei flow into these papillae (Figure 25-10b).

Now cell walls form between each pair of sexually compatible nuclei (Figure 25-10c).

Subsequently, the two nuclei fuse; the resultant cell is the diploid ascus (Figure 25-10d). (Consequently, the ascus is a zygote.)

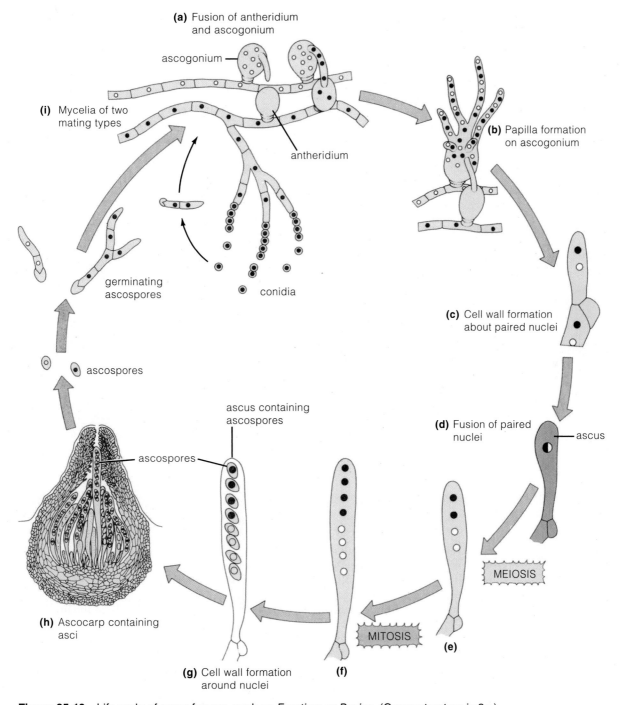

Figure 25-10 Life cycle of a sac fungus, such as *Eurotium* or *Peziza*. (Green structure is 2n.)

The nucleus of the ascus undergoes meiosis to form four nuclei (Figure 25-10e).

Mitosis then produces eight nuclei from these four (Figure 25-10f). (Notice that cytokinesis—cytoplasmic division—does not follow meiosis.)

Next, a cell wall forms about each nucleus, some of the cytoplasm of the ascus being included within the cell wall (Figure 25-10g).

Thus, eight haploid uninucleate ascospores (Figure 25-10g) have been formed.

While asci and ascospores are forming, surrounding hyphae proliferate to form the ascocarp (Figure 25-10h) around the asci.

As you see, two different types of ascospores are produced during meiosis. Each ascospore gives rise to a mycelium having only one mating type (Figure 25-10i).

B. *Peziza:* A Cup Fungus

The cup fungi are commonly found on soil during cool early spring and fall weather. Their sexual spore-containing structures are cup shaped, hence their common name.

PROCEDURE

1. Observe the preserved specimen of a cup fungus (Figure 25-11).
2. Actually, the structure we identify as a cup fungus is the "fruiting body," produced as a result of sexual reproduction by the fungus. Most of the organism is present within the soil as an extensive mycelium. Specifically, the fruiting body is called an **ascocarp**.
3. Obtain a prepared slide of the ascocarp of *Peziza* or a related cup fungus. Examine the slide with the medium-power and high-dry objectives of your compound microscope.
4. Identify the elongated fingerlike **asci**, which contain dark-colored, spherical **ascospores** (Figure 25-12).

Figure 25-11 Cup fungi (0.25×).

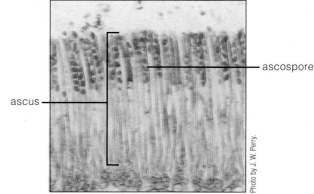

Figure 25-12 Cross section of an ascocarp from a cup fungus (186×).

25.5   Phylum Basidiomycota: Club Fungi *(About 30 min.)*

Members of this group of fungi are probably that first come to mind when we think of fungi, because this phylum contains those organisms commonly called "mushrooms." Actually the mushroom is only a portion of the fungus—it's the fruiting body, specifically a **basidiocarp**, containing the sexually produced haploid **basidiospores**. These basidiospores are produced by a club-shaped **basidium** for which the group is named. Much (if not most) of the fungal mycelium grows out of sight, within the substrate on which the basidiocarp is found.

MATERIALS

Per student:

■ commercial mushroom
■ prepared slide of mushroom pileus (cap), c.s. (*Coprinus*)
■ compound microscope

Per lab room:

■ demonstration specimens of various club fungi

A. Gill Fungi: The Mushrooms

Mushrooms called "gill fungi" have their sexual spores produced on sheets of hyphae that look like the gills of fish.

PROCEDURE

1. Obtain a fresh fruiting body, more properly called a **basidiocarp** (Figure 25-13). Identify the **stalk** and **cap**.

2. Look at the bottom surface of the cap, noting the numerous **gills**. It is on the surface of these gills that the haploid basidiospores are produced. Remember that all the structures you are examining are composed of aggregations of fungal hyphae.

3. Obtain a prepared slide of a cross section of the cap of a mushroom (Figure 25-14). Observe the slide first with the low-power objective of your compound microscope.

4. In the center of the cap, identify the **stalk**. The **gills** radiate from the stalk to the edge of the cap, much as spokes of a bicycle wheel radiate from the hub to the rim.

5. Switch to the high-dry objective to study a single gill (Figure 25-15). Note that the component hyphae produce club-shaped structures at the edge. These are the **basidia** (singular, *basidium*).

6. Each basidium produces four haploid **basidiospores**. Find them. (All four may not be in the same plane of section.)

Each basidiospore is attached to the basidium by a tiny hornlike projection. As the basidiospore matures, it is shot off the projection due to a buildup of turgor pressure within the basidium.

The life cycle of a typical mushroom is shown in Figure 25-16.

Because of genetic differences, basidiospores are of two different types. Mycelia produced by the two types of basidiospores are of two mating strains. Figure 25-16 shows these two different nuclei as open and closed (darkened) circles.

When a haploid basidiospore (Figure 25-16a) germinates, it produces a haploid *primary mycelium* (Figure 25-16b). The primary mycelium is incapable of producing a fruiting body.

Fusion between two sexually compatible mycelia (Figure 25-16b) must occur to continue the life cycle.

Surprisingly, the nuclei of the two mycelia don't fuse immediately; thus, each cell of this so-called *secondary mycelium* (Figure 25-16c) contains two genetically *different* nuclei. This condition is called "dikaryotic."

The secondary mycelium forms an extensive network within the substrate. An environmental or genetic trigger eventually stimulates the formation of the aerial basidiocarp (Figure 25-16d).

Each cell of the basidiocarp has two genetically different nuclei, including the basidia (Figure 25-16e) on the gills.

Within the basidia, the two nuclei fuse; the basidia are now diploid (Figure 25-16f).

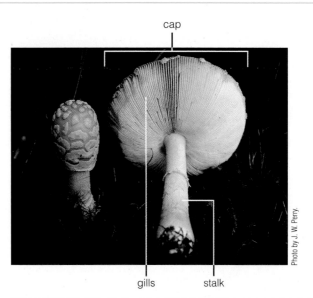

Figure 25-13 Mushroom basidiocarps. The one on the left is younger than that on the right (0.25×).

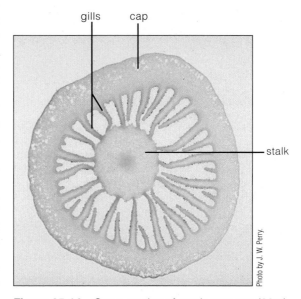

Figure 25-14 Cross section of mushroom cap (23×).

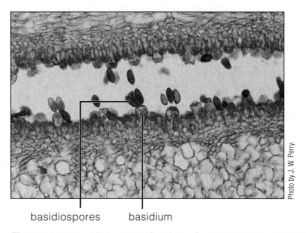

Figure 25-15 High magnification of a mushroom gill (287×).

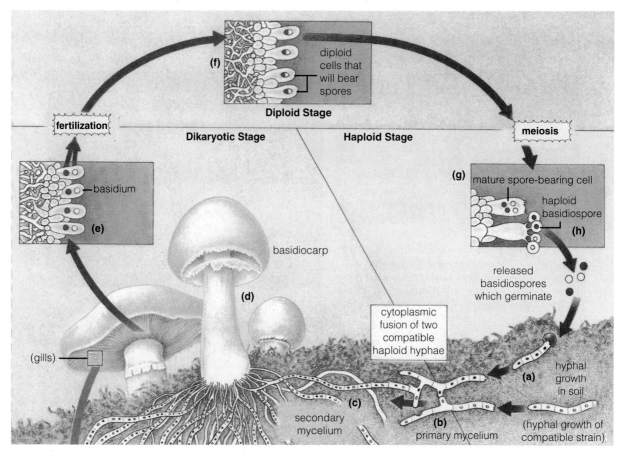

Figure 25-16   Life cycle of a mushroom.

Subsequently, these diploid nuclei undergo meiosis, forming genetically distinct nuclei (Figure 25-16g).

Each nucleus flows with a small amount of cytoplasm through the hornlike projections at the tip of the basidium to form a haploid basidiospore (Figure 25-16h).

Note that the gilled mushrooms do *not* reproduce by means of asexual conidia.

B. Other Club Fungi

A wide variety of basidiomycetes are not mushrooms. Let's examine a few representatives that you're likely to see in the field.

PROCEDURE

1. Examine the representatives of fruiting bodies of other members of the club fungi available in the laboratory.
2. Observe "puffballs" (Figure 25-17).The basidiospores of puffballs are contained within a spherical basidiocarp that develops a pore at the apex. Basidiospores are released when the puffball is crushed or hit by driving rain.
3. Examine "shelf fungi" (Figure 25-18). What you are looking at is actually the basidiocarp of the fungus that has been removed from the surface of a tree.

The presence of a basidiocarp of the familiar shelf fungi indicates that an extensive network of fungal hyphae is growing within a tree, digesting the cells of the wood. As a forester assesses a woodlot to determine the potential yield of usable wood, one of the things noted is the presence of shelf fungi, which indicates low-value (diseased) trees.

The most common shelf fungi are called polypores because of the numerous holes ("pores") on the lower surface of the fruiting body. These pores are lined with basidia bearing basidiospores.

4. Observe the undersurface of the basidiocarp that is on demonstration at a dissecting microscope and note the pores.

Figure 25-17 Puffballs. Note pore for spore escape (0.31×).

Figure 25-18 Shelf fungus growing around a fence post (0.12×).

25.6 "Imperfect Fungi" *(About 15 min.)*

This group consists of fungi for which no sexual stage is known. Someone once decided that in order for a life to be complete—to be "perfect"—sex was necessary. Thus, these fungi are "imperfect." Reproduction takes place primarily by means of asexual conidia.

These fungi are among the most economically important, producing antibiotics (for example, one species of *Penicillium* produces penicillin). Others produce the citric acid used in the soft-drink industry; still others are used in the manufacture of cheese. Some are important pathogens of both plants and animals.

MATERIALS

Per student:

- dissecting needle
- glass microscope slide
- coverslip
- prepared slide of *Penicillium* conidia (optional)
- compound microscope

Per student pair:

- dH₂O in dropping bottle
- culture of *Alternaria*

Per lab room:

- demonstration of *Penicillium*-covered food *and/or* plate cultures

A. Penicillium*

The genus *Penicillium* has numerous species. The common feature of all species in this genus is the distinctive shape of the hypha that produce the conidia (Figure 25-19).

PROCEDURE

1. Examine demonstration specimens of moldy oranges or other foods. The blue color is attributable to a pigment in the numerous conidia produced by this fungus, *Penicillium*.
2. With a dissecting needle, scrape some of the conidia from the surface of the moldy specimen (or from a petri dish culture plate containing *Penicillium*) and

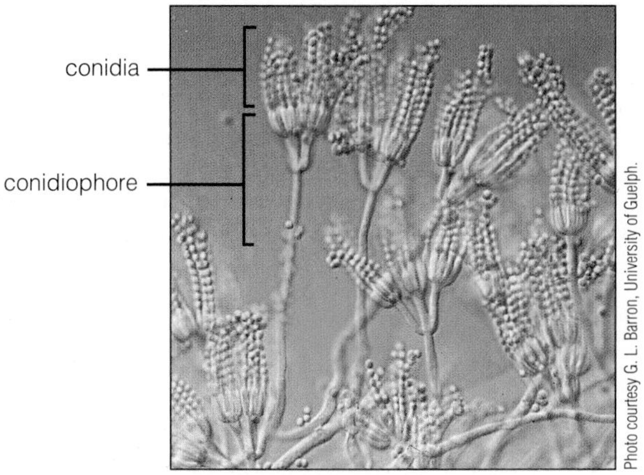

conidia

conidiophore

Figure 25-19 *Penicillium* (300×).

*Some species of *Penicillium* reproduce sexually, forming ascocarps. These species are classified in the phylum Ascomycota. However, not *all* fungi producing the conidiophore form called *Penicillium* reproduce sexually. Those that reproduce only by asexual means are considered "imperfect fungi."

prepare a wet mount. Observe your preparation using the high-dry objective. (Prepared slides may also be available.)

3. Identify the **conidiophore** and the numerous tiny, spherical **conidia** (Figure 25-19). The name *Penicillium* comes from the Latin word *penicillus*, meaning "a brush." (Appropriate, isn't it?)

B. *Alternaria*

Perhaps no other fungus causes more widespread human irritation than *Alternaria,* an allergy-causing organism. During the summer, many weather programs announce the daily pollen (from flowering plants, Exercise 24) and *Alternaria* spore counts as an index of air quality for allergy sufferers.

PROCEDURE

1. From the petri dish culture plate provided, remove a small portion of the mycelium and prepare a wet mount.

2. Examine with the high-dry objective of your compound microscope to find the **conidia** (Figure 25-20). Produced in chains, *Alternaria* conidia are multicellular, unlike those of *Penicillium* or *Aspergillus*. This makes identification very easy, since the conidia are quite distinct from all others produced by the fungi.

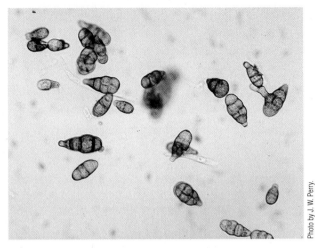

Figure 25-20   Conidia of *Alternaria* (332×).

25.7   Mutualistic Fungi *(About 20 min.)*

Fungi team up with plants or members of the bacterial or protist kingdoms to produce some remarkable mutualistic relationships. The best way to envision such relationships is to think of them as good interpersonal relationships, in which both members benefit and are made richer than would be possible for each individual on its own.

MATERIALS

Per lab room:

- demonstration specimens of crustose, foliose, and fruticose lichens
- demonstration slide of lichen, c.s. on compound microscope
- demonstration slide of mycorrhizal root, c.s. on compound microscope

A. Lichens

Lichens are organisms made up of fungi and either green algae (kingdom Protista) or cyanobacteria (kingdom Eubacteria). The algal or cyanobacterial cells photosynthesize, and the fungus absorbs a portion of the carbohydrates produced. The fungal mycelium provides a protective, moist shelter for the photosynthesizing cells. Hence, both partners benefit from the relationship.

PROCEDURE

1. Examine the demonstration specimen of a **crustose lichen** (Figure 25-21).

2. What is it growing on? _____

_____

_____

3. Write a sentence describing the crustose lichen.

_____

_____

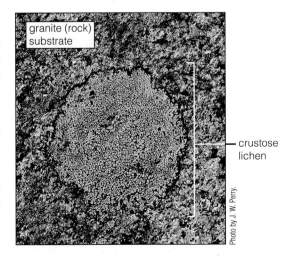

granite (rock) substrate

crustose lichen

Figure 25-21   Crustose lichen (0.5×).

4. Observe the demonstration specimen of a **foliose lichen** (Figure 25-22).
5. What is it growing on? ______________________
6. Write a sentence describing the foliose lichen.

____

____

____

____

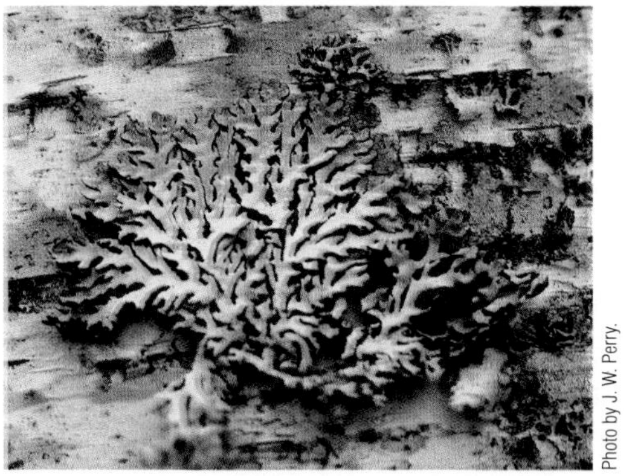
Photo by J. W. Perry.

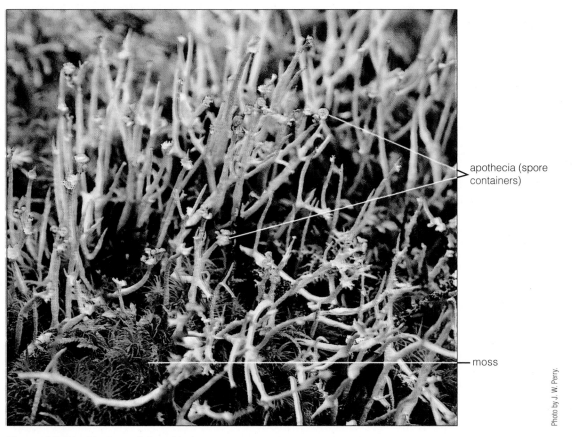

Figure 25-22 Foliose lichen (0.5×).

7. Examine the demonstration specimen of a **fruticose lichen** (Figure 25-23).
8. What is it growing on?

____

____

____

9. Write a sentence describing the fruticose lichen.

____

____

____

apothecia (spore containers)
moss
Photo by J. W. Perry.

Figure 25-23 Fruticose lichen (1×).

10. Observe the cross section of a lichen at the demonstration microscope. This particular lichen is composed of fungal and green algae cells (Figure 25-24).
11. Locate the **fungal mycelium** and then the **algae cells**.
12. It's likely that this slide has a cup-shaped fruiting body. To which fungal phylum does this fungus belong?

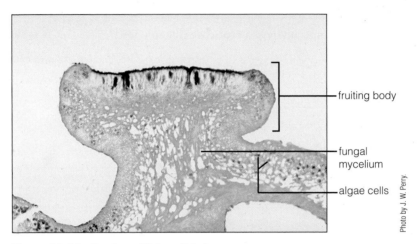

Figure 25-24 Section of lichen (84×).

B. Mycorrhizae

Mycorrhiza literally means "fungus root." A mycorrhiza is a mutualistic association between plant roots and certain species of fungi. In fact, many of the mushroom species (see Section 25.5.A.) you find in forests (especially those that appear to grow right from the ground) are part of a mycorrhizal association.

The fungus absorbs carbohydrates from the plant roots. The fungal hyphae penetrate much farther into the soil than plant root hairs can reach, absorbing water and dissolved mineral ions that are then released to the plant. Again, both partners benefit. In fact, many plants don't grow well in the absence of mycorrhizal associations.

There are two different types of mycorrhizal roots, **ectomycorrhizae** and **endomycorrhizae.** "Ectos" produce a fungal sheath around the root, and the hyphae penetrate between the cell walls of the root's cortex, but not into the root cells themselves. By contrast, "endos" do not form a covering sheath, and the fungi are found within the cells. We'll look at both types.

MATERIALS

Per student:

- ectomyccorhizal pine root, prep. slide, c.s.
- endomyccorhizal root, prep. slide, c.s.
- compound microscope

PROCEDURE

1. Obtain a prepared slide of an ectomycorrhizal pine root. Examine it with the medium- and high-power objective of your compound microscope.
2. Identify the sheath of fungal hyphae that surround the root.
3. Now look for hyphae between the cell walls within the cortex of the root.
4. Draw what you see in Figure 25-25.
5. Obtain a prepared slide of an endomycorrhizal root (Figure 25-26) and examine it as you did above.
6. Identify the hyphae within the plant root cells.

Figure 25-25 Drawing of ectomycorrhizal pine root (_____×).

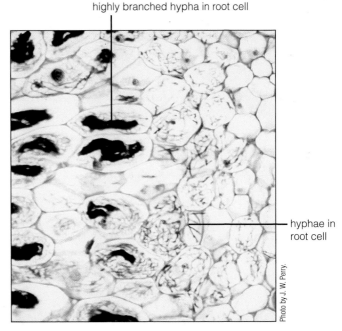

highly branched hypha in root cell

hyphae in root cell

Photo by J. W. Perry.

Figure 25-26 Cross section of endomycorrhizal root (110×).

| 25.8 | Experiment: Environmental Factors and Fungal Growth (About 45 min. to set up) |

Like all living organisms, environmental cues are instrumental in causing growth responses in fungi. Intuitively, we know that light is important for plant growth. Light also affects animals in a myriad of ways, from triggering reproductive events in deer to affecting the complex psyche of humans. What effect might light have on fungi? These experiments enable you to determine whether this environmental factor has any effect on fungal growth.

A. Light and Darkness

This experiment addresses the hypothesis that _light triggers spore production of certain fungi._

MATERIALS

Per experimental group (student pair):
- transfer loop
- bunsen burner (or alcohol lamp)
- matches or striker
- three petri plates containing potato dextrose agar (PDA)
- grease pencil or other marker

Per lab bench:
- test tubes with spore suspensions of _Trichoderma viride_, _Penicillium claviforme_, and _Aspergillus ornatus_

PROCEDURE
Work in pairs, and refer to Figure 25-27, which shows how to inoculate the cultures.

1. Receive instructions from your lab instructor as to which fungal culture you should use.

   Note: Do not open the petri plates until step 3.

2. Use a grease pencil or other lab marker to label each cover with your name and the name of the fungus you have been assigned. Write "CONTINUOUS LIGHT" on one petri plate, "CONTINUOUS DARKNESS" on a second plate, and "ALTERNATING LIGHT AND DARKNESS" on the third one.
3. Following the directions in Figure 25-27, inoculate each plate by touching the loop gently on the agar surface approximately in the middle of the plate.

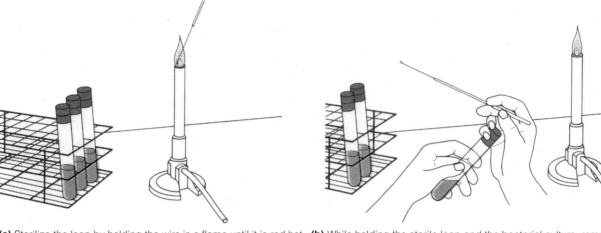

(a) Sterilize the loop by holding the wire in a flame until it is red hot. Allow it to cool before proceeding.

(b) While holding the sterile loop and the bacterial culture, remove the cap as shown.

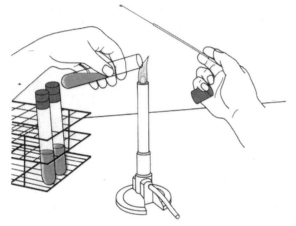

(c) Briefly heat the mouth of the tube in a burner flame before inserting the loop for an inoculum.

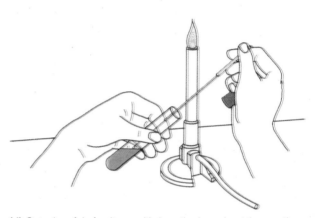

(d) Get a loopful of culture, withdraw the loop, heat the mouth of the tube, and replace the cap.

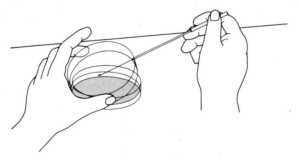

(e) To inoculate a solid medium in a petri plate, place the plate on a table and lift one edge of the cover.

Figure 25-27 Procedure for inoculating a culture plate.

After Case and Johnson, 1984.

4. Place one plate in each condition:
   - continuous light
   - continuous darkness
   - alternating light and dark
5. In Table 25-3, write your prediction of what will occur in the different environmental conditions.
6. Check the results after 5–7 days and make conclusions concerning the effect of light on the growth of the cultures. Record your results in Table 25-3. Draw what you observe in Figure 25-28.
7. Compare the results you obtained with the results of other experimenters who used different species. Record your conclusions in Table 25-3.

TABLE 25-3	Effect of Light on Fungal Sporulation in _____ (insert the fungus name)

Prediction:

Conditions:

Continuous light

Continuous darkness

Alternating light and darkness

Conclusions:

Trichoderma viride

Penicillium claviforme

Aspergillus ornatus

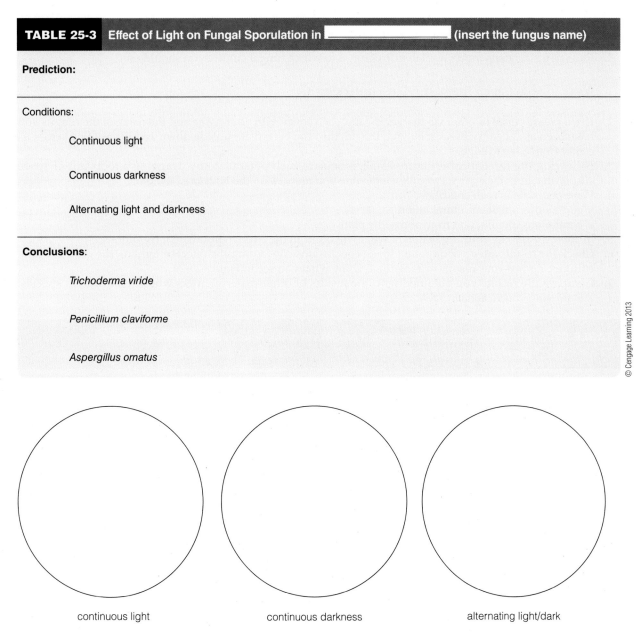

continuous light   continuous darkness   alternating light/dark

Figure 25-28   Drawing of effect of alternating light and darkness on sporulation of _____ (insert name of fungus).

© Cengage Learning 2013

B. Directional Illumination

This experiment addresses the hypothesis that *light from one direction causes the spore-bearing hyphae of certain fungi to grow in the direction of the light.*

MATERIALS

Per experimental group (student pair):

- transfer loop
- bunsen burner (or alcohol lamp)
- matches or striker
- two petri plates containing either potato dextrose agar (PDA) or cornmeal (CM) agar
- grease pencil or other marker

Per lab bench:

- test tubes with spore suspensions of
  Phycomyces blakesleeanus (inoculate onto PDA)
  Penicillium isariforme (inoculate onto PDA)
  Aspergillus giganteus (inoculate onto CM)

PROCEDURE

Work in pairs and refer to Figure 25-27, which shows how to inoculate cultures.

1. Receive instructions from your lab instructor as to which fungal culture you should use.

   Note: **Do not open the petri plates until step 3.**

2. Use a grease pencil or other lab marker to label each cover with your name and the name of the fungus you have been assigned. Without removing the lid, turn the petri plate upside down and place a line across its center from edge to edge.
3. Following the directions in Figure 25-27, inoculate each petri plate by dragging a loop full of spore suspension over the agar surface above the line you marked on the bottom of the plate.
4. Place one plate in each condition:
   - continuous unilateral illumination
   - continuous illumination from above the plate
5. In Table 25-4, write your prediction of what will occur in the two environmental conditions.
6. Check the results after 5–7 days and make conclusions concerning the effect of light on the growth of the cultures. Record your results in Table 25-4. Draw what you see in Figure 25-29.
7. Compare the results you obtained with the results of other experimenters who used different species. Record your conclusions in Table 25-4.

TABLE 25-4	Effect of Directional Illumination on Growth of the Fungus _____ (insert fungus name)

Prediction:

Conditions:

       Unilateral illumination

       Illumination from above

Conclusions:

       Phycomyces blakesleeanus

       Penicillium isariforme

       Aspergillus giganteus

© Cengage Learning 2013

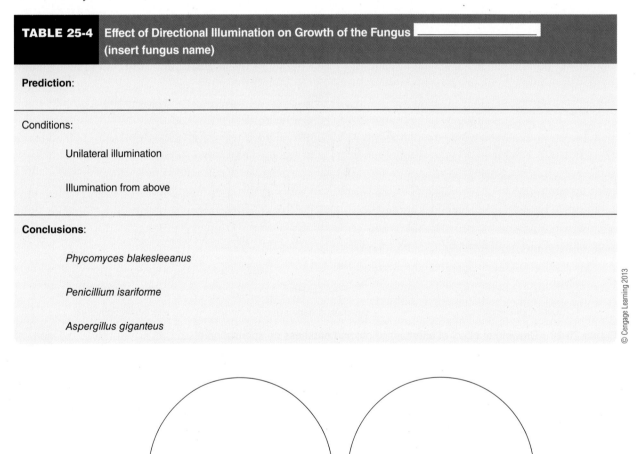

unilateral illumination   light from top of plate

Figure 25-29 Drawing of effect of unilateral illumination on growth of _____ (insert name of fungus).

_____ 1. An organism that grows specifically on nonliving organic material is called
   (a) an autotroph.
   (b) a heterotroph.
   (c) a parasite.
   (d) a saprophyte.

_____ 2. Taxonomic separation into fungal phyla is based on
   (a) sexual reproduction, or lack thereof.
   (b) whether the fungus is a parasite or saprophyte.
   (c) the production of certain metabolites, like citric acid.
   (d) the edibility of the fungus.

_____ 3. Fungi is to fungus as _____ is to _____.
   (a) mycelium/mycelia
   (b) hypha/hyphae
   (c) sporangia/sporangium
   (d) ascus/asci

_____ 4. Which statement is not true of the zygospore-forming fungi?
   (a) They are in the phylum Zygomycota.
   (b) Ascospores are found in an ascus.
   (c) *Rhizopus* is a representative genus.
   (d) A zygospore is formed after fertilization.

_____ 5. Which structures would you find in a sac fungus?
   (a) ascogonium, antheridium, zygospores
   (b) ascospores, oogonia, asci, ascocarps
   (c) basidia, basidiospores, basidiocarps
   (d) ascogonia, asci, ascocarps, ascospores

_____ 6. The club fungi are placed in the phylum Basidiomycota
   (a) because of their social nature.
   (b) because they form basidia.
   (c) because of the presence of an ascocarp.
   (d) because they are dikaryotic.

_____ 7. Which statement is not true of the imperfect fungi?
   (a) They reproduce sexually by means of conidia.
   (b) They form an ascocarp.
   (c) Sex organs are present in the form of oogonia and antheridia.
   (d) All of the above are false.

_____ 8. A relationship between two organisms in which both members benefit is said to be
   (a) parasitic.
   (b) saprotrophic.
   (c) mutualistic.
   (d) heterotrophic.

_____ 9. An organism that is made up of a fungus and an associated green alga or cyanobacterium is known as a
   (a) sac fungus.
   (b) lichen.
   (c) mycorrhizal root.
   (d) club fungus.

_____ 10. A mutualistic association between a plant root and a fungus is known as a
   (a) sac fungus.
   (b) lichen.
   (c) mycorrhizal root.
   (d) club fungus.

EXERCISE **25**

Fungi

25.2 Phylum Zygomycota: Zygosporangium-Forming Fungi

1. Observe this photo at the right of a portion of a fungus you examined in lab.
   (a) What is structure A?

   (b) Are the contents of structure A haploid or diploid?

2. Distinguish between a *hypha* and a *mycelium.*

3. Examine the below photomicrographs of a fungal structure you studied in lab. Is this structure labeled B in the photomicrograph below the product of sexual or asexual reproduction?

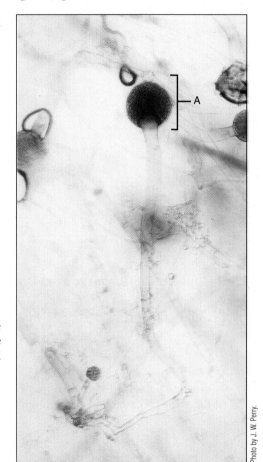

(130×)

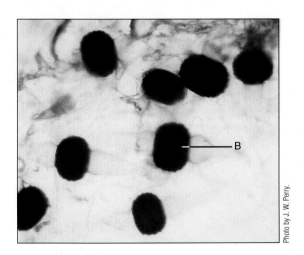

(120×)

25.4 Phylum Ascomycota: Sac Fungi

4. Distinguish among an *ascus*, an *ascospore*, and an *ascocarp*.

5. Walking in the woods, you find a cup-shaped fungus. Back in the lab, you remove a small portion from what appears to be its fertile surface and crush it on a microscope slide, preparing the wet mount that appears at the right. Identify the fingerlike structures present on the slide.

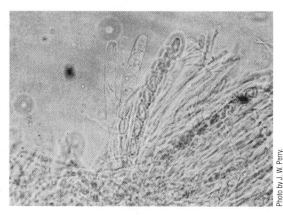

(287×)

25.5 Phylum Basidiomycota: Club Fungi

6. What type of spores are produced by the fungus pictured at the right?

(0.5×)

7. The photo below shows a fungus growing on a dead hemlock tree. What is the common name of a fungus of this sort? Be specific.

(0.1×)

25.6 "Imperfect Fungi"

8. Explain the name "imperfect fungi."

Food for Thought

9. Give the correct singular or plural form of the following words in the blanks provided.

	Singular	Plural
(a)	hypha	
(b)		mycelia
(c)	zygospore	
(d)		asci
(e)	basidium	
(f)		conidia

10. Lichens are frequently the first colonizers of hostile growing sites, including sunbaked or frozen rock, recently hardened lava, and even gravestones. How can lichens survive in habitats so seemingly devoid of nutrients and under such harsh physical conditions?

Sponges and Cnidarians

OBJECTIVES

After completing this exercise, you will be able to

1. define *larva, asymmetry, radial symmetry, bilateral symmetry, cephalization, coelom, pseudocoelom, monoecious, dioecious, invertebrates, spongocoel, osculum, spicules, choanocyte (collar cell), budding, spongin, polyp, tentacles, medusa, nematocyst, gastrovascular cavity, epidermis, gastrodermis, mesoglea;*

2. list the characteristics of animals;

3. describe how the phyla Porifera and Cnidaria respectively show the cell-specialization and tissue levels of organization;

4. explain the basic body plan of members of the phyla Porifera and Cnidaria;

5. describe the natural history of the phyla Porifera and Cnidaria;

6. identify representatives of the classes Calcarea and Demospongiae of the phylum Porifera and classes Hydrozoa, Scyphozoa, and Anthozoa of the phylum Cnidaria;

7. identify structures (and indicate associated functions) of representatives of these classes.

Introduction

Animals are multicellular, oxygen-consuming, heterotrophic (heterotrophs feed on other organisms or on organic waste) *organisms that exhibit considerable motility (spontaneous movement).* Most animals are *diploid* and *reproduce sexually*, although asexual reproduction is also common. The animal life cycle includes a *period of embryonic development, often with a* **larva**—sexually immature, free-living form that grows and transforms into an adult— or equivalent stage. During the embryonic development of most animals, *three primary germ layers—ectoderm, mesoderm,* and *endoderm*—form and give rise to the adult organs and tissues. Evidence accumulated from centuries of study supports the notion that animals evolved from a group of protistan-like ancestors distinct from those that gave rise to the plants and fungi.

Animal diversity is usually studied in the approximate order of a group's first appearance on Earth. Table 26-1 summarizes the trends evident in this sequence for the phyla considered in this and subsequent diversity exercises:

■ In terms of habitat, marine animals precede freshwater and terrestrial forms.

■ The organization of animal bodies is initially **asymmetrical**—cannot cut the body into two like halves—but rapidly becomes symmetrical, exhibiting first **radial symmetry**—any cut from the oral (mouth) to the other end through the center of the animal yields two like halves (Figure 26-1a) —and then bilateral symmetry. In **bilateral symmetry**, a cut in only one plane parallel to the main axis yields like right and left halves (Figure 26-1b).

Bilateral symmetry accompanies the movement of sensory structures and the mouth to the front end of the body to form a head, a process called **cephalization**.

Spaces within the body are first "baskets" used to filter seawater, then incomplete guts (no anus) appear, and finally complete guts (with an anus) are suspended in an entirely

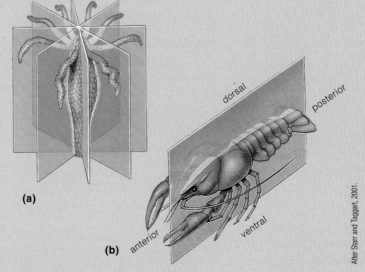

Figure 26-1 Examples of body symmetry. **(a)** Radial symmetry in a cnidarian polyp. **(b)** Bilateral symmetry in a crayfish, an arthropod.

After Starr and Taggart, 2001.

TABLE 26-1 Trends in the Evolution of Animals

Phylum	Common Name	Habitat	Body Organization	Level of Organization
Porifera	Sponges	Aquatic, mostly marine	None or crude radial symmetry	Cell specialization
Cnidaria	Cnidarians	Aquatic, mostly marine	Radial symmetry, incomplete gut	Tissue (ectoderm and endoderm only)
Platyhelminthes	Flatworms	Aquatic, moist terrestrial, or parasitic	Bilateral symmetry, head, incomplete gut	Organ system (all three primary germ layers)
Rotifera	Rotifers	Aquatic (mostly freshwater) or moist terrestrial	Bilateral symmetry, head, complete gut, pseudocoelom	Organ system
Annelida	Annelids some parasitic	Aquatic or moist terrestrial, bilateral symmetry, some parasitic	Bilateral symmetry, head, segmented, complete gut, coelom	Organ system
Mollusca	Molluscans	Aquatic or terrestrial, mostly marine	Unique body plan, complete gut, coelom	Organ system
Nematoda	Roundworms	Aquatic, moist terrestrial, or parasitic	Bilateral symmetry, head, complete gut, pseudocoelom	Organ system
Arthropod	Arthropods	Aquatic, terrestrial (land and air), some parasitic	Bilateral symmetry, head, complete gut, segmented, coelom	Organ system
Echinodermata	Echinoderms	Aquatic, marine	Radial symmetry (larvae are bilaterally symmetrical), no well-defined head, no segmentation, usually complete gut, coelom	Organ system
Chordata	Chordates	Aquatic, terrestrial (land and air)	Bilateral symmetry, head and postanal tail, segmentation, complete gut, coelom	Organ system

enclosed ventral body space. This ventral cavity is called a **coelom** if it is lined by tissue derived from mesoderm, and a **pseudocoelom** if the mesodermal lining is incomplete (Figure 26-2).

■ The level of organization of the animal body proceeds through aggregates of specialized cells, tissues formed by associations of similarly specialized cells, organs comprised of different tissues, and systems of functionally related organs.

■ Body support moves from self-associating scaffolds in sponges to hydrostatic bodies to shells and exoskeletons or endoskeletons comprised of internal support structures.

■ Sexual reproduction predominates in animals, but earlier phyla are asexual as well. Earlier forms are often **monoecious** or *hermaphroditic*—individuals have both male and female reproductive structures. Later forms are **dioecious** with male and female reproductive structures in separate sexes. Fertilization of the egg by sperm is usually external in *sessile* (immobile and attached to a surface such as the sea bottom) animals and internal in most mobile forms, and *parthenogenesis* (the egg develops without fertilization) occurs sporadically.

Body Support	Life Cycle	Special Characteristics
Spicules, spongin	Sexual reproduction (mostly monoecious), larva (amphiblastula), asexual reproduction (budding)	Spongocoel, osculum, gemmules, choanocytes (collar cells)
Hydrostatic gut	Sexual reproduction (monoecious or dioecious), larva (planula), asexual reproduction (budding and fragmentation)	Nerve net, nematocysts, two adult body forms—polyp (sessile) and medusa (free swimming)
Muscular	Sexual reproduction (mostly monoecious with internal fertilization); parasites have very complicated life cycles; some asexual reproduction (mostly transverse fission)	Brain, flame cells
Hydrostatic pseudocoelom	Sexual (dioecious with internal fertilization), some parthenogenesis, males much smaller than females or absent	Brain, flame cells, wheel organ, mastax, microscopic animals
Hydrostatic coelom	Sexual reproduction (monoecious or dioecious with mostly internal fertilization), larvae in some, asexual reproduction (mostly transverse fission)	Brain, nephridia, gills in some
Shells, floats (aquatic)	Sexual (monoecious or dioecious with external or internal fertilization), larvae in some	Brain, nephridium or nephridia, most have gills or lungs, external body divided into head, mantle, and foot
Hydrostatic pseudocoelom	Sexual reproduction (mostly dioecious with internal fertilization)	Brain, molt to grow
Exoskeleton of chitin	Sexual reproduction (mostly dioecious with internal fertilization), larvae, metamorphosis, some parthenogenesis	Brain, green glands or Malphigian tubules, gills, tracheae or book lungs, body regions (head, thorax, and abdomen), specialized segments form mouthparts and jointed appendages, molt to grow
Dermal endoskeleton of calcium carbonate (mesoderm)	Sexual reproduction (mostly dioecious with external fertilization), larvae, metamorphosis, high regenerative potential	Radial symmetry, water-vascular system
Most have an endoskeleton of cartilage or bone	Sexual reproduction (mostly dioecious with internal or external fertilization), some larvae, some metamorphosis	Brain, ventral heart, notochord, gill slits, and dorsal hollow nerve cord at some stage of life cycle

© Cengage Learning 2013

CAUTION

Specimens are kept in preservative solutions. Use lab gloves and safety goggles, especially if you wear contact lenses, whenever you or your lab mates handle a specimen. If safety goggles do not fit, your glasses should suffice. Wash any part of your body exposed to preservative with copious amounts of water. If preservative solution is splashed into your eyes, wash them with the safety eyewash bottle or eye irrigation apparatus for 15 minutes.

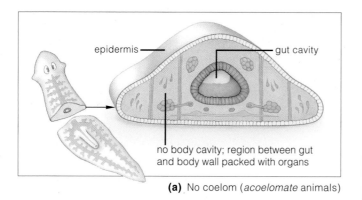

epidermis — — gut cavity

no body cavity; region between gut and body wall packed with organs

(a) No coelom (*acoelomate* animals)

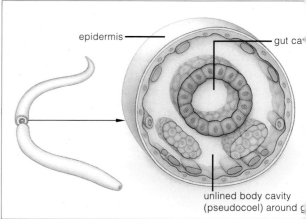

epidermis — — gut ca

unlined body cavity (pseudocoel) around g

(b) Pseudocoel (*pseudocoelomate* anir

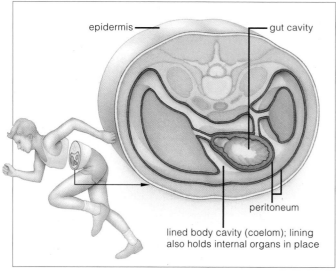

epidermis — — gut cavity

peritoneum

lined body cavity (coelom); lining also holds internal organs in place

(c) Coelom (coelomate animals)

Figure 26-2 Development of ventral body cavity in animals.

26.1 Sponges (Phylum Porifera) *(45 min.)*

Habitat	aquatic, mostly marine
Body Arrangement	asymmetrical or crude radial symmetry
Level of Organization	cell specialization
Body Support	spicules and/or spongin
Life Cycle	sexual reproduction (mostly monoecious), larva (amphiblastula), asexual reproduction (budding and fragmentation)
Special Characteristics	spongocoel, osculum, gemmules, choanocytes (collar cells)

Aristotle in his *Scala Naturae* said of sponges, "In the sea there are things which it is hard to label as either animal or vegetable." This was a reference to their sensitivity to stimulation and their relative lack of motility.

Sponges arose early from the main line of animal evolution (Figure 26-3). Most sponges are marine, and all forms are *aquatic*. As adults, sponges are *sessile,* but many disperse as free-swimming larvae. Although sponges possess a variety of different types of cells, they are atypical animals in their lack of definite tissues. They exhibit the *cell-specialization level of organization,* as there is a division of labor among the different cell types. Cells are organized into layers, but these associations of cells do not show all the characteristics of tissues. Sponges have a *crude radial symmetry or are asymmetrical.* Sponges along with the animals of most phyla are **invertebrates**—animals that do not posses a backbone or vertebral column.

Three body plans exist in sponges (Figure 26-4). They are referred to as *asconoid, syconoid,* and *leuconoid* and increase in the complexity of their structure in this order.

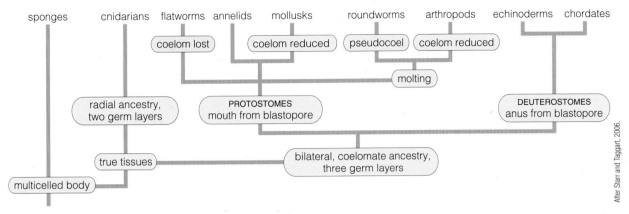

Figure 26-3 Presumed family tree for animals.

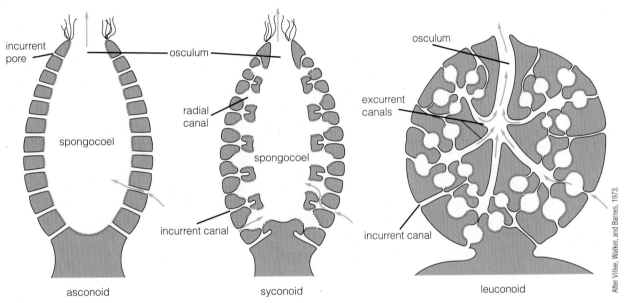

Figure 26-4 Three body plans of sponges. The blue arrows indicate the direction of water flow. Pink areas are lined by collar cells.

Water enters a vase-shaped asconoid sponge through *incurrent pores*; flows directly to a large central internal cavity, the **spongocoel**; and exits through a larger opening, the **osculum**. A syconoid sponge is similar, except that the inner side of the body wall pockets in to form *radial canals* into which branches of the *incurrent canals* drain. In a leuconoid sponge, the radial canals are internalized as chambers within the wall of the sponge. These chambers empty into *excurrent canals* that in turn lead to the osculum. Concurrent with the development of increasingly complex patterns of water flow, the area lined with *collar cells*, which in asconoid sponges is the inner wall of the spongocoel, also becomes internalized, first lining the radial canals of syconoid sponges and then the chambers of leuconoid sponges.

MATERIALS

Per student:

- dissection microscope
- hand lens
- compound microscope, lens paper, a bottle of lens-cleaning solution (optional), a lint-free cloth (optional)
- clean microscope slide and coverslip (optional)

Per student pair:

- preserved specimen of *Scypha (Grantia)*
- prepared slides of
  Scypha spicules (optional)
  a longitudinal section of *Scypha*
  a whole mount of *Leucosolenia*

- a whole mount of teased commercial sponge fibers
- a whole mount of *Spongilla* spicules
  freshwater sponge gemmules

Per student group (4):

- 100-mL beaker of heatproof glass (optional)
- bottle of 5% sodium hydroxide (NaOH) or potassium hydroxide (KOH) (optional)
- waste container for 5% hydroxide solution (optional)
- hot plate (optional)
- squeeze bottle of dH₂O (optional)

Per lab room:

■ demonstration collection of commercial sponges
■ squeeze bottle of 50% vinegar and 50% water
■ demonstration of living or dried freshwater sponges
■ boxes of different sizes of lab gloves
■ box of safety goggles

PROCEDURE

A. Class Calcarea: Calcareous Sponges

Members of this class contain **spicules** (skeletal elements) composed of calcium carbonate ($CaCO_3$). All three body plans are represented in this group. You will study members of the genera *Scypha*, which all have the syconoid body plan, and *Leukosolenia*, which all have asconoid body plans.

1. Dried cluster of *Scypha*.
   (a) With a hand lens or your dissection microscope, examine the general morphology of *Scypha* (Figure 26-5). Identify the osculum. Note the pores in the body wall. Observe the long spicules surrounding the osculum and the shorter spicules protruding from the surface of the sponge.
   (b) Asexual reproduction occurs by **budding**. Budding is a process whereby parental cells differentiate and grow outward from the parent to form a bud, sometimes breaking off to become a new individual. *Scypha* often buds from its base. If the buds don't separate, a cluster of individuals results. Look for buds on your specimen.

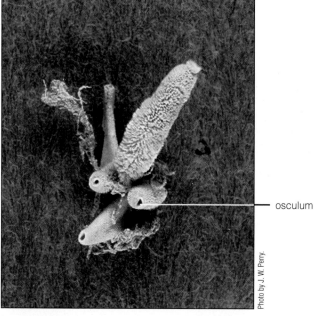

osculum

Figure 26-5 A cluster of dried specimens of the syconoid sponge *Scypha* (3×).

Photo by J. W. Perry.

CAUTION

Hot 5% hydroxide solutions are corrosive. Handle with extreme care. Wear lab gloves and safety goggles. If splashed, wash exposed surface for 15 minutes in running tap water; then flood the area with a 50% vinegar-and-water solution.

---

2. Examine a wet mount (w.m.) or a prepared slide of *Scypha* spicules with your compound microscope, using the medium-power objective. Draw what you observe in Figure 26-6. Follow these instructions to prepare the wet mount:
   (a) Place a small piece of a specimen of *Scypha* in a 100-mL heatproof beaker.
   (b) Carefully add enough 5% sodium hydroxide or potassium hydroxide solution to the beaker to cover the piece and bring this to a boil on the hot plate. Allow the beaker to cool.
   (c) Allow the remains of the sponge to settle to the bottom of the beaker and carefully pour off the liquid into the waste container while retaining the residue.
   (d) Wash the residue by adding distilled water. After the remains of the sponge settle, carefully pour off the water into the waste container.
   (e) Make a wet mount of the residue and draw a few of them in Figure 26-6.
3. Longitudinal section of *Scypha*.
   (a) Orient yourself to the general body plan and other features of *Scypha* (Figure 26-7). Examine a longitudinal section of this sponge with a compound microscope. In your section, trace the path of water as it flowed into and through the sponge in nature (Figure 26-8a). Identify the following structures and place them in the correct order for the water's circulation through the sponge by writing a 1st, 2nd, and 3rd, and so on in the second column of Table 26-2.

Figure 26-6 Calcium carbonate spicules of *Scypha*, w.m. (_____×).

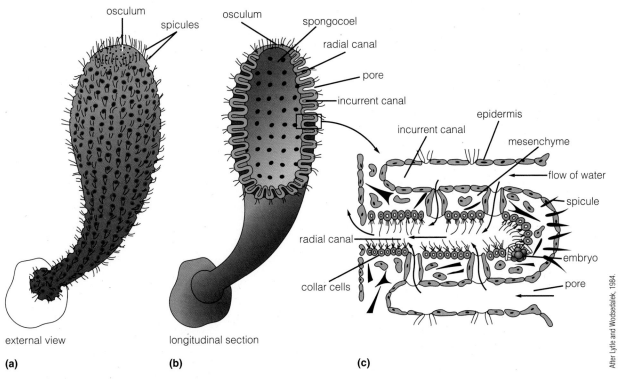

After Lytle and Wodsedalek, 1984.

(a) external view

(b) longitudinal section

(c)

Figure 26-7 *Scypha.* (**a**) External view. (**b**) Longitudinal section. (**c**) Magnified to show detail.

TABLE 26-2	Circulation of Water through the Syconoid Sponge *Scypha*
Structure	**Order for Water Circulation**
Spongocoel	
Incurrent canals	
Radial canals	
Incurrent pores	
Osculum	

© Cengage Learning 2013

(b) Examine the **collar cells**, which appear as numerous small, dark bodies lining the radial canals (Figure 26-8b). Each bears a flagellum that projects into the radial canal, but it is unlikely that you will be able to see the flagella, even with the high-dry objective. Examine the structure of a collar cell in Figure 26-9. The beating of the flagella of collar cells likely creates currents that help move water through the sponge. The microvilli—fingerlike projections that comprise their collars—play a major role in filtering fine food particles out of the water. Collar cells and perhaps other cells engulf and digest food fragments carried into the sponge by the flow of water.

(c) Identify the *epidermis* with its flat protective cells. It covers the outer surface and incurrent canals, and lines the spongocoel.

(d) Note that between the outer epidermis and inner epidermis lies a region of gelatinous material sometimes called the *mesohyle,* embedded in which are amoeboid cells and spicules.

(e) Sexual reproduction occurs in sponges. Although it's unlikely that you will observe gametes, look for early developmental stages (embryos) of *amphiblastula larvae* in the wall, perhaps protruding into the radial canals. These appear as relatively large, dark structures (Figures 26-7 and 26-8a). They would have soon finished development, broken loose, and exited the sponge through the osculum.

4. With your compound microscope, look for spicules and buds on the prepared slide of a whole mount of the asconoid sponge *Leucosolenia* (Figure 26-10).

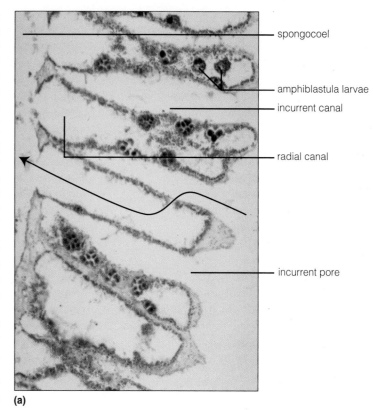

(a)

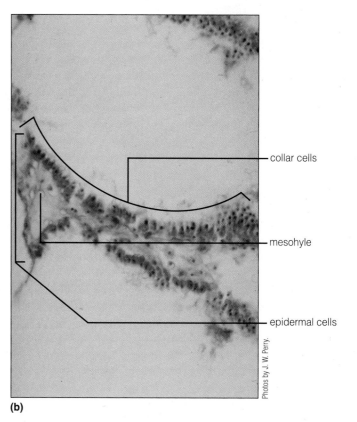

(b)

Figure 26-8   Longitudinal sections of *Scypha.* (**a**) The arrow indicates the path of water flow (96×). (**b**) Radial canals lined with collar cells (384×).

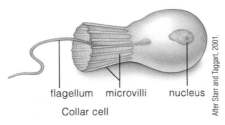

Figure 26-9   Choanocyte (collar cell).

flagellum   microvilli   nucleus

Collar cell

After Starr and Taggart, 2001.

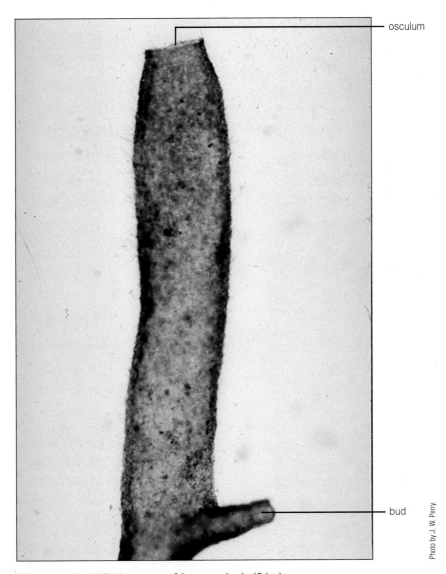

osculum

bud

Photo by J. W. Perry.

Figure 26-10   Whole mount of *Leucosolenia* (34×).

B. Class Demospongiae: Commercial and Freshwater Sponges

This class contains the majority of sponge species. Most sponges in this class are supported by **spongin** (Figure 26-11b, a flexible substance chemically similar to human hair), by spicules of silica, or both. Silica is silicon dioxide, a major component of sand and of many rocks, such as flint and quartz. All sponges in this class are leuconoid and are able to hold and pass large volumes of water in and through their bodies because of their complex canal systems. Some sponges with skeletons composed only of spongin fibers were once commercially valuable as bath sponges, but synthetic sponges have largely replaced them in today's marketplace.

1. Look over the demonstration specimens of commercial sponges available in your lab room (Figure 26-11a).

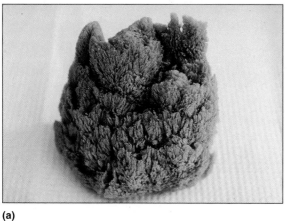 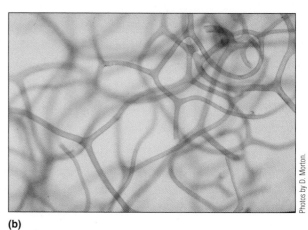

(a)   (b)

Figure 26-11 (a) Bath sponge. (b) Spongin (74×).

2. With your compound microscope, examine a prepared slide of teased commercial sponge fibers and draw them in Figure 26-12.
3. Examine a prepared slide of silica spicules obtained from a common freshwater sponge of the genus *Spongilla* and draw several of them in Figure 26-13.

Figure 26-12 Teased spongin fibers of a commercial sponge, w.m. (_____×).

Figure 26-13 Silica spicules of *Spongilla*, w.m. (_____×).

4. Freshwater and some marine sponges in this class reproduce asexually by forming *gemmules*, resistant internal buds. In the freshwater varieties, a shell of silica and spicules surrounds groups of cells. During the unfavorable conditions of winter, disintegration of the sponge releases the gemmules. Under the favorable conditions of spring, cells emerge from the gemmules and develop into new individuals. Find and examine a gemmule on the prepared slide of freshwater sponge gemmules (Figure 26-14).

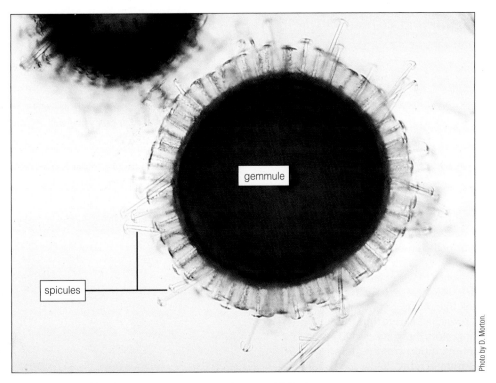

Figure 26-14 Freshwater sponge gemmule, w.m. (150×).

5. Observe the demonstration of living freshwater sponges available in your lab room (Figure 26-15). They look like yellowish-brown, wrinkled scum with perforations and sometimes have a greenish tint due to the presence of symbiotic algae. They are fairly easy to find in nature; search for them in clear, well-oxygenated water, encrusting submerged plant stems and debris such as a waterlogged twig. Slow streams and the edges of ponds where there is some wave action are good places to find freshwater sponges.

Figure 26-15 Dried freshwater sponges, *Sporilla fragilis* with gemmules (0.5×).

Habitat	aquatic, mostly marine
Body Arrangement	radial symmetry, gut (incomplete)
Level of Organization	tissue (derived from ectoderm and endoderm)
Body Support	hydrostatic (gastrovascular cavity and epitheliomuscular cells)
Life Cycle	sexual reproduction (monoecious or dioecious), larva (planula), asexual reproduction (budding)
Special Characteristics	tissues, two adult body forms—polyp (sessile) and medusa (free swimming)—nematocysts, nerve net

This phylum contains some of the most beautiful organisms in the seas. Adult bodies are sac- or bell-shaped and their mouths are surrounded with **tentacles** that move in response to water currents (Figure 26-16). Many are brightly colored and look like plants or flowers.

Cnidarians branched early from the protistan-like ancestors that gave rise to the main line of evolution from which other animals sprang (Figure 26-3).

These animals are *aquatic* and found mostly in shallow marine environments, with the notable exception of the freshwater hydras. Cnidarians have definite tissues and thus show a *tissue level of organization*. They are mostly *radially symmetrical*. Members of this phylum have *guts* and definite *nerve nets*.

Two main body forms exist in cnidarians (Figure 26-17): a **polyp** that is cylindrical in shape, with the oral end and **tentacles** directed upward and the aboral (away from the mouth) surface attached to the sea floor or freshwater bed and a free-swimming **medusa** (jellyfish) that is bell or umbrella shaped, with the oral end and **tentacles** directed downward. Either or both body forms may be present in the Cnidarian life cycle. Also found throughout the group is a larval form called a *planula*.

Cnidarians are efficient predators. All are carnivores that use their tentacles to capture unwary invertebrate and vertebrate prey. Food is captured with the aid of stinging elements called **nematocysts** (Figure 26-18). The nematocysts are discharged by a combination of mechanical and chemical stimuli. The prey is either pierced by the nematocyst or entangled in its filament and pulled toward the mouth by the tentacles. In some species, nematocysts discharge a toxin that paralyzes the prey. The mouth opens to receive the food, which is deposited in the gut. Digestion is started in the gut cavity by enzymes secreted by its lining. It is completed inside the lining cells after they engulf the partially digested food.

Photo by W. A. Yoder.

Figure 26-16   Pink-tipped sea anemone with mutualistic spotted cleaning shrimp (1×).

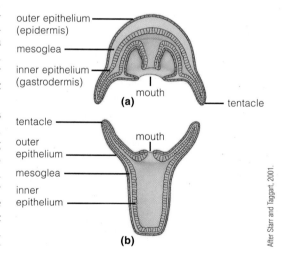

After Starr and Taggart, 2001.

Figure 26-17   Two body forms of Cnidarians. (**a**) Medusa. (**b**) Polyp.

MATERIALS

Per student:

- dissection microscope
- compound microscope, lens paper, a bottle of lens cleaning solution (optional), a lint-free cloth (optional)
- clean microscope slide and coverslip

Per student pair:

- prepared slides of
  whole mounts of *Hydra* (showing testes, ovaries, buds, and embryos)
  cross section of *Hydra*
- whole mounts of *Obelia* (showing colonial polyps and medusa)

Per student group (4):

■ small finger bowl
■ scalpel or razor blade
■ preserved specimen of *Gonionemus*
■ prepared slides of
    a whole mount of the medusa (*ephyra*) of
        Aurelia
    a whole mount of planula larvae of *Aurelia*
    a whole mount of scyphistoma of *Aurelia*
    a whole mount of the strobila of *Aurelia*

Per lab room:

■ demonstration specimens of *Physalia* and *Metridium*
■ demonstration collection of corals
■ boxes of different sizes of lab gloves
■ box of safety goggles

Per lab section:

■ culture of live *Hydra*
■ container of fresh or artificial stream water
■ dropper bottle of glutathione
■ culture of live copepods or cladocerans
■ dropper bottle of vinegar

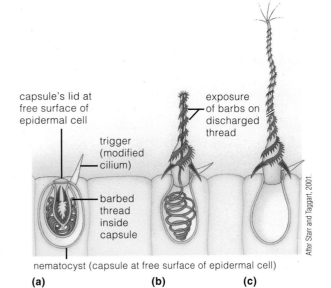

nematocyst (capsule at free surface of epidermal cell)

(a)   (b)   (c)

Figure 26-18   One type of nematocyst, a capsule with an inverted thread inside. This one has a bristlelike trigger (**a**). When prey (or predators) touch the trigger, the capsule becomes more "leaky" to water. As water diffuses inward, pressure inside the capsule increases and the thread is forced to turn inside out (**b**). The thread's tip extends to penetrate the prey, releasing a toxin as it does this (**c**).

PROCEDURE

A. Class Hydrozoa: Hydrozoans

This is mostly a marine group, except for the hydras, which live in freshwater. A polyp, medusa, or both may be present in the life cycle of a hydrozoan. Marine polyps tend to be colonial.

1. *Hydra* is a solitary polyp, and it does not have a medusae stage in its life cycle.
    (a) Place a live specimen of *Hydra* in a finger bowl of fresh stream water. Examine the live *Hydra* with a dissection microscope, along with prepared slides of whole mounts with the compound microscope, noting its polyp body form and other structural details (Figure 26-19).
    (b) Observe the tentacles at the oral end (Figure 26-19a). Find the elevation of the body at the base of the tentacles, in the center of which lies a mouth (Figure 26-19b). Note the swellings on the tentacles (Figure 26-19a); each is a stinging cell (*cnidocytes*) that contains a nematocyst. The clear area in the center of the body represents the gut or gastrovascular cavity (Figure 26-19d).
    (c) Look for evidence of sexual reproduction: one to several *testes* or an *ovary* on the body, both of which may occur on the same individual. Testes (Figure 26-19a) are conical and located more orally than the broad, flattened ovary (Figure 26-19b). You may be able to see a zygote or embryo marginally attached to or detached from the body of the *Hydra* (Figure 26-19c). Is *Hydra* monoecious or dioecious?

_____

    (d) *Hydra* also reproduces asexually by budding (Figure 26-19d). Look for buds on your specimens.
    (e) Examine a cross section of *Hydra* (Figure 26-20) and locate the epidermis. The epidermis is derived from ectoderm, one of the primary germ layers. The **epidermis** contains *epitheliomuscular cells, interstitial cells,* and a few stinging cells. Locate the lining of the gastrovascular cavity, or **gastrodermis**, which is derived from the endoderm and composed of epitheliomuscular cells (sometimes called *nutritive muscular cells*), gland cells, and interstitial cells. Between these two layers is a thin gelatinous membrane called the **mesoglea**. In *Hydra*, the mesoglea is without cells.

The two layers of epitheliomuscular cells contract in opposite directions, although in the *Hydra* epidermis the longitudinally arranged contractile filaments are more developed than the circular filaments of the gastrodermis. How might two layers of epitheliomuscular cells function to maintain and change the body shape of a Cnidarian?

_____

_____

_____

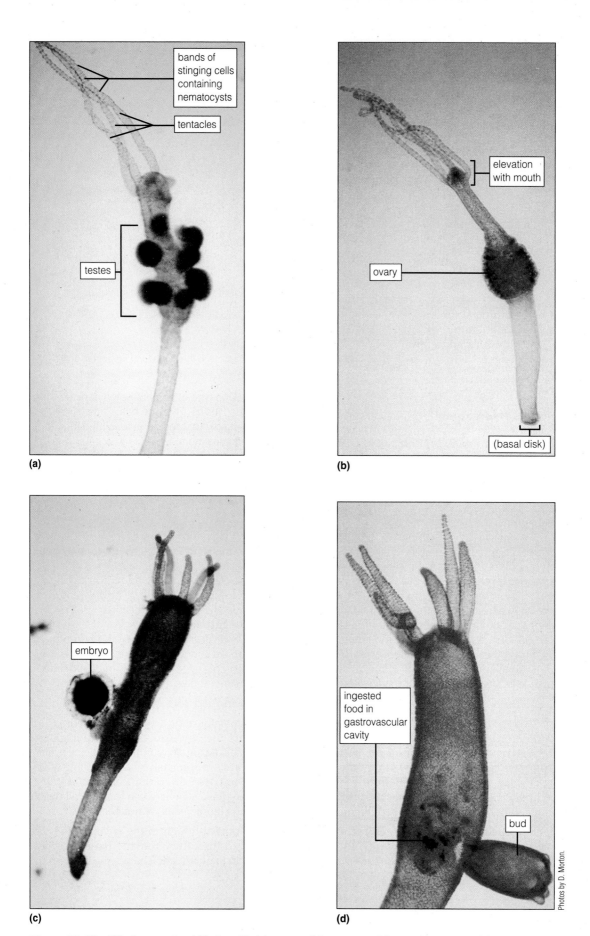

Figure 26-19 Whole mounts of *Hydra* with (**a**) testes, (**b**) an ovary, (**c**) an embryo, and (**d**) a bud (30×).

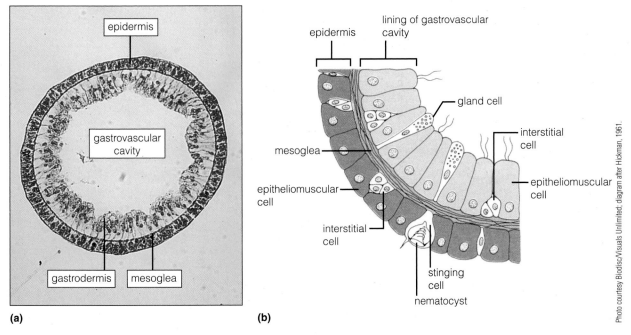

Figure 26-20 *Hydra.* (**a**) Cross section (100×). (**b**) Part of the wall.

What additional function can you suggest for the epitheliomuscular cells of the gastrodermis? (*Hint:* What happens to partially digested food?)

_____

_____

_____

_____

Interstitial cells can give rise to sperm, eggs, and any other type of cell in the body.

(**f**) If it is not already there, place the live *Hydra* in a small finger bowl to the dissection microscope. Add a drop of glutathione and a number of live copepodan or cladoceran crustaceans under the water. The glutathione stimulates the feeding response of *Hydra.* Look for feeding behavior by the *Hydra.* Note the action of the tentacles and the nematocysts. If you do not observe the discharge of nematocysts (Figure 26-21), place a drop of vinegar in the finger bowl and observe carefully. Describe what you observe.

_____

_____

_____

(**g**) Are any of your *Hydra* green? If so, use a scalpel or razor blade to remove a tentacle. Make a wet mount of the tentacle, being certain to crush it with the coverslip. What do you observe with your compound microscope? Can you explain the green color of your *Hydra?*

_____

_____

_____

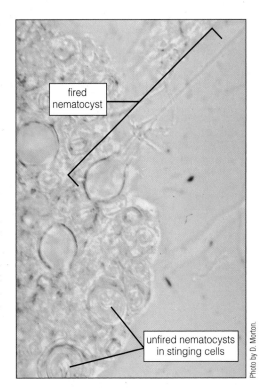

Figure 26-21 Portion of a tentacle of *Hydra* with fired and unfired nematocysts, w.m. (800×).

2. *Obelia* and *Gonionemus* are common examples of hydrozoans with both a polyp and a medusae stage in their life cycles (Figure 26-22).

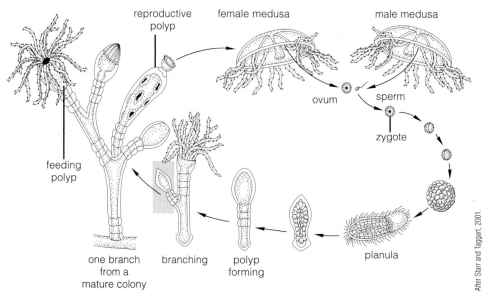

Figure 26-22 Life cycle of the marine hydrozoan *Obelia* that includes medusa and polyp stages. The medusa stage, produced asexually by a reproductive polyp, is free swimming. Fusion of gametes from male and female medusae leads to a zygote, which develops into a planula. The swimming or crawling planula develops into a polyp, which branches to start a new colony. As growth continues, new polyps are formed.

(a) With your compound microscope, examine a prepared slide of a whole mount of a colony of *Obelia* polyps (Figure 26-23a). Identify *feeding* and *reproductive polyps*. The reproductive polyps contain developing medusae.

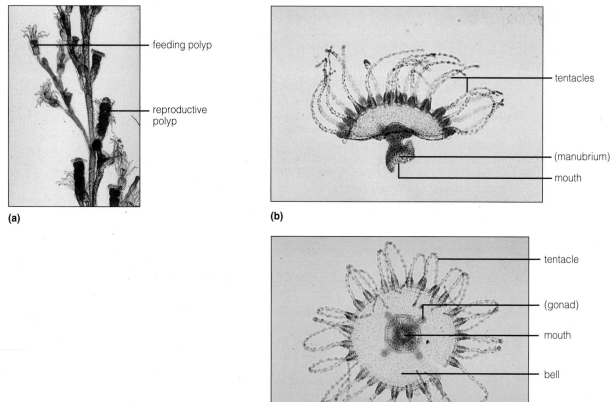

Figure 26-23 Whole mounts of *Obelia*.
(a) Colonial polyps (15×). (b) Side view of a medusa (92×). (c) Top view of a medusa (86×).

(b) Similarly observe a whole mount of a medusa and identify the *mouth*, the umbrella-like *bell*, and *tentacles* bearing bands of stinging cells (Figure 26-23b and c).

(c) Switch to a slide of a whole mount of a *Gonionemus* medusa. Compare your preserved specimen to the live medusa seen in Figure 26-24. In addition to the structures listed in Figure 26-23c, identify the muscular *velum* and the *adhesive pads* on the tentacles. Between the bases of the tentacles are balancing organs called *statocysts*. Medusa move by "jet propulsion." Contractions of the body bring about a pulsing motion that alternately fills and empties the cavity of the bell.

3. Look at the demonstration specimen of a preserved Portuguese man-of-war (*Physalia*) (Figure 26-25).

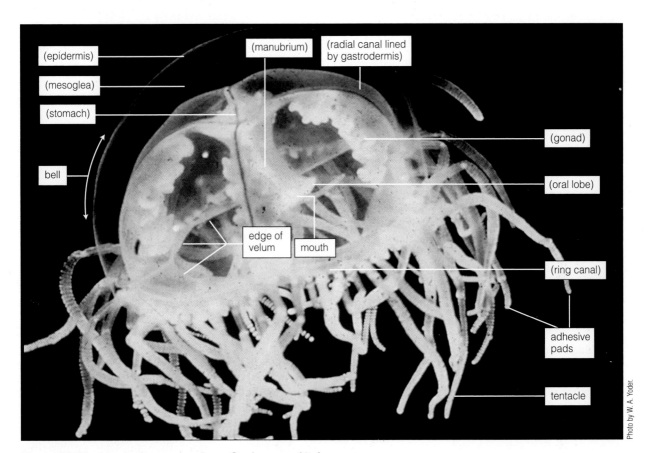

Figure 26-24   Live hydrozoan medusa, *Gonionemus* (4×).

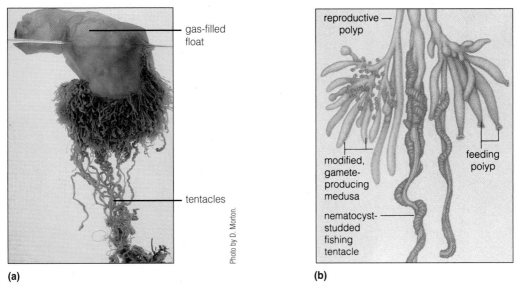

(a)   **(b)**

Figure 26-25   The Portuguese man-of-war. (**a**) Preserved specimen. (**b**) The tentacles contain several types of modified polyps and medusae.

This organism is really a colonial hydrozoan composed of several types of modified polyps and medusae. Note the conspicuous gas-filled *float*. Its sting is very painful to humans, and a possibly fatal neurotoxin is associated with the nematocysts of the tentacles.

B. Class Scyphozoa: True Jellyfish

This marine class contains the true jellyfish. The medusa is the predominant stage of the life cycle and is generally larger than hydrozoan medusae. The nematocysts in the tentacles and oral arms of jellyfish such as the sea nettle can produce painful stings. A tropical group called the sea wasps is more dangerous to humans than the hydrozoan Portuguese man-of-war. Some forms are bioluminescent, producing flashes of light by chemical reactions. Bioluminescence may lure prey toward the jellyfish or scare potential predators.

1. Look at a prepared slide of a whole mount of the medusa (ephyra) of *Aurelia* with your compound microscope (Figure 26-26). The structure of this medusa is similar to that of the hydrozoan, *Gonionemus*. However, note that the *tentacles* are short and that four *oral arms* surround the square mouth. In this case, prey is captured with the oral arms instead of the tentacles. Also, a velum is absent.

2. Examine prepared slides of planula larvae, a scyphistoma, and a strobila (Figure 26-27). Sexes are separate, and fertilization is internal. The zygote develops into a planula larva, which swims for a while before becoming a sessile polyp called a *scyphistoma*. This is an active, feeding stage, equipped with tentacles and a mouth. At certain times of the year, young medusae are produced asexually by transverse budding. This stage is called a *strobila*, and the process is called strobilization.

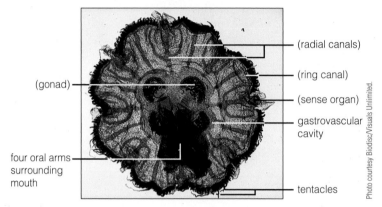

Figure 26-26   Scyphozoan jellyfish, *Aurelia*, w.m. (5×).

Photo courtesy Biodisc/Visuals Unlimited.

C. Class Anthozoa: Corals and Sea Anemones

Class Anthozoa is a marine group in which all members are heavy-bodied polyps that feed on small fish and invertebrates. The largest organisms in this class are the anemones. Both corals (Figure 26-28) and sea anemones attach to hard surfaces or burrow in the sand or mud.

1. Examine the demonstration exoskeletons of a variety of corals (Figure 26-29).

The epidermis of each colonial coral polyp secretes an exoskeleton of calcium carbonate. After a polyp dies, another polyp uses the remaining skeleton of the dead polyp as a foundation and secretes its own skeleton. In this fashion, corals have built up various types of reefs, atolls, and islands.

2. Examine the demonstration of a preserved or living specimen of *Metridium*, the common sea anemone (Figure 26-30). Note its stout, stumplike body, with the mouth and surrounding tentacles at the oral end and the *pedal disk* at the other end. What do you think is the function of the pedal disk?

_____

_____

Some species of anemones form symbiotic relationships with certain small fishes and invertebrates that live among their tentacles. These animals receive protection from the tentacles, which for some reason do not discharge their nematocysts into the animals' bodies. In return for the protective cover of the tentacles, the anemones benefit—for example, by ingesting food particles not eaten by the predatory fish.

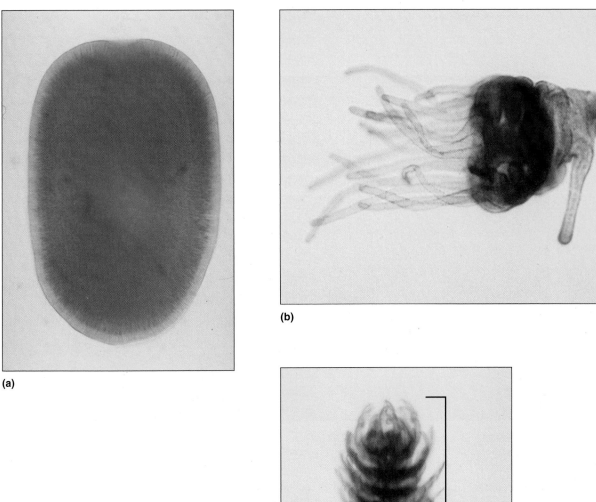

(a)

(b)

developing
medusae
(ephyrae)

Photos by D. Morton.

(c)

Figure 26-27 Stages in the life cycle of *Aurelia*. (**a**) Planular larvae, w.m. (400×). (**b**) Scyphistoma, w.m. (50×). (**c**) Strobila, w.m. (30×).

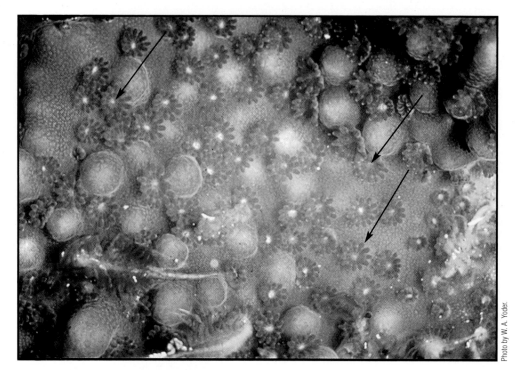

Photo by W. A. Yoder.

Figure 26-28   Live coral polyps (arrows) (1.5×).

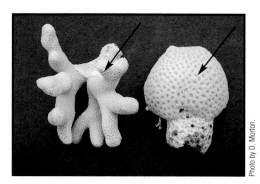

Photo by D. Morton.

Figure 26-29   Hard coral (0.30×). The arrows indicate sites that were occupied by individual polyps.

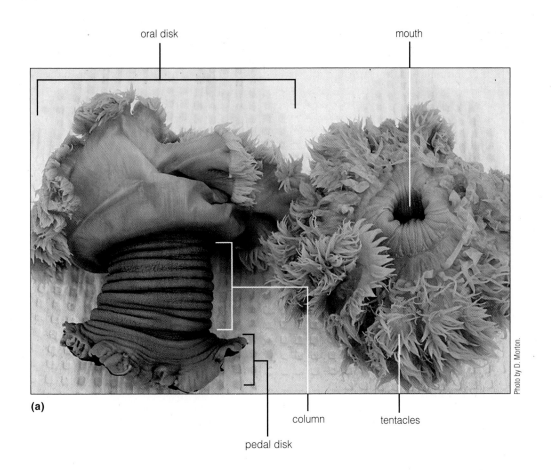

oral disk

mouth

column

tentacles

pedal disk

(a)

Photo by D. Morton.

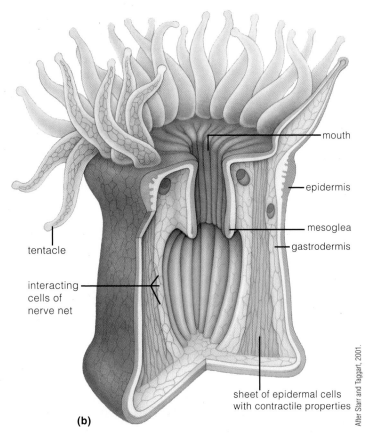

mouth

epidermis

mesoglea

gastrodermis

tentacle

interacting cells of nerve net

sheet of epidermal cells with contractile properties

After Starr and Taggart, 2001.

(b)

Figure 26-30 (a) Top and side views of preserved specimens of the common sea anemone, *Metridium* (1.5×). (b) Organization of a sea anemone.

_____ 1. Which of the following is characteristic of animals?
(a) heterotrophy
(b) autotrophy
(c) photosynthesis
(d) both b and c

_____ 2. Sponges are atypical animals because they
(a) reproduce sexually.
(b) are heterotrophic.
(c) lack definite tissues.
(d) are multicellular.

_____ 3. The "skeleton" of a typical sponge is composed of
(a) bones.
(b) cartilages.
(c) spicules.
(d) scales.

_____ 4. Many sponges produce both eggs and sperm and thus are
(a) monoecious.
(b) dioecious.
(c) hermaphroditic.
(d) both a and c

_____ 5. Freshwater sponges reproduce asexually by forming resistant internal buds called
(a) spicules.
(b) larvae.
(c) ovaries.
(d) gemmules.

_____ 6. Cnidarians evolved from an ancestral group of
(a) sponges.
(b) protests.
(c) fungi.
(d) plants.

_____ 7. A free-swimming jellyfish has a body form known as a
(a) polyp.
(b) strobila.
(c) medusa.
(d) hydra.

_____ 8. Cnidarians capture their prey with the use of
(a) poison claws.
(b) nematocysts.
(c) oral teeth.
(d) pedal disks.

_____ 9. Which statement is true for scyphozoan medusae?
(a) They are the dominant body form.
(b) They are generally smaller than hydrozoan medusae.
(c) Most live in freshwater.
(d) They are scavengers.

_____ 10. Coral reefs, atolls, and islands are the result of the buildup of coral
(a) polyps.
(b) medusae.
(c) exoskeletons.
(d) wastes.

EXERCISE **26**

Sponges and Cnidarians

Post-Lab Questions

Introduction

1. List six characteristics of the animal kingdom.

2. Define *asymmetry* and *radial symmetry*.

3. Identify the symmetry (asymmetry, radial symmetry, or bilateral symmetry) of the following objects.
   (a) fork

   (b) jagged piece of rock

   (c) rubber ball

26.1 Sponges (Phylum Porifera)

4. Define *cell specialization* and indicate how sponges exhibit this phenomenon.

5. Describe three ways that sponges reproduce.

6. Identify this photo; it is magnified 186×.

_____

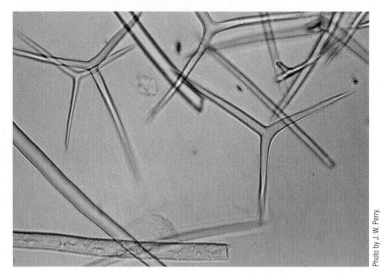

Photo by J. W. Perry.

26.2 Cnidarians (Phylum Cnidaria)

7. Define the tissue level of organization and indicate how cnidarians exhibit this phenomenon.

8. Name and draw the two body forms of cnidarians.

9. Describe the means by which the cnidarians seize and eat faster organisms.

Food for Thought

10. What adaptations of sponges and cnidarians have helped them to endure for so long on Earth?

Flatworms and Rotifers

After completing this exercise, you will be able to

1. define *acoelomate, regeneration, host, intermediate host, cuticle, scolex, proglottid, pseudocoelomate, cloaca, bladder worm;*

2. describe the tissue and organ-system levels of organization;

3. explain the basic body plan of members of the phyla Platyhelminthes and Rotifera;

4. describe the natural history of members of the phyla Platyhelminthes and Rotifera;

5. identify representatives of the classes Turbellaria, Trematoda, and Cestoda of the phylum Platyhelminthes; and representatives of phylum Rotifera;

6. identify structures (and indicate associated functions) of representatives of these groups;

7. outline the life cycles of a human liver fluke;

8. outline the life cycle of *Ascaris;*

9. compare and contrast the anatomy and life cycles of free-living and parasitic flatworms.

Introduction

If you've ever turned over rocks and leaves in a clear, cool, shallow stream or roadside ditch, you may have observed representatives of phylum Platyhelminthes without realizing it. Do you recall any dark, smooth, soft-bodied, flattened or ovoid masses approximately 1 cm or less in length attached to the bottom of the rocks or leaves? This description fits that of a free-living flatworm or planarian. Parasitic flukes and tapeworms also belong to this phylum.

Rotifers are microscopic, the smallest animals. They are mostly bottom-dwelling, freshwater forms, and they are an important prey of larger animals in aquatic food chains.

CAUTION

Specimens are kept in preservative solutions. Use lab gloves and safety goggles, especially if you wear contact lenses, whenever you or your lab mates handle a specimen or other fluids. If safety goggles do not fit, your glasses should suffice. Wash any part of your body exposed to preservative with copious amounts of water. If preservative solution or other fluids are splashed into your eyes, wash them with the safety eyewash bottle or eye irrigation apparatus for 15 minutes.

27.1 Flatworms (Phylum Platyhelminthes) *(About 75 min.)*

Habitat	aquatic, moist terrestrial, or parasitic
Body Arrangement	bilateral symmetry, acoelomate, head (cephalization), gut (incomplete)
Level of Organization	organ system (tissues derived from all three primary germ layers); nervous system with brain; excretory system with flame cells; no circulatory respiratory, or skeletal systems
Body Support	muscles
Life Cycle	sexual reproduction (mostly monoecious with internal fertilization); parasites have very complicated life cycles; some asexual reproduction (mostly transverse fission)
Special Characteristics	bilateral symmetry, head, organ systems

Flatworms have soft, flat, wormlike bodies with *bilateral symmetry* (Figure 26-1). This is the same type of symmetry we exhibit. Flatworms are **acoelomates** in that they lack a body cavity between the gut and body wall (Figure 26-2). The *tissues* of flatworms are *derived from all three primary germ layers:* ectoderm, mesoderm, and endoderm.

Organizationally, these animals represent an important evolutionary transition because they are the simplest forms to exhibit an organ-system level of organization (Figure 27-1) and *cephalization* (a definite head with sense organs). In an evolutionary sense, cephalization goes hand in hand with bilateral symmetry. When an animal with bilateral symmetry moves, the head generally leads, its sensors quickly detecting environmental stimuli that indicate the presence of food or danger ahead. Although there are definite organs, the *digestive system* is still of the *gastrovascular type, incomplete with a mouth and no anus.*

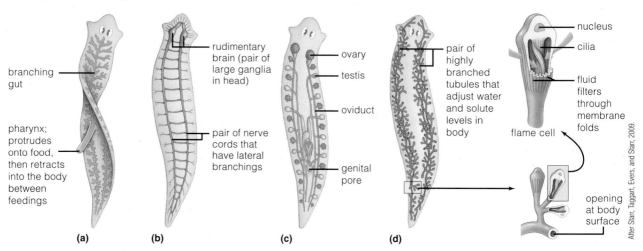

Figure 27-1 Organ systems of one type of flatworm: (**a**) digestive, (**b**) nervous, (**c**) reproductive, (**d**) water-regulating (excretory).

MATERIALS

Per student:

- dissection microscope
- compound microscope, lens paper, a bottle of lens-cleaning solution (optional), a lint-free cloth (optional)
- scalpel
- blunt probe

Per student pair:

- prepared slides of
    whole mount of *Dugesia* (planaria)
    whole mount of planaria with the digestive system stained
    cross section of planaria (pharyngeal region)
    whole mount of *Fasciola hepatica*
    whole mount of *Clonorchis sinensis* (if available)

- whole mount(s) of *Taenia pisiformis*, with scolex and different types of proglottids
- a whole mount of a bladder worm (cysticercus) of *Taenia solium*

Per student group (4):

- penlight
- glass petri dish (2)

Per lab room:

- chunk of beef liver
- container of fresh stream water
- container of live planarians
- demonstration collection of preserved tapeworms
- boxes of different sizes of latex gloves
- box of safety goggles

PROCEDURE

A. Class Turbellaria: Planarians

Most species in this class are free-living flatworms, although a few species are parasitic or symbiotic with other organisms. Free-living flatworms are found in marine and fresh waters on the underside of submerged rocks, leaves, and sticks. They are especially abundant in cool freshwater streams and along the ocean shoreline. Some live in wet or moist terrestrial habitats.

Planarians of the genus *Dugesia* are the representatives of this class typically studied in biology courses. They are relatively small flatworms, usually 2 cm or less in length. Planarians are monoecious—an individual has both male and female reproductive structures—but apparently do not self-fertilize but rather copulate with each other. Planarians exhibit a high degree of **regeneration**, in which lost body parts are gradually replaced.

1. Microscopically examine a whole mount of a planarian. Refer to Figure 27-2 and locate the **head**, **eyespots**, and **auricles**, lateral projections of the head that are organs sensitive to touch and to molecules dissolved in the water. The **eyespots** are photoreceptors. They are sensitive to light but do not form images.

2. Obtain and similarly examine a whole mount of a planarian with an injected digestive system. Identify the muscular **pharynx** within the **pharyngeal pouch**. The *mouth* is located at the free end of the pharynx. The mouth opens into the pharyngeal cavity, which in turn is continuous with the cavity of the **intestine**. To feed, a free planarian protrudes the pharynx from its ventral (belly) side and sucks up organic morsels (insect larvae, small crustaceans, and other small living and dead animals). The intestine is elaborately branched and occupies much of the animal's body. What function does the extensive branching of the intestine facilitate?

   _____

   _____

   _____

   _____

   _____

   _____

3. To observe feeding behavior, place a small piece of beef liver in a clean glass petri dish containing fresh stream water and add several live planarians. Describe what you see with the aid of a dissection microscope.

   _____

   _____

   _____

   _____

   _____

You may observe the elimination of waste from the gastrovascular cavity. If you do, describe what you see; if not, state the route by which this must occur.

_____

_____

_____

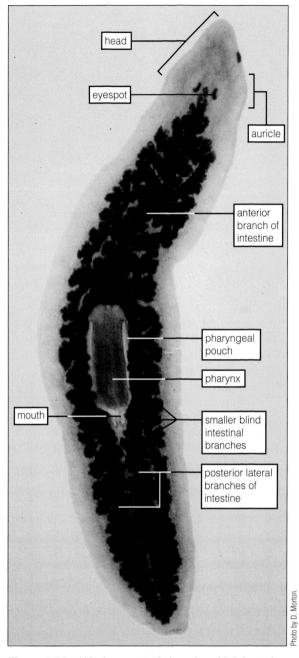

Figure 27-2 Whole mount of planaria, with injected digestive system (45×).

Photo by D. Morton.

Set the dish aside and examine it later, especially if you did not observe feeding behavior.

4. Examine a cross section through the pharynx of a planarian with your compound microscope (Figure 27-3). Find the following structures: *pharynx, pharyngeal pouch,* **nerve cords, epidermis, portions of the intestine**, and the several layers of muscles in the body wall. Note that the bottom epidermis is lined by **cilia**. What is the function of these cilia?

_____

_____

_____

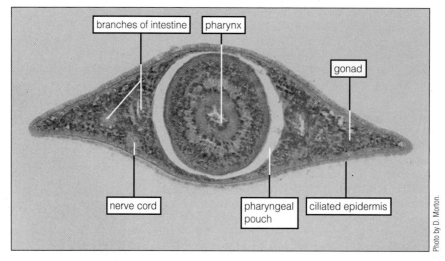

Figure 27-3 Cross section of planaria (150×).

Much of the rest of the body wall contains three layers of muscle: an outer circular layer, a middle diagonal layer, and an inner longitudinal layer. It is often difficult to distinguish all three layers.

5. Working in groups of four, use a dissection microscope to observe a living planarian in a clean glass petri dish filled with fresh stream water. Note its general size and shape. Touch the animal with a probe. What is its reaction? How does its shape change? Which muscle layers likely contract to produce this shape change?

_____

_____

_____

B. Class Trematoda: Flukes

Flukes are small, leaf-shaped, parasitic flatworms with complex life cycles. They are generally internal parasites of vertebrates, including humans. A protoplasmic but noncellular cuticle covers the surface of their bodies. What is a likely explanation for the presence of this tough, resistant cuticle?

_____

_____

_____

1. Examine a whole mount of a sheep liver fluke, *Fasciola hepatica*, noting its general morphology. Identify the structures indicated in Figure 27-4.
2. Examine a whole mount of an adult *Clonorchis sinensis*, a human liver fluke, if available, or Figure 27-5. This animal is a common parasite in Asia, and humans are the final **host**.

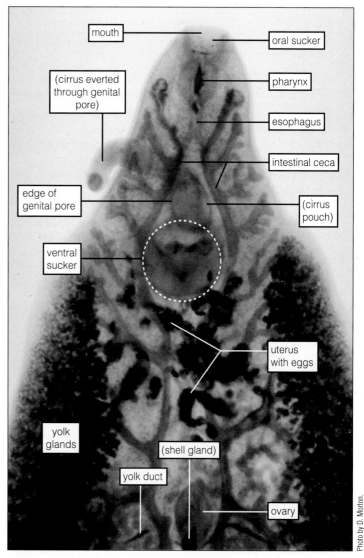

Figure 27-4 Anterior portion of *Fasciola hepatica*, the sheep liver fluke, w.m. (20×).

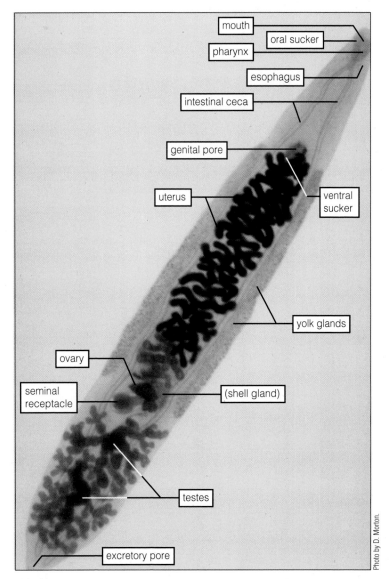

Figure 27-5 Photomicrograph of a whole mount of *Clonorchis sinensis*, a human liver fluke (20×).

(a) Identify the structures labeled in Figure 27-5. *Clonorchis* has two suckers for attachment to its host: an **oral sucker** surrounding the mouth and a **ventral sucker**. The digestive tract begins with the **mouth** and continues with a muscular **pharynx**, a short **esophagus**, and two blind pouches or **intestinal ceca**. Nitrogenous wastes are excreted at the tip of the back end through the **excretory pore**.

(b) *Clonorchis* is monoecious. Find the following female reproductive structures: **ovary, yolk glands, seminal receptacle,** and **uterus**. Find the following male reproductive structures: **testes** and **seminal vesicle**, which may be visible in your specimen. During copulation, sperm is conveyed from the testes to the *genital pore* by paired vasa efferentia, which fuse to form the vas deferens. These ducts are difficult to see in most specimens. Copulating individuals exchange sperm at the genital pore, and the sperm is stored temporarily in the seminal receptacle. The yolk glands provide yolk for nourishment of the developing eggs. *Eggs* can be seen as black dots in the uterus, where they mature.

(c) The life cycle of *Clonorchis sinensis* (Figure 27-6) involves two **intermediate hosts**, a snail and the golden carp. Intermediate hosts harbor sexually immature stages of a parasite. Fertilized eggs are expelled from the adult fluke into the human host's bile duct and are eventually voided with the feces of the host. If the eggs are shed in the appropriate moist or aquatic environment, they may be ingested by snails (the first intermediate host).

A larva emerges from the egg inside the snail. The larva passes through several developmental stages and gives rise to a number of tadpole-like larvae. These free-swimming larvae leave the snail,

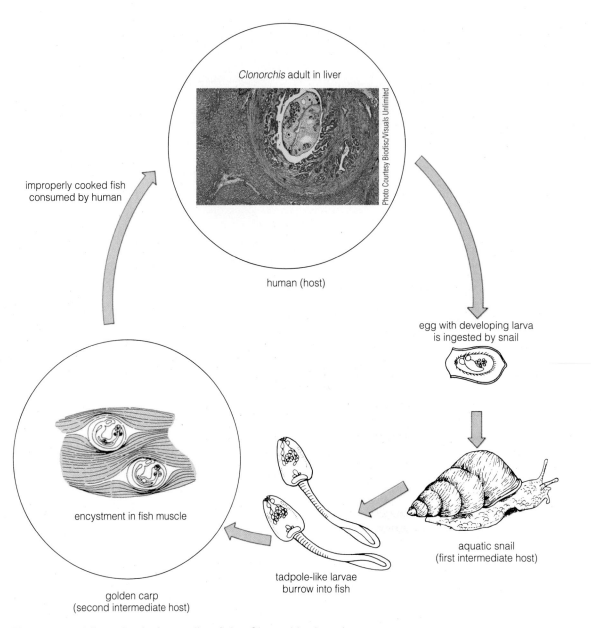

Figure 27-6 Life cycle of a human liver fluke, *Clonorchis sinensis*.

encounter a golden carp (the second intermediate host), and burrow into this fish to encyst in muscle tissue. If raw or improperly cooked fish is eaten by humans, cats, dogs, or some other mammal, the larvae emerge from the cysts and make their way up the host's bile duct into the liver, where they mature into adult flukes. The cycle is then repeated. What is the infectious stage (for humans) of this parasite?

_____

What two main precautions can humans take to reduce exposure to this parasite?

_____
_____
_____
_____
_____
_____

C. Class Cestoda: Tapeworms

Tapeworms are internal parasites that inhabit the intestine of vertebrates. Their flattened, multiunit bodies have no digestive system; instead, they absorb across their body walls predigested nutrients provided by the host. Their bodies are essentially reproductive machines, with extensive, well-developed reproductive organs occupying much of each mature body unit. A thick cuticle covers the body.

1. Look over the demonstration collection of preserved tapeworms taken from a variety of vertebrate hosts.
2. Examine a preserved specimen of *Taenia pisiformis*, a tapeworm of dogs and cats (Figure 27-7). Note that it is divided into three general regions: specialized head or **scolex**, **neck**, and **body**. The body is composed of successive units, the **proglottids**. New immature proglottids are added at the neck. As they move farther from the scolex, proglottids mature. The oldest, hindmost segments are full of eggs and ready to burst and release their zygotes (gravid).
3. Look at a prepared slide of a whole mount of a *Taenia pisiformis* scolex (Figure 27-8). Observe that the scolex is covered with hooks and suckers. For what task does the tapeworm use these **hooks** and **suckers**?

_____

_____

_____

Note the neck and the new proglottids forming at its base.

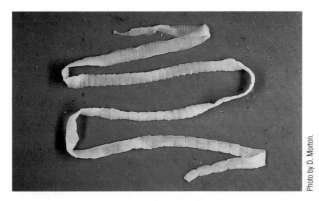

Figure 27-7   Preserved *Taenia* tapeworm (0.67×).

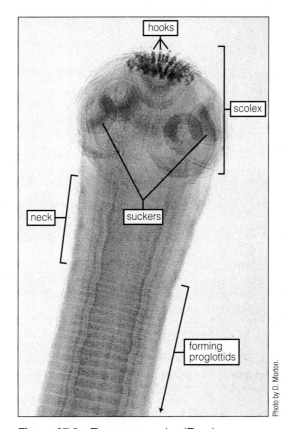

Figure 27-8   Tapeworm scolex (*Taenia pisiformis*), w.m. (40×).

4. Find a whole mount of mature proglottids of *Taenia pisiformis* and examine them with a dissection microscope. Identify the structures indicated in Figure 27-9. The lateral **genital pore** opens into the **vagina** and receives sperm from the **vas deferens**. The *uterus* is usually visible as a line running top to bottom in the middle of the proglottid. The *ovaries* are represented by dark, diffuse masses of tissue lateral to the uterus at the vaginal end of the proglottid. The **testes** are paler, more diffuse masses of tissue lateral to the uterus at the other end. Tapeworms are monoecious. They are one of the few animals that can self-fertilize, perhaps a result of their isolation from each other. Flanking most of the reproductive organs are the **excretory canals**.

5. Locate a gravid proglottid of *Taenia pisiformis* (Figure 27-10). Gravid proglottids are almost completely filled with zygotes in an expanded uterus ready for release as eggs.

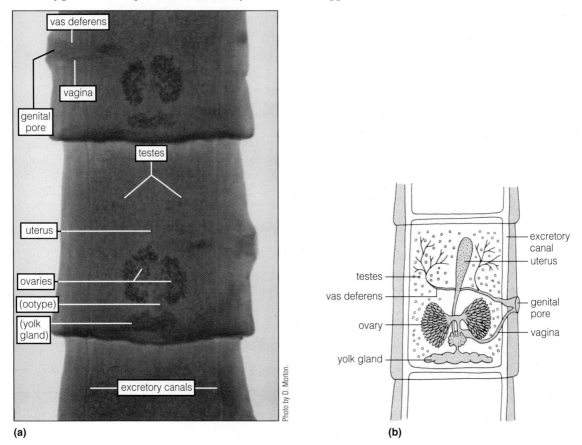

(a)   (b)

Figure 27-9   Tapeworm mature proglottids (*Taenia pisiformis*). (**a**) Whole mount (30×). (**b**) Diagram.

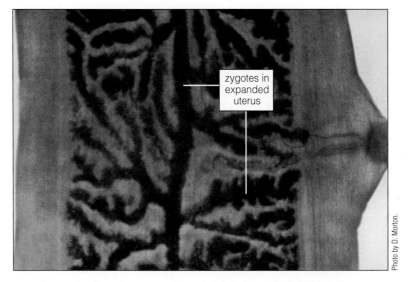

Figure 27-10   Tapeworm gravid proglottid (*Taenia pisiformis*), w.m. (30×).

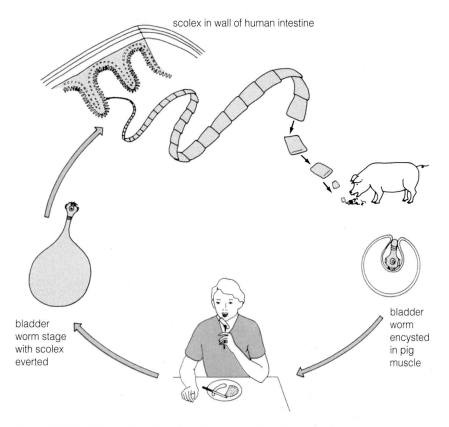

scolex in wall of human intestine

bladder worm stage with scolex everted

bladder worm encysted in pig muscle

Figure 27-11 Life cycle of the pork tapeworm, *Taenia solium*.

6. The life cycle of the pork tapeworm, *Taenia solium,* involves two host organisms, the pig and a human (Figure 27-11). Gravid proglottids containing embryos are voided with the feces of a human. If a pig ingests a proglottid, the embryos escape from the proglottid and bore into the intestinal wall and enter a blood or lymphatic vessel. From there they make their way to muscle, where they encyst as larvae known as **bladder worms**. If a human ingests improperly cooked pork, a bladder worm "pops" out of its encasement (bladder) and attaches its scolex to the intestinal wall. Here it matures and adds proglottids to complete the life cycle.

Which mammal is the host?

_____

Which mammal is the intermediate host?

_____

Examine the bladder worm stage of *Taenia solium* with the dissection microscope. Draw what you see in Figure 27-12.

Figure 27-12 Drawing of the bladder worm of *Taenia solium* (_____×).

27.2	**Rotifers (Phylum Rotifera) *(About 15 min.)***

Habitat	aquatic or moist terrestrial, mostly fresh water
Body Arrangement	bilateral symmetry, pseudocoelomate, head, gut (complete)
Level of Organization	organ system (tissues derived from all three primary germ layers); nervous system with brain; excretory system with flame cells; no circulatory, respiratory, or skeletal systems

Body Support	hydrostatic (pseudocoelom)
Life Cycle	sexual (dioecious with internal fertilization), some parthenogenesis, males much smaller than females or absent
Special Characteristics	microscopic animals, wheel organ, mastax

Rotifers are easy to find in rain gutters and spouts, and in the slimy material around the bases of buildings. Their thin, usually transparent body wall is covered by a cuticle composed of protein. Also like the nematodes to be studied later, rotifers are pseudocoelomates. Rotifers are dioecious; however, many species have no males, and the eggs develop parthenogenetically (without fertilization).

MATERIALS

Per student:

- clean microscope slide and coverslip
- compound microscope, lens paper, a bottle of lens-cleaning solution (optional), a lint-free cloth (optional)

Per lab room:

- live culture of rotifers
- disposable transfer pipet

PROCEDURE

1. With a disposable transfer pipet, place a drop from the rotifer culture on a clean slide and make a wet mount.
2. Observe the microscopic rotifers with the compound microscope. The elongate, cylindrical body is divisible into the **trunk** and the **foot** (Figure 27-13).

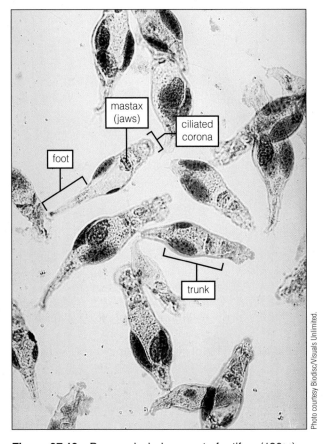

Figure 27-13   Prepared whole mount of rotifers (130×).

The foot bears two to several "sticky" *toes* for adhering to objects. The upper end of the trunk bears a **corona** of cilia. These cilia beat in such a fashion as to give the appearance of a wheel (or two) turning, hence the common references to this structure as the wheel organ and to these animals as wheel animals. The cilia create currents that sweep organic food particles into the mouth at the center of the wheel organ. The cilia also enable the organism to move. Inside the trunk, below the wheel organ, look for the movement of a grinding organ, the **mastax**. This structure is equipped with jaws made hard by a substance called chitin. Considering its location, what is the likely function of the mastax?

_____

_____

_____

_____ 1. Cephalization is
   (a) the division of the trunk into a scolex, neck, and body.
   (b) the presence of a head with sense organs.
   (c) sexual reproduction involving self-fertilization.
   (d) the ability to replace lost body parts.

_____ 2. Flatworms
   (a) are radially symmetrical.
   (b) possess a body cavity.
   (c) reproduce by parthenogenesis.
   (d) have tissues derived from all three primary germ layers.

_____ 3. Free-living flatworms (planaria) move by using
   (a) cilia.
   (b) flagella.
   (c) pseudopodia.
   (d) a muscular foot.

_____ 4. The flukes and tapeworms are covered by a protective
   (a) cuticle.
   (b) shell.
   (c) scales.
   (d) slime layer.

_____ 5. The life cycle of the human liver fluke has _____ intermediate hosts.
   (a) one
   (b) two
   (c) three
   (d) four

_____ 6. Tapeworms do not have a
   (a) reproductive system.
   (b) head.
   (c) excretory system.
   (d) digestive system.

_____ 7. The oldest proglottid of a tapeworm is
   (a) the scolex.
   (b) the one nearest its neck.
   (c) the one in its middle.
   (d) the one at its end.

_____ 8. Bladder worms are found in
   (a) proglottids in human feces.
   (b) proglottids in the human intestine.
   (c) improperly cooked pork.
   (d) none of the above

_____ 9. Nematodes and rotifers have a body cavity that is
   (a) called a coelom.
   (b) called a pseudocoelom.
   (c) completely lined by tissues derived from mesoderm.
   (d) both a and c

_____ 10. Rotifers obtain their food by the action of
   (a) tentacles.
   (b) cilia on the wheel organ.
   (c) a muscular pharynx.
   (d) pseudopodia.

EXERCISE **27**

Flatworms and Rotifers

Post-Lab Questions

27.1 Flatworms (Phylum Platyhelminthes)

1. How is bilateral symmetry different from radial symmetry?

2. List the characteristics of flatworms.

3. Distinguish between a host and an intermediate host.

4. Identify and label this illustration. It is magnified 9×.

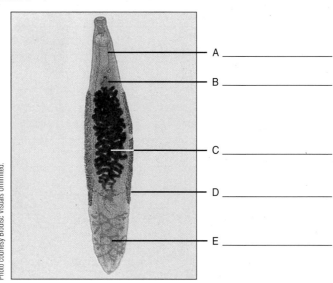

A _____

B _____

C _____

D _____

E _____

Photo courtesy Biodisc Visuals Unlimited.

5. Describe several adaptations of parasitic flatworms to their external environment.

27.2 Rotifers (Phylum Rotifera)

6. Pseudocoel means "false body cavity." Why is this term applied to the body cavity of rotifers?

Food for Thought

7. Contrast the digestive systems of free-living and parasitic flatworms.

8. Explain the relationship between the tremendous numbers of eggs produced by flukes and tapeworms and the complexity of their life cycles.

9. The relatively simple animals *Hydra* and *Dugesia* can regenerate lost body parts, but humans generally cannot. Discuss this in terms of tissue differentiation and the comparative levels of structural complexity of these organisms.

10. Search the Internet for sites that include references to flatworms or rotifers. List two addresses and briefly describe their contents.

    http:// _____

    _____

    _____

    _____

    http:// _____

    _____

    _____

    _____

Segmented Worms and Mollusks

After completing this exercise, you will be able to

1. define *coelom, coelomate, protostome, deuterostome, segmentation, peritoneum, setae, clitellum, cocoon, foot, mantle;*

2. differentiate between protostomes and deuterostomes;

3. describe the natural history of members of the phyla Annelida and Mollusca;

4. identify representatives of the classes Oligochaeta, Polychaeta, and Hirudinea of the phylum Annelida;

5. outline the life cycle of an earthworm;

6. identify representatives of the classes Bivalvia, Gastropoda, and Cephalopoda of the phylum Mollusca;

7. outline the life cycle of a freshwater clam;

8. identify structures (and indicate their associated functions) of the representatives of these phyla and classes.

Introduction

Some time after the origin of bilateral symmetry, the animals that subsequently evolved developed a true ventral body cavity or **coelom**. A true body cavity is lined by tissue derived entirely from mesoderm (Figure 26-2). The appearance of the coelom has great evolutionary significance, for it accommodates, cushions, and protects organs; allows an increase in overall body size and flexibility; and contributes to the development of systems primarily derived from mesoderm.

Coelomates, animals possessing a coelom (Figure 26-3), took two divergent evolutionary paths based on whether the *blastopore*—the first opening in a developing embryo—became the mouth (**protostomes**) or anus (**deuterostomes**) (Table 28-1). During subsequent evolutionary change, the coelom was reduced (mollusks and joint legged animals), lost (flatworms), or modified to a pseudocoel (rotifers and roundworms).

TABLE 28-1	**Evolution of Coelomates**	
Path	**Blastopore Becomes**	**Phyla**
Protostomes	Mouth	Mollusca, Annelida, and Arthropoda
Deuterostomes	Anus	Echinodermata, Hemichordata, and Chordata

© Cengage Learning 2013

CAUTION

Specimens are kept in preservative solutions. Use lab gloves and safety goggles, especially if you wear contact lenses, whenever you or your lab mates handle a specimen or other fluids. If safety goggles do not fit, your glasses should suffice. Wash any part of your body exposed to preservative with copious amounts of water. If preservative solution or other fluids are splashed into your eyes, wash them with the safety eyewash bottle or eye irrigation apparatus for 15 minutes.

Habitat	aquatic or moist terrestrial, some parasitic
Body Arrangement	bilateral symmetry, coelomate, head, body segmented, gut (complete)
Level of Organization	organ system (tissues derived from all three primary germ layers), nervous system with brain, circulatory system (closed), excretory system (nephridia), respiratory system (gills in some), no skeletal system
Body Support	hydrostatic (coelom)
Life Cycle	sexual reproduction (monoecious or dioecious with mostly internal fertilization), larvae in some, asexual reproduction (mostly transverse fission)
Special Characteristics	segmented, closed circulatory system

The most striking annelid characteristic is the division of the cylindrical trunk into a series of similar segments. *Annelids* were the first animals to evolve the condition of **segmentation**. Segmentation is internal as well as external, with segmentally arranged components of various organ systems and the body cavity. The coelom is more or less divided by *septa* into compartments within each segment, each lined by **peritoneum**. Unlike the proglottids of the cestodes, segments cannot function independently of each other. The *circulatory system* is *closed,* with the blood entirely contained in vessels. There is a well-developed *nervous system* with a central nervous system composed of two fused dorsal ganglia (brain) and ventral nerve cords. The circulatory system is not segmented. Segmentation is significant in evolutionary history because each segment or group of segments can become specialized. Although this is an important characteristic of most of the phyla we have yet to study, it is barely evident in the earthworm.

MATERIALS

Per student:

- pins
- blunt probe or dissecting needle

Per student pair:

- preserved earthworm
- dissection microscope
- dissection pan
- fine dissecting scissors
- *Eisenia foetida* cocoons

Per lab room:

- labeled demonstration dissection of the internal anatomy of the earthworm (optional)
- demonstration collection of living and preserved earthworms and polychaetes
- several preserved leeches
- live freshwater leeches in an aquarium
- large plastic bag for disposal of dissected specimens
- boxes of different sizes of lab gloves
- box of safety goggles

A. Oligochaetes (Class Oligochaeta)

PROCEDURE

1. Refer to Figure 28-1 as you examine the living earthworms on demonstration in the lab. You probably already know several characteristics of oligochaetes because of past observations with the earthworm (*Lumbricus terrestris*), especially if you fish. Recalling past experiences with this organism, list as many features of the earthworm as you can.

_____

_____

_____

_____

2. Let's see how you did! Obtain a preserved earthworm and have a dissection microscope handy. Is its body bilaterally or radially symmetrical?_____

_____

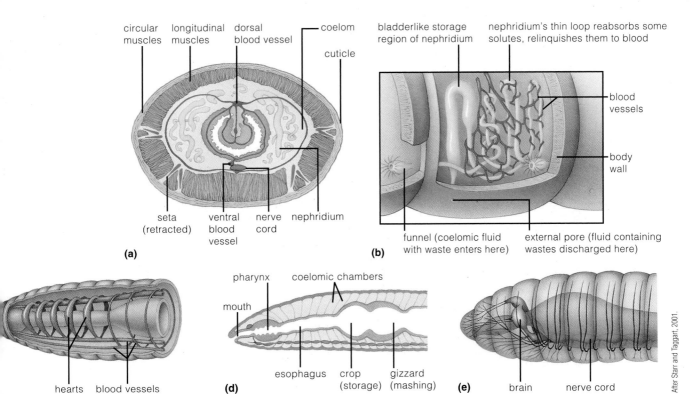

circular muscles · longitudinal muscles · dorsal blood vessel · coelom · cuticle

seta (retracted) · ventral blood vessel · nerve cord · nephridium

(a)

bladderlike storage region of nephridium · nephridium's thin loop reabsorbs some solutes, relinquishes them to blood

blood vessels · body wall

funnel (coelomic fluid with waste enters here) · external pore (fluid containing wastes discharged here)

(b)

hearts · blood vessels

pharynx · coelomic chambers · mouth

esophagus · crop (storage) · gizzard (mashing)

(d)

brain · nerve cord

(e)

After Starr and Taggart, 2001.

re 28-1 Earthworm body plan. (**a**) Midbody, transverse section. (**b**) A nephridium, one of many functional units that help maintain olume and the composition of body fluids. (**c**) Portion of the closed circulatory system. The system is linked in its functioning with ridia. (**d**) Part of the digestive system, near the worm's head end. (**e**) Part of the nervous system.

3. Examine the external anatomy (Figure 28-2). Note the segments.
   Count them and record the number:_____

   (a) Very few segments are added after hatching. Note the **prostomium**, a fleshy lobe that hangs over the **mouth**; both structures are part of the first segment.

   (b) Look for **setae**, tiny bristle-like structures on the ventral surface, with the dissection microscope. Take the specimen between your fingers and feel for them. What is their likely function?

   _____

   _____

   _____

   _____

   (c) Use the dissection microscope to help you find the following. A pair of small **excretory pores** is found on the lateral or ventral surfaces of each segment, except the first few and the last. On the sides of segment 14, the openings (*female pores*) of the oviducts can be seen. The openings (*male pores*) of the sperm ducts, with their swollen lips, can be found on segment 15. The **clitellum** is the enlarged ring beginning at segment 31 or 32 and ending with 37. This glandular structure secretes a slimy band around two copulating individuals. The **anus** is the vertical slit in the terminal segment.

4. *Dissection of earthworm internal anatomy.*
   (a) Place your worm in a dissection pan and stick a pin through each end of the specimen. With fine scissors, make an incision just to the right of the dorsal black line (blood vessel), 10 segments anterior to the anus, and cut superficially to the mouth. (You can use the prostomium as an indicator of the dorsal side of the animal if the black line cannot be seen.) Gently pull the incision apart and note that the coelom is divided into a series of compartments by septa. Carefully cut the septa free of the body wall with a scalpel and pin both sides to the pan at 5- to 10-segment intervals. Cover the specimen with about 0.5–1.0 cm of tap water and examine the internal anatomy (Figure 28-3).

   (b) Use your blunt probe or dissection needle to find the male reproductive organs, the light-colored **seminal vesicles**. These can be found in segments 9–12. They are composed of three sacs, within which are the *testes*. A sperm duct leads from the *seminal* vesicles to the male pore.

   (c) Find two pairs of female reproductive organs—the white, spherical **seminal receptacles** in segments 9 and 10. There are ovaries in segment 13. An oviduct runs from each ovary to the female pore.

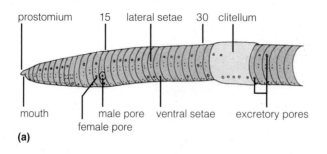

(a)

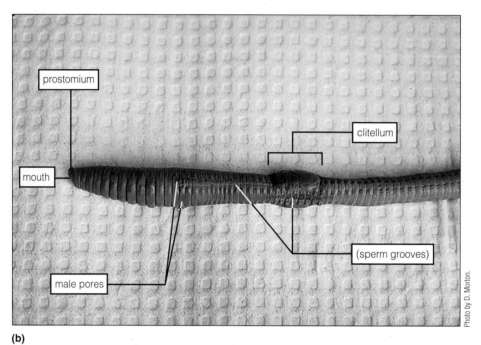

(b)

Figure 28-2 (**a**) External anatomy of the earthworm (numbers refer to the segment numbers). (**b**) External view of preserved earthworm.

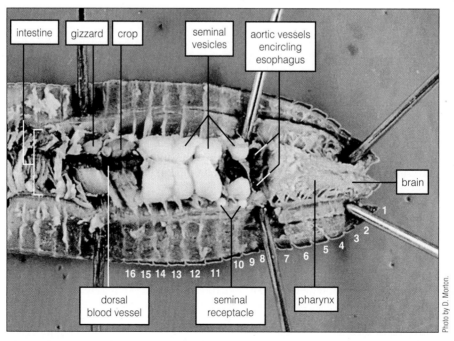

Figure 28-3 Dissection of the anterior end of an earthworm. The first 16 segments are numbered.

(d) Locate the organs of the digestive tract. The *buccal cavity* occupies the first three segments. The muscular **pharynx** is the swelling in the digestive tract just behind the mouth. This organ is responsible for sucking soil particles containing food into the mouth. The *esophagus* extends from the pharynx through segment 14. The *crop* is a large, thin-walled sac in segments 15 and 16 that serves as a temporary storage organ for food. The **gizzard**, a grinding organ, is posterior to the crop. The **intestine**, the site of digestion and absorption, extends from the gizzard to the anus.

(e) Trace the parts of the circulatory system. The **dorsal blood vessel** runs along the dorsal side of the digestive tract and contains blood that flows toward the head. Five pairs of **aortic vessels** ("hearts") encircle the esophagus in segments 7–11, connecting the dorsal blood vessel with the **ventral blood vessel** (Figure 28-4). The ventral vessel runs along the ventral side of the digestive tract, carrying blood backward. The primary pumps of the circulatory system are the dorsal and ventral vessels. The dorsal vessel contains valves to prevent blood from flowing backward.

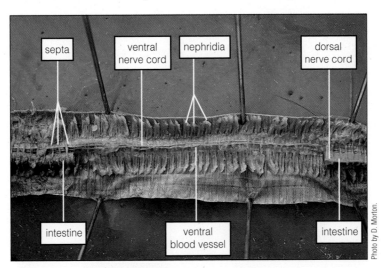

Figure 28-4 Dissection of the posterior end of an earthworm. The intestine has been partially removed to expose the ventral nerve cord.

(f) Find the two-lobed **brain** (cerebral ganglia) on the anterior, dorsal surface of the pharynx in segment 3. The **ventral nerve cord** arises from the brain and extends along the floor of the body cavity to the last segment. Although they are difficult to locate, paired lateral nerves branch from the segmental ganglia along the nerve cord.

(g) Locate a pair of **nephridia**—coiled white tubes—at the base of every segment but the first three and last one. These structures constitute the "kidneys" of the earthworm (Figure 28-4).

(h) Dispose of your dissected specimen in the large plastic bag provided for this purpose.

5. Use a dissection microscope to examine cocoons from *Eisenia foetida,* another species of earthworm. Identify eggs and embryos. Earthworms are monoecious—individuals have both male and female reproductive tracts. Copulation usually takes 2–3 hours with individuals facing belly to belly in opposite directions, joined together at their clitellums. These organs secrete a slimy substance that forms a band around the worms. Sperm are transferred between the participants, and the sperm are stored temporarily in the seminal receptacles.

   A few days after the individuals separate, each worm secretes a **cocoon**. The cocoon receives eggs and stored sperm from the seminal receptacles. The eggs are fertilized, and the cocoon moves forward as the worm backs out. The cocoon then slips off the front end of the worm and is deposited in the soil. In *Lumbricus terrestris*, a single young earthworm completes development using the other fertilized eggs as food, eventually breaks free of the cocoon, and becomes an adult in several weeks. As long as the stored sperm lasts, the earthworm continues to form cocoons.

6. Use your dissection microscope to examine a prepared slide of a cross section of the earthworm (Figure 28-5). First look at the body wall. A thin **cuticle** covers the body. The **epidermis**, which lies beneath it, secretes the cuticle, and internal to this are an **outer circular layer** and an **inner longitudinal layer** of muscle. Identify the other labeled structures in Figure 28-5.

B. Polychaetes (Class Polychaeta)

The polychaetes are abundant marine annelids with dorso-ventrally flattened bodies. They are most common at shallow depths in the intertidal zone at the seashore and are prey for a variety of marine invertebrates.

   Examine the polychaetes on demonstration in the lab. Although they resemble oligochaetes in many ways, notice that the polychaetes have fleshy, segmental outgrowths called **parapodia** (Figure 28-6). These structures

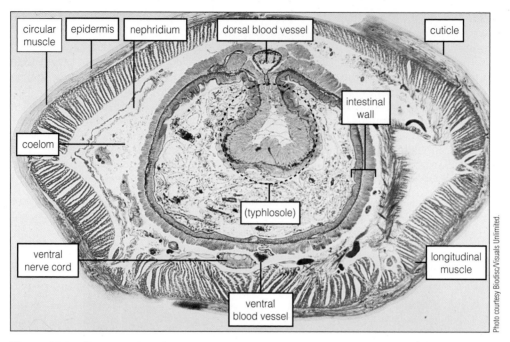

Figure 28-5 Cross section of an earthworm (20×).

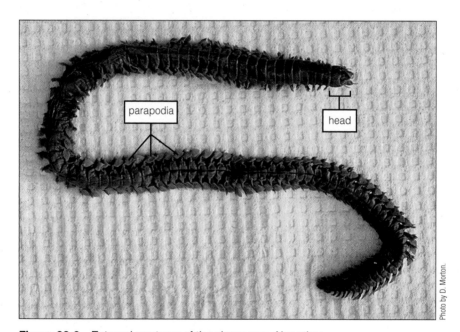

Figure 28-6 External anatomy of the clamworm, *Neanthes*.

are equipped with numerous setae, the characteristic for which they get their class name, *Polychaeta,* meaning "many bristles." Notice also the relatively complex **head**.

C. Leeches (Class Hirudinea)

The leeches are dorso-ventrally flattened annelids that are usually found in freshwater.

1. Examine the preserved specimens of leeches (Figure 28-7a). They lack parapodia and setae. They are predaceous or parasitic; parasitic species suck blood with the aid of a muscular pharynx. The segments surrounding the mouth are modified into a small **anterior sucker**. The mouth is supplied with jaws, made hard with the substance chitin, used to wound the host and supply the leech with blood. Parasitic leeches secrete an anticoagulant into the wound to prevent the blood from clotting. A **posterior sucker** helps the leech attach to the host or prey.

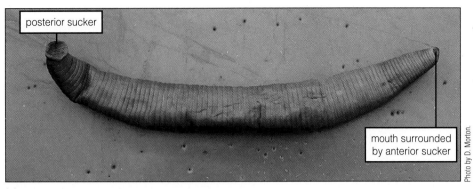

posterior sucker

mouth surrounded by anterior sucker

Photo by D. Morton.

(a)

J.A.L. Cooke/Oxford Scientific Films.

(b)

Figure 28-7 Leeches. **(a)** Preserved specimen. **(b)** Live, feeding on a human.

2. Observe the movement of living freshwater leeches (Figure 28-7b) as they swim through the water in an aquarium. Explain their movements in terms of the contraction and relaxation of circular and longitudinal layers of muscle that comprise part of their body wall.

_____

_____

_____

_____

28.2	Mollusks (Phylum Mollusca) *(40–60 min.)*

Habitat	aquatic or terrestrial, mostly marine
Body Arrangement	head and unique body plan, gut (complete)
Level of Organization	organ system (tissues derived from all three primary germ layers), nervous system with brain, excretory system (nephridium or nephridia), circulatory system (open), respiratory system (most have gills or lungs), no skeletal system
Body Support	shells, floats (aquatic)
Life Cycle	sexual (monoecious or dioecious with external or internal fertilization), larvae in some
Special Characteristics	coelomate; open circulatory system; respiratory system; external body divided into head, mantle, and foot

This is the second largest animal phylum and includes snails and nudibranchs, slugs, bivalves—clams, mussels, and oysters—octopuses, cuttlefish, and squids. Bodies are typically soft (*molluscus* in Latin means "soft") and often covered by protective shells. Apart from the squids and some octopuses (and the land snails and slugs, which have adopted a terrestrial existence), *mollusks inhabit shallow marine and fresh waters,* where they crawl along or burrow into the soft substrate. Many species are valued foods for humans.

Two characteristics distinguish the mollusks from other animals. They possess a ventral, muscular **foot** for movement and a dorsal integument, the **mantle**, which in different groups of mollusks secretes shells, functions in gaseous exchange, and facilitates movement. Even forms without shells usually have mantles.

MATERIALS

Per student:

- scalpel
- blunt probe or dissecting needle
- compound microscope, lens paper, a bottle of lens-cleaning solution (optional), a lint-free cloth (optional)

Per student pair:

- preserved freshwater clam
- dissection pan
- prepared slide of a cross section of clam gill
- dissection microscope
- glass petri dish
- prepared slide of glochidia

Per lab room:

- labeled demonstration dissection of a freshwater clam (optional) and the internal anatomy of a squid
- demonstration collection of bivalve shells, and living and preserved specimens
- freshwater aquarium with snails (for example, *Physa*)
- demonstration collection of gastropod shells, and living and preserved specimens
- several preserved squids, octopuses, and chambered nautiluses
- large plastic bag for disposal of dissected specimens
- boxes of different sizes of lab gloves
- box of safety goggles

PROCEDURE

A. Bivalves (Class Bivalvia)

Bivalves are so named because they have shells composed of two halves or *valves.* Included in this class are the clams, mussels, oysters, and scallops. They have a hatchet-shaped (laterally flattened) body and a foot used for burrowing in soft substrates.

1. *Study the external features of a freshwater clam.*
   (a) Obtain and examine a freshwater clam in a dissection pan (Figure 28-8). Each **valve** has a hump on the dorsal surface (back), the **umbo**, which points toward the front of the organism. The umbo is the oldest part of the valve.
   (b) Encircling the umbo are concentric rings of annual growth; the raised portions (*growth ridges*) represent periods of restricted winter growth. Determine the age of your clam by counting the number of ridges formed by the growth rings. Age of clam: _____

umbo   growth ridges

Figure 28-8   External view of a freshwater clam.

2. *Dissect a freshwater clam.*
   (a) Remove the left valve by cutting the *anterior* and *posterior adductor muscles* that hold the valves together. To do this, first identify the anterior adductor muscle in Figure 28-9 and then slip a scalpel between the valves at its approximate location feeling for a firm structure with the scalpel edge. Once you find it, cut back and forth. Repeat this process for the posterior adductor muscle. Once these muscles are severed, slide the scalpel between the valve to be removed and the tissues that line its inner surface. With the flat side of the scalpel, hold the tissues down while you remove the valve.
   (b) The membrane that lines the inner surface of each valve is the **mantle**, which secretes the calcium carbonate shell. Pearls form as a result of irritations (usually grains of sand or small pebbles) in the mantle and sometimes are found embedded in the *pearly layer,* the inner of three layers that comprise the shell. Note that the mantle forms an **incurrent siphon** and an **excurrent siphon** at the rear end.

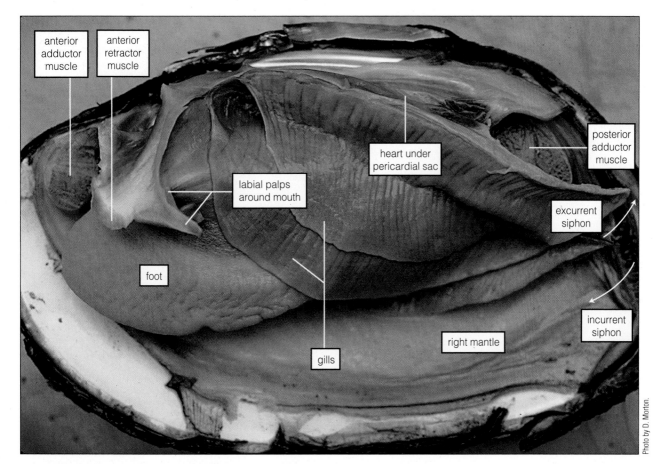

anterior adductor muscle

anterior retractor muscle

posterior adductor muscle

heart under pericardial sac

labial palps around mouth

excurrent siphon

foot

incurrent siphon

right mantle

gills

Photo by D. Morton.

Figure 28-9   Dissection of a freshwater clam. The left valve and mantle have been removed. Arrows indicate the path of water flow into the incurrent siphon and out of the excurrent siphon.

(c)   Remove the mantle on the left side by cutting it free with a scalpel. The relatively simple body plan includes a ventral foot and a dorsal visceral mass. Identify the **foot** and two sets of **gills**, one on each side of the *visceral mass*. The gills function in gaseous exchange. Water enters the incurrent siphon and flows over the gills, ultimately leaving through the excurrent siphon. Gills also trap food particles contained in incoming water; these are transported to the mouth by the cilia on the gills. Find the **labial palps**, which are situated around the mouth.

(d)   The circulatory system is *open,* so the blood passes from arteries leading from the heart through body spaces called sinuses. Blood is returned to the heart by veins draining the mantle and gills. Find the **heart** in the pericardial sac between the visceral mass and the hinge between the valves.

(e)   Dispose of your dissected specimen in the large plastic bag provided for this purpose.

(f)   Examine the labeled dissection prepared by your instructor. Identify the **mouth, esophagus, stomach, intestine, digestive gland** (liver), **anus, nephridium** (kidney), and **gonad** (Figure 28-10).

3.   Examine the demonstration slide of a transverse section of the gill with your compound microscope. Identify the structures labeled in Figure 28-11.

4.   Fertilization is internal in bivalves. Sperm taken in through the female's incurrent siphon fertilize eggs in the gills. In many bivalves, *glochidia* larvae develop from the fertilized eggs and after several months leave the gill area via the excurrent siphons to parasitize fish by clamping onto the gills or fins with their jaw-like valves. The fish grows tissue over the parasite, which remains attached for 2–3 months. The glochidium then breaks out of the fish to develop into an adult on the lake or river bottom. Microscopically look at glochidia in a prepared slide (Figure 28-12).

5.   *Study bilvalve diversity.* Examine the assortment of bivalve shells and preserved specimens on demonstration. List the features they have in common.

_____

_____

_____

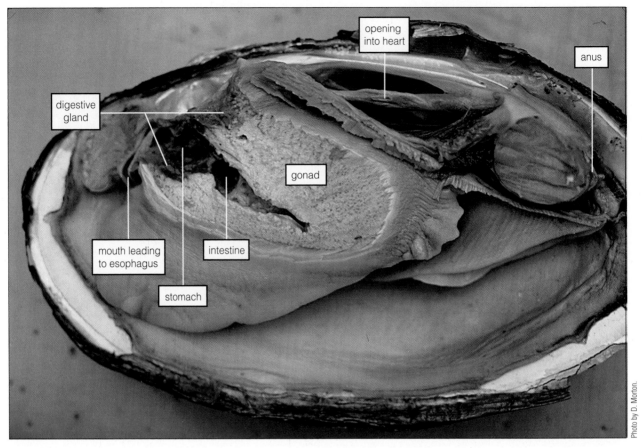

Figure 28-10 Further dissection of freshwater clam. The left gills and part of the visceral mass have been removed.

List some of the differences.

_____

_____

_____

How can you tell which live on sandy bottoms?

_____

B. Snails, Slugs, and Nudibranchs (Class Gastropoda)

Slugs and nudribranchs (Figure 28-13) have no shell; the former two are terrestrial, the latter marine. Nudibranchs are very ornate gastropods, often brightly colored, and frequently adorned with numerous fleshy projections.

1. Place a live freshwater snail (for example, *Physa*) in a glass petri dish. Once the snail has attached itself to the glass, invert the petri dish and use the dissection microscope to observe it moving along a slimy path.
   A *slime gland* in the front of the foot secretes mucus. Briefly describe the muscular contractions of the foot that allow the animal to glide over this mucus. _____

   _____

   Looking at the foot through the glass, you can see the **mouth**. Describe its location. _____

   _____

2. In snails with spirally coiled shells, the shell either spirals to the snail's right or left as it grows. Which way does your snail's shell spiral? _____

   _____

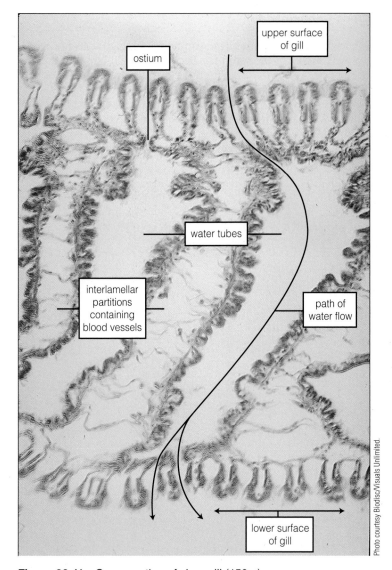

Photo courtesy Biodisc/Visuals Unlimited.

upper surface of gill

ostium

water tubes

interlamellar partitions containing blood vessels

path of water flow

lower surface of gill

Figure 28-11 Cross section of clam gill (150×).

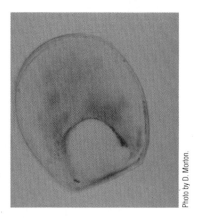

Photo by D. Morton.

Figure 28-12 Whole mount of glochidia (250×).

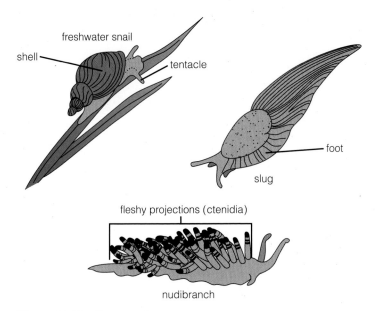

freshwater snail

shell

tentacle

foot

slug

fleshy projections (ctenidia)

nudibranch

Figure 28-13 Gastropods.

C. Squids, Octopuses, and Nautiluses (Class Cephalopoda)

Several significant modifications of the basic molluskan body plan occur in cephalopods (Figure 28-14), which are adapted for rapid swimming and catching prey. The foot of the squid has become divided into four pairs of arms and two tentacles, each with suckers, and a siphon. The tentacles, with their terminal suckers, shoot out to grasp prey. The muscular mantle draws water into its chamber when it relaxes.

Subsequent contraction of the mantle wall seals the cavity and forces water out of the siphon, resulting in jet propulsion propelling the animal through the water. Two lateral fins provide guidance. In addition, there are two large eyes that superficially resemble those of vertebrates. A relatively large brain coordinates all of these activities.

Nautiluses have shells, but shells are reduced in squids and absent in octopuses. The shell of the squid lies dorsally beneath the mantle and is called the pen.

1. Examine the external morphology of a squid. Identify the structures labeled in Figure 28-15.
2. Identify those structures indicated in Figure 28-16 on the demonstration dissection of the internal anatomy of the squid.
3. Look at the variety of cephalopod shells and preserved specimens on display in your lab room.

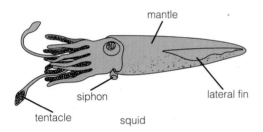

squid

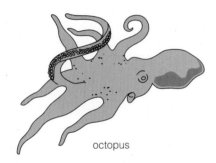

octopus

chambered nautilus

Figure 28-14   Cephalopods.

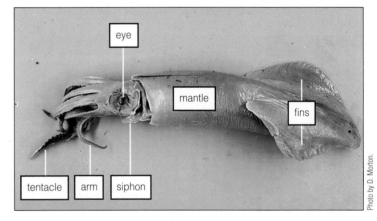

Figure 28-15   Lateral view of a preserved squid.

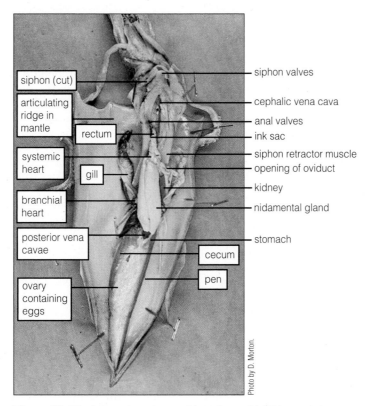

siphon (cut)

articulating
ridge in
mantle

rectum

systemic
heart

gill

branchial
heart

posterior vena
cavae

ovary
containing
eggs

cecum

pen

siphon valves

cephalic vena cava

anal valves

ink sac

siphon retractor muscle

opening of oviduct

kidney

nidamental gland

stomach

Photo by D. Morton.

Figure 28-16 Internal anatomy of female squid. Nidamental glands and the ovary are absent in male specimens, and a penis occupies the same location as the oviduct.

_____ 1. The protostomes are animals whose
   (a) stomach is in front of the crop.
   (b) mouth is covered by a fleshy lip.
   (c) blastopore becomes a mouth.
   (d) digestive tract is lined by mesoderm.

_____ 2. The deuterostomes are animals whose
   (a) stomach is in front of the crop.
   (b) mouth is covered by a fleshy lip.
   (c) blastopore becomes the anus.
   (d) digestive tract is lined by mesoderm.

_____ 3. Further evolution of the protostomes
   changed the coelom by
   (a) losing it.
   (b) reducing it.
   (c) it becoming a pseudocoel.
   (d) all of the above

_____ 4. Earthworms exhibit segmentation,
   defined as the
   (a) division of the body into a series of
        similar segments.
   (b) presence of a true coelom.
   (c) difference in size of the male and
        female.
   (d) presence of a "head" equipped with
        sensory organs.

_____ 5. The copulatory organ of the earthworm
   is the
   (a) penis.
   (b) clitellum.
   (c) gonopodium.
   (d) vestibule.

_____ 6. Leeches belong to the phylum
   (a) Arthropoda.
   (b) Mollusca.
   (c) Annelida.
   (d) Cnidaria.

_____ 7. One of the two major distinguishing characteristics of mollusks is
   (a) the presence of three body regions.
   (b) the mantle.
   (c) segmentation of the body.
   (d) jointed appendages.

_____ 8. Functions of the mantle in clams include
   (a) secreting the shell.
   (b) forming pearls around irritating
        grains of sand.
   (c) forming the incurrent and excurrent
        siphons.
   (d) all of the above

_____ 9. Which of the following is a gastropod?
   (a) clam
   (b) snail
   (c) squid
   (d) octopus

_____ 10. Which of the following is a cephalopod?
   (a) clam
   (b) snail
   (c) slug
   (d) octopus

EXERCISE **28**

Segmented Worms and Mollusks

Post-Lab Questions

Introduction

1. Indicate the differences between protostomes and deuterostomes.

2. List the animal phyla that are protostomes and those that are deuterostomes.

3. What happens to the coelom as each phyla of protostomes evolve?

28.1 Annelids (Phylum Annelida)

4. Briefly describe the life cycle of an earthworm.

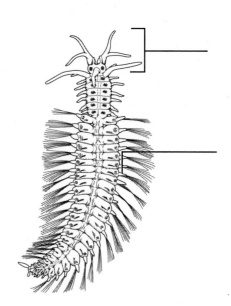

5. Label this diagram of an annelid and circle its class (Oligochaeta, Polychaeta, or Hirudinea).

28.2 Mollusks (Phylum Mollusca)

6. Explain what causes the growth ridges found on the shells of bivalves.

7. List some structures that distinguish bivalves from the other molluskan classes.

8. What characteristics distinguish slugs from nudibranchs?

Food for Thought

9. Is it possible that some of the monsters reported by ancient mariners were giant squid or octopi? Search the Internet before giving and explaining your answer.

10. Define *segmentation* and indicate how the annelids and your body exhibit this phenomenon.

Roundworms and Joint-Legged Animals

OBJECTIVES

After completing this exercise, you will be able to

1. define *pseudocoelomate, cloaca, bladder worm exoskeleton, hemolymph, chitin, carapace, sexual dimorphism, gills, tracheae;*

2. explain the term *ecdysis;*

3. identify representatives of the phylum Nematoda;

4. outline the life cycle of *Ascaris;*

5. identify representatives of the subphyla Crustacea, Hexapoda, Myriapoda, and Chelicerata of the phylum Arthropoda;

6. identify structures (and indicate their associated functions) of the representatives of these phyla and classes.

Introduction

Although joint-legged animals and roundworms at first appear quite different from each other (e.g., segmentation compared to none, legs compared to a legless wormlike body, and a reduced coelom compared to a pseudocoel), these two groups of animals share one characteristic: they molt. Molting or **ecdysis** is the periodic loss of the outer covering of the integument during growth. Unlike our epidermis, which in two to three weeks renews itself one cell at a time, the exoskeleton of joint-legged animals and the cuticle of roundworms are shed all at once. Actually, our hairs, which are grown in follicles derived from the epidermis, are likewise shed as a whole.

CAUTION

Specimens are kept in preservative solutions. Use lab gloves and safety goggles, especially if you wear contact lenses, whenever you or your lab mates handle a specimen or other fluids. If safety goggles do not fit, your glasses should suffice. Wash any part of your body exposed to preservative with copious amounts of water. If preservative solution or other fluids are splashed into your eyes, wash them with the safety eyewash bottle or eye irrigation apparatus for 15 minutes.

29.1   Roundworms (Phylum Nematoda) *(About 50 min.)*

Habitat	aquatic, moist terrestrial, or parasitic
Body Arrangement	bilateral symmetry, pseudocoelomate, head, gut (complete)
Level of Organization	organ system (tissues derived from all three primary germ layers); nervous system with brain; no circulatory, respiratory, or skeletal systems
Body Support	hydrostatic (pseudocoelom)
Life Cycle	sexual reproduction (mostly dioecious with internal fertilization)
Special Characteristics	pseudocoelomate, complete gut

Nematodes also called *roundworms*, like rotifers, are **pseudocoelomates**, which literally means they have a false body cavity or *pseudocoel* (Figure 26-2). This term is somewhat unfortunate, because there *is* a body cavity. It is not considered a true body cavity because tissues derived from mesoderm incompletely line it. Also, in an evolutionary sense it is likely derived from the body cavity of ancestral coelomates. Unlike flatworms, roundworms have a complete gut with both a mouth and an anus. Together, the pseudocoel and the complete digestive system comprise a tube-within-a-tube arrangement, a body plan that is seen in most of the phyla yet to be studied. Roundworms have slender, cylindrical bodies that taper at both ends. A tough, nonliving cuticle of protein covers them.

Parasitic roundworms of vertebrates, including humans, are ascarids, hookworms, *Trichinella* (Figure 29-1), pinworms, and the filarial worms of humans that cause elephantiasis by obstructing lymphatic vessels. Deworming of puppies often results in the expulsion of dog ascarids from the digestive tract. The most studied ascarid is *Ascaris lumbricoides*, which parasitizes humans.

MATERIALS

Per student:
- dissection microscope
- dissecting needle

Per student pair:
- finger bowl

Per lab room:
- collection of free-living roundworms
- tray of moist soil
- container of fresh stream water
- demonstration dissections of
    a male *Ascaris* and
    a female *Ascaris*
- demonstration microscope of a cross section of
    a male *Ascaris* and
    a female *Ascaris*
- boxes of different sizes of lab gloves
- box of safety goggles

A. Free-Living Roundworms

PROCEDURE

1. Examine the demonstration collection of free-living roundworms.
2. Place some moist soil in a finger bowl with some fresh stream water. Using a dissection microscope, look for translucent roundworms in the soil sample (Figure 29-2).

    Describe the behavior of any roundworms you find and make a note of any visible anatomical features they possess.

    _____
    _____
    _____
    _____
    _____
    _____
    _____
    _____
    _____
    _____
    _____

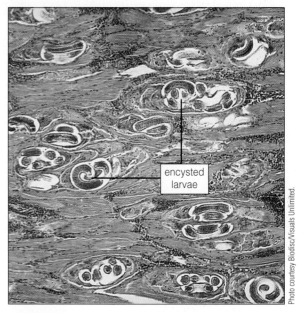

Figure 29-1 *Trichinella spiralis* larvae encysted in muscle (46×). This organism causes the disease trichinosis, which is usually contracted by eating undercooked pork or wild game.

encysted larvae

Photo courtesy Biodisc/Visuals Unlimited.

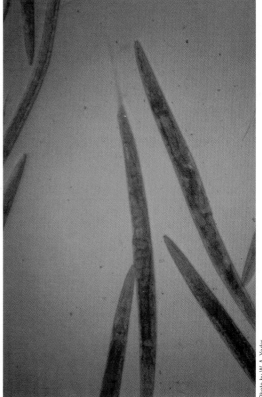

Figure 29-2 Free-living nematodes, *Rhabditella* females, w.m. (350×).

Photo by W. A. Yoder.

B. *Ascaris*

Ascaris is a common parasite of humans, pigs, horses, and other mammals. Adults range from 15 to 40 cm in length, with males being smaller and having a hook shape to the posterior part of the body.

CAUTION

With your lab gloves on, wash your hands and instruments thoroughly after examining *Ascaris* roundworms; although unlikely, some eggs may still be alive, even though the specimen has been preserved.

---

1. Examine a female *Ascaris* (Figure 29-3a). At the front end, the triangular *mouth* is surrounded by three lips. The opening on the ventral surface near the other end of the body is the *anus*. A *genital pore* is located about one-third the distance from the mouth.
2. Look at the demonstration dissection and demonstration cross section of a female *Ascaris* and identify the structures labeled in Figure 29-4a and b. The paired **lateral lines** each contain an *excretory tube*. These tubes open to the outside through a single **excretory pore** located near the mouth on the ventral surface. The **ovaries** are the threadlike ends of a Y-shaped reproductive tract. The ovary leads to a larger **oviduct** and a still larger **uterus**. The two uteri converge to form a short, muscular **vagina** that opens to the outside through the **genital pore**.
3. Examine a male *Ascaris* (Figure 29-3b). The lips and mouth are the same as that of the female. However, the male has a *cloaca*—an exit for both the reproductive and digestive systems—instead of an anus. A pair of *penial spicules* protrudes from this opening.

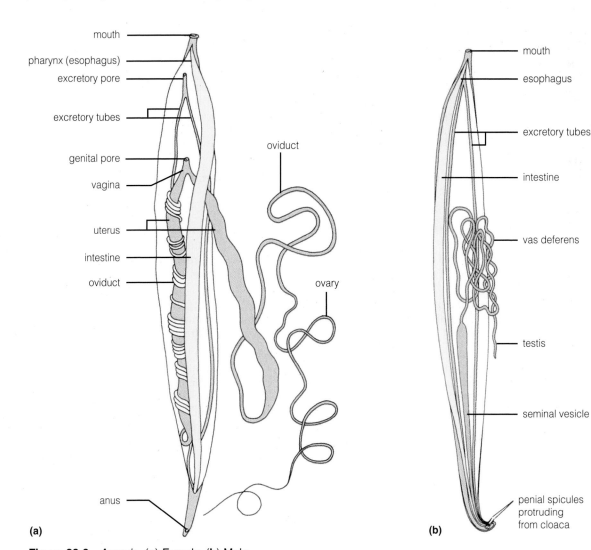

Figure 29-3 *Ascaris.* (**a**) Female. (**b**) Male.

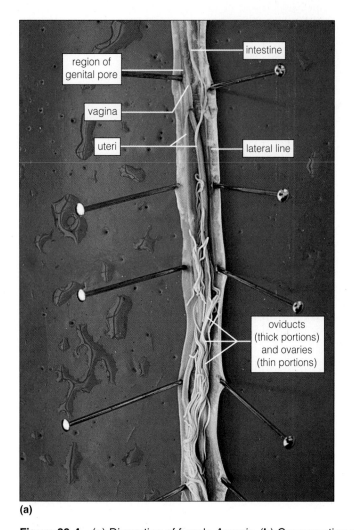

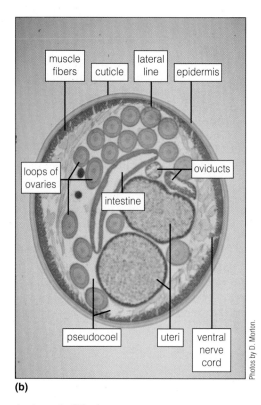

(a)

(b)

Photos by D. Morton.

Figure 29-4 (a) Dissection of female *Ascaris*. (b) Cross section of female *Ascaris* (26×).

4. Look at the demonstration dissection and demonstration cross section of a male *Ascaris* and identify the structures labeled in Figure 29-5a and b. The reproductive system superficially resembles that of the female. It consists of a single tubular structure that includes a threadlike **testis** joined to a larger **vas deferens**, an even larger **seminal vesicle**, and terminally a short, muscular ejaculatory duct that empties into the cloaca.

5. After copulation, the eggs are fertilized in the female oviduct. The eggs, surrounded by a thick shell, are expelled through the genital pore to the outside; as many as 200,000 may be shed into the host's intestinal tract per day. Examine a prepared slide with whole *Ascaris* eggs, and draw what you see in Figure 29-6.

The eggs are eliminated in the host's feces. They are very resistant and can live for years under adverse conditions. Infection with *Ascaris* results from ingesting eggs in food or water contaminated by feces or soil. The eggs pass through the stomach and hatch in the small intestine. The small larvae migrate through the bloodstream, exit the circulatory system into the respiratory system, and crawl up the respiratory tubes and out the trachea into the throat. There they are swallowed to reinfect the intestinal tract and mature. Sometimes larvae exit the nasal passages rather than being swallowed. The entire journey of the larva from intestine back to intestine takes about 10 days.

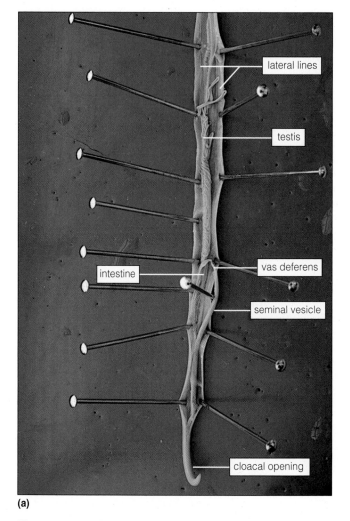

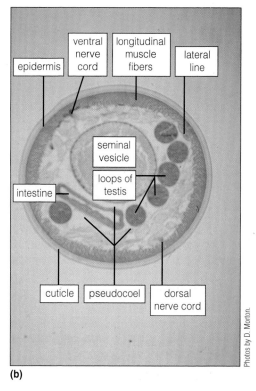

(a)

(b)

Photos by D. Morton.

Figure 29-5 (**a**) Dissection of male *Ascaris.* (**b**) Cross section of male *Ascaris* (35×).

Figure 29-6 Drawing of the *Ascaris* eggs
(_____×).

Habitat	aquatic, terrestrial (land and air), some parasitic
Body Arrangement	bilateral symmetry, coelomate, body regions (head, thorax, and abdomen), segmented (specialized segments form mouthparts and jointed appendages), gut (complete)
Level of Organization	organ system (tissues derived from all three primary germ layers), nervous system with brain, excretory system (green glands or Malphigian tubules), circulatory system (open), respiratory system (gills, trachae, or book lungs)
Body Support	exoskeleton of chitin
Life Cycle	sexual reproduction (mostly dioecious with internal fertilization), larvae, metamorphosis, some parthenogenesis
Special Characteristics	specialized segments, exoskeleton of chitin, metamorphosis, flight in some, complex behavior

In numbers of both species and individuals, this is the largest animal phylum. There are more than 1 million species of arthropods, more than all of the other animal phyla combined. Arthropods have both segmentation and an **exoskeleton** (external skeleton) of **chitin**—a structural polysaccharide. Arthropod ancestors had one pair of appendages (legs) per segment. In modern arthropods, some of these appendages evolved into other useful structures.

The body is typically divided into three regions: the *head, thorax,* and *abdomen.* The mouthparts are modified legs. The coelom is filled with the blood, or **hemolymph**, and connected to an *open circulatory system.* In addition to the exoskeleton, the evolutionary success of arthropods is due in large part to their *highly developed central nervous system, well-developed sense organs,* and *complex behavior.*

MATERIALS

Per student:

- compound microscope, lens paper, a bottle of lens-cleaning solution (optional), a lint-free cloth (optional)
- dissecting scissors
- blunt probe or dissecting needle
- scalpel

Per student pair:

- dissection microscope
- preserved crayfish
- dissection pan
- grasshopper (*Romalea*)

- prepared slide of a whole mount of grasshopper mouthparts

Per lab room:

- demonstration collections of
  living and preserved and mounted crustaceans
  living and preserved and mounted insects
  living and preserved centipedes
  living and preserved millipedes
  living and preserved and mounted spiders, horseshoe crabs, scorpions, ticks, and mites
- large plastic bag for disposal of dissected specimens
- boxes of different sizes of lab gloves
- box of safety goggles

PROCEDURE

A. Crustaceans (Subphylum Crustacea)

Some of our most prized seafoods are crustaceans. This group contains shrimps, lobsters, crabs, and crayfish—and animals of considerably less culinary interest, including water fleas, sand fleas, isopods (sow or pill bugs), ostracods, copepods, and barnacles. Most species are marine, but some live in freshwater. Most isopods occupy moist areas on land. For the most part, crustaceans are carnivorous, scavenging, or parasitic.

Crustaceans have two pairs of antennae on the head, and their exoskeleton is hardened on top and sometimes on the sides to form a protective **carapace**.

1. Examine the demonstration specimens of crustaceans on display in the lab.
2. Check out the external features of a crayfish.
   (a) Obtain a crayfish, a fairly typical crustacean, and rinse it in water at the sink. Place the rinsed crayfish in a dissection pan.
   (b) Recognize the two major body regions (Figure 29-7): an anterior **cephalothorax**, composed of a fused head and thorax, and a posterior *abdomen.* Note the **carapace**. The portion of the carapace that extends between the eyes is the **rostrum**.

(c) Identify the pair of large, image-forming **compound eyes** (comprised of many closely packed photosensory units) each one atop a movable stalk, a pair of **antennae**, and two pairs of smaller and shorter **antennules**. The first pair of antennules are considered short antennae.

(d) Find the **mouth**, which is comprised of three pairs of mouthparts—one pair of **mandibles** and two pairs of **maxillae**. Three pairs of **maxillapeds** (Figure 29-8) surround the mouthparts.

(a) **(b)** exopodite

gure 29-7   (**a**) Dorsal and (**b**) ventral views of the external anatomy of the crayfish. The five walking legs are numbered in (**a**).

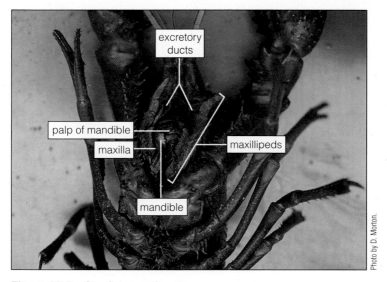

Figure 29-8   Crayfish mouthparts.

TABLE 29-1	Function of Crayfish Legs and Modified Appendages
Structure	Function
Antennule	Contains a statocyst, an organ of balance, in the basal segment
Antenna	Vision and chemoreception
Mandible	Jaw
Maxilla (first)	Food handling
Maxilla (second)	Creates current for gaseous exchange in the gills
Maxillipeds (first, second, and third)	Food handling
Walking legs (1–5)	Defense (chelipeds) and movement
Swimmerets (1–5)	Copulation (males) and brooding young (female)
Uropod (combines with telson and other uropod to form tail fan)	Swimming backward

© Cengage Learning 2013

(e) Locate the five pairs of large **walking legs**; the first, called the **chelipeds**, bear large pincers called **chelae**. Continuing toward the tail, count the five pairs of abdominal **swimmerets**. Near the end of the abdomen, locate the **uropods**, a pair of large, flattened lateral appendages located on both sides of the *telson*, which is an extension of the last abdominal segment. The **anus** opens onto the ventral surface of the telson.

(f) Table 29-1 summarizes in order from front to back down one side of the body those structures that are legs or modified appendages. Review them in order and familiarize yourself with their functions.

(g) Crayfish exhibit **sexual dimorphism** (morphological differences between male and female). Look at a number of specimens of both sexes, observe any differences in the characteristics listed in Table 29-2 (and shown in Figure 29-9), and fill in any blanks.

3. Now let's dissect a crayfish.

(a) Locate the **gills** within the branchial chambers by carefully cutting away the lateral flaps of the carapace with your scissors (Figure 29-10). The gills are feathery structures containing blood channels that function in gaseous exchange.

(b) With scissors, superficially cut forward from the rear of the carapace to just behind the eyes. With your scalpel, carefully separate the hard carapace from the thin, soft, underlying tissue. Next, remove the gills to reveal the internal organs (Figure 29-11).

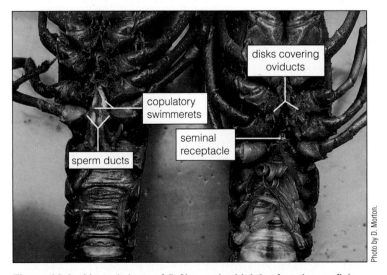

Photo by D. Morton.

Figure 29-9 Ventral views of (left) a male, (right) a female crayfish.

© Cengage Learning 2013

Characteristic	Male	Female
Swimmerets	Front two pairs of swimmerets enlarged for copulation and transferring sperm to the female	No modifications
Walking legs	Openings of the *sperm ducts* at the bases of 5th pair	*Seminal receptacle* in the middle between the bases of the 4th and 5th pairs and discs covering the oviducts at the bases of 3rd pair
Width of abdomen	Narrow	Broad

TABLE 29-2 Sexual Dimorphisms in Crayfish (Figure 29-9)

Figure 29-10 Internal anatomy of the crayfish with side of carapace removed.

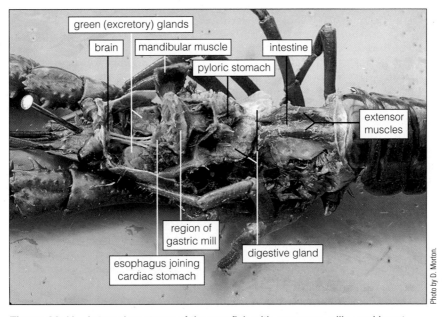

Figure 29-11 Internal anatomy of the crayfish with carapace, gills, and heart removed.

(c) Note the two longitudinal bands of *extensor muscles* that run dorsally through the thorax and abdomen. In the abdomen, find the large *flexor muscles* lying ventrally below the extensor muscles and intestine. What is the function of these two sets of muscles?

_____

_____

_____

(d) Locate the small, membranous heart or its location just posterior to the stomach. Blood flows from the heart into arteries, which open into *sinuses* (open spaces). Blood returns to the heart through holes in its wall.

(e) Identify the organs of the upper digestive tract. The mouth leads to the tubular **esophagus**, which empties into the **cardiac stomach**. At the border between the cardiac stomach and the **pyloric stomach** is the **gastric mill**, a grinding apparatus composed of three chitinous teeth. What role in digestion do you think this structure plays?

_____

_____

_____

(f) Note the large **digestive gland**. This organ secretes digestive enzymes into the cardiac stomach and takes up nutrients from the pyloric stomach. Absorption of nutrients from the tract continues in the intestine, which runs from the abdomen to the *anus*.

(g) Locate the **green glands**, the excretory organs situated ventrally in the head region near the base of the antennae. A duct leads to the outside from each green gland.

(h) The gonads lie beneath the heart. They will be obvious if your specimen was obtained during the reproductive season. If not, they will be small and difficult to locate. In a female, try to find the *ovaries* just beneath the heart. In the male, two white *testes* will occupy a similar location.

(i) Remove the organs in the cephalothorax to expose the *ventral nerve cord*. Observe the segmental ganglia and their paired lateral nerves. Trace the nerve cord forward to locate the **brain**. Note the nerves leading from the brain to the eyes, antennules, antennae, and mouthparts.

(j) Dispose of your dissected specimen in the large plastic bag provided for this purpose.

B. Insects (Subphylum Hexapoda)

Of the more than 1 million species of arthropods, more than 850,000 of them are insects. These animals occupy virtually every kind of terrestrial and freshwater habitat. This extraordinary proliferation of species is almost certainly due to the evolution of wings. Insects were the first organisms to fly and thus were able to exploit a variety of opportunities not available to other animals.

1. Examine insects on demonstration in the lab. Groups of insects you are probably familiar with—dragonflies and damselflies; grasshoppers, crickets, and katydids; cockroaches; "praying" mantids; termites; cicadas, aphids, leafhoppers, and spittlebugs; "true" bugs; beetles; "true" flies; fleas; butterflies and moths; and wasps, bees, and ants—each belongs to one of the orders of insects.

2. Examine the external features of a typical insect.
    (a) Obtain a preserved grasshopper of the genus *Romalea* and place it in a dissection pan (Figure 29-12).
    (b) Observe that the body consists of a **head, thorax,** and **abdomen**. The thorax is composed of three segments, each of which bears a pair of legs. The middle and back segments also each bear a pair of wings. The abdomen, unlike the situation in crustaceans, has no appendages.
    (c) On the sides of the thorax and abdomen find the **spiracles**, tiny openings into the **tracheae** (or breathing tubes), which course through the body and function in gaseous exchange.
    (d) Note that the **head** of the grasshopper contains a pair of **compound eyes**, which form images. Between these eyes, find three simple eyes, the **ocelli**, which although light sensitive, do not form images. A single pair of antennae characterizes the insects from the crustaceans (which have two) and the arachnids (which have none).
    (e) Locate the feeding appendages (Figure 29-13)—a pair of **mandibles** and two pairs of **maxillae**, the second pair fused together to form the lower lip, the **labium**. The upper lip, the **labrum**, covers the mandibles. Near the base of the labium is a tongue-like process called the **hypopharynx** (Figure 29-14).

3. Look at the demonstration slide of a whole mount of grasshopper mouthparts. Locate and label a mandible, a maxilla, the labium, and the labrum in Figure 29-14.

Mouthparts are extremely variable in insects, with modifications for chewing, piercing, siphoning, and sponging (Figure 29-15).

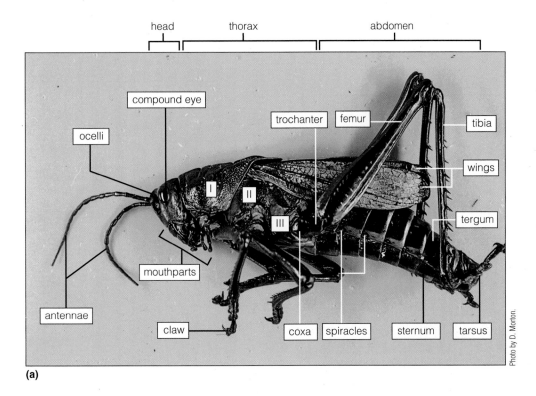

(a)

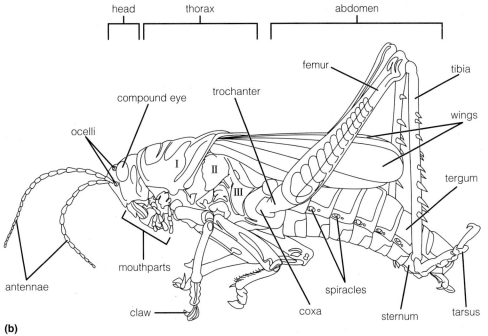

(b)

Figure 29-12 Photograph (**a**) and diagram (**b**) of the external anatomy and body form of a grasshopper, a typical insect.

C. Centipedes and Millipedes (Subphylum Myriapoda)

Centipedes are swift, predaceous arthropods adapted for running. The body is flattened with one pair of legs per segment, except for the head and the rear two segments (Figure 29-16a). The first segment of the body bears a pair of legs modified as *poison claws* for seizing and killing prey. In some tropical species (which can get as long as 20 cm), the poison can be dangerous to humans. Centipedes live under stones or the bark of logs. Their prey consists of other arthropods, worms, and mollusks. The common house centipede eats roaches, bedbugs, and other insects.

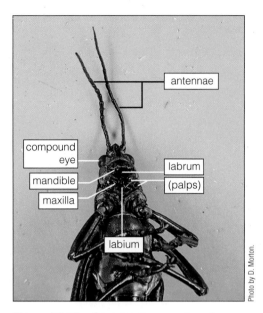

Figure 29-13   Grasshopper mouthparts.

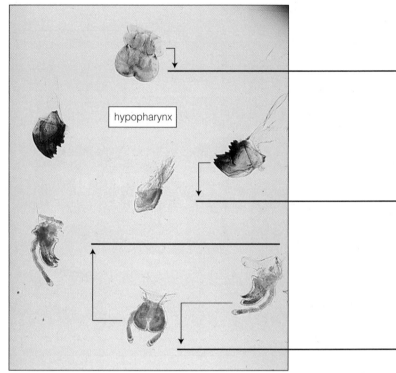

hypopharynx

Figure 29-14   Isolated grasshopper mouthparts, w.m. (7×).

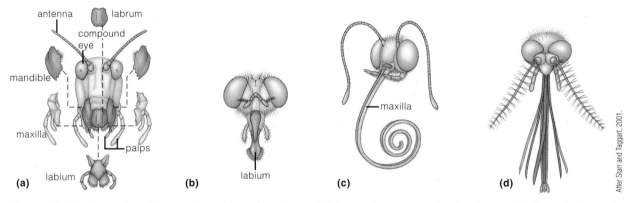

antenna   labrum
          compound
          eye
mandible
maxilla
          palps
labium
(a)   (b)   labium   (c)   maxilla   (d)

After Starr and Taggart, 2001.

Figure 29-15   Examples of insect appendages. Headparts of (a) grasshoppers, a chewing insect; (b) flies, which sponge up nutrients with a specialized labium; (c) butterflies, which siphon up nectar with a specialized maxilla; and (d) mosquitoes, with piercing and sucking appendages.

Although superficially similar, the millipedes and centipedes differ in many ways. The millipedes typically have cylindrical bodies. All segments, except those of the short thorax, bear two pairs of legs. There are no poison claws (Figure 29-16b). Millipedes are very slow arthropods that live in dark, moist places. They scavenge on decaying organic matter. When handled or disturbed, they often curl up into a ball and secrete a noxious fluid from their scent glands as a means of defense.

1. Examine the centipedes on demonstration in the lab.
2. Similarly observe the millipedes on display in the lab.

D. Subphylum Chelicerata: Spiders, Horseshoe Crabs, Scorpions, Ticks, and Mites

Today most species are terrestrial, but the early chelicerates were marine. Horseshoe crabs represent these ancient marine forms. Their bodies and five pairs of walking legs are concealed under a tough carapace (Figure 29-17a).

The body of spiders and scorpions consists of a *cephalothorax* and *abdomen*. In ticks and mites, these parts are fused. The cephalothorax bears six pairs of appendages, the rear four of which are walking legs (Figure 29-17b). How many pairs of walking legs do insects have? _____

(a)

(b)

Figure 29-16 (a) Live centipede.
(b) Live millipede.

(a)

(b)

Figure 29-17 (a) Ventral view of a horseshoe crab. (b) Live spider. *Continues*

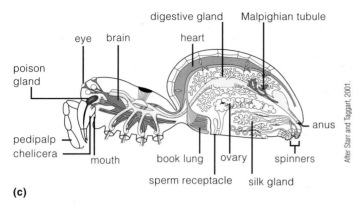

eye brain
digestive gland Malpighian tubule
poison gland
heart
pedipalp
chelicera
mouth
book lung
sperm receptacle
ovary
silk gland
spinners
anus

After Starr and Taggart, 2001.

Figure 29-17   (**c**) Internal anatomy of a spider. *Continued*

(c)

Chelicerates have no true mandibles or antennae, and their eyes are simple. In spiders, the first pair of appendages bear fangs. Each fang has a duct that connects to a poison gland. The second pair of appendages is used to chew and squeeze their prey. This second pair is also sensory, and males use them to store and transfer sperm to females. Most, but not all, spiders spin webs to catch their prey. They secrete digestive enzymes into the prey's (usually an insect's) body, and the liquefied remains are sucked up. Other chelicerates have front appendages specialized to their particular method of feeding. Ticks suck the blood of vertebrates and most mites are scavengers—for example, dust mites eat the shed skin cells that are a major component of house dust.

Examine the assortment of chelicerates on display in the lab.

_____ **1.** Structures that are shed in one piece include
   (a) the cuticle of roundworms.
   (b) the exoskeleton of arthropods.
   (c) human hair.
   (d) all of the above

_____ **2.** Nematodes and rotifers have a body cavity that is
   (a) called a coelom.
   (b) called a pseudocoelom.
   (c) completely lined by tissues derived from mesoderm.
   (d) both a and c

_____ **3.** Filarial roundworms cause elephantiasis by
   (a) encysting in muscle tissue.
   (b) promoting the growth of fatty tumors.
   (c) obstructing lymphatic vessels.
   (d) laying numerous eggs in the joints of their hosts.

_____ **4.** Digestive wastes of the male *Ascaris* roundworm exit the body through the
   (a) anus.
   (b) cloaca.
   (c) mouth.
   (d) excretory pore.

_____ **5.** The animal phylum with the most species is
   (a) Mollusca.
   (b) Annelida.
   (c) Arthropoda.
   (d) Platyhelminthes.

_____ **6.** The body of arthropods includes
   (a) a head.
   (b) a thorax.
   (c) an abdomen.
   (d) all of the above

_____ **7.** Sexual dimorphism, as seen in the crayfish, is
   (a) the presence of male and female individuals.
   (b) the production of eggs and sperm by the same individual.
   (c) another term for copulation.
   (d) the presence of observable differences between males and females.

_____ **8.** The grinding apparatus of the digestive system of the crayfish is/are the
   (a) oral teeth.
   (b) gizzard.
   (c) pharyngeal jaw.
   (d) gastric mill.

_____ **9.** The insects were the first organisms to
   (a) show bilateral symmetry.
   (b) exhibit segmentation.
   (c) fly.
   (d) develop lungs.

_____ **10.** Like the insects, the arachnids (spiders and so on) have
   (a) three pairs of walking legs.
   (b) one pair of antennae.
   (c) true mandibles.
   (d) an abdomen.

Name _____   Section Number _____

EXERCISE **29**

Roundworms and Joint-Legged Animals

Post-Lab Questions

Introduction

1. Indicate the differences between protostomes and deuterostomes, and list the phyla of animals in each group.

2. Describe ecdysis in roundworms and joint-legged animals.

29.1 Roundworms (Phylum Nematoda)

3. *Pseudocoel* means "false body cavity." Why is this term applied to the body cavity of roundworms?

29.2 Joint-Legged Animals (Phylum Arthropoda)

4. Define *sexual dimorphism,* describe it in the crayfish, and list two other animals that exhibit this phenomenon.

5. Identify and label the structures noted in this diagram.

A. _____   E. _____   H. _____
B. _____   F. _____   I. _____
C. _____   G. _____   J. _____
D. _____

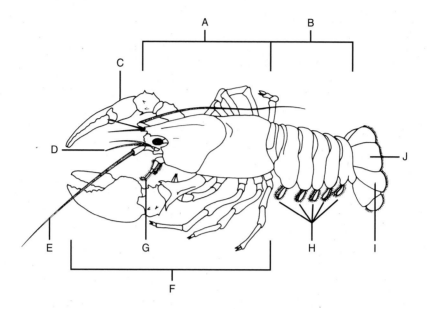

6. How has flight been at least partly responsible for the success of the insects?

7. List several differences between insects and spiders.

Food for Thought

8. List the roundworm parasites and use the Internet to learn how they gain entry into the body. Describe that process here.

9. Write the phylum and choose the appropriate description for each animal (or group of animals) listed in the following table.

Animal	Phylum	Symmetry (none, radial, or bilateral)	Level of Organization (cell, tissue, organ system)	Body Cavity (none, pseudocoelomate, or coelomate)	Gut (none, incomplete, or complete)
Sponge					
Hydra and Jellyfish					
Flatworm					
Rotifer					
Segmented worm					
Clam, snail, and octopus					
Roundworm					
Crayfish, insect, and spider					

10. Define *segmentation* and indicate how the annelids, arthropods, and your body exhibit this phenomenon.

Echinoderms and Invertebrate Chordates

OBJECTIVES

After completing this exercise, you will be able to

1. define *dermal endoskeleton, water-vascular system, notochord, pharyngeal gill slits, dorsal hollow nerve cord, post-anal tail, invertebrate;*

2. describe the natural history of members of phylum Echinodermata and the invertebrate members of phylum Chordata;

3. identify some representatives of the echinoderm classes (sea stars, brittle stars, sea urchins, and sea cucumbers);

4. compare and contrast the invertebrate chordates (subphyla Urochordata and Cephalochordata);

5. identify structures (and indicate associated functions) of the representatives of these phyla and subphyla.

Introduction

The ancestors of echinoderms and chordates, the phylum to which we belong, had a coelom (body cavity) and were deuterostomes (first embryonic opening becomes anus). Although the adult forms of these two phyla look quite different, chordates are thought to have evolved from the bilaterally symmetrical larvae of ancestral echinoderms some 600 million years ago.

There is an additional small phylum, the Hemichordata, whose members (acorn worms) have both echinoderm and chordate characteristics. Acorn worms are *long, slender, wormlike* animals (Figure 30-1a). Most *live in shallow seawater.* They *obtain food by burrowing in mud and sand or filter feeding.* Evidence of their former activity can be seen on exposed tidal flats where mud or sand that has passed through their bodies is left in numerous coiled, ropelike piles (Figure 30-1b).

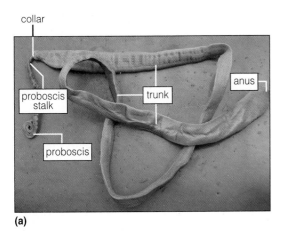

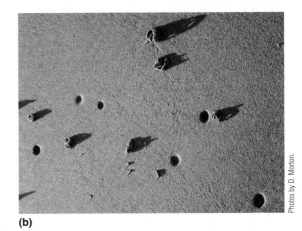

(a)   (b)

Figure 30-1   (a) Preserved acorn worm. (b) Ropelike piles of sand formed by burrowing acorn worms.

30.1   Echinoderms (Phylum Echinodermata) *(60 min.)*

Habitat	aquatic, marine
Body Arrangement	radial symmetry (larvae are bilaterally symmetrical), coelomate, no well-defined head, no segmentation, gut (usually complete)

Level of Organization	organ system (tissues derived from all three primary germ layers), circular nervous system with radial nerves, reduced circulatory system, no respiratory and excretory system, dermal endoskeleton (mesoderm)
Body Support	endoskeleton of calcium carbonate
Life Cycle	sexual reproduction (mostly dioecious with external fertilization), larvae, metamorphosis, high regenerative potential
Special Characteristics	radial symmetry, no segmentation, water-vascular system, dermal endoskeleton

CAUTION

Specimens are kept in preservative solutions. Use lab gloves and safety goggles, especially if you wear contact lenses, whenever you or your lab mates handle a specimen or other fluids. If safety goggles do not fit, your glasses should suffice. Wash any part of your body exposed to preservative with copious amounts of water. If preservative solution or other fluids are splashed into your eyes, wash them with the safety eyewash bottle or eye irrigation apparatus for 15 minutes.

Echinoderm means "spiny skin." These are marine animals; they live on the bottom of both shallow and deep seas. Their feeding methods range from trapping organic particles and plankton (e.g., sea lilies and feather stars) to scavenging (sea urchins) and predatory behavior (sea stars).

The echinoderms exhibit five-part *radial symmetry* and a calcareous (contains calcium carbonate) **dermal endoskeleton** (internal skeleton) composed of many small plates. Much of the *coelom* is taken up by a unique *water-vascular system,* which functions in movement, attachment, respiration, excretion, food handling, and sensory perception.

MATERIALS

Per student:

- dissecting scissors
- blunt probe or dissecting needle

Per student pair:

- dissection microscope
- preserved sea star
- dissection pan
- dissection pins
- prepared slide of
  a cross section of a sea star arm (ray)
  a whole mount of bipinnaria larvae
- compound microscope, lens paper, a bottle of lens-cleaning solution (optional), and a lint-free cloth (optional)

Per lab room:

- collections of
  preserved or mounted brittle stars, sea urchins, sea biscuits, and sand dollars (preserved specimens and skeletons), preserved sea cucumbers
- live specimens (optional)
- collection of preserved or mounted feather stars and sea lilies
- large plastic bag for disposal of dissected specimens
- boxes of different sizes of lab gloves
- box of safety goggles

A. Sea Stars (Class Asteroidea)

The sea stars are familiar occupants of the sea along shores and coral reefs. They are slow-moving animals that sometimes gather in large numbers on bare areas of rock. Many are brightly colored. Some sea stars are more than 1 m in diameter.

PROCEDURE

1. Obtain a preserved specimen of a sea star and keep your specimen moist with water in a dissection pan. Note the **central disk** on the *aboral* side—the side without the mouth and the **five arms** attached to it. Find the **sieve plate** (madreporite), a light-colored calcareous roundish structure near the edge of the disk between two arms (Figure 30-2a). The pores in the sieve plate open into the water-vascular system.
2. With a dissection microscope, note the many protective **spines** scattered over the surface of the body. Near the base of the spines are many small pincerlike structures, the **pedicellariae** (Figure 30-2b). The *valves* of these structures grasp objects that land on the surface of the body, such as potential parasites. Also among the spines are many soft, hollow **skin gills** (dermal branchiae) that communicate directly with the coelom and function in the exchange of gases and excretion of ammonia (a nitrogen-containing metabolic waste).
3. Locate the **mouth** on the oral side (Figure 30-3a). Note that an **ambulacral groove** extends from the mouth down the middle of the oral side of each arm. Numerous **tube feet** extend from the water-vascular system

(a)

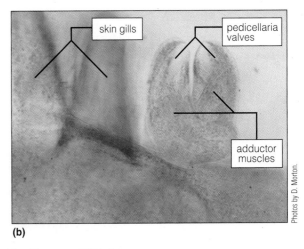

(b)

Figure 30-2 (a) Aboral view of sea star. (b) Pedicellaria and skin gills, w.m. (205×).

and occupy this groove. Each tube foot consists of a bulblike structure attached to a **sucker** (Figure 30-3b). The animal moves by alternating the suction and release mechanisms of the tube feet. Tube feet are also used to adhere to the shells of bivalves and to pry them open to get at their soft insides. The tube feet, along with the skin gills, also function in the exchange of gases and excretion of ammonia.

4. With your dissecting scissors, cut across the top of an arm about 1 cm from the tip. Next, cut out a rectangle of spiny skin by carefully cutting along each side of the arm to the central disk and then across the top of the arm at the edge of the disk. Observe the hard, calcareous plates of the dermal endoskeleton as you cut. Remove the rectangle of skin to uncover the large *coelom*, which contains the internal organs (Figure 30-4).

5. Cut around the *sieve plate* to remove the upper portion of the body wall of the *central disk*. The *mouth* connects via an extremely short esophagus to the clearly visible pouchlike **cardiac stomach**. The cardiac stomach opens into the upper *pyloric stomach*. A slender, short *intestine* leads from the upper side of the stomach to the *anus*. Find the two green, fingerlike **digestive glands** in each arm, which produce digestive enzymes and deliver them to the pyloric stomach.

6. Identify the dark-colored **gonads** that are located near the base of each arm. The sexes are separate but difficult to distinguish, except by microscopic examination. In nature, sperm and eggs are released into the seawater, and the fertilized eggs develop into bipinnaria larvae.

7. The water-vascular system is unique to echinoderms (Figure 30-5). Note that the *sieve plate* leads to a short **stone canal**, which in turn leads to the *circular canal* surrounding the mouth. Five *radial canals* lead from the circular canal into the ambulacral grooves. Each radial canal connects by short side branches with many pairs of tube feet.

8. The nervous system (not shown in Figure 30-3) is simple. A circular *nerve ring* surrounds the mouth, and a *radial nerve* extends from this into each arm, ending at a light-sensitive eyespot. There are no specific excretory organs. Dispose of your dissected specimen in the large plastic bag provided for this purpose.

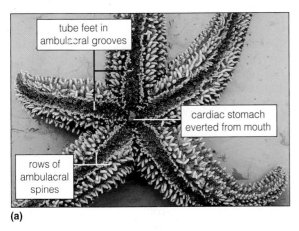

(a)

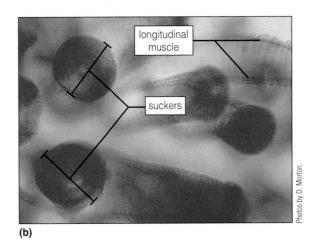

(b)

Figure 30-3 (a) Oral view of sea star. (b) Close-up of tube feet, w.m. (78×).

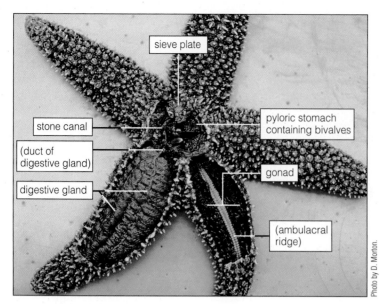

Figure 30-4   Internal anatomy of a sea star.

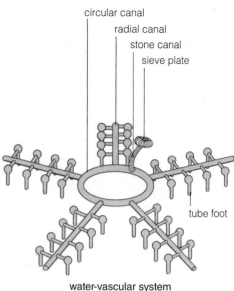

Figure 30-5   The water-vascular system of a five-armed sea star. The ring canal is located in the central disk, and each radial canal extends into one of the arms.

9. Examine a prepared slide of a cross section of an arm with your compound microscope. Identify the structures labeled in Figure 30-6.
10. Examine a prepared slide of a whole mount of bipinnaria larvae. The digestive tract, including the *mouth, esophagus, stomach, intestine*, and *anus*, is clearly visible (Figure 30-7).
11. Observe any available preserved, plastic-mounted, or living specimens of other sea stars.

B. Other Echinoderms

PROCEDURE

1. *Class Ophiuroidea* includes the brittle stars (Figure 30-8). These echinoderms have slender arms and move relatively rapidly, for echinoderms! In life, the rays are much more flexible than those of the sea stars. Examine preserved, plastic-mounted, or living specimens of brittle stars.
2. *Class Echinoidea* contains the sea urchins, animals that lack arms. They have long, movable spines, which surround the top and sides of a compact skeleton called the **test** (Figure 30-9). Sea urchins ingest food by means of a complex structure, *Aristotle's lantern*, which contains teeth. The sea biscuits and sand dollars are also members of this class. They have flat, disk-shaped bodies. Sea urchins, sea biscuits, and sand dollars, despite their lack of arms, all exhibit five-part radial symmetry and move by means of tube feet. Study preserved or living specimens and the tests (Figure 30-10) of members of this class.
3. *Class Holothuroidea* are the sea cucumbers (Figure 30-11). No spines are present. A collection of tentacles at one end surrounds the mouth. In some forms, the tube feet occur over the entire surface of the body. Sea cucumbers move in wormlike fashion as a result of the contractions of two layers of circular and longitudinal muscles. Examine preserved or living sea cucumbers.
4. *Class Crinoidea* contains the graceful, flowerlike feather stars and sea lilies (Figure 30-12). These stationary animals attach to the sea floor by a long stalk. The arms are featherlike; each contains a ciliated groove and tentacles (derived from tube feet) to direct food to the mouth. There can be from 5 to over 200 arms, some of which reach 35 cm. *Class Concentricycloidea* contains the sea daisies. Examine any available specimens and illustrations of these classes.

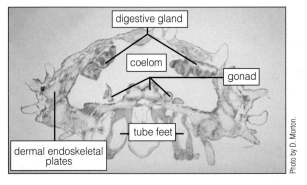

Figure 30-6 Cross section of a sea star arm (23×).

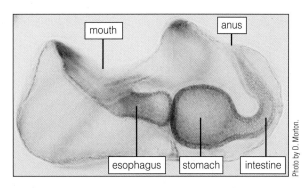

Figure 30-7 Bipinnaria larva (196×).

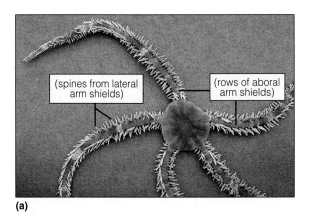

(a)

(b)

Figure 30-8 (a) Dried and (b) live brittle star.

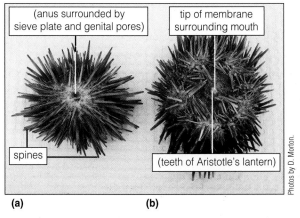

(a)   (b)

Figure 30-9 (a) Aboral and (b) oral views of a sea urchin.

(a)   (b)

Figure 30-10 (a) Sea biscuit and (b) sand dollar tests.

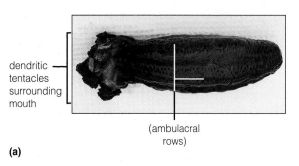

dendritic tentacles surrounding mouth

(ambulacral rows)

(a)

(b)

Photos by D. Morton and W. A. Yoder.

Figure 30-11 (a) Preserved and (b) live sea cucumber.

30.2 Chordates (Phylum Chordata) (50 min.)

Habitat	aquatic, terrestrial (land and air)
Body Arrangement	bilateral symmetry, coelomate, head and post-anal tail, segmentation, gut (complete)
Level of Organization	organ system (tissues derived from all three primary germ layers), nervous system with brain, circulatory system (closed with a ventral heart), organ systems highly developed
Body Support	most have a mesoderm-derived endoskeleton of cartilage or bone
Life Cycle	sexual reproduction (mostly dioecious with internal or external fertilization), some larvae, some metamorphosis
Special Characteristics	notochord, pharyngeal gill slits, dorsal hollow nerve cord, and a post-anal tail at some stage of life cycle

JWAlker/Jupiter Images

Figure 30-12 Feather star.

The word *chordate* refers to one of the four major diagnostic characteristics shared by all members of this phylum at least some time in their life span: a **notochord, pharyngeal gill slits,** a **dorsal hollow nerve cord,** and a **post-anal tail** (Table 30-1). Most chordates are also vertebrates—animals that possess a backbone or vertebral column—but a few are not. The latter animals belong to the subphlya Urochordata and Cephalochordata. They are **invertebrates** because they lack a vertebral column. All of the animal phyla you have studied so far are also invertebrates.

TABLE 30-1	**Chordate Characteristics**	
Characteristic	**Definition**	**Developmental Fate in Humans**
Notochord	A dorsal, flexible rod that provides support for most of the length of the body	Center of intervertebral disks
Pharyngeal gill slits	Paired, lateral outpockets of pharynx that perforate through its wall	Auditory tubes and endocrine glands of neck
Dorsal hollow nerve cord	A tube of nervous tissue dorsal to notochord	Brain and spinal cord
Post-anal tail	Posterior tail that extends past the anus	Present in embryo

© Cengage Learning 2013

MATERIALS

Per student:

■ blunt probe

Per student pair:

■ dissection microscope
■ compound microscope, lens paper, a bottle of lens-cleaning solution (optional), and a lint-free cloth (optional)
■ dissection pan

■ preserved or mounted sea tunicates (sea squirts)
■ preserved and plastic mounted lancelet specimens
■ prepared slide of a whole mount of a lancelet
■ prepared slide of a cross section of a lancelet in the region of the pharynx
■ boxes of different sizes of lab gloves
■ box of safety goggles

A. Tunicates (Subphylum Urochordata)

Tunicates are also called sea squirts. Both common names are descriptive of obvious features of these animals. A leather-like "tunic" covers the adult, and water is squirted out of an excurrent siphon (Figure 30-13).

Sea squirt larvae resemble tadpoles in general body form and have a notochord confined to the post-anal tail. The notochord and tail degenerate when the larva becomes an adult tunicate. The larva also has a primitive brain and sense organs. These structures undergo reorganization into a nerve ganglion and nerve net in the adult. The number of pharyngeal gill slits increases in the adult.

1. Examine a preserved or plastic mounted specimen of a sea squirt with a dissection microscope. These animals are small, inactive, and very common marine organisms that inhabit coastal areas of all oceans. The adult is a filter feeder, capturing organic particles into an incurrent siphon by ciliary action. Individuals either float freely in the water (singly or in groups) or are sessile, attached to the bottom as branching individuals or colonies.

2. Water enters a sea squirt through the *incurrent siphon*. It travels into the *pharynx*, where it seeps through *gill slits* to reach a chamber, the *atrium*. The water eventually exits through the *excurrent siphon*. Back in the pharynx, food particles are trapped in sticky mucus and then passed to the digestive tract. Undigested materials are discharged from the anus into the atrium. Identify as many of these structures as possible.

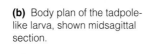

(b) Body plan of the tadpole-like larva, shown midsagittal section.

(a)

(c) A new larva swims about for a brief period. Metamorphosis begins when its head attaches to a substrate. The notochord, tail, and most of the nervous system are resorbed (recycled to form new tissues). Slits in the pharynx wall multiply. Organs rotate until the openings through which water enters and leaves the pharynx are directed away from the substrate.

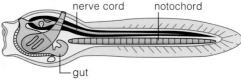
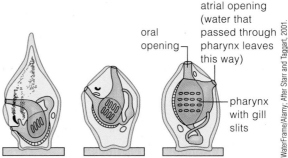

Figure 30-13 **(a)** Adult form of a sea squirt. **(b)** A sea squirt larva. **(c)** Metamorphosis of a larva into the adult.

B. Lancelets (Subphylum Cephalochordata)

The lancelets are distributed worldwide and are especially abundant in coastal areas with *warm, shallow waters*. *Amphioxus* (Figure 30-14) is the commonly studied representative of this subphylum. The word *Amphioxus* means "sharp at both ends," and this certainly describes its *elongate, fishlike body*. Lancelets reach 50–75 mm in length as adults. Despite their streamlined appearance, these organisms are not very active. They spend most of their time buried in the sand with their heads projecting while they filter organic particles from the water.

1. Examine a preserved lancelet with the dissection microscope (Figure 30-15). Note that the body is translucent, with a low, continuous dorsal and caudal (tail) **fin**. Find the mouth surrounded by tentacles and the

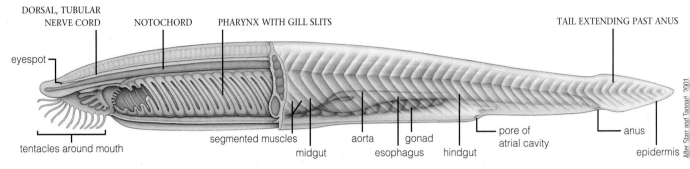

After Starr and Tannart 2001

Figure 30-14 Cutaway view of a lancelet. The four Chordate characteristics are capitalized.

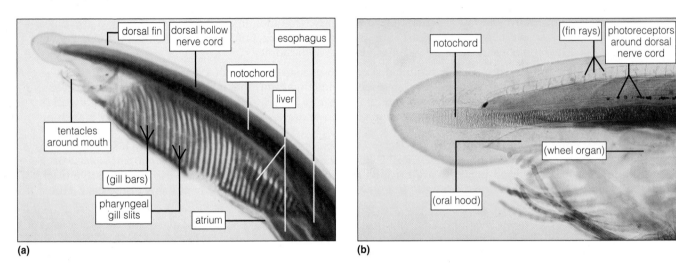

Photos by D. Morton.

(a)

(b)

Figure 30-15 Whole mounts of (**a**) anterior end (30×) and (**b**) head (100×) of a lancelet.

roughly 150 **pharyngeal gill slits**. Trace below the flow of water from the **mouth** to the pore of the atrial cavity.

_____

_____

_____

2. Describe the capture of food and the path it takes through the digestive system. Unlike the tunicates, the *anus* in a lancelet opens externally.

_____

_____

_____

3. Find the flexible **notochord** extending nearly the full length of the individual; this structure persists into the adult stage. There is a small brain, with a **dorsal hollow nerve cord** that bears light-sensitive cells (the *eyespot*) at its anterior end.

4. Examine a cross section of a lancelet with a compound microscope and identify the structures indicated in Figure 30-16.

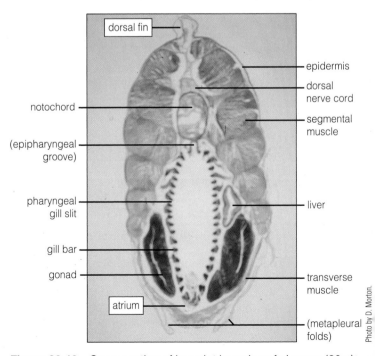

Photo by D. Morton.

Figure 30-16 Cross section of lancelet in region of pharynx (20×).

_____ 1. Echinoderms and chordates are
   (a) deuterostomes.
   (b) protostomes.
   (c) animals whose blastopore becomes a mouth.
   (d) both b and c

_____ 2. The skin gills of sea stars are structures
   (a) for the exchange of gases and excretion of ammonia.
   (b) of defense.
   (c) for movement.
   (d) that produce the skeletal elements.

_____ 3. The tube feet of a sea star
   (a) bear tiny toes.
   (b) are located only at the tips of the arms.
   (c) function in movement.
   (d) protect the organism from predatory fish.

_____ 4. Aristotle's lantern is a tooth-bearing structure of the
   (a) sea stars.
   (b) tunicates.
   (c) acorn worms.
   (d) sea urchins.

_____ 5. The sieve plate, stone canal, circular canal, and radial canals of a sea star are structures of the
   (a) nervous system.
   (b) water-vascular system.
   (c) digestive system.
   (d) excretory system.

_____ 6. The dermal endoskeleton of echinoderms is made of
   (a) cartilage.
   (b) bone.
   (c) calcium carbonate.
   (d) chitin.

_____ 7. Which structure is **not** a major diagnostic feature of the chordates?
   (a) dorsal hollow nerve cord
   (b) notochord
   (c) pharyngeal gill slits
   (d) vertebral column

_____ 8. The sea squirts and lancelets have an inner chamber that expels water. This chamber is called the
   (a) intestine.
   (b) bladder.
   (c) atrium.
   (d) nephridium.

_____ 9. Invertebrates
   (a) do not have a vertebral column.
   (b) have a vertebral column.
   (c) are all members of the phylum Chordata.
   (d) do not include animals that are members of the phylum Chordata.

_____ 10. Which animal is a chordate?
   (a) sea star
   (b) sea cucumber
   (c) acorn worm
   (d) lancelet

EXERCISE **30**

Echinoderms and Invertebrate Chordates

30.1 Echinoderms (Phylum Echinodermata)

1. Describe the various structures on the aboral "spiny skin" of a sea star and discuss their functions.

2. Describe the various functions of tube feet.

3. Identify the following structures on this aboral view of the tip of a sea star arm.

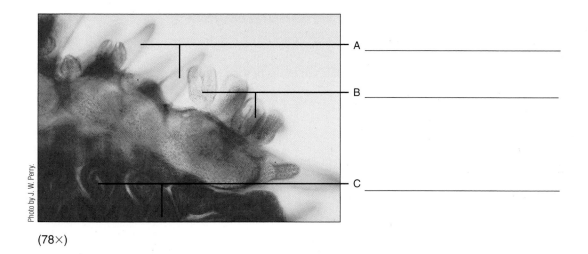

A _____

B _____

C _____

Photo by J. W. Perry.

(78×)

4. Describe reproduction in the sea star.

5. List in order the structures in the water-vascular system of a sea star.

   6. List the four major characteristics of chordates.

   7. Why are sea squirts and lancelets referred to as invertebrate chordates?

   8. Compare the presence or absence of the four chordate characteristics in adult sea squirts and lancelets.

   9. Compare arthropods, echinoderms, and chordates in terms of their similarities and differences.

Food for Thought

   10. Hypothesize an evolutionary relationship among the bipinnaria larva of echinoderms, larval and adult tunicates, and lancelets.

Vertebrates

OBJECTIVES

After completing this exercise, you will be able to

1. define *vertebrate, vertebral column, cranium, vertebrae, cloaca, lateral line, placoid scale, operculum, atrium, ventricle, artery, vein, amniotic egg, ectothermic, endothermic, viviparous;*

2. describe the basic characteristics of members of the subphylum Vertebrata;

3. identify representatives of the vertebrate classes, Cephalaspidomorphi, Chondrichthyes, Osteichthyes, Amphibia, Reptilia, Aves, and Mammalia;

4. identify structures (and indicate associated functions) of representatives of the vertebrate classes;

5. construct a dichotomous key to the animals (optional).

Introduction

In terms of the numbers of individuals and species, chordates do not constitute a large phylum; however, chordates known as vertebrates have had a disproportionately large ecological impact. The **vertebrates** have the *four basic characteristics of chordates* listed in the previous exercise, plus a cranium and a **vertebral column**. The dorsal hollow nerve cord has differentiated into a *brain* and a *spinal cord*. Bones protect both of these structures, the brain by the bones of the **cranium** (braincase) and the spinal cord by the **vertebrae**, the bones that make up the vertebral column.

Like echinoderm larvae and sea squirt larvae and lancelets, vertebrates are *cephalized, bilaterally symmetrical,* and *segmented*. In adults, segmentation is most easily seen in the musculature, vertebrae, and ribs. The body is typically divided into a *head, neck, trunk,* and *tail*. If appendages are present, they are paired, lateral *thoracic appendages* (pectoral fins, forelimbs, wings, and arms) and *pelvic appendages* (pelvic fins, hindlimbs, and legs), which are linked to the vertebral column and function to support and to help move the body. Seven classes of vertebrates have representatives alive today (lampreys, cartilaginous fishes, bony fishes, amphibians, reptiles, birds, and mammals), and one class is completely extinct (armored fishes).

CAUTION

Specimens are kept in preservative solutions. Use lab gloves and safety goggles, especially if you wear contact lenses, whenever you or your lab mates handle a specimen or other fluids. If safety goggles do not fit, your glasses should suffice. Wash any part of your body exposed to preservative with copious amounts of water. If preservative solution or other fluids are splashed into your eyes, wash them with the safety eyewash bottle or eye irrigation apparatus for 15 minutes.

31.1   Lampreys (Class Cephalaspidomorphi) *(12 min.)*

The ancestors of lampreys were the first vertebrates to evolve. Lampreys have a cartilaginous (made of cartilage), primitive skeleton.

Lampreys are represented by both marine and freshwater species and by some species that use seawater and fresh water at some point in their lives. Most lampreys feed on the blood and tissue of fishes by rasping wounds in their sides. A landlocked population of the sea lamprey is infamous for having nearly decimated the commercial fish industry in the Great Lakes. Only a vigorous control program that targeted lamprey larvae restored this industry.

MATERIALS

Per student:

■ scalpel
■ blunt probe or dissecting needle

Per student pair:

■ dissection microscope
■ dissection pan

Per student group:

■ prepared slide of a whole mount of a lamprey larva or ammocoete
■ preserved specimen of sea lamprey

Per lab room:

■ boxes of different sizes of lab gloves
■ box of safety goggles

PROCEDURE

1. With a dissection microscope, observe a prepared slide of a whole mount of a lamprey larva or ammocoete. Sea lampreys spawn in freshwater streams. Ammocoetes hatch from their eggs and, after a period of development, burrow into the sand and mud. They are long-lived, metamorphosing into adults after as long as 7 years. Because of their longevity and dissimilar appearance, adult lampreys and ammocoetes were long thought to be separate species.

The ammocoete is considered by many to be the closest living form to ancestral chordates. Unlike a lancelet, it has an eye (actually two median eyes), internal ears (otic vesicles), and a heart, as well as several other typical vertebrate organs. Identify the structures labeled in Figure 31-1.

2. Examine a preserved sea lamprey (Figure 31-2). Note its slender, rounded body. The skin of the lamprey is soft and lacks scales. Look at the round, suckerlike **mouth**, inside of which are circular rows of horny, rasping **teeth** and a deep, rasping **tongue**. Although there are no lateral, paired appendages, there are two **dorsal fins** and a **caudal fin** (tail fin). Count and record below the number of pairs of external **gill slits**.

_____

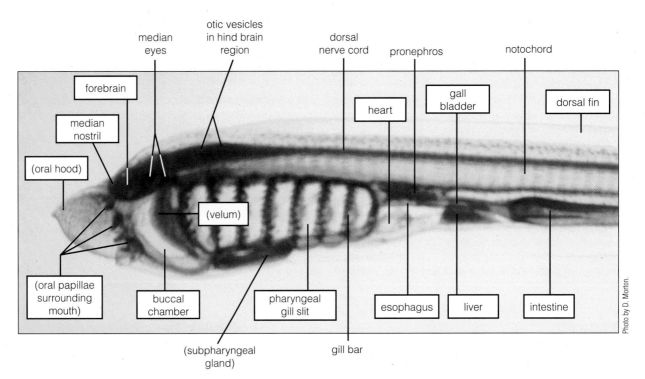

Figure 31-1 Whole mount of ammocoete (40×).

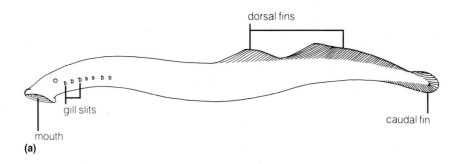

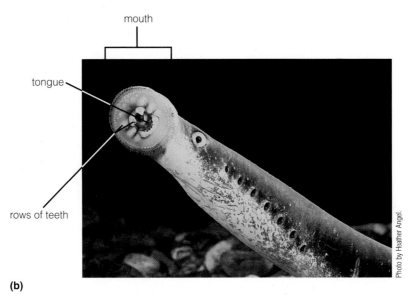

Figure 31-2 **(a)** Side view and **(b)** oral disk of a sea lamprey.

31.2   Cartilaginous Fishes (Class Chondrichthyes) *(12 min.)*

The first vertebrate fossils to show jaws and paired appendages were the heavily armored *placoderms* (class Placodermi). These fishes arose about 425 million years ago. Current hypotheses suggest that the ancestors of placoderms relatively quickly gave rise to the cartilaginous fishes.

Members of the class Chondrichthyes are *jawed fishes* with *cartilaginous skeletons* and *paired appendages.* The evolution of jaws and paired appendages and the evolution of more efficient respiration and a better developed nervous system, including sensory structures, were critical events in the history of vertebrate evolution. Together these adaptations allowed cartilaginous fishes to chase, catch, and eat larger and more active prey. This mostly marine group includes the carnivorous skates, rays, and sharks.

MATERIALS

Per student:

- scalpel
- forceps
- dissecting scissors

Per student pair:

- dissection microscope
- dissection pan

Per student group:

- preserved specimen of dog fish (shark)

Per lab room:

- collection of
    preserved cartilaginous fishes
    shark jaws with teeth
- boxes of different sizes of lab gloves
- box of safety goggles

PROCEDURE

1. Look at the assortment of cartilaginous fishes on display.
2. Refer to Figure 31-3a as you examine the external anatomy of a preserved dog fish (shark).
   - **(a)** How many dorsal fins are there? _____
   - **(b)** Notice the front pair of lateral appendages, the **pectoral fins,** and the back pair, the **pelvic fins**. The **tail fin** has a *dorsal lobe* larger than the *ventral* one. In the male, the pelvic fins bear **claspers,** thin processes for transferring sperm to the oviducts of the female. Examine the pelvic fins of a female and a male.
   - **(c)** Identify the opening of the **cloaca** just in front of the pelvic fins. The cloaca is the terminal organ that receives the products of the digestive, excretory, and reproductive systems.
   - **(d)** There are five to seven pairs of gill slits in cartilaginous fishes, six in the dog fish. Observe the most anterior gill slit, which is called the *spiracle,* located just behind the eye. Trace the flow of seawater in nature through the mouth, over the pharyngeal gills, and out of the gill slits.

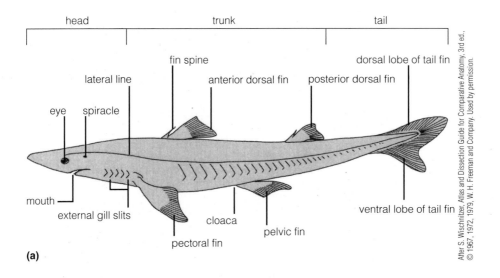

(a)

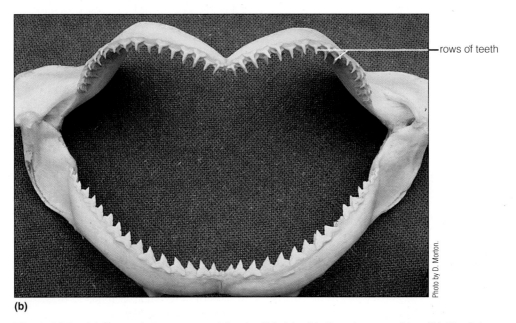

(b)

Figure 31-3 (a) External appearance of the dogfish (shark), *Squalus acanthias*. (b) Shark jaw.

(e) Locate the **nostrils,** which open into blind olfactory sacs; in cartilaginous fishes, they don't connect with the pharynx, as your nostrils do. In cartilaginous fishes, they function solely in olfaction (smell). The **eyes** are effective visual organs at short range and in dim light. There are no eyelids.

(f) Find the dashed line that runs along each side of the body. Each is called a **lateral line** and together function to detect vibrations in the water. Lateral lines consist of a series of minute canals perpendicular to the surface that contain sensory hair cells. When the hairs are disturbed, nerve impulses are initiated; their frequency enables the fish to locate the disturbance.

(g) **Placoid scales,** toothlike outgrowths of the skin, cover the body. Run your hand from head to tail along the length of the animal. How does it feel?

_____

Now run your hand in the opposite direction along the animal. How does it feel this time?

_____

Cut out a small piece of skin with your scissors, pick it up with your forceps, and examine its surface with the dissection microscope. Draw what you see in Figure 31-4.

Figure 31-4 Drawing of shark skin (_____×).

3. Examine the demonstration of a shark jaw with its multiple rows of teeth (Figure 31-3b).

31.3   Bony Fishes (Class Osteichthyes) *(12 min.)*

The bony fishes, the fishes with which you are most familiar, inhabit virtually all the waters of the world and are the largest vertebrate group. They are economically important, commercially and as game species. The ancestors of cartilaginous fish gave rise to the first bony fishes about 415 million years ago. As their name suggests, the *skeleton* is at least *partly ossified (bony),* and the flat *scales* that cover at least some of the surface of most bony fishes are *bony* as well. Gill slits are housed in a common chamber covered by a bony movable flap, the **operculum**.

MATERIALS

Per student:

- scalpel
- compound microscope, lens paper, a bottle of lens-cleaning solution (optional), and a lint-free cloth (optional)
- clean microscope slide and coverslip
- dH₂O in dropping bottle
- forceps
- blunt probe or dissecting needle

Per student pair:

- dissection pan

Per student group:

- preserved specimens of yellow perch

Per lab room:

- collection of preserved bony fishes
- boxes of different sizes of lab gloves
- box of safety goggles

PROCEDURE

1. Examine the assortment of bony fishes on display.
2. Examine a yellow perch as a representative advanced bony fish (Figure 31-5).
   (a) Identify the *spiny* **dorsal fin** located in front of the *soft* **dorsal fin.** Paired **pectoral fins** and **pelvic fins** are also present, and behind the **anus** is an unpaired **anal fin.** The **tail fin** consists of *dorsal* and *ventral lobes* of approximately equal size. The fins, as in the cartilaginous fishes, are used to brake, steer, and maintain an upright position in the water.

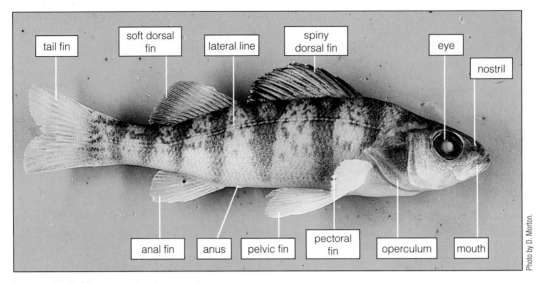

Figure 31-5 Preserved yellow perch.

(b) Note the double **nostrils**, leading into and out of *olfactory sacs*. The eyes resemble those in sharks, with no eyelids. With forceps, pry open the operculum to see the organs of respiration, the **gills.**

(c) Find the **lateral line.** This functions similarly to that of the shark.

(d) Bony scales cover the body. Using forceps, remove a scale, make a wet mount, and examine it with the compound microscope. Locate the *annual rings*, which indicate the approximate age of the fish. How are the annual rings of the fish scale analogous to the annual ridges of the clam shell?

_____

_____

_____

Draw the scale in Figure 31-6.

Figure 31-6 Drawing of yellow perch scale (_____×).

31.4 Amphibians (Class Amphibia) *(30 min.)*

Amphibians include the first vertebrates to partially live on land, having evolved about 370 million years ago from a group of lobe-finned fishes. The paired appendages are modified as legs, which support the individual during movement on land. Respiration by one species or another or by the different developmental stages of one species is by lungs, gills, and the highly vascularized skin and lining of the mouth. There is a *three-chambered heart,* compared to the two-chambered heart of fishes, with two *circuits for the blood circulation,* compared to one circuit in fishes. Reproduction requires water, or at least moist conditions on land. Fertilized eggs hatch in water, and larvae generally live in water. The skeleton is bonier than that of the bony fishes, but a considerable proportion of it remains cartilaginous. The *skin* is usually *smooth and moist*, with mucous glands; scales are usually absent. This group of vertebrates includes the frogs, toads, salamanders, and tropical, limbless, burrowing, mostly blind caecilians.

MATERIALS

Per student:

- forceps
- dissecting scissors
- blunt probe or dissecting needle
- dissection pins

Per student pair:

- preserved farm-raised leopard frog

Per lab room:

- collection of preserved bony fishes
- collection of preserved amphibians
- skeleton of frog
- skeleton of human
- boxes of different sizes of lab gloves
- box of safety glasses

PROCEDURE

1. Examine preserved specimens of a variety of amphibians.
2. The leopard frog illustrates well the general features of the vertebrates and the specific characteristics of the amphibians. Obtain a preserved specimen and after rinsing it in tap water, examine its external anatomy (Figure 31-7).

   (a) Find the two *nostrils* at the tip of the head. These are used for inspiration and expiration of air. Just behind the eye locate a disk-like structure, the *tympanum,* the outer wall of the middle ear. There is no external ear. The tympanum is larger in the male than in the female. Examine the frogs of other students in your lab.

   Is your frog a male or female?

   _____

   (b) At the back end of the body, locate the *cloacal opening.*

   (c) The forelimbs are divided into three main parts: the **upper arm, forearm,** and **hand.** The hand is divided into a **wrist, palm,** and **fingers** (digits). The three divisions of the hindlimbs are the **thigh, shank** (lower leg), and **foot.** The foot is further divided into three parts: the **ankle, sole,** and **toes** (digits).

3. Examine the internal anatomy of the frog.

   (a) With the scissors, cut through the joint at both angles of the jaw. Identify the structures labeled in Figure 31-8. The region containing the opening into the **esophagus** and the **glottis** is the **pharynx** (throat). The esophagus leads to the stomach and the rest of the digestive tract; the glottis leads into the blind respiratory tract.

   (b) Fasten the frog, ventral side up, with pins to the wax or rubbery surface of a dissection pan. Lift the skin with forceps.

Figure 31-7 External anatomy of the frog.

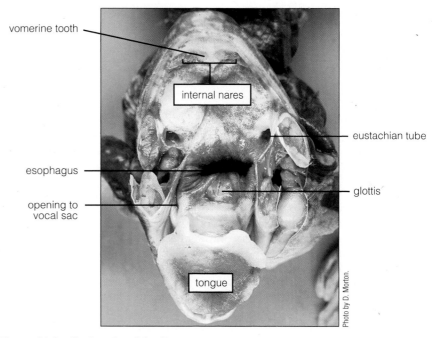

Figure 31-8 Oral cavity of the frog.

Then make a superficial cut with your scissors just left or right of center from the lower abdomen forward to the tip of the lower jaw. Pin back the skin on both sides to expose the large abdominal muscles. Lift these muscles with your forceps, and cut through the body wall with your scissors again from the lower abdomen forward to the tip of the lower jaw, cutting through the sternum (breastbone) but not damaging the internal organs. Pin back the body wall as you did the skin to expose the internal organs (Figure 31-9).

(c) Use Figure 31-9 to locate the *spleen* (Figure 31-9c) and the following organs of the digestive system: **stomach, small intestine, large intestine** (colon), **liver, gall bladder,** and **pancreas.** Trace the esophagus from its opening in the pharynx to the stomach. Find the **bronchi** (the singular is *bronchus*), which lead toward the **lungs.** Are the bronchi dorsal or ventral to the esophagus?

_____

(d) Identify the **heart** with two thin-walled **atria** and the thick-walled **ventricle** and the **conus arteriosus,** along with the right and left branches of the **truncus arteriosus** (Figure 31-9a). The latter structure carries blood to the paired **pulmocutaneous arches** and **systemic arches.** In frogs and other amphibians a *pulmocutaneous circuit* circulates blood to and from the lungs and the skin, and a *systemic circuit* services the rest of the body. Even though there is only one ventricle, because of the structure of the heart and its large arteries, most of the deoxygenated blood returning from the body is pumped into the pulmocutaneous circuit before the oxygenated blood is pumped into the systemic circuit. In comparison, a fish has a two-chambered heart with only one atrium and a one-circuit circulation. A two-circuit system is more efficient because blood pressure drops less after the blood is oxygenated. Partition of the ventricle of the heart and the further separation of the two circuits will continue in subsequent vertebrate classes.

(e) Refer to Figures 31-9 and 31-10 and find the major *arteries* listed in Table 31-1.

TABLE 31-1	Major Arteries of the Frog	
Arteries	**Location**	**Function Is to Convey Blood to**
Pulmocutaneous arches	First pair of vessels branching off truncus arteriosus	Lungs and skin
Systemic arches	Second pair of vessels branching off truncus arteriosus	The rest of the body
Common carotids (carotid arches)	Branch from the systemic arches	Head
Subclavians	Branch from the systemic arches	Arms
Dorsal aorta	Formed by the fusion of the systemic arches along dorsal body wall in mid abdomen	Lower half of the body
Common iliacs	Division of dorsal aorta in lower abdomen	Legs

© Cengage Learning 2013

Which of these arteries carries deoxygenated blood? _____

(f) Locate the **anterior** and **posterior vena cavae,** which return deoxygenated blood from the systemic circuit to the right atrium of the heart (Figures 31-9c and 31-10). Other veins return blood from the head, arms, and legs to one of the vena cavae. Right and left **pulmonary veins** return oxygenated blood from the lungs to the left atrium.

(g) The excretory and reproductive organs together comprise the *urogenital system* (Figure 31-11). Locate the urine-producing excretory organs, the pair of **kidneys** located dorsally in the body cavity. The **ureter,** a duct that leads from the kidney to the **urinary bladder** (Figure 31-9a), an organ that stores urine for the recovery of water back into the circulatory system. The bladder empties into the *cloaca.*

(h) Refer to Figure 31-11a and, in the male, locate the **testes** (Figure 31-9c); these organs produce sperm that are carried to the kidneys through tiny tubules, the *vasa efferentia.* The ureters serve a dual function in male frogs, transporting both urine and sperm to the cloaca. Find the vestigial female oviducts (Figure 31-9b), which are located lateral to the urogenital system.

(i) Refer to Figure 31-11b and, in the female, find the *ovaries* (Figure 31-9d); these organs expel eggs into the *oviducts.* The oviducts lead into the *uterus.* As in the male, the reproductive tract ends in the cloaca.

(j) Locate the *fat bodies* (Figures 31-9c and d), many yellowish, branched structures just above the kidneys. They store food reserves for hibernation and reproduction.

(k) The nervous system of vertebrates is composed of (1) the *central nervous system,* the brain and spinal cord, and (2) the *peripheral nervous system,* nerves extending from the central nervous system. Turn your frog over and remove the skin from the dorsal surface of the head between the eyes and along the vertebral column. With your scalpel, shave thin sections of bone from the skull, noting the shape and size of the *cranium,* until you expose the *brain.* Pick away small pieces of bone with your forceps to expose the entire brain. Use the same procedure to expose the vertebrae and *spinal cord.* Note the *cranial nerves* coming from the brain and the *spinal nerves* coming from the spinal cord. Spinal nerves are also easy to see running along the dorsal wall of the abdomen.

(l) Dispose of your dissected specimen in the large plastic bag provided for this purpose.

4. Compare the frog and human skeleton.

(a) The skeleton of vertebrates consists of the following: (1) the **axial skeleton,** consisting of the bones of the **skull, sternum, hyoid bone, vertebral column,** and rib cage (when present); and (2) the **appendicular skeleton,** with the bones of its **girdles** and their appendages. The **pectoral girdle** consists of the paired **clavicle** and **scapula** bones (**suprascapular** in the bullfrog). The latter articulates forms a joint with the **humerus.** The **pubis, ischium,** and **ilium** bones form the pelvic girdle, which articulates with the femur. Compare the skeleton of the bullfrog and human (Figure 31-12), and identify the axial and appendicular portions of the skeleton.

(b) There are differences in the two skeletons, but their basic plan is remarkably similar. To a large extent, the size and shape of the different parts of the skeleton of a vertebrate correlate with body specializations and behavior. List some reasons for the differences in the girdles, appendages, and cranium of the frog and human skeletons.

_____

_____

_____

31.5 Reptiles (Class Reptilia) *(12 min.)*

Reptiles are the oldest group of vertebrates *adapted to life primarily on land,* although many do live in fresh water or seawater. They arose from the ancestors of amphibians about 300 million years ago. Their *skeleton* is bonier than that of amphibians. The *dry skin* is covered by *epidermal scales* and there are almost no skin glands. Reptiles lay **amniotic eggs** (Figure 31-13). Inside these eggs, embryos are suspended by *fetal membranes* in an internal aquatic environment surrounded by a shell to prevent their drying out.

There is no larval stage. The amniotic egg (characteristic of birds and some mammals as well as reptiles) and lack of larvae are adaptations to living an entire life cycle on land. The *three* or *four-chambered hearts* of reptiles consist of two atria and a partially divided ventricle or in crocodiles, two ventricles. The nervous system, especially the brain, is more highly developed than that of amphibians.

Reptiles, like the fishes and amphibians, are largely **ectothermic,** without the capability to maintain a relatively high core body temperature physiologically and with a greater dependence on gaining or losing heat from or to their external environment. Reptiles and other ectotherms have an adaptive advantage over endothermic mammals in warm and humid areas of Earth because they expend minimal energy on maintaining body temperature, which leaves more energy for reproduction. This advantage is reflected in the greater numbers of individuals and species of reptiles compared to mammals in the tropics. Mammals have the advantage in moderate to cold areas.

MATERIALS

Per student pair:

- dissection pan

Per lab room:

- collection of preserved reptiles
- skeleton of snake

- turtle skeleton or shell
- boxes of different sizes of lab gloves
- box of safety goggles

PROCEDURE

1. Examine an assortment of preserved reptiles and note their diversity as a group. This diverse group of vertebrates includes the turtles, lizards, snakes, crocodiles, and alligators.

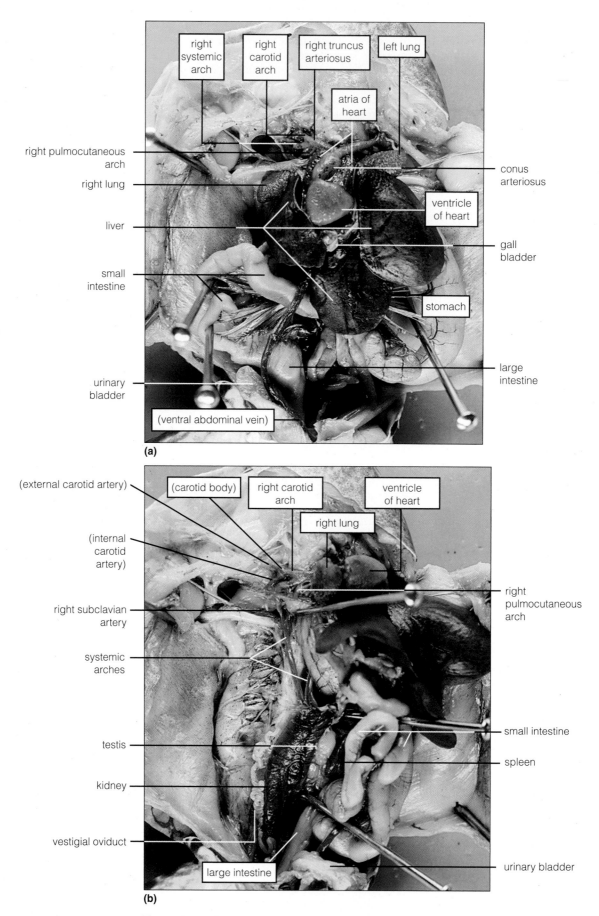

Figure 31-9 Abdominal views of dissected frogs. (**a–c**) Male; (**d**) female. *Continues.*

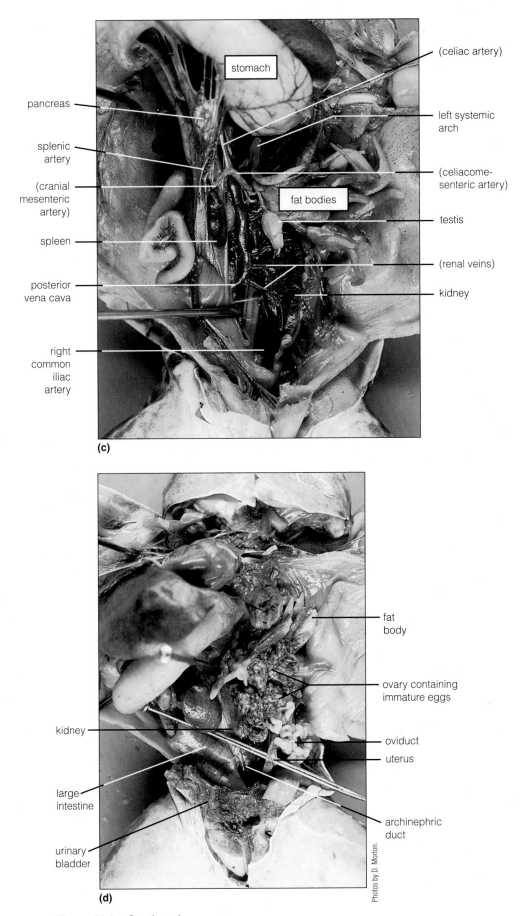

stomach

(celiac artery)

pancreas

left systemic arch

splenic artery

(celiacomesenteric artery)

(cranial mesenteric artery)

fat bodies

testis

spleen

(renal veins)

posterior vena cava

kidney

right common iliac artery

(c)

fat body

ovary containing immature eggs

kidney

oviduct

uterus

large intestine

archinephric duct

urinary bladder

Photos by D. Morton.

(d)

Figure 31-9 *Continued.*

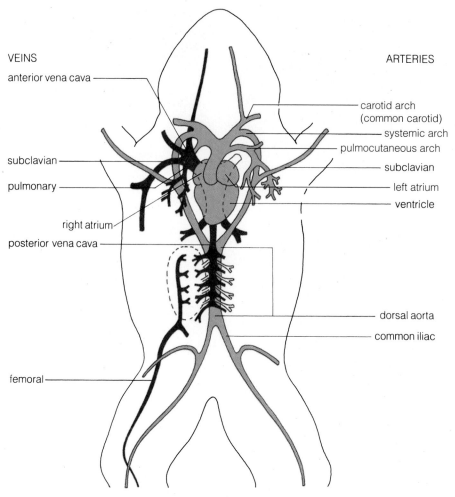

VEINS

anterior vena cava

subclavian

pulmonary

right atrium

posterior vena cava

femoral

ARTERIES

carotid arch
(common carotid)
systemic arch
pulmocutaneous arch
subclavian
left atrium
ventricle

dorsal aorta
common iliac

Figure 31-10 Ventral view of the major blood vessels of the frog.

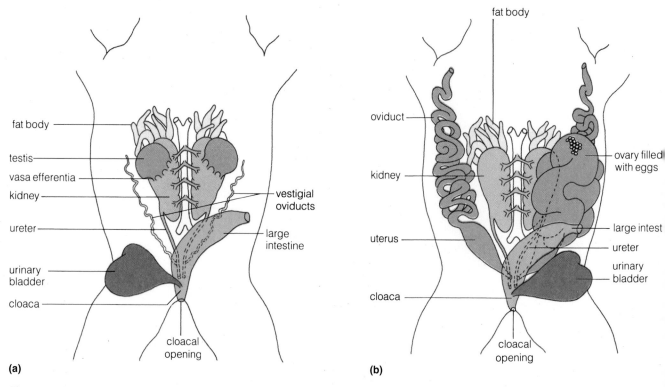

fat body

oviduct

kidney

uterus

cloaca

ovary filled
with eggs

large intest

ureter

urinary
bladder

cloacal
opening

fat body

testis

vasa efferentia

kidney

ureter

urinary
bladder

cloaca

vestigial
oviducts

large
intestine

cloacal
opening

(a)

(b)

Figure 31-11 Ventral views of the urogenital systems of (**a**) male and (**b**) female frogs.

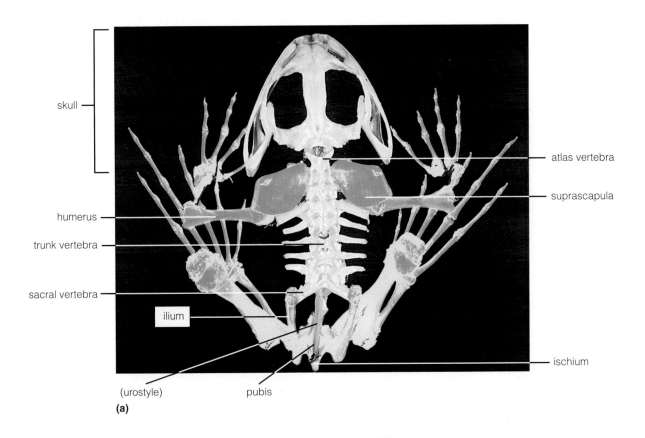

(a)

skull

atlas vertebra

suprascapula

humerus

trunk vertebra

sacral vertebra

ilium

ischium

(urostyle)

pubis

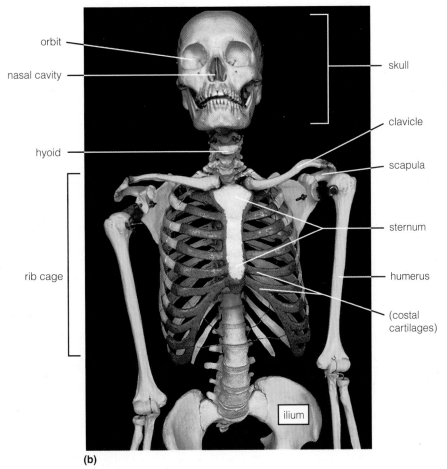

(b)

orbit

nasal cavity

hyoid

rib cage

skull

clavicle

scapula

sternum

humerus

(costal cartilages)

ilium

Figure 31-12 Skeletons of (**a**) a bullfrog and (**b, c**) a human. The appendicular skeleton is colored yellow and the axial skeleton is uncolored. *Continues.*

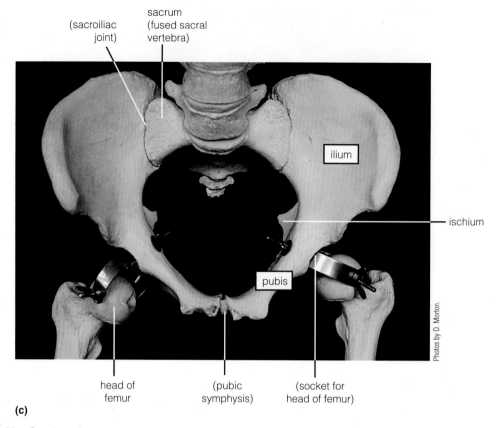

Figure 31-12 *Continued.*

2. Compare the skeleton of a snake (Figure 31-14) with that of the frog. Describe any differences you see.

_____
_____
_____
_____
_____

Identify the axial and appendicular portions of the snake skeleton.

3. Look at the inside of the *carapace* (the upper portion) of a turtle shell. What portions of the skeleton are incorporated into the shell?

_____
_____
_____

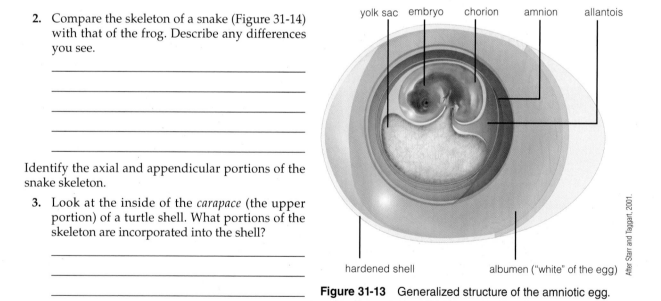

Figure 31-13 Generalized structure of the amniotic egg.

31.6   Birds (Class Aves) *(12 min.)*

Birds are likely descended from small bipedal (walked on two legs) reptiles about 160 million years ago and are thought to be closely related to dinosaurs. Their bodies are covered with *feathers*; and scales, reminiscent of their reptilian heritage, are present on the feet. The front limbs in most birds are modified as *wings* for flight. An additional internal adaptation for flight is the well-developed *sternum* (breastbone), with a *keel* for the attachment of powerful muscles for flight (Figure 31-15). Birds are **endothermic** vertebrates, capable of maintaining high core body temperatures physiologically.

numerous
free ribs
(no sternum)

Photo by D. Morton.

unfused lower jaw and
other adaptations allow for
a flexible jaw mechanism

Figure 31-14 Snake skeleton.

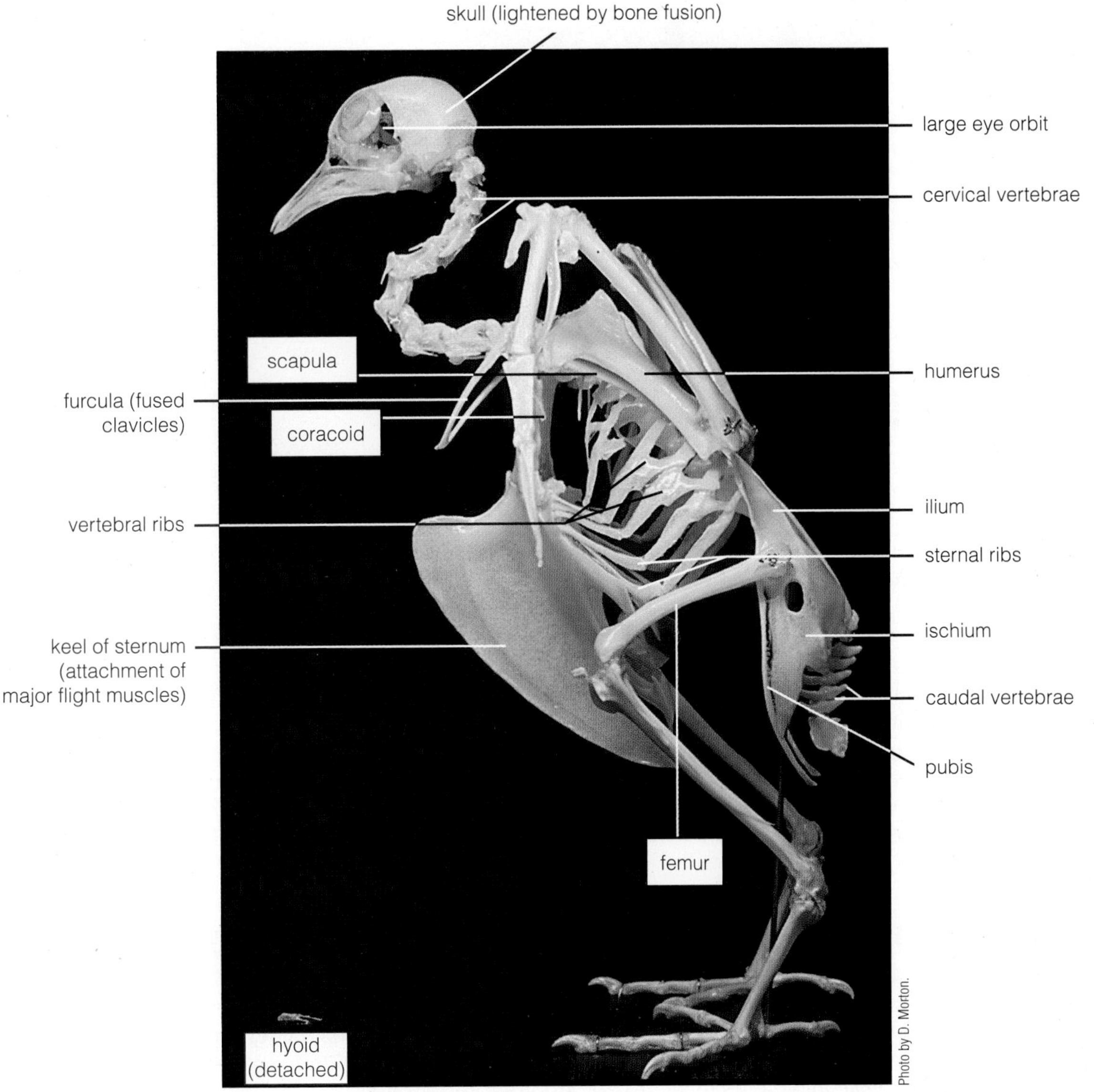

skull (lightened by bone fusion)

large eye orbit

cervical vertebrae

scapula

humerus

furcula (fused
clavicles)

coracoid

ilium

vertebral ribs

sternal ribs

ischium

keel of sternum
(attachment of
major flight muscles)

caudal vertebrae

pubis

femur

hyoid
(detached)

Photo by D. Morton.

Figure 31-15 Bird skeleton.

The major *bones* of birds are *hollow* and contain *air sacs connected to the lungs.* Birds have a *four-chambered heart,* with two atria and two ventricles (Figure 31-16). This permits the complete separation of oxygenated and deoxygenated blood. Why is this circulatory arrangement an advantage for birds, as opposed to that found in amphibians and reptiles? (*Hint:* Recall that birds are endothermic and most can fly.)

_____

_____

_____

The nervous system resembles that of reptiles. However, the *larger brain* permits more muscular coordination, resulting in advanced behavior; the optic lobes are especially well developed in association with a keen sense of sight.

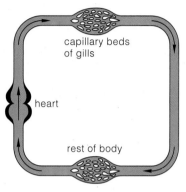 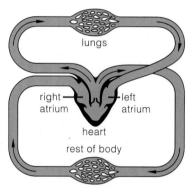 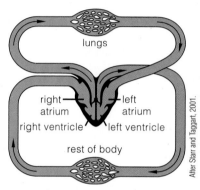

(a) In fishes, a two-chambered heart (atrium, ventricle) pumps blood in one circuit. Blood picks up oxygen in gills and delivers it to rest of body. Oxygen-poor blood flows back to heart.

(b) In amphibians, a heart pumps blood through two partially separate circuits. Blood flows to lungs, picks up oxygen, and returns to heart. It mixes with oxygen-poor blood still in heart, flows to rest of body, and returns to heart.

(c) In birds and mammals, the heart is fu partitioned into two halves. Blood circula in two circuits: from the heart's right half lungs and back, then from the heart's lef half to oxygen-requiring tissues and bacl

Figure 31-16 Circulatory systems of (**a**) fishes, (**b**) amphibians, and (**c**) birds and mammals.

MATERIALS

Per student pair:
- dissection microscope

Per lab room:
- an assortment of feathers, stuffed birds, or bird illustrations
- several field guides to birds

PROCEDURE

1. Examine a feather and identify the **rachis** (shaft) and **vane** (Figure 31-17). Examine the vane with the dissection microscope and note the **barbs** and **barbules.** What else do birds use their feathers for besides flight?

_____

_____

2. Identify the axial and appendicular portions of the skeleton (Figure 31-15). Note the various adaptations for flight, including the elongated wrists and ankles.

3. The upper and lower jaws are modified as variously shaped *beaks* or *bills,* with the shapes reflecting the feeding habits of the species. No teeth are present in adults. Examine the assortment of birds or bird illustrations and speculate on their food preferences by studying the configurations of their bills. Check your conclusions with a field guide or other suitable source that describes the food habits.

31.7   Mammals (Class Mammalia) *(12 min.)*

Mammals arose from an ancient branch of reptiles over 200 million years ago. The term *mammal* stems from the *mammary glands,* which all adult female mammals possess. They produce milk that is fed to mammalian young (Figure 31-18). The *mammalian brain* is the *largest* of all other vertebrates, and its surface area is increased by grooves and folds (Figure 31-19). Behavior is more complex and flexible. *Hair* is present during some portion of the life span. Mammals are equipped with a variety of modifications and outgrowths of the skin in addition to hair, all made of the protein keratin. These include horns, spines, nails, claws, and hoofs. Mammalian dentition is intricate, with teeth specialized for cutting, grooming, wounding, tearing, slicing, and grinding. The teeth in the upper and lower jaws match up.

Like birds, mammals are *endothermic* and the *heart* is *four-chambered*. The ear frequently has a cartilaginous outer portion called the *pinna,* which functions to collect sound waves. In addition to mammary glands, there are three other types of skin glands: *sebaceous, sweat,* and *scent glands.*

MATERIALS

Per lab room:

- stuffed mammals or mammal illustrations
- mammalian placentas and embryos (in utero)

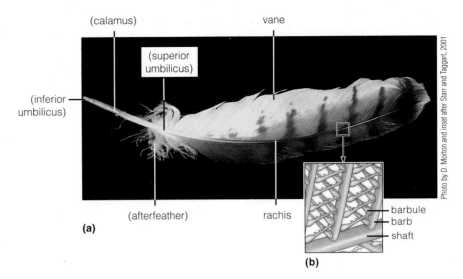

Figure 31-17 (a) Feather and (b) expanded view of the vane.

Figure 31-18 Female dog with nursing puppies.

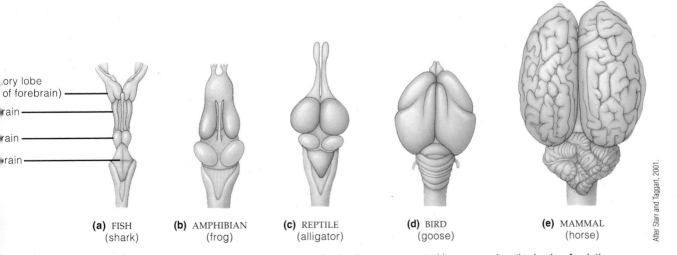

re 31-19 Evolutionary trend toward an expanded, more complex brain, as suggested by comparing the brain of existing brates: (a) shark; (b) frog; (c) alligator; (d) goose; and (e) horse. These dorsal views are not to the same scale.

PROCEDURE

1. Study the assortment of mammals on display and list as many functions for hair as you can.

_____

_____

2. Except for two species of *monotremes* that lay eggs (platypus and spiny anteater), most mammals are **viviparous**—bear live young. Before birth, an embryo (later a fetus) develops in the wall of the uterus along with a *placenta,* a connection between the mother and the embryo. The placenta exchanges gases with, delivers nutrients to, and removes wastes from the embryo, among other functions.

Some viviparous mammals (*marsupials*—kangaroos, opossums, and so on) give birth early, and along with spiny anteater hatchlings, continue development suckling milk in a skin pouch. *Eutherian* mammals—those you are most likely to be familiar with, give birth later, but the young of all mammals receive extended parental care. Examine the placentas and embryos in uteri of mammals on display in the lab. Can you suggest a relationship between the early development of mammals and the comparative sophistication of the nervous system and behavior of adult mammals?

_____

3. Look at Figure 31-20 and identify the teeth present in the skull of the human skeleton. Fill in the numbers of each type in the upper and lower jaws and their functions in Table 31-2.

TABLE 31-2	Human Dentition		
Type of Tooth	**Number in Upper Jaw**	**Number in Lower Jaw**	**Function**
Incisors			
Canines			
Premolars			
Molars			

© Cengage Learning 2013

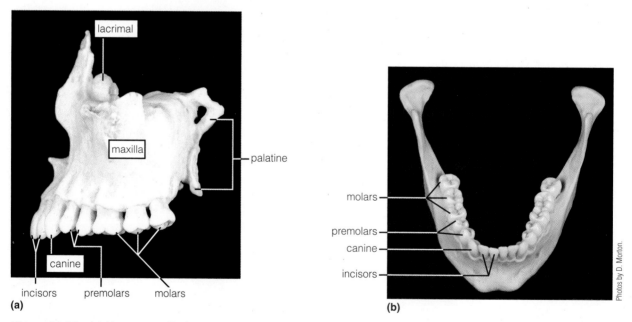

Figure 31-20 (**a**) Human maxilla in side view. (**b**) Top view of the mandible.

31.8 Construction of a Dichotomous Key to the Animals (optional) *(50 min.)*

Your instructor will provide you with instructions and perhaps other pertinent materials to use with your textbook and lab manual.

_____ 1. Animals that have the four basic characteristics of chordates plus a vertebral column are
(a) invertebrates.
(b) hemichordates.
(c) cephalochordates.
(d) vertebrates.

_____ 2. The ammocoete is the larva of
(a) lampreys.
(b) sharks.
(c) bony fishes.
(d) amphibians.

_____ 3. Placoid scales are characteristic of class
(a) Cephalaspidomorphi.
(b) Chondrichthyes.
(c) Osteichthyes.
(d) Reptilia.

_____ 4. The bony movable flap that covers the gills and gill slits is called the
(a) operculum.
(b) cranium.
(c) tympanum.
(d) colon.

_____ 5. Which group of animals is endothermic?
(a) fishes
(b) mammals
(c) reptiles
(d) amphibians

_____ 6. The bones that surround and protect the brain are collectively called the
(a) spinal cord.
(b) vertebral column.
(c) pelvic girdle.
(d) cranium.

_____ 7. The lampreys are unusual vertebrates in that they have no
(a) eyes.
(b) jaws.
(c) gill slits.
(d) mouth.

_____ 8. The structure of the cartilaginous and bony fishes that detects vibrations in the water is the
(a) anal fin.
(b) operculum.
(c) nostrils.
(d) lateral line.

_____ 9. Amphibians have
(a) a two-chambered heart.
(b) a three-chambered heart.
(c) a four-chambered heart.
(d) none of the above

_____ 10. The amniotic egg of reptiles, birds, and mammals is an adaptation to
(a) thwart carnivorous predators.
(b) allow life on land.
(c) compensate for the short period of development of the young.
(d) protect the young from the nitrogenous wastes of the mother during embryonic development.

EXERCISE 31

Vertebrates

Post-Lab Questions

Introduction

1. What characteristic of vertebrates is missing in all invertebrates?

2. List the four basic characteristics of chordates.

31.1 Lampreys (Class Cephalaspidomorphi)

3. What is an ammocoete? What is its possible significance to the evolution of vertebrates?

4. How do lampreys differ from other vertebrates?

31.2 Cartilaginous Fishes (Class Chondrichthyes)

5. Explain how the adaptations of cartilaginous fishes make them better predators.

31.3 Bony Fishes (Class Osteichthyes)

6. Identify this vertebrate structure. What do the rings present?

(24×)

Photo courtesy of Biodisc/Visuals Unlimited.

31.5 Reptiles (Class Reptilia)

7. Describe an amniotic egg. What is its evolutionary significance?

31.7 Mammals (Class Mammalia)

8. What are the unique characteristics of mammals?

9. Describe the evolution of the vertebrate heart.

Food for Thought

10. About 65 million years ago, an asteroid impact likely caused a mass extinction, which led to the demise of dinosaurs and created new opportunities for surviving plants and animals. Create a short evolutionary scenario for the next human-made or natural mass extinction event.

Plant Organization: Vegetative Organs of Flowering Plants

OBJECTIVES

After completing this exercise, you will be able to

1. define *vegetative, morphology, dicotyledon (dicot), monocotyledon (monocot), cotyledon, taproot system, node, internode, adventitious root, herb (herbaceous plant), woody (woody plant), heartwood, sapwood, growth increment (annual ring)*;

2. identify and give the function of the external structures of flowering plants (those in **boldface**);

3. identify and give the functions of the tissues and cell types of roots, stems, and leaves (those in **boldface**);

4. determine the age of woody branches.

Introduction

It's difficult to overstate the importance of plant life. Are you sitting on a wooden chair? If so, you're perched on part of a tree. No, you say? Perhaps your chair is covered with fabric. If it's natural fabric other than wool, it's a plant product. If the chair is covered with plastic, that cover was made from petroleum products derived from plant material that lived millions of years ago. The energy used to create the chair was probably derived from burning petroleum products or coal, also derived from plant material.

Obviously, plants are an important part of our lives. This exercise introduces you to the external and internal structure of the **vegetative organs**—those not associated with sexual reproduction—of flowering plants. The organs are the roots, stems, and leaves. Flowers are the sexual reproductive organs, and were considered in detail in Exercise 24.

Each organ is usually distinguished by its shape and form, its **morphology**. But the cells and tissues comprising these three organs are remarkably similar. Each organ is covered by the protective **dermal tissue**; each possesses **vascular tissue** that transports water, minerals, and the products of photosynthesis; and each contains **ground tissue**, that which is covered by the dermal tissue and in which the vascular tissue is embedded. Examine Figure 32-1, which illustrates these relationships.

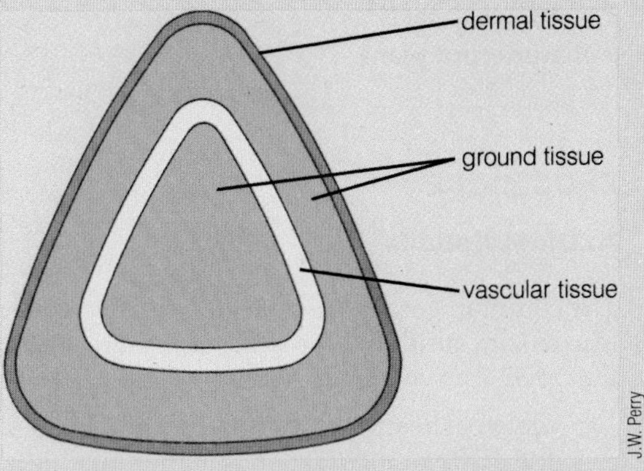

The organs of the plant body are more similar than they are dissimilar. In fact, for this reason the differences between a root and a stem or leaf are said to be *quantitative* rather than *qualitative*. That is, these differences are in the number and arrangement of cells and tissues, not the type. Consequently, the plant body is a continuous unit from one organ to the next.

In this exercise, we'll study the organs of two rather large assemblages of flowering plants, **dicotyledons** (dicots)

Figure 32-1 Model plant organ in cross section.

and **monocotyledons** (monocots). **Cotyledons** are the leaves that are formed within the seed. Dicots have two seed leaves, monocots one. In addition to this fundamental difference, dicots and monocots have other dissimilarities, as shown in Figure 32-2. Refer to Figure 32-2 as you proceed through each section.

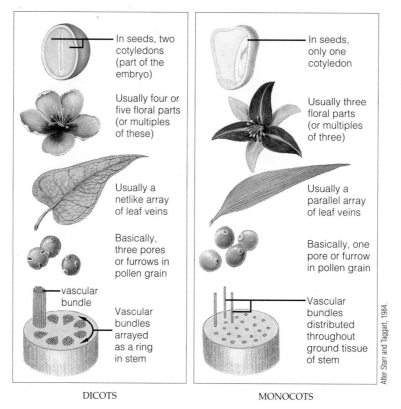

In seeds, two cotyledons (part of the embryo)

Usually four or five floral parts (or multiples of these)

Usually a netlike array of leaf veins

Basically, three pores or furrows in pollen grain

vascular bundle

Vascular bundles arrayed as a ring in stem

In seeds, only one cotyledon

Usually three floral parts (or multiples of three)

Usually a parallel array of leaf veins

Basically, one pore or furrow in pollen grain

Vascular bundles distributed throughout ground tissue of stem

After Starr and Taggart, 1984.

DICOTS

MONOCOTS

Figure 32-2   Comparison of dicot and monocot structure.

32.1   External Structure of the Flowering Plant *(About 30 min.)*

MATERIALS

Per student group (table):

■ mature corn plant

Per lab room:

■ living bean and corn plants in flats
■ potted geranium and dumbcane plants
■ dishpan half-filled with water

PROCEDURE

A. Dicotyledons

The common garden bean, *Phaseolus vulgaris,* is a representative dicot. Other familiar examples of dicotyledons (dicots) are sunflowers, roses, cucumbers, peas, maples, and oaks. Let's look at its external morphology. Label Figure 32-3 as you study the bean plant.

1. Obtain a bean plant by gently removing it from the medium in which it is growing. Wash the root system in the dishpan provided, not in the sink.
2. The plant consists of a **root system**, that portion usually below ground, and a typically aerial **shoot system**. Examine the root system first. The root system of the bean is an example of a **taproot system**, that is, one consisting of one large **primary root** (the **taproot**) from which **lateral roots** arise. Identify the taproot and lateral roots.
3. Now turn your attention to the shoot system, consisting of the stem and the leaves.
4. Identify the points of attachment of the leaves to the stem, called **nodes;** the regions between nodes are **internodes**.
5. Look in the upper angle created by the junction of the stem and leaf stalk to find the **axillary bud**. These buds give rise to branches and/or flowers.
6. Find the **terminal bud** at the very tip of the shoot system. The terminal bud contains an apical meristem that accounts for increases in length of the shoot system.

If lateral branches are produced from axillary buds, each lateral branch is terminated by a terminal bud and possesses nodes, internodes, and leaves, complete with axillary buds. As you see, the shoot system can be a highly branched structure.

7. Look several centimeters above the soil line for the lowermost node on the stem. If the plant is relatively young, you should find the **cotyledons** attached to this node. The cotyledons shrivel as food stored in them is used for the early growth of the seedling. Eventually the cotyledons fall off.

The cotyledons are sometimes called *seed leaves* because they are fully formed (although unexpanded) in the seed. By contrast, most of the leaves you're observing on the bean plant were immature or not present at all in the seed.

Now let's examine the other component of the shoot system, the foliage leaves (also called *true leaves* in contrast to the cotyledons). The first-formed foliage leaves (those nearest the cotyledons) are **simple leaves**, each leaf having one undivided blade.

8. Identify the **petiole** (leaf stalk) and **blade** on a simple leaf, and label it on Figure 32-3.

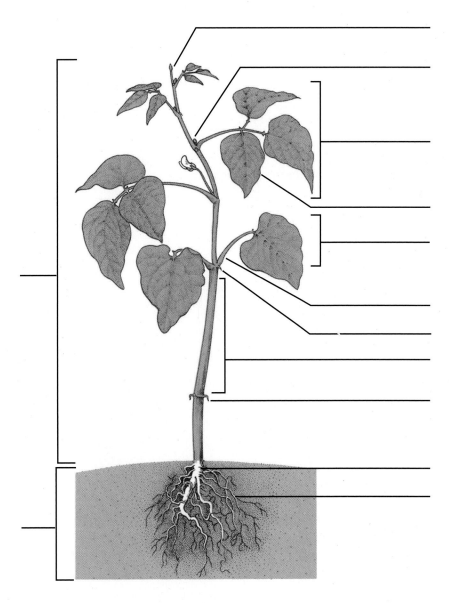

Figure 32-3 External structure of the bean plant.
Labels: root system, shoot system, primary root (taproot), lateral root, node, internode, axillary bud, terminal bud, cotyledon remnant, petiole, blade, simple leaf, compound leaf, leaflet

In bean plants, subsequently formed leaves are **compound leaves**, consisting of three **leaflets** per petiole. Each leaflet has its own short stalk and blade.

9. Identify a compound leaf, and label the petiole and leaflets in Figure 32-3.
10. Note the netted arrangement of veins in the blades. Veins contain vascular tissues—the xylem and phloem.
11. Identify the **midvein**; it is the largest vein and runs down the center of the blade, giving rise to numerous lateral veins.

Now that you have a general idea of the external structure of a typical dicot, let's look at another one, so that you can use your knowledge of a particular plant to make some generalizations.

12. Obtain a potted geranium plant and examine the external structure of the shoot. Identify the following parts, checking off each as you go along.

_____ node   _____ terminal bud

_____ internode   _____ leaf petiole

_____ axillary bud   _____ leaf blade

What color are the stems of the plants you've examined? _____

What structure in the cytoplasm of the cells making up the stem is responsible for this color?

_____

What is the function of this structure? _____

What, then, is one function of the stem? _____

13. Compare the leaves of the bean with those of the geranium.

Are the geranium leaves simple or compound? (You may wish to refer to some of the figures in Exercise 18, Taxonomy: Classifying and Naming Organisms, if you have difficulty deciding.) _____

Is there a single midvein in the geranium, or are there many large veins? _____

List all the features shared by the leaves of beans and geraniums.

_____

List any differences you observe in the leaves.

_____

B. Monocotyledons

Corn (*Zea mays*) is a **monocotyledon**. These plants have only one cotyledon (seed leaf). You're probably familiar with a number of monocots: lilies, onions, orchids, coconuts, bananas, and the grasses. (Did you realize that corn is actually a grass?)

Label Figure 32-4 as you study the corn plant.

1. Remove a single young plant from its growing medium and wash its root system in the dishpan. Note that the seed (technically this is a fruit called a "grain") is still attached to the plant.
2. Identify the **root system**. Note that there is no one particularly prominent root. In most monocots, the primary root is short lived and is replaced by numerous **adventitious roots**. Adventitious roots are roots that arise from places other than existing roots.

soil line

soil line

fibrous
adventitious
root system

Photos by J. W. Perry

(a)   (b)

Figure 32-4 External structure of a corn plant. (**a**) Seedling (0.5×). (**b**) Lower portion of mature plant (0.1×).
Labels: root system, prop root, leaf sheath

3. Trace the adventitious roots back to the corn grain.

Where do they originate?

_____

As the roots branch, they develop into a **fibrous root system**, one particularly well suited to prevent soil erosion.

4. Examine the mature corn plant. Identify the large **prop roots** at the base of the plant.

Where do prop roots arise from?

_____

Would you classify these as adventitious roots? (yes or no) _____

5. The shoot system of a young corn plant appears somewhat less complex than that of the bean. There seem to be no nodes or internodes. Look at the mature corn plant again. You should see that nodes and internodes do indeed exist. In your young plant, elongation of the stem has not yet taken place to any appreciable extent.

6. Strip off the leaves of the young corn plant. Keep doing so until you find the shoot apex. (It's deeply embedded.)

7. Examine the leaves of the corn plant in more detail. Note the absence of a petiole and the presence of a **sheath** that extends down the stem. The leaf sheath adds strength to the stem. Look at the veins, which have a parallel arrangement. (Contrast this to the petioled, netted-vein arrangement of the bean leaves.)

You may wonder how representative corn is of all monocotyledons. As it turns out, it's quite representative of most grasses, but not particularly so of monocots as a whole. Let's look at another monocot, a common horticultural plant found in many homes, called dumbcane (_Diffenbachia_).*

8. Obtain a potted specimen of the dumbcane plant. Observe its external morphology, comparing it with the corn plant.

Does the dumbcane have sheathing leaves like corn, or does each leaf have a petiole?

_____

Are the veins in the leaves parallel, or is netted venation present? _____
(_Hint:_ Look on the lower surface of the leaves, where it is most obvious.)

Is there a midvein? (yes or no) _____

Is the terminal bud obvious or is it deeply embedded, as in the corn plant? _____

Are prop roots present on the dumbcane plant? _____

32.2   The Root System _(About 30 min.)_

MATERIALS

Per student:

- single-edged razor blade
- clean microscope slide
- coverslip
- prepared slide of buttercup (_Ranunculus_) root, c.s.
- prepared slide of corn (_Zea_) root, c.s.
- compound microscope

Per student pair:

- dH$_2$O in dropping bottles

Per lab room:

- germinating radish seeds in large petri dishes
- demonstration slide of Casparian strip in endodermal cell walls

PROCEDURE

A. Living Root Tip

1. Obtain a germinating radish seed. Identify the **primary root**. Its fuzzy appearance is due to the numerous tiny **root hairs** (Figure 32-5). Root hairs increase the absorptive surface of the root tremendously.

2. Using a razor blade, cut off the seed and discard it, then make a wet mount of the primary root. (Add enough water so that no air surrounds the root. _Do not_ squash the root.)

---

* The common name _dumbcane_ has its origin in a use by the ancient Greeks. When they tired of long orations by their senators, Greeks sometimes ground up parts of the shoot and added it to a drink. Because certain cells contain needle-shaped crystals, consumption caused a temporary paralysis of the larynx (voice box), ending the oration. Today, dumbcane is a hazard to young children. Ingestion causes throat swelling that can lead to suffocation.

3. Examine your preparation with the low-power objective of your compound microscope. (If you're having difficulty seeing the root clearly, increase its contrast by closing the microscope's diaphragm somewhat.) Locate the conical root tip.

4. At the very end of the root tip, identify the protective **root cap** that covers the tip. As a root grows through the soil, the tip is thrust between soil particles. Were it not for the root cap, the apical meristem containing the dividing cells would be damaged.

5. Find the root hairs.

Do they originate all the way down to the root cap? (yes or no) _____

6. Examine the root hairs carefully.

What happens to their length as you observe them at increasing distance from the root tip?

_____

_____

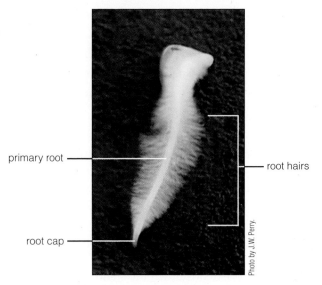

Figure 32-5   Living primary root (4×).

The youngest root hairs are the shortest. What does this imply regarding their point of origin and pattern of maturation?

_____

_____

B. Dicot Root Anatomy

Now let's see what the internal architecture of the root looks like.

1. Obtain a prepared slide containing a mature buttercup (*Ranunculus*) root in cross section. Refer to Figure 32-6 as you study this slide. Examine the slide first with the low-power objective of your compound microscope to gain an overall impression of the organization of the tissues present.

2. Starting at the edge of the root, identify the **epidermis**.

3. Moving inward, locate the **cortex** and the central **vascular column**. These regions represent the dermal, ground, and vascular tissue systems, respectively.

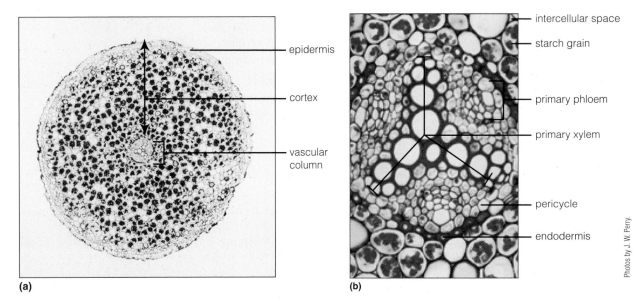

(a)   (b)

Figure 32-6   Cross section of buttercup (*Ranunculus*) root. (**a**) 45×. (**b**) Portion of cortex and vascular column (75×).

4. Switch to the medium-power objective for further study. Look at the outermost layer of cells, the **epidermis**.
5. Beneath the epidermis find the relatively wide **cortex**, consisting of parenchyma cells that contain numerous **starch grains**.

Based on the presence of starch grains, what would you suspect one function of this root might be?

_____

_____

6. Switch to the high-dry objective. Between the cells of the cortex, find numerous **intercellular spaces**.

The innermost layer of the cortex is given a special name. This cylinder, a single cell thick, is called the **endodermis**.

7. Locate the endodermis on your slide and in Figure 32-6b.

Unlike the rest of the cortical cells, endodermal cells *do not* have intercellular spaces between them. The endodermis regulates the movement of water and dissolved substances into the vascular column. Each endodermal cell possesses a **Casparian strip** within its radial and transverse walls. To visualize this arrangement, imagine a rectangular box (the endodermal cell) that has a rubber band (the Casparian strip) around its long dimension. Now imagine that the rubber band is actually part of the wall of the box. Figure 32-7 diagrams this arrangement.

The Casparian strip consists of waxy material. Water and substances dissolved in the water normally move through the cell *walls* as they flow from the edge of the root toward the vascular column. Because the Casparian strip consists of a waxy material, it effectively waterproofs the cell wall of the endodermal cells. Consequently, substances moving from the cortex toward the vascular cylinder (or vice versa) must flow through the cytoplasm of the endodermal cell.

As you may recall, the cytoplasm is bounded by the differentially permeable plasma membrane. Thus, dissolved substances are "filtered" through endodermal cells, which regulate what goes into or comes out of the vascular column.

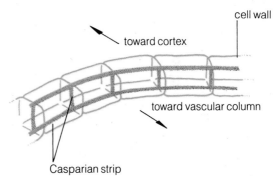

Casparian strip
(imagine that the cell wall is transparent; no cytoplasmic components are shown)

Figure 32-7 Diagrammatic representation of endodermal cells with Casparian strip.

8. In many cases, the Casparian strip is difficult to distinguish. With the highest magnification available, look for a red "dot" on the radial wall of the endodermal cells. If you cannot distinguish it on your own slide, examine the demonstration slide (Figure 32-8), which has been selected to show the Casparian strip. On this slide, you should be able to see its bandlike nature.
9. Return to your own slide, switch back to the medium-power objective, and focus your attention on the central vascular column. The cell layer immediately beneath the endodermis is the **pericycle**. Cells of the pericycle have the capacity to divide and produce lateral roots, like those you saw in the bean plant (Figure 32-3).
10. Finally, find the **primary xylem**, consisting of three or four ridges of thick-walled cells. (The stain used by most slide manufacturers stains the xylem cell walls red.) The xylem is the principal water-conducting tissue of the plant.
11. Between the "arms" of the xylem find the **primary phloem**, the tissue responsible for long-distance transport of carbohydrates produced by photosynthesis (known as photosynthates).

Primary xylem and phloem make up the *primary* vascular tissues, which are those produced by the apical meristems at the tips of the root and shoot. Later, we'll examine a *secondary* vascular tissue.

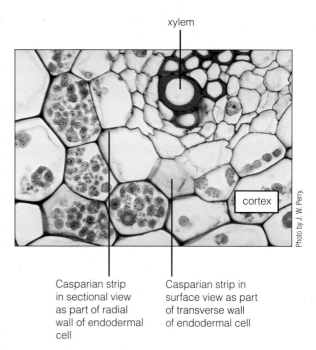

Casparian strip in sectional view as part of radial wall of endodermal cell

Casparian strip in surface view as part of transverse wall of endodermal cell

Figure 32-8 Endodermal cells showing Casparian strip (296×).

C. Monocot Root Anatomy

Now that you have an understanding of the internal structure of a dicot root, let's compare and contrast it with that of a monocot. Draw and label the monocot root in Figure 32-9 as you proceed with your study.

1. Obtain a prepared slide of a cross section of a corn (*Zea mays*) root. Examine it with the low-power objective of your compound microscope. Draw the general shape of the corn root in Figure 32-9.
2. Identify and draw the **epidermis**.
3. Next locate the **cortex**. Count the number of cell layers and draw the cortex.

Do you find any starch grains in the cortex of the corn stem? _____

4. Now locate the **endodermis**. If the root you have is a mature one, you will see that the endodermis has very thick walls on three sides.
5. The vascular column of a monocot is quite different from a dicot in several ways. First note that the center is occupied by thin-walled cells. These cells make up a region called the **pith**.
6. Draw the pith.
7. Next note that the **xylem** of this monocot root consists of scattered cells. Look at Figure 32-10, a high magnification photomicrograph of the xylem and phloem of this root.
8. Draw the xylem.

Figure 32-9   Drawing of a corn (monocot) root, c.s. (_____×).
Labels: epidermis, cortex, endodermis, pith, xylem, phloem

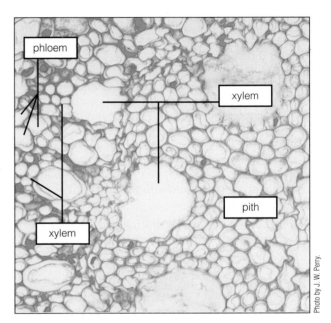

Figure 32-10   Portion of a cross section of a corn (monocot) root showing vascular tissue (140×).

32.3   The Shoot System: Stems *(About 50 min.)*

MATERIALS

Per student:

- prepared slides of
  herbaceous dicot stem, c.s. (flax, *Linum;* or alfalfa, *Medicago*)
  monocot stem, c.s. (corn, *Zea*)
  woody stem, c.s. (basswood, *Tilia*)
- woody twig (hickory, *Carya;* or horse chestnut, *Aesculus*)
- metric ruler or meter stick
- compound microscope

Per student pair:

- cross section of woody branch (tree trunk)
- dissecting microscope

Per lab room:

- demonstration slide of lenticel

PROCEDURE

A. Dicot Stem: Primary Structure

Remember that the root and shoot are basically similar in structure; only the arrangement of tissues differs. Dicot stems have their vascular tissues arranged in a more or less complete ring of individual bundles of vascular tissue (called vascular bundles). Moreover, the ground tissue of dicots can be differentiated into two regions: **pith** and **cortex**.

You've probably heard the term *herb*. An *herb* is an **herbaceous plant**, one that develops no or very little wood. Herbaceous plants have only primary tissues. Woody plants develop secondary tissues—wood and bark.

Beans, flax, and alfalfa are examples of herbaceous dicots. Maples and oaks are woody dicots. Let's look at an herbaceous stem first.

1. Obtain a slide of an herbaceous dicot stem (flax or alfalfa). This slide may be labeled "herbaceous dicot stem." As you study this slide, refer to Figure 32-11, of a partial section of an herbaceous dicot stem.
2. Using the low- and medium-power objectives, identify the single-layered **epidermis** covering the stem, a multilayered **cortex** between the epidermis and **vascular bundles**, and the **pith** in the center of the stem.
3. Observe a vascular bundle with the high-dry objective (Figure 32-12). Adjacent to the cortex, find the **primary phloem**.
4. Just to the inside of the primary phloem, locate the thin-walled cells of the **vascular cambium**. (The vascular cambium is the lateral meristem that produces wood and secondary phloem, both secondary tissues. Despite your specimen being an herbaceous plant, a vascular cambium may be present. Generally, this meristem does not produce enough secondary tissue to result in the plant being considered woody.)
5. Locate the thick-walled cells of the **primary xylem**. The primary xylem is that vascular tissue closest to the pith. (The wall of the xylem cells is probably stained red.)

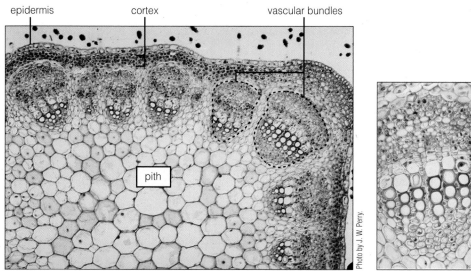

Figure 32-11 Portion of a cross section of an herbaceous dicot stem (75×).

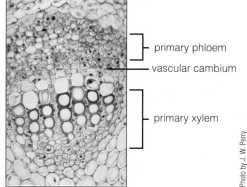

Figure 32-12 Vascular bundle of an herbaceous dicot stem (192×).

B. External Features of Woody Dicot Stems

Woody plants are those that undergo secondary growth, producing the secondary tissues. Both roots and shoots may have secondary growth. This growth occurs because of activity of the two meristems near the edge of the plant—the vascular cambium and the cork cambium. The vascular cambium produces secondary xylem (wood) and secondary phloem, while the cork cambium produces a portion of the bark called the **periderm**.

1. Examine a twig of hickory (*Carya*) or buckeye (*Aesculus*) that has lost its leaves. Label Figure 32-13 as you study the twig.
2. Find the large **terminal bud** at the tip of the twig. If your twig is branched, each branch has its own terminal bud.
3. Identify the shield-shaped **leaf scars** at each **node**. (Remember, a node is the region where a leaf attaches to the stem.) Leaf scars represent the point at which the leaf petiole was attached on the stem.
4. Within each leaf scar, note the numerous dots. These are the **vascular bundle scars**. Immediately above and adjacent to most leaf scars should be an **axillary bud**.

5. In the **internode** regions of the twig, locate the small raised bumps on the surface; these are **lenticels,** the regions of the periderm that allow for exchange of gases.

6. Return to the terminal bud. Note that the bud is surrounded by **bud scales**. When these scales fall off during spring growth, they leave **terminal bud scale scars**. Because a terminal bud is produced at the end of each growing season, you can use groups of terminal bud scale scars to determine the age of a twig.

   If the most recent growth took place during the last growing season (summer), when was the portion of the twig immediately adjacent to the cut end produced? _____

(Remember to date all regions between successive bud scale scars.)

Now that you've got some idea of the structure of a particular woody plant, let's see what amount of growth takes place in several different species of plants.

7. Measure the distance between a number of successive terminal bud scale scars in your specimen. Average the results to get an approximate idea of how much growth is produced annually. Record your results in Table 32-1.

8. Now obtain twigs of several other species, including the tree of heaven (*Ailanthus*), and do the same.

Would you say the growth rate is similar or quite variable among species that grow in your area? _____

_____

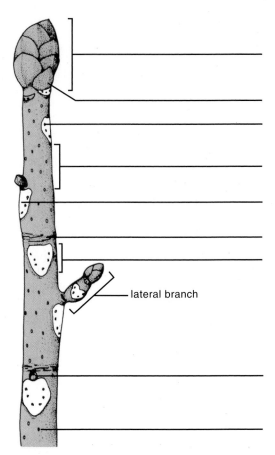

— lateral branch

Figure 32-13 External structure of a woody stem. **Labels:** terminal bud, leaf scar, node, vascular bundle scar, axillary bud, internode, lenticel, bud scale, terminal bud scale scars

TABLE 32-1	Average Annual Terminal Growth in Woody Stems	
Species	**Average Distance between Terminal Bud Scale Scars (cm)**	
Carya (or *Aesculus*)		
Ailanthus		

© Cengage Learning 2013

C. Secondary Growth: Gross Anatomy of a Woody Dicot Stem

1. Examine a cross section of a tree trunk (Figure 32-14; *trunk* is the nonscientific term for any large, woody stem). A tiny region in the center of the stem is the **pith**. Most of the trunk is made up of **wood** (also called *secondary xylem*). Locate the pith and the wood.

2. Now identify the **bark.**

As a woody stem grows larger, it requires additional structural support and more tissue for transport of water. These functions are accomplished by the wood. However, it is only the outer few years' growth of wood that is involved in water conduction. In some tree species, like that pictured in Figure 32-14, the distinction between conducting and nonconducting wood is obvious because of the color differences in each. The nonconducting wood, very dark in Figure 32-14, is called the **heartwood**.

3. Locate the heartwood in your stem.

The heartwood of many species, like black walnut or cherry, is highly prized for furniture making. Although virtually all trees develop heartwood, not all species have heartwood that is *visibly* distinct from the conducting wood.

4. The conducting wood, significantly lighter in color in Figure 32-14, is the **sapwood**. Examine the photo closely, because the color difference between the sapwood and the inner layers of the bark is subtle. Attempt to identify the sapwood on the stem you are examining.

5. Count the **growth increments (annual rings)** within the wood to estimate the age of the stem when the section was cut. (In most woods, a growth increment includes *both* a light and a dark layer of cells.)

How old would you estimate your section to be? _____ years

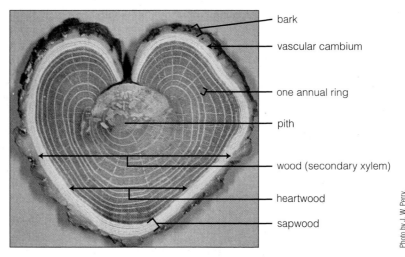

Figure 32-14   Cross section of a woody locust stem. This stem was injured several years into its growth, causing the unusual shape (0.5×).

6. Now find the **vascular cambium** located *between* the most recently formed wood (secondary xylem) and the **bark**. The bark is everything *external* to the vascular cambium.

7. Identify the bark, consisting of **secondary phloem** and **periderm**. (As the vascular cambium produces secondary phloem to the outside, the primary phloem, cortex, and epidermis are sloughed off, much as dead skin on your body is shed.)

The periderm performs the same function as the epidermis before the epidermis ruptured as a result of the stem's increase in girth. What is the function of the periderm?

_____

8. Within the wood, find the **rays**, which appear as lines running from the center toward the edge of the stem.
9. Examine the wood with a dissecting microscope and locate the numerous holes in the wood. These are the cut ends of the water-conducting cells, often called **pores**.

D. Secondary Growth: Microscopic Anatomy of a Woody Dicot Stem

Now that you've got an idea of the composition of a woody stem, let's examine one with the microscope.

1. Obtain a prepared slide of a cross section of a woody stem (basswood, *Tilia,* or another stem). Examine it with the various magnifications available on the dissecting microscope. Use Figure 32-15 as a reference.

2. Starting at the edge, identify the **periderm** (darkly stained cells). Depending on the age of the stem, **cortex** (thin-walled cells with few contents) may be present just beneath the periderm. Now find the broad band of **secondary phloem** (consisting of cells in pie-shaped wedges). Identify the **vascular cambium** (a narrow band of cells separating the secondary phloem

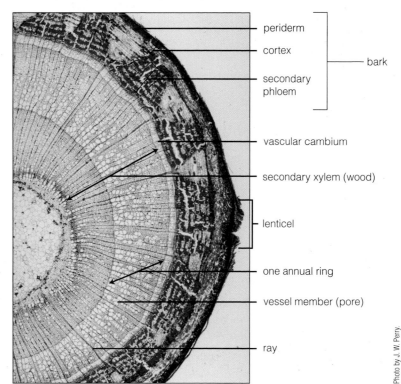

Figure 32-15   Cross section of a woody dicot stem (15×).

from secondary xylem), **secondary xylem (wood)**, and pith (large, thin-walled cells in the center). Count the number of growth increments (annual rings). Note the largest, thick-walled cells in the wood. These are the pores.

How old is this section? _____ years

3. Find the **rays** running through the wood. Rays are parenchyma cells that carry water and photosynthates *laterally* in the stem. (For the most part, the xylem and phloem carry substances *vertically* in the plant.)

Recall that the periderm replaces the epidermis as secondary growth takes place. The epidermis had stomata, which allowed the exchanging of gases between the plant and the environment. When the epidermis was shed, so were the stomata. But the need for exchange of gases still exists because the living cells require oxygen for respiration (Exercise 10). The plant has solved this problem by having special regions, **lenticels**, in the periderm. Lenticels are groups of cells with lots of intercellular space, in contrast to the tightly packed cells in the rest of the periderm.

4. Identify a lenticel on your slide. If none is found, examine the *demonstration* slide, specifically chosen to demonstrate this feature.

E. Structure of a Monocot Stem

Monocot and dicot stems differ in several ways. Monocot stems lack differentiation of the ground tissue into cortex and pith. Unlike the dicot stem whose vascular tissues are arranged in a ring, the vascular bundles of monocots are scattered throughout the ground tissue.

As you study the monocot stem, draw what you see in Figure 32-16.

1. Obtain a prepared slide of a monocot (corn, *Zea mays*) stem cross section. Observe the slide first using the low-power objective of your compound microscope, or perhaps even a dissecting microscope.
2. Draw its general appearance in Figure 32-16, labeling the **epidermis** on the outside, the thin-walled cells comprising the **ground tissue**, and the numerous **vascular bundles** scattered within the vascular tissue.
3. Switch to the medium-power objective of your compound microscope to study a single vascular bundle more closely. Draw the detail of one vascular bundle in Figure 32-17. (The vascular bundles of monocots look a lot like the face of a human or other primate.)
4. Observe the thick-walled cells surrounding the vascular bundle. These cells make up the bundle sheath. (The walls of these cells are often stained red.)
5. Note the very large cells that give the impression of the vascular bundle's "eyes, nose, and mouth." These cells make up the water-conducting xylem. (The cell walls are often stained red.) Draw and label these xylem.
6. Finally, find the phloem, located on the vascular bundle's "forehead." (Phloem cell walls are typically stained green or blue.)

Figure 32-16   Drawing of a corn (monocot) stem, c.s. (_____×).
Labels: epidermis, ground tissue, vascular bundles

Figure 32-17   Drawing of the vascular bundle of a corn (monocot) stem, c.s. (_____×).
Labels: bundle sheath, xylem, phloem

Leaves make up the second part of the shoot system. You examined the external morphology of the leaves in Section 32.1. Now let's look inside leaves, starting with a typical dicot leaf.

MATERIALS

Per student:

■ prepared slide of dicot leaf, c.s. (lilac, *Syringa*)
■ prepared slide of monocot leaf, c.s. (corn, *Zea*)
■ compound microscope

PROCEDURE

A. Dicot Leaf Anatomy

1. Obtain a prepared slide of a cross section of a dicot leaf (lilac or other). Refer to Figure 32-18 as you examine the leaf.

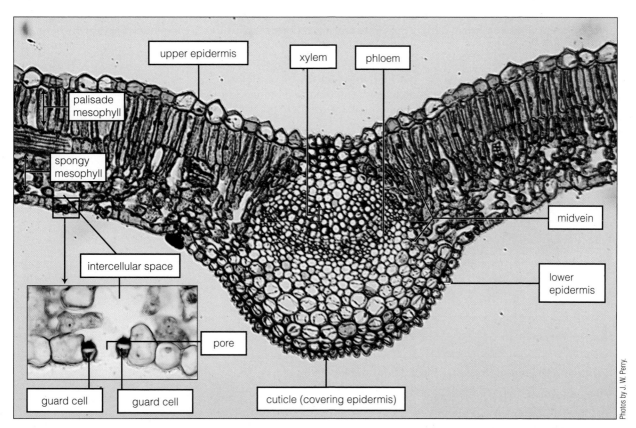

Figure 32-18   Cross section of a dicot leaf with midrib (105×); inset of a stoma (480×).

2. Examine the leaf first with the low-power objective to gain an overall impression of its morphology. Note the size and orientation of the **veins** within the leaf. The veins contain the xylem and phloem.
3. Find the centrally located **midvein** within the **midrib**, the midvein-supporting tissue. You might think of the midvein as the major pipeline of the leaf, carrying water and minerals to the leaf and materials produced during photosynthesis to sites where they will be used during respiration.
4. Use the high-dry objective to examine a portion of the blade to one side of the midvein. Starting at the top surface of the leaf, find the **cuticle**, a waxy, water-impervious substance covering the **upper epidermis**. The epidermis is a single layer of tightly appressed cells that are tightly bound together, without any intercellular spaces.

5. The ground tissue of the leaf is represented by the **mesophyll** (literally "middle leaf"). In dicot leaves, the mesophyll is usually divided into two distinct regions; immediately below the upper epidermis, find the two layers of **palisade mesophyll**. These columnar-shaped cells are rich in chloroplasts.
6. Below the palisade mesophyll, find the loosely arranged **spongy mesophyll**. Note the large volume of **intercellular air space** within the spongy mesophyll.

Does the spongy mesophyll contain any chloroplasts? (yes or no) _____

What is one function that occurs within the spongy mesophyll? _____

7. In the **lower epidermis**, find a **stoma** (plural: *stomata*) with its **guard cells** and the **pore** (inset, Figure 32-18). Large **epidermal hairs** shaped somewhat like mushrooms are usually found on the lower epidermis.

Is the lower epidermal layer covered by a cuticle? _____

8. Compare the abundance of stomata within the lower epidermis with that in the upper epidermis.

Which epidermal surface has more stomata? _____

9. Examine the midvein in greater detail. Find the thick-walled xylem cells (often stained red).
10. Below the xylem, locate the **phloem** (usually stained green).
11. Now identify and examine the smaller veins within the lamina (blade). Note that these, too, contain both xylem and phloem.

B. Monocot Leaf Anatomy

Let's compare the structure of a monocot grass leaf with that of a dicot. Draw the leaf in Figure 32-19 as you proceed with your study.

1. Obtain a prepared slide of a corn (*Zea mays*) leaf. Examine it first with the low-power objective of your compound microscope to gain an overall impression of its structure. Draw what you see in Figure 32-19.
2. Identify the **upper** and **lower epidermis**. Label them in Figure 32-19.
3. Search on both epidermal layers for **stomata**. Count them on each layer.

Are there more, fewer, or about the same number on the upper epidermis versus the lower?

_____

4. Think about the orientation of the leaf blade on the plant. (If necessary, examine a living or dried corn plant to gain an impression of the orientation.)

Is the orientation of the corn leaves the same or different from that of a dicot, such as the bean plant?

_____

If you determined it is different, in what way is it different?

_____

5. Make a hypothesis relating the orientation of the leaf blade and the number of stomata on each surface of a leaf blade.

_____

_____

_____

6. Look at the **mesophyll** of the leaf. Draw and label it in Figure 32-19.

Is the mesophyll divided into palisade and spongy layers? (yes or no) _____

7. Note the **intercellular air spaces** within the mesophyll. Draw and label the intercellular air spaces.
8. Examine a single vascular bundle. Note the larger cells that surround the vascular bundle. This "wreath" of cells is the **bundle sheath**, a layer that is exceptionally conspicuous in grasses. Draw and label the bundle sheath.
9. Note that the bundle sheath cells are tightly packed, with no intercellular spaces separating them. The bundle sheath serves a function like that of the endodermis in the root.
10. Finally, identify the larger **xylem** cells and **phloem** cells within the vascular bundle. Draw and label these tissues in Figure 32-19.

Figure 32-19 Drawing of a corn (monocot) leaf, c.s. (_____×).
Labels: cuticle, epidermis, stoma, guard cells, mesophyll, bundle sheath, xylem, phloem

_____ 1. The study of a plant's structure is
   (a) physiology.
   (b) morphology.
   (c) taxonomy.
   (d) botany.

_____ 2. A plant with two seed leaves is
   (a) a monocotyledon.
   (b) a dicotyledon.
   (c) exemplified by corn.
   (d) a dihybrid.

_____ 3. A taproot system lacks
   (a) lateral roots.
   (b) a taproot.
   (c) both of the above
   (d) none of the above

_____ 4. Which structure is **not** part of the shoot system?
   (a) stems
   (b) leaves
   (c) lateral roots
   (d) axillary buds

_____ 5. An axillary bud
   (a) is found along internodes.
   (b) produces new roots.
   (c) is the structure from which branches and flowers arise.
   (d) is the same as a terminal bud.

_____ 6. The endodermis
   (a) is the outer covering of the root.
   (b) is part of the vascular tissue.
   (c) contains the Casparian strip, which regulates the movement of substances.
   (d) is none of the above.

_____ 7. Meristems are
   (a) located at the tips of stems.
   (b) located at the tips of roots.
   (c) regions of active growth.
   (d) all of the above

_____ 8. To determine the age of a woody twig, one counts the
   (a) nodes.
   (b) leaf scars.
   (c) lenticels.
   (d) regions between sets of terminal bud scale scars.

_____ 9. The midrib of a leaf
   (a) contains the midvein.
   (b) contains only xylem.
   (c) is part of the spongy mesophyll.
   (d) contains only phloem.

_____ 10. The bundle sheath in a monocot leaf
   (a) is filled with intercellular space.
   (b) is the location of stomata.
   (c) is covered with a cuticle to prevent water loss.
   (d) functions in a manner somewhat similar to the root endodermis.

EXERCISE **3 2**

Plant Organization: Vegetative Organs of Flowering Plants

Post-Lab Questions

32.1 External Structure of the Flowering Plant

1. What type of root system do you see on the dandelion at the right?

32.2 The Root System

2. Describe the location, structure, and importance of the Casparian strip.

(0.25×)

32.3 The Shoot System: Stems

3. On the figure below, identify structures **A**, **B**, and **C**.

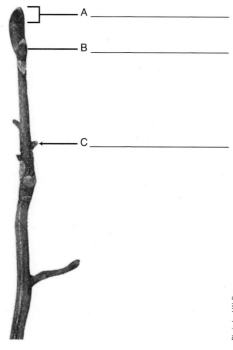

A _____

B _____

C _____

(0.5×)

4. Identify the structures labeled **A** and **B** on the figure below.

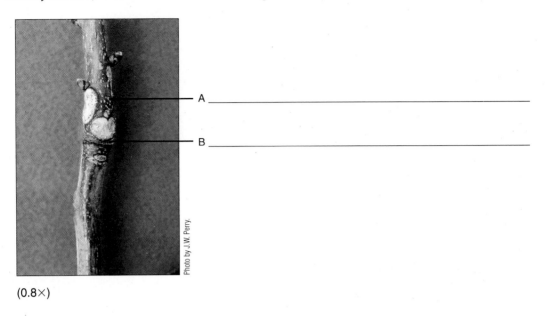

A _____

B _____

(0.8×)

5. What feature(s) would you use to determine the age of a woody twig?

6. Label the diagram using the following terms: bark; heartwood; sapwood; secondary phloem; vascular cambium.

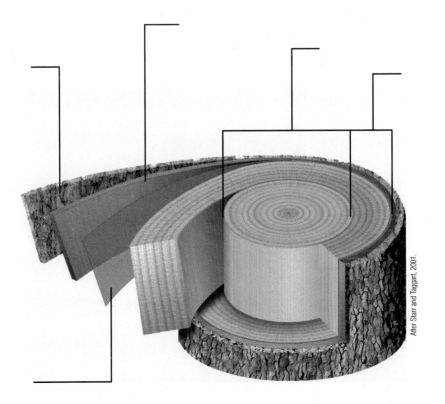

7. The photo shows the microscopic appearance of maple wood. Using your knowledge of the woody stem section, identify cell type **A** and the "line" of cells at **B**. (Note: The outside of the tree from which this section was taken is shown in the photo that accompanies question 8.)

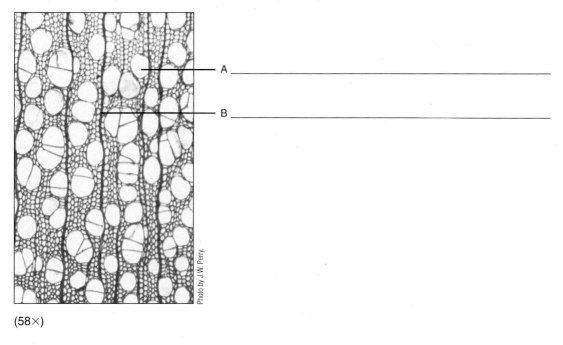

A _____

B _____

(58×)

8. A section cut from an ash branch is shown below.
   (a) Which meristem is located at **A**?

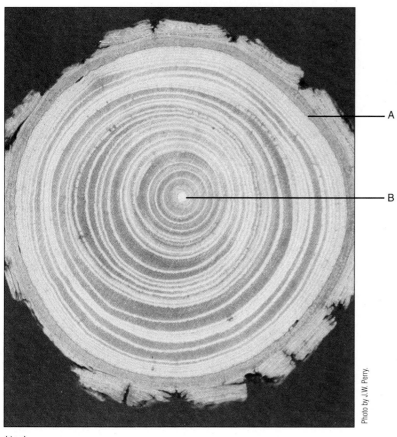

A

B

(1×)

(b) Identify region **B**.

(c) Within ±3 years, how old was this branch when cut?

32.4 The Shoot System: Leaves

9. The following photo shows a section of a leaf from a dicot that is adapted to a dry environment. Its lower epidermis has depressions, and the stomata are located in the cavities. Even though it's different from the leaf you studied in lab, identify the regions labeled **A**, **B**, and **C**.

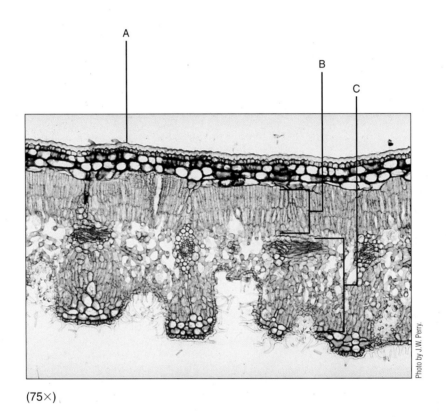

(75×)

Food for Thought

10. A major problem for land plants is water conservation. Most water is lost through stomata due to evaporation at the surface of the leaf. Many plants, including lilac (the leaf section you examined), orient their leaves perpendicular to the drying force of the sun's rays. What did you observe about the relative abundance of stomata in the lower epidermis versus the upper epidermis? Why do you think this distribution has evolved?

Animal Organization

OBJECTIVES

After completing this exercise, you will be able to

1. define *tissue, organ, system, organism, histology, basement membrane, goblet cells, cilia, brush border of microvilli, keratinization, keratin, collagen fibers, elastic fibers, fibroblast, fat cells, lumen, chondrocytes, lacunae, Haversian systems, osteocytes, actin filaments, myosin filaments, intercalated disk, neuron, cell body, axon, dendrites, neuroglia, body cavities—thoracic, abdominal, pleural, pericardial, pelvic, cranial, spinal;*

2. discuss the high degree of organization present in animal structure and explain its significance;

3. recognize the four basic tissues and their common mammalian subtypes;

4. list the functions of the four basic tissues and their common mammalian subtypes;

5. explain how the four basic tissues combine to make organs;

6. list each system of a mammal and its vital functions;

7. describe the basic plan of the mammalian body;

8. explain the layout of the body's major cavities;

9. locate the major organs in a mammal's body.

Introduction

Each animal cell is specialized to emphasize certain activities, although most continue to carry on basic functions such as cellular respiration. Groups of similarly specialized cells along with any extracellular material, associate together to form **tissues**. Different subtypes from each of the four basic tissue types are combined much like quadruple-decker sandwiches to make the body's **organs**. Groups of related organs are strung together functionally, and usually structurally, to form **systems**. Taken together and arranged correctly, systems form the entire body. The levels of organization of an animal with systems can be summarized as follows:

Animal organism
↑
Systems (e.g., digestive, respiratory)
↑
Organs (e.g., esophagus, stomach, trachea, lungs)
↑
Epithelial, connective, muscular, and nervous tissues
↑
Specialized cells
↑
Organelles

33.1   Tissues and Organs *(50 min.)*

The study of normal tissues is called **histology**. The four basic tissue types are **epithelial tissue, connective tissue, muscle tissue**, and **nervous tissue**. Each type has subtypes. In your study of these subtypes, you will examine a variety of tissues and organs, all permanently mounted on glass microscope slides. Most slides have sections of tissues and organs, usually 6 to 10 μm thick. Other slides have whole pieces of organs that are either transparent or are gently pulled apart (teased) and spread on the surface of the slide until they are thin enough to see through. The tissues of some organs are simply smeared onto the surface of slides. Differences in tissue structure go with the particular functions they perform.

MATERIALS

Per student:

- compound microscope, lens paper, a bottle of lens-cleaning solution (optional), a lint-free cloth (optional), a dropper bottle of immersion oil (optional)
- prepared slide of a whole mount of mesentery (simple squamous)
- prepared slide of a section of the cortex of the mammalian kidney
- prepared slide of a section of trachea
- prepared slide of a small intestine (preferably a cross section of the ileum)
- prepared slide of a section of esophagus
- prepared slide of a section of mammalian skin
- prepared slide of sections of contracted and distended urinary bladders
- prepared slide of a teased spread of loose (areolar) connective tissue

- prepared slide of a tendon, longitudinal section
- prepared slide of a compact bone ground cross section
- prepared slide of a section of white adipose tissue
- prepared slide of sections of the three muscle types
- prepared slide of a smear of the spinal cord of an ox (neurons)
- colored pencils
- computer with Internet access (optional)

Per lab room:

- demonstration of intercalated disks in cardiac muscle tissue
- 50-mL beaker three-fourths full of water
- 50-mL beaker three-fourths full of immersion oil
- two small glass rods

PROCEDURE

A. Epithelial Tissues

Epithelial tissues are widespread throughout the body, covering both the body's outer surface (epidermis of skin) and inner surfaces (e.g., ventral body cavities, inner surface of the small intestine, etc.). Their main functions are *protection* and *transport*—secretion, absorption, and excretion. Specialized functions include sensory reception and the maintenance of the body's gametes (eggs and sperm).

Epithelial cells carry on rapid cell division in adults as well as children, and various stages of mitosis are often seen in this basic tissue type. As a consequence, epithelia have the highest rates of cell turnover among tissues. For example, the lining of the small intestine replaces itself every 3 to 4 days. Epithelial tissues do not have blood vessels and are attached to the underlying connective tissue by an extracellular **basement membrane** that is difficult to see if it is not specially stained. Use the following questions to identify most subtypes of epithelial tissue:

QUESTION 1. What is the shape of the outermost cells?

There are three possible answers: *squamous* (scale-like or flat), *cuboidal*, or *columnar*. These choices describe the shape of the cells when they are viewed in a section perpendicular to the surface. In surface view, all three types are multisided (polygonal).

QUESTION 2. How many layers of cells are there?

Epithelial tissues are *simple* if there is one layer of cells; they are *stratified* if there are two or more. Figure 33-1 applies these definitions to squamous epithelia. *Pseudostratified* epithelial subtypes appear stratified but really are not because their cells all rest on the basement membrane (Figure 33-2). As only the tallest columnar-shaped cells reach the free surface, cell nuclei lie at different levels, giving the false appearance of several layers of cells.

In addition, the epithelium that lines the inside of the urinary bladder (and parts of other nearby urinary system

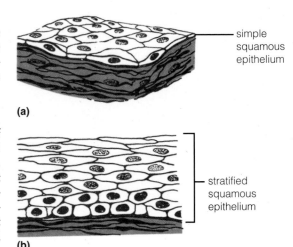

(a)

(b)

Figure 33-1 Squamous epithelia, (**a**) simple and (**b**) stratified.

organs) changes its subtype as it fills or empties with urine. This epithelium is called *transitional* to avoid confusion.

There are six common subtypes of epithelia: *simple squamous, simple cuboidal, simple columnar, pseudostratified columnar, stratified squamous,* and *transitional.*

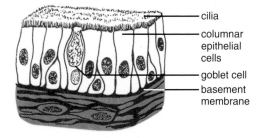

Figure 33-2 Pseudostratified columnar epithelium. It is columnar because of the shape of the cells that reach the free surface.

1. Examine the whole mount of *mesentery* with your compound microscope (Figure 33-3). The mesentery is a fold of the abdominal wall and holds the intestines in place. This slide is a surface view of the **simple squamous epithelium** (Figure 33-4), which lines the inside of the ventral body cavities and the organs within.

What is the shape of the surface cells? _____

Draw what you see in Figure 33-5 and fill in the total magnification of the compound microscope you are using. Repeat this procedure for each subsequent drawing. Alternatively, if computers and Internet access are available, construct an atlas of links to images of sections, and so on, of the various tissues and organs under study.

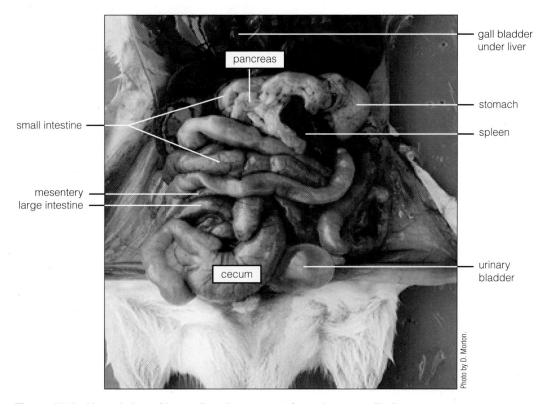

Figure 33-3 Ventral view of lower digestive system of a male mouse (2×).

2. Look at a section of the *kidney* (Figure 33-6). With the help of Figure 33-7a, find a *renal corpuscle.* A renal corpuscle is composed of a tuft of capillaries surrounded by a hollow capsule formed from the cup-shaped end of one of the kidney's functional units, the nephron. Identify a side view of the simple squamous epithelium that comprises the outer wall of the capsule (Figure 33-7b). Around the renal corpuscle, you can see a number of transverse and oblique sections of the tubular portion of the nephron. Their walls are composed of **simple cuboidal epithelium.** The main function of the nephrons and the entire kidney is the production of urine. Draw simple squamous epithelium in Figure 33-8. In Figure 33-9, draw simple cuboidal epithelium.

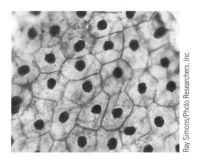

Figure 33-4 Surface view of simple squamous epithelium of the mesentery (300×).

Figure 33-5 Drawing of the surface of simple squamous epithelium (_____×).

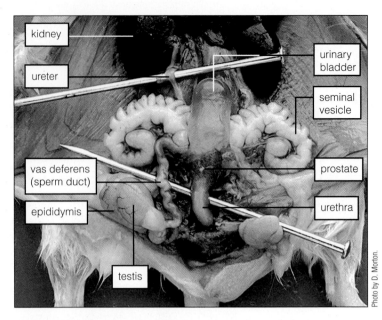

Photo by D. Morton.

Figure 33-6 Ventral view of the urinary and reproductive systems of a male mouse after removal of the digestive system (3×).

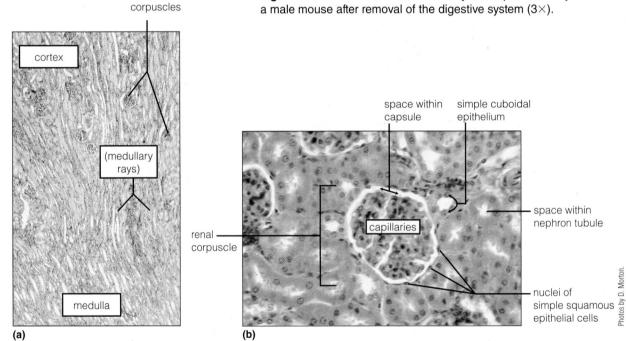

Photos by D. Morton.

Figure 33-7 Sections of the cortex of the kidney, (**a**) low power (25×) and (**b**) medium power (183×) views.

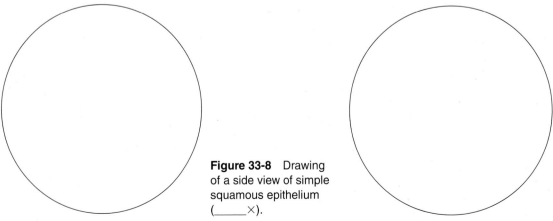

Figure 33-8 Drawing of a side view of simple squamous epithelium (_____×).

Figure 33-9 Dra of a side view of cuboidal epitheliu (_____×).

3. Check out the section of the *trachea* in Figure 33-10. The trachea is a tubular organ that conveys air to and from the lungs. Note that its inner surface is lined by **pseudostratified columnar epithelium** (Figure 33-11). Locate the unicellular glands, which are called **goblet cells** because of their shape. They secrete mucus that is difficult to see unless it is specifically stained. Using high power, do you see the

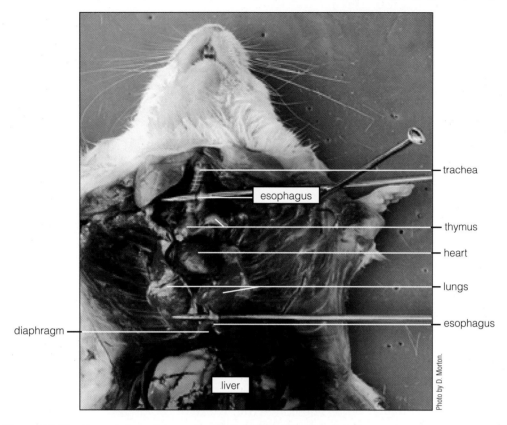

Figure 33-10 Ventral view of the organs in the thoracic cavity of a dissected mouse (2×).

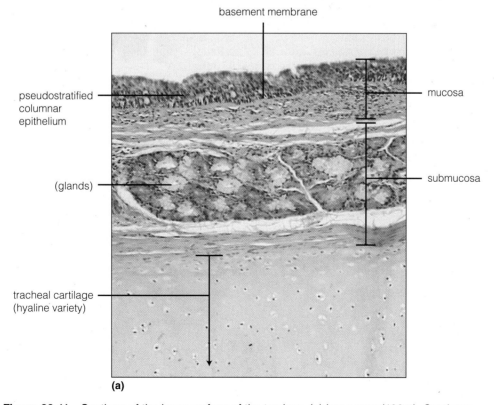

(a)

Figure 33-11 Sections of the inner surface of the trachea, **(a)** low power (100×). *Continues.*

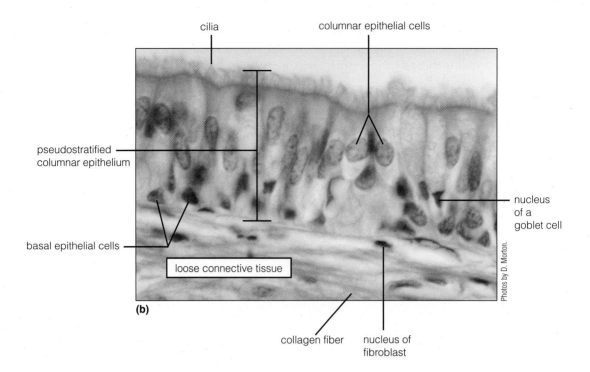

cila   columnar epithelial cells

pseudostratified
columnar epithelium

basal epithelial cells

loose connective tissue

nucleus
of a
goblet cell

Photos by D. Morton.

(b)

collagen fiber   nucleus of
                   fibroblast

Figure 33-11 *Continued.* Sections of the inner surface of the trachea, **(b)** oil immersion (1000×).

numerous hair-like structures that project from the surface of the columnar epithelial cells? These are **cilia**, which in life move together, sweeping mucus-trapped bacteria and debris up the trachea to the throat. When you clear your throat, you collect this mucus and swallow it (a form of recycling). Draw pseudostratified columnar epithelium in Figure 33-12.

4. Observe a section of the *small intestine* (Figure 33-3). The small intestine is a tubular organ that connects the stomach to the large intestine. Its inner surface is lined with **simple columnar epithelium** (Figure 33-13). Do not miss the *goblet cells* present in the epithelium. Locate the **brush border of microvilli** situated along the free surface of the epithelium. Do not confuse this relatively thin structure with the cilia from the previous slide. Microvilli are primarily responsible for the large surface area of the small intestine. Individual microvilli can best be seen at the electron-microscopic level (Figure 33-14). In Figure 33-15, draw simple columnar epithelium.

5. Replace the slide on your compound light microscope with a section of the *esophagus* (Figure 33-10). The tubular esophagus connects the throat to the stomach. Examine the epithelium lining the inner surface of the esophagus (Figure 33-16). Is the shape of the outermost cells of the epithelium squamous, cuboidal, or columnar? _____

Figure 33-12 Drawing of a side view of pseudostratified columnar epithelium (_____×).

Are there one or many layers of cells in this tissue? _____

Name this subtype of epithelial tissue. _____

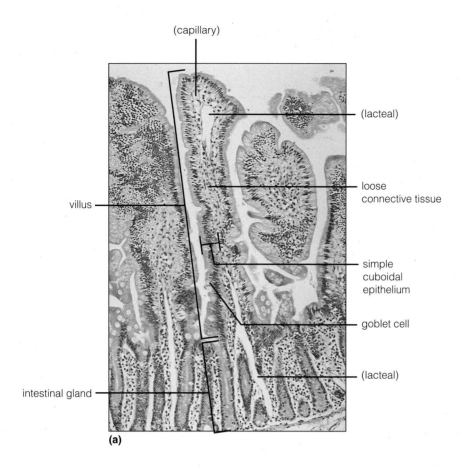

(capillary)

(lacteal)

loose
connective tissue

villus

simple
cuboidal
epithelium

goblet cell

(lacteal)

intestinal gland

(a)

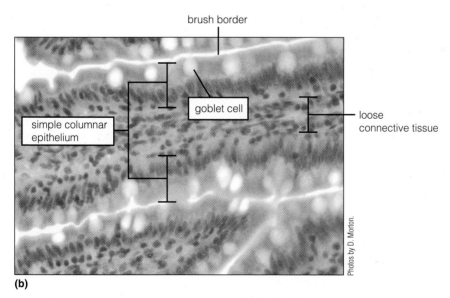

brush border

simple columnar
epithelium

goblet cell

loose
connective tissue

Photos by D. Morton.

(b)

Figure 33-13 Sections of the inner surface of the small intestine,
(**a**) medium-power (250×) and (**b**) high-power (400×) views.

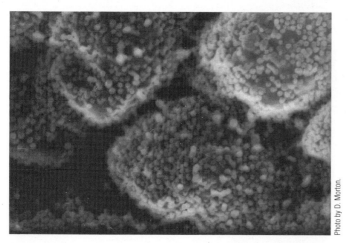

Figure 33-14 Scanning electron micrograph of a surface view of microvilli of simple columnar epithelial cells of the fundic cecum (stomach) of a vampire bat (10,000×).

Figure 33-15 Drawing of a side view of simple columnar epithelium (_____×).

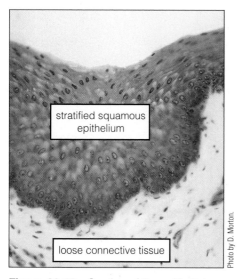

stratified squamous epithelium

loose connective tissue

Figure 33-16 Section of the epithelial lining of the inner surface of the esophagus (300×).

Figure 33-17 Drawing of a side view of stratified squamous epithelium (_____×).

In Figure 33-17, draw stratified squamous epithelium.

6. Now look at a section of the *skin*. The skin is divided into three layers (Figure 33-18). The *epidermis* is composed of stratified squamous epithelium. The *dermis* and *hypodermis* (or subcutaneous layer) are connective tissue and will be studied later.

The epidermis is the most extreme example of a protective epithelium. The process of keratinization, whereby the cells transform themselves into bags of the protein keratin, causes strata in the epidermis. This substance greatly contributes to the skin's toughness, flexibility, and water-resistant surface.

At one of the free edges of your section of mammalian skin, locate the **keratinized stratified squamous epithelium**. It will be stained bluer than the predominately pink connective tissue layers. Hair follicles and multicellular sweat glands may be present in the connective tissue layers. These structures grow into the connective tissue layers from the epidermis during the development of the skin. Draw keratinized stratified squamous epithelium in Figure 33-19.

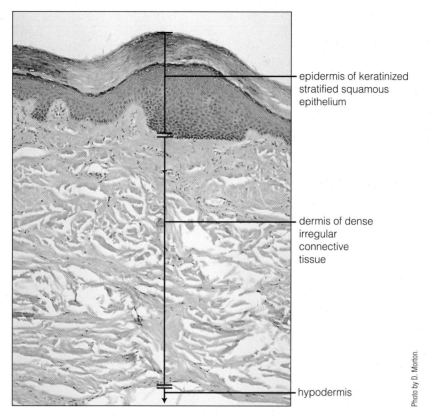

Figure 33-18   Section of the skin (30×).

epidermis of keratinized stratified squamous epithelium

dermis of dense irregular connective tissue

hypodermis

Photo by D. Morton.

Figure 33-19   Drawing of a side view of keratinized stratified squamous epithelium (_____×).

7. The *urinary bladder* (Figure 33-6) slide has two sections, one from a contracted bladder and the other from a distended bladder. Locate the **transitional epithelium** at the surface of each section (Figure 33-20). The transitional epithelium from the contracted bladder appears like stratified cuboidal epithelium. In the distended bladder, the transitional epithelium is thinner and looks like stratified squamous epithelium. Draw a high-power view of these two extremes in Figure 33-21.

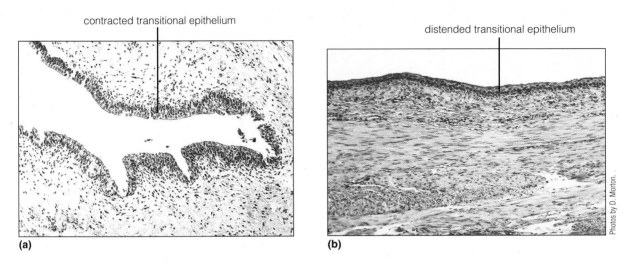

contracted transitional epithelium

distended transitional epithelium

(a)

(b)

Photos by D. Morton.

Figure 33-20 Sections of (**a**) contracted and (**b**) distended urinary bladders (89×).

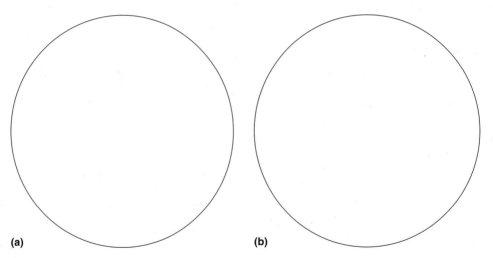

(a)

(b)

Figure 33-21 Drawings of high-power views of transitional epithelia from (**a**) contracted and (**b**) distended urinary bladders (_____×).

B. Connective Tissue

Connective tissues occur in all parts of the body. They contain a large amount of material external to the cells, called the *extracellular matrix*. This matrix consists of *fibers embedded in ground substance*. Two basic kinds of fibers exist: collagen and elastic. Whereas collagen fibers are tough, flexible, and inelastic, elastic fibers stretch when pulled, returning to their original length when the pull is removed. Except for cartilage, connective tissues contain blood vessels. Similar to epithelia, connective tissue cells are capable of cell division in adults but at a reduced rate. Table 33-1 lists most connective tissue subtypes along with their matrix characteristics.

The hard and special groups are sometimes combined as a specialized category of connective tissue subtypes. Blood is studied in Exercise 40. Find and examine subtypes in the following slides.

1. First look at a section of *teased spread of loose or areolar connective tissue*. **Loose connective tissue** forms much of the packing material of the body and fills in the spaces between other tissues. Identify **collagen fibers** and **elastic fibers** (Figure 33-22). Collagen fibers are stained light pink and are variable in diameter, but they are generally wider than the elastic fibers. Elastic fibers are darkly stained, thin, and branched. Areolar connective tissue contains a number of different cell types. To see a **fibroblast**—the cell that produces the matrix of loose and other soft connective tissues—look for an elongated, oval-shaped nucleus associated

TABLE 33-1	Connective Tissue Groups and Common Subtypes	
Group	**Subtype**	**Matrix Contains**
Soft	Loose (also called areolar)	Few collagen and elastic fibers arranged apparently randomly
	Dense irregular	Many collagen or elastic fibers arranged apparently randomly
	Dense regular	Many collagen or elastic fibers arranged in a parallel fashion
Hard	Cartilage	Many collagen or elastic fibers and polymerized (or jellylike) ground substance
	Bone	Many collagen fibers and mineralized ground substance
Special	Adipose	Delicate collagen fibers
	Blood	Plasma

© Cengage Learning 2013

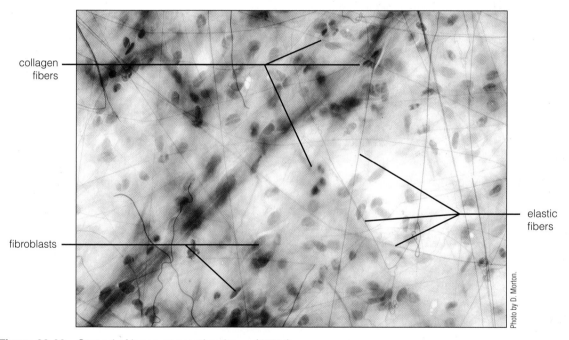

collagen fibers

fibroblasts

elastic fibers

Photo by D. Morton.

Figure 33-22 Spread of loose connective tissue (450×).

with a fiber. Many kinds of cells that play important roles in the body's immune system live in this tissue. The spaces between fibers and cells contain ground substance in the living tissue. The ground substance of all soft connective tissues consists of a viscous soup of carbohydrate/protein molecules and is usually extracted from tissue sections during processing. Draw loose connective tissue in Figure 33-23.

2. Reexamine the section of *skin*. Look at the second layer of the skin, the dermis, which is primarily composed of **dense irregular collagenous connective tissue** (Figure 33-18). The dermis cushions the body from everyday stresses and strains. Note that the looser irregular connective tissue of the hypodermis has a lower concentration of collagen fibers and islands of fat cells. It functions as a shock absorber, an insulating layer, and a site for storing water and energy. In mammals that move their skin independently of the rest of the body (e.g., cats), skeletal muscle tissue is also found in the hypodermis. In Figure 33-24, draw dense irregular collagenous connective tissue.

Figure 33-23 Drawing of a spread of loose connective tissue (_____×).

Figure 33-24 Drawing of dense irregular collagenous connective tissue (_____×).

3. A *tendon* connects a skeletal muscle organ to a bone organ. Tendons are composed largely of **dense regular collagenous connective tissue**. Examine a longitudinal section of a tendon (Figure 33-25).

Describe the density and arrangement of the fibers.

_____

_____

_____

This design makes tendons very strong, much like a rope composed of braided strings, which in turn are made of even smaller fibers. How are the fibroblasts oriented relative to the arrangement of the fibers?

_____

_____

Draw dense regular collagenous connective tissue in Figure 33-26.

4. Reexamine the section of the trachea (Figures 33-10 and 33-11a). Find a portion of one of the supporting cartilage, which prevents the wall from collapsing and closing the **lumen** (the space within any hollow organ). These cartilaginous structures are composed of **hyaline cartilage** tissue (Figure 33-27). Look for the **chondrocytes** (cartilage cells) that are located in small holes in the matrix called **lacunae** (singular, *lacuna*).

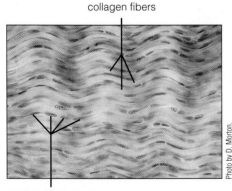

collagen fibers

nuclei of fibroblasts
aligned in a row
between collagen fibers

Photo by D. Morton.

Figure 33-25 Dense collagenous connective tissue from a longitudinal section of a tendon (250×).

Figure 33-26 Drawing of dense regular collagenous connective tissue (_____×).

Although invisible, many collagen fibers are embedded in the polymerized ground substance. They cannot be seen because the indices of refraction of these two matrix components are similar. Your instructor has set up a demonstration of this phenomenon. Observe the two labeled beakers, one filled with immersion oil and the other with water. Look at the glass rod placed in each of them. The index of refraction is about 1.58 for glass, about 1.33 for water, and about 1.52 for immersion oil. In which fluid is it easier to see the glass rod?

The difference between the index of refraction of the glass rod and immersion oil is less than that between the glass rod and water. This is exactly why collagen fibers are hard to see in the ground substance of hyaline cartilage.

In locations where cartilage has to be more durable (intervertebral disks, for example), the collagen content is higher and the fibers are visible. In other sites (such as outer ear flaps), large numbers of elastic fibers are present. Elastic cartilage is deformed by a small force and returns to its original shape when the force is removed. Draw hyaline cartilage in Figure 33-28.

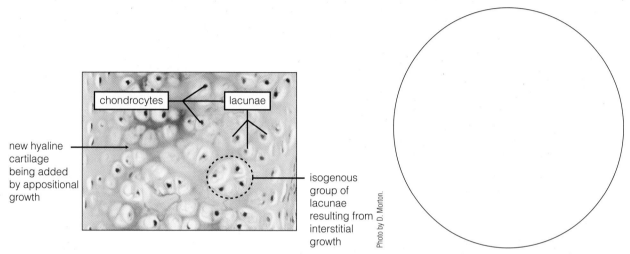

Figure 33-27 Section of hyaline cartilage in the wall of the trachea (130×).

Figure 33-28 Drawing of hyaline cartilage (_____×).

5. Living bones are amazingly strong. Bone organs are mostly composed of hard yet flexible **bone** tissue. Its hardness is attributable to minerals (predominantly calcium–phosphate salts called *hydroxyapatites*) deposited in the matrix. Its flexibility comes from having the highest collagen content of all connective tissues.

Examine the cross section of compact bone tissue (Figure 33-29). Grinding a piece of the shaft of a long bone with coarse and then finer stones until a thin wafer remains has produced this preparation. Although only the mineral part of the matrix is present, the basic architecture has been preserved. Compact bone tissue is primarily composed of longitudinally arranged **Haversian systems** (also called *osteons*). Locate a Haversian system and identify its *central canal, lamellae* (singular, *lamella), lacunae,* and *canaliculi*—little canals that you see connecting lacunae with each other and with the central canal.

In living bone, blood vessels and nerves are present in central canals, concentric layers of matrix form the lamellae, and the intervening rings of lacunae contain bone cells called **osteocytes**. In young living bone, the canaliculi contain the cytoplasmic processes of osteocytes. Thus, the osteocytes can easily exchange nutrients, wastes, and other molecules with the blood. By comparison, substances in cartilage have to diffuse across the matrix between chondrocytes and blood vessels in the surrounding soft connective tissue. Which tissue, bone or cartilage, will heal quicker? Explain why.

In Figure 33-30, draw a cross section of compact bone tissue.

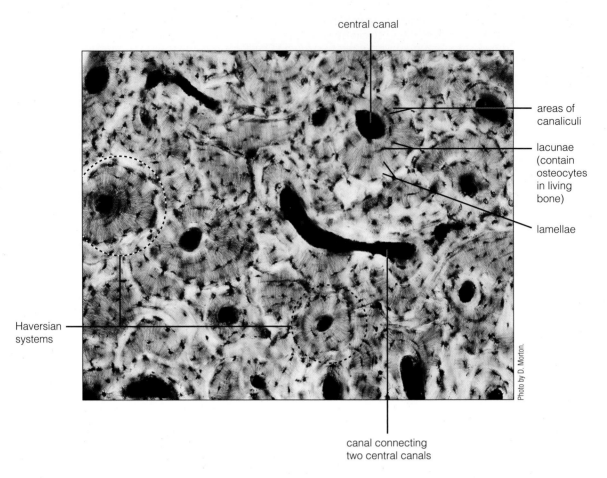

central canal

areas of
canaliculi

lacunae
(contain
osteocytes
in living
bone)

lamellae

Haversian
systems

canal connecting
two central canals

Photo by D. Morton.

Figure 33-29　Ground cross section of a long bone showing compact bone tissue (100×).

Figure 33-30　Drawing of compact bone
tissue (_____×).

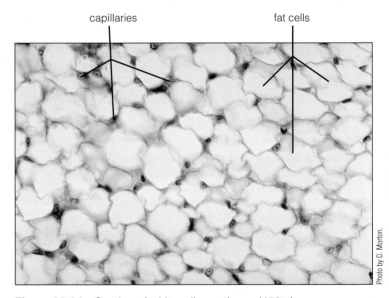

capillaries

fat cells

Photo by D. Morton.

Figure 33-31　Section of white adipose tissue (450×).

6. As you see in Figure 33-31, **white adipose tissue** consists mainly of **fat cells**. However, careful examination of the section using high power shows them to be surrounded by delicate collagen fibers, fibroblasts, and capillaries. Because the fat has been lost during the slide preparation, the fat cells look empty. The primary function of this tissue is energy storage. Draw white adipose tissue in Figure 33-32.

C. Muscle Tissue

Muscle tissue is contractile in that its cells (fibers) can shorten. If shortening occurs, it produces changes in the position of body parts. If shortening does not occur, contractions will produce changes in tension. Contraction results from interactions between two types of protein filaments: **actin** and **myosin**. Similar to epithelial tissue, muscle tissue is primarily cellular; but unlike both epithelial and connective tissues, its cells do not normally divide in adults. Therefore, the repair of damaged and dead fibers is limited. The three subtypes of muscle tissue are *skeletal, cardiac,* and *smooth.*

Figure 33-32 Drawing of white adipose tissue (_____×).

Skeletal muscle tissue is under voluntary control (i.e., pursing your lips for a kiss). This means that your conscious mind can order skeletal muscles to contract, although most of their contractions are actually involuntary (e.g., during complex movements like catching a ball). Cardiac muscle and smooth muscle tissues are always under *involuntary control.* Involuntary control means that contractions are directed at the unconscious level.

1. Examine your slide with sections of all three subtypes (Figure 33-33). Identify their characteristics (Table 33-2). In skeletal and cardiac muscle fibers, actin and myosin filaments overlap to produce the alternating pattern of light and dark bands (cross- or transverse striations) seen in these tissues. Only cardiac muscle cells are branched. Where the branch of one fiber joins another, the cells are stuck together

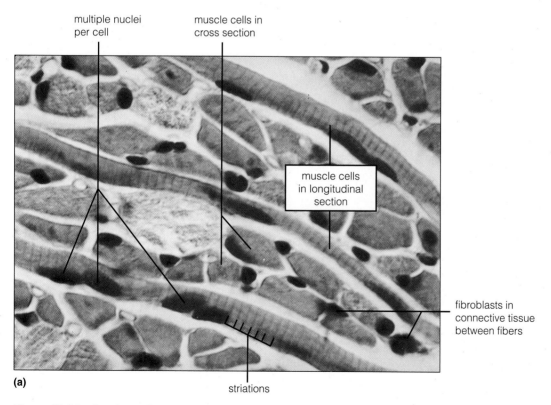

multiple nuclei per cell

muscle cells in cross section

muscle cells in longitudinal section

fibroblasts in connective tissue between fibers

(a)

striations

Figure 33-33 Sections of muscle tissue: **(a)** skeletal (400×). *Continues.*

intercalated disks

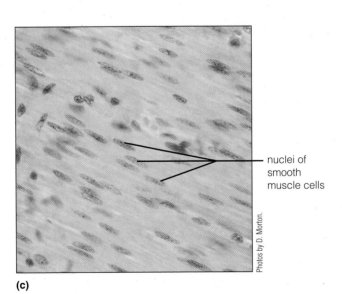

(b)

nucleus of cardiac
muscle cell

branch

nuclei of
smooth
muscle cells

Photos by D. Morton.

(c)

Figure 33-33 *Continued.* Sections of muscle tissue:
(**b**) cardiac (1000×), (**c**) smooth (400×).

TABLE 33-2	Characteristics of Muscle Tissue Subtypes			
Subtypes	Organ Location	Position and Number of Nuclei in Fibers	Cross-Striated Fibers	Special Fiber Features
Skeletal	Skeletal muscles	Peripheral and many	Yes	Long, cylindrical shape
Cardiac	Heart	Central and usually one	Yes	Branches and intercalated disks
Smooth	Walls of internal organs	Central and one	No	Shape tapers at both ends

© Cengage Learning 2013

and in direct communication through a complex of cell-to-cell junctions. The complex is called an **intercalated disk**. Find an intercalated disk in your section, but if you have trouble seeing it, look at the demonstration set up by your instructor. Draw the three muscle tissue subtypes in Figure 33-34.

D. Nervous Tissue

Nervous tissue is found in the brain, spinal cord, nerves, and all of the body's organs. Its function is the point-to-point transmission of information. Messages are carried by impulses that travel along the functional unit of the nervous system, the **neuron** (or nerve cell). Similar to muscle cells, neurons do not divide in adults, so natural replacement is impossible.

1. The largest cells in Figure 33-35 are *motor neurons*, which connect the spinal cord to muscle fibers or glands. Find a similar cell in your smear of an ox spinal cord. The motor neuron has a **cell body** and a number of slender cytoplasmic extensions called *neuron processes*, including one long **axon** and several shorter **dendrites**. The dendrites and cell body are stimulated within the spinal cord, and the axon conducts

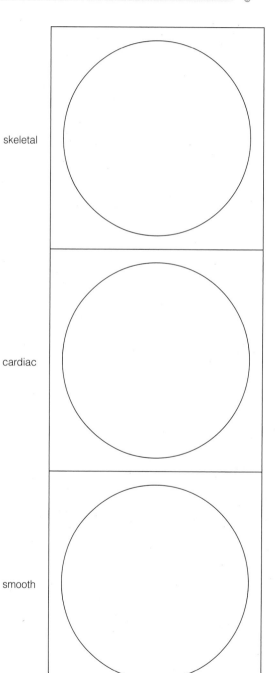

skeletal

cardiac

smooth

Figure 33-34 Drawings of muscle tissues (_____×).

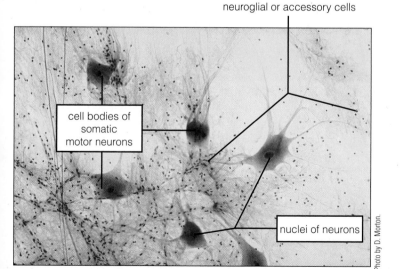

smaller nuclei of numerous neuroglial or accessory cells

cell bodies of somatic motor neurons

nuclei of neurons

Photo by D. Morton.

Figure 33-35 Smear of spinal cord, showing nervous tissue (100×).

impulses out of it. The cells with smaller nuclei are accessory cells called **neuroglia**. Accessory cells help neurons function and make up about half the mass of nervous tissue. In Figure 33-36, draw a smear of nervous tissue.

Figure 33-36 Drawing of nervous tissue (_____ ×).

33.2   Systems

There are 11 systems in mammals, and each one contains a number of organs. In this portion of the exercise, you will identify the vital functions of these systems and many of their constituent organs. Many of these organs are located in or near the major **body cavities**. The *ventral* (located toward the belly surface) and *dorsal* (located toward the back surface) body cavities of humans, which are typical of mammals, are shown in Figure 33-37. A muscular *diaphragm* separates the **thoracic cavity** and **abdominal** cavities. The thoracic cavity is further subdivided into two lateral (away from the midline) **pleural cavities** and a medial (at or near the midline) **pericardial cavity**. The abdominal cavity is continuous with a lower pelvic cavity, and together they are alternatively referred to as the *abdominopelvic cavity*. The continuous **cranial** and **spinal cavities** make up the dorsal body cavity. Table 33-3 summarizes the layout of the body cavities.

MATERIALS

Per student:

- textbook
- colored pencils—green, yellow, black, red, brown, pink, and blue

Per lab room:

- labeled demonstration dissection of a mouse (optional)
- demonstration dissection of a sheep brain

PROCEDURE

1. Using your text, list the vital functions of each system in Table 33-4.

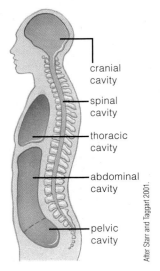

Figure 33-37 Major body cavities of humans.

After Starr and Taggart 2001.

TABLE 33-3	Layout of Mammalian Body Cavities		
Cavity			**Contains**
Ventral	Thoracic	Pleural (right and left)	Lungs
		Pericardial	Heart
	Abdominopelvic	Abdominal	Stomach, small intestine, liver, and so on
		Pelvic	Urinary bladder, and so on
Dorsal	Cranial		Brain
	Spinal		Spinal cord

© Cengage Learning 2013

TABLE 33-4	The Organ Systems of Mammals	
Systems	**Vital Functions**	
Integumentary		
Nervous		
Endocrine		
Skeletal		
Muscular		
Circulatory		
Lymphatic		
Respiratory		
Digestive		
Urinary		
Reproductive		

© Cengage Learning 2013

2. Look at the photos of dissected mice (Figures 33-3, 33-6, and 33-10), which show the organs of the ventral body cavities. In addition, your instructor may have prepared a labeled demonstration dissection of a mouse. Use your text to identify the systems the labeled organs belong to, underlining each name with a colored pencil: green for the respiratory system, yellow for the digestive system, black for the lymphatic system, red for the circulatory system, brown for the urinary system, pink for the female reproductive system, and blue for the male reproductive system. Use Figure 33-38 to see the urinary and reproductive systems of a dissected female mouse.
3. The organs of the nervous system are located in the dorsal body cavities—the brain in the cranial cavity and the spinal cord in the spinal cavity. Examine the demonstration dissection of a sheep brain and identify the structures labeled in Figure 33-39.

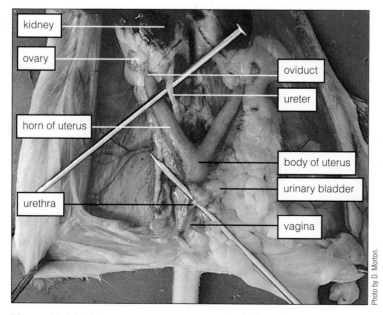

Figure 33-38 Ventral view of the urinary and reproductive systems of a female mouse (3×).

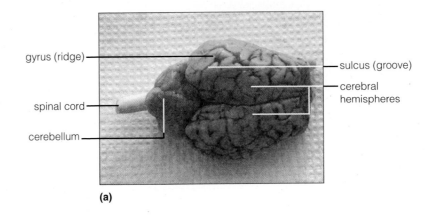

gyrus (ridge)

spinal cord

cerebellum

sulcus (groove)

cerebral hemispheres

(a)

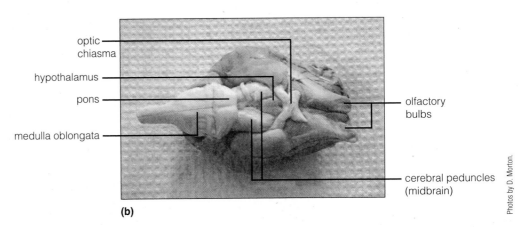

optic chiasma

hypothalamus

pons

medulla oblongata

olfactory bulbs

cerebral peduncles (midbrain)

(b)

Photos by D. Morton.

Figure 33-39   Sheep brain: dorsal (**a**) and ventral (**b**) views.

____ 1. Histology is the study of
(a) cells.
(b) organelles.
(c) tissues.
(d) organisms.

____ 2. A collection of similarly specialized cells and any extracellular material they secrete and maintain describes
(a) an organ.
(b) a system.
(c) a tissue.
(d) organelles.

____ 3. Organs strung together functionally and usually structurally form
(a) organs.
(b) systems.
(c) tissues.
(d) organelles.

____ 4. _____ are constructed of all four basic tissue types.
(a) Organs
(b) Systems
(c) Cells
(d) Organelles

____ 5. An epithelial tissue formed by more than one layer of cells and with columnlike cells at the surface is called
(a) simple squamous.
(b) stratified squamous.
(c) simple columnar.
(d) stratified columnar.

____ 6. The middle layer of the skin
(a) is called the dermis.
(b) is primarily connective tissue.
(c) contains collagen fibers.
(d) is all of the above.

____ 7. To which subtype of muscle tissue does a fiber with cross-striations and many peripherally located nuclei belong?
(a) skeletal
(b) cardiac
(c) smooth
(d) none of the above

____ 8. In which tissue would you look for cells that function in point-to-point communication?
(a) connective
(b) epithelial
(c) muscle
(d) nervous

____ 9. The ventral body cavities include
(a) the thoracic cavity.
(b) the cranial cavity.
(c) the abdominopelvic cavity.
(d) both a and c

____ 10. The cranial cavity contains
(a) the lungs and heart.
(b) the spinal cord.
(c) the brain.
(d) both b and c

EXERCISE 33

Animal Organization

Post-Lab Questions

Introduction

1. In the correct order from smallest to largest, list the levels of organization present in most animals.

33.1 Tissues and Organs

2. Describe the main structural characteristics of the four basic tissues.

   (a) Epithelial tissue

   (b) Connective tissue

   (c) Muscle tissue

   (d) Nervous tissue

3. Describe the main functions of the four basic tissues.

   (a) Epithelial tissue

   (b) Connective tissue

   (c) Muscle tissue

   (d) Nervous tissue

4. Identify the following tissues.

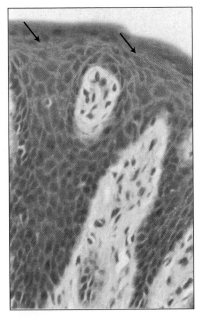

(186×)

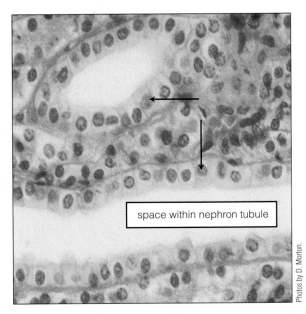

space within nephron tubule

Photos by D. Morton.

(480×)

A. _____

B. _____

5. Choose any organ. Describe how its construction allows for its specific function.

6. Briefly describe the ventral body cavities and the major organs they contain.

7. Briefly describe the dorsal body cavities and the major organs they contain.

8. Identify the following structures:

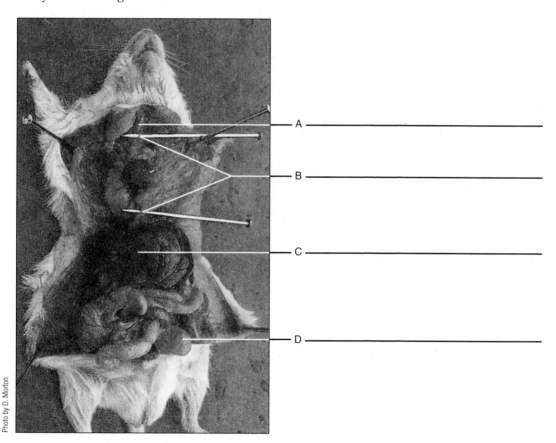

A —————————————————

B —————————————————

C —————————————————

D —————————————————

Photo by D. Morton.

Food for Thought

9. Bone is a subtype of connective tissue, and bones are organs. How are the two related yet different from each other? Can you think of another tissue/organ pair that potentially creates a similar confusing situation?

10. Search for the phrases "tissue engineering" and "organ engineering" on the Internet. List two websites and briefly summarize their content.

    http:// _____

    _____

    _____

    http:// _____

    _____

    _____

Dissection of the Fetal Pig: Introduction, External Anatomy, and the Muscular System

OBJECTIVES

After completing this exercise, you will be able to

1. define *fetus, digitigrade locomotion, plantigrade locomotion, biped, antagonistic muscles;*
2. locate and describe the external features of a fetal pig;
3. determine the sex of a fetal pig;
4. describe the function of the umbilical cord;
5. define the origin, insertion, and action of a skeletal muscle;
6. identify some of the major skeletal muscles and muscle groups of a mammal.

Introduction

In this and the following two exercises, you will examine in some detail the external and internal anatomy of an embalmed fetal (from **fetus,** an unborn mammal) pig. Many aspects of fetal structural and functional organization are identical with those of other mammals, including humans. Thus, a study of the fetal pig is in a very real sense a study of ourselves.

The specimens you will use in the next three exercises were purchased from a biological supply house, which obtained them from a plant where pregnant sows are slaughtered for food. On average, a sow produces 7–12 offspring per litter. The period of development in the uterus (*gestation period*) is approximately 112–115 days. Generally, lab specimens are approximately 20–30 cm (8–12 in.) long, and their age is between 100 days and nearly full term.

The arterial and venous systems are injected under pressure with latex or a rubberlike compound. Red latex is injected into the arteries through the umbilical artery. Then an incision is made in the throat and blue latex is injected into the jugular and other veins.

While dissecting the fetal pig, keep several points in mind. First, be aware that *to dissect* does not mean "to cut up," but rather primarily "to expose to view." Thus, proceed carefully and never cut or move more than is necessary to expose a given part. Second, for each structure or organ that you identify in the pig, ask yourself if an equivalent one is present in your body. If so, where is it located, and is its function similar to that in the fetal pig? Finally, pay particular attention to the spatial relationships of organs and other structures as you expose them. Carefully identify each structure and determine its organ system, its relationship to that organ system, and its general function. Then determine how it relates to the other organ systems in the body. By proceeding in this manner, you will greatly enhance your understanding of the structure and function of the mammalian body.

To understand the dissection directions, you need to be familiar with the terms used in virtually all anatomical work (Appendix 3). Spend a few minutes relating each of these terms to the fetal pig body and to your body as well. Figure 34-1 will aid you in this activity.

CAUTION

Specimens are kept in preservative solutions. Use lab gloves and safety goggles, especially if you wear contact lenses, whenever you or your lab mates handle a specimen or other fluids. If safety goggles do not fit, your glasses should suffice. Wash any part of your body exposed to preservative with copious amounts of water. If preservative solution or other fluids are splashed into your eyes, wash them with the safety eyewash bottle or eye irrigation apparatus for 15 minutes.

Before you begin your examination of the internal structure of the fetal pig, you will examine the external features of its body. This gives you an opportunity to compare the body of the pig with other mammals. Remember, it is the external surface of an organism that has the greatest amount of contact with the environment. Thus, the greatest differences between two organisms may be their external rather than their internal features.

MATERIALS

Per student pair:

- one preserved fetal pig injected with red and blue latex
- plastic bag to store fetal pig
- dissection pan

Per lab room:

- liquid waste disposal bottle
- box of nametags (tags may be provided with fetal pigs)
- permanent marking pens or pencils
- boxes of different sizes of lab gloves
- box of safety goggles

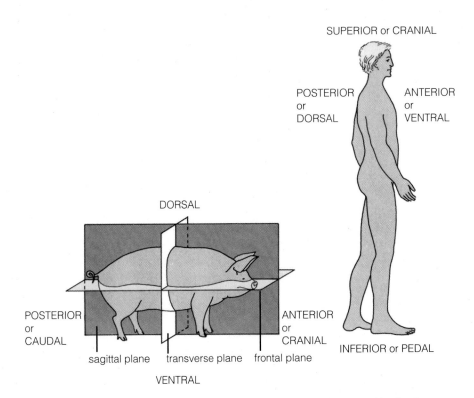

Figure 34-1　Lateral view of pig and human with terms of direction and body planes.

PROCEDURE

A. Preparing the Fetal Pig for Examination

1. The fluid used to preserve fetal pigs may be irritating to hands and eyes. Therefore, wear protective gloves during dissection and put on protective goggles.
2. Obtain a fetal pig and place it on a dissection pan lined with paper towels. If the plastic bag supplied with your pig contains any excess preservative, pour the liquid into the waste bottle provided by your instructor, and save the bag. As you will be using the same fetal pig for several days, you should place it in the plastic bag at the end of each day's exercise so it doesn't dry out.
3. For easy identification, tie a nametag to a hindleg, the bag, or both, *according to the instructor's wishes.* Use a marking pen or pencil to fill in the tag.

B. Body Regions and Their Features

1. Identify the four regions of the fetal pig body: the large, compact **head;** the **neck;** the **trunk** with four **limbs** (or appendages); and the *tail* (Figure 34-2).

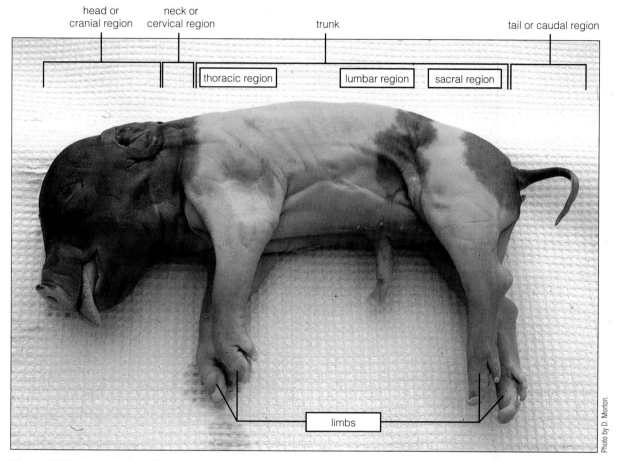

Figure 34-2 Lateral view of a fetal pig with the four major body regions indicated.

2. Examine the head in more detail (Figure 34-3) and identify the **eyes** with **upper** and **lower lids**, the **external ears**, the **mouth**, and the characteristic **nose** or for pigs, *snout*. Note the position of the **nostrils** or *external nares* on the snout. Feel the texture of the snout. It is composed of bone, cartilage, and other tough connective tissue and as such allows the pig to root and push soil and debris in its search for food.

3. Open the pig's mouth and note the **tongue** with its covering of **papillae**, which contain *taste buds*. Papillae are especially concentrated and prominent along the posterior edges and tip of the tongue. Also notice if any *baby teeth* are present. Like humans, pigs are omnivores; that is, they eat both animal and plant matter.

4. Note that the trunk is divided further into the **thoracic region**, **lumbar region**, and **sacral region**. These regions, along with the cervical region in the neck, also describe the corresponding regions of the vertebral column or spine.

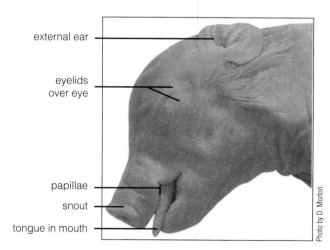

Figure 34-3 External features of the fetal pig head.

5. Place the pig on its back (dorsal surface) and examine its *abdomen* (belly). The most prominent feature of the underside (ventral surface) of the fetal pig is the **umbilical cord** seen near the posterior end of the abdomen (see Figure 34-4).

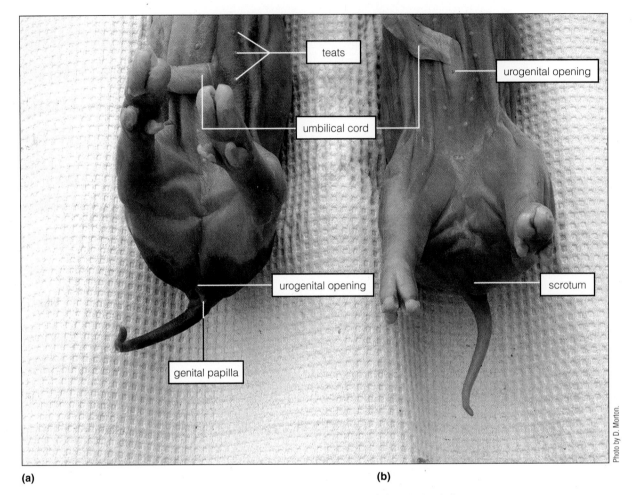

Figure 34-4   Ventral view of lower half of the body of (**a**) female and (**b**) male fetal pigs.

Is an umbilical cord present in the adult pig or human? _____ (yes or no)

During its development, the fetus was connected to the placenta on the uterine wall of its mother's reproductive system by way of the umbilical cord. The cord contains two *arteries* (red latex), a large *vein* (blue latex), and a fourth vessel, usually collapsed, the *allantoic duct*. The blood in the umbilical vein carries nutrients and oxygen from the mother to the fetus, and blood in the umbilical arteries carries waste materials and carbon dioxide from the fetus to the mother.

6. Note on the ventral surface of the pig the pairs of **nipples** or *teats* (Figure 34-4). Both male and female pigs may have from five to eight pairs of these structures situated in two parallel rows on the **thoracic region** (chest) and **abdominal region** of the body. Finally, locate the **anus**, the posterior opening of the digestive tract. The anus is situated immediately under (ventral to) the tail.

7. Locate and identify the following: the **wrist, lower forelimb, elbow, upper forelimb, shoulder, ankle, shank, knee, thigh,** and **hip** (Figure 34-5).

8. Examine the toes. The first *toe*, or *digit*, which corresponds to your big toe or thumb, is absent in both forelimbs and hindlimbs of the pig. Furthermore, the second and fifth digits are reduced in size, and the middle two digits, the third and fourth, are flattened or *hoofed*.

9. Pigs and other hoofed animals walk with the weight of their body borne on the tips of the digits. This type of walking is referred to as **digitigrade locomotion**. By contrast, humans use the entire foot for walking and have **plantigrade locomotion**. Compare the structure of your hands and feet with the foot of a pig.

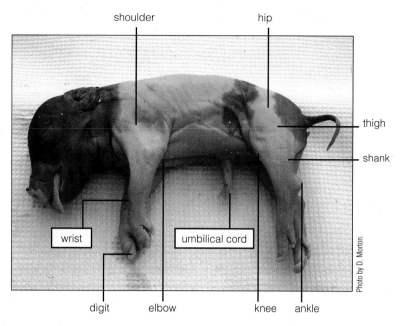

shoulder   hip

thigh

shank

wrist

umbilical cord

digit   elbow   knee   ankle

Photo by D. Morton.

Figure 34-5 Lateral view of external features of the fetal pig.

C. Determining the Sex of Your Fetal Pig

1. Find the *urogenital opening* immediately ventral to the anus in females (Figure 34-4a) and just behind the umbilical cord in males (Figure 34-4b). A small fleshy *genital papilla* projects from the urogenital opening of female fetal pigs. All fetal pigs have a common urogenital opening shared by the urinary system and the genital (reproductive) system. This situation persists in adult male pigs and humans. In adult female pigs and humans, however, there are separate openings to the urinary and reproductive systems.

2. In your or another group's male fetal pig, find a swelling on the posterior portion of the abdomen between the upper ends of the hindlimbs. The swelling is the **scrotum**, which contains the sperm producing *testes*, a pair of small, oval structures that are part of the male reproductive system. These are generally easy to locate in older fetuses because during late development they descend into the scrotum. Identify the *penis*, a large, tubular structure immediately under the skin just posterior to the urogenital opening.

34.2   Muscular System *(About 90 min.)*

The contractions of *skeletal muscle organs* enable the body and its parts to move. Through the contractions of skeletal muscles, you maintain your posture and can thus wink your eye, wave at a friend, or tap your foot in time to the beat of music. In this section, you will examine the structure of skeletal muscles and learn to identify some of them and describe their functions. The movement or movements produced by a particular skeletal muscle is called its **action**.

Skeletal muscles are attached to the various bones of the skeleton by tough strips of fibrous connective tissue called *tendons*. The three parts of a typical muscle are the **origin** (the end attached to the less mobile portion of the skeleton), the **insertion** (the end attached to the portion of the skeleton that likely moves when the muscle contracts), and the **belly** (the middle portion between the points of attachment).

Realize that the attachments of the muscles can only be understood with reference to the skeleton and its movable joints. Although we won't examine the skeleton in detail, it will be necessary to refer to its various parts (Figure 34-6). The skeletal muscles and their attachments to the skeleton of pigs and humans are remarkably similar. Most differences relate to how pigs and humans walk, humans being **bipeds** (walk on the hindlimbs only).

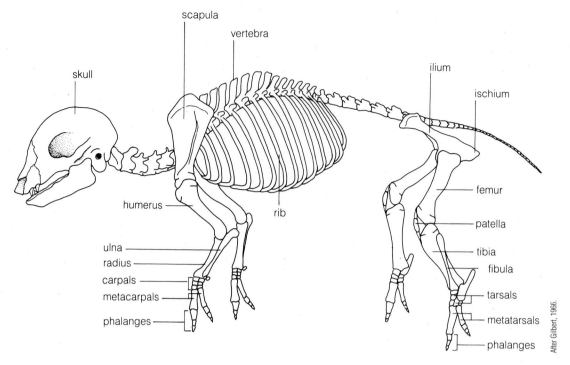

Figure 34-6 Skeleton of the fetal pig. The hyoid bone, located in the upper neck, and the sternum, to which many of the ribs are attached midventrally, are not included in this illustration.

A skeletal muscle that moves an appendage one way usually has one or more opposing muscles that move it in the opposite direction. Muscles with such opposite actions are called **antagonistic muscles**. For example, the biceps brachii is responsible for bending (flexing) the lower forelimb (or foreleg in the pig), at the elbow and the triceps brachii (its antagonist) brings about straightening or extending the forelimb. The forearm can be held in any position by the simultaneous contraction of both of these opposing muscles.

MATERIALS

Per student pair:

- one preserved fetal pig injected with red and blue latex
- plastic bag to store fetal pig
- dissection pan
- one dissecting kit including the following: scalpel, blunt probe, dissecting needle, scissors, and forceps
- two pieces of string 60 cm long or four large rubber bands

Per lab room:

- boxes of different sizes of lab gloves
- box of safety goggles

PROCEDURE

A. Preparing the Fetal Pig for Dissection of the Skeletal Muscles

1. Place your specimen on its dorsal side in a dissection pan lined with paper towels.
2. Tie one end of a string to the left forelimb at the wrist, pass the string underneath the dissection pan, then tie the other end of the string to the right wrist so that the legs are spread apart under tension. Repeat with the hindlimbs, using the second piece of string. If rubber bands are available, tie two rubber bands together to make them longer. Then loop one end of the band around the right forelimb close to the foot. Bring the rubber band under the dissecting pan and loop it around the other forelimb to anchor the feet securely. Repeat this procedure with another set of rubber bands and anchor the hindlimbs to the pan.
3. For the following dissection, refer to Figure 34-7. The numbers in the drawing refer to the order in which the incisions are made in the dissection.

Using a scalpel, make an incision on the ventral side of the pig at the base of the neck (1). *Make sure you cut only through the skin and not the underlying tissues.* With your scissors, continue the incision posteriorly to the umbilical cord, then around the cord on the left side (the pig's left side) to the region between the hindlegs. From this point, carry on the cut along the medial surface of the left hindleg toward the foot (2) and around the pig's ankle (3). Make a similar cut from the midventral incision in the thorax down the medial surface of the left forelimb (4) and around the pig's wrist (5).

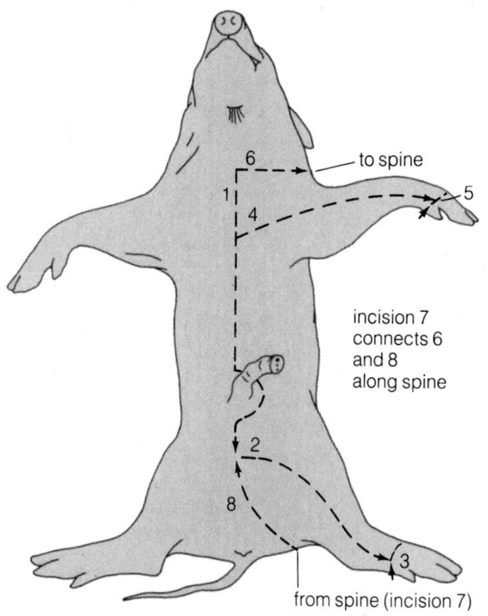

Figure 34-7 Cuts needed to expose muscles.

4. Return to the original incision on the ventral surface of the neck and extend it around the left side of the pig's neck dorsally toward the spine (6). To do this, you'll need to remove the string or rubber bands and place the pig on its right side in the dissecting pan. When you reach the spine, continue the incision posteriorly along the spine to the base of the tail (7). Finally, connect the posterior ends of incisions 7 and 1 (8). After completing the incisions, place the pig on its back and secure the legs with string or rubber bands as you did earlier.

5. To skin your specimen, grasp the cut edge of the skin at the base of the throat and begin easing the skin loose from the underlying tissues. Use your blunt probe between the skin and underlying connective tissues, working slowly until you have removed the skin from the ventral portion of the trunk and the limbs. Then, turn your pig on its right side and remove the skin along the lateral (left side) and dorsal surfaces to the vertebral region (skin over the spine).

6. In the region of the neck, shoulder, and trunk, you may notice a layer of light brown muscle fibers adhering to the skin. These fibers comprise the *cutaneous maximus* muscle, which is responsible for the twitching of the skin that rids the pig of insects and other irritants. Humans don't have this layer of muscle.

7. After your specimen is skinned, the muscles will not appear as clearly defined as in the illustrations. This is because adipose tissue (fat) and a layer of relatively loose connective tissue (*superficial fascia*) cover a denser, tougher layer of *deep fascia*. This deep fascia connects one muscle to another and maintains them in their proper position relative to one another. It will be necessary to break through the deep fascia as you proceed. Remember, however, that you are working with a fetus and as such its structures are immature and can be easily torn. Therefore, proceed with care.

8. As you attempt to identify the various muscles, you may find that the boundary of a given muscle is easily seen, whereas in others it seems to blend with those around it. In order to define the limits of a muscle, use a blunt probe to tease away the overlying adipose and connective tissues until you can see the direction of the muscle fibers. Look for changes in the direction of the muscle fibers and attempt to slip the blunt probe or flat edge of your scalpel handle between the two separate layers at this point. If the two layers separate readily from one another, you have located the boundary between two different muscles. Now proceed with the exercise and attempt to identify some of the major muscles in the fetal pig as *directed by your instructor*. Obviously, not all the skeletal muscles have been included, but only those that are relatively easy to identify and that will illustrate the general principles of skeletal muscular action.

B. Muscles of the Shoulder and Back

1. The **latissimus dorsi** (Figure 34-8) is a broad muscle wrapped around the sides of the thoracic region and chest. Carefully pick away the adipose tissue and loose connective tissue from the sides of the chest until you can see its fibers. The origin of this muscle is the **lumbodorsal fascia** and nearby bones (Figure 34-8). It inserts by a tendon into the medial side of the proximal end of the humerus (the major bone of the upper forelimb). The action of the latissimus dorsi is to move the upper forelimb dorsally and posteriorly (or extend it at the shoulder joint).

2. The **trapezius** (Figure 34-8) is a broad muscle anterior to the latissimus dorsi. Its origin is from the base of the skull to the tenth thoracic vertebrae, and it inserts on the spine of the scapula (the shoulder blade). Its action is to draw or pull the shoulder medially (or elevates it).

3. The **deltoid** (Figure 34-8) is a relatively broad muscle that covers the shoulder region. It originates from the scapula and inserts on the humerus. Its action raises and pulls the upper forelimb anteriorly (or abducts and flexes it at the shoulder).

4. If you have not done so, carefully remove the cutaneous muscle and jellylike connective tissues covering the side of the neck. You should now be able to observe a broad, flat strip of muscle, the **brachiocephalic** (Figures 34-8 and 34-9), running from its origin on the back of the skull to its insertion on the anterior surface of the distal end of the humerus. Its action pulls the upper forelimb anteriorly (or flexes it at the shoulder).

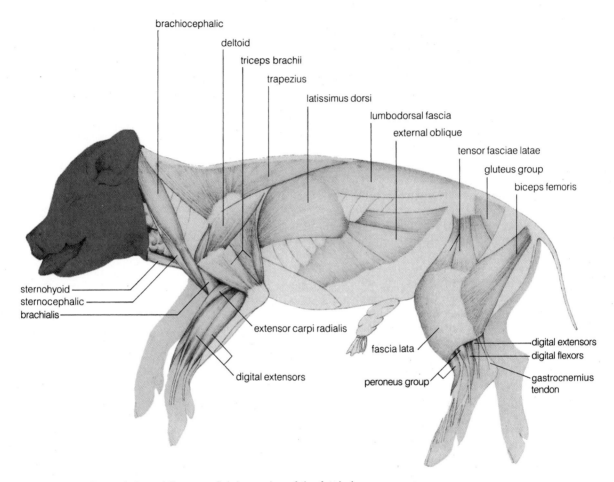

Figure 34-8 Lateral view of the superficial muscles of the fetal pig.

C. Muscles of the Forelimb

1. First find the **triceps brachii** (Figures 34-8 and 34-9), a large muscle that virtually covers the entire outer and posterior surface of the upper forelimb. It originates from the scapula and proximal third of the humerus and inserts on the proximal end of the ulna (one of the two bones of the lower forelimb). Its action is to extend the lower forelimb at the elbow.
2. To locate the **biceps brachii** (Figure 34-9), untie your pig, place it on its dorsal side, and again secure the legs with string or rubber bands. Look for a rather small, spindle-shaped muscle covering the anterior surface of the humerus. This muscle originates on the scapula and inserts on the radius (the other bone of the lower forelimb) and ulna. Its action is to flex the lower forelimb and to act antagonistically with the triceps. In order to see it clearly, you'll need to cut through the overlying muscle (the **brachialis**; see Figure 34-9) and pull the cut edges back. The brachialis is a small muscle that also flexes the elbow. Its origin is the proximal humerus, and it inserts on the ulna.
3. There are numerous other muscles in the lower forelimb, most of which are concerned with extending or flexing the foot and digits (see Figures 34-8 and 34-9).

D. Muscles of the Throat and Chest

1. Now locate the **sternocephalic** muscle (Figures 34-8 and 34-9), a long muscle lying ventral to the brachiocephalic. Its origin is the sternum, and it inserts into the lateral portion of the skull just behind the external ear known as the mastoid process. Its action draws the snout toward the chest (or flexes the head).

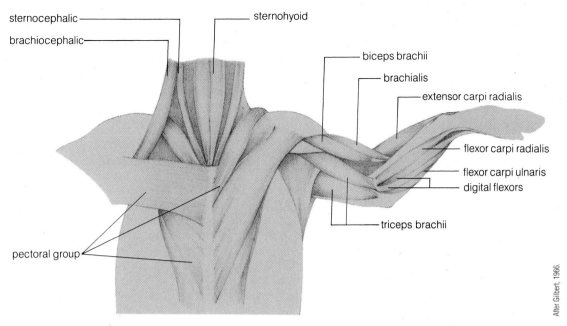

Figure 34-9 Muscles of the ventral thoracic region and forelimb. Some of the superficial muscles of the pig's left side have been removed.

2. Next find the **sternohyoid** muscle (Figures 34-8 and 34-9), which consists of two long flat strips running from the sternum (origin) to the hyoid (insertion). When this muscle contracts, it retracts and depresses the base of the tongue and the larynx, as, for example, when swallowing.

3. Note the large **pectoral group** of muscles (Figure 34-9) that originate on the ventral side of the sternum and insert on the humerus. The pectoral group moves the upper forelimb medially toward the chest (or adducts it at the shoulder). If you wish, you may carefully cut the belly of the superficial pectoral and bend it back to examine the underlying muscles more closely.

E. Muscles of the Abdominal Region

The major lateral abdominal muscles consist of the outer **external oblique**, the **internal oblique** (the midlayer), and the **transversus** (the inner layer). To locate these muscles, turn your specimen on its side. Now find the fibers of the external oblique (Figures 34-8 and 34-10) and observe that they run diagonally, so that their ventral ends are posterior to their dorsal ends. Carefully cut a "window" approximately 1 cm square in the external

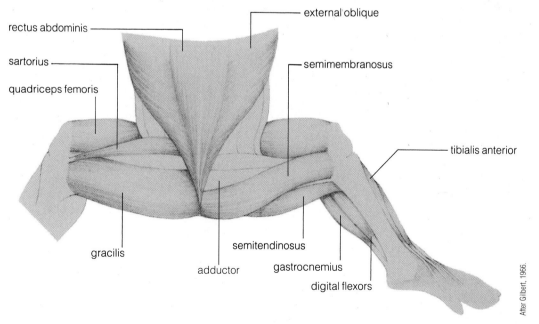

Figure 34-10 Ventral view of muscles of the abdominal region and hindlimb. Some superficial muscles on the pig's left side have been removed.

Dissection of the Fetal Pig: Introduction, External Anatomy, and the Muscular System **539**

oblique to expose a portion of the internal oblique. (Remember that the muscles of the fetus are extremely thin. Take care not to cut into the body cavity at this time, as you will release a large amount of fluid.) Notice that the fibers of the internal oblique run at nearly right angles to the direction of those of the external oblique. Now, using the same careful technique as previously, remove a small portion of the internal oblique and attempt to reveal the fibers of the transversus, which as the name suggests run transversely across the abdomen. Collectively, the action of these abdominal muscles, together with the **rectus abdominus** (Figure 34-10), is to flex the trunk and compress the abdominal organs (for example, to aid expiration or defecation).

F. Muscles of the Hip and Thigh

1. Begin your dissection of the muscles of the hip and thigh by locating the **biceps femoris** (Figure 34-8). This conspicuous superficial muscle covers most of the caudal half of the lateral surface of the thigh. It originates from the ischium of the pelvic girdle and inserts on the lateral fascia attached to the femur and tibia near the knee. Its action pulls the thigh posteriorly and laterally (or extends and abducts it at the hip).
2. Next, locate the **tensor fasciae latae** (Figure 34-8), the most anterior of the thigh muscles. Its origin is the ilium of the pelvic girdle and this short, triangular-shaped muscle continues down the anterior and lateral surface of the thigh as a sheet of connective tissue, the *fascia lata*, which attaches to the patellar ligament. Its actions are to tense the fascia lata, draw the thigh anteriorly (or flex it at the hip), and extend the shank at the knee.
3. Now carefully free the tensor fasciae latae at its insertion, but leave the origin and its medial portion intact. Cut the biceps femoris at its insertion near the tibia, and peel it back to expose the deeper muscles of the thigh. Identify the *vastus lateralis*, the most lateral of the four quadriceps femoris muscles, and the **semitendinosus** (Figures 34-10 and 34-11). Carefully expose the lateral portion of the **semimembranosus**, which is situated under the semitendinosus (Figure 34-11). The biceps femoris, semitendinosus, and semimembranosus are collectively called the **hamstring muscles**. In general, the hamstrings originate on the ischium of the pelvic girdle and insert around the back of the knee. Their collective actions are to pull the thigh posteriorly (or extend it at the hip) and flex the shank at the knee.
4. To locate other muscles of the thigh, place your pig on its dorsal side and tie back the legs with string or rubber bands. Referring to Figures 34-10 and 34-11, locate the *vastus medialis*—the most medial of the quadriceps femoris muscles, the **sartorius,** the **gracilis,** and the medial side of the **semimembranosus.** The sartorius and gracilis originate from the bones of the pelvic girdle and insert on the medial side of the tibia. These muscles draw the thigh medially (or adduct it at the hip) and the sartorius helps pull the thigh anteriorly (or flex it at the hip).

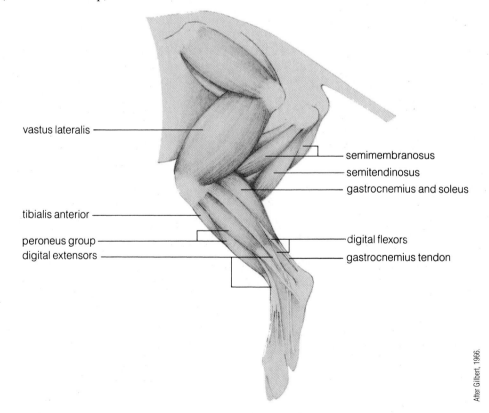

Figure 34-11   Lateral view of the muscles of the hindlimb.

5. Separate the anterior edge of the vastus medialis and find between it and the vastus lateralis muscle a third quadriceps femoris muscle, the *rectus femoris*. A much-reduced fourth quadriceps femoris muscle, the *vastus intermedius*, lies beneath the rectus femoris. Collectively, the **quadriceps femoris** muscles originate on the upper surface of the femur and on the ilium of the pelvic girdle (rectus femoris) and insert via the patellar ligament onto the anterior surface of the tibia. Their general actions are to pull the thigh anteriorly (or flex it at the hip) and extend the shank at the knee. If you wish, you may carefully cut the belly of the sartorius and gracilis and bend their bellies back to examine the underlying muscles more closely.

G. Muscles of the Hindlimb

1. The **gastrocnemius** (Figures 34-10 and 34-11) and underlying *soleus* (Figure 34-11) originate, respectively, from the distal end of the femur and the head of the fibula. The Achilles tendon inserts them both to the calcaneus (heel bone). These muscles extend the foot at the ankle, and the gastrocnemius helps flex the shank at the knee.
2. Other muscles of the lower hindlimb flex the foot (*tibialis anterior*, Figures 34-10 and 34-11; *peroneus group*, Figures 34-8 and 34-11), while others flex and extend the digits (*digital flexors*, Figures 34-8 and 34-11; *digital extensors*, Figures 34-8 and 34-11). Spend a few minutes examining the lower hindlimb and attempt to identify several of these muscles.
3. Place your pig back in the plastic bag. Tie it shut to prevent your pig from drying out. Dispose of any paper towels that contain preservative *as directed by your instructor*. Clean the tray, dissecting tools, and the lab table.

_____ 1. A fetus is
   (a) a newborn pig.
   (b) a newborn human.
   (c) an unborn mammal.
   (d) all of the above

_____ 2. *To dissect* means primarily
   (a) to cut open.
   (b) to remove all internal organs.
   (c) to expose to view.
   (d) all of the above

_____ 3. When the directions for a fetal pig dissection refer to the left, they are referring to
   (a) your left.
   (b) the pig's left.
   (c) the pig's right.
   (d) both a and c

_____ 4. In a fetal pig, *dorsal* and *ventral* refer respectively to
   (a) the head and tail regions of the body.
   (b) the tail and the head regions of the body.
   (c) the upper (back) portion and the lower (underside) portion of the body.
   (d) the lower (underside) portion and the upper (back) portion of the body.

_____ 5. The umbilical cord functions to
   (a) carry waste products in the blood from the fetus to the mother.
   (b) carry waste products in the blood from the mother to the fetus.
   (c) carry oxygen in the blood from the mother to the fetus.
   (d) do both a and c.

_____ 6. The female *fetal* pig is similar to the male *fetal* pig in that its body
   (a) has separate openings for the urinary system and the reproductive system.
   (b) has a common opening for the urinary system and the reproductive system.
   (c) has an opening for the urinary system but none for the reproductive system.
   (d) has an opening for the reproductive system but none for the excretory system.

_____ 7. Pigs have digitigrade locomotion because they walk on
   (a) their ankles.
   (b) the soles of their feet.
   (c) the tips of their toes, which are modified as hooves.
   (d) their hands and knees.

_____ 8. The origin of a skeletal muscle is
   (a) attached to the less mobile portion of the skeleton.
   (b) attached to the portion of the skeleton that moves when the muscle contracts.
   (c) never attached to the skeleton.
   (d) its belly.

_____ 9. The insertion of a skeletal muscle is
   (a) attached to the less mobile portion of the skeleton.
   (b) attached to the portion of the skeleton that moves when the muscle contracts.
   (c) never attached to the skeleton.
   (d) its belly.

_____ 10. The biceps brachii is responsible for flexing the forearm while the triceps brachii extends the forearm. Muscles with such opposite actions are called
   (a) cooperative.
   (b) antagonistic.
   (c) involuntary.
   (d) sensory.

EXERCISE **34**

Dissection of the Fetal Pig: Introduction, External Anatomy, and the Muscular System

Post-Lab Questions

34.1 External Anatomy of the Fetal Pig

1. Identify the external features of a fetal pig noted on the drawing.

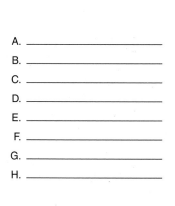

A. _____

B. _____

C. _____

D. _____

E. _____

F. _____

G. _____

H. _____

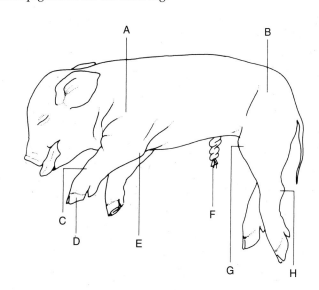

2. What is the function of the umbilical cord?

3. Briefly describe how your feet differ from those of the pig.

4. Using external features, briefly describe how you can determine the difference between a male and a female fetal pig.

34.2 Muscular System

5. Define in general terms the *origin*, *insertion*, and *action* of a skeletal muscle organ.

6. What does the phrase *antagonistic* muscles mean?

7. Complete the table.

Muscle Group	Skeletal Muscles in Group	General Actions of Group
Hamstrings	_____ _____ _____	_____ _____ _____
Quadriceps femoris	_____ _____	_____ _____

Food for Thought

8. The right and left sternocleidomastoid muscles of humans originate on the upper surfaces of breast bone and collar bones, pass along both sides of the neck, and insert just behind the ears on the mastoid processes. Describe the action that occurs if both sides contract together.

9. Explain the basic difference between digitigrade and plantigrade locomotion.

10. Search the Internet for sites that describe human body movements. List two sites and briefly summarize their contents.

http:// _____

_____

_____

http:// _____

_____

_____

Dissection of the Fetal Pig:
Digestive, Respiratory, and Circulatory Systems

After completing this exercise, you will be able to

1. define *diaphragm, thoracic cavity, abdominopelvic cavity, exocrine gland, endocrine gland, digestive tract, blood vessels—arteries, veins,* and *capillaries, pulmonary* and *systemic circuits of circulatory system, portal vein;*

2. locate the organs of the digestive, respiratory, and circulatory systems in a fetal pig;

3. describe and give the functions of the organs of the digestive, respiratory, and circulatory systems in a living mammal;

4. explain the importance of the digestive, respiratory, and circulatory systems to a living mammal;

5. trace the pathway of ingested food through the digestive tract;

6. trace the pathway of oxygen and carbon dioxide into and out of the lungs of a mammal;

7. identify the major blood vessels of a fetal pig;

8. locate, name, and describe the functions of the chambers of the heart;

9. name the internal structures of the heart.

Introduction

The ventral body cavities of your fetal pig contain most of the organs of the digestive, respiratory, and urogenital systems, as well as the heart and major vessels of the circulatory system. Remember that *dissecting* does not primarily mean "cutting up" but rather "exposing to view." Thus, proceed carefully, as the internal organs are fragile. Do not remove any structure unless directed to do so. Work closely with your partner, making sure that each organ and structure is fully identified and studied by both of you before you proceed to the next step.

CAUTION

Specimens are kept in preservative solutions. Use lab gloves and safety goggles, especially if you wear contact lenses, whenever you or your lab mates handle a specimen or other fluids. If safety goggles do not fit, your glasses should suffice. Wash any part of your body exposed to preservative with copious amounts of water. If preservative solution or other fluids are splashed into your eyes, wash them with the safety eyewash bottle or eye irrigation apparatus for 15 minutes.

35.1   Ventral Body Cavities *(10 min.)*

For the following dissection, use Figure 35-1 as a guide for making the various incisions. The numbers in the figure correspond to the numbers in the following directions.

MATERIALS

Per student pair:
- one preserved fetal pig injected with red and blue latex
- plastic bag to store fetal pig
- dissection pan
- dissecting kit
- dissecting pins

- bone shears
- two pieces of string each 60 cm long *or* four large rubber bands
- piece of string 20 cm long

Per lab room:
- liquid waste disposal bottle
- boxes of different sizes of lab gloves
- box of safety goggles

PROCEDURE

1. Place the pig, ventral side up, in the dissection pan and restrain it using string or rubber bands as you did in the previous exercise. Begin an incision at the small tuft of hair on the upper portion of the neck (1), and continue it posteriorly to approximately 1.5 cm anterior to the umbilical cord. Cut through the muscle layer but not too deeply, or you may damage the internal organs.

2. *If your pig is a male, move on to step 3.* If your pig is a *female,* make a second incision (2F) completely around the umbilical cord and continue it posteriorly for about 3 cm, stopping at a point between the hindlimbs.

3. If your fetal pig is a *male,* make the second incision (2M) as a half circle anterior to the umbilical cord and then proceed with two parallel incisions posteriorly to a region between the hindlimbs. The two incisions are necessary to avoid cutting the *penis,* which lies under the skin just posterior to the umbilical cord. The incisions made in the region of the *scrotum* should be made carefully so as not to damage the testes, which are lying just under the skin.

4. Carefully deepen incisions 1 and 2 until the body cavity is exposed. In order to make lateral flaps of the muscle tissue, which can be folded out of the way, make a third (3) and fourth (4) incision as illustrated in Figure 35-1. Now, carefully open the body cavity. If it is filled with fluid, pour the fluid into the waste container provided (not into the sink!) and carefully rinse out the cavity with a little water.

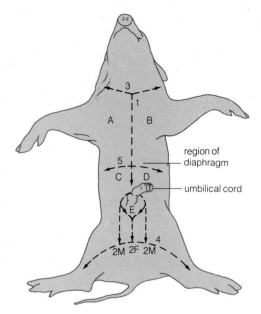

Figure 35-1 Ventral view of the fetal pig, with the position of incisions indicated. Specific incisions for male and female specimens are marked M and F, respectively.

5. Use your fingers to locate the lower margin of the *rib cage.* Just below it, make a fifth (5) incision laterally in both directions from the first incision (1). In this region is the **diaphragm,** a skeletal muscular sheet connected to the body wall and separating the two major ventral body cavities: an anterior **thoracic cavity** and a posterior **abdominopelvic cavity.** Use your scalpel to free the diaphragm from the body wall (do not remove it, however).

6. Carefully peel back flaps A, B, C, and D (see Figure 35-1) and pin them beneath your pig. It may be necessary to cut through the ventral part of the rib cage with a pair of bone shears to separate body wall flaps A and B. Do so carefully, so as not to damage the heart and lungs, which are located in the thoracic cavity.

7. To free the umbilical cord and the flesh immediately surrounding it (flap E), first locate the umbilical vein—a dark, tubular structure extending from the umbilical cord forward (anteriorly) to the liver. Then tie small pieces of string around the vein in two places (approximately 1.5 cm apart) and with your scissors cut through the vein between the pieces of string. Now pull back the umbilical cord to a position between the hindlegs and pin the flesh surrounding it to the body. The pieces of string around the umbilical vein will aid in identifying this structure during the section on the circulatory system.

8. When the body cavities are fully exposed, carefully remove any excess red or blue latex that may be present. (This occurs when some veins and arteries burst when injected with latex.) Remove large pieces with forceps. Remove your pig from the dissection pan and rinse out any smaller pieces of latex in a sink equipped to screen out debris.

9. *If you are continuing with the dissection, move on to the next section.* Otherwise, place your pig back in the plastic bag. Tie it shut to prevent your pig from drying out. Dispose of any paper towels that contain preservative *as directed by your instructor.* Clean the tray, dissecting tools, and the laboratory table.

35.2   Digestive System (*About 40 min.*)

The digestive system of a vertebrate consists, in order, of the mouth, oral cavity, pharynx, esophagus, stomach, small intestine, large intestine or colon, cecum, rectum, anal canal, and anus and associated structures and glands (salivary glands, gallbladder, liver, and pancreas). In addition to these digestive system organs, you will locate and identify the thymus and thyroid, two endocrine system glands. The digestive system digests the complex molecules in food, absorbs the useful end products of digestion along with water and most everything else that the body adds to digesting food, and processes indigestible remnants for defecation.

MATERIALS

Per student pair:

- one preserved fetal pig injected with red and blue latex
- plastic bag to store fetal pig
- dissection pan
- dissecting kit
- dissecting pins
- bone shears
- two pieces of string each 60 cm long *or* four large rubber bands
- dissection microscope
- meter stick (optional)

Per lab room:

- liquid waste disposal bottle
- boxes of different sizes of lab gloves
- box of safety goggles

PROCEDURE

A. Mouth

1. Place your pig in the dissection pan. Observe the area between the **lips** and **gums**; this is called the **vestibule**. The larger area behind the gums is referred to as the **oral cavity**.

2. Carefully cut through the corners of the mouth and back toward the ears with bone shears until the lower jaw can be dropped and the oral cavity exposed (Figure 35-2).

3. If teeth are present, carefully extract a **tooth** and examine it. A tooth consists of the *crown*, the *neck* (surrounded by the gum), and the *root* (embedded in the jawbone). If your specimen does not have exposed teeth, cut into the gums and look for developing teeth.

4. Feel the roof of the oral cavity and determine the position of the **hard palate** and **soft palate** (Figure 35-2). What is the difference between the two regions?

   _____

   _____

   _____

   _____

   _____

5. Posterior to the soft palate is the **pharynx** (Figure 35-2). In humans the portion of the pharynx just posterior to the oral cavity is also referred to as the *throat*. Note that unlike humans, the pig does not have a fingerlike piece of tissue, the **uvula**, projecting from the posterior region of the soft palate. Confirm its presence in humans by looking into the throat of your lab partner.

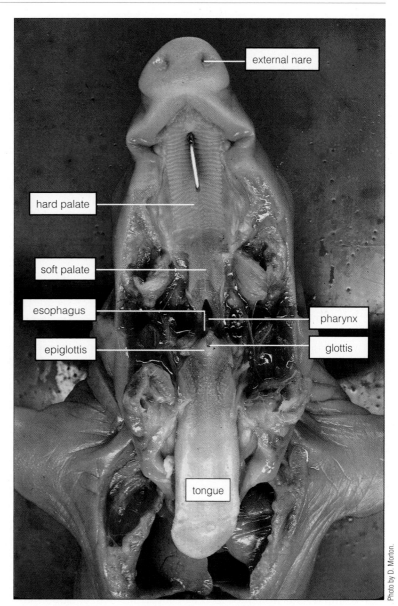

Figure 35-2   Structures of the oral cavity and pharynx.

Dissection of the Fetal Pig: Digestive, Respiratory, and Circulatory Systems   **547**

6. Carefully close the pig's jaws. Now, in the neck locate the **trachea** (Figure 35-4), a tube that is kept open throughout its length by a series of cartilaginous supports. Although the trachea is actually a part of the respiratory system, its identification aids in finding the **esophagus** (Figure 35-4), which lies behind it on its dorsal surface. Carefully slit the esophagus and insert a blunt probe into it and run it back toward the mouth. Open the mouth and note where the probe emerges. This is the opening of the esophagus (Figure 35-2).
7. Continue your study of the pharynx by locating the opening to the *larynx*, the **glottis**. It can be identified by the presence of a small white cartilaginous flap, the **epiglottis**, on its ventral surface (Figure 35-2). The epiglottis covers the glottis when a mammal swallows.

B. Salivary Glands

1. Place your pig on its right side and, proceeding from the base of the ear, carefully cut through the skin to the corner of the eye, then ventrally toward the chin, and, finally, continue the incision posteriorly toward the forelimb. Carefully remove the skin.
2. Tease away the muscle tissue below the ear to reveal a large, relatively dark, triangular **parotid gland**. This salivary gland extends from the edge of the ear posteriorly to halfway down the neck (Figure 35-3). Note that the parotid appears to be composed of many small nodules compared to the fibrous large masseter muscle lying underneath and anterior to it.
3. Cut through the middle of the parotid gland to expose the somewhat lobed **mandibular gland** (Figure 35-3). Do not confuse this second salivary gland with the small, oval lymph nodes present in the head and neck region.
4. The third salivary gland is the **sublingual gland**. The fluids secreted by the mandibular and sublingual glands are more viscous than that secreted by the parotid. Collectively, the secretions by the three salivary glands maintain the oral cavity in a moist condition, ease the mixing and swallowing of food, and contain enzymes that begin the breakdown of starch to sugars.

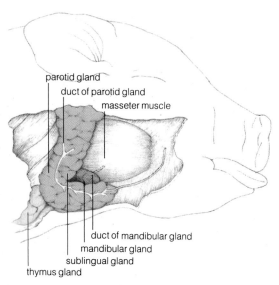

Figure 35-3 Lateral view of the fetal pig head with salivary and thymus glands exposed.

C. Thymus and Thyroid Glands

1. Work from the ventral side of your pig with the legs secured by string or rubber bands and the body wall flaps pinned to the sides of the body.
2. Identify the **thymus gland**, a whitish structure that is divided into two lobes. It extends from the neck where it covers the trachea and larynx to the upper thoracic cavity where it partially covers the anterior portion of the heart (Figures 35-3, 35-4). The thymus plays important roles in the development and maintenance of the body's immune system.
3. Immediately beneath the thymus in the neck region find the **thyroid gland**, a small reddish, oval mass with a relatively solid consistency (Figure 35-4). Thyroid hormones function in the regulation of metabolism, growth, and development.

D. Liver, Gallbladder, and Pancreas

1. Identify the brownish colored **liver** (Figure 35-4), which is the largest organ of the abdominopelvic cavity. Count and carefully determine the extent of all of its four lobes. The liver has many functions, including secreting *bile*.

The liver also plays a very important role in maintaining the stable composition of the blood. The nutrients from a digested meal are absorbed into the blood of the small intestine. This blood, which contains high concentrations of sugars like glucose and amino acids, is transported from the small intestine to the liver (via the hepatic portal vein), and there the excess glucose is converted to glycogen for storage. If the liver has stored a full capacity of glycogen, it converts the glucose into fat, which is stored in other parts of the body. The liver also removes excess amino acids from the blood by converting them to carbohydrates and fats. During this process, an amino group ($-NH_2$) is removed from the amino acid and converted into ammonia (NH_3). Ammonia is a very toxic substance. The liver combines it with carbon dioxide to form less toxic urea, which is then eliminated from the body in urine.

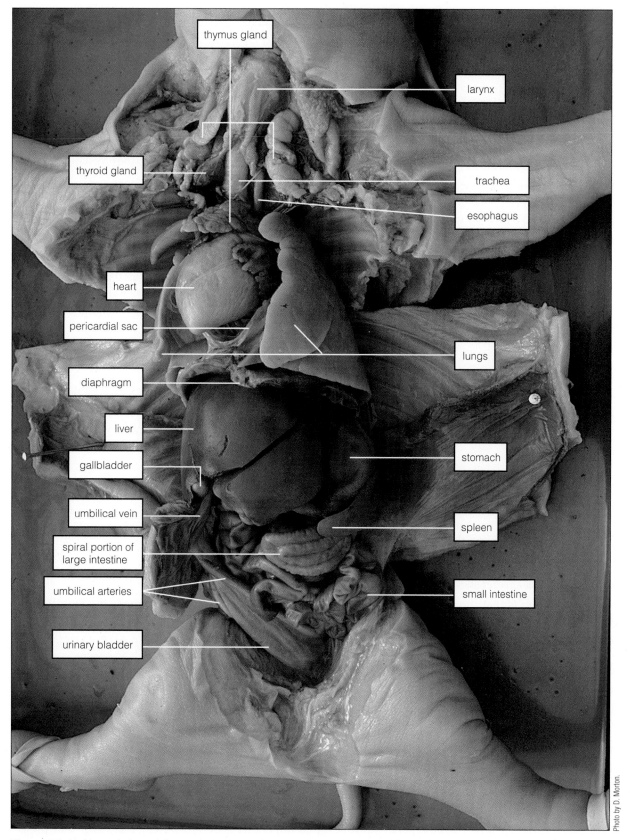

thymus gland

larynx

thyroid gland

trachea

esophagus

heart

pericardial sac

lungs

diaphragm

liver

stomach

gallbladder

umbilical vein

spiral portion of large intestine

spleen

umbilical arteries

small intestine

urinary bladder

Photo by D. Morton.

Figure 35-4 Ventral view of the general internal anatomy of the fetal pig.

2. Lift the right central lobe of the liver to expose the **gallbladder**. This saclike organ stores the bile secreted by the liver. The *cystic duct* from the gallbladder unites with the *hepatic duct* from the liver to form the common *bile duct* which empties into the first portion of the *small intestine* (Figure 35-4). *If your instructor wishes, attempt to locate the hepatic and common bile ducts and trace them from the liver to the small intestine. Be careful not to injure the hepatic portal vein*, which parallels these ducts.

3. Carefully move the small intestine and locate the **pancreas**, an elongated globular mass lying between the *stomach* and small intestine. The *pancreatic duct* carries digestive enzymes and other substances produced by the pancreas to the duodenum. (Do not attempt to find the pancreatic duct, however, as it is too small to be dissected satisfactorily.) The pancreas is both an **exocrine gland** (whose secretions are released into a duct) and an **endocrine gland** (whose hormones are released into the blood). The endocrine portion of the pancreas secretes insulin and other hormones involved with controlling the levels of glucose in the blood of mammals.

E. Stomach, Small Intestine, Large Intestine or Colon, Rectum, and Anus

1. Earlier in this section, you made a small slit in the esophagus. Return to this incision and insert a blunt probe through the slit in the esophagus, only this time posteriorly until its tip enters the **stomach**, a bean-shaped organ dorsal and to the left of the liver (Figure 35-4). Push the lobes of the liver to one side to fully expose the stomach. Note that the esophagus penetrates the diaphragm before joining the *cardiac end* (near the heart) of the stomach. The other end of the stomach, which empties into the small intestine, is called the pyloric end. Two muscular rings, the *cardiac sphincter* and the *pyloric sphincter,* control the movement of food through the stomach. Feel for these sphincters by gently squeezing these rings of smooth muscle tissue situated at the entrance and exit of the stomach, between your index finger and thumb.

2. Cut the stomach lengthwise with your scissors. Describe any contents of the stomach.

_____

_____

The contents of the fetal pig's digestive tract are called *meconium* and are composed of a variety of substances, including amniotic fluid swallowed by the fetus, epithelial cells sloughed off from the digestive tract, and hair.

3. Clean out the stomach and note the folds or *rugae* on its internal surface. What role might the rugae play in digestion?

_____

_____

4. The **small intestine** (Figure 35-4) is divided into three regions: the *duodenum,* the *jejunum,* and the *ileum.* The first portion, the duodenum, leaves the pyloric end of the stomach and runs along the edge of the pancreas. The junctions of the duodenum and ileum with the jejunum cannot easily be distinguished.

5. A thin membrane, the **mesentery**, holds the coils of the small intestine together. Cut the mesentery from the dorsal body wall and from between the coils of the small intestine. Uncoil the small intestine. Measure the length with a meter stick and record it: _____ cm

A rule of thumb is that the small intestine in both pigs and humans is about five times the length of the individual.

6. Using your scissors, cut out a small section of the intestine, slit it lengthwise, and place it in a clear, shallow dish filled with water. Now examine it using a dissection microscope.

How does the inner surface appear?

_____

_____

_____

Locate the *villi.* Most of the nutrients provided by the digestive process are absorbed by these small projections from the wall of the small intestine.

7. Locate the juncture of the **large intestine** (Figure 35-4), or *colon,* and the ileum. This may be more difficult in a pig than in a human because in the former there is not such a noticeable difference in the size of the small and large intestines. However, this juncture is marked by the presence of a blind pouch, the *cecum,* which in the pig is relatively large. In humans, the cecum is very short and bears a small fingerlike projection called the *appendix.*

8. As with other junctures in the digestive tract, the region where the ileum joins the large intestine is the site of a sphincter of smooth muscle, sometimes called the *ileocecal valve.* Feel for it by rolling the junction between your index finger and thumb.

9. The coiled large intestine stretches from the cecum to the straight **rectum**, which opens to the outside at the **anus**. The **anal canal** is the site of the final muscle in the alimentary canal, the *anal sphincter.* Locate the rectum, anus, and anal sphincter, but do not dissect these structures at this time. You may, however, *at the direction of your instructor,* remove a piece of the colon and examine it with a dissecting microscope.

10. Review the digestive system by tracing the pathway of an ingested indigestible fiber in and out of the body.

11. *If you are continuing with the dissection, move on to the next section.* Otherwise, place your pig and any excised organs back in the plastic bag. Tie it shut to prevent your pig from drying out. Dispose of any paper towels that contain preservative *as directed by your instructor.* Clean the tray, dissecting tools, and the laboratory table.

35.3 | Respiratory System *(10 min.)*

The respiratory system of a mammal consists of various organs and structures associated with the exchange of gasses between the internal and external environment. Air rich in oxygen is inhaled into the air sacs of the lungs. Oxygen diffuses from this air into the capillaries of the lungs while carbon dioxide moves in the other direction. Carbon dioxide–rich air is then exhaled from the body.

MATERIALS

Per student pair:

- one preserved fetal pig injected with red and blue latex
- plastic bag to store fetal pig
- dissection pan
- dissecting kit
- dissecting pins
- two pieces of string each 60 cm long *or* four large rubber bands

Per lab room:

- liquid waste disposal bottle
- boxes of different sizes of lab gloves
- box of safety goggles

PROCEDURE

A. Nose

1. Untie your pig from the dissection pan. In the pig and other mammals, molecules of air enter the body through the *nostrils* and pass through a pair of **nasal cavities** dorsal to the hard palate and into the nasal portion of the pharynx or *nasopharynx* (Figure 35-2). Examine the nostrils and hard and soft palates, and then carefully cut the soft palate longitudinally to examine the nasopharynx of your fetal pig.

2. From the nasopharynx, air passes through the *glottis* into the **larynx** (Figures 35-4 and 35-5a and b). In humans, the front of the larynx is often referred to as the *Adam's apple* or *voice box*.

B. Trachea, Bronchial Tubes, and Lungs

1. Place the pig, ventral side up, in the dissection pan and restrain it using string or rubber bands as you did in the first section of this exercise. Pin back the body flaps to the sides of the body.

2. Slit the larynx longitudinally to expose the *vocal cords* (Figure 35-5c).

3. Locate the trachea (Figure 35-5). The trachea extends from the larynx and divides into two major branches, the **bronchi** (singular, *bronchus*), to the lungs. Note again the series of cartilaginous structures that prevent the trachea from collapsing. These apparent rings of cartilage are actually incomplete on their dorsal side.

4. Note that the thoracic cavity is divided into two **pleural cavities**, which contain the lungs, and a **pericardial sac** that is located between them. Inside this sac is the *pericardial cavity* and the *heart.* Carefully examine the lungs and note the thin, transparent **pleural membrane** that lines the inner surface of the thoracic cavity and the outer surface of the lungs. The right lung consists of four lobes and the left of two or three. Are the lungs of the fetal pig filled with air? _____ (yes or no)

5. Carefully push the *heart* to one side and gently tease away some of the lung tissue to expose the bronchi. Notice that the bronchi divide into smaller and smaller branches. These branches continue to divide and branch into finer and finer structures, eventually ending as microscopic air sacs called **alveoli** (singular, *alveolus*). The thin walls of the alveoli are well supplied with blood capillaries, and it is here that the exchange of carbon dioxide and oxygen occurs after birth.
Where does this exchange occur in the fetus? _____

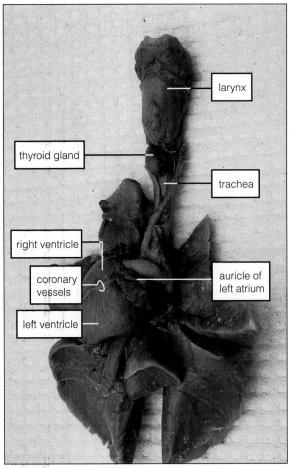

(a)

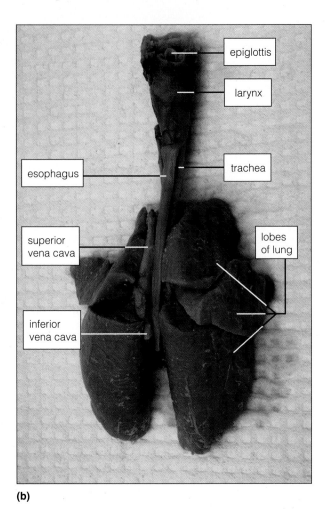

(b)

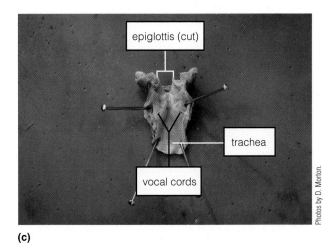

Photos by D. Morton.

(c)

Figure 35-5   (a) Ventral and (b) dorsal views of the respiratory system of the fetal pig. In (c), the larynx has been slit open to show the vocal cords.

6. Now relocate the diaphragm and note its position in relation to the lungs. Contraction of this skeletal muscle in part results in inhalation.

7. Complete your study of the respiratory system of the fetal pig by tracing the pathway of a carbon dioxide molecule from an alveolus to the nostrils.

8. *If you are continuing with the dissection, move on to the next section.* Otherwise, place your pig and any excised organs back in the plastic bag. Tie it shut to prevent your pig from drying out. Dispose of any paper towels that contain preservative *as directed by your instructor.* Clean the tray, dissecting tools, and the laboratory table.

The mammalian body contains a vast network of **blood vessels** that transport blood to and from every living cell. Blood contains cells and plasma. Plasma contains water, oxygen, carbon dioxide, nutrients, metabolic wastes, hormones, and other substances. In mammals, blood is pumped through arteries and arterioles into the **capillaries** of the tissues, from where blood flows into **veins** that transport it back to the heart. The exchange of substances between the blood and tissues occurs across the thin walls of **capillaries.**

The circulatory system is divided into two circuits: the **pulmonary circuit**, which involves blood flow to and from the lungs, and the **systemic circuit**, which allows the flow of blood to and from the rest of the body. In this section, you will study these two circuits and examine how the heart directs the flow of blood through them both in a fetus and in an adult.

MATERIALS

Per student pair:

- one preserved fetal pig injected with red and blue latex
- plastic bag to store fetal pig
- dissection pan
- dissecting kit
- dissecting pins
- two pieces of string each 60 cm long *or* four large rubber bands

Per lab room:

- liquid waste disposal bottle
- boxes of different sizes of lab gloves
- box of safety goggles

PROCEDURE

A. Pulmonary Circuit and Surface Anatomy of the Heart

1. Place the pig, ventral side up, in the dissection pan and restrain it using string or rubber bands as you did in the first section of this exercise. Pin back the body flaps to the sides of the body.
2. If it is not already torn, open the pericardial sac. Similar to the situation in the pleural cavities, the inside of the pericardial sac and the outside of the heart is lined by the **pericardial membrane**.
3. Identify the four chambers of the heart (Figure 35-6)—the **right** and **left atria** (singular, *atrium*), and the larger **right** and **left ventricles**. On the surface of the heart locate the **coronary vessels** lying in the diagonal groove between the two ventricles (Figure 35-6). The coronary arteries and their branches supply blood directly to the heart. (The heart is a muscle and as such has the same requirements as any other organ.) When these vessels become severely occluded, a heart attack may occur. It is the coronary arteries and their branches that are replaced or "bypassed" in coronary bypass surgery.
4. Gently push the heart to the left and identify two relatively large blue veins (Figure 35-6), called the **superior vena cava** and the **inferior vena cava** in humans and the *anterior vena cava* and the *posterior vena cava* in pigs. After birth, oxygen-poor (or carbon dioxide–rich) blood from all of the body except the lungs and heart returns from the systemic circuit to the right atrium of the heart through these large veins. Trace the inferior vena cava a short distance from the heart.

Through what structure does the inferior vena cava pass? _____

5. The blood that enters the right atrium passes to the right ventricle and then is forced into the pulmonary circuit as the heart contracts. Identify the **pulmonary trunk**, which lies between the left and right atria and extends dorsally and to the left (Figure 35-6a). It branches to form the **left** and **right pulmonary arteries** (Figure 35-6b).

What color latex do these pulmonary arteries contain? _____ (red or blue)
After birth, do these vessels carry blood rich or poor in oxygen? _____ (rich or poor)
Carefully move the heart aside and follow the pulmonary arteries to the lungs.

6. After birth, once the blood is oxygenated (and the carbon dioxide removed) in the lungs, it returns to the left atrium of the heart in the **left** and **right pulmonary veins**, which complete the pulmonary circuit. Carefully move the lungs and heart aside and locate these large vessels (Figure 35-6b).
7. From the left atrium of the adult, the oxygenated blood passes to the left ventricle. Blood is forced into the **aorta** as the heart contracts, starting its trip through the systemic circuit. Locate the aorta (Figure 35-6a and b), which leads dorsally out of the left ventricle of the heart. Note that its base is partially covered by the pulmonary trunk coming from the right ventricle.
8. Examine how blood circulation is different in fetal mammals (Figure 35-7). In the fetus, most of the pulmonary circuit is bypassed twice. First, most blood from the right ventricle enters the aorta directly from

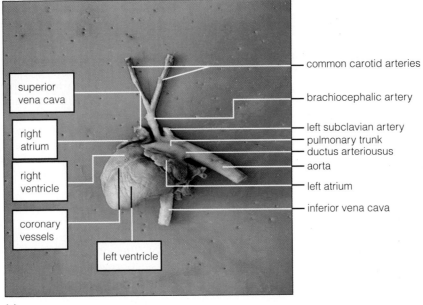

(a)

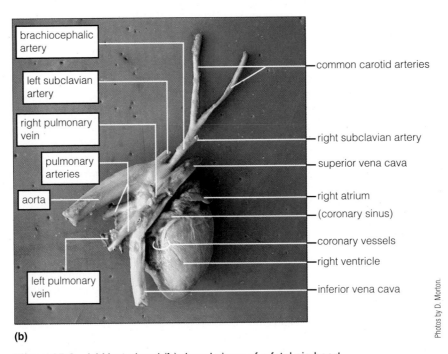

(b)

Figure 35-6 **(a)** Ventral and **(b)** dorsal views of a fetal pig heart.

Photos by D. Morton.

the pulmonary trunk through the **ductus arteriosus**, a large but short vessel connecting the pulmonary trunk directly to the aorta. The second bypass occurs when most of the blood delivered to the right atrium, mostly from the posterior portions of the body by the inferior vena cava, passes directly into the left atrium through a temporary opening in the wall separating the right and left atria (*foramen ovale*). Thus, this blood entirely bypasses the pulmonary circuit. Identify the ductus arteriosus (Figures 35-6 and 35-7).

Why is it not necessary for large quantities of blood to enter the pulmonary system of a fetus?

With the first breath of the newborn, the ductus arteriosus contracts and the foramen ovale closes. Circulation through the pulmonary circuit is increased dramatically. Then, during the eight weeks following birth, the ductus arteriosus forms a fibrous strand of connective tissue, the *ligamentum arteriosum*, and the foramen ovale permanently fuses shut (*fossa ovalis*).

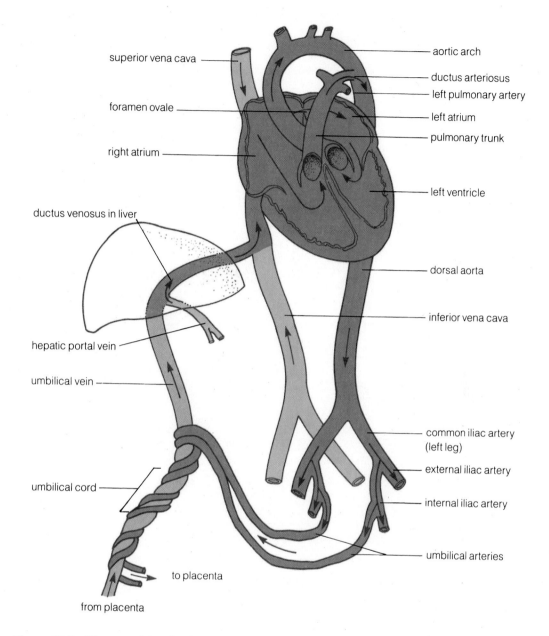

superior vena cava

foramen ovale

right atrium

ductus venosus in liver

hepatic portal vein

umbilical vein

umbilical cord

to placenta

from placenta

aortic arch

ductus arteriosus

left pulmonary artery

left atrium

pulmonary trunk

left ventricle

dorsal aorta

inferior vena cava

common iliac artery
(left leg)

external iliac artery

internal iliac artery

umbilical arteries

Figure 35-7 Diagram of the circulatory system of a fetal mammal. Arrows indicate the flow of blood. Pink represents fully oxygenated blood. The blue indicates oxygen-depleted blood.

B. Systemic Circuit—Major Arteries and Veins Anterior to the Heart

1. The systemic circuit begins with the aorta. This large vessel leads anteriorly out of the left ventricle of the heart and makes a sharp turn to the left (the so-called **aortic arch**, Figure 35-7) and proceeds posteriorly through the body as the **dorsal aorta** (Figure 35-8a). All of the major arteries of the body arise from these two regions (the aortic arch and dorsal aorta) of the aorta.

2. Locate the first visible vessel, the **brachiocephalic artery** (Figure 35-8a), to branch from the aortic arch. The first vessels to branch from the aorta are the coronary arteries; these cannot be seen without dissecting the heart. The brachiocephalic artery branches to give rise to the **right subclavian artery** (Figure 35-8a), going to the right forelimb, and the **carotid trunk** (Figure 35-8a), whose branches course anteriorly through the neck and head. Trace the right subclavian artery and its branches through the shoulder region to the right forelimb. As it passes through the shoulder region, the name of the right subclavian changes to the *axillary artery* and then to the *brachial artery* when it enters the upper forelimb.

3. Return to the aorta and locate the second visible vessel to branch from the aortic arch, the **left subclavian artery** (Figure 35-8a). The left subclavian artery and its branches pass through the shoulder and left

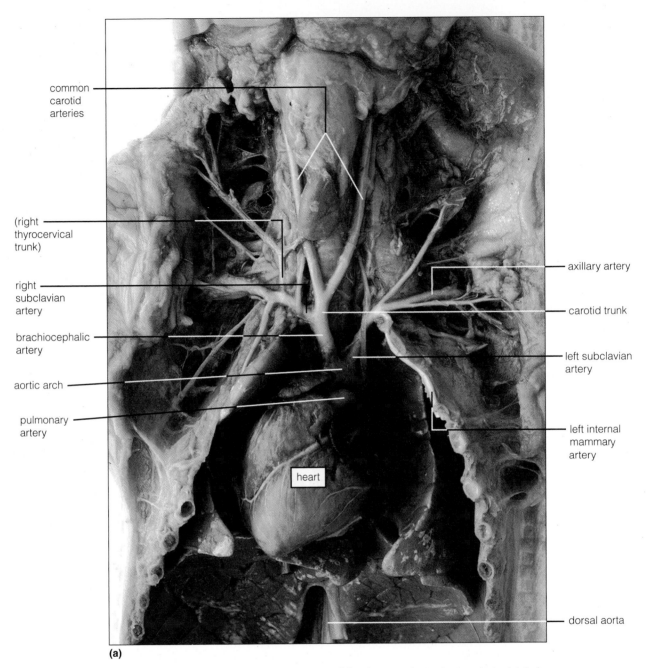

common
carotid
arteries

(right
thyrocervical
trunk)

right
subclavian
artery

brachiocephalic
artery

aortic arch

pulmonary
artery

axillary artery

carotid trunk

left subclavian
artery

left internal
mammary
artery

heart

dorsal aorta

(a)

Figure 35-8 Ventral views of (**a**) arteries and (**b**) veins anterior to the heart of a fetal pig.

forelimb or arm in the same manner as the right subclavian artery, described above. As you trace the course of the left subclavian, notice that some of its branches feed the muscles of the chest and back.

4. Return to the right forelimb and locate the venous system that passes through this appendage. Because the veins are relatively thin-walled, this may be very difficult. Also, some of them may not be injected with blue latex and will appear a brownish color. If possible, follow the *brachial vein* to the *axillary* and the **subclavian vein** (Figure 35-8b) until the latter becomes the **brachiocephalic vein** (Figure 35-8b). It should be relatively easy to follow the brachiocephalic to its juncture with the previously identified superior vena cava (Figure 35-8a; it forms a prominent V), which returns blood to the right atrium of the heart.

5. In order to examine the arterial system that serves the throat and head, locate the carotid trunk (Figure 35-8a; a branch of the brachiocephalic artery; see above). This short branch of the brachiocephalic artery immediately splits into the **left** and **right common carotid arteries**. Each of these vessels divides into the *internal* and *external common carotid arteries*. Remove the thymus and thyroid glands and considerable muscle tissue in the throat to locate the anterior portions of the common carotid arteries. As you locate and trace the carotid arteries, look for a white "fiber" that parallels them. This is the *vagus nerve.*

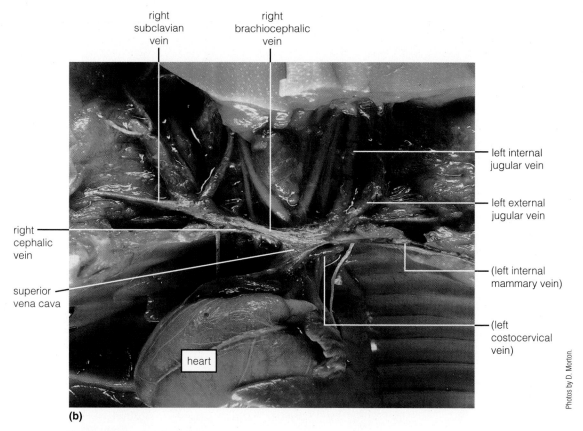

right subclavian vein

right brachiocephalic vein

left internal jugular vein

left external jugular vein

right cephalic vein

(left internal mammary vein)

superior vena cava

(left costocervical vein)

heart

Photos by D. Morton.

(b)

Figure 35-8 *Continued.*

6. On either side of the neck are the major veins that drain the head and throat region. The **internal** and **external jugular veins** (Figure 35-8b) join the subclavian veins (from the forelimbs) to form the previously identified brachiocephalic vein. The latter leads into the superior vena cava.

C. Systemic Circuit—Major Arteries and Veins Posterior to the Heart

1. The posterior extension of the aortic arch is the dorsal aorta. As the name implies, the dorsal aorta lies near the dorsal body wall running parallel to the spine. From this large vessel arise all of the arterial branches that feed the organs, glands, and muscles of the abdominal region and the muscles of the hindlimbs and tail.

2. Follow the dorsal aorta posteriorly. Carefully move the liver and stomach of the pig and use a dissection needle to scrape away the sheet of tissue that connects the dorsal aorta to the pig's back. Locate the **celiac artery** (Figure 35-9a), and its branches that deliver blood to the stomach, spleen, and liver. Continue to follow the dorsal aorta posteriorly and locate the **superior mesenteric artery** (Figure 35-9a). This vessel, just posterior to the celiac artery, branches to the pancreas and duodenum of the small intestine.

3. Posterior to the superior mesenteric artery are the **renal arteries** (Figure 35-9a), relatively short vessels that connect the dorsal aorta and the kidneys. At this time, it's easy to locate the **renal veins** (Figure 35-9b), which drain blood from the kidneys to the inferior vena cava.

4. As you follow the dorsal aorta posteriorly beyond the kidneys, the **external iliac arteries** (Figure 35-7) branch, one into each hindlimb. Each leg is also drained with a major vein, the **common iliac vein** (Figure 35-9a), which joins the inferior vena cava.

5. Follow the branches of the dorsal aorta into the tail region, being careful not to cut the two intervening branches. The small extension toward the tail region is called the *sacral artery* as it leaves the dorsal aorta and the *caudal artery* when it enters the tail.

6. Just anterior to the sacral artery, the **internal iliac arteries** (Figure 35-7) branch from the dorsal aorta. These enlarge and form the two **umbilical arteries** (Figure 35-7), which run through the umbilical cord to the placenta. Cut the umbilical cord transversely and note the arrangement of the umbilical arteries within it. Consider the composition of the blood as it travels through the umbilical arteries to the placenta. In what gas is its blood rich? _____ (oxygen or carbon dioxide)

7. Locate the two pieces of vein that you tied with string in Section 35.1. This is the **umbilical vein** (Figure 35-7), through which blood rich in nutrients and oxygen flows from the placenta of the mother back to the fetus. Locate the umbilical vein in the umbilical cord and follow it anteriorly toward the liver. When the

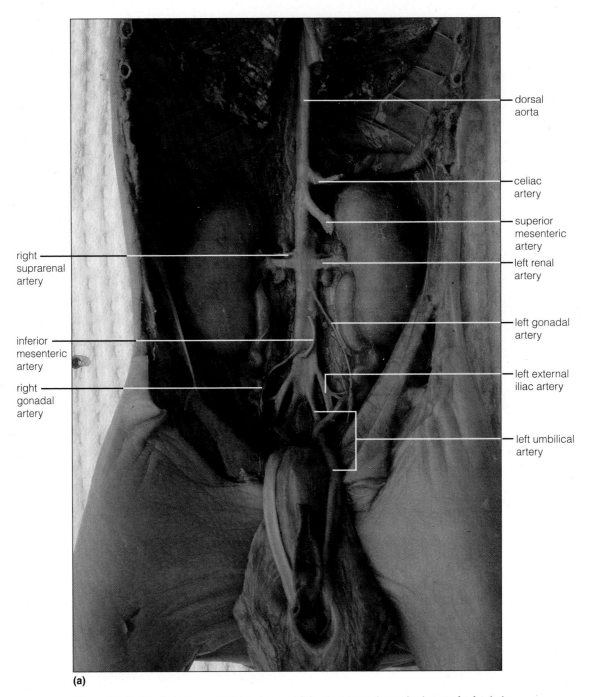

right suprarenal artery

inferior mesenteric artery

right gonadal artery

dorsal aorta

celiac artery

superior mesenteric artery

left renal artery

left gonadal artery

left external iliac artery

left umbilical artery

(a)

Figure 35-9   Ventral views of (**a**) arteries and (**b**) veins posterior to the heart of a fetal pig.

umbilical vein reaches the liver, it becomes the **ductus venosus** (Figure 35-7), which continues anteriorly within the substance of the liver and joins the inferior vena cava. The umbilical arteries, the umbilical veins, and ductus venosus become modified into ligaments following the birth of the fetus.
What is the relationship between the navel and the umbilical cord?

After birth the *hepatic portal system,* which consists of a network of veins, collects all the blood from the lower digestive tract and associated organs (stomach, small intestine, pancreas, and spleen). This blood is carried to the liver via the *hepatic portal vein.* In general, a **portal vein** is one that collects blood from the capillaries of one organ and transfers it to the capillaries of another organ. After birth, the *hepatic veins* drain the blood of the liver. These join the inferior vena cava just anterior to the point where the ductus venosus joins the inferior vena cava (Figure 35-10).

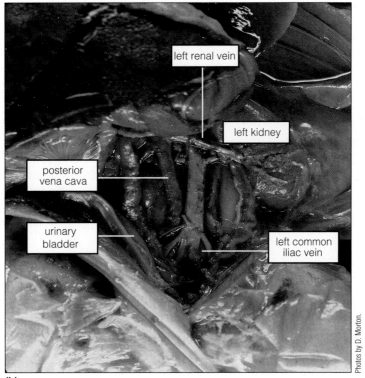

Photos by D. Morton.

(b)

Figure 35-9 *Continued.*

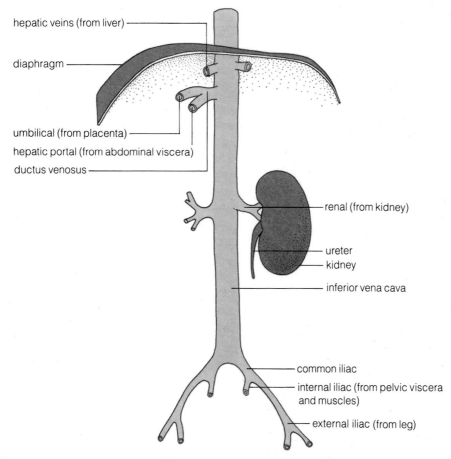

Figure 35-10 Relationship of umbilical vein, hepatic portal vein, and hepatic veins to the inferior vena cava in a fetal pig.

8. Complete your dissection of the systemic circulatory system by tracing the inferior vena cava from the abdominal cavity through the **diaphragm** and to the thoracic cavity, where it joins the superior vena cava before entering the right atrium of the heart.

9. *If you are continuing with the dissection, move on to the next section.* Otherwise, place your pig and any excised organs back in the plastic bag. Tie it shut to prevent your pig from drying out. Dispose of any paper towels that contain preservative *as directed by your instructor.* Clean the tray, dissecting tools, and the laboratory table.

35.5 Internal Structure of the Heart *(15 min.)*

The hearts of a fetal pig and a human are quite similar. For example, the heart is the primary pump of the circulatory system, it is four-chambered, and it is primarily composed of cardiac muscle tissue. The same can be said of all mammals and birds and, except for the number of chambers, of all vertebrates and some invertebrates as well. Although the fetal pig heart is small, keep in mind as you dissect it that the same structures are present in your own heart.

MATERIALS

Per student pair:
- one preserved fetal pig injected with red and blue latex
- plastic bag to store fetal pig
- dissection pan
- dissecting kit

Per lab room:
- liquid waste disposal bottle
- boxes of different sizes of lab gloves
- box of safety goggles

PROCEDURE

1. Carefully free the heart from the fetal body by cutting through the superior and inferior venae cavae, the subclavians, the common carotids, and the dorsal aorta just posterior to the heart. Cut through the left and right pulmonary veins and the pulmonary arteries at their juncture with the lungs. Remove the heart from the fetus and place it on paper towels with its ventral surface facing you, as it was in the thoracic cavity. If any remnants of the pericardial sac are still present, carefully remove them from around the heart.

2. Review the location of the four heart chambers: the left and right atria and the left and right ventricles. Locate the coronary artery and vein in the longitudinal groove running between the left and right ventricles.

3. Place the heart dorsal-side up and locate the inferior and superior venae cavae. Cut through these vessels with your scissors and expose the interior chamber of the right atrium (see Figure 35-11b, incision 1). Carefully remove any latex and coagulated blood in the right atrium. Between the atrium and the right ventricular cavity are three membranous cusps attached to the wall of the right atrium. This is the **tricuspid valve**. The open ends of its cusps face downward like open parachutes into the cavity of the right ventricle.

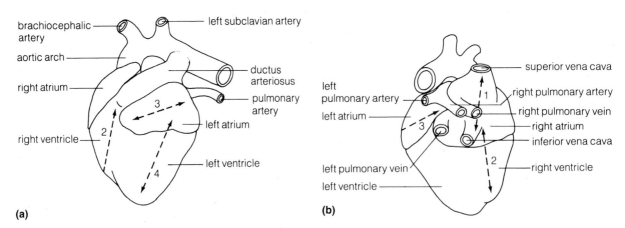

Figure 35-11 (a) Ventral and (b) dorsal views of fetal pig heart, showing numbered incisions for dissection.

4. Continue working from the dorsal side and cut into the right ventricle (incision 2). With your forceps and needle, remove any latex that obstructs your view. You should be able to see the three cusps of the **pulmonary semilunar valve** at the juncture of the pulmonary artery and the right ventricle. The open ends of the cusps face into the pulmonary trunk and thus prevent a backward flow of blood into the ventricle.

5. Examine the internal walls of the ventricle. If you wish, you may extend incision 2 to the ventral side of the right ventricle. Look for muscular ridges on the inside wall. These are the **papillary muscles**, and arising from them are relatively fine fibers, the **chordae tendinae**. The chordae tendinae are attached to the edges of the tricuspid valve and they are commonly called the heartstrings.

6. Next, with the heart's ventral surface facing you, locate the ductus arteriosus and the aorta. (Remember that the ductus arteriosus is a shunt between the pulmonary trunk and the aorta.)

Cut open the left atrium (incision 3) and the left ventricle (incision 4). Remove any latex. On the dorsal wall of the heart, find the pulmonary veins from the inside of the left atrium. Next, locate the *bicuspid valve* (consisting of two cusps) between the left atrium and left ventricle.

Do the cusps appear similar to the tricuspid valve? _____ (yes or no)

7. Turn the heart so that the ventral surface is facing you and examine the cavity of the left ventricle. Note the papillary muscles and the chordae tendinae in the left ventricle.

Do they appear similar to those in the right ventricle? _____ (yes or no)

8. Insert a probe into the aorta from the exterior of the heart and note where it enters the cavity of the left ventricle. At this point there is another valve, the **aortic semilunar valve**, with three cusps. Is the orientation of the aortic semilunar valve the same as that of the semilunar valve between the pulmonary trunk and the right ventricle? _____ (yes or no)

9. Recall that in the fetus a temporary opening, the foramen ovale (Figure 35-7), exists between the right and left atria. Look for it in your specimen.

10. Complete your study by looking for differences in the thickness of the walls of the atria and those of the ventricles. Why is the wall of the left ventricle thicker than that of the right ventricle?

_____

_____

_____

11. *If you are continuing with the dissection, move on to the next exercise.* Otherwise, place your pig and any excised organs back in the plastic bag. Tie it shut to prevent your pig from drying out. Dispose of any paper towels that contain preservative *as directed by your instructor.* Clean the tray, dissecting tools, and the laboratory table.

_____ 1. The two *major ventral* body cavities of a fetal pig are the
  (a) thoracic and pleural.
  (b) thoracic and pericardial.
  (c) abdominopelvic and thoracic.
  (d) abdominopelvic and pericardial.

_____ 2. The diaphragm is a sheetlike skeletal muscle that separates the
  (a) thoracic and pleural cavities.
  (b) thoracic and pericardial cavities.
  (c) thoracic and abdominopelvic cavities.
  (d) pleural and pericardial cavities.

_____ 3. The digestive system is concerned with
  (a) blood circulation.
  (b) digestion and the absorption of nutrients.
  (c) reproduction.
  (d) excretion of urine.

_____ 4. The liver functions to
  (a) produce bile.
  (b) pump blood.
  (c) form urea.
  (d) do both a and c.

_____ 5. As a general rule, the small intestine of a pig or human is
  (a) about 60 cm long.
  (b) about 1.5 m long.
  (c) about as long as the individual is tall (or long in the case of the pig).
  (d) about five times the height of the individual.

_____ 6. The cardiac, pyloric, anal, and ileocecal sphincters are all part of the
  (a) digestive tract.
  (b) respiratory tract.
  (c) circulatory system.
  (d) muscular system.

_____ 7. In humans, the front of the larynx is commonly referred to as
  (a) the voice box.
  (b) the Adam's apple.
  (c) the food pipe.
  (d) both a and b

_____ 8. The microscopic air sacs or alveoli are the sites where blood
  (a) picks up oxygen.
  (b) gives up carbon dioxide.
  (c) gives up oxygen.
  (d) both a and b

_____ 9. A vein is a blood vessel that always carries
  (a) blood toward the heart.
  (b) blood away from the heart.
  (c) oxygen-rich blood.
  (d) oxygen-poor blood.

_____ 10. The hearts of a fetal pig and a human are similar in that they are
  (a) the primary pump of the circulatory system of the body.
  (b) both four chambered.
  (c) composed of cardiac muscle tissue.
  (d) all of the above

EXERCISE **3 5**

Dissection of the Fetal Pig: Digestive, Respiratory, and Circulatory Systems

Post-Lab Questions

35.1 Ventral Body Cavities

1. Describe the location of the two *major* ventral body cavities.

35.2 Digestive System

2. Describe and give the general function of the digestive tract.

3. List in order the organs through which food and other substances pass in their journey into, through, and out the digestive tract.

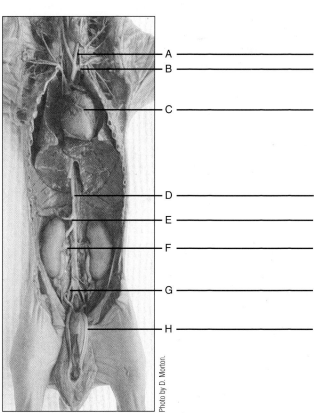

35.3 Respiratory System

4. Describe the major similarities and differences in the location, structure, and function of the trachea and esophagus.

35.4 Blood Vessels and the Surface Anatomy of the Heart

5. Identify the structures labeled in the photo on the right.

A _____
B _____
C _____
D _____
E _____
F _____
G _____
H _____

Photo by D. Morton.

6. What is the main difference between the pulmonary and the systemic circuits of the circulatory system?

7. What is the foramen ovale? What is its fate after birth?

8. With regard to blood circulation, what is the difference between an artery and a vein?

9. Briefly describe the function of a portal vein system (for example, the hepatic portal vein).

Food for Thought

10. Search the Internet for sites that describe artificial hearts. List two sites and briefly summarize their contents.

http:// _____

_____

_____

http:// _____

_____

_____

Dissection of the Fetal Pig: Urogenital and Nervous Systems

After completing this exercise, you will be able to

1. define *urea, peritoneum, urine, urinary bladder, homologous, ovulation, semen, inguinal hernia, vasectomy, nephron, meningitis;*

2. locate the organs of the urinary, reproductive, and nervous systems in a fetal pig;

3. describe and give the functions of the organs of the urinary, reproductive, and nervous systems;

4. explain the importance of the urinary, reproductive, and nervous systems to a living mammal;

5. locate, name, and describe the function of the internal structures of the kidney;

6. list in order the three meninges;

7. explain how the spinal nerves are connected to the spinal cord;

8. identify and give the general functions of the structures of the brain and spinal cord.

Introduction

In this exercise you will complete your study of the internal anatomy of the fetal pig. To do this, we will examine the system that removes metabolic wastes from the bloodstream (*urinary system*), the system responsible for the production of new individuals or offspring (*reproductive system*), and the system responsible for the integration and control of all of the body (*nervous system*).

CAUTION

Specimens are kept in preservative solutions. Use lab gloves and safety goggles, especially if you wear contact lenses, whenever you or your lab mates handle a specimen or other fluids. If safety goggles do not fit, your glasses should suffice. Wash any part of your body exposed to preservative with copious amounts of water. If preservative solution or other fluids are splashed into your eyes, wash them with the safety eyewash bottle or eye irrigation apparatus for 15 minutes.

36.1 Urogenital System *(40 min.)*

The urinary and reproductive systems are traditionally dissected together (as the *urogenital system*) because they share several anatomical features.

MATERIALS

Per student pair:

- one preserved fetal pig injected with red and blue latex
- plastic bag to store fetal pig
- dissection pan
- dissecting kit
- dissecting pins
- four large rubber bands *or* two pieces of string, each 60 cm long

Per lab room:

- liquid waste disposal bottle
- boxes of different sizes of lab gloves
- box of safety goggles

PROCEDURE

A. Urinary System

Like humans, the pig is a terrestrial organism and, as such, must conserve water. At the same time, metabolic wastes must be continuously removed from the blood. Furthermore, the composition of the blood must be constantly monitored and adjusted so that the cells of the body are bathed in a fluid of constant composition.

Much of the potentially poisonous waste occurs in the form of **urea** and results from the metabolism of amino acids in the liver. Urea is filtered from the bloodstream in the kidneys, which also regulate water and salt balance.

1. Place your pig on its back in the dissection pan and use string or rubber bands to secure the legs, as you did in the preceding exercises. Pin the lateral body-wall flaps to the dorsal side of your specimen and pull the umbilical cord and surrounding tissue back between the hindlimbs.
2. Find the paired **kidneys** in the lumbar region of the body cavity pressed against the dorsal body wall (Figure 36-1). They are covered by the **peritoneum**, the smooth, rather shiny membrane that lines the abdominopelvic cavity. (You may have already removed much of this during the dissection of the circulatory system in the preceding exercise.) The main function of the kidneys is the production of **urine**, a fluid containing urea and other waste substances dissolved in water.
3. To expose the right kidney, carefully lift up the abdominal organs and move them anteriorly and to the left. Using a dissecting needle, carefully scrape away the peritoneum so that the kidney bean-shaped kidney and the **ureter**—the duct that connects it to the bladder—are easily seen. Note the central depression in the surface of the kidney. This is the *hilus,* the region where the ureter leaves, the *renal vein* leaves, and the *renal artery* enters the kidney.
4. Carefully follow the ureter from the hilus to the **allantoic bladder**. Then lift the bladder and find the **urethra**. The latter is the structure through which urine passes from the bladder to the outside of the animal. In the male, the urethra is very long and passes through the *penis* to the outside of the body (Figure 36-1). Notice that the urethra passes posteriorly for a distance of approximately 2 cm and then turns sharply anteriorly and ventrally before entering the penis. In the female fetal pig, the urethra is short and passes posteriorly to join with the *vagina* to form the *urogenital sinus* (Figure 36-2).
5. Locate the **allantoic duct**, which leads from the allantoic bladder into the umbilical cord. The allantoic duct is largely a vestigial structure, for most of the wastes produced in the kidneys of the fetus are carried in the bloodstream through the umbilical vein to the placenta, where they are removed in the body of the mother. Following birth, the allantoic duct collapses, and the allantoic bladder is incorporated into the **urinary bladder**, which functions to store urine.

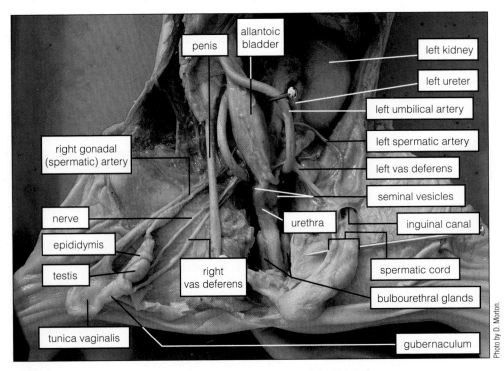

Figure 36-1   Ventral view of the male urogenital system of the fetal pig.

B. Female Reproductive System

In terrestrial organisms, fertilization (the fusion of the nuclei of male and female gametes) occurs internally, where a relatively stable aquaticlike environment is maintained. Once fertilization has occurred, the zygote divides to form an embryo and eventually a fetus. In mammals, all of this growth and development occurs within the female's uterus, which nourishes the developing offspring until it is born.

1. Examine the **vulva**, the collective term for the external genitalia of the female. In the pig, the vulva includes the *genital papilla* on the outside of the body, the *labia* or lips found on either side of the urogenital sinus, and the **clitoris**, a small body of erectile tissue on the ventral portion of the urogenital sinus. Also included in the vulva is the opening of the urogenital sinus itself (Figure 36-2).

2. In the female fetal pig, the **urogenital sinus** is the common passage for the urethra and the **vagina**. To locate these structures, carefully insert your scissors into the opening of the urogenital sinus and cut this structure from the side. Locate where the ducts of the vagina and urethra enter to form the urogenital sinus.

3. The urogenital sinus is not present in the adult female pig. During the subsequent development of the fetus, the sinus is reduced in size until the vagina and the urethra each develop

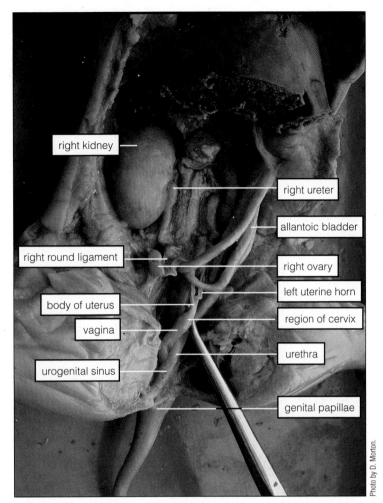

Figure 36-2   Ventral view of the female urogenital system of the fetal pig.

their own, separate opening to the outside. Thus, in the adult female pig, urine exits through the urinary opening. This is the situation in most adult female mammals.

How does this compare with the structure of the reproductive system of most male mammals?

_____

_____

_____

4. Relocate the *clitoris*. This small, rounded region on the inner ventral surface of the urogenital sinus is **homologous** (that is, similar in developmental origin) to that part of the erectile tissue of the male penis, which encloses the urethra. In the female, note that the urethra opens posteriorly to the clitoris.

5. Follow the thick-walled muscular vagina anteriorly to the **uterus**. In the pig, the uterus consists of three structures or regions: the **cervix** at the entrance to the uterus, the **uterine body**, and the two **uterine horns** (Figure 36-2). Note that the uterine horns unite to form the body of the uterus. The pig has a *bicornuate uterus*, in which the fetuses develop within the uterine horns. In the human female there are no uterine horns, and the fetus develops within the body of the *simplex uterus.*

6. From the uterine horns, follow the **oviducts** to the **ovaries** (female gonads). The ovaries are small, yellowish kidney bean–shaped structures that lie just posterior to the kidneys. They are the sites of egg production and the source of the female sex hormones, estrogen and progesterone.

All of the eggs that a female produces during her lifetime are present in the ovaries at birth. After puberty, eggs will mature, rupture from the surface of the ovaries, and enter the oviducts. This process is called **ovulation**. If viable sperm are present in the upper third of an oviduct when it contains eggs, fertilization may occur. In this case, the fertilized egg or zygote will develop into an embryo and pass down the oviduct to become implanted in the wall of the uterine horn (pig) or the uterine body (human).

7. Identify the membranous broad and round ligaments. The **broad ligament**, which originates from the **dorsal body wall** (Figure 36-2), supports the ovaries, oviducts, and uterine horns. The **round ligament**, which also supports the ovaries, extends from the lateral wall and crosses the broad ligament diagonally.

C. Male Reproductive System

The male reproductive system of a mammal consists of organs and structures that primarily function in the production of sperm, their transit to the base of the penis during sexual excitement, and their subsequent ejaculation in **semen**—sperm plus the secretions of the *sex accessory glands.*

1. Locate the **testes** (male gonads; Figure 36-1), the site of **sperm** production and source of *testosterone,* the male sex hormone. In older fetuses they are located in the *scrotum,* but in younger fetuses they can be found anywhere between the kidneys and the scrotum.

For viable sperm to be produced in adult males, the testes must be situated outside of the abdominopelvic cavity, where body temperatures are slightly lower than within. Thus, during normal development, the testes undergo a posterior migration, or descent, into the scrotum.

2. Locate the scrotum. Make a midline incision through this structure, cutting through the muscle tissue. Pull out the two elongated bulbous structures covered with a transparent membrane. This membrane is the *tunica vaginalis* and is actually an outpocketing of the abdominal wall. Notice the tough white cord that connects the posterior end of the testes to the inner face of the sac (Figure 36-1). This cord, the **gubernaculum**, is homologous to the round ligament in the female reproductive system. It grows more slowly than the surrounding tissues and thus aids in pulling the testes posteriorly into the scrotal sacs.
3. Cut through the tunica vaginalis to expose a single testis and find the **epididymis**, the tightly coiled tube that lies along one side of the testis. Sperm produced in the testes are stored in the epididymides until ejaculation.
4. Note the slender, elongated **spermatic cord** that emerges anteriorly from each testis (Figure 36-1). Gently pull the cord and note that it passes through an opening, the **inguinal canal**, which is actually an opening from the abdominopelvic cavity into the scrotum. It is through this opening that the testes descend during their migration into the scrotum.

Some human males develop an **inguinal hernia,** a condition in which part of the intestine drops through the inguinal canal into the scrotum. Pigs and other four-legged mammals (hint) do not develop inguinal hernias. Why do you think this is so?

_____

_____

_____

5. Using the tips of your scissors, slit open the spermatic cord attached to the dissected testis. Note that it contains the *sperm duct* or **vas deferens** (plural, *vasa deferentia*), the spermatic vein, the **spermatic artery**, and the spermatic nerve. It is the vas deferens that is severed when a human male has a **vasectomy**. Follow the vas deferens to the base of the bladder, where it loops up and over the ureter and then continues posteriorly to enter the urethra.
6. Expose the full length of the **penis** and its juncture with the urethra. To find the latter, make an incision with your scalpel through the muscles in the midventral line between the hindlegs. When the cut is deep enough, they will lie flat. Carefully remove the muscle tissue and pubic bone on each side of the cut until the urethra is exposed. With a blunt probe, tear the connective tissue connecting the urethra to the rectum, which lies dorsal to it.
7. Locate a pair of small glands, the **seminal vesicles**, on the dorsal surface of the urethra where the two vasa deferentia enter. Situated between the bases of the seminal vesicles is the **prostate gland**. The other sex accessory glands are the **bulbourethral glands**, two elongated structures lying on either side of the juncture of the penis and urethra.

The seminal vesicles, the prostate gland, and the bulbourethral glands all secrete fluids that together with sperm, form semen, which is ejaculated during sexual intercourse. In addition to sperm, semen is mostly water, sugar, and other molecules that provide a supportive environment for the sperm.

8. *If you are continuing with the dissection, move on to the next section.* Otherwise, place your pig back in the plastic bag. Tie it shut to prevent your pig from drying out. Dispose of any paper towels that contain preservative *as directed by your instructor.* Clean the tray, dissecting tools, and the laboratory table.

In the first part of this exercise, you located the kidneys, a pair of kidney bean–shaped structures lying on either side of the spine in the lumbar region of the body. Although the kidneys are situated below the diaphragm, they are actually located outside the peritoneum, the membrane that lines the abdominal cavity. During this procedure, you will examine the internal anatomy of the kidney, including its functional unit, the **nephron**.

MATERIALS

Per student pair:

- one preserved fetal pig injected with red and blue latex
- plastic bag to store fetal pig
- dissection pan
- dissecting kit
- dissecting pins
- two pieces of string, each 60 cm long *or* four large rubber bands
- compound light microscope
- prepared section of kidney

Per lab room:

- liquid waste disposal bottle
- boxes of different sizes of lab gloves
- box of safety goggles

PROCEDURE

A. Anatomy

Each kidney consists of numerous nephrons, which function to filter the blood and start the process of urine formation. A system of ducts completes urine formation and transports it to a space, the *renal pelvis*, which is drained by the ureter. The ureter transports the urine to the urinary bladder.

1. Return to the right kidney and attempt to identify the *adrenal gland.* Look for a tiny, cream colored, comma-shaped body located on the medial, anterior side of the kidney. This is another of the body's endocrine organs. The paired adrenal glands secrete several hormones, including adrenaline (epinephrine). Free the right kidney by severing the renal vein, renal artery, and ureter. Remove the kidney from the body cavity and place it on a paper towel with the central depression to the right.
2. With your scalpel, carefully cut the kidney in half lengthwise, as you would separate the two halves of a peanut. Examine the cut surface of one of the halves and locate the three major regions of the kidney—the outer **cortex**, the inner **medulla**, and the **renal pelvis** (Figure 36-3a). The cortex and medulla contain different portions of the nephrons and their associated blood vessels.

B. Microscopic Structure of the Nephrons

Each **nephron** is a little tube or tubule that is closed at one end. The closed end is expanded and collapsed upon itself much like a deflated basketball pushed in by your fist, only the inside space is connected to the space within a tube attached to the other side. The expanded, collapsed end of the nephron is called the *capsule* and in the kidney the space within its inner wall (where your fist would be) contains a network of capillaries, the *glomerulus* (plural, glomeruli; Figure 36-3b). The combination of the capsule and the glomerulus is called a *renal corpuscle.*

Urine formation begins with the ultrafiltration of blood across the capillary and inner capsule walls into the space within the capsule. After ultrafiltration, the filtrate travels through the rest of the nephron—*proximal convoluted tubule,* the *loop of Henle,* and the *distal convoluted tubule.* Much of the water, ions, sugars, and other useful substances are reabsorbed into the blood in the *peritubular capillaries.* While the reabsorption process recaptures many useful molecules, substances such as ammonia, potassium, and hydrogen ions are secreted to join the urea in the forming urine. Thus, the nephron carries out its excretory and salt balance functions in three steps: filtration, reabsorption, and tubular secretion.

Although all the activities of the nephron are extremely vital to the health of a mammal, the importance of the reabsorption function is easily illustrated with a few numbers. The renal corpuscles of human kidneys produce approximately 180 L (approximately 180 quarts) of filtrate each day. About 99% of this ultrafiltrate is reabsorbed as water, primarily by the nephrons. If they were not so efficient, we would have to drink constantly just to replenish the fluid lost.

Examine Figure 36-3b, which shows a nephron, its named portions, and associated blood vessels.

1. Examine a prepared section of the kidney with your compound microscope. Identify the *cortex* and *medulla.* In the cortex locate *renal corpuscles, glomeruli,* and *capsules* (Figure 36-4).

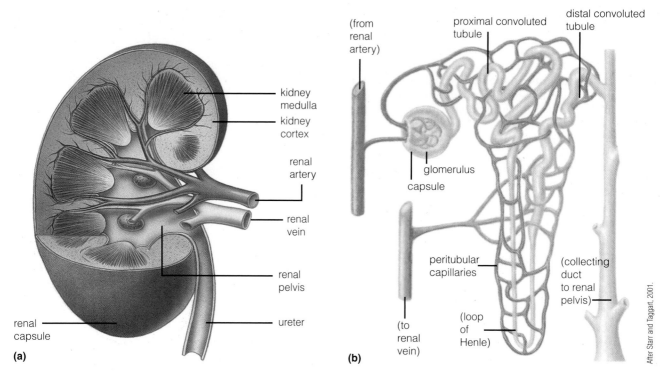

Figure 36-3 (a) Longitudinal section of human kidney. (b) A nephron, the functional unit of the kidney.

(a)

kidney medulla

kidney cortex

renal artery

renal vein

renal pelvis

renal capsule

ureter

(b)

(from renal artery)

proximal convoluted tubule

distal convoluted tubule

glomerulus

capsule

peritubular capillaries

(collecting duct to renal pelvis)

(to renal vein)

(loop of Henle)

After Starr and Taggart, 2001.

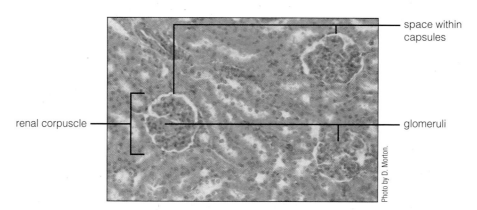

space within capsules

renal corpuscle

glomeruli

Photo by D. Morton.

Figure 36-4 Section of the cortex of a kidney (186×).

2. Identify cross sections of the tubular portion of nephrons.
3. *If you are continuing with the dissection, move on to the next section.* Otherwise, place your pig and any excised organs back in the plastic bag. Tie it shut to prevent your pig from drying out. Dispose of any paper towels that contain preservative *as directed by your instructor.* Clean the tray, dissecting tools, and the laboratory table.

36.3 | **Nervous System** *(40 min.)*

The nervous system of the pig, and in fact of all vertebrates, can be divided into two major components: the *central nervous system* (CNS) and the *peripheral nervous system* (PNS). The CNS consists of the brain and the spinal cord, which serves as the primary link between the brain and much of the PNS. The PNS primarily includes the large network of *nerves* outside the CNS. Through the cranial and spinal nerves of the peripheral system, impulses enter and leave the central nervous system.

MATERIALS

Per student group (4):

- one preserved fetal pig injected with red and blue latex
- plastic bag to store fetal pig
- dissection pan
- dissecting kit
- dissecting pins
- two pieces of string, each 60 cm long *or* four large rubber bands
- dissection microscope

Per lab room:

- liquid waste disposal bottle
- boxes of different sizes of lab gloves
- box of safety goggles

PROCEDURE

As it is very time consuming to expose the full length of the spinal cord, work in groups of four. One pair in the group exposes the anterior portion of the spinal cord and the other pair its posterior region.

A. Spinal Cord

1. Turn the body-wall flaps of the fetal pig inward and place your specimen ventral-side down on the dissection pan. *Proceed carefully with this portion of the dissection*, as the nervous tissues of a fetus are extremely delicate and can be easily destroyed.

2. Carefully remove a strip of muscle about 1.5 cm wide from the base of the neck posteriorly along the spinal column to the tail. This will expose the **spines** and the **vertebral arches** of the vertebrae (Figure 36-5). With a sharp scalpel, gradually cut away the spines and the neural arches of the vertebrae until the **spinal cord** is exposed (Figure 36-6).

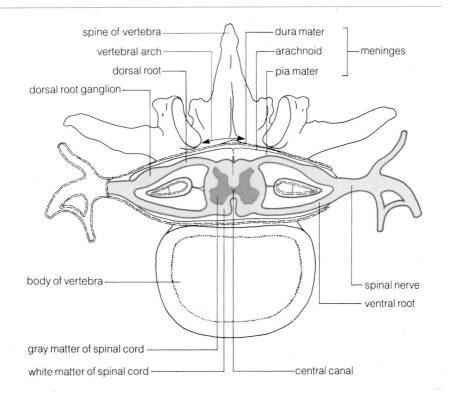

Figure 36-5 Cross section of a vertebra and spinal cord of a mammal.

3. Note the enlargements of the spinal cord at the level of the forelimbs and hindlimbs. These are the **cervical** and **lumbar enlargements**, respectively; they result from the presence of a large number of nerve cells, or *neurons*, supplying the appendages in these regions.

4. At the anterior end of the body, the spinal cord widens to become the **medulla oblongata**, the most posterior portion of the brain. At its caudal end, the spinal cord narrows to a relatively thin strand of tissue called the *filum terminale.*

5. Surrounding the spinal cord and the brain are a set of three membranes, the **meninges** (Figure 36-5). The outermost layer, the **dura mater**, is the most apparent and adheres to the underside of the cranial and spinal bones. The dura mater is a tough, fibrous connective tissue sheath that you cut through to expose the spinal cord. The middle layer, the **arachnoid**, is very delicate and collapses when the spinal canal is opened. The innermost layer, the **pia mater**, adheres closely to the surface of the spinal cord and brain. If you can't identify the outer and inner meninges, attempt to locate them when you dissect the brain.

In certain severe viral or bacterial infections, the meninges around the spinal cord and/or brain can become inflamed. This serious condition is known as **meningitis**.

6. Note the origin of the **spinal nerves** on either side of the spinal cord (Figures 36-5 and 36-6). Each spinal nerve is composed of a **dorsal** and a **ventral root** (Figure 36-5). The dorsal root, carrying sensory impulses into the spinal cord, can be easily identified by the presence of a distinct swelling called the **dorsal root ganglion**. The ventral root, which carries motor impulses from the cord to some type of effector (a skeletal muscle, for example), has no ganglion and is not as easily seen as the dorsal root in a dorsal view.

7. Remove a short cross section (0.5 cm long) of the spinal cord and examine it with a dissection microscope. Observe the cut end and note the prominent H-shaped area in the center (Figure 36-5). This is the **gray matter**, which is composed of the cell bodies of neurons, their unmyelinated processes, and accessory cells. The **white matter** around the "H" is primarily made up of myelinated (surrounded by a white, fatty insulating material) neuron processes that conduct messages to and from the brain.

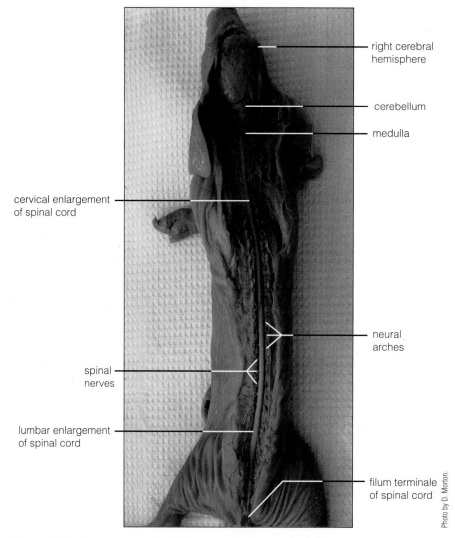

Figure 36-6   Dorsal view of the brain, spinal cord, and spinal nerves of a fetal pig.

B. Brain

1. Using your scalpel and scissors, make a longitudinal cut through the skin and muscle tissue of the dorsal portion of the head, beginning at the base of the snout and ending at the base of the skull. From the anterior portion of this incision, make a transverse cut to the angle of the jaws and another transverse incision from the base of the skull to a level just ventral to the ears. Peel the skin and muscle back to expose the skull.

2. Penetrate the skull with the lower blade of your scissors and make a shallow cut along its mid-dorsal line. Now make two cuts, about 2 cm apart, at right angles to the first cut. With forceps, carefully break off pieces of the skull until the entire dorsal and lateral areas of the brain are exposed (Figure 36-6).

3. If you were not able to identify the meninges covering the spinal cord, locate the dura mater lining the inside of the skull and the pia mater on the surface of the brain. As with the spinal cord, the arachnoid layer (middle layer) will not be apparent.
4. Cut the spinal cord at the base of the brain and carefully remove it from the skull. The brain is composed of the right and left *cerebral hemispheres* (collectively, the largest part of the brain, the **cerebrum**; Figure 36-6), separated by a prominent *longitudinal groove*; a smaller mass posterior to the cerebrum, the **cerebellum**; and the medulla oblongata or more simply, the *medulla*, poking out from under the cerebellum. The **pons** lies just anterior to the medulla on the ventral side of the brain.

In general, the more posterior portions of the brain direct most involuntary, unconscious, and mechanical processes. For example, the cerebellum unconsciously controls posture and contains motor programs (like computer programs) for many complex body movements. The cerebrum is responsible for such activities as reasoning, memory, conscious thought, language, and sensory decoding, activities that are generally associated with intelligence.

5. Place your pig and any excised organs back in the plastic bag. Tie it shut to prevent your pig from drying out. Dispose of any paper towels that contain preservative *as directed by your instructor.* Clean the tray, dissecting tools, and the laboratory table.

_____ **1.** The urogenital system refers to the
  (a) urinary and reproductive systems.
  (b) urinary and excretory systems.
  (c) reproductive system.
  (d) external genitalia.

_____ **2.** The ureters drain urine into the
  (a) renal pelvis.
  (b) cecum.
  (c) urinary bladder.
  (d) small intestine.

_____ **3.** The clitoris of the female and a portion
  of the penis of the male are homologous
  structures. This means they have a
  similar
  (a) function.
  (b) structure.
  (c) developmental origin.
  (d) origin and structure.

_____ **4.** The testes of a male differ from the ovaries
  of a female in that the testes
  (a) develop in the body cavity and
      migrate to a position outside of the
      body cavity.
  (b) require a slightly higher temperature
      than that of the body to produce
      viable gametes.
  (c) produce zygotes.
  (d) do both a and b.

_____ **5.** When a human male has a vasectomy, the
  operation involves
  (a) removal of the male gonads or testes.
  (b) removal of the urethra.
  (c) the severing of the vas deferens.
  (d) removal of the prostate gland.

_____ **6.** Semen contains
  (a) sperm.
  (b) the secretions of sex accessory glands.
  (c) eggs.
  (d) both a and b

_____ **7.** The functional unit of the kidney is the
  (a) renal pelvis.
  (b) ureter.
  (c) cortex.
  (d) nephron.

_____ **8.** The central nervous system of a mammal
  includes the
  (a) brain.
  (b) spinal cord.
  (c) brain and spinal cord.
  (d) brain, spinal cord, and every major
      nerve in the body.

_____ **9.** The brain is surrounded by a set of
  membranes called the
  (a) pleural membranes.
  (b) peritoneum.
  (c) pericardial membranes.
  (d) meninges.

_____ **10.** The largest part of the brain of a
  mammal is the
  (a) cerebrum.
  (b) cerebellum.
  (c) pons.
  (d) medulla oblongata.

EXERCISE **3 6**

Dissection of the Fetal Pig: Urogenital and Nervous Systems

Post-Lab Questions

36.1 Urogenital System

1. What is the urogenital system?

2. Briefly describe the functions of the kidney, ureters, bladder, and urethra in the adult male pig.

3. What is the vulva? Include in your answer a description of its components.

4. How do the uteruses of female pigs and humans differ? Include in your discussion the site of embryo implantation.

5. What is the inguinal canal in males and how does it form during development?

6. Identify the structures in the photo on the right.

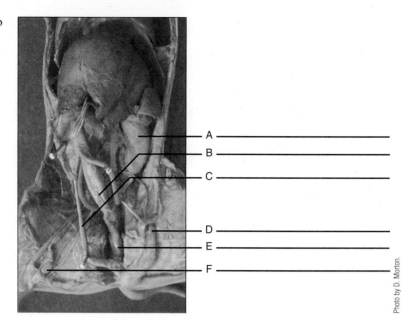

A —————————————————

B —————————————————

C —————————————————

D —————————————————

E —————————————————

F —————————————————

Photo by D. Morton.

36.2 Kidney

7. Name the functional unit of the kidney. Briefly describe how it operates.

36.3 Nervous System

8. Describe the basic organization of the nervous system of a mammal.

9. List the three meninges in order from the surface of the brain to the inside of the skull.

Food for Thought

10. Search the Internet for sites that describe kidney transplants and dialysis treatments. List two sites and briefly summarize their contents.

http:// ————————————————————————————

————————————————————————————

————————————————————————————

http:// ————————————————————————————

————————————————————————————

————————————————————————————

Human Sensations, Reflexes, and Reactions

OBJECTIVES

After completing this exercise, you will be able to

1. define *sensory receptors, stimuli, sensory neurons, motor neurons, effectors, interneurons, integration, sensations, perception, proprioception, modality, free neuron endings, encapsulated neuron endings (Meissner's and Pacinian corpuscles), sensory projection, phantom pain, sensory adaptation, reflex, reflex arc, stretch reflexes, patella reflex, muscle spindle, pupillary reflex, swallowing reflex, reaction;*

2. describe the flow of information through the nervous system;

3. state the nature and function of sensations;

4. describe a stretch reflex;

5. describe the pupillary reflex;

6. distinguish between a reflex and a reaction;

7. measure visual reaction time.

Introduction

Interactions between parts of your body and between your body and the external environment depend on the uninterrupted flow of information through the nervous system (Figure 37-1).

Sensory receptors detect changes in the energy of variables (**stimuli**) such as an increase in blood pressure or a decrease in skin temperature. The sensory receptors stimulated are typically those most sensitive to the particular stimulus type. **Sensory neurons** transmit this information to the brain and spinal cord. **Motor neurons** transmit information from the brain and spinal cord to the **effectors**, which bring about the appropriate response. The brain and spinal cord together make up the central nervous system (CNS).

Effectors are muscles or glands. *Somatic motor neurons* control skeletal muscles that usually respond to the external environment. *Autonomic motor neurons* control smooth muscles, cardiac muscle, and glands that usually respond within the body. The processes of sensory and motor neurons along with connective tissue are located in the *nerves* of the peripheral nervous system (PNS).

Within the CNS, sensory neurons can directly stimulate motor neurons; more frequently though, the sensory neurons stimulate one or more **interneurons**, which lie entirely within the CNS. In turn, these interneurons stimulate the motor neurons, other interneurons, or both. The sum of all these interconnections allows for **integration**, a process during which sensory and other information is processed and appropriate actions taken.

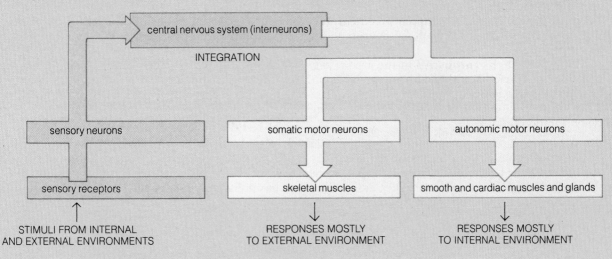

Figure 37-1 Flow of information through the nervous system.

The *conscious mind*, which is largely located in the forebrain, feels some of the sensory information as **sensations**. The brain understands this information (**perception**) and uses it to initiate responses (e.g., avoids a source of pain or finds water and drink when thirsty).

Our bodies have sensory receptors for light, sound, chemicals, temperature, tissue damage, and mechanical displacement. Sensations we feel include sight, hearing, taste, smell, hot, cold, pain, touch, pressure, vibration, equilibrium, and **proprioception** (knowledge of the position of the body and movement of the various body parts). There are also a number of complex sensations such as thirst, hunger, and nausea.

Sensory receptors and the sensations they produce have three characteristics: *modality, sensory projection*, and *sensory adaptation*. These characteristics can be easily demonstrated by investigating the skin's sensory receptors.

MATERIALS

Per student pair:

- compound microscope
- prepared slide of mammalian skin stained with hematoxylin and eosin (containing Meissner's corpuscles) and ideally, Pacinian corpuscles (optional)
- felt-tip, nonpermanent pen
- bristle (those from a moderately sized house paint brush will do)
- dissecting needle
- scientific calculator
- two blunt probes in a 250-mL beaker of ice water

- two blunt probes in a 250-mL beaker of hot tap water (the hot water will have to be replenished every 5 minutes)
- ice bag
- camel-hair brush
- reflex hammer
- three 1000-mL beakers, one containing ice water, one 45°C water, and one room temperature water
- tissue paper

Per lab room:

- demonstration slide of a Pacinian corpuscle

PROCEDURE

A. Modality

Modality is the particular sensation that results from the stimulation of a particular sensory receptor. For example, the modalities of taste—bitter, sour, salty, sweet, and umami (savory)—are associated with five different types of taste buds. However, although every sensory receptor is most sensitive to one type of stimulus, the modality actually depends on where in the brain the sensory neurons from the sensory receptor (or the interneurons to which they connect) terminate in the brain. Modality cannot be encoded in the messages carried by sensory neurons, because every impulse (action potential) in that message is identical. The only information neurons transmit is the absence or presence of a stimulus and (if one is present) its intensity—low stimulus intensities produce a low frequency of impulses, and high stimulus intensities produce a high frequency of impulses.

1. Examine a prepared section of skin (Figure 37-2). Two categories of sensory receptors are present: **free neuron endings** and **encapsulated neuron endings**.

Free neuron endings are almost impossible to see in typically stained sections, but note their distribution in Figure 37-2. Stimulating different free neuron endings produces sensations of pain, crude touch, and perhaps cold and hot. Encapsulated neuron endings consist of neuron tips surrounded by a connective tissue capsule.

Find **Meissner's corpuscles** in the *dermal papillae* (Figure 37-2). Meissner's corpuscles are sensory receptors for fine touch and low frequency vibration. Now look for **Pacinian corpuscles** between the dermis and hypodermis. Pacinian corpuscles look like a cut onion and are sensory receptors for pressure and high frequency vibration. Not all skin sections contain a Pacinian corpuscle. If you cannot locate one, look at the demonstration slide.

2. With a felt-tip, nonpermanent ink pen, draw a 25-cell, 0.5-cm grid as shown in Figure 37-3 on the inside of your lab partner's forearm, just above the wrist.
3. You are now the investigator, and your lab partner is the subject. At this point, ask your lab partner to close his or her eyes. Using a bristle, touch the center of each box in the grid, if the bristle bends, you are pressing too hard. Ask your lab partner to announce when a touch is felt. Do not count responses given when you remove the bristle. Just count those that coincide with the initial touch. Mark each positive response with a T in the upper left-hand corner of the corresponding grid in Figure 37-3.

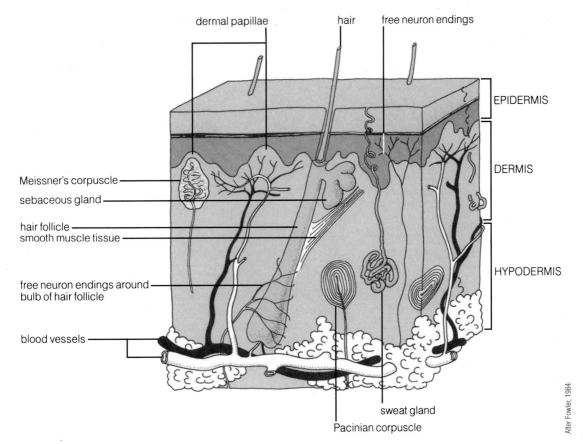

dermal papillae   hair   free neuron endings

EPIDERMIS

DERMIS

Meissner's corpuscle

sebaceous gland

hair follicle
smooth muscle tissue

HYPODERMIS

free neuron endings around
bulb of hair follicle

blood vessels

sweat gland
Pacinian corpuscle

After Fowler, 1984.

Figure 37-2 Diagram of skin.

4. Repeat the above with a clean dissecting needle. This time, if you feel a prick, mark P for "pain" in the upper right-hand corner of the corresponding box in Figure 37-3.

> **CAUTION**
>
> Do not press; simply let the tip of the dissecting needle rest on the surface of the skin.

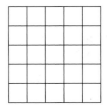

Figure 37-3 Grid for testing skin stimuli and recording modality data.

5. Repeat the above with a chilled blunt probe. Before using the blunt probe, dry it with tissue paper. The blunt probe will warm up over time, so switch it with the second chilled blunt probe every five trials. This time, mark each positive response with a C for "cold" in the lower left-hand corner of the corresponding box in Figure 37-3.

6. Repeat the above with a heated blunt probe. Before using the blunt probe, dry it with tissue paper. Use the two blunt probes alternately every five trials. This time, mark each positive response with an H for "hot" in the lower right-hand corner of the corresponding box in Figure 37-3.

7. Record the total number of positive responses for each stimulus in Table 37-1. Calculate the density of sensory receptors for each modality (multiply the number of positive responses for each stimulus by 4) and record them in the third column of Table 37-1.

8. Repeat steps 2–7, only this time you are the subject and your lab partner is the investigator.

9. Does each cell in the grid contain a sensory receptor for all four modalities studied? _____ (yes or no)

10. Can you see a pattern or patterns in the distribution of positive responses marked in Figure 37-3? _____ (yes or no)

11. Are the densities for the sensory receptors for each modality the same? _____ (yes or no)

TABLE 37-1	Positive Identifications to Stimuli Applied to 25 Cells in a 0.25-cm² Patch of Skin	
Stimulus	Number of Responses	Density (Responses/cm²)
Touch		
Pain		
Cold		
Hot		

© Cengage Learning 2013

B. Sensory Projection

All sensations are felt in the brain. However, before the conscious mind receives a sensation, it is assigned back to its source, the sensory receptor. This phenomenon is called **sensory projection**. This is a very important characteristic of sensations because it allows the conscious mind to perceive the body as part of the world around it. You have probably experienced sensory projection. A common example is the "pins and needles" you feel in your hand and forearm when you accidentally jar the nerve that passes over the inside of the elbow (so-called funny bone). The sensory neurons in the nerve are stimulated, and your brain projects the sensation back to the sensory receptors. Another example is the **phantom pain** and other sensations that recent amputees sometimes "feel" in missing limbs. This occurs because the sensory neurons that once served the missing body part are activated by the trauma of the amputation.

1. Obtain an ice bag from the freezer.
2. Hold the ice bag against the inside of your elbow for 2 to 5 minutes.
3. Describe any sensations felt in the hand or forearm for your lab partner to record them in Figure 37-4.
4. Continue to hold the ice bag on your elbow and check for any loss of sensation by your lab partner gently stroking your forearm and hand with a camel-hair brush. Sensations may also be felt after the ice bag is removed.
5. If no results are obtained, try tapping the inside of the elbow just above the funny bone with the reflex hammer.
6. Similarly test your lab partner.
7. What can you conclude about sensory projection and the sensory receptors on the surface of the hand and forearm?

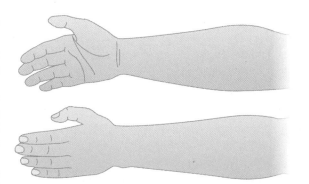

Figure 37-4 Front and back views of forearm and hand for recording projection data.

_____

_____

_____

C. Sensory Adaptation

The intensity of the signal produced by a sensory receptor depends in part on the strength of the stimulus and sometimes on the degree to which the sensory receptor was stimulated before the current stimulus. Most sensory receptors undergo **sensory adaptation** to a constant stimulus over time. For example, when you first enter a dark room after being in bright light, you cannot see. After a while, your photoreceptors adapt to the new light conditions, and your vision improves.

1. Partially fill each of three 1000-mL beakers with ice water, water at room temperature, and water at about 45°C.
2. Place one hand in the ice water and one in the warm water. After 1 minute, place both hands *simultaneously* in the water at room temperature.
3. Describe the sensation of temperature in each hand to your lab partner, who should record these descriptions in Table 37-2.

TABLE 37-2	Sensations Felt When Preadapted Hands Are Placed in Room-Temperature Water
Relative Temperature of Preadaption	Result
Cold	
Warm	

© Cengage Learning 2013

4. What can you conclude about the skin sensory receptors for temperature and their capacity for sensory adaptation?

_____

_____

_____

What about the ability of other kinds of sensory receptors to adapt? Use your own experiences to answer question 4 above for touch, smell, and pain. (Hint for touch: compare how your clothes feel after a morning shower with how they feel later.)

_____

_____

_____

_____

_____

37.2 Reflexes

A **reflex** is an involuntary response to the reception of a stimulus. The simplest **reflex arc** consists of a sensory receptor, sensory neuron, motor neuron, and effector. *Involuntary* means that your conscious mind does not decide the response to the stimulus. However, the conscious mind may be aware after the reflex has taken place. Reflexes of which we are unaware occur most often in the internal environment (e.g., reflexes involved in adjustments of blood pressure).

MATERIALS

Per student pair:

- reflex hammer
- penlight

PROCEDURE

A. Stretch Reflexes

Stretch reflexes are an example of the simplest type because the interneurons are not directly involved (Figure 37-5). The sensory neuron connects directly with the motor neuron in the spinal cord. Stretch reflexes are important in controlling balance and complex skeletal muscular movements such as walking. Physicians often test these reflexes during physical examinations to check for spinal nerve damage. You have probably experienced one of these tests, the **patella reflex**. In this test, the sensory receptor is the **muscle spindle** in the quadriceps femoris muscle group located on the front of the thigh, which is attached through its tendon and the patellar ligament to the top of the front surface of the tibia. The tibia is the larger of the two lower leg bones. The patella (kneecap) is embedded in the middle of the combined tendon/ligament. The muscle spindle detects any stretching of the muscle. The effector is the muscle itself.

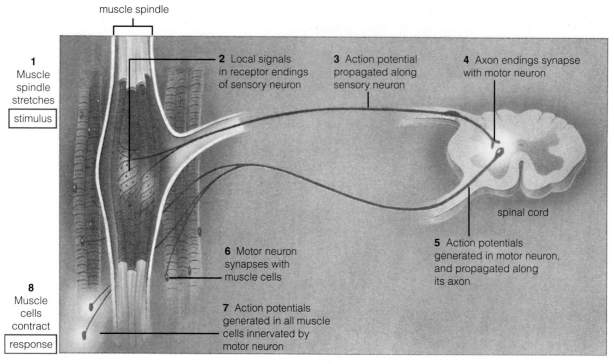

muscle spindle

1 Muscle spindle stretches

stimulus

2 Local signals in receptor endings of sensory neuron

3 Action potential propagated along sensory neuron

4 Axon endings synapse with motor neuron

spinal cord

5 Action potentials generated in motor neuron, and propagated along its axon

6 Motor neuron synapses with muscle cells

7 Action potentials generated in all muscle cells innervated by motor neuron

8 Muscle cells contract

response

Figure 37-5 Parts of a stretch reflex. Numbers indicate the sequence of events of this simple reflex, starting with the stimulus—the stretching of the muscle spindle.

1. Sit on a clean lab bench and shut your eyes.
2. When you are not expecting it, your lab partner's role is to tap the patella ligament with a reflex hammer (Figure 37-6). Describe the response.

_____

_____

If you have trouble producing a response, repeat steps 1 and 2, but this time distract yourself by counting backward from 10.

3. Even with your eyes shut, are you aware of the stimulus and the response? _____ (yes or no) This is because of pressure sensory receptors that sense the tap and because of proprioceptors that sense movement of the leg.
4. Stretch reflexes are *somatic reflexes* because they involve somatic motor neurons and skeletal muscles. Can you willfully inhibit a stretch reflex? _____ (yes or no)
5. Similarly test your lab partner.

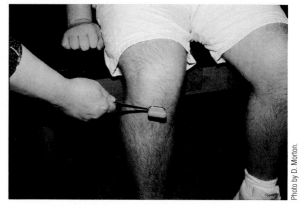

Photo by D. Morton.

Figure 37-6 Area to tap to produce patella reflex.

B. Pupillary Reflex

1. Shine the penlight into your lab partner's eyes. How is the size of the pupil (the diameter of the opening into the eye that is surrounded by the pigmented iris) affected? It gets _____ (smaller, larger).

2. Now turn off the penlight. How is the size of the pupil affected this time? It gets _____ (smaller, larger).

3. Repeat steps 1 and 2. Which is faster, constriction of the iris (which makes the pupil smaller) or dilation of the iris (which makes the pupil larger)? _____

4. Without seeing it, are you aware of the pupil's changing diameter? _____ (yes or no)

5. The **pupillary reflex** is an autonomic reflex because it involves an autonomic motor neuron and in this case, smooth muscle. Can you willfully inhibit the pupillary reflex? _____ (yes or no)

6. Similarly test your lab partner.

C. Complex Reflexes

Complex reflexes involve many reflex arcs and interneurons. A good example is swallowing. The stimulus in the **swallowing reflex** is the movement of saliva, food, or drink into the posterior oral cavity. The response is swallowing.

1. Cup your hand around your neck and swallow. Feel the complex skeletal muscular movements involved in swallowing. Do you consciously control all these muscles? _____ (yes or no)
2. Test whether it is possible to swallow several times in quick succession. Can you do this? _____ (yes or no)
3. Explain this result. (Hint: it has something to do with the stimulus.)

_____

_____

_____

_____

_____

4. What part of swallowing does your conscious mind control, and what part is a reflex?

_____

_____

_____

37.3 Reactions

A **reaction** is a voluntary response to the reception of a stimulus. *Voluntary* means that your conscious mind initiates the reaction. An example is swatting a fly when it has landed in an accessible spot. Because neurons must carry the sensory message to the cerebral cortex and the message to react back to the motor neuron, a reaction takes more time than a reflex. *Reaction time* is the sum of the time it takes for

- The stimulus to reach the sensory receptor
- The sensory receptor to process the message
- A sensory neuron to carry the message to the integration center
- The integration center to process the information
- A motor neuron to carry the response to the effector
- The effector to respond

Visual reaction time can easily be measured with a reaction-time ruler. This device makes use of the principle of progressive acceleration of a falling object.

MATERIALS

Per student pair:

- reaction-time ruler (Reaction Time Kit available from Carolina Biological Supply Company)
- chair or stool
- scientific calculator

PROCEDURE

1. Sit on a chair or stool (Figure 37-7).
2. Your lab partner stands facing you and holds the *release end* of the reaction-time ruler with the thumb and forefinger of the dominant hand at eye level or higher.

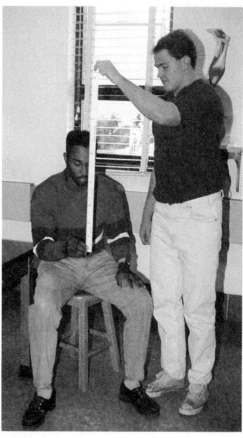

Figure 37-7 Two students measuring visual reaction time.

3. Position the thumb and forefinger of your dominant hand around the *thumb line* on the ruler. The space between the thumb and forefinger should be about 1 inch.
4. Tell your lab partner when you are ready to be tested.
5. Any time during the next 10 seconds, your lab partner releases the ruler.
6. Catch the ruler between the thumb and forefinger as soon as it starts to fall. The line under your thumb represents visual reaction time in milliseconds.
7. Your lab partner reads your reaction time from the ruler and records it in Table 37-3.
8. Repeat steps 2 to 7 10 times and calculate the average reaction time from the 10 trials.
9. Similarly test your lab partner.
10. The reaction times of most of the 10 trials should be similar, but perhaps the first few or one at random may be relatively different from the others. If this is true for your own or your lab partner's data, suggest some reasons for this variability.

_____

_____

_____

_____

_____

11. If opportunity and interest allow, the *Reaction Time Kit Instructions* booklet has a number of suggestions for other experiments you can easily do with the reaction-time ruler.

TABLE 37-3 Reaction-Time Data		
Trial	**Subject 1**	**Subject 2**
1		
2		
3		
4		
5		
6		
7		
8		
9		
10		
Total		
Average (Total/10)		

© Cengage Learning 2013

_____ 1. Neurons that carry messages from sensory receptors to the CNS are
(a) sensory.
(b) motor.
(c) interneurons.
(d) both a and b

_____ 2. Neurons that carry messages from the CNS to effectors are
(a) sensory.
(b) motor.
(c) interneurons.
(d) both a and b

_____ 3. Neurons that carry messages within the CNS are
(a) sensory.
(b) motor.
(c) interneurons.
(d) autonomic.

_____ 4. Knowledge of the position and movement of the various body parts is
(a) modality.
(b) sensory projection.
(c) sensory adaptation.
(d) proprioception.

_____ 5. Skin contains
(a) free neuron endings.
(b) encapsulated neuron endings.
(c) no nervous tissue.
(d) both a and b

_____ 6. Which characteristic of sensory receptors does phantom pain illustrate?
(a) modality
(b) sensory projection
(c) sensory adaptation
(d) proprioception

_____ 7. A simple reflex arc is made up of a sensory receptor and
(a) a sensory neuron.
(b) a motor neuron.
(c) an effector.
(d) all of the above

_____ 8. A stretch reflex is
(a) somatic.
(b) autonomic.
(c) both a and b
(d) none of the above

_____ 9. A pupillary reflex is
(a) somatic.
(b) autonomic.
(c) both a and b
(d) none of the above

_____ 10. A reaction is
(a) a reflex.
(b) involuntary.
(c) voluntary.
(d) both a and b

EXERCISE **37**

Human Sensations, Reflexes, and Reactions

Introduction

1. Where in the brain does your consciousness reside?

37.1 Sensations

2. In your own words, define the following terms:

   (a) modality

   (b) sensory projection

   (c) sensory adaptation

3. Identify the structures indicated in these photos.

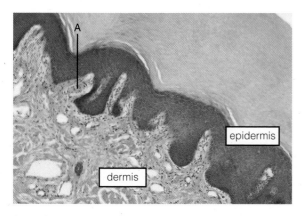

(100×)

A. _____
B. _____

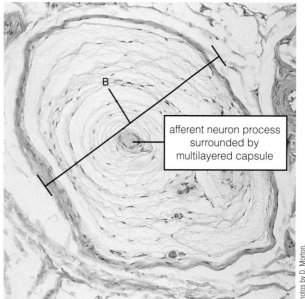

afferent neuron process surrounded by multilayered capsule

(110×)

37.2 Reflexes

4. Diagram the basic steps of a stretch reflex arc.

5. How does the patella reflex differ from the pupillary reflex?

37.3 Reactions

6. Indicate whether the following actions are caused by reactions or reflexes by putting a check in the correct column in the table.

Action	Reaction	Reflex
A baby wetting a diaper		
Braking a car to avoid an accident		
Withdrawing your hand from a hot stove surface		
Sneezing		
Waving to a friend across the street		

Food for Thought

7. What are the advantages and disadvantages to an organism of sensory receptor adaptation?

8. All animals do not perceive the external environment in exactly the same way. List some examples from your own knowledge and readings in the textbook.

9. To survive, an animal needs all its sensory receptors working, and even then, it cannot fully sense the external environment. What extra sensory receptors do you think would be advantageous to the survival of humans in this modern world?

10. Why is it advantageous for organisms *not* to be consciously aware of all the activity in their nervous systems?

Structure and Function of the Sensory Organs

OBJECTIVES

After completing this exercise, you will be able to

1. define *visual acuity, myopia, hyperopia, astigmatism, near point of eye, presbyopia;*
2. explain the differences among photoreceptors, mechanoreceptors, and chemoreceptors;
3. recognize and describe the structures of the external eye, eyeball, and ear;
4. determine visual acuity and the presence or absence of astigmatism;
5. describe the various reflexes important to proper vision;
6. recognize the retina, organ of Corti, taste buds, and olfactory epithelium;
7. explain briefly the function of rods and cones;
8. describe afterimages;
9. explain how we sense the source of a sound;
10. discuss the similarities and differences between the senses of taste and smell.

Introduction

What we know about the world around us is the direct result of the activity of our sensory receptors in sensory organs and structures on the surface of our bodies, especially our heads. These sensory organs and structures contain sensory receptors that receive information about the form and amount of light, atmospheric sound waves, movements of our heads, chemicals that enter our mouths and nostrils, and other stimuli. They contain specialized neurons and sometimes accessory cells that process this information and transform it into bioelectrical energy. This bioelectrical energy produces impulses in sensory neurons that extend from the sensory receptors along neuron pathways to particular parts of our brains, where the impulses are received and perceived as particular sensations. Typically, the frequency of the impulses is interpreted as intensity. Being aware of the nature and limitations of our senses is crucial to understanding ourselves.

38.1   The Eyes and Vision

Of all our senses, vision is the most developed. The retina of the eye contains **photoreceptors** (sensory receptors sensitive not only to light intensity but also to particular groups of wavelengths of light, which the brain interprets as color). You will discover that the eye is similar to a digital camera in structure. For example, our eyes use a lens to focus an image on an array of light-sensitive receptors.

MATERIALS

Per student:

- compound light microscope
- section of eye with retina and optic nerve

Per student pair:

- two sheets of white paper, one blank and one with dime-sized black dot; black and colored construction paper; a white paper dot cut-out and cut-outs of differently colored paper shapes (heart, star, and so on)
- mirror

- preserved sheep eye
- dissecting scissors
- blunt probe or dissecting needle
- forceps
- scalpel
- measuring tape
- penlight

Per student group (4):

- eye model

Per lab room:

- liquid waste disposal bottle
- boxes of different sizes of lab gloves
- safety goggles
- Snellen chart

- astigmatism chart
- place to stand with a taped line 10 ft away and a red taped line 20 ft away from charts

PROCEDURE

A. External Eye Structures

The surface of the eyes is kept moist by *lacrimal fluid* (tears) secreted by the *lacrimal gland*, which is located under the skin above and just lateral to the center of the visible portion of the eyeball (Figure 38-1). Its ducts open under the lateral third of the upper *eyelid*. This fluid is swept across the visible surface of the eyeball when we blink. Excess fluid drains into the openings (*lacrimal puncta*) of ducts located at the tip of small elevations (*lacrimal papillae*) situated on the medial rims of the eyelids. These canals join to the *nasolacrimal sac*, which connects via the *nasolacrimal duct* to the *nasal cavity*.

1. *Wash your hands thoroughly.* Consult Figure 38-1 and examine one of your eyes with a mirror. Identify the **eyelids, eyelashes, lacrimal papillae, lacrimal punta,** and **lacrimal caruncle** (a reddish body located at the inner corner of each eye that contains lubricating glands). From your own experience, suggest an additional function for eyelids and eyelashes other than moving tears over the eyeball.

   _____

   _____

   _____

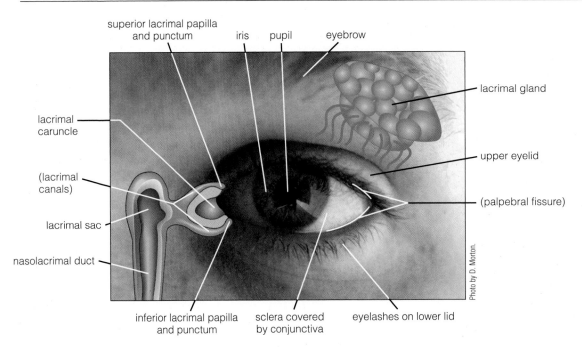

Figure 38-1 Human eye.

2. Which eyelid moves when you open and close your eyes?

   _____

3. Why do your eyes water when you have a cold?

   _____

   _____

4. Open your eyes wide. Note that only a small part of the eyeball can be seen through the fissure in the skin located between the eyelids. The "white of the eye" is the **sclera**. It is actually covered by an unseen transparent epithelial membrane, the **conjunctiva**, which is continuous with the epithelium lining the insides of the eyelids. Identify the equally transparent **cornea**, a dome of sclera at the center of the visible eyeball. Through it you can see the pigmented **iris**, which functions as a diaphragm that opens (dilates) and closes (constricts) the **pupil** (the hole through which light passes into the eyeball).

B. Structure and Function of the Eyeball

Many human eye models are available. Figure 38-2 shows two such models.

1. Examine a model of the eye and, consulting Figure 38-2, identify the following features:
   - **Outer eye muscles:** skeletal muscles that move the eyeball
   - Sclera and cornea
   - **Choroid:** the black-pigmented middle layer of the eye. This layer is continuous with the blue, brown, or other-colored iris, which surrounds the pupil. Just behind the iris and also part of the choroid is the ring-like **ciliary body**. Both the iris and ciliary body have *inner eye muscles* composed of smooth muscle tissue.

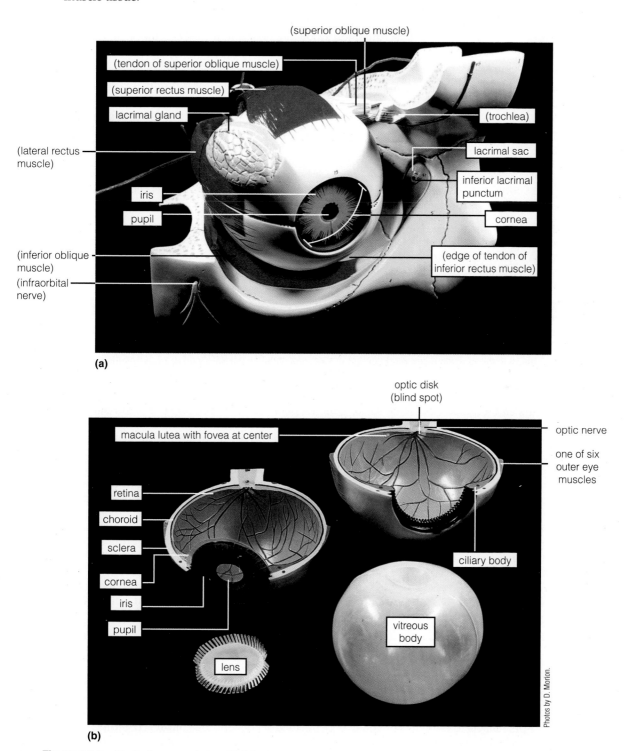

Figure 38-2 Typical eye models, with labels.

- **Retina:** the inner layer of the eye, which lines the back two-thirds of the eyeball. The retina contains photoreceptors. At the optical center of the retina is a slight depression called the **fovea**, which contains the greatest concentration of photoreceptors for color and is surrounded by a yellowish spot, the **macula lutea**. Neuron pathways from the photoreceptors converge to form the root of the optic nerve (sometimes called the optic disk). Because there are no photoreceptors in this region, its presence creates a hole in the image perceived by the brain and therefore it is also called the **blind spot**.
- **Optic nerve:** connects the retinal neuron pathways to the brain. Note that its outer layer is continuous with the sclera.
- **Lens:** a transparent, convex structure suspended by threadlike ligaments connected to the ciliary body and located just behind the pupil.
- **Spaces:** the space behind the lens is called the **posterior cavity** (Figure 38-3a), which is filled with a thick, gooey fluid or humor, forming a transparent **vitreous body**. The space between the lens and the cornea, including the pupil, is called the **anterior cavity** and is filled by the clear, watery **aqueous humor**. The iris divides the anterior cavity into an anterior chamber in front of it and a posterior chamber behind.

2. Use your compound light microscope to look at a prepared slide (or slides) with a section (or sections) of the optic nerve and retina. Note the blind spot (Figure 38-3a). From the inside toward the outside of the eyeball, identify the three layers of neurons and the pigment epithelium that comprise the retina (Figure 38-3b). Photoreceptors comprise the layer of neurons closest to the pigment epithelium. There are two types of photoreceptors, **cones** and **rods**. Cones are concentrated in the center of the retina, in and around the fovea, and are stimulated only by bright light. Different levels of stimulation of the three subtypes of cones—red, green, and blue—allow for color vision. Color blindness results when one or two of the cone subtypes are impaired or missing. Rods are found away from the center of the retina and are more sensitive to light, functioning only in dim conditions.

C. Dissection of the Sheep Eye

Sheep eyes are a byproduct of the slaughtering of sheep for food. They are purchased from biological supply companies.

CAUTION

Sheep eyes are kept in preservative solutions. Use lab gloves whenever you handle a specimen. Wash any part of your body exposed to this solution with lots of water. If preservative solution is splashed into your eyes, wash them with the safety eyewash for 15 minutes. Even if you wear contact lenses, you should wear safety goggles during dissection and when observing or studying a dissection. Eyeglasses should suffice in a situation when the goggles do not fit over them.

---

1. Work in pairs and wear lab gloves and safety goggles. Obtain a preserved sheep eye and place it on a stack of several paper towels. Identify its external features (Figure 38-4a).
2. Use dissecting scissors to remove the whitish connective tissue obscuring the pinkish *eye muscles*. This should also expose the tough, white, fibrous connective tissue of the sclera (Figure 38-4b). In life, the outer eye muscles move the eyeball.
3. With a scalpel, make an incision about 1/4 inch in back of and parallel to the edge of the cornea into the *anterior cavity*. Use scissors to extend the incision completely around the eyeball and gently pull apart the two pieces. Identify the internal features of the eyeball illustrated in Figure 38-5.
4. Allow the *lens* and *vitreous body* of the posterior cavity to flow out of the posterior portion of the eyeball. Fill the posterior portion of the eyeball with water so as to flatten the delicate, whitish retina for examination. Identify the *blind spot* (Figure 38-5b) and, depending on the state of preservation, perhaps the macula lutea (Figure 38-2).
5. Examine the lens (Figure 38-5c) and vitreous body. One after the other, place them on an index card bearing a small letter *e*. If the lens is cloudy, do not use it. What is the effect of the lens and vitreous body on the image you see?

_____

_____

D. Vision

Now that you know the anatomy and function of the structures of the eye, let's investigate some of the important concepts related to the sense of vision—visual acuity, the lens and pupil accommodation reflexes, depth perception, the blind spot, and afterimages. **Visual acuity** refers to the "sharpness" of your vision, or how well you can see detail.

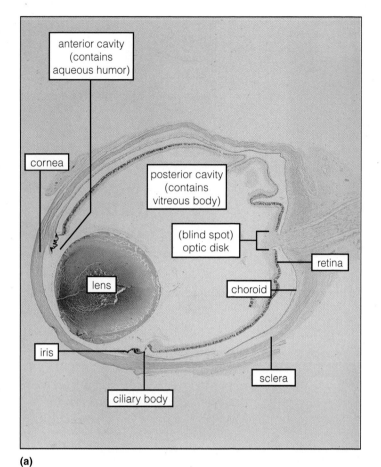

(a)

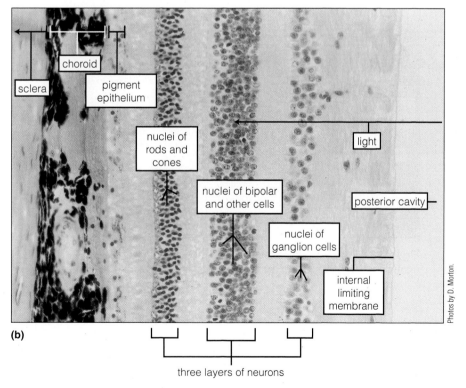

(b)

three layers of neurons

Figure 38-3 Sections of (**a**) back of the eyeball of a fetal eye (15×) and (**b**) a higher magnification view of the retina of a monkey (300×). The pigment epithelium and three layers of neurons comprise the retina.

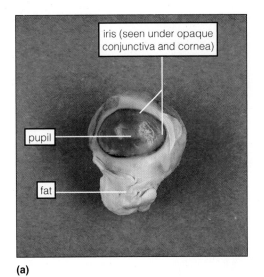

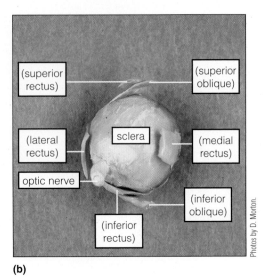

(a)   **(b)**

Photos by D. Morton.

Figure 38-4 (**a**) Anterior view of the external features and (**b**) posterior view with dissected muscles of a preserved sheep eye.

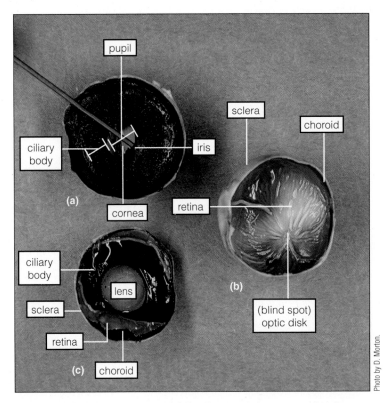

Photo by D. Morton.

Figure 38-5 Internal structures of a dissected sheep eye showing (**a**) inside of front of eyeball, (**b**) inside of back of eyeball, and (**c**) retina.

1. Determine your visual acuity using a Snellen chart as directed by your instructor. If you wear eyeglasses, perform this test both with and without them. Record your results in Table 38-1. A normal eye can read the line of letters marked 20 at 20 ft (red line) and is designated 20/20. If the eye can read only the letter marked 200, it is designated 20/200, and so on. Such an eye has **myopia** and is nearsighted. It is also possible for an eye to be farsighted (**hyperopia**). These conditions are usually the result of an elongated (myopia) or shortened (hyperopia) eyeball.

TABLE 38-1	Visual Acuity	
Eye	With Corrective Glasses or Contacts	Without Corrective Lenses
Right		
Left		

© Cengage Learning 2013

2. Visual acuity also can be reduced by **astigmatism**. Astigmatism is caused when one of the transparent surfaces (i.e., the cornea or lens) of the eye is not uniformly curved in all planes. Check for astigmatism by viewing a series of radiating lines on an astigmatism chart from a distance of 10 ft. To astigmatic individuals, some lines will look different, appearing sharper, thicker, or darker than other lines. Determine whether your right or left eye (or both) is astigmatic. If you wear eyeglasses, perform this test with them and then without them. Note the numbers of any lines that appear sharper, thicker, or darker in Table 38-2. If you wear glasses, detect the correction for astigmatism in your lenses by looking at the chart after rotating your lenses 90°.

TABLE 38-2	Astigmatism	
Eye	With Corrective Glasses or Contacts	Without Corrective Lenses
Right		
Left		

© Cengage Learning 2013

3. Your eyes accommodate focusing on objects at varying distances from the eye. Light rays are virtually parallel from an object 20 ft and farther away. Light rays from closer objects diverge, and these divergent rays must be bent at a sharper angle to focus them on the retina. **Lens accommodation** is accomplished by contracting the ciliary muscles to take tension off the lens, thus allowing it to bulge and become more convex.

Have your lab partner measure the distance from your eyes to the nearest point at which you can sharply focus on an object such as the words in this lab manual. Record this value in Table 38-3. Repeat this procedure with and without glasses, if you wear them. This distance is called the **near point** and is a relative measure of the ability of the lens to accommodate for close vision. Lens accommodation requires an elastic lens. As an individual grows older, the lens becomes more inelastic, and to focus on an object, it must be held farther and farther from the eye. However, the object then appears smaller and smaller. This condition is called **presbyopia**. To get letters large enough to read, older people use bifocals (or similar eyeglasses) or extra big print.

TABLE 38-3	Near Point	
Eye	With Corrective Glasses or Contacts	Without Corrective Lenses
Right		
Left		

© Cengage Learning 2013

4. To demonstrate pupil accommodation for changes in light intensity, shine a penlight intermittently into your lab partner's eye and note any changes in the diameter of the pupil. These responses are caused by the **photopupil reflex**. Describe these changes.

_____

_____

_____

Which occurs more rapidly? _____ (constriction or dilation)

Given changing light conditions, what advantage might this difference confer?

_____

_____

_____

   Note that when your lab partner switches focus from a distant to a nearby object, the pupils also constrict (**pupil accommodation for distance reflex**). This is done to block out the most divergent (most difficult to focus) rays of light.

5. Observe that when your lab partner switches focus from a distant to a nearby object, the eyeballs converge (**eye convergence reflex**). Convergence allows for binocular vision (overlapped images), which is necessary for depth perception. Gently press one eyeball out of line by pressing on an eyelid while viewing an object.

How many objects do you see? _____

6. Hold Figure 38-6 about 20 inches from your right eye. Close your left eye and place your right eye directly over the dot. Stare at the dot while bringing the page closer to the eye until the dot disappears. At this point, the image of the dot is falling on the blind spot of the retina.

7. Place a sheet of white paper with a black spot about the size of a dime onto another sheet of blank white paper. Under a bright light, stare at the black spot for 20 seconds, trying to move your head and eyes as little as possible. Slide off the top sheet, continuing to look at the blank sheet. Describe the **afterimage**. _____

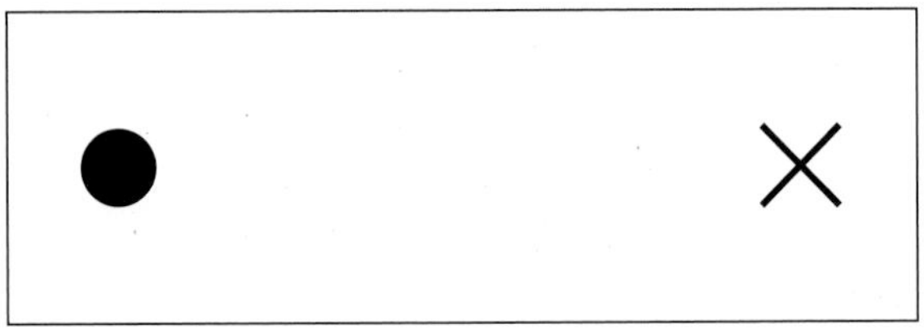

Figure 38-6   Demonstration of blind spot.

_____

_____

Repeat this test, substituting a top sheet of black construction paper with a white dot on it. Describe the afterimage.

_____

_____

Bits of bright red, blue, green, and yellow paper can also be used in this activity. Just as the afterimage of black is a more intense white and vice versa, color vision has afterimages that are essentially the "opposite" color of the original color. To demonstrate this, place a red heart on a green sheet of construction paper; then place this combination over the sheet of blank white paper. Under a bright light, stare at the red heart for 20 seconds, trying to move your head and eyes as little as possible. Slide off the top sheet, continuing to look at the blank white sheet. Describe the afterimage.

_____

_____

Repeat the experiment using a green heart and a piece of red construction paper and describe the results.

_____

_____

38.2   The Ears and Hearing

The ear contains several groups of mechanoreceptors (sensory receptors for mechanical energy). Depending on their specific location, their stimulation is interpreted by the brain as the sense of hearing or the sense of equilibrium.

MATERIALS

Per student:

- compound light microscope
- section of cochlea

Per student pair:

- mug-sized container stuffed to overflowing with cotton batting

Per student group (4):

- ear model

A. Structure and Function of the Ear

Except for the visible flaps (*auricles*) located on either side of the head, the paired ears are embedded in the temporal bones of the skull.

1. Examine a model of the ear and, consulting Figure 38-7, identify the following features:
   - **Outer ear:** includes **auricles** and **outer ear canal**, which funnel sound waves in air to the **tympanic membrane**.

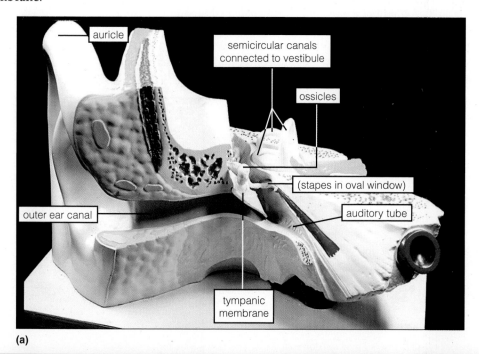

(a)

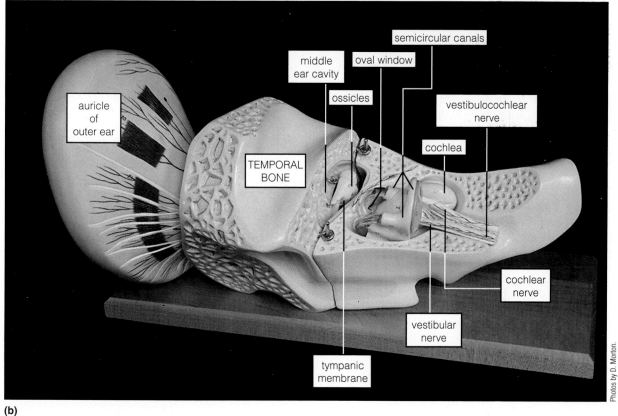

(b)

Figure 38-7   Typical ear models, with labels.

- **Middle ear:** contains inner ear **ossicles** (three small bones shaped like a hammer, anvil, and stirrup), which conduct vibrations in bone tissue from the tympanic membrane to the **oval window** of the cochlea. The middle ear space is connected to the upper pharynx (space behind the nasal and oral cavities) by the **auditory tube**, which functions to equalize air pressure on both sides of the tympanic membrane. You experience this when your ears "pop" in a car when driving quickly through steep changes in elevation.
- **Inner ear:** composed of the snail shell-like cochlea, vestibule, and semicircular canals. The **cochlea** contains fluid-filled spaces that conduct vibrations from the end plate of the stirrup-shaped ossicle, which is lodged in the **oval window**, to mechanoreceptors for hearing in the **organ of Corti** (or *spiral organ*; Figure 38-8). A membrane-covered *round window* dissipates old vibrations from the cochlea into the air in the middle ear space. The **semicircular canals** and **vestibule** contain fluid-filled sacs with structures containing mechanoreceptors that aid in equilibrium.

2. Observe a section of the cochlea with your compound light microscope. Consult Figure 38-8 and identify the *scala vestibuli, vestibular membrane, cochlear duct, organ of Corti, basilar membrane,* and *scala tympani.*

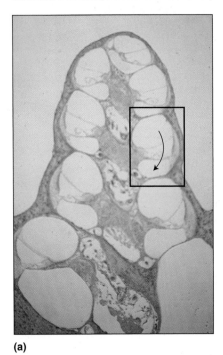

(a)

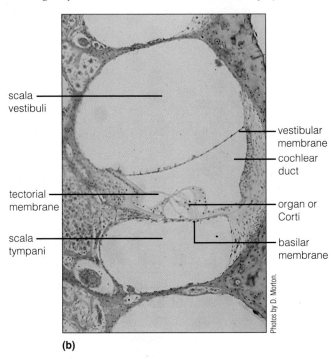

scala vestibuli

vestibular membrane

cochlear duct

tectorial membrane

organ or Corti

scala tympani

basilar membrane

Photos by D. Morton.

(b)

Figure 38-8 Cross section of the cochlea. The arrow shows direction taken by sound waves (25×). The boxed area in (**a**) is magnified in (**b**) 107×.

B. Sense of Direction of Sound

Why do we have two ears? Two ears enable our brains to perceive the direction of sound waves.

1. To demonstrate this, work in pairs. One of you sits on a stool with eyes closed. The other circles the stool quietly, stops, and calls the sitter's name in a high-pitched voice. (This might seem funny, but high-pitched sound does not penetrate bone as well as low-pitched sound.) Each time the sitter's name is called, the sitter tries to point to the caller. If a straight line from the sitter's finger "hits" the caller, place a check mark after trial 1 in the second column of Table 38-4. Repeat this test for a total of 10 trials.

2. Now cover one ear with a mug-sized container stuffed to overflowing with cotton batting and repeat step 1 ten more times. Use the third column of Table 38-4 to record "hits."

TABLE 38-4	Sense of Sound Direction	
Trial	**Ears Uncovered**	**One Ear Covered**
1		
2		
3		
4		
5		
6		
7		
8		
9		
10		
Total		

© Cengage Learning 2013

3. Total the number of check marks in both columns.
4. Consider the time of arrival and the relative intensity of the waves at each ear and suggest two ideas as to how the brain determines the direction of the sound source.

_____

_____

_____

_____

38.3   Taste and Smell

Chemoreceptors (sensory receptors for chemicals) are located in the taste buds of the tongue and in the olfactory epithelium of the nasal cavity.

MATERIALS

Per student:

- compound light microscope
- section of rabbit tongue
- section of nasal cavity

Per student group (4):

- box of tissues
- permanent marker
- four 250-mL Erlenmeyer flasks, one each with
    10% sucrose solution marked SW
    1% acetic acid solution marked SR
    5% NaCl solution marked SL
    0.5% quinine sulfate solution or tonic water marked BT
- three dishes containing cubes of a different vegetable or fruit (e.g., apple, onion, and potato)

Per lab room:

- box of applicator sticks
- container of small disposable beakers
- box of paper cups
- source of drinking water

PROCEDURE

A. Location of Chemoreceptors

Five types of taste buds function to detect sweet, salty, savory, acidic, and bitter substances in food and drink. Savory taste buds are also called umami after the name of the scientist who discovered them. Taste buds are probably sensitive to all of these substances but are particularly sensitive to one. A much larger number of different types of chemoreceptors in the olfactory epithelium detect a larger number of more vaporizable substances.

1. Examine a section of rabbit tongue with your compound light microscope. Look at the outer edge of the organ between the projections (foliate papillae) and identify the **taste buds** (Figure 38-9a). They consist of neuron endings surrounded by *taste cells* (chemoreceptors) with *microvilli* projecting through a *taste pore* and supportive epithelial cells.
2. Likewise, examine a section of the nasal cavity. Identify the **olfactory epithelium** lining the roof of the nasal cavity (Figure 38-10). Similar to taste buds, the olfactory epithelium contains neuron endings, chemoreceptors, and supportive epithelial cells.

B. Interaction between Taste and Smell

On an everyday level, we tend to mix up the sense of taste and smell. For example, how many times have you heard a statement like, "This pizza tastes great"? How much of this "great taste" is actually taste, and how much of it is smell?

1. Rinse your mouth with drinking water. Shut your eyes, pinch your nostrils closed, and have your lab partner place a small piece of vegetable or fruit on your tongue. Attempt to identify it. Record its identity or, if you cannot identify it, put a question mark in Table 38-5. With your nostrils still pinched closed, chew the piece thoroughly but do not swallow it. Again attempt to identify it. Record this and subsequent results in Table 38-5. Open your nostrils and identify the vegetable or fruit.

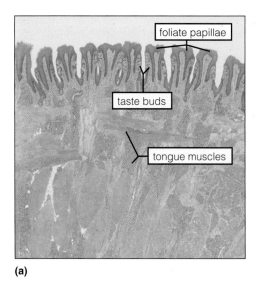

(a)

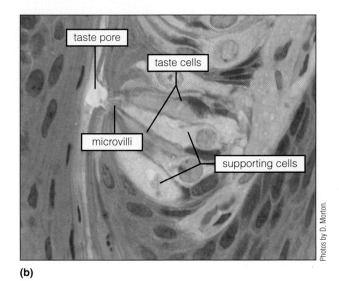

(b)

Figure 38-9 (a) Foliate papillae in rabbit tongue (25×) and (b) magnified taste bud (1000×).

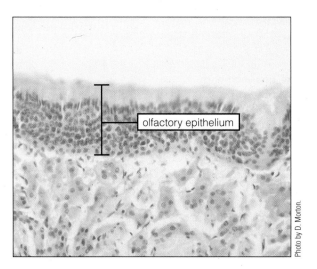

Figure 38-10 Olfactory epithelium (250×).

2. Repeat step 1 twice, using two more different vegetables or fruits.
3. In a strictly biological sense, when your nostrils are blocked by congestion, can you taste and smell as well as when they are open? Explain your answer.

_____

_____

_____

TABLE 38-5	Identity of Vegetable/Fruit		
Trial	**Placed on Tongue**	**After Chewing**	**Nostrils Open**
1			
2			
3			

_____ **1.** The cornea is part of the
   (a) sclera.
   (b) choroid.
   (c) retina.
   (d) lens.

_____ **2.** Parts of the choroid form
   (a) the lens.
   (b) the ciliary body.
   (c) the iris.
   (d) both b and c

_____ **3.** Photoreceptors are found in the
   (a) lens.
   (b) ciliary body.
   (c) iris.
   (d) retina.

_____ **4.** The space between the lens and iris is
   (a) the posterior cavity.
   (b) the posterior chamber.
   (c) the anterior chamber.
   (d) none of the above

_____ **5.** Nearsightedness is also called
   (a) myopia.
   (b) hyperopia.
   (c) presbyopia.
   (d) none of the above

_____ **6.** Ossicles are found in
   (a) the outer ear.
   (b) the middle ear.
   (c) the inner ear.
   (d) both a and c

_____ **7.** Sensory receptors found in the ear are
   (a) chemoreceptors.
   (b) mechanoreceptors.
   (c) photoreceptors.
   (d) none of the above

_____ **8.** Taste buds detect
   (a) light.
   (b) atmospheric sound waves.
   (c) chemicals in food and drink.
   (d) movements of the head.

_____ **9.** The olfactory epithelium is found in the
   (a) ear.
   (b) nasal cavity.
   (c) eye.
   (d) tongue.

_____ **10.** The most developed sense in humans is
   (a) hearing.
   (b) vision.
   (c) taste.
   (d) smell.

EXERCISE **38**

Structure and Function of the Sensory Organs

Post-Lab Questions

38.1 The Eyes and Vision

1. Define the following terms:
   (a) myopia

   (b) hyperopia

   (c) presbyopia

2. How does uncorrected myopia and hyperopia affect the near point?

3. What types and subtypes of photoreceptors are present in the retina? Briefly describe their function.

4. What is an afterimage? How is it formed?

5. List the eye structures that light passes through as it travels from outside the body to the retina.

38.2 The Ears and Hearing

6. List the ear structures that sound waves and resulting vibrations pass through as they travel from outside the body to the organ of Corti.

38.3 Taste and Smell

7. Why do humans and other animals have two nostrils?

8. When your nasal passages are blocked because of a heavy cold, food seems to lose its flavor. For example, pizza tastes like salt. Explain why this is so.

Food for Thought

9. According to your textbook and the Internet, in what part of the eye are cataracts located?

10. According to your textbook and the Internet, what is the relationship between the fluid in the anterior cavity and the disease glaucoma?

Human Skeletal and Muscular Systems

OBJECTIVES

After completing this exercise, you will be able to

1. define *bones, ligaments, joints, skeletal muscles, tendons, muscle tone, posture, sutures, synovial joints, diaphysis, epiphyses, compact bone, spongy bone, marrow cavity, lever;*

2. identify the major bones of the human skeleton;

3. describe the structure of a typical bone;

4. define *origin, insertion,* and *action* as these terms apply to skeletal muscles and their tendons;

5. distinguish between isometric and isotonic contractions of skeletal muscles;

6. give everyday and anatomical examples of the three classes of levers;

7. present a simple biomechanical analysis of walking.

Introduction

The skeletal system and muscular system are often considered together to stress their close structural and functional ties. These two systems often are referred to as the musculoskeletal system. They determine the basic shape of your body, support your other systems, and provide the means by which you move.

Bones are the main organs of the skeletal system. They are largely constructed of bone tissue, although all four basic tissue types are present. The places in the body where two or more bones are connected are called **joints**. The joints you are most familiar with are the shoulder, elbow, wrist, hip, knee, and ankle. However, there are many others. Around many joints, bones are held together by structures called **ligaments**. Ligaments are primarily dense connective tissue that is more or less elastic. Elastic ligaments around mobile joints stretch to allow movement.

Skeletal muscles are the main organs of the muscular system and are composed primarily of skeletal muscle tissue. Skeletal muscles are connected to bones by straplike dense fibrous connective tissue structures called **tendons**. Tendons are inelastic, so all of the force of skeletal muscle contraction is transferred to the skeleton.

When a skeletal muscle contracts, movement may or may not occur. If the skeletal muscle is allowed to shorten, the bone moves, and in doing so it moves some body part. On the other hand, if the skeletal muscle does not shorten, the tension in that muscle and in its tendons increases. All skeletal muscles exhibit tension or **muscle tone** except when you are asleep. This tension maintains **posture**—the ability to hold the body erect and to keep the position of its parts, all against the pull of gravity.

The organs of the skeletal and muscular systems have other functions. Bones protect internal organs (for example, the skull protects the brain, eyes, and ears). Bones also store minerals and produce blood cells in the bone marrow. When body temperature drops below a certain level, skeletal muscles produce heat by shivering.

39.1 Adult Human Skeleton *(About 45 min.)*

An articulated human skeleton is prepared by joining the degreased and bleached bones of an individual so that many of the bones can be moved as they were in life. Often, plastic casts of the original bones are used.

When fresh bones are prepared for study, their organic portion is lost. These bones consist only of the mineral portion of bone tissue. Original details remain, but the bones are brittle. *Therefore, you must handle bones gently. Use a pipe cleaner to point out details and never use a pencil or pen because it is very difficult to remove marks.*

MATERIALS

Per student:

- pipe cleaner
- compound microscope, lens paper, a bottle of lens-cleaning solution, a lint-free cloth
- prepared slides of
    a synovial joint
    a ground cross section of compact bone (optional)
    a cross section of a skeletal muscle

Per student group:

- articulated adult human skeleton (natural bone or plastic)
- femur
- femur that has been sawed into two halves lengthwise

Per lab room:

- labeled chart and illustrations of the adult human skeleton

PROCEDURE

A. Identification of Some Bones

There are 206 separate bones in the adult human skeleton. Using the labeled chart and illustrations of the human skeleton, identify the following bones on the articulated human skeleton and label them in Figure 39-1.

1. Axial skeleton

    - skull (28 separate bones, including middle ear bones) (Figure 39-2)
    - vertebrae (singular, *vertebra;* 26 separate bones, including the sacrum, which is composed of five fused vertebrae, and the coccyx, which is usually composed of four fused vertebrae) (Figure 39-3)
    - ribs (12 pairs of ribs for a total of 24 separate bones) (Figure 39-4)
    - sternum (three fused bones) (Figure 39-4)
    - hyoid (only bone that does not form a joint with another bone) (Figure 39-4)

2. Appendicular skeleton (these are all paired bones found on the right and left sides of the body)

    pectoral girdle (shoulder) (Figure 39-5)

    - scapula
    - clavicle

    arm (Figures 39-5, 39-6)

    - humerus
    - radius
    - ulna
    - carpals (8)
    - metacarpals (5)
    - phalanges (14)

    pelvic girdle (hip) (Figure 39-7)

    - coxal bone (three fused bones—pubis, ischium, ilium)

    leg (Figure 39-8)

    - femur
    - tibia
    - fibula
    - patella
    - tarsals (7)
    - metatarsals (5)
    - phalanges (14)

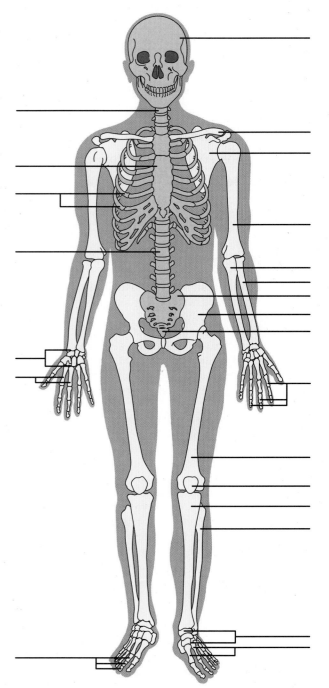

Figure 39-1   Label this front view of the adult human skeleton (axial portion shaded gray and appendicular portion colored yellow).
Labels: bones listed in section A, wrist, elbow, shoulder, hip, knee, ankle

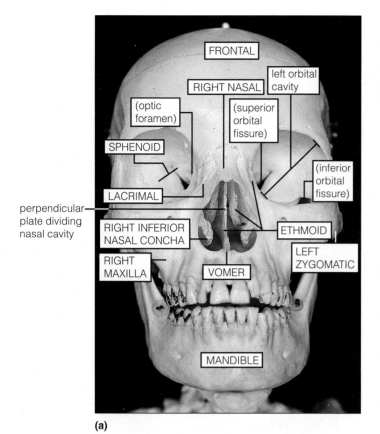

(a)

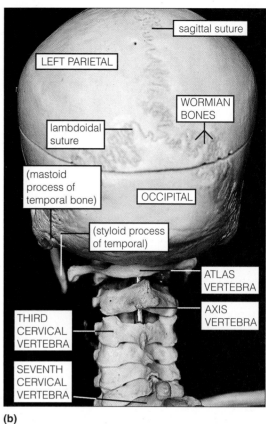

(b)

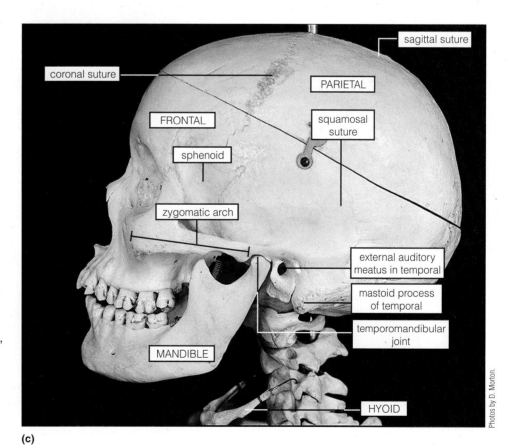

Figure 39-2 (a) Front, (b) back, and (c) side views of the skull. The names of bones are capitalized in this and subsequent figures to clearly separate them from other features.

(c)

Photos by D. Morton.

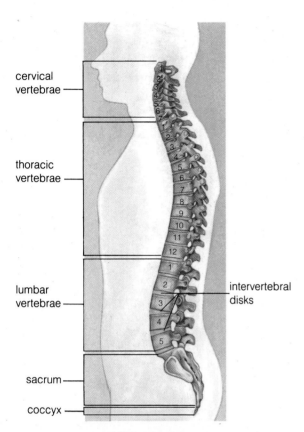

cervical vertebrae

thoracic vertebrae

lumbar vertebrae

intervertebral disks

sacrum

coccyx

Figure 39-3 Side of the vertebral column.

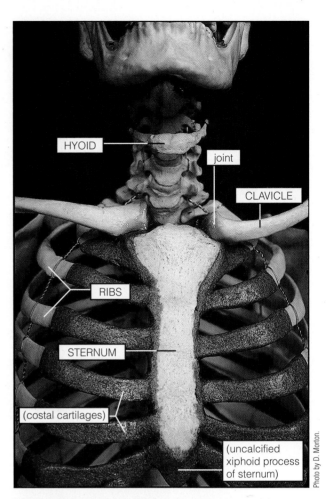

HYOID

joint

CLAVICLE

RIBS

STERNUM

(costal cartilages)

(uncalcified xiphoid process of sternum)

Photo by D. Morton.

Figure 39-4 Anterior view of skeleton of upper trunk.

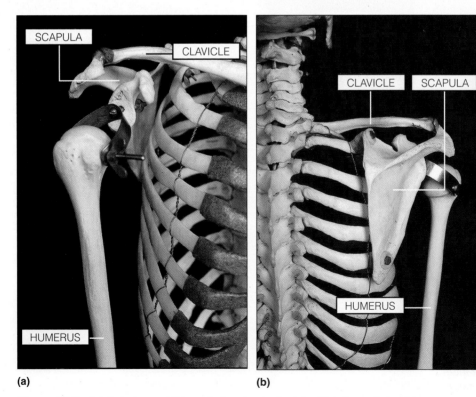

SCAPULA

CLAVICLE

HUMERUS

(a)

CLAVICLE

SCAPULA

HUMERUS

(b)

Photos by D. Morton.

Figure 39-5 (a) Anterior and (b) posterior views of pectoral girdle and shoulder.

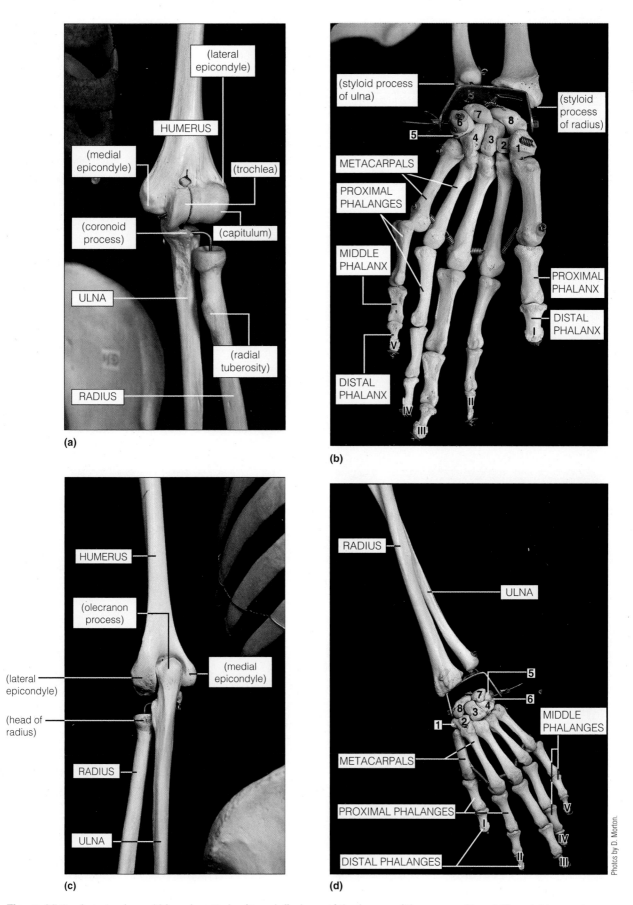

(a)

(lateral epicondyle)

HUMERUS

(medial epicondyle)

(trochlea)

(coronoid process)

(capitulum)

ULNA

RADIUS

(radial tuberosity)

(b)

(styloid process of ulna)

(styloid process of radius)

METACARPALS

PROXIMAL PHALANGES

MIDDLE PHALANX

PROXIMAL PHALANX

DISTAL PHALANX

DISTAL PHALANX

(c)

HUMERUS

(olecranon process)

(medial epicondyle)

(lateral epicondyle)

(head of radius)

RADIUS

ULNA

(d)

RADIUS

ULNA

MIDDLE PHALANGES

METACARPALS

PROXIMAL PHALANGES

DISTAL PHALANGES

Photos by D. Morton.

Figure 39-6 Anterior (**a** and **b**) and posterior (**c** and **d**) views of the bones of the arm and hand. The eight carpal bones in **b** and **d** are numbered and the five digits in **b** and **d** are labeled with Roman numerals.

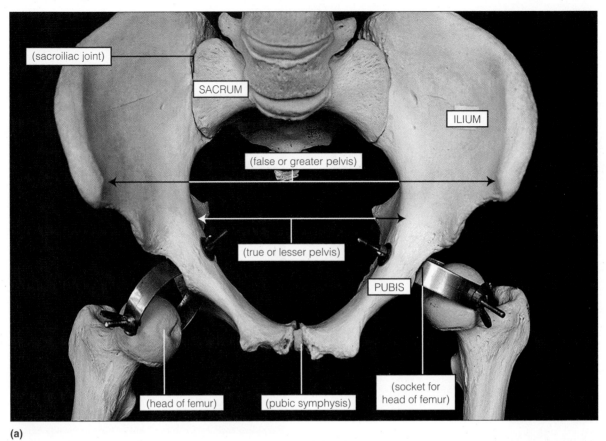

(a)

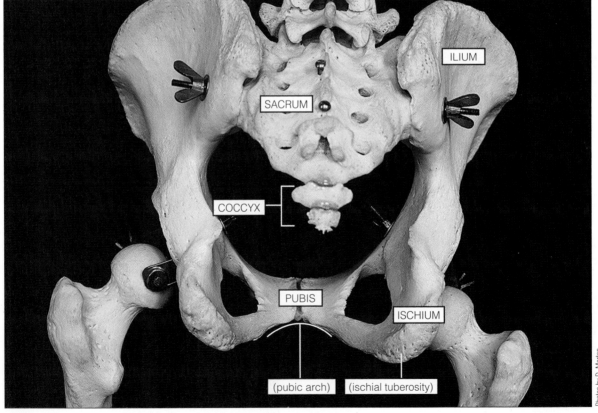

(b)

Figure 39-7 (a) Anterior and (b) posterior views of pelvic girdle and hip.

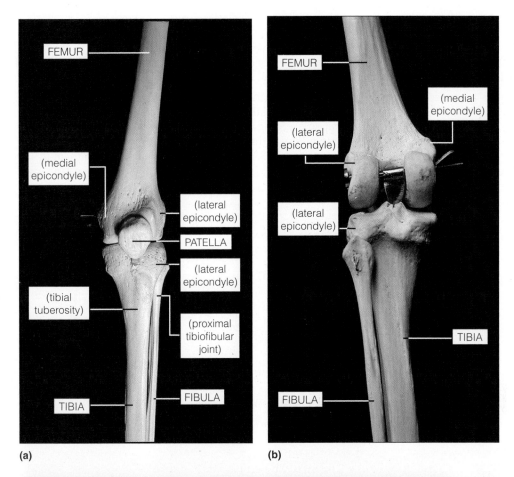

(a)

(b)

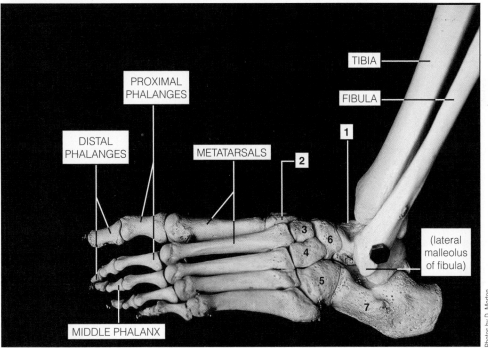

(c)

Figure 39-8 (**a**) Anterior and (**b**) posterior views of knee and lateral view of foot. (**c**) The seven tarsal bones are numbered.

B. Joints

The degree of movements allowed at different joints ranges from none to freely movable (Figure 39-9). Examples of immovable joints are the **sutures** that connect the bones of the roof of the skull of young adults. Joints that allow the freest movement are **synovial joints,** such as the ones listed in Table 39-1.

1. Examine a prepared section of a synovial joint with your compound microscope. Identify the structures labeled in Figure 39-9. The *fibrous capsule* of synovial joints is lined by a *synovial membrane,* which secretes lubricating *synovial fluid.* The fibrous capsule and ligaments function to stabilize synovial joints. Ligaments can be located outside and inside the capsule, and they may be thickenings of its wall.
2. Identify the synovial joints listed in Figure 39-1 and list the adjacent bones that form them in Table 39-1.

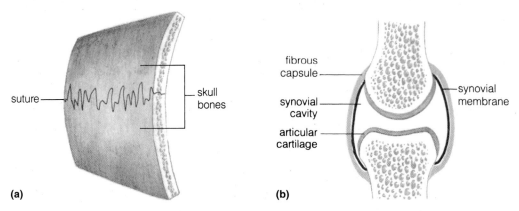

(a) (b)

Figure 39-9 Diagrams of (**a**) a suture and (**b**) a synovial joint.

TABLE 39-1	Bones that Form the Major Synovial Joints
Joint	**Adjacent Bones**
Wrist	
Elbow	
Shoulder	
Hip	
Knee	
Ankle	

© Cengage Learning 2013

C. Surface Features

There are many places on your body surface where bones can be felt. However, it is often difficult to tell specifically which bone you are feeling. Some are easy.

1. Feel the bone supporting your lower jaw, the *mandible.* This is the only bone of the skull that forms a synovial joint with another skull bone.
2. Let's try a harder example, the piece of bone that projects from the point of the elbow joint. Touch it and alternately extend and flex the forearm, increasing and decreasing the angle between the forearm and upper arm, respectively.

Which part of the arm does the projection move with, forearm or upper arm?

_____

While still touching this projection, alternately turn the hand palm down and up. Does the projection move? _____ (yes or no)

Which bone belongs to this projection?

In general, to identify a portion of a bone near a joint, move the body parts adjacent to the joint while touching the bone.

3. Identify the bones that have the surface features listed in Table 39-2.

TABLE 39-2	Surface Features and Bones	
Surface Feature	**Bone**
Knuckles	
Bump next to the wrist and on the same side of the upper appendage as the little finger	
Smaller bump next to the wrist and on the same side of the upper appendage as the thumb	
Bump next to and outside the ankle	
Bump next to and inside the ankle	

© Cengage Learning 2013

D. Structure of a Bone

1. Look at a femur, the longest bone of the skeleton (Figure 39-10). It consists of a shaft, or **diaphysis**, with two knobby ends, or **epiphyses** (singular, *epiphysis*). One end has a narrow neck and a round head.

Which bone does the femur join? _____

To which bone of the skeleton does the other end join? _____

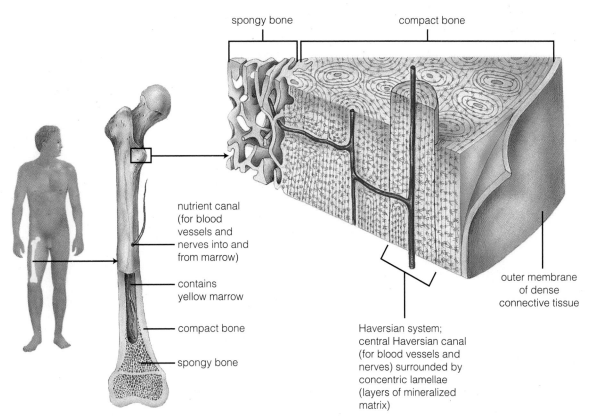

spongy bone   compact bone

nutrient canal (for blood vessels and nerves into and from marrow)

contains yellow marrow

compact bone

spongy bone

Haversian system; central Haversian canal (for blood vessels and nerves) surrounded by concentric lamellae (layers of mineralized matrix)

outer membrane of dense connective tissue

After Starr, 1991.

Figure 39-10 Structure of the femur.

Note the other surface features on the femur, such as projections of various sizes and lines. These surface features are attachment sites for tendons and ligaments.

Are there small tunnels opening onto the surface of the femur? _____ (yes or no)

In life, these nutrient canals serve as routes for blood vessels and nerves.

2. Examine a femur that has been sawed in half lengthwise. There are two kinds of adult bone tissue: **compact bone** and **spongy bone**. Compact bone is solid and dense and is found on the surface of the femur. Spongy bone is latticelike and is found on the inside of the femur, primarily in the epiphyses and surrounding the **marrow cavity**.

Which kind of bone tissue looks denser? _____

Comparing pieces of equal size, which kind of bone tissue looks lighter? _____

3. *Optional.* Instructions for the study of a transverse section of ground compact bone are located in Exercise 33.

E. Structure of a Skeletal Muscle

Skeletal muscle organs mostly have their skeletal muscle fibers arranged parallel to the axis along which the muscle shortens when contracting. A substantial amount of connective tissue surrounds the fibers and connects them to the tendons.

Use your compound microscope to examine a section of a skeletal muscle. Look for fibers, fiber bundles, the more or less loose connective tissue located between fibers and between bundles of fibers, and the fibrous connective tissue that surrounds the entire organ (Figure 39-11).

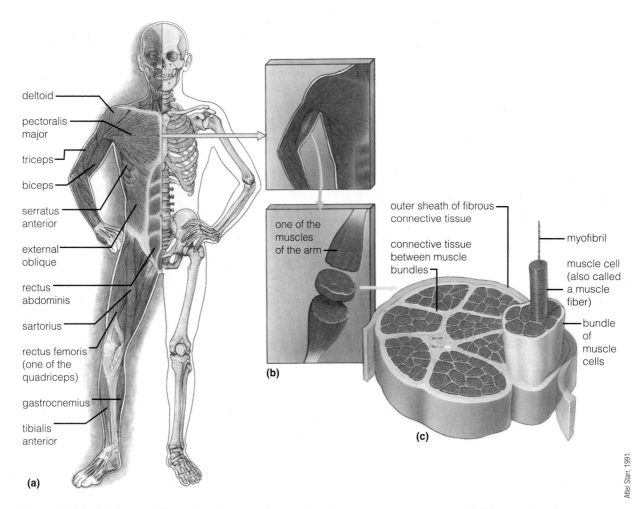

Figure 39-11 (a) Some of the major skeletal muscles of the human muscular system. (b) Closer look at the structure of a skeletal muscle organ. (c) Transverse section of a skeletal muscle organ.

Much of the skeletal system is a system of levers, in which each bone is a lever and the joints are fulcrums. During a typical movement, one end of a skeletal muscle, the **origin**, remains stationary. The other end, the **insertion**, moves along with the bone and surrounding body part. The movement produced by the contraction is the **action** of the skeletal muscle. Most insertions are close to their joints, and the advantage gained by this is that the muscle has to shorten a small distance to produce a large movement of the corresponding body part.

MATERIALS

Per student pair:

- pair of scissors
- toggle switch mounted on a board (Alternatively, you can use any light switches present in the room.)
- pair of forceps
- pencil
- textbook

PROCEDURE

A. Classes of Lever

Levers are simple machines. When a pulling force or effort is applied to a lever, it moves about its fulcrum, overcoming a resistance or moving a load.

1. There are three classes of levers (Figure 39-12):
   - Class I. The fulcrum is located between the effort and the load.
   - Class II. The load is located between the fulcrum and the effort.
   - Class III. The effort is located between the fulcrum and the load. Class III levers are the most common in the skeletal system.

Test your understanding of the three classes of levers by examining the common objects listed in Table 39-3 and then matching them with the appropriate class of lever.

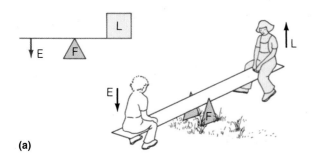

(a)

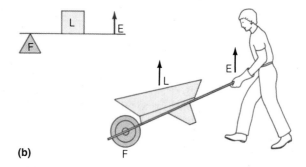

(b)

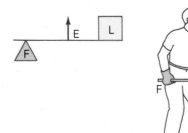

(c)

Figure 39-12 **(a)** Seesaw, an example of a class I lever. **(b)** Wheelbarrow, an example of a class II lever. **(c)** Lifting a spade with one hand while holding the handle stationary with the other hand, an example of a class III lever; E—effort, F—fulcrum, and L—load.

TABLE 39-3	**Lever Class of Common Objects**
Lever	**Object**
Class I _____	**a.** scissors
Class II _____	**b.** toggle switch or light switch
Class III _____	**c.** forceps

© Cengage Learning 2013

2. To remember the relative position of the fulcrum, load, and effort for each class of lever, use this mnemonic (memory device): "1, 2, 3; F, L, E." For example, because 2 has the same relative position as L in Figure 39-12b, the mnemonic tells you that a class II lever has the Load in the center, the Effort at one end, and the Fulcrum at the other. Try it for the other two classes of levers.

B. Analysis of Simple Movements

Let's analyze three simple movements: flexion of the forearm, extension of the forearm, and plantar flexion of the foot (Figure 39-13).

1. *Flexion of forearm.* While sitting, turn your hand so the palm is up and place it under the lab bench. Try to flex the forearm (decrease the angle between the forearm and upper arm). Because the skeletal muscle that is attempting to flex the forearm cannot shorten, the tension in it will increase. A contraction of a skeletal muscle in which tension increases but no movement results is called an **isometric contraction**. Feel with your other hand the front surface of the upper arm. The large tense muscle is the biceps brachii. Its origin is the scapula, and its insertion is the radius. Which joint is the fulcrum?

_____

Now place a pencil in the palm of your hand and flex the forearm. A contraction of a skeletal muscle that results in movement is called an **isotonic contraction**. There is no increase in tension during the movement. Feel the tension in the biceps brachii as you make this movement. Repeat this procedure, but replace the pencil with a textbook. Both the pencil and the book are adding to the load being lifted, the forearm.

In which case—lifting the pencil or the textbook—was the tension in the biceps brachii the greatest?

_____

When you lift any object, the tension in the muscle must equal the weight of that object before movement can occur. Therefore, normal movements have an isometric phase followed by an isotonic phase.

Where is the pulling force applied? (insertion, origin, or both the insertion and origin)

_____

Even simple movements require the coordination of a group of muscles. For example, the origin does not move because other skeletal muscles hold the scapula stationary. What class of lever (I, II, or III) is illustrated by the preceding example?

_____

2. *Extension of forearm.* Place the hand, still palm up, on the top of the lab bench and try to extend the forearm (increase the angle between the forearm and the upper arm). Feel for a tense muscle on the back surface of the upper arm. This is the triceps brachii. The origin of the triceps brachii is the scapula and the upper humerus; its insertion is the *olecranon process* of the ulna (Figure 39-13). The fulcrum is the same as the previous example, except that it has shifted position relative to the effort and the load.

What class of lever is illustrated by this movement? _____

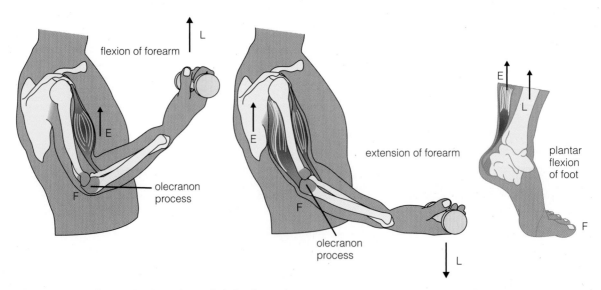

Figure 39-13 Some simple actions of skeletal muscles.

Extension of the forearm is the opposite movement to flexion of the forearm. Hold the textbook, palm still up, halfway between full flexion and full extension. Feel the tension in the biceps brachii and triceps brachii. Repeat this procedure without the book. Is the tension in the biceps brachii greater with or without the book?

_____

The state of contraction of a group of skeletal muscles has to be coordinated to accomplish a particular movement or element of posture. Both the tendons of the biceps brachii and the triceps brachii are pulling on their insertions on the bones of the forearm to keep the forearm stationary. Other muscles are keeping the shoulder stationary.

3. *Plantar flexion of foot.* You need to stand up to do this movement. A lab partner should stand behind you and watch that you do not fall during this procedure. With one hand on the lab bench to steady your balance, stand on the tips of your toes. With your other hand, feel one of the very large tense muscles on the back of each calf. This is the gastrocnemius. The origin of the gastrocnemius is the femur, and its insertion is a tarsal—the calcaneus or heel bone. The fulcrum is the metatarsal–phalangeal joints, and the weight is the weight of the body transmitted through the tibia. What class of lever does this movement demonstrate?

_____

39.3   Walking *(20 min.)*

Walking is a complex activity that requires many movements and the coordinated contractions of several groups of skeletal muscles. For each leg, walking involves two phases, which together make up the *step cycle*. The *stance phase* occurs when the leg bears weight, and the *swing phase* occurs when the leg is in the air.

MATERIALS

Per lab room:
- safe place to walk

PROCEDURE

1. *Follow your instructor's directions as to where to walk safely.* Walk a few normal steps, concentrating on one leg.

What part of the foot (toe or heel) strikes the ground first? _____

What part of your foot leaves the ground last? _____

Does it leave passively, or does it push off? _____

2. Now put your hands on your hips and concentrate on what your pelvic girdle is doing while you walk. First take short strides and then long ones.

Does the pelvic girdle rotate more during short or long strides? _____

_____

Rotation of the pelvic girdle can be demonstrated in a different way. Find a lab partner of about equal height. Walk right next to each other but out of step, that is, with opposite feet leading. First take short steps and then long ones. What happens?

_____

This sideways movement is called *lateral displacement*. Incidentally, females in general have to rotate their pelvic girdles a little more than males for a given length of stride. This is due to differences in the proportions of the female and male pelvic girdles.

3. *Vertical displacement* also occurs during walking. From the side, observe two individuals of equal height walking out of step and next to each other.

Do their heads remain at the same level, or do they bob up and down? _____

_____ 1. Ligaments connect
   (a) bones to bones.
   (b) skeletal muscles to bones.
   (c) tendons to bones.
   (d) skeletal muscles to tendons.

_____ 2. Tendons connect
   (a) bones to bones.
   (b) skeletal muscles to bones.
   (c) ligaments to bones.
   (d) skeletal muscles to tendons.

_____ 3. Which bone is part of the axial skeleton?
   (a) clavicle
   (b) radius
   (c) coxal bone
   (d) sternum

_____ 4. The two kinds of bone tissue are
   (a) compact and loose.
   (b) compact and spongy.
   (c) dense and spongy.
   (d) loose and dense.

_____ 5. There are _____ classes of levers.
   (a) two
   (b) three
   (c) four
   (d) more than four

_____ 6. The class of lever in which the effort is located between the fulcrum and the load is called
   (a) class I.
   (b) class II.
   (c) class III.
   (d) class IV.

_____ 7. The end of the skeletal muscle that remains stationary during a movement is
   (a) the action.
   (b) the origin.
   (c) the insertion.
   (d) none of the above

_____ 8. In an isotonic contraction of a skeletal muscle,
   (a) the tension in the muscle increases.
   (b) movement occurs.
   (c) no movement occurs.
   (d) both a and c occur.

_____ 9. In an isometric contraction of a skeletal muscle,
   (a) the tension in the muscle increases.
   (b) movement occurs.
   (c) no movement occurs.
   (d) both a and c occur.

_____ 10. The step cycle of walking consists of
   (a) a stance phase.
   (b) a swing phase.
   (c) both a and b
   (d) none of the above

EXERCISE **39**

Human Skeletal and Muscular Systems

Post-Lab Questions

39.1 Adult Human Skeleton

1. Match the following bones to their location in the body.

    Bone   **Location**
    _____ radius   a. pectoral girdle
    _____ coxal bone   b. leg
    _____ ribs   c. axial skeleton
    _____ scapula   d. arm
    _____ fibula   e. pelvic girdle

2. This photo below is of a femur that has been sawed in half lengthwise. Draw arrows from the labels below to the two types of bone tissue seen here.

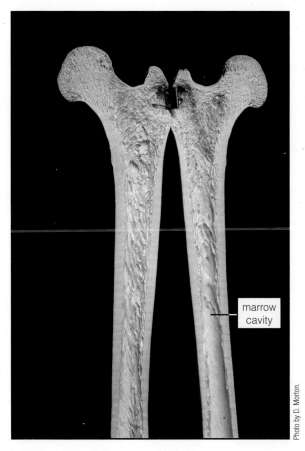

marrow cavity

Photo by D. Morton.

Labels: compact bone, spongy bone

3. Identify the bones indicated in this photo.

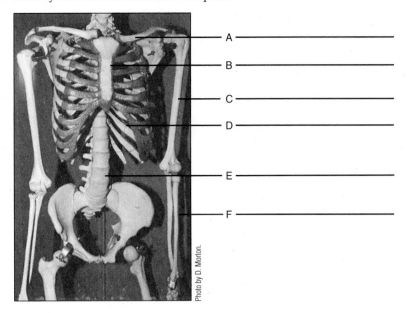

A —————————————————————————

B —————————————————————————

C —————————————————————————

D —————————————————————————

E —————————————————————————

F —————————————————————————

Photo by D. Morton.

4. Label the fibrous capsule and synovial membrane of this joint.

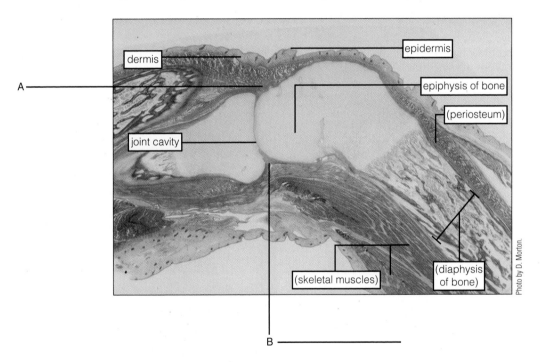

dermis

epidermis

A ————————————

epiphysis of bone

(periosteum)

joint cavity

(skeletal muscles)

(diaphysis of bone)

Photo by D. Morton.

B —————————————————

39.2 Leverage and Movement

5. Define the following terms:
   (a) the *insertion* of a skeletal muscle

   (b) the *origin* of a skeletal muscle

   (c) the *action* of a skeletal muscle

6. Explain the difference between isometric and isotonic contractions. How are both important to normal body movements?

7. Draw and label the structures of a typical skeletal muscle organ.

39.3 Walking

8. In your own words, describe one step in the walking cycle.

Food for Thought

9. The skeletal muscle that flexes (bends) the forearm after pronation (palm down position as in a pull-up) is the brachialis. Its origin is the humerus, and the insertion is the upper front of the ulna. Identify the class of lever involved and explain why you made this choice.

10. Search the Internet for sites that describe diseases of bones (for example, osteoporosis) and of skeletal muscles (such as muscular dystrophy). List two sites and briefly summarize their contents:

    http:// _____

    _____

    _____

    http:// _____

    _____

    _____

Human Blood and Circulation

OBJECTIVES

After completing this exercise, you will be able to

1. define *different types of blood vessels, blood, heart, pulmonary circuit, systemic circuit, homeostasis, plasma, hematocrit, agglutination, blood pressure, elastic membranes, valves, sinoatrial node, acetylcholine, epinephrine;*

2. identify and give the characteristics and functions of the different types of blood cells;

3. explain the ABO and Rh blood group systems;

4. describe how to perform a hematocrit and blood typing;

5. give the structure and function of the different types of blood vessels;

6. describe how blood flows through the circulatory system;

7. name the four chambers and four valves of the heart and describe the route blood takes through them;

8. describe how the heart contracts;

9. explain how the heart is controlled.

Introduction

Circulation—the bulk transport of fluid around the body—connects the specialized cells of multicellular organisms, even though they are separated physically. A circulatory system is a necessary step in the evolution of complex, larger organisms.

Most coelomates (animals with a true body cavity) have a circulatory system with a branching network of pipes (**blood vessels**) that contain fluid (**blood**). Blood is pumped through the blood vessels by one or more *hearts*. Some invertebrates, like the clam, crayfish, and insect, have an *open circulatory system*, in which the blood percolates directly through the body tissues (Figure 40-1). Vertebrates and some invertebrates, like the earthworm, have a *closed circulatory system*, in which the blood flows solely within the blood vessels (Figure 40-2).

Humans and other mammals and birds have a single four-chambered heart with two completely separate networks of blood vessels, the pulmonary and systemic circuits (Figure 40-3). The **heart** receives blood from both circuits and pumps it back into them, alternating the blood between each one. The pattern of blood vessels is the same in each circuit:

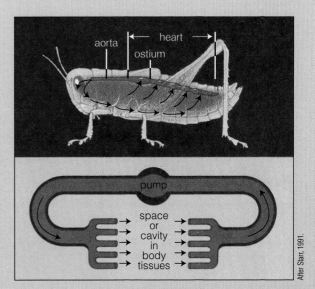

Figure 40-1 Open circulatory system.

heart → arteries → arterioles → capillaries → venules → veins → heart

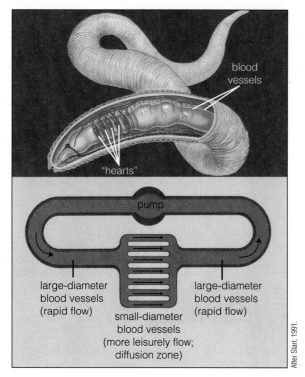

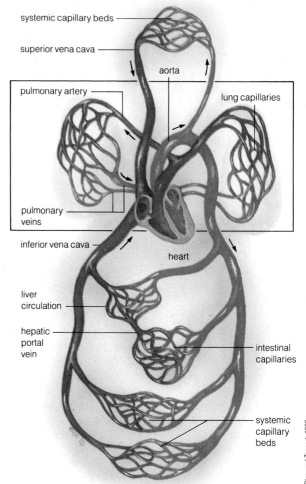

Figure 40-2 Closed circulatory system.

See Table 40-1 for the functions of each of these vessel types.

The **pulmonary circuit** carries oxygen-depleted blood to the capillary beds of the lungs, where oxygen is loaded and where excess carbon dioxide is unloaded. Pulmonary veins drain the oxygen-rich blood back to the heart. The **systemic circuit** takes the oxygen-rich blood from the heart and conveys it to the rest of the body's capillary beds, where oxygen is unloaded and excess carbon dioxide is picked up. Systemic veins drain the oxygen-depleted blood back to the heart.

Figure 40-3 The human circulatory system. The pulmonary circuit is enclosed by the box. Red indicates oxygenated blood; blue indicates oxygen-depleted blood.

TABLE 40-1	Types of Blood Vessels and Their Major Functions
Types	**Functions**
Elastic arteries	1. Receive blood from heart 2. Deliver blood to more numerous muscular arteries 3. Maintain blood pressure between contractions of the heart
Muscular arteries	1. Deliver blood to more numerous and smaller arterioles 2. Regulate blood flow to organs
Arterioles	1. Deliver blood to more numerous and smaller capillaries 2. Regulate peripheral resistance 3. Precapillary sphincters of smooth muscle regulate blood flow into particular capillary beds
Capillaries	1. Exchange dissolved gases, nutrients, wastes, etc. with fluid surrounding cells (interstitial fluid) 2. Form interstitial fluid
Venules	1. Drain blood into fewer and larger veins 2. Serve as a blood reservoir
Veins	1. Drain blood into fewer and larger veins, and finally back to heart 2. Serve as a blood reservoir

© Cengage Learning 2013

In a few cases, this pattern of blood flow is interrupted by a **portal vein**, which connects two sets of capillary beds. The most prominent example is the *hepatic portal vein*, which transports blood from capillary beds in the intestines, stomach, and spleen to beds of large capillaries in the liver.

The circulatory system plays a central role in the maintenance of **homeostasis**—a stable internal environment (Exercise 4). Homeostatic mechanisms throughout the body function to keep the physical and chemical properties of the blood within physiological limits. By means of the fast-circulating blood, the effects of homeostatic mechanisms are spread first to the interstitial fluid and then to the intracellular fluid throughout the body.

40.1 Blood *(40 min. or more)*

Human blood is about 45% cells by volume, although it is only slightly thicker than water. Blood cells are suspended in a straw-colored fluid called **plasma** (about 55% of total blood volume). Plasma is mostly water but contains many dissolved substances, including gases, nutrients, wastes, ions, hormones, enzymes, and antibodies.

MATERIALS

Per student:

- compound microscope, lens paper, bottle of lens-cleaning solution (optional), lint-free cloth (optional), dropper bottle of immersion oil (optional)
- prepared slide of a Wright- or Giemsa-stained smear of human blood

Per lab room:

- eosinophil on demonstration (compound microscope)
- basophil on demonstration (compound microscope)
- dropper bottle(s) of aseptic or simulated blood
- dropper bottles of anti-A, anti-B, anti-D (Rh) sera

- box of depression or stirring sticks
- glass slides and wax pencil or plastic blood typing trays

Per student group (4):

- plain capillary tubes
- microhematocrit centrifuge
- ruler or hematocrit reader
- blood kits (optional)

Per lab section:

- blood waste disposal jar

PROCEDURE

A. Formed Elements (Cells and Platelets) of Blood

The abundance and size of the various blood cells are presented in Table 40-2. Table 40-3 lists the major functions of the blood cells.

1. Use your compound microscope to examine the prepared slide of a stained smear of blood with medium power. Note the numerous, small pink-stained red blood cells, or **erythrocytes** (Figure 40-4a). Each erythrocyte is a biconcave disk without a nucleus. Scattered among the erythrocytes are a much smaller number of blue/purple-stained cells. These are white blood cells, or **leukocytes**. Center a leukocyte and rotate the nosepiece to the high-dry objective. What part of the cell, nucleus or cytoplasm, is stained blue/purple?

TABLE 40-2	**Characteristics of Formed Elements of Blood**		
Cell or Fragment	Number/mm³ in Peripheral Blood	Percent of Leukocytes	Size (µm)
Erythrocytes	4.5–5.5 million	—	7 by 2
Platelets	250,000–300,000	—	2–5
Neutrophils	3000–6750	65	10–12
Eosinophils	100–360	3	10–12
Basophils	25–90	1	8–10
Lymphocytes	1000–2700	25	5–8
Monocytes	150–750	6	9–15

© Cengage Learning 2013

TABLE 40-3 Functions of Formed Elements of Blood

Cell or Fragment	Functions
Erythrocytes	Contain hemoglobin, which transports oxygen, and carbonic anhydrase, which promotes transport of carbon dioxide by the blood
Platelets	Source of substances that aid in blood clotting
Neutrophils	Leave the blood early in an inflammation to become phagocytes (cells that eat bacteria and debris)
Eosinophils	Phagocytosis of antigen–antibody complexes; numbers are elevated during allergic reactions
Basophils	Granules contain a substance (histamine) that makes blood vessels leaky and a substance (heparin) that inhibits blood clotting
Lymphocytes	Perform many functions central to immunity
Monocytes	Leave the blood to form phagocytic cells called macrophages

© Cengage Learning 2013

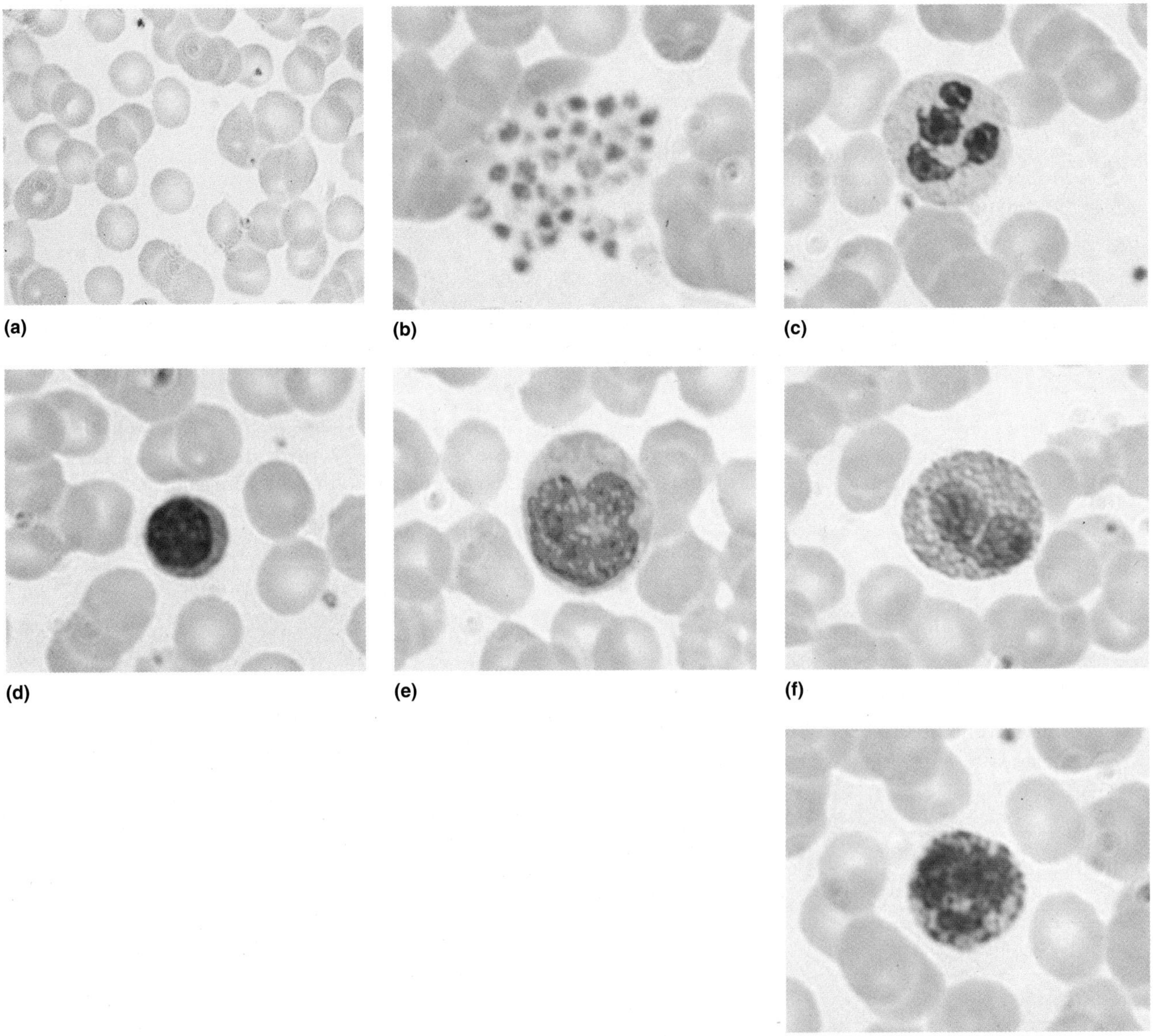

(a)

(b)

(c)

(d)

(e)

(f)

(g)

Photos by D. Morton.

Figure 40-4 Formed elements of blood (**a–g**, 1439×).

2. Using high power, preferably the oil-immersion objective, move the slide slowly and look for cell fragments between the erythrocytes and leukocytes. They usually have one small, blue-stained granule in them and are often clumped together. These cell fragments are **platelets**, or thrombocytes (Figure 40-4b).
3. Locate at least three of the five leukocytes—**neutrophils, lymphocytes**, and **monocytes** (Figure 40-4c, d, and e). Search for them with the high-dry objective. When you find one, center and examine it. If your microscope has an oil-immersion objective (and with the permission of the instructor), use it to look at each blood cell. You may also find **eosinophils** and **basophils** (Figure 40-4f and g), which are normally the rarest leukocyte types. If you have not found an eosinophil or a basophil by the time you have identified the three common leukocyte types, look at one or both of the demonstrations of these cells that have been set up by your instructor.

The three types of leukocytes with the suffix -*phil* (for *philic*, meaning "to like") have large *specific granules*. The prefix in their names refers to the staining characteristics of the specific granules: *neutro-* for neutral (that is, little staining by either of the two dyes in typical blood stains—eosin and methylene blue); *eosino-* because the specific granules stain with the pink dye eosin; and *baso-* because the specific granules stain with the *basic* dye methylene blue.

B. Hematocrit

If you have donated blood, you've probably had your **hematocrit** or *percent packed red blood cell volume* taken. A lower than normal hematocrit is one indicator of anemia—a lower than normal hemoglobin concentration.

1. Work in groups of four. Fill a plain capillary tube to the mark with aseptic or simulated blood and seal it with clay. Fresh blood requires a capillary tube with the inside surface coated with the anticoagulant heparin to prevent clotting.
2. Place the sealed capillary tube in a numbered groove of a microhematocrit centrifuge along with those from other groups. The clay-stopped end of the capillary tube should point towards the outside rim of the centrifuge. Balance your tube with another group's tube or another tube like yours with water instead of blood.
3. After your instructor has secured and spun the tubes, and opened the centrifuge, recover your tube. Determine the percent packed red blood cell volume using a ruler (Figure 40-5) or mechanical measuring device and record it: _____ %
4. Dispose of your capillary tube in the blood waste disposal jar.

C. Blood Typing

On the surfaces of your red blood cells are one or more *antigens* that will cause their agglutination if exposed to the complementary *antibodies*. **Agglutination** is the clumping of erythrocytes. This could theoretically occur during a blood transfusion. The transfusion of

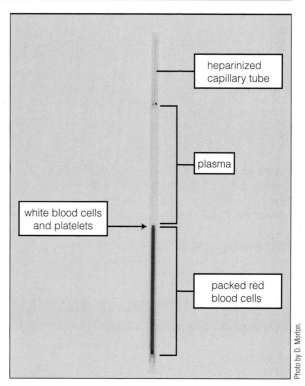

Figure 40-5 Hematocrit. The percent packed volume of red blood cells is calculated by dividing the height of the column of packed blood cells by the length of the total column and multiplying by 100. In this case, it is 53%.

incompatible blood causes the destruction of donor erythrocytes and perhaps the death of the patient when the clumped cells block blood vessels. For these reasons, blood used for transfusion is very carefully matched for compatibility with the patient's blood.

In the **ABO blood typing system**, erythrocytes can have A antigen alone or the B antigen alone, both, or neither. If one or both antigens are not present, the plasma contains the antibody or antibodies for the missing antigen (Table 40-4). For example, if an individual's blood is type A, then their plasma contains anti-B antibodies (blue in Figure 40-6a). Therefore, a type A individual cannot safely receive blood from type B and AB donors, because the anti-B antibodies in their plasma will agglutinate the donor erythrocytes (Figure 40-6b).

Individuals with which blood type (sometimes called the *universal donor*) can theoretically give blood to all other blood types? _____ (A, B, AB, or O)

Explain why this is so. _____

_____

_____

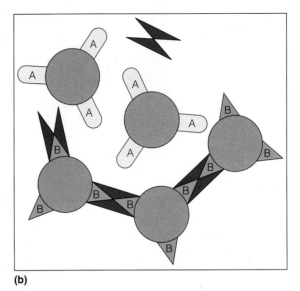

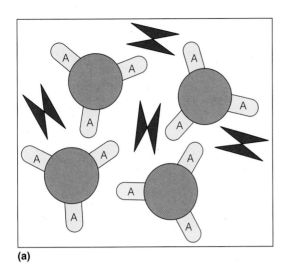

(a)

(b)

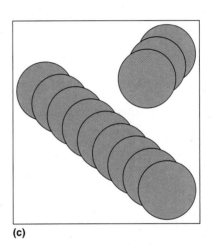

(c)

Figure 40-6 (a) Red blood cells (RBCs) from a type A individual. The RBCs have A antigen on their surfaces and anti-B antibody (blue) in the plasma. (b) If type B RBCs are introduced into the circulation, agglutination of these cells will occur. (c) The round surfaces of the RBCs have more antigen, so they tend to stack up in a "slipped stack of poker chips" pattern.

TABLE 40-4	ABO Blood Types		
Blood Types	Antigens Present on Erythrocytes	Antibodies Present in Plasma	Plasma Agglutinates
A	A	Anti-B	B and AB
B	B	Anti-A	A and AB
AB	A and B	None	None
O	None	Anti-A and B	A, B, and AB

© Cengage Learning 2013

Individuals with which blood type (sometimes called the *universal recipient*) can theoretically receive blood from any other type? _____ (A, B, AB, or O)

Explain why this is so. _____

_____

_____

Usually a standard blood typing procedure includes a test for the Rh factor or D antigen. For example, people with A+ blood have both the A and D antigens on the surface of their erythrocytes. An A− individual has only the A antigen. About 86% of the population in the United States is Rh+. However, Rh− individuals don't have the anti-D antibody unless they have been exposed to Rh+ erythrocytes. This could happen during a transfusion of Rh+ blood or during the birth of an Rh+ child if not prevented by injecting antibody (RhoGAM) during the pregnancy and shortly after birth. The injected antibody ties up any D antigen and prevents the mother's body from making anti-D antibody, which would otherwise attack a subsequent Rh+ fetus.

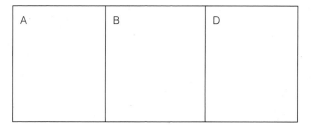

Figure 40-7 An ABO/D grid made with a glass slide.

1. Work in groups of four. Gather a microscope slide, several mixing sticks, and a wax pencil. Alternatively, use a plastic blood typing tray. If you use the latter, skip step 2.

2. With the wax pencil, divide each microscope slide into thirds with two lines perpendicular to the long axis of the slide (see Figure 40-7). Label the upper left-hand corner of the left-hand box or depression with an A. Similarly, label the next box B and the last box D.

3. Place one drop of aseptic or simulated blood in the middle of each box on the slide or each depression in the tray. Then place one drop of anti-A serum next to the box A blood cells, one drop of anti-B serum next to the box B blood cells, and one drop of anti-D serum next to box D blood cells. To mix the drops, stir each set with an unused end of a mixing stick, or rock the tray back and forth.

4. If the antigen is present, the erythrocytes in that mixture will clump or the mixture will be cloudier. If it isn't present, the mixture will not change (Figure 40-8).

Record the blood type. _____ (A+, A−, B+, B−, AB+, AB−, or O+ or O−)

5. Your instructor may ask you to type several samples of aseptic or simulated blood. If so instructed, repeat steps 1–4 above.

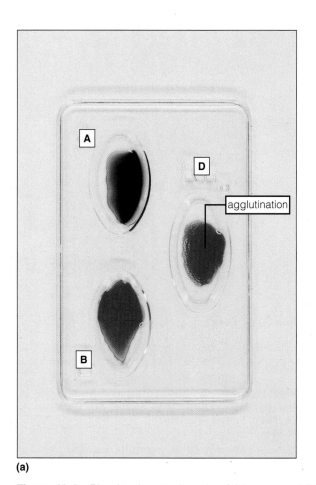

(a)

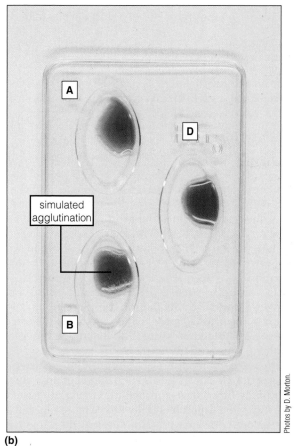

(b)

Figure 40-8 Blood typing results using (a) human and (b) simulated blood.

D. Other Blood Activities

A number of blood kits are commercially available. If you are asked to use one of these kits, follow the instructions carefully.

40.2 Blood Vessels *(40 min. or more)*

The basic structure of blood vessels is shown in Figure 40-9.

MATERIALS

Per student:

- compound microscope
- prepared slide of a companion artery and vein, c.s.
- prepared slide of a whole mount of mesentery

Per student pair:

- fish net
- small fish (3–4 cm long) in an aquarium
- 3 × 7-cm piece of absorbent cotton
- petri dish
- coverslip
- dissecting needle

Per student group (4):

- container of anesthetic dissolved in dechlorinated water
- squeeze bottle of dechlorinated water

Per lab room:

- safe area to run in place
- several meter sticks taped vertically to the walls
- clock with a second hand

PROCEDURE

A. Arteries

Each contraction of the heart pumps blood into the space within the arteries. The rate of flow of blood out of the heart and into the arteries per minute is called *cardiac output* (CO). The arterial space is fairly constant, and it is somewhat difficult for blood to flow through the blood vessels, especially out of the arterioles. This resistance to the flow of blood is called *peripheral resistance* (PR). As cardiac output or peripheral resistance or both, increase, more blood has to fit into the arterial space, which increases the force that the blood exerts on the walls of the arteries. This force is called **blood pressure** (BP). Blood pressure is directly proportional to the product of cardiac output and peripheral resistance (BP ∝ CO × PR).

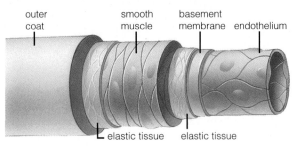

(a) ARTERY

outer coat · smooth muscle · basement membrane · endothelium · elastic tissue · elastic tissue

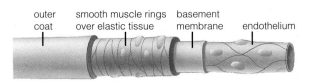

(b) ARTERIOLE

outer coat · smooth muscle rings over elastic tissue · basement membrane · endothelium

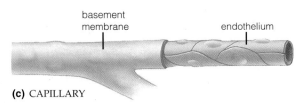

(c) CAPILLARY

basement membrane · endothelium

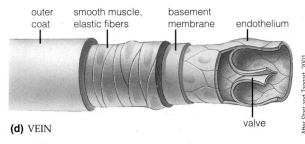

(d) VEIN

outer coat · smooth muscle, elastic fibers · basement membrane · endothelium · valve

After Starr and Taggart, 2001.

Figure 40-9 Structure of blood vessels.

Pressure in blood vessels is highest in the arteries leaving the heart; gradually decreasing the farther a vessel is from the heart (Table 40-5). Blood or any other liquid or gas always flows from high to low pressure. So accordingly, blood flows through the circulatory system down this pressure gradient.

Would you expect blood flow to be more rapid in arteries or veins? _____

Explain your answer. (*Hint:* look at the pressure differences between arteries, capillaries, and veins in Table 40-5.)

_____

_____

_____

TABLE 40-5	Blood Pressure in Different Parts of the Circulatory System of a Young Man at Rest
Locations	**Blood Pressures (mm Hg)**
Right atrium of heart	5/0 (systolic/diastolic)
Right ventricle of heart	25/5
Pulmonary arteries	20
Arterioles and capillaries of lung	20–10
Pulmonary veins	10
Left atrium of heart	10/0
Left ventricle of heart	120/10
Brachial artery	120/80
Arterioles	100–50
Capillaries	50–20
Veins	20–0

© Cengage Learning 2013

The walls of the largest arteries contain many **elastic membranes**, which are stretched during contraction (*systole*) of the heart. When the heart is relaxing (*diastole*), these membranes rebound and squeeze the blood, maintaining blood pressure and flow. Valves at the point where the aorta and the trunk of the pulmonary arteries leave the heart prevent the backflow of blood.

When a physician takes your blood pressure, it's usually of the brachial artery of the upper arm and with the body at rest. A blood pressure of 120/80 means the systolic pressure is 120 mm of mercury (Hg) and the diastolic pressure is 80 mm Hg. The difference between systolic and diastolic pressures (*pulse pressure*) produces a pulse that you can feel in arteries that pass close to the skin.

Blood pressure changes with health, emotional state, activity, and other factors.

1. Get a prepared slide of a companion artery and vein. With your compound microscope, locate and examine the cross section of an artery (Figure 40-10).

Arteries have thick walls compared to other blood vessels. They have an outer coat of connective tissue, a middle coat of smooth muscle tissue, and an inner coat of simple squamous epithelium (*endothelium*). Elastic membranes separate the three coats. The middle coat is the thickest.

2. Find your *radial pulse* in the radial artery (Figure 40-11). Use the index and middle fingers of your other hand. A pulse occurs every time the heart contracts. The strength of the pulse is an estimate of the difference between the systolic and diastolic blood pressure.

3. Sit down and count the number of pulses in 15 seconds. To calculate your heart rate, multiply by 4. Record these numbers in Table 40-6.

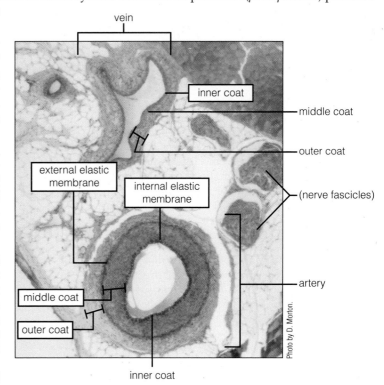

Figure 40-10 Photomicrograph of cross section of an artery and vein (35×).

TABLE 40-6	Heart Rate under Different Conditions	
Condition	**15-sec. Counts**	**Heart Rate (beats per minute)**
Sitting at rest	_____ × 4 =	
Holding breath	_____ × 4 =	
Running in place	_____ × 4 =	

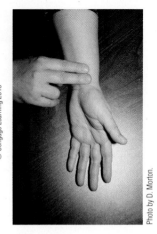

Figure 40-11 Feeling the radial pulse.

© Cengage Learning 2013

Photo by D. Morton.

CAUTION

Do not do the following procedures if you have any medical problems with your lungs or heart. All subjects should be seated except where otherwise indicated and should stop immediately if they feel faint.

---

4. Hold your breath. After 10 seconds have passed, count the number of pulses, then calculate your heart rate as in step 3. Record these numbers in Table 40-6.

Compared to your resting heart rate, does your heart rate increase, decrease, or remain the same? _____

   When you hold your breath, you decrease the return of blood to the heart. This reduces pulse pressure. Homeostatic mechanisms increase heart rate to compensate for reduced blood pressure.

5. Now run in place for 2 minutes in the area designated by your lab instructor. Immediately after sitting down, count the number of pulses, calculate your heart rate, and record these numbers. After running in place, does your heart rate increase, decrease, or remain the same?

_____

Explain these results.

_____

_____

_____

6. *Optional.* Instructions for measuring arterial blood pressure are provided in Exercise 4.

B. Capillaries

Capillaries have a very thin wall that consists of endothelium.

1. Obtain a prepared slide of a whole mount of mesentery and examine it with your compound microscope. Look for groups of blood vessels running through the connective tissue (Figure 40-12). The smallest vessels, which branch and join with each other, are capillaries. Can you see red blood cells inside the capillaries? _____ (yes, no)

2. Use the net to catch a small fish from the aquarium and place it in the anesthetic fluid. Treat the fish gently, and it will not be harmed by this procedure.

3. After the fish turns belly up, wrap its body in cotton made soaking wet with dechlorinated water. Place the fish in half a petri dish so that the tail is in the center.

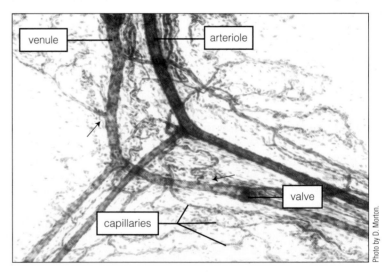

Figure 40-12 Arteriole, venule, and capillary bed in a whole mount of mesentery. Arrows indicate capillaries joining venule (100×).

Photo by D. Morton.

4. Using dechlorinated water, make a wet mount of the posterior two-thirds of the fish's tail and examine it with the low-power, medium-power, and high-dry objectives of the compound microscope. Use the lowest illumination that still allows you to see the blood flowing in the vessels. If necessary, you can temporarily close the condenser iris diaphragm to create more contrast. The smallest blood vessels are capillaries.

Can you see red blood cells moving through them? _____ (yes or no)

What other vessels can you identify?

_____

_____

Is the blood flowing at the same speed in all of the capillaries? _____(yes or no)

Describe blood flow in the fish's tail.

_____

_____

5. Return the fish to the aquarium, wash the half petri dish, and squeeze out the cotton in the sink before dropping it in the trash can.

C. Veins

For the blood to return to the heart after passing through capillary beds below the heart, it must overcome the force of gravity. Veins have **valves** to prevent the backflow of blood away from the heart. Primarily muscular and breathing movements move blood from one segment between valves to another.

1. Again look at a prepared slide of a companion artery and vein. Find and examine the cross section of a vein (Figure 40-12).

The vein has thinner walls and a larger lumen compared to its companion artery. Veins have an outer coat of connective tissue, a middle coat of smooth muscle tissue, and an inner coat of endothelium. Elastic membranes may be present. The outer coat is the thickest layer of the three coats. Compared to arteries, the walls of veins appear more disorganized.

2. Work in pairs. Notice the veins as the subject's arm hangs down at the side of the body. You can easily see the veins because they are full of blood. This is usually best seen on the back of the hand. Now raise the arm above the head. Describe and explain any changes that take place.

_____

_____

_____

3. Using one of the meter sticks vertically taped to the wall, determine the venous pressure in the veins of the hand. Hold the subject's arm straight out at the level of the heart. Record the reading in millimeters where a recognizable point on the hand crosses the meter stick in the second column of the second row in Table 40-7 (measurement 1).

TABLE 40-7	Measurement of Venous Blood Pressure
Measurement 1	_____ mm
Measurement 2	_____ mm
Measurement 2 – measurement 1	_____ mm H$_2$O
_____ mm H$_2$O 0.074 mm Hg/mm H$_2$O	_____ mm Hg

© Cengage Learning 2013

4. Raise the arm slowly until the veins in the hand collapse (be sure that most of the muscles in the arm are relaxed). Read where the same point on the hand crosses the meter stick and record it in millimeters in the second column of the third row of Table 40-7 (measurement 2). The difference between the two readings gives you the venous pressure expressed in millimeters of water. Calculate the venous pressure in millimeters of mercury (mm Hg) by multiplying by 0.074 mm Hg/mm H$_2$O.

How does the venous pressure compare to arterial pressures?

_____

_____

5. Look at the veins of the subject's forearm and hand. The swellings that occur at various intervals are valves. Figure 40-12 shows a smaller valve in a venule. Choose a section between two swellings that doesn't have any side branches. Place one finger on the swelling away from the heart and with another finger press the blood forward (toward the heart) beyond the next swelling.

Does the vein fill up with blood again? _____ (yes or no)

Now remove the finger and observe what happens. Try this again, but press blood in the opposite direction. Discuss your observations.

_____

_____

_____

_____

40.3 The Heart *(30 min. or more)*

Normally the heart beats over 100,000 times a day, pumping the blood around the circulatory system (Figure 40-13).

MATERIALS

Per lab group bench:

■ human heart model

Per lab section:

■ demonstration of the effects of acetylcholine and epinephrine on the heart of a doubly pithed frog (no brain or spinal cord) kept moist with amphibian Ringer's solution (balanced salts solution). Because of environmental concerns, frogs should be farm-raised.

PROCEDURE

A. Human Heart Model

Examine a model of the human heart and identify and trace the flow through its four chambers and four valves (Figure 40-14).

1. The **right atrium** (Figures 40-13 and 40-14) receives blood from the systemic circuit via the *superior vena cava*, the *inferior vena cava*, and the *coronary sinus* (which drains blood from capillary beds in the heart itself).
2. When the right atrium contracts, blood is pushed through the **right AV valve** (*atrioventricular* or *tricuspid valve*) into the **right ventricle**.
3. Contraction of the right ventricle forces this blood into the trunk of the *pulmonary arteries* to start its journey through the pulmonary circuit.
4. The **left atrium** receives blood from pulmonary circuit via the *pulmonary veins*.
5. Now, blood is forced through the **left AV valve** (*bicuspid valve*) into the **left ventricle** by the contraction of the left atrium.
6. Contraction of the left ventricle pushes blood into the *aorta* to begin its travel through the systemic circuit. The *aortic semilunar valves* prevent the backflow of blood into the left ventricle.

B. Demonstration of Frog Heart Function

1. Your instructor has set up a demonstration of a frog heart in place in the opened thorax of a frog. Although the frog has a three-chambered heart—two atria and one ventricle—the heart's function and control are essentially the same as those of humans. The nervous system of the frog has been destroyed, so it does not feel pain or control heart action.

   Is the heart beating? _____ (yes or no)
2. Carefully observe the beating heart.

Does the entire heart muscle contract simultaneously or do some parts contract before others? _____

_____

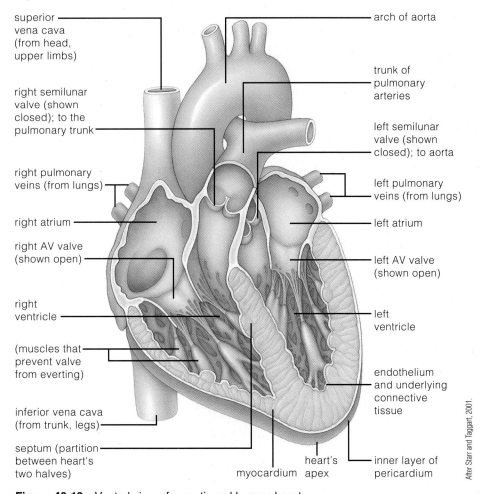

superior vena cava (from head, upper limbs)

right semilunar valve (shown closed); to the pulmonary trunk

right pulmonary veins (from lungs)

right atrium

right AV valve (shown open)

right ventricle

(muscles that prevent valve from everting)

inferior vena cava (from trunk, legs)

septum (partition between heart's two halves)

myocardium

heart's apex

arch of aorta

trunk of pulmonary arteries

left semilunar valve (shown closed); to aorta

left pulmonary veins (from lungs)

left atrium

left AV valve (shown open)

left ventricle

endothelium and underlying connective tissue

inner layer of pericardium

After Starr and Taggart, 2001.

Figure 40-13   Ventral view of a sectioned human heart.

Is there a pattern in the way each contraction sweeps across the heart or is it totally random as to when a particular part contracts? _____

If you look carefully, you can see the order in which the chambers contract. Record your observations.

_____

_____

_____

The primary pacemaker of the heart, the **sinoatrial node**, is located in the right atrium. It functions independently of the nervous system, firing rhythmically. Each time the sinoatrial node fires, it initiates a message to contract. This message spreads over the atria. Then special heart cells amplify and conduct the message throughout the ventricle.

3. Count how many times the heart contracts in 15 seconds, record it in Table 40-8 as the control 1 rate, then calculate the heart rate.

TABLE 40-8	Effect of Acetylcholine and Epinephrine on the Heart Rate of a Frog	
Conditions	**15-sec. Counts**	**Heart Rate (beats per minute)**
Control 1	_____ × 4 =	
After acetylcholine	_____ × 4 =	
Control 2	_____ × 4 =	
After epinephrine	_____ × 4 =	

© Cengage Learning 2013

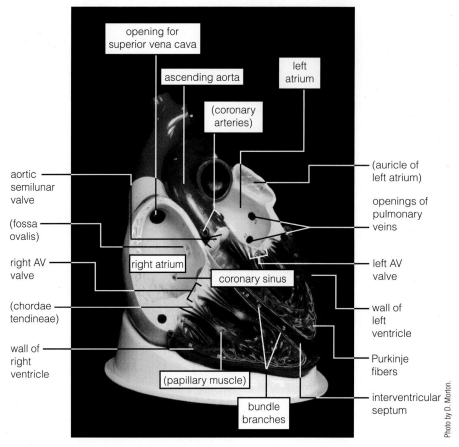

Figure 40-14 Model of a human heart.

4. Note when your instructor places several drops of an acetylcholine solution on the heart. After a minute passes, repeat step 3, recording the count in row 2.
5. Watch your instructor thoroughly flush the thoracic cavity with balanced salt solution, wait 3 minutes, then determine the control 2 heart rate.
6. Now observe as your instructor places several drops of an epinephrine solution on the heart. After 1 minute, repeat step 3, recording the count in row 4.
7. Plot your results on Figure 40-15.
8. Describe the effects of acetylcholine and epinephrine on the heart rate.

_____

_____

_____

_____

9. In an intact frog, the heart rate is modified by input from the central nervous system. The heart rate is affected by the amount of **acetylcholine** and **norepinephrine** secreted by neurons around the sinoatrial node. During restful activities, acetylcholine slows the heart rate and thus acts as a brake on the sinoatrial node. Pain, strong emotions, the anticipation of exercise, and the fight-or-flight response can all increase the secretion of norepinephrine. Norepinephrine, which has a molecular structure and action similar to epinephrine, speeds the heart rate and thus acts as an accelerator on the sinoatrial node and can override the parasympathetic brake.

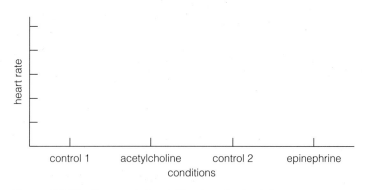

Figure 40-15 Bar graph for plotting heart rate data.

_____ **1.** The number of circuits in the circulatory systems of humans is
(a) one.
(b) two.
(c) three.
(d) four.

_____ **2.** Blood contains
(a) dissolved gases.
(b) dissolved nutrients.
(c) dissolved hormones.
(d) all of the above

_____ **3.** Red blood cells are
(a) erythrocytes.
(b) leukocytes.
(c) platelets.
(d) all of the above

_____ **4.** The most common leukocyte in the blood is
(a) a lymphocyte.
(b) an eosinophil.
(c) a basophil.
(d) a neutrophil.

_____ **5.** The cellular fragments in the blood that function in blood clotting are
(a) erythrocytes.
(b) leukocytes.
(c) platelets.
(d) none of the above

_____ **6.** Blood vessels that return blood from capillaries back to the heart are
(a) arteries.
(b) veins.
(c) portal veins.
(d) arterioles.

_____ **7.** Blood vessels that connect networks of capillary beds are
(a) arteries.
(b) veins.
(c) portal veins.
(d) both b and c

_____ **8.** From which chamber of the heart does the right ventricle receive blood?
(a) right atrium
(b) left atrium
(c) left ventricle
(d) none of the above

_____ **9.** How many chambers does the frog heart have?
(a) one
(b) two
(c) three
(d) four

_____ **10.** The primary pacemaker of the heart is the
(a) bicuspid valve.
(b) tricuspid valve.
(c) aorta.
(d) sinoatrial node.

Name _____   Section Number _____

EXERCISE **4 0**

Human Blood and Circulation

Post-Lab Questions

40.1 Blood

1. Name and give the staining characteristics and functions of the three leukocytes with specific granules.

    (a)

    (b)

    (c)

2. Describe the shape, content, and function of an erythrocyte.

3. Why can't you give type A blood to a type B patient?

40.2 Blood Vessels

4. Describe how to take someone's hematocrit.

5. Identify the indicated blood vessel.

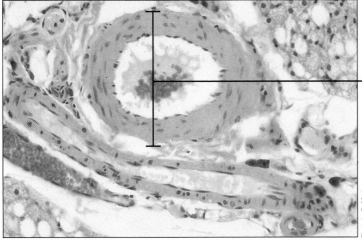

_____

(250×)

6. Pretend you are an erythrocyte in the right atrium of the heart. Describe one trip through the human circulatory system, ending back where you started.

40.3 The Heart

7. Explain how the heart of a double pithed frog can continue to contract in an organized manner after the nervous system is destroyed.

8. How is the heart rate controlled by the nervous system in an intact organism?

Food for Thought

9. Considering its relationship with the other systems of the body, what is the importance of the circulatory system?

10. Search the Internet for sites about arteriosclerosis. List two sites and briefly summarize their contents.

http:// _____

_____

_____

http:// _____

_____

_____

Human Respiration

OBJECTIVES

After completing this exercise, you will be able to

1. define *ventilation, inhalation, exhalation, breathing, cohesion, tidal volume, inspiratory reserve volume, expiratory reserve volume, residual volume, vital capacity, chemoreceptor;*

2. list the skeletal muscles used in breathing and give the specific function of each;

3. trace the flow of air through the organs and structures of the respiratory system;

4. explain how air moves in and out of the lungs during respiration in the human;

5. explain how air moves in and out of the lungs during respiration in the frog;

6. distinguish among negative pressure inhalation, positive pressure exhalation, and positive pressure inhalation;

7. describe the relationship between vital capacity and lung volumes and the interrelationships among lung volumes;

8. explain the importance of CO_2 concentration in the blood and other body fluids to the control of respiration.

Introduction

Carbohydrate metabolism and cellular respiration break down glucose, consume oxygen (O_2), and produce carbon dioxide (CO_2) and water (H_2O) to provide energy (adenosine triphosphate or ATP) to fuel cellular activities.

For cellular respiration to continue, O_2 must be replenished and CO_2 removed from cells by the process of diffusion (Exercise 7). The efficiency of diffusion to transport substances is great over short distances but decreases rapidly at longer distances. Evolution, however, has selected for organisms of different sizes—from one-celled species to the blue whale, the largest living animal. Because diffusion works well only over short distances, animals about the size of earthworms and larger have circulatory systems that move dissolved gases around the body, and animals a little larger have respiratory systems. For example, the clam and crayfish move water across gills to facilitate the body's gain of O_2 and loss of CO_2.

In humans and other air-breathing vertebrates, O_2 uptake and CO_2 elimination occur by diffusion across the moistened thin membranes of millions of tiny sacs and their surrounding capillaries located in the lungs (Figure 41-1). These animals are protected from excessive water loss via evaporation from the very large, moist respiratory surface by having the lungs positioned inside the body.

The main function of the rest of the respiratory system is **ventilation**—the exchange of gases between the lungs and the atmosphere. The movement of gases in (**inhalation**) and out (**exhalation**) of the respiratory system requires the rhythmical contraction of skeletal muscles (**breathing**).

41.1 Breathing *(50 min.)*

Let's review some anatomical terms. The trunk of the body is divided into an upper *thorax,* which is supported by the rib cage and contains the *thoracic cavity,* and a lower *abdomen.* The thoracic and abdominal cavities are separated by a partition of skeletal muscle called the *diaphragm.* The thoracic cavity contains two *pleural cavities,* which contain the lungs and the *pericardial cavity* surrounding the heart.

The muscles of breathing and their roles in inhalation and exhalation are listed in Table 41-1. The external and internal intercostal muscles are located between the ribs. Contraction of the diaphragm increases the size of the thoracic cavity by lowering its floor. Relaxation of the diaphragm allows it to spring back to its original position. When they are contracted, the abdominal muscles can squeeze the internal organs, pushing up the diaphragm to decrease the size of the thoracic cavity.

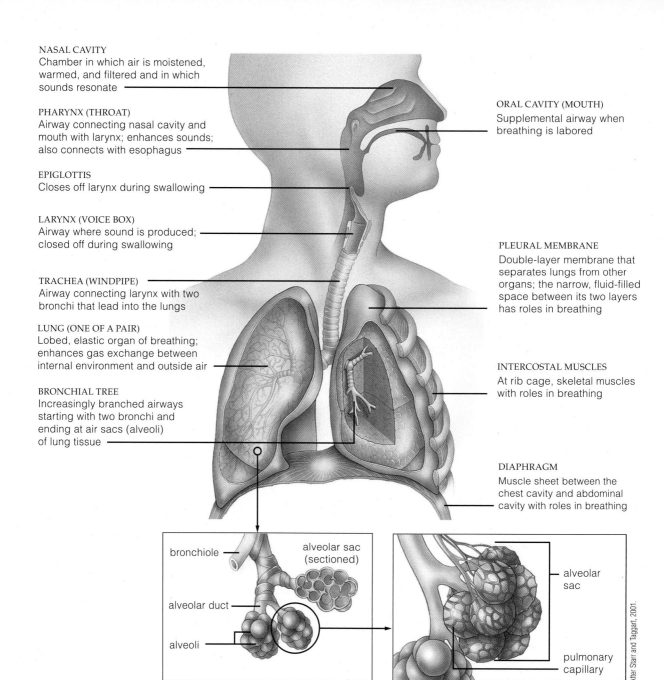

NASAL CAVITY
Chamber in which air is moistened, warmed, and filtered and in which sounds resonate

PHARYNX (THROAT)
Airway connecting nasal cavity and mouth with larynx; enhances sounds; also connects with esophagus

EPIGLOTTIS
Closes off larynx during swallowing

LARYNX (VOICE BOX)
Airway where sound is produced; closed off during swallowing

TRACHEA (WINDPIPE)
Airway connecting larynx with two bronchi that lead into the lungs

LUNG (ONE OF A PAIR)
Lobed, elastic organ of breathing; enhances gas exchange between internal environment and outside air

BRONCHIAL TREE
Increasingly branched airways starting with two bronchi and ending at air sacs (alveoli) of lung tissue

ORAL CAVITY (MOUTH)
Supplemental airway when breathing is labored

PLEURAL MEMBRANE
Double-layer membrane that separates lungs from other organs; the narrow, fluid-filled space between its two layers has roles in breathing

INTERCOSTAL MUSCLES
At rib cage, skeletal muscles with roles in breathing

DIAPHRAGM
Muscle sheet between the chest cavity and abdominal cavity with roles in breathing

bronchiole

alveolar sac (sectioned)

alveolar duct

alveoli

alveolar sac

pulmonary capillary

After Starr and Taggart, 2001.

Figure 41-1 Human respiratory system.

	Muscles				
Stage of Respiratory Cycle	External Intercostals	Internal Intercostals	Diaphragm	Neck and Shoulder	Abdominal
Restful inhalation	Relaxed	Relaxed	Contracted	Relaxed	Relaxed
Inspiratory reserve	Contracted	Relaxed	Contracted	Contracted	Relaxed
Restful exhalation	Relaxed	Relaxed	Relaxed	Relaxed	Relaxed
Expiratory reserve	Relaxed	Contracted	Relaxed	Relaxed	Contracted

TABLE 41-1 Muscles Used to Breathe

© Cengage Learning 2013

MATERIALS

Per student:

■ two pieces of paper (each 14 × 21.5 cm—half a sheet of notebook paper)

Per group (2):

■ metric tape measure
■ large caliper with linear scale (for example, Collyer pelvimeter)

Per group (4):

■ functional model of lung
■ models of the organs and structures of the respiratory system
■ prepared slide of a section of mammalian lung

Per lab room:

■ frogs in terrarium or video of breathing frog
■ clock with second hand

PROCEDURE

A. Ventilatory Ducts and Lungs

The respiratory system consists of the lungs and the ducts that shuttle air between the atmosphere and the lungs. Its organs and structures from the outside in are the two **external nares** (nostrils), the **nasal cavity** (divided into right and left sides by a partition called the **nasal septum**), the **pharynx** (shared with the digestive tract), the **glottis** (and its cover the **epiglottis**), the **larynx** (voice box), the **trachea** (windpipe), two **bronchi** (singular, *bronchus*), and its branches, which terminally open into myriad tiny sacs called **alveoli** (singular, *alveolus*) in the **lungs**. Along with respiration, sound production is a major function of the respiratory tract.

1. Look at the various models of respiratory tract organs and structures and identify as many of the preceding boldfaced terms as possible.
2. Get your compound light microscope and examine a lung section at high power; identify the alveoli (Figure 41-2).

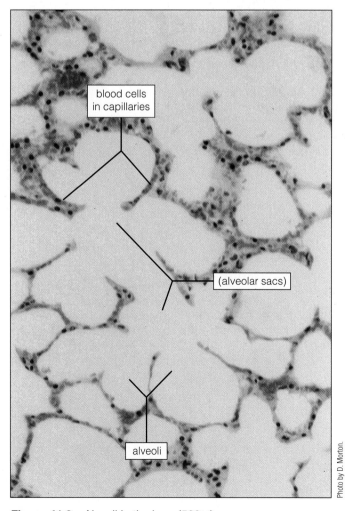

Figure 41-2 Alveoli in the lung (500×).

B. Ventilation

All flow occurs down a pressure gradient. When you let go of an untied inflated balloon, it flies away, propelled by the jet of air flowing out of it. The air flows out because the pressure is higher inside the balloon. The high pressure inside the balloon is maintained by the energy stored in its stretched elastic wall.

When the thoracic cavity expands during inspiration, first the pressure in the pleural sacs decreases, and then the pressure within the lungs decreases. Because the pressure outside the body is now higher than that in the lungs, and assuming the connecting *ventilatory ducts* (trachea and so on) are not blocked, air flows into the lungs (Figure 41-3). This is called **negative pressure inhalation**.

The opposite occurs during expiration. The size of the thorax and pleural sacs decreases, the pressure in the lungs increases, and air flows out of the body down its pressure gradient. This is called **positive pressure exhalation**.

The pressure in the pleural sacs is actually always below atmospheric pressure, which means the lungs are always partially inflated after birth. Thus, a hole in a pleural sac or lung will result in a collapsed lung.

Cohesion (sticking together) of the wet pleural membranes lining the outsides of the lungs and insides of the body walls of the pleural cavities aids inhalation. Exhalation depends in part on the *elastic recoil* (like letting go of a stretched rubber band) of lung tissue.

1. Work in groups of four. Look at the functional lung model. The "Y" tube is analogous to the ventilatory ducts. The balloons represent the lungs. The space within the transparent chamber represents the thoracic spaces and its rubber floor, the muscular diaphragm.
2. Pull down the rubber diaphragm. Describe what happens to the balloons.

_____

As you pull down the rubber diaphragm, does the volume of the space in the container increase or decrease?

_____

As the volume changes, does the pressure in the container increase or decrease?

_____

As the balloons inflate, does the volume of air in the balloons increase or decrease?

_____

Why do the balloons inflate?

_____

_____

_____

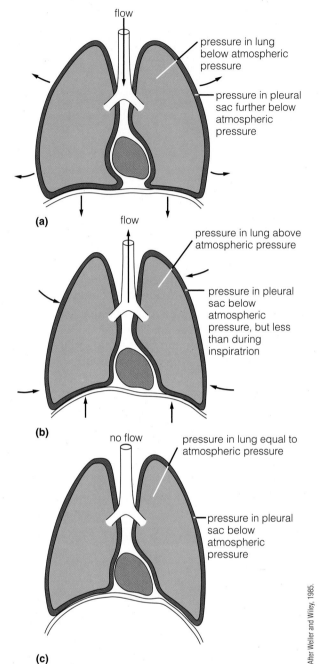

(a)

pressure in lung below atmospheric pressure

pressure in pleural sac further below atmospheric pressure

(b)

pressure in lung above atmospheric pressure

pressure in pleural sac below atmospheric pressure, but less than during inspiratrion

(c)

pressure in lung equal to atmospheric pressure

pressure in pleural sac below atmospheric pressure

After Weller and Wiley, 1985.

Figure 41-3 Changes in the thoracic cavity during (**a**) negative pressure inhalation and (**b**) positive pressure exhalation, and corresponding movements of air. (**c**) The thoracic cavity at the end of an expiration.

3. Push up on the rubber diaphragm. Describe what happens to the balloons and why it takes place.

_____

_____

_____

4. Pull the rubber diaphragm down and push it up several times in succession to simulate breathing.
5. Pucker up your lips and inhale. As you inhale, place one piece of paper directly over your lips. What occurs?

_____

_____

The negative pressure created in your lungs by the contraction of the muscles of inhalation causes this suction.

6. Fold the narrow ends of the two pieces of paper to produce 2- to 3-cm flaps. Open the flaps and use them as handles. Hold a piece of paper with each hand and touch the papers' flat surfaces together in front of you (Figure 41-4). Pull them apart. Now, thoroughly wet both pieces of paper with water and again touch their flat surfaces together in front of you. Pull them apart. What difference did the water make?

_____

_____

7. Some vertebrates, such as the frog, inhale by pushing air into the lungs. This is the positive pressure inhalation. Observe a frog out of water or watch a video of a breathing frog. The frog inhales by sucking in air through the nostrils by lowering the floor of the mouth. Valves in the nostrils are then closed and the floor of the mouth raised, thus increasing the pressure and forcing the air into the lungs. The upper portion of the ventilatory duct can be closed to keep the air in the lungs. Exhalation occurs by elastic recoil of the lungs with the ventilatory duct open. In the frog, both inhalation and exhalation are the result of positive pressure.

What is the frog's respiratory rate (breaths per minute)? Count and record how many times the frog lowers and raises the floor of the mouth (one breath) in 3 minutes. _____ breaths/3 minutes

Divide by 3 to calculate the average respiratory rate and record it: _____ breaths/minute

Respiration in the frog is supplemented by gas exchange across the moist skin. Also, as they are ectotherms (do not maintain a high body temperature using physiological means), frogs generally have a lower metabolic rate and, therefore, a lesser demand for O_2 than a mammal of the same size.

C. Breathing Movements

1. Place your hands on your abdomen and take three breaths—three deep inspirations followed by three forced expirations. What do you feel your body doing during each inspiration?

_____

_____

What do you feel your body doing during each expiration?

_____

_____

2. Place your hands on your chest and repeat step 1. What do you feel your body doing during each inspiration?

_____

_____

What do you feel your body doing during each expiration?

_____

_____

Figure 41-4 Use of two pieces of paper to demonstrate cohesion.

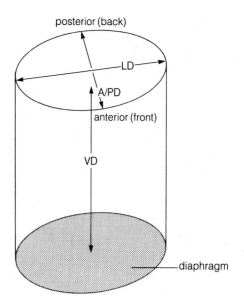

Figure 41-5 Thoracic diameters: LD, lateral diameter; A/PD, anteroposterior diameter; and VD, vertical diameter.

D. Measurements of the Thorax

The size of the thorax can be described by three so-called diameters: the lateral diameter (LD), the anteroposterior diameter (A/PD), and the vertical diameter (VD; Figure 41-5). The vertical diameter is the only one that cannot be measured easily.

Make the following observations and record them in Table 41-2.

1. Work in pairs. Take turns measuring the circumference of each other's chest with a tape measure at two levels, under the armpits (axillae, C_{AX}) and at the lower tip of the sternum (xiphoid process, C_{XP}) for the conditions listed in Table 41-2. While the measurements are being taken, it is extremely important not to tense muscles other than those used for respiration. For example, do not raise the arms.
2. With calipers, also measure the A/PD and LD at the nipple line for these same conditions. The distance between the tips of the calipers is read off the scale in centimeters.

TABLE 41-2	Chest Measurements (cm)				
Subject	**Condition**	**C_{AX}**	**C_{XP}**	**A/PD**	**LD**
You	At the end of a restful inhalation				
	At the end of a restful (passive) exhalation				
	At the end of a forced (maximum) inhalation				
	At the end of a forced (maximum) exhalation				
Your lab partner	At the end of a restful inhalation				
	At the end of a restful (passive) expiration				
	At the end of a forced (maximum) inhalation				
	At the end of a forced (maximum) exhalation				

© Cengage Learning 2013

3. About two-thirds of the air inhaled during a restful inhalation is due to contraction of the diaphragm. Interpret the data in Table 41-2 and in your own words describe changes in the size of the thorax during

(a) a restful inhalation:

C_{AX} _____

C_{XP} _____

A/PD _____

LD _____

(b) a passive exhalation:

C_{AX} _____

C_{XP} _____

A/PD _____

LD _____

(c) an inspiratory reserve:

C_{AX} _____

C_{XP} _____

A/PD _____

LD _____

(d) an expiratory reserve:

C_{AX} _____

C_{XP} _____

A/PD _____

LD _____

4. Does the size of the thorax change significantly during a restful inhalation or a passive exhalation? _____ (yes or no)

How does the shape of the thorax change during an inspiratory reserve?

_____

_____

_____

How does the shape of the thorax change during a subsequent expiratory reserve?

_____

_____

_____

41.2　Spirometry *(30 min.)*

Air in the lungs is divided into four mutually exclusive volumes (Figure 41-6): tidal volume (TV), inspiratory reserve volume (IRV), expiratory reserve volume (ERV), and residual volume (RV).

Tidal volume is the volume of air inhaled or exhaled during a breath. It normally varies from a minimum at rest to a maximum during strenuous exercise.

Inspiratory reserve volume is the volume of air you can voluntarily inhale after inhalation of the tidal volume. **Expiratory reserve volume** is the volume of air you can voluntarily exhale after an exhalation of the tidal volume. IRV and ERV both decrease as TV increases.

Residual volume is the volume of air that cannot be exhaled from the lungs. That is, normal lungs are always partially inflated.

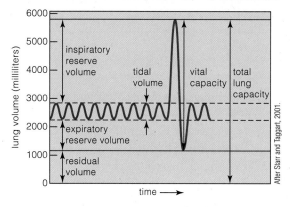

Figure 41-6 Spirogram shows the defined lung volumes and capacities.

There are four capacities derived from the four volumes:

$$\text{inspiratory capacity (IC)} = \text{TV} + \text{IRV}$$

$$\text{functional residual capacity (FRC)} = \text{ERV} + \text{RV}$$

$$\text{vital capacity (VC)} = \text{TV} + \text{IRV} + \text{ERV}$$

$$\text{total lung capacity (TLC)} = \text{total of all four lung volumes}$$

All the lung volumes except the residual volume can be measured or calculated from measurements obtained using a simple spirometer or lung volume bag. A more sophisticated recording spirometer plots respiration over time (a spirogram). Figure 41-6 shows a spirogram and the relationships of the lung volumes and capacities.

CAUTION

Always use a sterile mouthpiece and do not inhale air from either a simple spirometer or a lung volume bag.

MATERIALS

Per group (4):

- noseclip (optional)
- simple spirometer or lung volume bag

PROCEDURE

1. Work in groups of four. Sit quietly and breathe restfully. Use a noseclip or hold your nose. After you feel comfortable, start counting as you inhale. After the fourth inhalation, exhale normally into the spirometer or lung volume bag. Read the volume indicated by the spirometer or squeeze the air to the end of the lung volume bag and read the volume from the wall of the bag. Record the volume below (trial 1). Reset the spirometer or squeeze the air out of the lung volume bag. Repeat this procedure two more times (trials 2 and 3) and calculate the total and average tidal volume at rest.

   trial 1 _____ mL   trial 3 _____ mL

   trial 2 _____ mL   total _____ mL

   Divide the total by 3 = _____ mL to calculate the average TV at rest. Record the TV in Table 41-3.

TABLE 41-3	Vital Capacity and Lung Volumes at Rest (mL)
Measure	**Volume (mL)**
Tidal volume	
Inspiratory reserve volume	
Expiratory reserve volume	
Vital capacity	

© Cengage Learning 2013

2. Determine the volume of air you can forcibly exhale after a restful inhalation (average of three trials).
   trial 1 _____ mL   trial 3 _____ mL

   trial 2 _____ mL   total _____ mL

   Divide the total by 3 = _____ mL. This is the average sum of ERV and TV at rest.

3. Determine the volume of air you can forcibly exhale after a forceful inhalation (average of three trials).
   trial 1 _____ mL   trial 3 _____ mL

   trial 2 _____ mL   total _____ mL

   Divide the total by 3 = _____ mL to calculate the average VC. Record VC in Table 41-3.

4. Calculate the ERV at rest by subtracting step 1's result from step 2's result.
   _____ mL (step 2) – _____ mL (step 1) = _____ mL.
   Record ERV in Table 41-3.

5. Calculate the IRV at rest by subtracting step 2's result from step 3's result.
   _____ mL (step 3) – _____ mL (step 2) = _____ mL.
   Record the IRV in Table 41-3.

6. Does **vital capacity** change as tidal volume increases or decreases? _____ (yes or no)

7. Measure and record your height in centimeters. _____ cm

Write your vital capacity/height on the board—your name isn't necessary.

Plot the vital capacity of each student in your lab section on the graph at right.

Is there a relationship between vital capacity and height? If so, describe it mathematically using a graphing calculator or with words.

_____

_____

_____

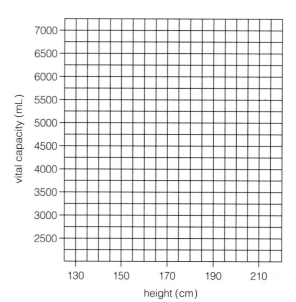

41.3 Control of Respiration *(20 min.)*

The control of respiration, both the rate and depth of breathing, is very complex. Simply stated, **chemoreceptors** (receptors for chemicals such as O_2, CO_2, and hydrogen ions, or H^+), stretch receptors in the ventilatory ducts, and centers in the brain stem (part of the brain that connects to the spinal cord) control respiration.

Your own experience has taught you that respiration is to some extent under the control of the conscious mind. We can decide to stop breathing or to breathe more rapidly and deeply. However, the unconscious mind can override voluntary control. The classic example of this is the inability to hold one's breath for more than a few minutes. In this section you will make further pertinent observations related to the hypothesis that CO_2 concentration in the blood and other body fluids is the most important stimulus for the control of respiration.

MATERIALS

Per student pair:
■ small mirror

Per lab room:
■ a safe place to exercise
■ clock with second hand

PROCEDURE

CAUTION

Do not do the following activities if you have any medical problems with your lungs or heart. All subjects should stop immediately if they feel faint.

1. Work in pairs. Write two predictions in Table 41-4. In prediction 1 forecast the effects on respiratory rate and the ability to hold one's breath after hyperventilation (overventilating the lungs by forced deep breathing). (For example, if I hyperventilate, then. . .) In prediction 2, forecast the effects on respiratory rate and the ability to hold one's breath after exercise.

2. Sit down and after you feel comfortable, have your lab partner count the number of times you breathe in 3 minutes. If it's difficult to see you breathe, have your partner place a small mirror under your nose. Record the number of breaths: _____ breaths. Divide this number by 3 to calculate the average respiratory rate *at rest* and write it in the third column of Table 41-4.

3. Determine how many minutes you can hold your breath after a restful inhalation and record the time in the fourth column of Table 41-4.

TABLE 41-4 Respiratory Data

Prediction 1:

Prediction 2:

	Condition	Respiratory Rate (breaths/minute)	Breath Holding (minutes)
You	At rest		
	After hyperventilation		
	After exercise		
Your lab partner	At rest		
	After hyperventilation		
	After exercise		

Conclusions:

© Cengage Learning 2013

4. Now breathe as deeply and as rapidly as possible (*hyperventilate*). Try to take at least 10 breaths but stop as soon as you can answer the following question. (*In any case, do not continue for more than 20 breaths.*) As time goes on, does it become easier or more difficult to continue rapid deep breathing?

   Immediately have your lab partner count and record the number of breaths you take in the next 3 minutes: _____ breaths.

5. How many minutes can you hold your breath *immediately* after hyperventilating? Record this time in the fourth column of Table 41-4.

6. Divide the number in step 4 by 3 to calculate the average respiratory rate *after hyperventilation* and write it in the third column of Table 41-4. Does hyperventilation increase, decrease, or have no effect on the CO_2 concentration of the blood?

7. When fully recovered, carefully run in place for 2 minutes in the area designated by your lab instructor. Immediately after sitting down, again have your lab partner count how many breaths you take in 3 minutes and record it: _____ breaths. Now determine how long you can hold your breath *immediately* after 2 minutes of exercise and record it in the fourth column of Table 41-4. As you did earlier, calculate the average respiratory rate *after exercise* and record it in the third column of Table 41-4. Does running in place increase or decrease the CO_2 concentration of the blood? What causes the CO_2 concentration to change while you are running in place?

8. Switch roles with your lab partner and repeat steps 1–7 recording this data in Table 41-4.

9. Look over the data in Table 41-4 and summarize your results:

Write a conclusion in Table 41-4 as to whether your results supported your predictions.

When an animal increases its activity, its muscles need more O_2 to support their higher level of cellular metabolism. Increasing both breathing and the circulation of blood through the lungs, heart, and skeletal muscles enhances oxygen delivery.

From observing our own bodies, we know that exercise is accompanied by increases in the rate of breathing, the depth of breathing, and the heart rate. Intuitively, we also expect a similar elevation in arterial blood pressure because this would increase the blood pressure difference between the elastic arteries near the heart and the arterioles near the capillaries. This steeper pressure gradient would result in a higher blood velocity—the speed at which blood flows—just as a boulder rolls faster down a steeper hill. Higher blood velocity along with increased activity of the respiratory system means more O_2 is transported to the heart and skeletal muscles.

So, how do heart rate, systolic blood pressure, respiratory rate, and tidal volume change as the intensity of exercise increases? This experiment addresses the hypothesis that *all these factors will increase as the intensity of exercise increases to contribute to delivering more and more O_2 to the skeletal muscles.*

MATERIALS

Per student:
- calculator

Per lab room:
- television or monitor
- video cassette or DVD player
- video cassette or DVD: "Experiment: Biology; The Physiology of Exercise"

PROCEDURE

As a class you will watch a video of a young man on a stationary bicycle. After recording his heart rate, systolic blood pressure, respiratory rate, and tidal volume at rest, you will record data for these same observations after four periods of work—pedaling against a constant resistance at 10, 15, 20, and 25 km/hour.

1. State a prediction for this experiment and write it in Table 41-5.
2. Watch the video and record the data in Table 41-5. Certain numbers need to be multiplied by 10 or 100 to convert them to the units used in the table. However, don't convert any of the values read off the various meters in the video, because the conversion factors are included in the table.
3. List the experimental variables.
   Independent variable _____
   Dependent variables _____
   Controlled variables (as many as you can identify)

   _____

   _____

TABLE 41-5 | Physiology of Exercise

Prediction:

Intensity of Exercise	Heat Rate (beats/min.)	Systolic Pressure (mm Hg)	Respiratory Rate (breaths/min.)	Tidal Volume (L)
0 km/hr (rest)	____ × 10	____	____ × 10	____
10 km/hr	____ × 10	____	____ × 10	____
15 km/hr	____ × 100	____	____ × 10	____
20 km/hr	____ × 100	____	____ × 10	____
25 km/hr	____ × 100	____	____ × 10	____

Conclusion:

© Cengage Learning 2013

4. Plot these results in the following graphs:

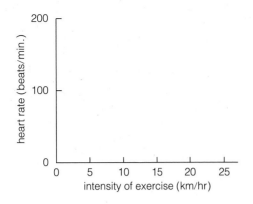

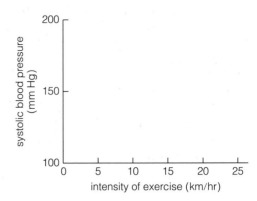

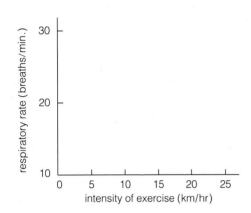

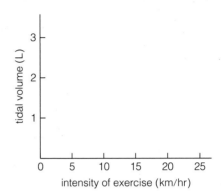

5. Let's analyze the respiratory data further.
   At 10 km/hr, did respiratory rate increase? _____ (yes or no)
   Did tidal volume increase? _____ (yes or no)
6. During which period of work did the respiratory rate first increase? _____ km/hr
7. Calculate the amount of air exhaled per minute during the most strenuous exercise period (25 km/hr).

$$\underline{\hspace{2cm}} \frac{\text{breaths}}{\text{min.}} \times \underline{\hspace{2cm}} \frac{\text{L}}{\text{breath}} \times \underline{\hspace{2cm}} \frac{\text{L}}{\text{min.}}$$

This is the respiratory minute volume. This value is very close to this subject's maximum respiratory minute volume.

_____ 1. Which muscles may contract during inspiration?
(a) external intercostals
(b) internal intercostals
(c) abdominal
(d) both b and c

_____ 2. Which muscles contract during a restful expiration?
(a) external intercostals
(b) internal intercostals
(c) diaphragm
(d) none of the above

_____ 3. Which muscles may contract during a more forceful expiration?
(a) external intercostals
(b) diaphragm
(c) abdominal
(d) both b and c

_____ 4. An untied inflated balloon flies because
(a) the pressure is higher inside than outside the balloon.
(b) the pressure is lower inside than outside the balloon.
(c) air flows down its pressure gradient.
(d) both a and c occur.

_____ 5. Human ventilation is
(a) negative pressure inhalation.
(b) positive pressure inhalation.
(c) negative pressure exhalation.
(d) both b and c

_____ 6. Frog ventilation is
(a) negative pressure inhalation.
(b) positive pressure inhalation.
(c) positive pressure exhalation.
(d) both b and c

_____ 7. Vital capacity is always equal to
(a) tidal volume.
(b) inspiratory reserve volume.
(c) expiratory reserve volume.
(d) a + b + c.

_____ 8. An instrument that measures lung volumes is a
(a) caliper.
(b) spirometer.
(c) barometer.
(d) stethoscope.

_____ 9. Respiration is controlled by
(a) chemoreceptors.
(b) stretch receptors.
(c) centers in the brain stem.
(d) all of the above

_____ 10. The most important stimulus in the control of respiration is the concentration in the blood and other body fluids of
(a) oxygen (O_2).
(b) carbon dioxide (CO_2).
(c) hydrogen ions (H^+).
(d) nitrogen (N_2).

EXERCISE **41**

Human Respiration

Post-Lab Questions

41.1 Breathing

1. Which skeletal muscles are contracted during
   (a) restful inhalation?

   (b) inspiratory reserve?

   (c) restful exhalation?

   (d) expiratory reserve?

2. How does the size of the thorax change during
   (a) inhalation?

   (b) exhalation?

3. How does the potential volume of the pleural sacs change during
   (a) inhalation?

   (b) exhalation?

4. Define these terms:
   (a) negative pressure inhalation

   (b) positive pressure exhalation

5. How does breathing in a human differ from that in a frog?

41.2 Spirometry

6. Explain the relationship among vital capacity, tidal volume, inspiratory reserve volume, and expiratory reserve volume.

41.3 Control of Respiration

7. What substance is the most important stimulus in the control of respiration? How is its production linked to changes in metabolic rate, such as occur during exercise?

41.4 Experiment: Physiology of Exercise

8. Give an explanation for the fact that athletes' resting heart rates are usually slower than the average rate for healthy humans.

Food for Thought

9. Explain why hyperventilation can prolong the time you can hold your breath. Can this be dangerous (for example, hyperventilation followed by swimming underwater)?

10. Search the Internet for sites about emphysema. List two sites below and briefly summarize their contents.

    http:// _____

    _____

    _____

    http:// _____

    _____

    _____

Animal Development

OBJECTIVES

After completing this exercise, you will be able to

1. define *sexual reproduction, gametes, ovum, sperm, zygote, fertilization, gonads, ovaries, testes, gametogenesis, oogenesis, spermatogenesis, meiosis, yolk, acrosome, blastodisc, vegetal pole, animal pole, seminiferous tubules, interstitial cells, testosterone, Sertoli cells, follicle, estrogen, progesterone, ovulation, corpus luteum, implantation, uterus, blastomere, morula, blastula, blastocoel, gray crescent, blastocyst, inner cell mass, trophoblast, placenta, archenteron, blastopore, primitive streak, notochord, neural tube, somites, amniotic cavity, embryonic disc;*

2. draw a mammalian sperm and label its segments;

3. describe the structure of chicken and frog ova and how they differ from a typical mammalian ovum;

4. recognize interstitial cells, seminiferous tubules, Sertoli cells, and sperm in a prepared section of a mammalian testis;

5. recognize primary oocytes, primordial and secondary follicles, and a corpus luteum in a prepared slide of a mammalian ovary;

6. describe spermatogenesis and oogenesis in mammals;

7. describe the events and consequences of sperm penetration and fertilization.

8. compare and contrast cleavage in the sea star, frog, chicken, and human;

9. explain differences in cleavage according to (a) the amount and distribution of yolk in the ovum and (b) the evolution of mammals;

10. describe gastrulation, the formation of the primary germ layers (ectoderm, mesoderm, endoderm), and their derivatives.

Introduction

Sexual reproduction typically involves the fusion of the nuclei of two **gametes**, called the **ovum** (plural, *ova*) in females and **sperm** in males, to form the first cell of a new individual, the **zygote**. This fusion is referred to as **fertilization**.

Gametes are produced in reproductive organs called **gonads**, usually in individuals of two separate sexes—**ovaries** in females and **testes** in males. **Gametogenesis**, gamete formation, reduces the number of chromosomes by half. Gametes have 22 *autosomes*, one of each of the 22 pairs found in most other cells of the body plus one *sex chromosome*, an X or Y. Gametes with a half set of chromosomes are referred to as *haploid*. Cells with a full set of chromosomes are *diploid*. Fertilization produces a diploid zygote, whose combination of genes is unlike those of either parent. Other than cases of identical twins, triplets, and so on, it is also very unlikely that the genes of one zygote will be the same as those of any other zygote, even those derived from the same parents. This variation is an advantage to a species in a changing and unpredictable environment. Why?

_____

_____

_____

Note: Depending on the timing of your lab and to prepare for the observation of a live frog zygote, your instructor may demonstrate the fertilization of frog eggs before you start this exercise.

Most sperm have at least one flagellum and are specialized for motility. They contribute little more than their chromosomes to the zygote.

Ova are specialized for storing nutrients, and they contain the molecules and organelles needed to fuel, direct, and maintain the early development of the embryo. Nutrients are stored as **yolk** in the cytoplasm. Consequently, mammalian ova are larger than body cells and in some species reach a diameter of 0.2 mm. In frogs, additional yolk increases the diameter of the ovum to 2 mm, and in chickens, it reaches about 3 cm.

MATERIALS

Per student:

- compound light microscope, lens paper, bottle of lens-cleaning solution (optional)
- lint-free cloth (optional)
- glass microscope slide
- coverslip
- dissecting needle
- one-piece plastic dropping pipet

Per student pair:

- unfertilized hen's egg
- several paper towels
- dissection pan
- Syracuse dish

- two camel-hair brushes
- dissection microscope

Per lab group (table or bench):

- model of a frog ovum (optional)

Per lab section:

- live frog sperm in pond water
- live frog ova in pond water

Per lab room:

- phase-contrast compound light microscope (optional)

PROCEDURE

A. Sperm

1. With the high-dry objective of your compound microscope, study a prepared slide of bull sperm. Draw several sperm in Figure 42-1. Each sperm has three major segments: the *head, midpiece,* and *tail.* Label the major segments of one sperm in your drawing. The tail is typically composed of a single *flagellum.*

2. *Skip this step if your compound microscope does not have an oil-immersion objective.* Switch to the oil-immersion objective. Find the **acrosome** located in the head of a sperm and *mitochondria,* in its midpiece. The acrosome partially covers the front of the nucleus. The function of the acrosome is to release enzymes to aid in the penetration of the egg. Why does a sperm have a lot of mitochondria relative to its cell size?

   _____

   _____

   _____

   _____

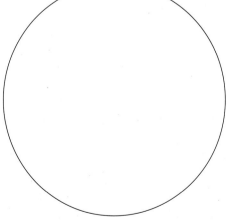

3. Place one drop of frog sperm suspension on a glass microscope slide and make a wet mount. Observe the movements of sperm flagella using either the phase-contrast compound microscope or your compound microscope with the iris diaphragm partially closed to increase contrast. Describe what you see.

   _____

   _____

   _____

   _____

Figure 42-1   Drawing of bull sperm (_____×).

B. Ova

1. Investigate the egg of a chicken.
   (a) Obtain an unfertilized chicken egg (Figure 42-2). Crack it open as you would in the kitchen and carefully spill the contents into a hollow made from paper towels placed in a dissection pan.
   (b) Only the yolk is the ovum. Look on its surface for the **blastodisc,** a small white spot just under the cell membrane. This area is free of yolk and contains the nucleus. This is where fertilization occurs if sperm are present in the hen's oviducts. The walls of the oviduct secrete the *albumin* (egg white).
   (c) Examine the *shell* and the two *shell membranes*. The shell membranes are fused except in the region of the air space at the blunt end of the egg. Note the two shock absorber–like *chalazae* (singular, *chalaza*), which suspend the ovum between the ends of the egg. They are made of thickened albumin and may help rotate the ovum to keep the blastodisc always on top of the yolk.

2. Compare a frog egg to a chicken egg.
   (a) Gently place a live frog egg in a Syracuse dish half filled with pond water and examine it with a dissection microscope. Use two clean camel-hair brushes to transfer the egg. *Do not let the egg dry out.* The ovum is enclosed in a protective jelly membrane. Which surface (light side or dark side) of the ovum floats up in the pond water?

   _____

   (b) Look at a model of a frog ovum. Ova from different species of animals vary in the amount and distribution of yolk. Frog ova have a moderate amount of yolk that is concentrated in the lower half of the ovum (Figure 42-3). This half of the ovum is called the **vegetal pole.** The nucleus is located in the upper, yolk-free half, the **animal pole.** Note that the animal pole is black. This is because it contains pigment granules. Why does a frog ovum have less yolk than a bird ovum?

   _____

   _____

Suggest one or more possible functions for the black pigment in the animal pole. (*Hint:* one function is the same as that for the pigment in your skin that increases when exposed to sunlight.)

_____

_____

_____

3. See Figure 42-8 for a diagram of a human egg as it would appear in the upper oviduct. The egg is surrounded by a membrane, the *zona pellucida*, and a layer of follicle cells. Why do most mammalian ova have very little yolk?

   _____

   _____

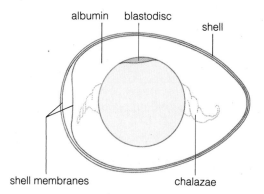

Figure 42-2 Unfertilized chicken egg.

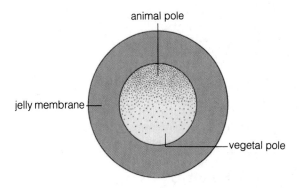

Figure 42-3 Frog egg.

Gamete formation (**gametogenesis**) is called **oogenesis** in females and **spermatogenesis** in males. The general scheme and terminology of gametogenesis are summarized in Table 42-1. **Meiosis** is a process by which events in the nucleus of a diploid cell use two cytoplasmic divisions to form haploid gametes or cells that will mature into gametes. The nuclear events associated with the first division are referred to as meiosis I, and those of the second division are called meiosis II.

TABLE 42-1   Gametogenesis

	Type of Cell	
Condition of Cell	**Male**	**Female**
Mitotically active	Spermatogonium	Oogonium
Before meiosis I	Primary spermatocyte	Primary oocyte
Before meiosis II	Secondary spermatocyte	Secondary oocyte and first polar body
After meiosis II	Spermatid	Ovum and three polar bodies
After differentiation	Sperm	—

© Cengage Learning 2013

MATERIALS

Per student:

- compound light microscope, lens paper, bottle of lens-cleaning solution (optional), lint-free cloth (optional)
- prepared slides of a section of
    adult mammalian testis stained with iron hematoxylin
                                               adult mammalian ovary with follicles
                                               adult mammalian ovary with a corpus luteum

PROCEDURE

A. Mammalian Spermatogenesis

1. With your compound microscope, look at a prepared slide of the testis (Figure 42-4). Most of the interior of a testis is filled with coiled **seminiferous tubules**. Transverse and oblique sections are present in your slide. Look for glandular **interstitial cells** between the seminiferous tubules. Interstitial cells secrete the male sex hormone **testosterone.**
2. Center a cross section of a seminiferous tubule in the field of view and increase the magnification until you see only a portion of the tubule's wall. In the wall of the seminiferous tubule, identify as many stages of spermatogenesis as possible, as well as **Sertoli cells**, which function to nurture the developing sperm (Figure 42-5).

B. Mammalian Oogenesis

After birth, in humans and most other mammals, oogonia are not present because all of them have started oogenesis. There is an excess supply of primary oocytes present at birth (~2 million in a newborn girl).

In the ovaries, primary oocytes are located within cellular balls called **follicles**. In addition to supporting the developing oocytes, follicle cells also secrete the female sex hormones, **estrogen** and **progesterone**. A typical mammalian ovary contains a number of stages of follicular development (Figure 42-6).

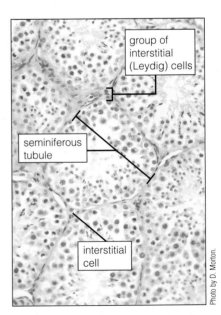

Figure 42-4   Transverse section of the testis (200×).

A primordial follicle whose thin walls are one cell thick initially surrounds each primary oocyte. Most follicles degenerate, and in humans, only about 300,000 primary oocytes remain at puberty. The majority of these oocytes also degenerate. With each turn of the female cycle, several follicles begin to mature, and one (rarely two or more) of each batch of oocytes finally bursts from the surface of the ovary and is swept into the oviduct. The release of oocytes from the ovary is called **ovulation** (Figure 42-7).

Around the time of ovulation, the primary oocyte divides to form a secondary oocyte and the first polar body. The zona pellucida and a capsule of follicle cells (Figure 42-8) surround an ovulated secondary oocyte. A sperm must first pass through this barrier before penetration of the egg membrane and fertilization can occur. Upon fertilization, the secondary oocyte divides again to form an ovum and another polar body. The first polar body may also divide, resulting in a total of four polar bodies.

Follicle cells left behind in the ovary after ovulation develop collectively into a large roundish structure called the **corpus luteum** (the name means "yellow body" and refers to its color in a fresh ovary). The corpus luteum continues to secrete female sex hormones, especially progesterone.

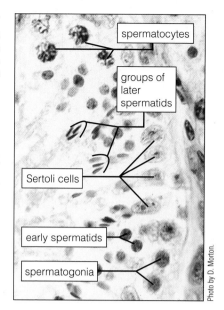

Figure 42-5 Spermatogenesis (800×).

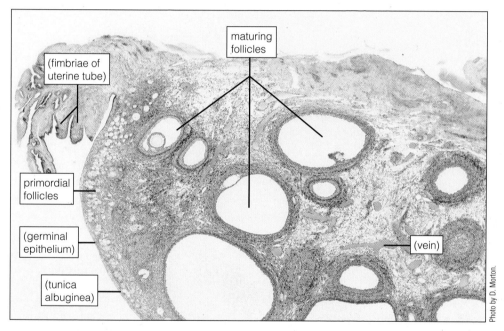

Figure 42-6 Mammalian ovary (25×).

1. With the compound microscope, examine the prepared slide of an ovary. Look for primordial follicles (Figure 42-9a).

Both the follicles and their primary oocytes are small compared with maturing follicles and their primary oocytes. Groups of primordial follicles tend to occur between maturing follicles or between maturing follicles and the ovarian wall. Note that a single layer of smaller cells compose the wall of the primordial follicle.

2. Now look for maturing follicles. In maturing follicles, the size of the cells in the follicle wall and of the primary oocyte itself increases. Also, the follicle cells divide, causing the wall to become first two cells thick and then multilayered. Then a space appears between the follicle cells (Figure 42-9b). This fluid-filled space increases in size until the primary oocyte and the follicle cells immediately around the primary oocyte are suspended in it (Figure 42-9c). The mass of cells is connected to the rest of the wall by a narrow stalk of follicle cells. Just before ovulation, the follicle reaches its maximum size and bulges from the surface of the ovary.

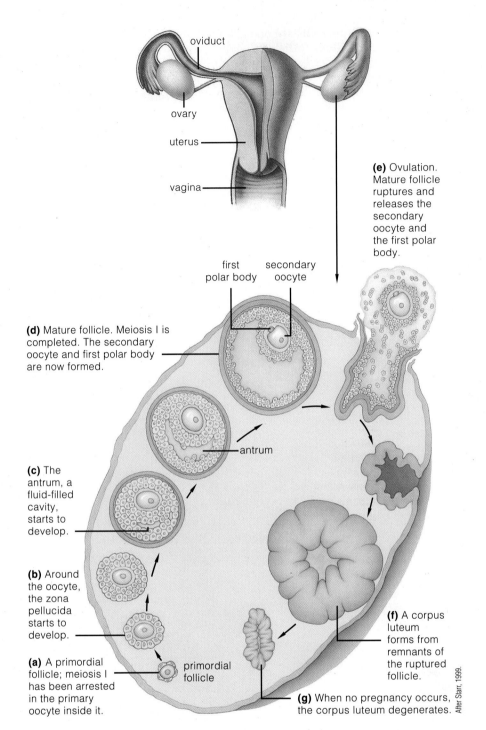

oviduct

ovary

uterus

vagina

(e) Ovulation. Mature follicle ruptures and releases the secondary oocyte and the first polar body.

first polar body   secondary oocyte

(d) Mature follicle. Meiosis I is completed. The secondary oocyte and first polar body are now formed.

antrum

(c) The antrum, a fluid-filled cavity, starts to develop.

(b) Around the oocyte, the zona pellucida starts to develop.

(a) A primordial follicle; meiosis I has been arrested in the primary oocyte inside it.

primordial follicle

(f) A corpus luteum forms from remnants of the ruptured follicle.

(g) When no pregnancy occurs, the corpus luteum degenerates.

After Starr, 1999.

Figure 42-7 Cyclic events in a human ovary, drawn as if sliced lengthwise through its midsection. Events in the ovarian cycle proceed from the growth and maturation of follicles, through ovulation (rupturing of a mature follicle with a concurrent release of a secondary oocyte). For illustrative purposes, the formation of these structures is drawn as if they move clockwise around the ovary. In reality, their maturation occurs at the same site.

3. Replace the slide on the microscope with one of a mammalian ovary with a corpus luteum. Locate a corpus luteum (Figure 42-10).
4. If an embryo does not successfully implant in the **uterus**, the corpus luteum degenerates and is replaced by scar tissue. This scar is now called the *corpus albicans* (Figure 42-11).

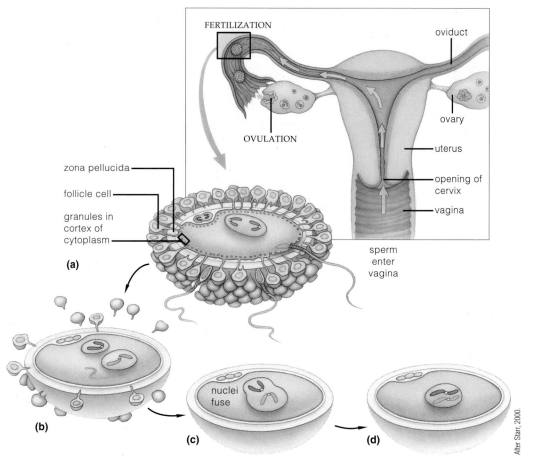

Figure 42-8 Fertilization. (**a**) Many sperm surround a secondary oocyte. Acrosomal enzymes clear a path through the zona pellucida. (**b**) When a sperm does penetrate the secondary oocyte, granules in the egg cortex release substances that make the zona pellucida impenetrable to other sperm. Penetration also stimulates meiosis II of the oocyte's nucleus. (**c**) The sperm's tail degenerates and its nucleus enlarges and fuses with the oocyte nucleus. (**d**) With fusion, fertilization is over. The zygote has formed.

42.3 Sperm Penetration and Fertilization in the Frog *(10 min.)*

In humans and most other mammals, both fertilization (in the upper oviduct) and **implantation** and subsequent development to a late stage (in the uterine wall) take place in the mother. Other animals, such as frogs, have external fertilization and development. At the beginning of the lab, or just before, your instructor induced ovulation in a female frog. Sperm obtained from the shredded testes of a male frog were then added to fertilize these ova.

MATERIALS

Per student pair:
- Syracuse dish
- two camel-hair brushes
- dissection microscope

Per lab group (table or bench):
- model of a frog zygote (optional)

Per lab section:
- live frog zygotes in pond water
- pond water
- video or link(s) to Internet site(s) with video of frog development (optional)

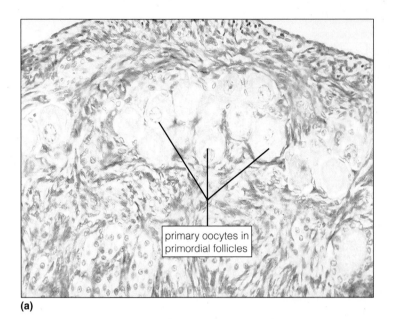

(a)

primary oocytes in
primordial follicles

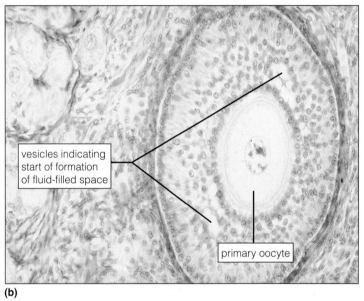

(b)

vesicles indicating
start of formation
of fluid-filled space

primary oocyte

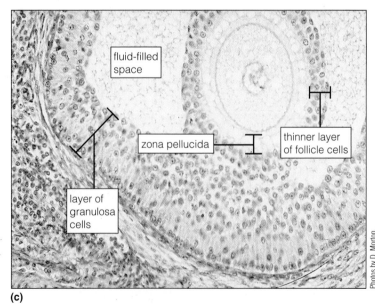

fluid-filled
space

zona pellucida

thinner layer
of follicle cells

layer of
granulosa
cells

Photos by D. Morton.

(c)

Figure 42-9 (a) Group of primordial
follicles (450×), (b) maturing follicles (450×),
and (c) portion of a mature secondary
follicle (40×).

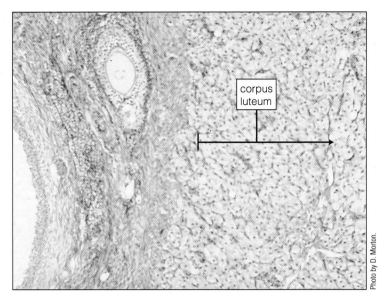

Figure 42-10 Corpus luteum (100×).

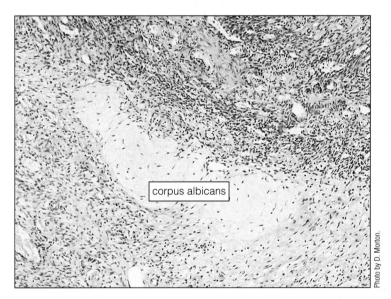

Figure 42-11 Corpus albicans (100×).

PROCEDURE

1. Using two clean camel-hair brushes, gently transfer to a Syracuse dish half-filled with pond water a live frog zygote that has not yet begun to cleave and examine it with a dissecting microscope. *Do not let the zygote dry out.*

Which surface of the zygote (light or dark side) floats up in the pond water? ＿＿＿＿＿＿＿＿＿＿＿＿＿
Compared with ova, do frog zygotes show a change in their pattern of coloration after fertilization? (*Hint:* look between the darkly pigmented and unpigmented areas.) ＿＿＿＿＿＿＿＿＿＿＿ (yes, no)
If so, describe what you observe.

＿＿＿
＿＿＿

2. If you cannot observe fertilization directly, study a model of a frog zygote, watch a video, or follow links provided by your instructor on early frog development.

Cleavage is a special type of cell division that occurs first in the zygote and then in the cells formed by successive cleavages, the **blastomeres**. Unlike typical cell division, there is no period of cytoplasmic growth between the cleavage divisions, causing the blastomeres to become smaller and smaller. After a number of cleavages, the blastomeres form a solid cluster of cells called the **morula**. Cleavage continues until a hollow ball of cells, called the **blastula**, is formed. This stage marks the end of cleavage. Because of the smaller blastomeres, the size of the blastula is only slightly larger than that of the zygote.

In many organisms whose ova have little yolk (e.g., sea stars and humans), cleavage is *complete* and nearly *equal*, resulting in separate blastomeres that are all about the same size. In frogs and other animals with moderate amounts of yolk, cleavage is complete but unequal with the larger yolk-laden cells at the vegetal pole and the smaller cells with little yolk at the animal pole. There is so much yolk in the ova of many animals (e.g., most fish, reptiles, birds, and the two mammals that lay eggs—platypuses and spiny anteaters) that complete cleavage of the entire zygote is impossible, so it forms a disclike embryo that sits on the surface of the yolk. This type of cleavage is incomplete and is called *discoidal*.

CAUTION

Preserved specimens are kept in preservative solutions. Use lab gloves whenever you handle a specimen. Wash any part of your body exposed to this solution with lots of water. If preservative solution is splashed into your eyes, wash them with the safety eyewash for 15 minutes. Even if you wear contact lenses, you should wear safety goggles during a procedure and when observing a procedure. Eyeglasses should suffice when goggles do not fit over them.

MATERIALS

Per student:

- compound light microscope, lens paper, bottle of lens-cleaning solution (optional), lint-free cloth (optional)
- prepared slide of a whole mount of sea star development through gastrulation

Per student pair:

- two Syracuse dishes
- two blue camel-hair brushes (optional)
- two red camel-hair brushes
- dissection microscope

Per lab section:

- frog embryos in pond water from eggs fertilized 1 hour before the lab (optional)

Per lab room:

- preserved specimens of two-, four-, and eight-cell cleavage stages; morulae (32-cell cleavage stage) and blastulae of frog in easily accessible screw-top containers
- source of dH_2O
- pond water (optional)
- models of early frog development
- models of human development
- boxes of different sizes of lab gloves
- safety goggles

PROCEDURE

A. Cleavage in Sea Stars

1. With your compound microscope, observe a prepared slide with whole mounts of early sea star embryos. Find a zygote; two-, four-, and eight-cell cleavage stages; a morula; and a blastula and draw them in Figure 42-12. Be sure to adjust the fine-focus knob to see the three-dimensional aspects of these stages. The blastula is a hollow ball of flagellated blastomeres surrounding a cavity called the **blastocoel**. Label the blastomeres and blastocoel in your drawing of the sea star blastula.
2. What is the orientation (parallel or perpendicular) relative to each other of the first two cleavage planes?

_____

What is the orientation (parallel or perpendicular) of the third cleavage plane compared with the first cleavage plane?

_____

B. Cleavage in Frogs

1. *Optional.* If your instructor prepared fertilized frog eggs 1 hour before the lab, you will be able to watch the first two cleavage divisions during this laboratory. At room temperature about 1 hour after addition of the sperm to the eggs, a region of less pigmented cytoplasm, the **gray crescent**, appears opposite the site

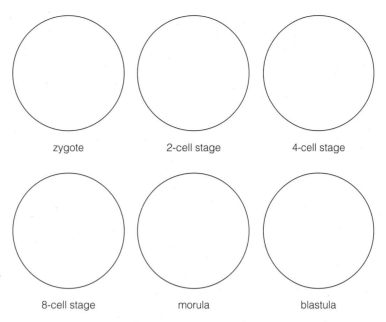

zygote   2-cell stage   4-cell stage

Figure 42-12 Drawings of early
sea star developmental stages
(_____×).

8-cell stage   morula   blastula

of sperm penetration (Figure 42-13). The first cleavage division occurs about 2 hours after fertilization, the second a half hour later, and the third after another 2 hours. Gently place several live frog zygotes in a Syracuse dish half filled with pond water and examine them with a dissection microscope. Use two blue camel-hair brushes to transfer the zygote. *Do not let the zygote dry out.*

pigmented cortex

yolk-rich cytoplasm

sperm penetrating frog egg

gray crescent

After Starr and Taggart, 2001.

Figure 42-13 Formation of the gray crescent.

2. Wear lab gloves and safety goggles. If living embryos are unavailable, look at preserved specimens under the dissecting microscope. Use two red camel-hair brushes to transfer each stage in turn to a Syracuse dish half filled with distilled water. When you are done, return the specimen to its container. *The embryo specimens are in a preservative solution, so make sure you do not use the red brushes to manipulate the live embryos because the residual preservative will harm them.*
3. Examine preserved specimens of eight-cell, morula, and blastula stages.
4. Study the models of the early stages of frog development.
5. Draw the stages of frog development in Figure 42-14.
6. Note that cleavage in the vegetal pole lags behind that of the animal pole. Why? (*Hint*: what is present in the vegetal pole that would hinder cleavage?)

_____

_____

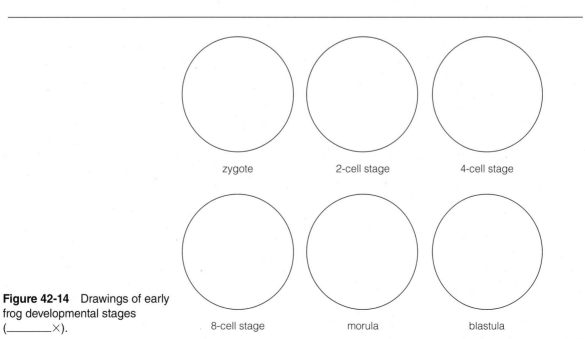

zygote   2-cell stage   4-cell stage

Figure 42-14 Drawings of early
frog developmental stages
(_____×).

8-cell stage   morula   blastula

C. Cleavage in Chickens

In chickens, cleavage of the blastodisc forms a layer of cells called the *blastoderm*, which in time becomes separated from the yolk by a cavity, the blastocoel. Further development of the embryo occurs only in the blastoderm (Figure 42-15).

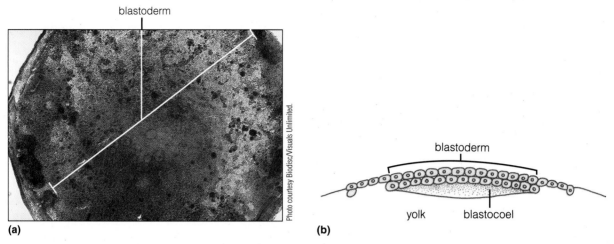

Figure 42-15   Cleavage in the chicken. (**a**) Surface view (42×). (**b**) Transverse section along diameter indicated in (**a**).

D. Cleavage in Humans

Examine models of the early stages of human development. Because the ovum has little yolk, cleavage is complete and nearly equal. At the end of cleavage, a hollow ball of cells is formed. However, this **blastocyst** differs from a blastula in that a group of cells aggregate at one pole of the inner surface of the blastocoel (Figure 42-16). These cells are called the **inner cell mass**, and similar to the blastoderm of chickens, further development of the embryo proceeds there only. The remaining cells that surround the blastocoel are called the **trophoblast**. The blastula stage coincides with implantation of the embryo in the uterus.

Current evolutionary thought is that modern reptiles, birds, and mammals evolved from earlier reptilian ancestors that had external development and ova rich in yolk. Internal development in mammals linked nourishment of the embryo directly to the mother and removed the need for large, yolky eggs. The cost of producing excessive yolk would have been a biological liability; in this case, natural selection (an evolutionary process) resulted in mammalian ova with little yolk. However, gastrulation and subsequent developmental events in reptiles, birds, and mammals are remarkably similar.

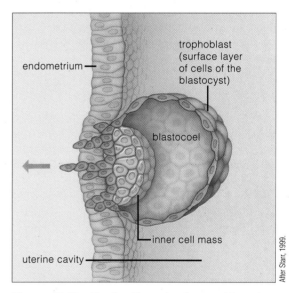

Figure 42-16   Section through a human blastocyst as it attaches to the inner lining (endometrium) of the uterus and starts implantation by burrowing into it (day 6–7 after fertilization).

42.5   Gastrulation *(15 min.)*

Gastrulation is the first time embryonic cells undergo typical cell division with periods of growth between divisions. These cells start the process of specialization and undertake migrations from one embryonic location to another. Groups of some cells even undergo programmed cell death (*apoptosis*) as part of the developmental process. The basic body plan is established. This body is typically made up of three **primary germ layers** or tissues and a front and back end, which can be divided along its longitudinal axis into near identical right and left sides. In animals having at least the organ level of organization, the primary germ layers are called the **ectoderm**, **mesoderm**, and **endoderm** and give rise to organs and, after continued cell specializations, subtypes of the four adult basic tissues (Table 42-2).

TABLE 42-2 — Tissue Derivatives of the Primary Germ Layers

Layer	Derivatives
Ectoderm	Nervous tissues, epidermis (keratinized stratified squamous epithelium) and the structures it forms
Mesoderm	Muscle tissues, connective tissues, and epithelia of the urinary and reproductive systems
Endoderm	Epithelial lining of most of the digestive tract and respiratory tract, and associated glands (e.g., the liver)

© Cengage Learning 2013

As development proceeds, four extraembryonic membranes (also called *fetal membranes* in mammals) form in support of further embryonic development both in the shells of birds (Figure 42-17) and in the uterine walls of mammals. In mammals, the exchange of gasses, nutrients, wastes, and so on takes place across the **placenta** (Figure 42-18), which is partly maternal and partly embryonic in origin.

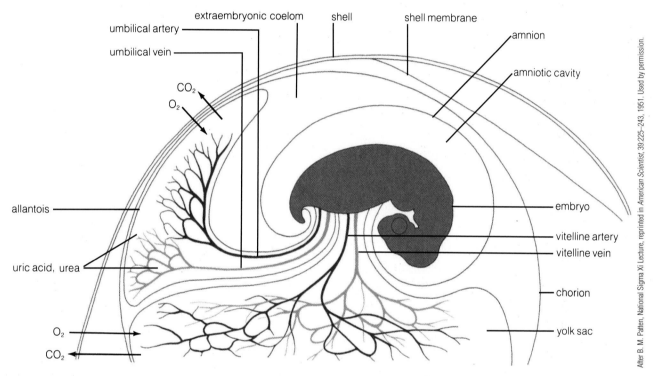

After B. M. Patten, National Sigma Xi Lecture, reprinted in *American Scientist*, 39:225–243, 1951. Used by permission.

Figure 42-17 Extraembryonic membranes of the chicken.

MATERIALS

Per student:

- compound light microscope
- lens paper
- bottle of lens-cleaning solution (optional)
- lint-free cloth (optional)
- prepared slide of a whole mount of sea star development through gastrulation
- prepared slides with whole mounts of chicken embryos at 18 and 24 hours of incubation

Per student pair:

- two Syracuse dishes
- two blue camel-hair brushes (optional)
- two red camel-hair brushes
- dissection microscope
- fertile hen's eggs incubated for 18 and 24 hours (optional)

Per student group (bench or table):

- models of human gastrulation

Per lab section:

- frog embryos (early gastrulae) in pond water from eggs fertilized 21 hours before the lab (optional)
- frog embryos (late gastrulae) in pond water from eggs fertilized 36 hours before the lab (optional)
- pond water

Per lab room:

- preserved specimens of early and late gastrulae of frog in easily accessible screw-top containers
- a source of dH$_2$O
- pond water (optional)
- models of early and late gastrulae of frog
- boxes of different sizes of lab gloves
- safety goggles

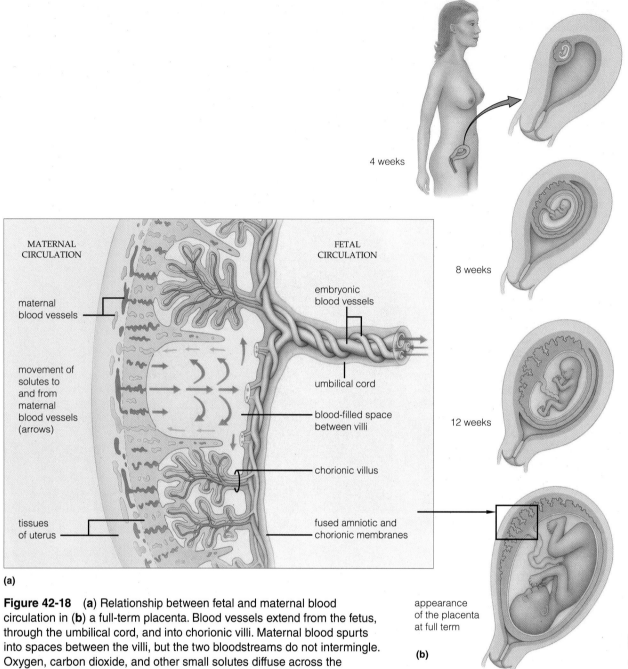

(a)

Figure 42-18 **(a)** Relationship between fetal and maternal blood circulation in **(b)** a full-term placenta. Blood vessels extend from the fetus, through the umbilical cord, and into chorionic villi. Maternal blood spurts into spaces between the villi, but the two bloodstreams do not intermingle. Oxygen, carbon dioxide, and other small solutes diffuse across the placental membrane surface.

4 weeks

8 weeks

12 weeks

appearance of the placenta at full term

(b)

After Starr, 2000.

PROCEDURE

A. Gastrulation in Sea Stars

1. Reexamine with the compound microscope the prepared slide bearing whole mounts of the early stages of sea star development. Using the low- and medium-power objectives, find a blastula (Figure 42-19a).
2. Locate an early gastrula (Figure 42-19b). As gastrulation starts, the vegetal pole flattens and folds in like a pocket to create a new cavity, the **archenteron** ("ancient gut").
3. Find a late gastrula (Figure 42-19c). The hole connecting the archenteron to the outside is called the **blastopore**. Gastrulation initially forms two layers of cells—the ectoderm covering the outside of the gastrula and the endoderm lining the archenteron. Then the mesoderm buds off from the inner tip of the archenteron, and its cells migrate to form a third layer of cells between the ectoderm and endoderm.

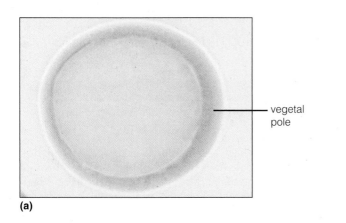

(a)

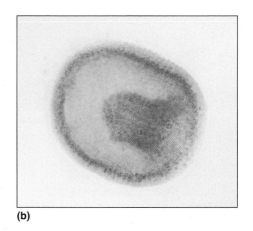

(b)

vegetal
pole

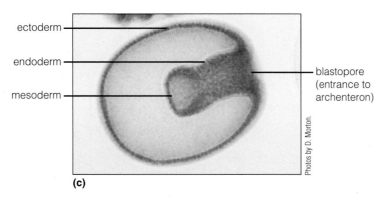

ectoderm

endoderm

mesoderm

blastopore
(entrance to
archenteron)

Photos by D. Morton.

(c)

Figure 42-19 Gastrulation in the sea star (320×). (**a**) Late blastula, (**b**) early gastrula, and (**c**) late gastrula.

B. Gastrulation in Frogs

1. Gastrulation in frogs is affected by the large amount of yolk in the vegetal hemisphere. Because the pigmented cells of the animal pole divide faster, they partially overgrow the yolk-laden cells of the vegetal pole. Examine a living or preserved gastrula using the same protocol previously described for earlier developmental stages.

2. Examine models of an early and a late gastrula. Note that gastrulation does not occur simultaneously over the surface of the vegetal pole. Rather, it starts at a point just under what will be the anus of the adult frog and continues, forming a crescent (Figure 42-20a) that will close to form a circle around a plug of yolk-laden cells, the *yolk plug* (Figure 42-20b). The initial point of the folding-in is referred to as the *dorsal lip of the blastopore.*

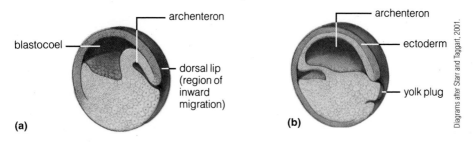

archenteron

blastocoel

dorsal lip
(region of
inward
migration)

(a)

archenteron

ectoderm

yolk plug

Diagrams after Starr and Taggart, 2001.

(b)

Figure 42-20 (**a**) Early and (**b**) late gastrulation in the frog.

C. Gastrulation in Chickens

1. Obtain a prepared slide bearing a whole mount of a chicken gastrula (18 hours of incubation) and one of an embryo incubated for 24 hours.

CAUTION

Do not use the high-power objectives when examining whole mounts of thick material. Its use will break the coverslip.

---

2. Observe the gastrula using only the low- and medium-power objectives. Similar to cleavage, gastrulation in chickens is influenced by the large amount of yolk. The gastrula does not fold in like a pocket through a blastopore; rather, cells move or migrate into a groove on the blastoderm called the **primitive streak** (Figure 42-21a). At one end of the primitive streak, the migrating cells pile up to form *Hensen's node*. This is thought to be equivalent to the dorsal lip of the blastopore.

3. Examine the slide of the embryo at 24 hours of incubation. The three-layered embryo forms in the same axis as the primitive streak but in front of Hensen's node (Figure 42-21b). The three layers from the top of the

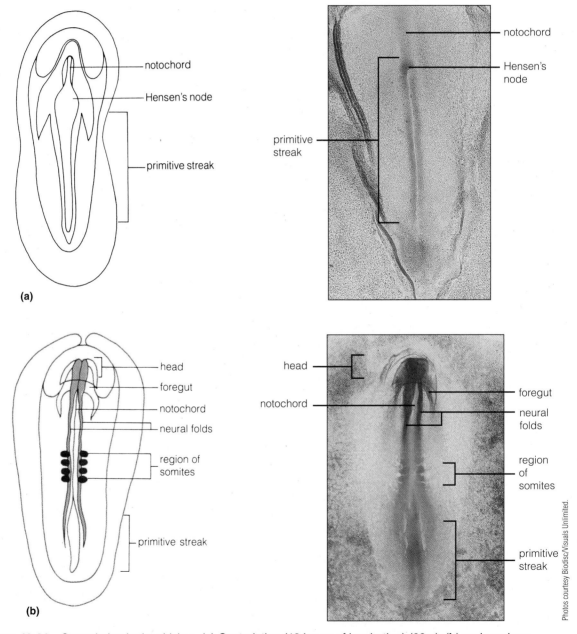

Figure 42-21 Gastrulation in the chicken. (**a**) Gastrulation (18 hours of incubation) (28×); (**b**) early embryo (24 hours of incubation) (16×). The embryos have been removed from the surface of the yolk.

embryo toward the yolk are the ectoderm, mesoderm, and endoderm. One of the first recognizable structures in the embryo is the mesodermal **notochord.** The notochord induces the formation of the embryo's nervous system from the ectoderm. Find *neural folds* on either side of the notochord. With time, the neural folds will fuse like a zipper, anterior to posterior, and in so doing will form the **neural tube** (Figure 42-22). Likewise, the head of the embryo has lifted off the surface of the yolk and has thereby formed the anterior portion of the digestive tract, the *foregut.*

4. Note the two rows of **somites** on either side of the neural folds (Figure 42-21b). These are segmental condensations of mesoderm that will later develop into the skeleton, the skeletal muscles of the trunk, and the dermis of the skin. Count the number of pairs of somites present in this embryo and record this number:

_____

5. *Optional.* If your instructor has fertile hen's eggs incubated for 18 and 24 hours, examine them under the dissection microscope.

D. Gastrulation in Humans

Examine the models of human gastrulation. Human gastrulation follows much the same scenario as that of chickens. However, before gastrulation occurs, a cavity forms between the inner cell mass and the trophoblast. This is the **amniotic cavity** of one of the fetal membranes, the *amnion* (Figure 42-23). Also, cells from the inner cell mass grow downward along the inner surface of the trophoblast, fuse, and form the **yolk sac cavity.** Between the amniotic and yolk sac cavities is the embryonic disc. The **embryonic disc** in mammals is the equivalent of the blastoderm in chickens. Thus, gastrulation commences in the embryonic disc with the formation of a primitive streak.

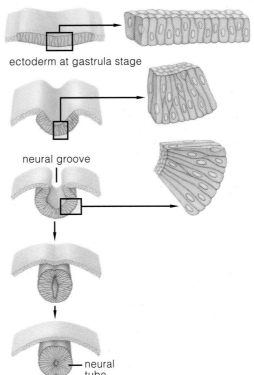

ectoderm at gastrula stage

neural groove

neural tube

Figure 42-22 Formation of the neural tube.

After Starr and Taggart, 2001.

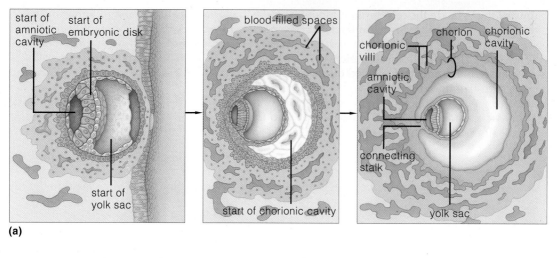

(a)

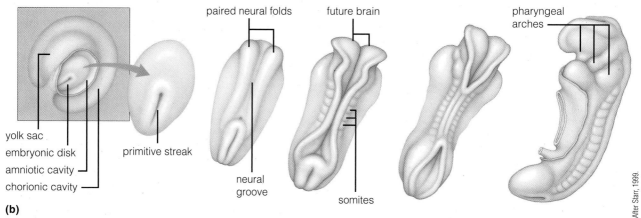

(b)

Figure 42-23 (**a**) Series of sections through a human embryo showing development just after implantation (days 10–14 after fertilization). (**b**) Formation of the primitive streak (day 15 after fertilization).

After Starr, 1999.

_____ 1. Sperm
   (a) are male gametes.
   (b) are female gametes.
   (c) contain yolk.
   (d) are produced in the ovaries.

_____ 2. Ova are
   (a) male gametes.
   (b) female gametes.
   (c) specialized for motility.
   (d) both a and c

_____ 3. The formation of gametes in the gonads can best be called
   (a) gametogenesis.
   (b) spermatogenesis.
   (c) oogenesis.
   (d) none of the above

_____ 4. Which structure is found in a section of the testis?
   (a) secondary spermatocytes
   (b) sperm
   (c) Sertoli cells
   (d) all of the above

_____ 5. Oocytes are found in an ovary in
   (a) seminiferous tubules.
   (b) follicles.
   (c) corpora lutea.
   (d) none of the above

_____ 6. One primary oocyte will form
   (a) one ovum.
   (b) four ova.
   (c) up to three polar bodies.
   (d) both a and c

_____ 7. Which process results in a zygote?
   (a) meiosis
   (b) mitosis
   (c) fertilization
   (d) none of the above

_____ 8. The entrance into the cavity formed during gastrulation in sea stars and frogs is called the
   (a) archenteron.
   (b) blastocoel.
   (c) blastocyst.
   (d) blastopore.

_____ 9. The three primary germ layers are formed during
   (a) fertilization.
   (b) cleavage.
   (c) organ formation.
   (d) gastrulation.

_____ 10. Muscle tissues, connective tissues, and the epithelia lining the urinary and reproductive systems arise from
   (a) mesoderm.
   (b) endoderm.
   (c) ectoderm.
   (d) two of the above.

EXERCISE **42**

Animal Development

Post-Lab Questions

42.1 Gametes

1. Define and characterize the following:
   (a) gametes

   (b) gonads

   (c) gametogenesis

2. Describe the similarities and differences between sperm and ova.

3. Compare chicken, frog, and mammalian ova as to the size and amount of yolk present.

42.2 Gametogenesis

4. Why do you think four sperm cells are produced as a result of gametogenesis but only one ovum is produced?

5. Are oogonia present in adult women? _____ (yes or no) Explain your answer.

42.3 Sperm Penetration and Fertilization in the Frog

6. Read about the gray crescent in your textbook. How does its location relate to the site of sperm penetration and the cytoplasmic events following it?

42.4 Cleavage

7. Identify the following sea star developmental stages (185×).

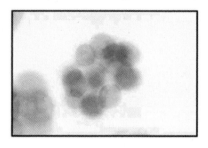

A. _____

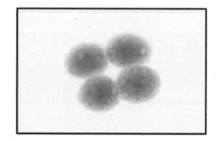

B. _____

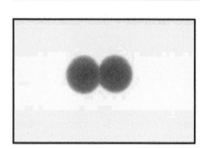

C. _____

D. _____

Photos by D. Morton.

8. How do the amount and distribution of yolk in animal zygotes affect cleavage?

42.5 Gastrulation

9. List the primary germ layers formed by gastrulation. What tissues will they form in the adult?

Food for Thought

10. As various newspaper articles, books, and movies suggest, the cloning of human beings—producing new individuals from activated somatic cells, perhaps followed by uterine implant—is now possible. Can you suggest any biological advantages or disadvantages to having the Earth populated with clones of a few of the best examples of our species?

The Natural Arsenal: An Experimental Study of the Relationships Between Plants and Animals

OBJECTIVES

After completing this exercise, you will be able to

1. define *herbivore, plant secondary compound, allelochemical, toxin, bioactive, cytotoxic, bioassay*;

2. explain the ecological role of many plant toxins and their potential effects on herbivores, pathogens, or competitors;

3. describe the nature of the chemical "arms race" between plants and herbivores;

4. state scientific hypotheses and predictions, design a simple experiment with proper controls and replication, evaluate the results, and reach conclusions based on the data provided by the experiment.

Introduction

"NO! Don't touch that leaf!" ... "Don't eat that berry!" Sound like something grownups said to you when you were young? Why shouldn't you eat the berries of American yew bushes, or any part of the foxglove plant? Why is one nightshade species called "deadly"? Why have so many of us learned to identify poison ivy ("leaves of three, let it be")? Because they're "poisonous," right?

So why do plants produce chemicals that are harmful to people or other animals? Recall that plants occupy the producer trophic level. They use solar energy to build organic molecules from water and carbon dioxide in the process of photosynthesis, and also incorporate into their tissues mineral nutrients absorbed via their roots. Plants compete with each other for crucial resources and occupy the base of most food webs, with animals and other **herbivore** consumers (herbivores are organisms that eat plants) eating them in order to acquire the nutrients and energy stored in the plant body.

Plants would seem to be at a distinct disadvantage in the predation and competition relationships we just described. The vast majority are rooted in the earth; certainly plants aren't able to evade predators by moving away. Nor can they physically move to more advantageous locations for gathering sunshine or minerals. Nonetheless, they are not defenseless against the vast array of herbivores and competitors. In addition to various physical defenses like thorns and tough tissues, most plants have highly effective chemical weapons.

All higher plants have **secondary compounds**. These are chemicals that do NOT fall into the major chemical groupings of carbohydrates, proteins, fats, or nucleic acids. They occur irregularly, with most secondary compounds occurring in a few kinds of plants but not in others. More importantly, most secondary substances have no known role in the metabolism of the plants in which they occur. Rather, they seem to perform a variety of functions, including that of chemical defense agent.

The secondary chemical compound signals that are active between different species are called **allelochemicals**. Caffeine, nicotine, strychnine, organic cyanide compounds, and curare are just a few of the "weapons" in the chemical arsenal that plants have evolved to defend themselves against animal consumers, plant competitors, and pathogenic (disease-causing) microbes. For example, tannins in the leaves of oaks and other plants combine with leaf proteins and digestive enzymes in an insect's gut and inhibit protein digestion by the insect. Thus tannins considerably slow the growth of most caterpillars and other herbivores.

Of course, herbivores may counter such toxic effects through the evolutionary modification of their own biochemistry or physiology. (**Toxins** react with specific cellular components to kill cells or alter growth or development in harmful ways.) Herbivore species that coevolve detoxification mechanisms of secondary chemicals may be able to specialize on plant hosts that are poisonous to most other species.

One familiar example is the Monarch butterfly, *Danaus plexippus*. The adult female lays eggs on milkweed (*Asclepias*) plants. The eggs hatch into an immature stage called a *larva*, which we know as the "caterpillar." The larvae eat the milkweed leaves and eventually complete their developmental cycle. What makes the Monarch/ milkweed relationship so intriguing is that milkweed plants contain potent heart poisons that deter nearly all herbivore feeding. Monarch butterflies have evolved metabolic pathways, however, to detoxify the milkweed poison. In fact, Monarch larvae store the milkweed toxin in their tissues and thereby protect themselves from predators such as birds!

Not all allelochemicals act as toxins that reduce herbivore survival. Some alter the development of insects by interfering with the production of crucial hormones, while others act as repellants that prevent feeding. Still others have antibiotic effects that inhibit infections by pathogens.

In this exercise, you will investigate the harmful effects of some plant compounds on animals and other plants, focusing especially on those bioactive compounds that may be **cytotoxic**, or poisonous to specific cell types. You'll work with a group to design and conduct bioassays to identify potentially **bioactive** compounds, that is, molecules that have a targeted biological effect on survival, growth, or reproduction. A **bioassay** is a method of screening for potentially active substances by exposing a test population of living organisms to the substance in a controlled environment. A bioassay provides an evaluation by measuring the quantity of a substance that causes a defined effective dose.

Bioassays are used for a variety of purposes. They enable researchers to find plant chemicals that show biological activity on selected research organisms. For example, botanists and biochemists use bioassays to screen unstudied plants for potentially useful, naturally occurring drugs. The National Institutes of Health, the National Cancer Institute, and several major U.S. pharmaceutical companies have launched intensive bioassay "chemical prospecting" efforts in tropical rainforests to find new sources of drugs in the fights against cancer, AIDS, heart disease, and many other health problems.

Bioassays are also used by the pharmaceutical industry in other ways. Some pharmaceuticals are cytotoxic agents; chemotherapy agents, for example, are toxic to cancer cells. Certain side effects of chemotherapy are due to the chemical's cytotoxic effect on other body cells as well. Cytotoxic effects are due to interactions with specific biochemical pathways. Therefore, these chemicals are toxic to cells that utilize that specific pathway, but have little to no effect on other cells that don't use the pathway.

While it's unlikely that you will discover an unidentified cancer-fighting chemical or natural herbicide in the plants in your lab, you might very well find cytotoxic agents. It's important to keep in mind that the reason those chemicals exist in the first place is the result of coevolution between plants and their herbivores, competitors, and pathogens. The more we know about what organisms make these chemicals, why they are made, how they are made, and the consequences of the secondary compounds to the parties involved, the more we will know about the evolutionary process and about potential practical uses of such chemicals for our own benefit.

You will use two test organisms in your bioassays. The first, *Artemia,* or brine shrimp, is a small crustacean inhabitant of salty lakes like the Great Salt Lake. *Artemia* are filter feeders that sweep algae and other tiny food particles into their mouths as they swim. They therefore usually occupy a primary consumer (herbivore) position in aquatic ecosystems. One bioassay will look for effects on *Artemia* cells as a general indication of the presence of cytotoxic compounds in a plant extract.

You may have used the second bioassay organism, Wisconsin Fast Plants (*Brassica rapa*), in a previous bioassay (see Exercise 1). Fast Plants are small, very fast-growing plants of the cabbage family. A bioassay with Fast Plants serves to identify potential herbicides (plant-killing compounds) or plant-growth stimulants by looking at the effect of a plant extract on seed germination and plant growth over a single day.

43.1   Experimental Design

Work in groups of four to explore the effects of plant secondary compounds on cell survival of *Artemia,* and/or the effects on seed germination and growth of Fast Plants. Your instructor will direct you to the plant materials that are available for testing.

Develop a hypothesis that you will test. Your hypothesis might relate to differences between extracts of different plant species with respect to effects on *Artemia* cell survival. Or you might wish to test the effect that a single plant extract has on *Artemia* cell survival versus its effect on Fast Plant germination and growth.

Remember, there is no *single* correct hypothesis.
The hypothesis we will test is:

_____

_____

_____

Design an experiment using the materials available in the lab and the following bioassay procedures to test your hypothesis.

Our experimental design is:

_____

_____

_____

_____

What will be your control treatment(s)? What is the purpose of the control treatment?

_____

_____

_____

How will you provide replication to improve the reliability of your results?

_____

_____

Write a prediction for your experiment.

_____

_____

List the possible general outcomes of your experiment. For each, indicate if that outcome would support _or_ falsify your prediction and hypothesis.

_____

_____

_____

_____

_____

Develop a work plan to carry out your experiment so that each group member has specific tasks to accomplish and contribute to the overall group effort. For example, if your experiment will test the differences between extracts of two plant species with respect to their ability to harm _Artemia,_ two group members might prepare the bioassay for one species, and the other two might run the bioassay for the second plant species. You also might divide up the work of collecting data for the duration of the bioassay.

43.2   Bioassay Procedures

A. Preparation of Plant Extract (About 20 min.)

MATERIALS

Per student group:
- 25-mL graduated cylinder
- two 20-mL beakers
- 50-mL beaker
- small mortar and pestle
- spatula
- plant material of your choice
- glass stirring rod

Per lab room:
- several electronic balances able to weigh in 0.1-g increments
- weighing papers or boats
- liquid nitrogen (optional)

PROCEDURE

1. Grind about 1 g of the plant tissue to a fine paste with the mortar and pestle. (If liquid nitrogen is available, carefully pour a small amount over the plant material to instantly freeze it.)

CAUTION
Liquid nitrogen is extremely cold and will cause cryogenic burns. DO NOT touch the liquid or allow any to splash onto your skin.

If using leaf tissue, try not to include large veins; if using root tissue, wash off all soil very well. Grind the tissue _for at least 5 minutes._ The goal is to rupture virtually all the plant cells to mix their contents in the bioassay liquid. When you think you've ground the tissue adequately, continue grinding for two more minutes.

2. Use the balance to measure out two 0.1-g samples of the paste onto two pieces of weighing paper or two weighing boats.

3. Measure 10 mL of prepared brine into one labeled 20-mL beaker, and 10 mL of distilled water into a second labeled 20-mL beaker. Mix a 0.1-g sample of plant paste into each beaker, being careful to scrape off ALL plant material, and stir to create uniform suspensions. This extract has a concentration of 0.1 g plant material/10 mL, or 10 mg/mL. (Recall that 1000 mg = 1 g.) You will use the brine mixture for brine shrimp bioassays, and the dH$_2$O mixture for Fast Plant bioassays.

B. *Artemia* Bioassay (About 1 hour)

MATERIALS

Per student group:

- glass microscope slides
- coverslips
- fine-pointed, water-resistant marker (Sharpie or similar) or wax pencil
- wide-mouth plastic graduated transfer pipets
- razor blade or scalpel
- trypan blue stain in dropping bottle
- acetic acid in dropping bottle
- dissecting needle
- compound microscope
- plant extract(s) prepared in brine in part A

Per lab room:

- live *Artemia* in aquarium or beaker
- plant extracts
- paper towels

PROCEDURE

1. Label one microscope slide "Negative control" and a second microscope slide "Positive control."
2. Obtain a number of brine shrimp and brine in the 50-mL beaker.
3. Place a brine shrimp onto each slide, transferring only a few drops of brine with the animal. Blot excess brine with a paper towel. The brine shrimp should be surrounded in a small brine droplet.
4. Swirl the trypan blue stain bottle to suspend the stain mixture evenly. Place a couple drops of stain atop the brine shrimp on each slide. The brine shrimp should not be able to swim, but should be covered by the liquid. Blot excess stain with a paper towel.
5. Place the negative control slide on the microscope stage and focus on the brine shrimp at low magnification. Observe the structure of the animal, locating the point where the "tail" emerges behind the main body. (See Figure 43-1.)
6. Using a sharp scalpel, cut about $^1/_2$ to $^2/_3$ of the tail off the brine shrimp.
7. Put a coverslip over the severed tail sections. Observe the cut ends of the tail. Amoebocytes, the small roundish cells that ooze out of the tail cavity, are excellent bioassay subjects. They respond to cytotoxic agents in much the same way that human cells do. Healthy cells exclude trypan blue stain and appear colorless, but damaged and dead cells will take up the stain and appear blue (Figure 43-2). Do not confuse clumps of dye crystals with the very round cells.
8. Find a field of view where the amoebocytes aren't being moved by liquid movements. Count the number of cells that are colorless in appearance, and the number that are blue for a total of about 25 cells total. Calculate the percentage of dead cells. This number represents the "background" rate for dead cells. Record your data in Table 43-1.
9. Repeat steps 4–5 for the positive control brine shrimp.
10. Add a drop of acetic acid to the severed tail section for the positive control treatment and mix with a dissecting needle. (Acetic acid is a potent cytotoxin.) Add a coverslip and wait a couple minutes before observing with the microscope.
11. Count unstained and stained cells (about 25 total), and calculate the percentage of dead cells. This number illustrates the effects of a strong cytotoxin on *Artemia* amoebocytes. If you see no stained cells, wait up to 30 minutes for a cytotoxic effect to become visible. Record your data in Table 43-1.
12. Test your plant extract(s) similarly by cutting off a brine shrimp tail in a drop of trypan blue stain on a clean slide, then adding 1–2 drops of plant extract. Add a coverslip, wait a few minutes, then count stained (dead) and unstained (living) cells (about 25 total). Record your data in Table 43-1.

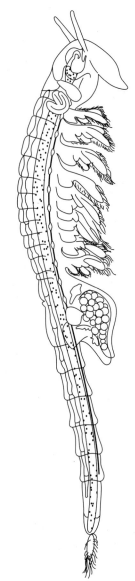

Figure 43-1 *Artemia*, brine shrimp.

 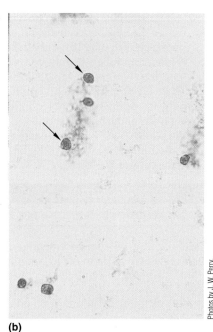

Figure 43-2 (a) Healthy (unstained) cells, (b) dead (stained) cells (500×).

(a)   (b)

Photos by J. W. Perry.

TABLE 43-1 | **Effect of Plant Extracts on *Artemia* Amoebocytes**

Treatment	Droplet Order on Slide	Number Unstained (Live) Cells	Number Stained (Dead) Cells	Percentage Dead Cells
Negative control	Brine + trypan blue			
Positive control	Brine + trypan blue + acetic acid			
Plant extract: _____	Brine + trypan blue + plant extract 1			
Plant extract: _____	Brine + trypan blue + plant extract 2			
Plant extract: _____	Brine + trypan blue + plant extract 3			

© Cengage Learning 2013

C. Fast Plant Bioassay (About 30 min.)

MATERIALS

Per student group:

- fine-pointed, water-resistant marker (Sharpie or similar)
- 35-mm film canister with lid prepunched with four holes
- four microcentrifuge tubes with caps
- four paper-towel wicks

- disposable pipets and bulb
- forceps
- eight Wisconsin Fast Plants seeds
- 15-cm plastic ruler
- dH₂O in dropping bottle
- plant extract(s) prepared in dH₂O in Part A

PROCEDURE

1. Using a fine-pointed, water-resistant marker, label the caps and/or sides of the four microcentrifuge tubes 10.0, 1.0, 0.1, and C (for Control).
2. Place 10 drops of the previously prepared plant extract into the tube labeled 10.0. This tube now contains 10 mg/mL of the test extract. You will now make *serial dilutions* of the solution, producing concentrations of 1.0 mg/mL and 0.1 mg/mL of the test substance:
   (a) Add 9 drops of dH₂O to tubes 1.0 and 0.1.
   (b) From tube 10.0, remove a small quantity of the solution with the disposable pipet and place *one* drop into tube 1.0. Mix the contents thoroughly by "thipping" the side of the microcentrifuge tube (flicking the

sides of the bottom of the tube with the index finger of your dominant hand while holding the sides of the top of the tube between the index finger and thumb of your other hand). Return any solution remaining in the pipet to tube 10.0. Tube 1.0 now contains a concentration 10% of that in tube 10.0, or 1.0 mg/mL.

   (c)   Now, from tube 1.0, remove a small quantity of the solution with *another* disposable pipet and place *one* drop in tube 0.1. Mix the contents thoroughly and return any solution left in the pipet to tube 1.0. Tube 0.1 now contains a 1% concentration of the original, or 0.1 mg/mL.

3. From the dropping bottle, place 10 drops of dH_2O into tube C. This is the control for the experiment, containing none of the test solution.

4. Insert a paper-towel wick into each microcentrifuge tube (Figure 43-3). After the solution has moistened the wick, use forceps to place two RCBr seeds near the top of each wick. The seeds should adhere to the wick. Close the caps of the tubes.

CAUTION

Do not allow any of the wick to protrude outside the cap.

---

5. Carefully insert the microcentrifuge assay tubes through the holes in the film canister lid (Figure 43-4). Set the experiment aside in the location indicated by your instructor.

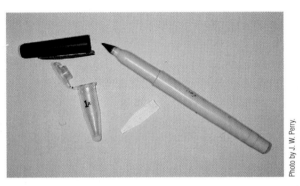

Figure 43-3   A labeled bioassay tube with paper wick.

Figure 43-4 Experimental apparatus for Fast Plant bioassay.

6. How can you provide replication for your experiment? Why would this be a good idea? Explain.

_____

_____

7. After 24 hours or more, examine your bioassay to see how many, and which, seeds germinated. Record your results in Table 43-2. In addition to recording the response as "germinated" (G) or "did not germinate" (DNG), also indicate the extent of any germination by measuring and recording the length in millimeters (mm) of the roots and shoots of seed 1 (S1) and seed 2 (S2) for each treatment. If you are testing a second plant extract, record those results in Table 43-3.

TABLE 43-2 Effect of _____ Extract on Germination of Fast Plant Seeds

Tube	Germination		Length of Roots (mm)		Length of Shoots (mm)	
	Seed 1	Seed 2	Seed 1	Seed 2	Seed 1	Seed 2
Control (dH₂O)						
0.1 mg/mL extract						
1.0 mg/mL extract						
10.0 mg/mL extract						

© Cengage Learning 2013

TABLE 43-3	Effect of _____ Extract on Germination of Fast Plant Seeds					
	Germination		Length of Roots (mm)		Length of Shoots (mm)	
Tube	Seed 1	Seed 2	Seed 1	Seed 2	Seed 1	Seed 2
Control (dH$_2$O)						
0.1 mg/mL extract						
1.0 mg/mL extract						
10.0 mg/mL extract						

© Cengage Learning 2013

43.3 Analysis of Results

Do your bioassay results indicate any cytotoxicity of the plant extract(s) you tested? _____

If yes, do your results indicate cytotoxicity toward plant cells, animal cells, or both? _____

Using your experimental results, write a conclusion regarding the effect of the plant extract(s) on your test organism(s).

_____

_____

Does your conclusion support your prediction *or* prove it unsound? Is your hypothesis supported or falsified by your conclusion? Explain.

_____

_____

_____

List possible sources of error in your experiment and explain why or how they might decrease the reliability of your results.

_____

_____

_____

_____ 1. Plant secondary compounds are chemicals that
  (a) have no known metabolic function in the plant.
  (b) are required for enzyme function.
  (c) are generally carbohydrates or proteins.
  (d) are important in cell division.

_____ 2. Allelochemicals are active in preventing
  (a) herbivore feeding.
  (b) germination of plant competitors.
  (c) infection by pathogens.
  (d) all of the above

_____ 3. Chemicals that are poisonous to specific cell types are said to be
  (a) hazardous.
  (b) cytotoxic.
  (c) allelochemicals.
  (d) bioactive.

_____ 4. Bioassays
  (a) are performed in a test tube using killed cells.
  (b) expose a population of living organisms to a test substance.
  (c) are used for chemotherapy.
  (d) are none of the above.

_____ 5. _Artemia_, brine shrimp,
  (a) are aquatic organisms.
  (b) live in salty ecosystems.
  (c) are filter-feeding primary consumers.
  (d) are all of the above.

_____ 6. Plant defenses against herbivore grazing include
  (a) thorns.
  (b) plant secondary compounds.
  (c) mobility.
  (d) both a and b

_____ 7. Trypan blue is
  (a) taken up by living cells.
  (b) taken up by dead cells.
  (c) taken up by both living and dead cells.
  (d) used to identify plant cells.

_____ 8. Dead _Artemia_ cells that have been treated with trypan blue appear
  (a) colorless.
  (b) blue.
  (c) red.
  (d) indistinguishable from living treated cells.

_____ 9. The dH_2O-only treatment in the Fast Plant bioassay provides
  (a) a control.
  (b) moisture for seed germination.
  (c) dilution of potential cytotoxins.
  (d) none of the above

_____ 10. _Artemia_ and Fast Plant bioassays can be used to identify
  (a) cytotoxins.
  (b) herbicides.
  (c) plant growth stimulators.
  (d) all of the above

EXERCISE **4 3**

The Natural Arsenal: An Experimental Study of the Relationships Between Plants and Animals

Post-Lab Questions

Introduction

1. Why have plants evolved secondary compounds?

2. Describe how a pharmaceutical company might use a bioassay to identify potential drugs in rainforest plants.

3. Would a chemical that is cytotoxic to brine shrimp cells necessarily have the same effect on Fast Plant cells? Explain.

43.1 Experimental Design

4. How can replication be increased in the *Artemia* bioassay to improve reliability of results?

43.2.B. Artemia *Bioassay*

5. What was the purpose of the *Artemia* slide prepared with droplets of
   (a) brine and trypan blue only?

   (b) brine, trypan blue, and acetic acid?

43.2.C. Fast Plant Bioassay

6. If seed germination took place in 100% of plants at the 1.0 mg/mL concentration, and no germination was observed at the 0.1 mg/mL concentration, how would you modify a second experiment to determine the actual effective concentration?

Food for Thought

7. If you found no cytotoxic effects with your plant extract, does this prove the extract is not cytotoxic? Explain.

8. If you found no cytotoxic effects with your plant extract, does this prove that the plant species from which the extract was made has no cytotoxic allelochemicals? Explain.

9. How might you alter the two bioassay procedures to provide more information? Describe and explain.
   (a)   *Artemia* bioassay:

   (b)   Fast Plant bioassay:

10. What other kinds of tests could you perform to provide more information about possible cytotoxicity of a plant extract?

Ecology: Living Organisms in Their Environment

OBJECTIVES

After completing this exercise, you will be able to

1. define *ecology, population, community, ecosystem, producer, consumer, trophic level, primary consumer, secondary consumer, tertiary consumer, decomposer, parasite, herbivore, carnivore, omnivore, survivorship curve;*

2. construct a food web;

3. determine an organism's trophic level and its relationships with other species in the community;

4. identify the three basic types of survivorship curves and describe the trends exhibited by each.

Introduction

Ecology. The word certainly brings to mind groups who advocate for a clean environment, who lobby the government to prevent loss of natural areas, or who protest the policies or actions of other groups. To the biologist, however, this is *not* what ecology as a science is all about. Rather, these activities are environmental concerns that have grown out of ecological studies.

The word *ecology* is derived from the Greek *oikos*, which means "house," so ecology can be thought of as the study of a "home"—be it the earth, a neighborhood, our house, or a pond. Less broadly defined, **ecology** is the study of the interactions between living organisms and their environment. Individuals, whether they are bacteria or humans, do not exist in a vacuum. Rather, they interact with one another in complex ways. As such, ecology is an extremely diverse and complex study. There are so many aspects of ecology that it's difficult to pick one, or even a few, to represent the entire science.

Let's look at the levels of organization considered in an ecological study.

A **population** of organisms is a group of individuals of the same species occupying a given area at a given time. For example, you and your colleagues in the lab at this moment are a population; you are all *Homo sapiens*. Your population consists of humans in biology class right now. The place you occupy, the lab room, is your *habitat*. Of course, you share your habitat with other organisms; unseen bacteria, fungi, and possibly insects are present in the room as well. Maybe plants decorate the lab room; if so, these too comprise yet another population in your habitat.

A **community** consists of *all* the populations of species that occupy a given area at a given time. To use our simple example, your lab community consists of humans, bacteria, fungi, insects, and plants.

Our community exists in a physical and chemical environment. The combination of the community and its environment comprise an **ecosystem**. Your lab ecosystem consists of humans, bacteria, insects, plants, an atmosphere, lab benches, walls, floors, light, heat, humidity, and so on. As you can see, an ecosystem has both living and nonliving components.

In this exercise, we will introduce some of the principles of ecology. You will study three concepts that demonstrate how ecological studies are performed: food webs, energy flow through ecosystems, and survivorship curves.

44.1 Food Webs *(About 25 min.)*

Consider the community of organisms in an environment. Each population of organisms plays a different role in the community, that is, each community has organisms that are producers and organisms that are consumers.

Producers are autotrophic organisms, so most plants and a few microorganisms are **producers** of biomass (the substance of living organisms). They utilize the energy of the sun or chemical energy to synthesize organic (carbon-containing) compounds from inorganic compounds.

Consumers are heterotrophic organisms that feed on preexisting biomass of other organisms, living and/or dead.

In ecological studies, organisms are usually classified into feeding levels (**trophic levels**) by what they eat, as follows:

- Photosynthetic **producers** form the base of nearly all food chains and food webs.
- **Primary consumers**, known as **herbivores**, are animals that eat plant material. Primary consumers thus consume producers.
- **Secondary consumers** are organisms that eat primary consumers. These organisms are called **carnivores**, animals that consume other animals.
- **Tertiary consumers** consume secondary consumers. They are also called carnivores.
- Sometimes feeding strategies cross trophic levels. **Omnivores** are organisms that eat either producers or consumers. They thus consume at various trophic levels depending on the particular food item eaten.
- **Parasites**—organisms that absorb nutrients from a living host organism for a period of time—might be considered primary, secondary, or tertiary consumers depending on whether they are living on a producer or another consumer.
- **Decomposers** include fungi, most bacteria, and some protistans that break down dead organic material from organisms of *all* trophic levels into smaller molecules. The decomposers then absorb the nutrients and ultimately play a crucial role in recycling the nutrients back into the ecosystem. Decomposers operate at all trophic levels.

How would you classify yourself with respect to your feeding strategy and trophic level?

_____

_____

PROCEDURE

Below you will find a partial listing of species in a forest community and some of their source(s) of energy. In the blanks provided, list the *most specific* trophic level of each species from the descriptions above.

- Human (black raspberries, hickory nuts, deer, rabbits) _____
- Black raspberry (sun) _____
- Deer (black raspberries) _____
- Bear (black raspberries, deer) _____
- Coyote (deer, rabbits, black raspberries) _____
- Caterpillar (living hickory leaves) _____
- Bacterial species 1 (living black raspberry leaves) _____
- Bacterial species 2 (dead skin cells of deer) _____
- Bacterial species 3 (dead hickory trees, dead black raspberries, dead humans, dead black bears, dead deer) _____
- Weasels (young rabbits) _____
- Mosquito (blood of living humans, deer, and bears) _____
- Hickory tree (sun) _____
- Fungal species 1 (living black raspberry stems) _____
- Fungal species 2 (dead black raspberries) _____
- Rabbit (black raspberries) _____

Now, in Table 44-1, construct a *food web* by placing the organisms in their appropriate trophic level(s). Leave space around the names if more than one organism occupies a trophic level. Then connect lines from the organisms to their energy source(s) to complete the food web.

Example: Secondary consumer (carnivore)   mountain lion
                                                    |
            Primary consumer (herbivore)   deer
                                                    |
            Producer   grass

TABLE 44-1	A Simplified Food Web
Trophic Level	**Organism(s)**
Decomposer:	
Secondary consumer (carnivore):	
Secondary consumer (parasite):	
Primary consumer (herbivore):	
Primary consumer (parasite):	
Omnivore:	
Producer:	

© Cengage Learning 2013

Suppose a new parasite species that kills black raspberry plants is introduced from another part of the world into this forest community. Speculate on the consequences of loss of black raspberries for the rabbit population.

_____

_____

How might the elimination of black raspberries affect the coyote populations? The weasel population?

_____

_____

What if pesticide spraying unintentionally kills both fungal species? What effect might the unintended consequences of these species losses have on both the raspberry and rabbit populations?

_____

_____

_____

44.2   Flow of Energy Through an Ecosystem *(About 15 min.)*

Ecologists have learned through careful measurement that the amount of energy captured decreases at each succeeding trophic level relative to that in the initial sunlight falling on the ecosystem. Of the total amount of sunlight falling on a green plant, the plant captures—*at best*—1% of this energy. That is, only a small fraction of the light energy falling on the surface of a leaf is converted to the chemical energy of ATP. A primary consumer on average only captures about 10% of the energy stored in a green plant. A secondary consumer usually captures only about 10% of the energy stored in the body of the animal that *it* eats, and so on.

What percentage of the energy contained in plant foods has been captured by the secondary consumer that feeds on an herbivore (primary consumer)? _____

A tertiary consumer eats the first (herbivore-eating) secondary consumer. Assuming a similar flow of energy, how much of the sun's *original* energy does this secondary consumer gain? _____

Suppose an omnivore can obtain all the nutritional requirements necessary for life by eating either plant material or animal material. From an energy standpoint, which route will allow the omnivore to increase efficiency and capture the greater amount of the sun's energy? _____

What other routes would enable this organism to obtain equal amounts of the sun's energy? _____

_____

_____

Let's apply this thinking to the human diet. If you were to consume plant products exclusively (i.e., were vegan), what trophic level would you occupy? _____ What proportion, on average, of the sun's energy that fell on the plants that you eat would you be capturing when you eat a pound of salad greens? _____

If you were to depart from your vegan diet and eat a 16 oz. steak from a beef steer fattened on corn, what proportion of the sun's energy that fell on the corn plants would you be capturing? _____

Which meal, the salad or the steak, allows you to gain the largest proportion of the sun's energy that fell on agricultural ecosystems? _____

44.3   Survivorship*

Within a population, some individuals die very young while others live into old age. To a large extent, the *pattern* of survivorship is species dependent. Generally, **survivorship** takes on one of three patterns, which we summarize using **survivorship curves,** graphs that plot the pattern of mortality (death) in a population.

Humans in developed countries with good health-care services are characterized by a Type I curve (see Figure 44-4) in which there is high survivorship until older age, then high mortality. The insurance industry uses this information to determine risk groups. The premiums they charge are based on an individual's risk group, with risk of death (and cost of premiums) highest for the oldest purchasers of insurance.

While survivorship curves for humans are relatively easy to generate, similar information on other species is much more difficult to determine. It can be a difficult and lengthy project to determine the age of an individual plant or animal, let alone track an entire population over a period of years. However, survivorship rates can be demonstrated in the laboratory using nonliving objects.

In this section, you will study the "populations" of dice and soap bubbles, using them to model real populations and to construct survivorship curves. You will subject these populations to different kinds of stress to determine their effects on survivorship curves.

MATERIALS

Per student:

■ 15-cm ruler

Per student group (4):

■ bucket containing 50 dice
■ soap bubble solution and wand
■ stopwatch or digital watch

Per lab room:

■ copy of Figures 44-2 and 44-3 plus equipment for projection
■ one set of red, blue, black, and green pens or fine-point markers
■ one or more survivorship frames (see Figure 44-1)

PROCEDURE

A. Dice Survivorship (About 30 min.)

Work in a group of four. One person should be assigned to dump the dice, another to record data, and the other two to count.

Population 1

1. Empty the bucket of 50 dice onto the floor or another space without obstructions.
2. Assume that individuals who come up as 1's die of heart disease. All others survive.
3. Pick up all the 1's, set them aside (in the cemetery), and count the number of individuals who have survived. Record the number of *survivors* in this generation (generation 1) in the left-hand column of Table 44-2.
4. Return the survivors to the bucket.
5. Dump those survivors onto the floor again and remove the deaths (1's) that occur during this second generation.
6. Count and record the number of survivors.
7. Continue this process until all the dice have "died" from heart disease.
8. Now determine the percentage of survivors for each generation using the following formula and record in Table 44-2:

$$\text{percentage surviving} = \frac{\text{number surviving}}{50} \times 100\%$$

---

*Adapted from an exercise courtesy J. Shepherd, Mercer University, personal communication.

Population 2

1. Start again with a full bucket of 50 dice. Assume that 1's die of heart disease and 2's die of cancer. Proceed as described for population 1, recording the counts in Table 44-2, until all dice are "dead."
2. Again determine the percentage of survivors for each generation with the following formula:

$$\text{percentage surviving} = \frac{\text{number surviving}}{50} \times 100\%$$

TABLE 44-2 Survival Rates of Dice Populations				
	Population 1 Survivors (Heart Disease Only)		Population 2 Survivors (Cancer and Heart Disease)	
Generation	Number	Percentage	Number	Percentage
0	50	100%	50	100%
1				
2				
3				
4				
5				
6				
7				
8				
9				
10				
11				
12				
13				
14				
15				
16				
17				
18				
19				
20				
21				
22				
23				
24				
25				

© Cengage Learning 2013

How does adding another cause of death affect the survivorship rate?

_____

_____

Do you think this activity models human survivorship well? Why or why not?

_____

_____

B. Soap Bubble Survivorship (About 30 min.)

Work in groups of four. One person blows bubbles, a second person times the "life" of the bubble, a third observes the bubble to determine the point at which it breaks (dies), and the fourth records data in Table 44-3.

Bubbles are "alive" as long as they are intact, but "die" when they burst. You will be following bubbles and timing how long they "live."

Different student groups in your lab section will create three different populations of soap bubbles. Your group will be assigned a population to work with during this exercise:

Population 1. Once the bubble leaves the wand, group members wave, blow, or fan in an effort to keep the bubble in the air and prevent it from breaking (dying) as long as possible.

Population 2. Group members do nothing to interfere with the bubbles or keep them up in the air once the bubbles leave the wand.

Population 3. This group uses a wand mounted on a wooden frame (Figure 44-1). The person blowing bubbles stands at the wand facing the frame, dips the wand in the soap solution, and tries to blow the bubbles through the opening in the frame. *Bubbles that break on the paper "hazard" or any other part of the frame rather than passing through the frame are timed and included in the data.* Bubbles that fall to the side without passing through or breaking on the frame are ignored (not counted in the data). ***Do not*** *attempt to manipulate the frame in any way so as to increase the chances that the bubbles will pass through it.*

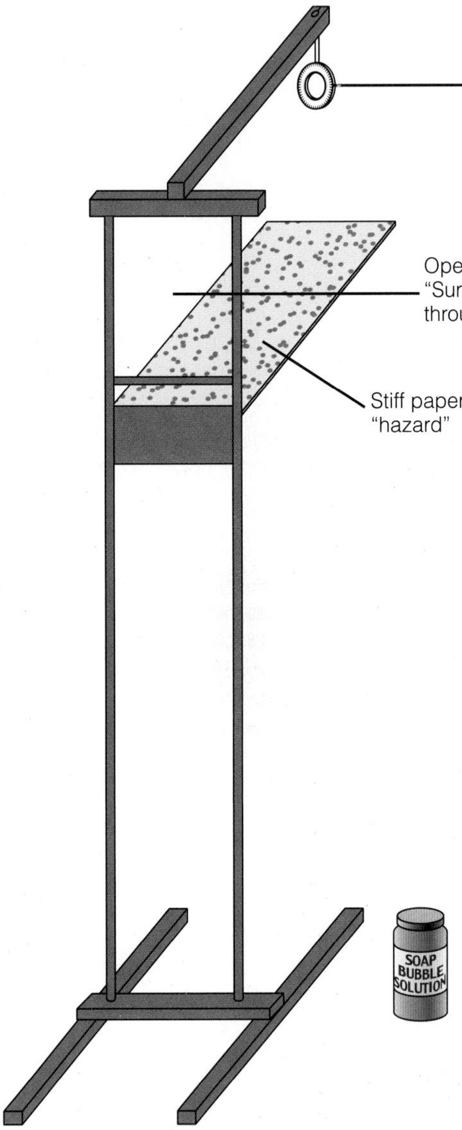

1. Practice blowing bubbles for a few minutes until you can generate them with the wand. Even though you may generate several bubbles, follow only one at a time to measure its life span.
2. The timer should start the watch once the bubble is free of the wand. When the bubble bursts, note the time, and place a check or tick mark at the corresponding line in the "Number" column under the "Bubbles Dying at This Age" section in Table 44-3.
3. Measure the life span of 50 bubbles.
4. Summarize your data as follows:
   (a) Count the number of checks (number of bubbles dying) at each age. Record the number in the column marked "Total."
   (b) To determine the *number* surviving at each age, subtract the number dying at each age from the number surviving at the previous age. So if 5 bubbles break (die) at age 1 second, then 50 − 5 = 45 survive at least 1 second. If 10 more bubbles die at age 2 seconds, then 45 − 10 = 35 survive at least 2 seconds. Continue until all 50 bubbles are accounted for.

Figure 44-1   Frame for soap bubble population 3.

(c) Calculate the *percentage* surviving at each age. Since at birth (the moment the bubble leaves the wand) 50 bubbles are alive, 100% are alive at age 0. Use the following formula:

$$\text{percentage surviving to this age} = \frac{\text{number surviving at age}}{50} \times 100\%$$

TABLE 44-3	Survivorship Data for Soap Bubble Population				
	Bubbles Dying at This Age			**Bubbles Surviving to This Age**	
Age at Death (sec)	**Number (✔ = 1)**	**Total**		**Number**	**Percentage**
0	0	0		50	100%
1					
2					
3					
4					
5					
6					
7					
8					
9					
10					
11					
12					
13					
14					
15					
16					
17					
18					
19					
20					
21					
22					
23					
24					
25					
26					
27					
28					
29					
30+					

© Cengage Learning 2013

PROCEDURE

Work alone for this activity.

1. Use Figure 44-2 to make an *arithmetic plo*t of survivorship of both the dice and the soap bubble populations. Note that the horizontal lines are the same distance apart on this graph.

Using open circles to indicate the *percentage* of *dice* surviving at each age, plot the percentage surviving data from Table 44-2. Using closed circles to indicate the *percentage* of *soap bubbles* surviving at each age, plot the percentage surviving data from Table 44-3.

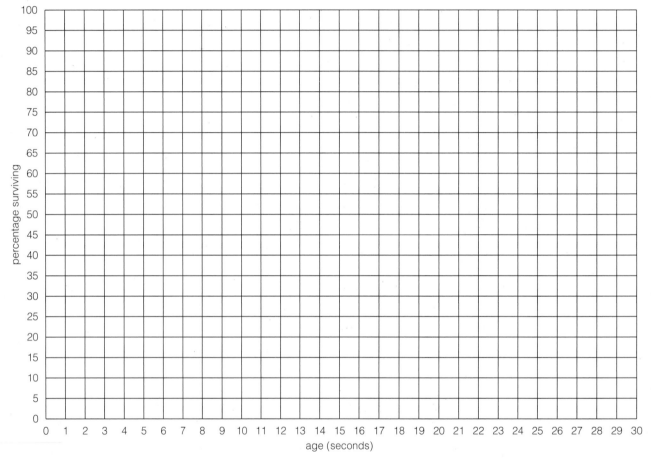

Figure 44-2 Arithmetic plot of survivorship.

2. Now you will use Figure 44-3 to make a *logarithmic plo*t of survivorship. Note that on this *semilog* graph, the horizontal lines become closer together toward the top.

If you've never used semilog paper, examine Figure 44-3 more closely. The scale on the vertical axis runs from 1 at the bottom to 9 near the middle, then from 1 to 9 near the top; there is a 1 at the topmost line. The lines are spaced according to the *logarithms* of the numbers.

The 1 at the top represents 100%, the 9 below it 90%, and so on. This means that the 1 in the middle of the page is 10%, while the 9 below it represents 9%; the 8, 8%; and so forth until the 1 at the bottom, which represents 1%.

Plot the survivorship of the dice and soap bubbles that your group generated, using the percentage surviving data from Tables 44-2 and 44-3. Open circles indicate the percentage of dice surviving and closed circles the percentage of soap bubbles surviving. Connect the points, forming a curve for each population.

After you've plotted your data, copy your plots onto the transparency or other page your instructor has for this purpose, using the pens provided. Different colored pens are available so that each population will have a separate color.

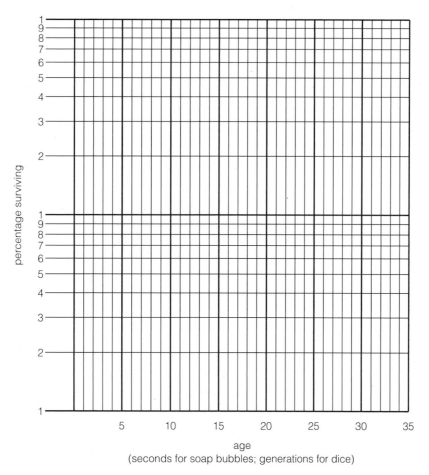

Figure 44-3 Logarithmic plot of survivorship.

44.5 Interpreting the Survivorship Curves *(About 15 min.)*

First examine the plots from the dice populations. In population 1 (heart disease only), typically a constant one-sixth, or 17%, of the population dies on average at each age. In population 2, usually two-sixths, or 33%, dies on average at each age. As you can see, on the *arithmetic plot* these data form a smooth curve, while on the *logarithmic plot* they form a straight line.

Both types of graphs provide useful information. A straight line on a logarithmic plot indicates the death rate is *constant*, but it's easier to see that more individuals die at a young age than at older ages in the arithmetic plot.

In natural populations, three basic trends of survivorship affecting population size have been identified. These are shown in Figure 44-4 and summarized in Table 44-4.

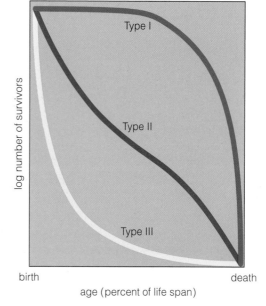

Figure 44-4 Three basic types of survivorship curves.

TABLE 44-4	**Survivorship Curves**
Type	**Trend**
I	Low mortality early in life, most deaths occurring in a narrow time span at maturity
II	Rate of mortality fairly constant at all ages
III	High mortality early in life, then long life for the few that remain

© Cengage Learning 2013

Now examine the logarithmic survivorship curves for the soap bubble population. How do they compare with the Type I, II, and III curves in Figure 44-4?

_____

_____

Do any of the soap bubble populations show a constant death rate for at least part of their life span? If so, for which population?

_____

_____

How did the treatments that populations 1, 2, and 3 were subjected to affect the shape of the curves?

_____

_____

_____

_____ **1.** The Greek word _oikos_ means
  (a) environment.
  (b) habitat.
  (c) house.
  (d) community.

_____ **2.** A population includes
  (a) all organisms of the same species in one place at a given moment.
  (b) the organisms and their physical environment.
  (c) the physical environment in which an organism exists.
  (d) all organisms within the environment at a given moment.

_____ **3.** The place that an organism resides is
  (a) its community.
  (b) its habitat.
  (c) its population.
  (d) all of the above

_____ **4.** An ecosystem
  (a) includes populations but not habitats.
  (b) is the same as a community.
  (c) consists of a community and the physical and chemical environment.
  (d) is none of the above.

_____ **5.** Which organism is **not** a consumer?
  (a) killer whale
  (b) human
  (c) bean plant
  (d) fungus

_____ **6.** The trophic level that an organism belongs to
  (a) is determined by what it uses as an energy source.
  (b) determines the population it belongs to.
  (c) can be thought of as its feeding level.
  (d) is both a and c.

_____ **7.** A decomposer
  (a) is exemplified by a fungus.
  (b) is the same as a parasite.
  (c) feeds on living organic material.
  (d) feeds on primary consumers only.

_____ **8.** An herbivore
  (a) eats plant material.
  (b) eats primary producers.
  (c) is exemplified by a deer.
  (d) is all of the above.

_____ **9.** Which term best describes a human being?
  (a) omnivore
  (b) herbivore
  (c) carnivore
  (d) parasite

_____ **10.** A survivorship curve shows
  (a) the number of individuals surviving to a particular age.
  (b) the cause of death of an individual.
  (c) the place an organism exists in its environment.
  (d) the organism's trophic level.

EXERCISE **44**

Ecology: Living Organisms in Their Environment

Introduction

1. Define the terms *herbivore, carnivore,* and *omnivore.*

2. Define the terms *producer, primary consumer,* and *secondary consumer.*

3. Characterize yourself, using the terms listed in questions 1 and 2. Explain your answer.

44.1 Food Webs

4. Examine the food web you constructed in Table 44-1. Suppose a severe outbreak of rabies occurs in the coyote population, resulting in the death of the entire population. Draw a rough graph that describes the impact of the loss of this predator on each of the rabbit, deer, and raspberry populations over time. Explain your reasoning.

5. You are probably aware of the so-called ozone "hole" over Antarctica and other areas of the planet. The destruction of Earth's protective ozone layer allows more harmful ultraviolet (UV) light to reach the surface. Already, a significant increase in skin cancer is being noticed, a disease known to be caused by excessive exposure to UV light (including that produced by tanning lights). Not as widely known is the damage UV light causes to the photosynthetic machinery of green plants and protists.

   Explain what would happen over time to the populations of organisms of your food web (Table 44-1) if ozone depletion were to become widespread.

44.2 Flow of Energy Through an Ecosystem

6. You may have heard the advice to "eat low on the food chain." Explain how choosing a soybean-protein "burger" rather than a beef burger for lunch is more energy efficient with respect to energy flow and utilization in trophic levels.

44.3 Survivorship

7. Would you expect a population in which most members survive for a long time to produce few, or many, offspring? Under which kinds of environmental conditions would this type of survivorship pattern be best adapted?

8. Suppose a human population exhibits a Type III survival curve. What would you expect to happen to the curve over time if further dramatic improvements in medical technology take place?

9. Return to the survivorship model you created using dice: you probably found that the chance of dying from heart disease is approximately one-sixth for each die, indicating that survivorship is essentially the same for each age group. Compare what happened in your model with a realistic projection showing at what ages most humans die of heart disease.

Food for Thought

10. Ecologist Aldo Leopold, the father of American conservation, said, "The last word in ignorance is the man who says of an animal or plant: 'What good is it?' ... If the biota, in the course of aeons, has built something we like but do not understand, then who but a fool would discard seeming useless parts? To keep every cog and wheel is the first precaution of intelligent tinkering."

    These are strong words indeed. How do they relate to what you've explored in this exercise?

Human Impact on the Environment

OBJECTIVES

After completing this exercise, you will be able to

1. define *runoff, evapotranspiration, fossil fuels, habitat loss, stream, pool, riffle, abiotic, biotic, watershed, point source pollution, nonpoint source pollution, pH, dissolved oxygen, nitrates, phosphates, chloride, fecal coliform bacteria, macroinvertebrate, larvae, Stream Monitoring Biotic Index;*

2. explain how human alterations of land may affect water resources, temperature, atmospheric composition, pollution, and maintenance of species;

3. describe the impacts of inorganic and nutrient pollution on stream organisms;

4. explain the importance of dissolved oxygen in stream water;

5. describe how the aquatic macroinvertebrates present in a stream can indicate water quality;

6. discuss limitations and strengths of standard chemical water tests and the Stream Quality Index;

7. discuss some of the ways in which human activity in a watershed can affect stream water quality.

Introduction

Global warming, extinction of species, damage to the stratospheric ozone layer, toxic chemical spills, ground water contamination, and deforestation. This partial list of environmental effects linked to our expanding human population is diverse and alarming. Each of these issues is a complex problem with scientific, economic, philosophical, sociological, and political aspects, and no single exercise (or entire course!) can fully address any of these issues. However, this exercise, which focuses on some of the ways humans alter terrestrial (land) habitats and affect water quality, will give you an idea of how scientists build the data base they use to assess environmental problems.

45.1   Impact of Land Use Changes

We depend completely on a healthy environment to provide clean air, water, and food; maintain the balance of the atmosphere and climate; decompose wastes; maintain wildlife and other organisms; and provide us with pharmaceuticals, building materials, and much more. Our economies and cultures rely on intact, healthy, functioning environments.

While every organism impacts the environment, our human numbers and technologies make us the most significant force of landscape alteration as we convert natural landscapes to human uses. In this activity you will analyze several locales typical in the United States to gauge some ways that human-caused changes affect their abilities to provide essential environmental services.

PROCEDURE

Study Figure 45-1, photos of a (a) suburban shopping area, (b) residential development, and (c) protected natural area, all located within two miles of each other in east-central Wisconsin. The mall and residential development occupy land that was similar to the natural area before their construction. (Alternatively, your instructor may provide information about three local areas.) As you read the questions and explanations below, fill in Table 45-1. Indicate briefly for each numbered statement how the characteristics of each site affect that feature.

(a)

(b)

(c)

Photos by Joy B. Perry

Figure 45-1 East-central Wisconsin (**a**) strip mall, (**b**) residential development, (**c**) protected natural area.

We rely on adequate fresh water from groundwater and surface sources (rivers and lakes) for irrigation water to grow crops, for industrial processes, and for personal and home usage.

1. Compacted, impenetrable or steeply sloped ground surfaces cause most rainwater and snow melt to run off, flowing over the ground surface by gravity toward low points. Most **runoff** eventually reaches streams, carrying with it any pollutants picked up along the way. In contrast, level, porous ground surfaces with many plant roots allow rainfall to soak into the soil and eventually replenish groundwater supplies. To what extent does each site affect runoff, and how?

2. Wetlands and soils with good water infiltration are able to release water to support plant growth during dry periods. Plants are the base of food chains, with all animals (including humans) depending directly or indirectly on plants for their nutrition. To what extent does each site affect water availability during a drought, and how?

   The ground surface and vegetation on a site can modify daily and annual temperature changes.

3. Plants release water from pores in their leaves in a process called **evapotranspiration**. The evaporation creates a local cooling effect. To what extent does each site have a cooling effect due to evapotranspiration?

4. Dark-colored solid surfaces such as building roofs, roads, and parking lots absorb solar radiation and release that energy as heat. These surfaces have a heating effect on their surroundings, in contrast to vegetated or light-colored surfaces. To what extent does each site have a heating effect on the surrounding area?

Fossil fuels (petroleum, natural gas, and coal) provide most of the energy needed for transportation and electricity production in the United States. The energy is released when the fuels are burned; combustion

TABLE 45-1 Impact of Land Use Changes

Feature	Site _____	Site _____	Site _____
Impact on water quantity:			
1. Runoff during rainfall			
2. Water availability during drought			
Impact on temperature fluctuations:			
3. Cooling by evapotranspiration			
4. More heat produced than absorbed from sun			
Impact on pollution and climate:			
5. Increases air pollutants from fossil fuel combustion			
6. Net producer *or* consumer of carbon dioxide			
7. Pollutants added to soil and/or water			
Impact on species diversity:			
8. Number of species greater than, similar to, *or* less than 150 years ago			
9. Changes in predominance of native species from 150 years ago			

© Cengage Learning 2013

products include carbon dioxide, carbon monoxide, soot, ozone, and other substances that damage the human respiratory system.

In addition, rising carbon dioxide levels in the atmosphere accelerate global warming. Plants remove carbon dioxide from the atmosphere in the process of photosynthesis; as all organisms undergo cellular respiration and when microbes decompose dead organic matter, the carbon dioxide is returned to the atmosphere.

Additional human contributions to pollution include trash and waste production and disposal, pesticide use, and overfertilization of lawns and agricultural areas that may pollute water resources.

5. To what extent does each site degrade air quality by causing increased fossil fuel combustion?
6. Is each site a net producer *or* consumer of carbon dioxide?
7. What kinds of pollutants does each area release into the soil and/or water?

Habitat loss is the single greatest cause for extinction today, followed closely by the impact of invasive species. Every species has specific requirements for the range of physical and chemical properties of its habitat, as well as for interactions with many other species. Some species have very narrow requirements; even small environmental disturbances can overstress these organisms and cause their local or global extinction. Conversely, other species are adaptable to a broad range of conditions and have become widely distributed geographically. Some of these species, such as English sparrows, Norway rats, German cockroaches, and fire ants, spread aggressively and become invasive, reducing native species.

8. Do you estimate the number of species living on each site to be greater than, similar to, or less than the number present 150 years ago? _____
9. What changes, if any, do you think there have there been in the types of species present over the past 150 years? _____

Which human activities or types of changes did you observe to have the greatest impact on the environment? Explain. _____

_____

_____

Which environmental services seem to have been least impacted by commercial and housing development? Which have been most impacted? Explain. _____

_____

_____

Which of the sites has the greatest ability to provide essential environmental services? _____
Which of the sites has the least ability to provide essential environmental services? _____

We will always need energy sources and homes, as well as other resources from the environment. Suggest three changes that would lessen the environmental impact of commercial and residential developments.

1. _____

2. _____

3. _____

45.2 Evaluation of Water Quality of a Stream

A **stream** is a body of running water flowing on the earth. Streams range from shallow, narrow brooks tumbling down mountainsides to broad, deep rivers like the Mississippi. Regardless of its size, every stream has distinct major habitat types that influence the types of organisms living there. Relatively slow-moving, deep waters are called **pools**, while **riffles** are shallow areas with fast-moving current bubbling over rocks (Figure 45-2).

Riffles are places where stream water tumbling over rocks and gravel bottom recharges the water with oxygen. Because of the higher dissolved oxygen levels and the rocks and gravel that provide habitat, most stream inhabitants live in riffles. The rapid current brings oxygen for cellular respiration as well as a steady supply of food from upstream. Pools, on the other hand, are still areas with lower oxygen content. Pools tend to have muddy beds, catch organic matter from upstream, and are areas of decomposition and nutrient recycling.

Gaseous, liquid, and solid substances enter the stream and either dissolve in the water or are carried along as suspended particles. The current thus transports these substances to where they may affect organisms downstream. The current speed, type of bottom structure, temperature, and dissolved substances are the **abiotic** components of the stream ecosystem. These abiotic components influence the life and habits of the organisms inhabiting the flowing water, the **biotic,** or living, components of the stream ecosystem.

The single most important factor influencing the abiotic and biotic components is the character of the **watershed**, all the surrounding land area that serves as drainage for a stream. A watershed may cover a few acres of land surrounding a very small stream, or include tens of thousands of square kilometers for a large river. You can think of a watershed as a bowl, with the stream flowing across the bottom. Water that falls on the higher elevations (the sides of the bowl) of a watershed eventually flows into the stream. Your instructor will provide you with information about the boundaries and terrain of the watershed of the stream you will study.

(a)   (b)

Figure 45-2   Stream habitats. (**a**) Pool leading into a riffle. (**b**) Closer look at a riffle.

The quality of a stream depends on its watershed and the human activities that occur there. In general, a stream is of the highest quality when the watershed and streamside vegetation remain in a natural state, while most human influences tend to be detrimental to stream quality. These human-induced effects can be lumped under the general term *pollution*. The most common types of water pollutants are:

- *Human pathogens,* disease-causing agents, such as bacteria, viruses, and protists.
- *Toxic chemicals,* both organic (carbon-containing) and inorganic. Toxic chemicals can harm humans, and fish and other aquatic life. They include acids, heavy metals, petroleum products, pesticides, and detergents.
- *Organic wastes,* such as sewage, food-processing and paper-making wastes, and animal wastes. Organic wastes can be decomposed by bacteria, but the process uses up oxygen dissolved in the water. The decomposition of large quantities of organic wastes can deplete the water of oxygen, causing fish, insects, and other aquatic oxygen-requiring organisms to die.
- *Plant nutrients,* such as nitrates and phosphates, can cause too much growth of algae and aquatic plants. When the masses of algae and plants die, the decay process can again deplete the critical dissolved oxygen (DO) in the water.
- *Sediments,* or suspended matter, the particles of soil or other solids that become suspended in the water. Sediment originates where soil has been exposed to rainfall, most often from construction sites, forestry operations, and agriculture. This is the biggest pollutant by weight, and it can reduce photosynthesis by clouding the water and can carry pesticides and other harmful substances with the soil particles. When sediment settles, it can destroy fish feeding and spawning areas and clog and fill stream channels, lakes, and harbors.
- *Heat,* or thermal pollution. Heated water holds less dissolved oxygen than colder water, so the addition of heated water by industries must be closely monitored to ensure that dissolved oxygen isn't driven to too low levels.

All these substances enter water as either "point" or "nonpoint" pollution. **Point source pollution** occurs at a specific location where the pollutants are released from a pipe, ditch, or sewer. Examples of point sources include sewage treatment plants, storm drains, and factory discharges. Point sources are relatively easy to identify, regulate, and monitor. Federal and state governments regulate the type and quantity of some types of pollutants for each point source discharge.

Nonpoint source pollution enters streams from a vast number of sources, with no specific point of entry contributing a large amount of pollutants. Most nonpoint source pollution results when rainfall and runoff carry pollutants from large, poorly defined areas into a stream. Examples of nonpoint source pollution include the washing of fertilizers and pesticides from lawns and agricultural fields, and soil erosion from construction sites. Because sources of nonpoint pollution are difficult to identify and define, it is much more difficult to regulate.

What human activities can you name that might be affecting your study stream? _____

_____

_____

45.2.A. Evaluation and Sampling at the Stream

In this section, you will take several measurements in the water and collect water samples for analysis. Your class will divide into groups, with each group performing different tasks. As a class you will measure the temperature and pH of the water, plus determine several nutrient/pollutant levels. Additionally, you will collect macroinvertebrates for classification and sample the water for the presence of potentially pathogenic (disease-causing) bacteria.

45.2.A.1. Physical and Chemical Sampling

MATERIALS

Per student:

- disposable plastic gloves (optional)

Per student group (4):

- nail clippers or scissors
- plastic beaker
- plastic sample bottles
- watch or clock

Per lab room:

- thermometer
- portable pH meter or pH indicator paper
- Hach or other dissolved oxygen test kit
- Hach or other nitrate test kit
- Hach or other chloride test kit
- Hach or other phosphate test kit
- dH_2O
- paper towels

PROCEDURE

(a) **Temperature.** Stream organisms are typically more sensitive to higher temperatures. In addition to affecting the metabolic rate of the cold-blooded aquatic organisms, cooler water holds more dissolved oxygen than warmer water.

Measure the water temperature holding a thermometer about 10 cm below the water surface if possible. Keep the thermometer in the water for approximately 1 min. so the temperature can equilibrate. Record the temperature measurements from three locations, then total them and calculate the average temperature. Enter the average temperature in Table 45-2.

Temperature: 1 _____ ; 2 _____ ; 3 _____ ; Average: _____

TABLE 45-2	Stream Measurements and Water Quality Standards			
Physical/Chemical Measurement	**Average Value**	**Water Quality Standards**		**Probable Cause of Degradation[a]**
		General[b]	**Trout Stream**	
Temperature		<32°C	<20°–23°C	Lack of shade
pH		6.0–8.5	6.5–8.5	Acid drainage
Dissolved oxygen (DO)		>5 mg/L	>6 mg/L	Organic and nutrient sources
Nitrate (NO_3^-)		<2 mg/L	<10 mg/L	Organic and nutrient sources; fertilizers
Phosphate (PO_4^-)		<0.1 mg/L		Organic and nutrient sources; fertilizers
Chloride (Cl^-)		<30 mg/L		Road salt, sewage, agricultural chemicals
Total coliform bacteria		200 colonies/100 mL[c]		Sewage, animal wastes

[a] Probable cause(s) of values failing to meet standards.
[b] Standards for swimming and to protect general aquatic life. (*Source*: Wisconsin Department of Natural Resources.)
[c] Standards for safe human contact. (*Source*: Wisconsin Department of Natural Resources.)

© Cengage Learning 2013

(b) **pH.** pH is a measure of the concentration of hydrogen ions in the water. pH signifies the degree of acidity or alkalinity of water. The pH scale ranges from 0 (most acidic) to 14 (most basic). A solution with a pH of 7 is neutral. A change of only 1 on the scale means a tenfold change in the hydrogen ion concentration of the water.

Most aquatic organisms thrive in water of pH 6–8. pH that is too low can damage fish gills. Low water pH can also cause some minerals, such as aluminum, to become toxic. The higher the pH, the more carbonates, bicarbonates, and salts are dissolved in the water. Streams with moderate concentrations of these substances are nutritionally rich and support abundant life.

Use the pH meter or indicator paper to determine the pH of your stream water. Try to sample both riffle and pool areas. Determine and record the average pH in Table 45-1.

pH: 1 _____ ; 2 _____ ; 3 _____ ; Average: _____

Perform each following chemical analyses on three sites in the stream, average your readings, and record the average concentration in Table 45-2. Follow the instructions provided with each test kit.

(c) **Dissolved oxygen (DO).** Aquatic animals, like most living organisms, require oxygen for their metabolic processes. This oxygen is present in dissolved form in water. If much organic matter (for example, sewage) is present in water, microorganisms use the dissolved oxygen in the process of decomposing the organic matter. In that case, the level of dissolved oxygen might fall too low to be available for aquatic animals. DO requirements vary by species and physical condition, but many studies suggest that 4–5 parts per million (ppm) is the minimum needed to support a diverse fish population. The DO of good fishing waters generally averages about 9 ppm. When DO levels drop below 3 ppm, even pollution-tolerant fish species may die.

Dissolved oxygen: 1 _____ ; 2 _____ ; 3 _____ ; Average: _____

(d) Nitrate (NO_3^-). Nitrogen (N) is a nutrient required for plant life. Nitrogen in the form of **nitrates** (NO_3^-) is a common component of the synthetic fertilizers used on lawns, golf courses, and farm fields to stimulate plant growth. Leaking sewage tanks, municipal wastewater treatment plants, manure from livestock and other animals, and discharges from car exhausts are all sources of stream nitrates. Excessive nitrates in streams can cause overgrowth of algae and other plant life. This plant growth may decrease the light available to other stream life, but eventually dies and decays. The decay process can deplete water of the vital dissolved oxygen.

Nitrate: 1 _____ ;   2 _____ ;   3 _____ ;   Average: _____

(e) Phosphate (PO_4^-). Phosphorous (P), another element crucial to plant life, is also a common component of commercial fertilizers in the form of **phosphates** (PO_4^-). When it rains, varying amounts of phosphates wash from farm fields and other fertilized areas into nearby waterways. Like nitrates, phosphate enrichment also causes nutrient problems in streams. Excessive levels indicate pollution by nutrient sources or organic substances (fertilizers, sewage, detergents, animal wastes).

Phosphate: 1 _____ ;   2 _____ ;   3 _____ ;   Average: _____

(f) Chloride (Cl^-). Chloride is one of the major ions in sewage; it also enters waterways when washed from roadways that have been treated with salt (NaCl, sometimes $CaCl_2$) in wintertime to melt ice and snow. Agricultural chemicals are another source of chloride in water. Water with high chloride content is toxic to plant life.

Chloride: 1 _____ ;   2 _____ ;   3 _____ ;   Average: _____

45.2.A.2. Biological Sampling

MATERIALS

Per student:

- hip boots or other footwear that can get wet
- forceps or tweezers

Per student group (4):

- sterile water sample collection bottles or containers
- kick-seine or D-net sampler
- white enamel pan
- bottle of alcohol
- sample collecting jars
- waterproof markers

Per lab room:

- one or more Coliscan dish plus Easygel bottle
- 1-mL sterile dropping pipets
- nail clipper

PROCEDURE

(a) Fecal coliform bacteria. Fecal coliform bacteria are tiny, rod-shaped bacteria that live in the intestines of humans and all warm-blooded animals and pass out of the body in feces. They can cause human disease if they enter a drinking water supply, and they are also associated with other pathogenic bacteria. You will sample the water of your stream to determine whether any fecal coliform bacteria are present.
   1. Use a marker to label the edge of a Coliscan plate with the date and other identifying information.
   2. Collect a water sample from beneath the surface of the stream.
   3. Use a sterile pipet and transfer 1 mL of the water into the Easygel bottle. Be careful not to touch the inside of the bottle or cap with anything to maintain sterility. Swirl the bottle to mix, then pour into the labeled petri dish. Place lid on dish and gently swirl until the entire dish is covered with liquid. Be careful not to splash over the side or on the lid.
   4. Place dishes right-side-up on a level spot, preferably warm, until solid.
   5. Incubate the sample at 35°–40°C for 24 hours, *or* at room temperature for 48 hours.

(b) Aquatic macroinvertebrates. Macroinvertebrates, animals that are large enough to be seen with the naked eye and that lack a backbone, are an important part of the stream ecosystem. Nearly all of the macroinvertebrates present in the water are immature insects, called **larvae**, which mature and then spend their adult lives on land or in the air. Most aquatic larvae feed on decaying organic matter, but some graze on algae or pursue other animals. Other aquatic macroinvertebrates include mollusks (shelled organisms such as snails and clams), arthropods (relatives of shrimp and lobsters such as crayfish and scuds), and various types of worms.

Most aquatic macroinvertebrates, especially those that live on or in the bottom of riffle areas are good indicators of water quality. Aquatic macroinvertebrates can be separated into four general groups on the basis of their tolerance to pollution. Some bottom dwellers like dobsonfly larvae need a high-quality, unpolluted environment and are placed in Group 1 (Figure 45-3). Dominance of Group 1 organisms in a stream signifies good water quality. Group 2 organisms, including caddisfly and dragonfly larvae, also require good water quality, with low levels of pollutants. Group 3 organisms include blackfly larvae and scuds. They can exist in a wide range of

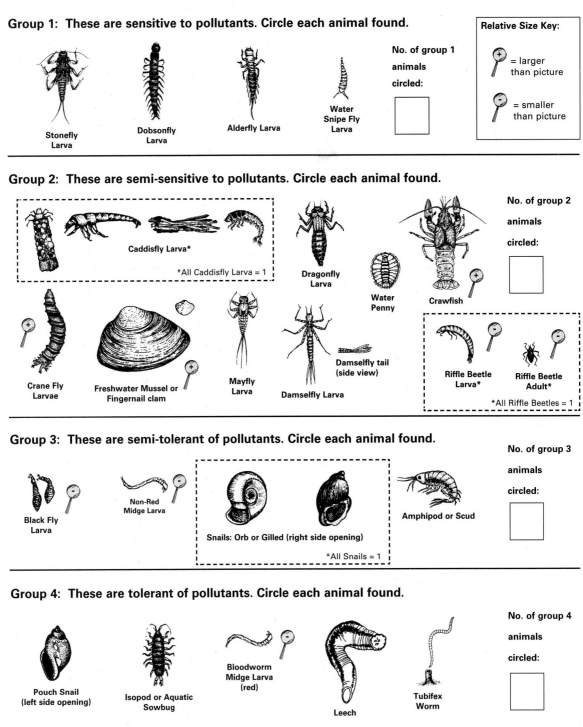

Group 1: These are sensitive to pollutants. Circle each animal found.

Stonefly Larva
Dobsonfly Larva
Alderfly Larva
Water Snipe Fly Larva

No. of group 1 animals circled:

Relative Size Key:
+ = larger than picture
- = smaller than picture

Group 2: These are semi-sensitive to pollutants. Circle each animal found.

Caddisfly Larva*
*All Caddisfly Larva = 1
Dragonfly Larva
Water Penny
Crawfish

No. of group 2 animals circled:

Crane Fly Larvae
Freshwater Mussel or Fingernail clam
Mayfly Larva
Damselfly Larva
Damselfly tail (side view)
Riffle Beetle Larva*
Riffle Beetle Adult*
*All Riffle Beetles = 1

Group 3: These are semi-tolerant of pollutants. Circle each animal found.

Black Fly Larva
Non-Red Midge Larva
Snails: Orb or Gilled (right side opening)
*All Snails = 1
Amphipod or Scud

No. of group 3 animals circled:

Group 4: These are tolerant of pollutants. Circle each animal found.

Pouch Snail (left side opening)
Isopod or Aquatic Sowbug
Bloodworm Midge Larva (red)
Leech
Tubifex Worm

No. of group 4 animals circled:

For more information, call (608) 265-3887 or (608) 264-8948.
Download and print data sheets from
http://www.watermonitoring.uwex.edu/wav/monitoring/sheets.html

Water Action Volunteers

© 2008 University of Wisconsin. This publication is part of a seven-series set, "Water Action Volunteers – Volunteer Monitoring Factsheet Series." All recording forms are free and available from the WAV coordinator. WAV is a cooperative program between the University of Wisconsin-Extension & the Wisconsin Department of Natural Resources. University of Wisconsin-Extension is an EEO/Affirmative Action employer and provides equal opportunities in employment and programming, including Title IX and ADA requirements.

© 2008 University of Wisconsin

Figure 45-3 Stream insects and other macroinvertebrates.

water quality conditions and include crayfish and dragonfly larvae. Group 4 organisms are tolerant of pollution and include red midge and aquatic sowbugs and leeches.

By looking at the diversity of macroinvertebrates present, you will be able to get a general idea of the health of a stream. Typically, waters of higher quality support a greater diversity (more kinds) of pollution-intolerant macroinvertebrates. As water becomes more polluted, fewer kinds of organisms thrive. You will collect samples of macroinvertebrates and separate them into general types to determine a Stream Quality Index Value.

Try to use the kick-seine or D-frame net to sample from riffle areas only, since these areas usually have a higher quality environment, with more dissolved oxygen, for macroinvertebrates. Sample from pools only if most of the stream is pool area.

Work in groups for this activity.

1. Select a riffle area with a depth of 7 to 30 cm, and stones that are 5 to 25 cm in diameter or larger if possible. Move into the sample area from the downstream side to minimize disturbance.
2. One student should hold the kick-seine or D-frame net at the downstream edge of the riffle. Be sure that the bottom edge of the screen is held tightly against the stream bottom. Don't allow any water to flow over the top of the seine or you might lose specimens.
3. Other group members should walk in front of the net and begin the collection procedure. Pick up all rocks fist-sized or larger and rub their surfaces thoroughly in front of the net to dislodge all organisms into the seine. Then disturb the streambed for a distance of 1 m upstream of the net. Stir up the bed with hands and feet until the entire 1-m^2 area has been worked over. Kick the streambed with a sideways motion to allow bottom-dwelling organisms to be carried into the net.
4. Remove the net from the water with a forward-scooping motion so that no specimens are washed from the surface. Carry the seine to a flat area of the stream bank for examination.
5. Rinse the organisms and debris collected to the bottom of the screen. Many aquatic macroinvertebrates are very small, so be certain to remove all of them from the screen. Use forceps to remove any that cling to the screen. Place the collection in a collecting jar; drain off as much water as possible and cover the material with alcohol to kill and preserve the organisms. Label each jar to identify the area sampled.

45.2.B. Laboratory Analysis—Evaluating Biological Tests and Samples

1. Fecal Coliform Bacteria
Inspect the dishes that have been incubated and count the *purple* colonies (*not* light blue, blue-green, white or pink). Multiply that number by 100 to determine the number of fecal coliform bacterial colonies per 100 mL of water. Record in Table 45-2.

Fecal coliform bacteria in excess of 200 colonies per 100 mL (2 colonies per plate) indicate contamination of water with animal or human wastes and are unsafe for human contact.

2. Macroinvertebrates: determining the Stream Monitoring Biotic Index Value
Identification of aquatic organisms can be difficult. However, biological sampling results can be rapidly translated into a water quality assessment without actually identifying each insect specimen, but by determining the number of kinds of certain macroinvertebrates that were collected.

MATERIALS

Per student:

- petri dishes
- forceps or tweezers
- white paper *or* white enamel pan
- dissection microscope *or* magnifying glass

PROCEDURE

(a) Pour a portion of a macroinvertebrate sample into a petri dish. Place a piece of white paper onto the stage of a dissecting microscope to provide an opaque white background, and then observe the specimens in the petri dish. Alternatively, you might pour a sample into a white enamel pan and observe the specimens with a magnifying glass.

(b) Use forceps or tweezers to sort all the organisms into "look-alike" groups, or taxa (Figure 45-3). Look primarily at body shape and number of legs and "tails" to aid in identification. Record each taxonomic group present in your sample under the appropriate group number in Table 45-3.

Group 1 are pollution-intolerant organisms; Group 2 are semi-sensitive to pollutants; Group 3 are semi-tolerant to pollutants; and Group 4 organisms are pollution-tolerant. *For purposes of this lab, disregard organisms you may find but that are not pictured in Figure 45-3.*

c. Determine the Stream Monitoring Biotic Index Value, a measure of stream ecosystem health, as follows (see example below):

1. Each pollution-tolerance group is assigned an importance value as follows:
   Group 1 = importance value 4
   Group 2 = importance value 3
   Group 3 = importance value 2
   Group 4 = importance value 1

2. Count the number of taxa (*types* of organisms) in each group and multiply that number by the importance value. Record the total number of taxa identified for each group, plus the resulting group values in Table 45-3. Also record the total number of taxa identified in Table 45-3.

3. Add the group values for all four groups; record in Table 45-3.

4. Divide the totaled group values by the total number of different taxa identified to yield the Stream Monitoring Biotic Index Value.

Example:

Taxa Identified in Stream Sample

Group 1 (Pollution Intolerant)	Group 2 (Semi-sensitive to Pollutants)	Group 3 (Semi-tolerant to Pollutants)	Group 4 (Pollution Tolerant)
stonefly	caddisfly	scud	pouch snail
	crawfish	gilled snail	aquatic sowbug
			leech
No. taxa **1** × Importance Value **4** = Group Value **4**	No. taxa **2** × Importance Value **3** = Group Value **6**	No. taxa **2** × Importance Value **2** = Group Value **4**	No. taxa **3** × Importance Value **1** = Group Value **3**

A. Total Number Taxa Identified = 1 + 2 + 2 + 3 = **8**
B. Sum of Group Values = 4 + 6 + 4 + 3 = **17**
C. Stream Monitoring Biotic Index Value = Sum of Group Values (B.) ÷ Total Number Taxa Identified (A.) = 17 ÷ 8 = **2.12**

TABLE 45-3 Stream Monitoring Biotic Index Worksheet for (stream name) ▭▭▭▭▭▭

Taxa Identified in Stream Samples

Group 1 (Pollution Intolerant)	Group 2 (Semi-sensitive to Pollutants)	Group 3 (Semi-tolerant to Pollutants)	Group 4 (Pollution Tolerant)
No. taxa _____ × Importance Value 4 = Group Value _____	No. taxa _____ × Importance Value 3 = Group Value _____	No. taxa _____ × Importance Value 2 = Group Value _____	No. taxa _____ × Importance Value 1 = Group Value _____

A. Total Number Taxa Identified = _____
B. Sum of Group Values = _____ + _____ + _____ + _____ = _____
C. Stream Monitoring Biotic Index Value = Sum of Group Values (B.) ÷ Total Number Taxa Identified (A.) = _____

© Cengage Learning 2013

TABLE 45-4	Interpreting the Stream Monitoring Biotic Index Value	
Stream Monitoring Biotic Index Value		**Water Quality**
3.6 and above		Excellent
2.6–3.5		Good
2.1–2.5		Fair
2.0 or less		Poor

© Cengage Learning 2013

What does this index number mean? Table 45-4 provides guidance in interpreting the Stream Monitoring Biotic Index Value.

In addition to indicating general water quality, the aquatic macroinvertebrate survey can provide you with information regarding the types of pollutants that may be affecting the stream. If you identified few taxa of pollution-tolerant invertebrates, but observed great numbers of individuals of these taxa, the stream water may be overly enriched with organic matter, or have severe organic pollution. Conversely, if you found relatively few taxa and only a few individuals of each kind (or if you found *no* macroinvertebrates but the stream appears clean), you might suspect the presence of toxic chemicals. A diversity of taxa, with an abundance of individuals for many of the taxa, indicates a healthy stream with good to excellent water quality.

45.2.C. Analysis of Results

Table 45-2 lists acceptable levels of a few chemicals in water to protect aquatic life, as well as the probable causes of test values that fail to meet those standards. Compare your test results with those values. Also consider your findings from biological observations.

Are your observations, chemical measurements, and biological observations consistent? (Refer to Tables 45-2 and 45-3.) That is, do they *all* indicate that your stream is healthy, with good water quality? Or conversely, do they *all* point to a stream in trouble, with poor water quality? Or do your results conflict, with some indicators of good quality and some indicators of poor water quality? Explain. _____

_____

_____

If your results aren't clear-cut, speculate on why this might be so.

_____

_____

What might you do to increase the consistency and reliability of your results?

_____

_____

Which of your observed or measured factors indicate(s) higher quality water?

_____

_____

Which of your observed or measured factors indicate(s) lower water quality?

_____

_____

Why might you find a low Stream Monitoring Biotic Index Value, even if the chemical and physical parameters you measured are within limits for "good" water quality?

_____

_____

If you found indications of organic or nutrient pollution, what might be some of the sources for this pollution in your stream's watershed?

_____

_____

_____

(Continues on p. 714.)

Do you think most of the pollutants in your stream result from point source or nonpoint source pollutants? Why?

_____

_____

Briefly describe three ways that the negative impacts on your stream could be reduced or eliminated.

_____

_____

_____

_____ 1. Runoff of rainwater is minimized on
   (a) hard surfaces like parking lots.
   (b) compacted ground.
   (c) level, spongy ground.
   (d) steep slopes.

_____ 2. Dark-colored solid surfaces
   (a) have a heating effect on their surroundings.
   (b) have a cooling effect on their surroundings.
   (c) release carbon dioxide.
   (d) increase evapotranspiration.

_____ 3. The greatest cause of extinction of species is
   (a) runoff.
   (b) habitat loss.
   (c) invasive species.
   (d) evapotranspiration.

_____ 4. Which of these is **not** a type of water pollutant?
   (a) dissolved oxygen
   (b) toxic chemicals
   (c) organic wastes
   (d) sediment

_____ 5. pH is
   (a) a measure of phosphorous concentration.
   (b) a measure of hydrogen ion concentration.
   (c) measured by counting the number of different kinds of aquatic insects.
   (d) measured on a scale from 1 to 100.

_____ 6. Dissolved oxygen content in stream water is affected by
   (a) organic matter in the water.
   (b) overgrowth of algae.
   (c) water nutrient levels.
   (d) all of the above

_____ 7. Which pollutant would most likely be the source of coliform bacteria in a stream?
   (a) fertilizers from a golf course
   (b) discharge from a factory
   (c) forestry activities
   (d) sewage or animal wastes

_____ 8. Most of the aquatic macroinvertebrates in a stream are
   (a) clams.
   (b) insect larvae.
   (c) worms.
   (d) arthropods such as crayfish.

_____ 9. Group 1 aquatic macroinvertebrates
   (a) require high-quality, unpolluted water.
   (b) require polluted water.
   (c) are tolerant of a wide range of water quality conditions.
   (d) include leeches.

_____ 10. Nonpoint source pollutants include
   (a) fertilizers from lawns.
   (b) pesticides from farm fields.
   (c) soil erosion from construction sites.
   (d) all of the above

Name _____ Section Number _____

EXERCISE **45**

Human Impact on the Environment

Post-Lab Questions

45.1 Impact of Land Use Changes

1. A wooded area is converted to a housing development and strip mall. Predict and briefly describe three ways this change in land use could affect atmospheric carbon dioxide levels.

2. How can plants reduce the temperature of their surroundings? Would this effect be as great during winter in northern temperature regions as during the summer?

3. Describe how changing the ways that land surfaces are used by humans cause changes in water resources.

45.2 Evaluation of Water Quality of a Stream

4. Why is it important for sewage treatment plants to remove nitrates and phosphates from sewage before discharging it into a river? Explain.

5. Trout are fish that require relatively low-temperature water. Suppose a shaded trout stream has its overhanging streamside trees removed. Explain the possible effects of this action on the trout population over the course of a summer.

45.2.A. Evaluation and Sampling at the Stream

6. At right are partial results of chemical sampling of one stream. What do these results tell you about the suitability of this stream as fish habitat? What are likely causes for these values?

Parameter	Measurement
Fecal coliform bacteria	1800/100 mL
Nitrate	8.5 mg/L
DO	3 mg/L

45.2.B. Laboratory Analysis—Evaluating Biological Tests and Samples

7. Here are the data from one stream's aquatic macroinvertebrate sample.

Taxa Present:

Group 1 (Pollution Intolerant)	Group 2 (Semi-sensitive to Pollutants)	Group 3 (Semi-tolerant to Pollutants)	Group 4 (Pollution Tolerant)
Stonefly larvae	Crane fly larvae	Non-Red Midge larvae	Sowbug
Dobsonfly larvae	Dragonfly larvae	Scud	
	Crayfish		
	Caddisfly larvae		
	Damselfly larvae		

Calculate the Stream Monitoring Biotic Index Value. What does this sample indicate about the water quality?

8. Referring to the stream in question 7, suppose you also know there are very few individuals of the Group 1 and 2 taxa present while scuds are extremely abundant. Does this information alter your assessment of the water quality of the stream? Why or why not?

Food for Thought

9. Describe two ways that stream water quality that is healthy for trout differs from that in which trout would be unlikely to survive.

10. A stream flows through marshes, farmland, and suburban housing developments before emptying into the ocean. Describe three ways that the water quality of the stream is affected by the different land areas along its course.

Animal Behavior

OBJECTIVES

After completing this exercise, you will be able to

1. define *instinctive behavior, learned behavior, phototaxis, geotaxis, thigmotaxis, posture;*

2. discuss the role of the genes in determining behavior;

3. create a list of major environmental stimuli;

4. describe what negative taxis and positive taxis mean;

5. list two aspects of frog behavior that do not require a forebrain and one that does not require a brain;

6. distinguish among the various environmental constituents that individuals can respond to.

Introduction

Animal behavior is the sum of the combined responses of individuals and groups to all of the stimuli they receive from external and internal environments. Receptors like the eyes, ears, olfactory epithelium, and taste buds gather information about various stimuli. Then, chemical messengers—hormones and the neurotransmitters of nervous tissues—direct glands and muscles as to how to respond, producing behavior. Responses result from reflexes, which occur below the level of consciousness, and reactions, which require a conscious decision.

Some behavior is in its entirety programmed in and directed by the genes. This **instinctive behavior**, initially at least, does not need any prior experience or learning. Other behavior does require some prior experience and learning. This **learned behavior** still has a genetic component, but the appropriate response or choice of responses must be learned. A clear example of learned behavior for us is the recognition, interpretation, and creation of spoken and written language. However, this behavior still requires adequate ears, a complex brain, and the correct vocal apparatus, the building of which is genetically programmed. Ultimately, it is the action of evolutionary agents over time that determines the genes animals possess and therefore the behaviors they will and can exhibit.

Not all learned behaviors require a conscious decision. Such behaviors include the imprinting of newborn goslings and many other animals on their mother and the classical conditioning of Pavlov's dogs to salivate when a bell is rung whether food is present or not. Other learned behaviors do require a conscious decision, such as braking your car after you observe a child's ball rolling into the street from between parked vehicles.

46.1   Instinctive Behavior *(40 min.)*

Examples of instinctive behavior include basic responses to major environmental stimuli such as gravity (**geotaxis**), light (**phototaxis**), and touch (**thigmotaxis**) and more complex actions such as the suckling, grasping, and smiling responses in human babies and food preference in newborn garter snakes. Responses can be positive (individuals are attracted to the stimulus) or negative (individuals avoid the stimulus). List up to five other major environmental stimuli that animals may respond to:

1. _____
2. _____
3. _____
4. _____
5. _____

MATERIALS

Per student pair:
- mealworm beetle
- plastic tray
- wooden or plastic blocks

Per lab room:
- snails in an algae-coated freshwater aquarium
- cover for the left half of the aquarium

PROCEDURE

A. Response to Gravity

1. Work in pairs. Describe the feeding behavior of snails in an algae-coated freshwater aquarium.

    _____

    _____

    _____

2. In the second column of Table 46-1, count and record the number of snails that orient with their heads pointed more upward than downward and those with their heads more downward than upward.

TABLE 46-1	Orientation of Snails in a Freshwater Aquarium	
Orientation	**Number**	**Percentage**
Heads upward		
Heads downward		
Total		

© Cengage Learning 2013

3. Total the number of upward and downward pointed snails and calculate the percentage of upwardly and downwardly pointing snails by dividing each group by the total and multiplying by 100, and record these percentages in the third column of Table 46-1.
4. Are most of the snails oriented with their heads directed downward (positive geotaxis) or upward (negative geotaxis)? _____

If the numbers are about equal, then the snails don't respond or are neutral in their response to gravity.

B. Response to Light

1. Count how many snails are in each half of the aquarium and record the numbers in the second column of Table 46-2.

TABLE 46-2	Snail Counts in the Light and Dark Halves of an Aquarium		
Side/Condition	**Initial Count**	**Final Count**	**Number Gained (+) or Lost (−)**
Right/Light			
Left/Dark			

© Cengage Learning 2013

2. At this time, your instructor will place a light over the right half of the aquarium and cover the left half.
3. At the end of the lab period, remove the cover and again count the snails in the right/light and left/dark sides of the aquarium. Record the numbers in the third column of Table 46-2.
4. Record the number gained or lost from each side of the aquarium (fourth column).
5. Are snails phototactic? _____ (yes or no)

    If so, what is their response? _____ (negative or positive)

C. Response to Touch

1. Obtain a plastic tray and some wooden or plastic blocks and construct a simple maze.
2. Place a mealworm beetle in the center.
3. Describe its reaction to touching (walking into) an obstacle.

   _____

   _____

   _____

   _____

   _____

   _____

4. Is your beetle thigmotactic? _____ (yes or no) If so, what is the response? _____ (negative or positive)
5. Start again and note if your beetle initially turns right or left in response to touching a wall by recording a check mark in the second or third column of Table 46-3. Continue to look until you have recorded a total of 10 observations.
6. Describe whether your beetle shows any possible dominance for turning to the right or left? _____

TABLE 46-3	Response of Beetle to Touch	
Observations	Right	Left
1		
2		
3		
4		
5		
6		
7		
8		
9		
10		
Totals		

© Cengage Learning 2013

46.2   Frog Behavior *(30 min.)*

It is often difficult to distinguish between instinctive and learned behavior in animals and for that matter, to recognize when a conscious decision is made. How other animals think and whether or not they think like humans are subjects of great controversy.

MATERIALS

Per student group (4):
- farm-raised frog
- plastic tray
- small rectangular plastic tank filled with room-temperature H_2O

Per lab room:
- farm-raised frog, scissors, dissecting needle for instructor demonstration (optional)
- boxes of different sizes of lab gloves
- box of safety goggles

CAUTION

The secretions of amphibian skin may be irritating, and the skin itself may harbor bacteria. Use lab gloves and safety goggles whenever you handle a specimen and do not touch exposed parts of your body, especially your eyes.

PROCEDURE

1. Work in groups of four students. Put on lab gloves and goggles and place a frog in the middle of a plastic tray. If the frog jumps off the tray, catch and replace it. Once it sits still, describe the typical frog **posture**—the position of the body and its parts.

   _____

   _____

   _____

2. Place the frog on its back and let go. Describe its response.

   _____

   _____

   _____

   Speculate as to the stimulus that initiates this righting response.

   _____

3. Put the frog in a tank of water and observe the frog's response to this stimulus. If the frog does not move, give a little push from the rear. Describe in some detail its response.

_____

_____

_____

4. Hold the frog in your hand and stretch out a leg. Give a firm pinch to the frog's toes but not so hard that you risk breaking them. Describe the frog's response.

_____

_____

_____

5. Consider how much of this behavior is instinctive or learned and how much requires conscious decisions on the part of the frog. Discuss your observations, come to a group consensus, and summarize it here.

_____

_____

_____

6. Which of these behaviors can be done by a frog without a forebrain (no consciousness), without a brain, and without a central nervous system or CNS (no brain or spinal cord)? Record your consensus predictions by placing a check mark in the Prediction columns of Table 46-4.

TABLE 46-4	Frog Behavior					
	Without a Forebrain		Without a Brain		Without a CNS	
Response	Prediction	Result	Prediction	Result	Prediction	Result
Righting						
Swimming						
Withdrawal						

© Cengage Learning 2013

7. Your instructor may demonstrate, show a videotape, ask you to watch video clips on a computer, or simply tell you about the correct results for a frog without a forebrain, without a brain, and without a CNS. The forebrain is removed by cutting off the upper jaw from the back of the eyes onward with a pair of scissors. The brain is removed by placing a needle between the cranium and skull and pushing it forward and scrambling the tissue. A frog without a CNS is produced by scrambling the brain and then turning the dissecting needle around and pushing it downwards, wiping out the spinal cord. Record the results for each condition in Table 46-4. How do the results summarized in Table 46-4 agree or contradict your predictions? _____

_____

_____

8. Do they affect what you thought about instinctive and learned behavior in humans?

_____

_____

_____

46.3 Tropical Fish Behavior (30 min. or more)

In nature, animals don't live in isolation from the physical and chemical makeup of the environment or from other species—animals and plants. Behavior occurs between individuals of the same species and individuals of different species, as well as between individuals and nonliving environmental constituents.

MATERIALS
- aquarium with five species of tropical fish
- sign that displays a picture, name, and code for each tropical species

PROCEDURE

1. Go to the tropical fish aquarium. Note the sign with a picture of each species of fish present, its name, a code, and the number of individuals of each species in the tank. Transfer this information to Table 46-5.

TABLE 46-5	Fish Present in Aquarium		
Common Name	Species Name	Code	Number of Individuals

© Cengage Learning 2013

2. Sit and watch one of the five species of fish in the aquarium. Write the species name in the title row of Table 46-6.
3. Also note in Table 46-6 any behaviors you observe, their frequency, and, if appropriate, the living (same species, another species; use codes), or nonliving environmental constituent (plant, rock, bottom of tank, sides of tank, shady or open areas of tank, and so on) at which it is directed. Observations could include a brief description of feeding, display (taking a particular body orientation), mating, egg laying, aggression (such as fin nipping), location in tank, schooling (fish swim in synchrony), jumping, or relative swimming speed.

TABLE 46-6	Behavioral Observations of _____			
Behavior	Frequency	Same Species	Another Species	Nonliving

© Cengage Learning 2013

4. Describe your chosen species' behavior under these conditions.

_____
_____
_____
_____
_____

46.4 Phototaxis Experiment *(60 min.)*

In this part of the exercise, you will test the hypothesis that *planarians exhibit negative phototaxis.*

MATERIALS

Per student group (4):

- petri dish
- aerated real or artificial stream water
- planaria
- penlight

PROCEDURE

1. Work in a group of four. In Table 46-7, predict the planarian's response to a light pointed at the front of its head, the side of its head, and its tail when the planarian is placed in a petri dish of fresh stream water. As done previously, use "If … then … " formats for your statements.

TABLE 46-7	Presence or Absence, and Direction of Phototaxis in Planarians		
Predictions:			
Trial	**Light Directed at Front of Head**	**Light Directed at Side of Head**	**Light Directed at Tail**
1			
2			
3			
4			
5			
Most Common Response:			
Conclusion:			

© Cengage Learning 2013

2. Cooperate with other groups and darken the room. Determine the orientation of a living specimen in a glass petri dish filled with fresh stream water. Point a penlight from in front directly at its head. Turn the penlight on. Describe briefly in the second column of Table 46-7 the specimen's response or lack of response (moves or turns toward the light, moves or turns away from the light, or no response). Repeat this trial for a total of five times, giving the specimen a 2-minute rest between each trial.

3. Repeat the procedure described in step 2, except shine the light directly at the side of the specimen's head and record the response in the third column of Table 46-7 (moves or turns away from the light, moves or turns toward the light, or no response).

4. Again, repeat the procedure described in step 2, except shine the light directly at the specimen's tail and record the response in the fourth column of Table 46-7 (moves or turns toward the light, moves or turns away from the light, or no response).

5. Note the most common response to each set of trials in Table 46-7 and write a conclusion as to whether the prediction is rejected or accepted.

6. Share your results with the other groups and collect their information. Are your results consistent with their results?

_____

_____

_____

7. If you were to look for planaria in a stream, where would you look for them? Base your answer on the results of this experiment.

_____

_____

_____

_____ 1. Behavior that is entirely controlled by genes and has no learning component is
   (a) learned.
   (b) habituated.
   (c) instinctive.
   (d) classically conditioned.

_____ 2. Learned behavior
   (a) has a learning component.
   (b) has a genetic component.
   (c) has no genetic component.
   (d) has both a and b.

_____ 3. All behavior
   (a) has a genetic component.
   (b) involves learning.
   (c) requires a conscious decision.
   (d) both a and c

_____ 4. Braking your car after you observe a child's ball rolling into the street from between parked vehicles
   (a) is instinctive.
   (b) requires a conscious decision.
   (c) is a reflex.
   (d) both a and b

_____ 5. An individual who avoids light is exhibiting
   (a) negative phototaxis.
   (b) positive geotaxis.
   (c) negative thigmotaxis.
   (d) positive thigmotaxis.

_____ 6. An individual who is attracted to light is exhibiting
   (a) negative phototaxis.
   (b) positive phototaxis.
   (c) negative thigmotaxis.
   (d) positive thigmotaxis.

_____ 7. An individual who responds to walking into an obstruction by pushing into it harder is exhibiting
   (a) negative phototaxis.
   (b) positive geotaxis.
   (c) negative thigmotaxis.
   (d) positive thigmotaxis.

_____ 8. An individual who avoids the pull of gravity is exhibiting
   (a) negative phototaxis.
   (b) positive geotaxis.
   (c) negative geotaxis.
   (d) positive thigmotaxis.

_____ 9. Schooling is
   (a) a feeding behavior.
   (b) fin nipping.
   (c) fish swimming in synchrony.
   (d) a mating behavior.

_____ 10. Fish can exhibit behaviors that are directed to
   (a) members of the same species.
   (b) members of different species.
   (c) nonliving constituents of the environment.
   (d) all of the above

EXERCISE 46

Animal Behavior

Post-Lab Questions

Introduction

1. Compare the contributions of genes and learning to instinctive behavior and learned behavior.

2. Does all learned behavior require a conscious decision before responding to a stimulus? Explain why or why not.

46.1 Instinctive Behavior

3. Match the terms in the left column to the definitions in the right column. Use each answer only once.

    _____ a. negative geotaxis   i.   avoidance of light

    _____ b. positive geotaxis   ii. attracted to gravity

    _____ c. negative phototaxis   iii. attracted to light

    _____ d. positive phototaxis   iv. avoidance of touch

    _____ e. negative thigmotaxis   v.   avoidance of gravity

    _____ f. positive thigmotaxis   vi. attracted to touch

46.2 Frog Behavior

4. Does a frog need a consciousness to behave like a frog? Explain.

46.3 Tropical Fish Behavior

5. Although a tropical fish aquarium is at best an approximation of a natural habitat, what constituents does the aquarium share with the natural habitat that are important to understanding the behavior of a particular species?

Food for Thought

6. What would you call an aquatic organism's avoidance of high concentrations of a chemical in the water?

7. Look up the fight–flight response in your textbook. Describe what it is and how it is still an advantageous part of human behavior.

8. You probably have had one if not several pet mammals—dogs, cats, and so on. Based on your interactions with them, hypothesize as to whether they can think like humans. If you believe that they can think, how and why might their thoughts be different from ours? (*Hint*: consider the differences between the language of humans and other mammals.)

9. Search the Internet for sites about animal thinking. List two sites and briefly summarize their contents.

   http:// _____

   _____

   _____

   http:// _____

   _____

   _____

10. Search the Internet for sites that include animal cams. List and briefly describe the contents of two sites that you would recommend.

    http:// _____

    _____

    _____

    http:// _____

    _____

    _____

Measurement Conversions

Metric to American Standard	American Standard to Metric
Length	**Length**
1 mm = 0.039 inch	1 inch = 2.54 cm
1 cm = 0.394 inch	1 foot = 0.305 m
1 m = 3.28 feet	1 yard = 0.914 m
1 m = 1.09 yards	1 mile = 1.61 km
1 km = 0.622 mile	
Volume	**Volume**
1 mL = 0.0338 fluid ounce	1 fluid ounce = 29.6 mL
1 L = 4.23 cups	1 cup = 237 mL
1 L = 2.11 pints	1 pint = 0.474 L
1 L = 1.06 quarts	1 quart = 0.947 L
1 L = 0.264 gallon	1 gallon = 3.79 L
Mass	**Mass**
1 mg = 0.0000353 ounce	1 ounce = 28.3 g
1 g = 0.0353 ounce	1 pound = 0.454 kg
1 kg = 2.21 pounds	

Genetics Problems

You may find it helpful to draw your own Punnett squares on a separate sheet of paper for the following problems.

Monohybrid Problems with Complete Dominance

1. In mice, black fur (*B*) is dominant over brown fur (*b*). Breeding a brown mouse and a homozygous black mouse produces all black offspring.
    (a) What is the genotype of the *gametes* produced by the brown-furred parent? _____
    (b) What genotype is the brown-furred parent? _____
    (c) What genotype is the black-furred parent? _____
    (d) What genotype is the black-furred offspring? _____
    (e) If two F_1 mice are bred with one another, what phenotype will the F_2 offspring be, and in what proportion?
    phenotype _____
    proportion _____

2. The presence of horns on Hereford cattle is controlled by a single gene. The hornless (*H*) condition is dominant over the horned (*h*) condition. A hornless cow was crossed repeatedly with the same horned bull. The following results were obtained in the F_1 offspring:
    eight hornless cattle
    seven horned cattle
    What are the parents' genotypes?
    cow _____
    bull _____

3. In fruit flies, red eyes (*R*) are dominant over purple eyes (*r*). Two red-eyed fruit flies were crossed, producing the following offspring:
    76 red-eyed flies
    24 purple-eyed flies
    (a) What is the approximate ratio of red-eyed to purple-eyed flies? _____
    (b) Based on your experience with previous problems, what two genotypes give rise to this ratio? _____
    (c) What are the parents' genotypes? _____
    (d) What is the genotypic ratio of the F_1 offspring? _____
    (e) What is the phenotypic ratio of the F_1 offspring? _____

Monohybrid Problems with Incomplete Dominance

4. Petunia flower color is governed by two alleles, but neither allele is truly dominant over the other. Petunias with the genotype R^1R^1 are red-flowered, those that are heterozygous (R^1R^2) are pink, and those with the R^2R^2 genotype are white. This is an example of **incomplete dominance**. (Note that superscripts are used rather than upper- and lowercase letters to describe the alleles.)
    (a) If a white-flowered plant is crossed with a red-flowered petunia, what is the genotypic ratio of the F_1 offspring? _____
    (b) What is the phenotypic ratio of the F_1 offspring? _____
    (c) If two of the F_1 offspring are crossed, what phenotypes will appear in the F_2 generation? _____
    (d) What will be the genotypic ratio in the F_2 generation? _____

Monohybrid Problems Illustrating Codominance

5. Another type of monohybrid inheritance involves the expression of *both* phenotypes in the heterozygous situation. This is called **codominance**.

   One well-known example of codominance occurs in the coat color of Shorthorn cattle. Those with reddish-gray roan coats are heterozygous (RR'), and result from a mating between a red (RR) Shorthorn and one that's white ($R'R'$). Roan cattle don't have roan-colored hairs, as would be expected with incomplete dominance, but rather appear roan as a result of having both red *and* white hairs. Thus, the roan coloration is not a consequence of pigments blending in each hair. Because the R and R' alleles are *both* fully expressed in the heterozygote, they are codominant.

   (a) If a roan Shorthorn cow is mated with a white bull, what will be the genotypic and phenotypic ratios in the F_1 generation?

   genotypic ratio ——————

   phenotypic ratio ——————

   (b) List the parental genotypes of crosses that could produce at least some

   white offspring ————————————————————————————————

   roan offspring ————————————————————————————————

Monohybrid, Sex-linked Problems

6. In humans, as well as in many other animals, sex is determined by special sex chromosomes. An individual containing two X chromosomes is a female, while an individual possessing an X and a Y chromosome is a male. (Rare exceptions of XY females and XX males have recently been discovered.)

   I am a male/female (circle one).

   (a) What sex chromosomes do you have? ——————

   (b) In terms of sex chromosomes, what type of gametes (ova) does a female produce? ——————

   (c) What are the possible sex chromosomes in a male's sperm cells? ——————

   (d) Which parent's gametes will determine the sex of the offspring? ——————

7. The sex chromosomes bear alleles for traits, just like the other chromosomes in our bodies. Genes that occur on the sex chromosomes are said to be sex-linked. More specifically, the genes present on the X chromosome are said to be X-linked. Many more genes are present on the X chromosome than are found on the Y chromosome. Nonetheless, those genes found on the Y chromosome are said to be Y-linked.

   The Y chromosome is smaller than its homologue, the X chromosome. Consequently, most of the loci present on the X chromosome are absent on the Y chromosome.

   In humans, color vision is X-linked; the gene for color vision is located on the X chromosome but is absent from the Y chromosome.

   Normal color vision (X^N) is dominant over color blindness (X^n). Suppose a color-blind man fathers the children of a woman with the genotype $X^N X^N$.

   (a) What genotype is the father? ——————

   (b) What proportion of daughters will be color-blind? ——————

   (c) What proportion of sons will be color-blind? ——————

8. One daughter from the preceding problem marries a color-blind man.

   (a) What proportion of their sons will be color-blind? (Another way to think of this is to ask: What are the *chances* that their sons will be color-blind?) ——————

   (b) Explain how a color-blind daughter might result from this couple.

   ——

   ——

Dihybrid Problems

Recall that pigmented eyes (P) are dominant to nonpigmented (p), and dimpled chins (D) are dominant to nondimpled chins (d).

9. A pigment-eyed, dimple-chinned man marries a blue-eyed woman without a dimpled chin. Their first-born child is blue-eyed and has a dimpled chin.

    (a) What are the possible genotypes of the father? _____

    (b) What genotype is the mother? _____

    (c) What alleles may have been carried by the father's sperm? _____

10. Suppose a dimple-chinned, blue-eyed man whose father lacked a dimple marries a woman who is homozygous recessive for both traits.

    (a) What is the expected genotypic ratio of children produced in this marriage? _____

    (b) What is the expected phenotypic ratio? _____

11. In his original work on the genetics of garden peas, Mendel found that yellow seed color (YY, Yy) is dominant over green seeds (yy) and that round seed shape (RR, Rr) is dominant over shrunken seeds (rr). Mendel crossed pure-breeding (homozygous) yellow, round-seeded plants with green, shrunken-seeded plants.

    (a) What will be the genotype and phenotype of the F_1 produced from such a cross?

        genotype _____

        phenotype _____

    (b) If the F_1 plants are crossed, what will be the expected phenotypic ratio of the F_2 generation?

    _____

Multiple Alleles

12. The major blood groups in humans are determined by **multiple alleles**; that is there are *more than* two possible alleles, any one of which can occupy a locus.

    In this ABO blood group system, a single gene can exist in any of three allelic forms: I^A, I^B, or i. The alleles A and B code for production of antigen A and antigen B (two proteins) on the surface of red blood cells. Alleles A and B are codominant, while allele i is recessive.

    Four blood groups (phenotypes) are possible from combinations of these alleles (Table 1).

TABLE 1	The ABO Blood Groups		
Blood Type	**Antigen Present**	**Antibody Present**	**Genotype**
O	Neither A nor B	A and B	ii
A	A	B	$I^A I^A$ or $I^A i$
B	B	A	$I^B I^B$ or $I^B i$
AB	AB	Neither A nor B	$I^A I^B$

    (a) Is it possible for a child with blood type O to be produced by two AB parents? _____ (yes or no) Explain.

    _____

    _____

    _____

    (b) In a case of disputed paternity, the child is type O, the mother type A. Could an individual of the following blood types be the father? _____ Explain each possibility.

    O_____

    A_____

    B_____

    AB_____

Chi-Square Analysis

13. In fruit flies, red eyes (R) are dominant over white eyes (r). A student performs a cross between a heterozygous red-eyed fly and a white-eyed fly. The student counts the offspring and finds 65 red-eyed flies and 49 white-eyed flies.

   (a) What is the expected phenotypic ratio of this cross? _____

   (b) Using a χ^2 test, determine whether the deviation between the observed and the expected is the result of chance.

   $\chi^2 = $ _____

   (c) Conclusion

   _____

   _____

14. In fruit flies, gray body (G) is dominant over ebony body (g).

   (a) A red-eyed, gray-bodied fly known to be heterozygous for both traits is mated with a white-eyed fly that is heterozygous for body color.

   What is the expected phenotypic ratio for this mating? _____

   (b) The observed offspring consist of 15 white-eyed, ebony-bodied flies; 31 white-eyed, gray-bodied flies; 12 red-eyed, ebony-bodied flies; and 38 red-eyed, gray-bodied flies.

   What is the χ^2 value for this cross? _____

   (c) Is it likely that the observed results "fit" the expected values? _____

Terms of Orientation in and Around the Animal Body

Body Shapes

Symmetry. The body can be divided into almost identical halves.

Asymmetry. The body cannot be divided into almost identical halves (for example, many sponges).

Radial symmetry. The body is shaped like a cylinder (for example, sea anemone) or wheel (for example, sea star).

Bilateral symmetry. The body is shaped like ours in that it can be divided into halves by only one symmetrical plane (midsagittal).

Directions in the Body

Dorsal. At or toward the back surface of the body.

Ventral. At or toward the belly surface of the body.

Anterior. At or toward the head of the body—ventral surface of humans.

Posterior. At or toward the tail or rear end of the body—dorsal surface of humans.

Medial. At or near the midline of a body. The prefix *mid-* is often used in combination with other terms (for example, midventral).

Lateral. Away from the midline of a body.

Superior. Over or placed above some point of reference—toward the head of humans.

Inferior. Under or placed below some point of reference—away from the head of humans.

Proximal. Close to some point of reference or close to a point of attachment of an appendage to the trunk of the body.

Distal. Away from some point of reference or away from a point of attachment of an appendage to the trunk of the body.

Longitudinal. Parallel to the midline of a body.

Axis. An imaginary line around which a body or structure can rotate. The midline or *longitudinal axis* is the central axis of a symmetrical body or structure.

Axial. Placed at or along an axis.

Radial. Arranged symmetrically around an axis like the spokes of a wheel.

Planes of the Body

Sagittal. Passes vertically to the ground and divides the body into right and left sides. The *midsagittal* or *median plane* passes through the longitudinal axis and divides the body into right and left halves.

Frontal. Passes at right angles to the sagittal plane and divides the body into dorsal and ventral parts.

Transverse. Passes from side to side at right angles to both the sagittal and frontal planes and divides the body into anterior and posterior parts—superior and inferior parts of humans. This plane of section is often referred to as a cross section.

Illustration References

Abramoff, P., and R. G. Thomson. 1982. *Laboratory Outlines in Biology III.* New York: W. H. Freeman.

Case, C. L., and T. R. Johnson. 1984. *Experiments in Microbiology.* Menlo Park, California: Benjamin/Cummings.

Fowler, I. 1984. *Human Anatomy.* Belmont, California: Wadsworth.

Gilbert, S. G. 1966. *Pictorial Anatomy of the Fetal Pig.* Second Edition. Seattle, Washington: University of Washington Press.

Hickman, C. P. 1961. *Integrated Principles of Zoology.* Second Edition. St. Louis, Missouri: C. V. Mosby.

Kessel, R. G., and C. Y. Shih. 1974. *Scanning Electron Microscopy in Biology.* New York: Springer-Verlag.

Lytle, C. F., and J. E. Wodsedalek. 1984. *General Zoology Laboratory Guide.* Complete Version. Ninth Edition. Dubuque, Iowa: Wm. C. Brown.

Patten, B. M. 1951. *American Scientist* 39: 225–243.

Scagel, R. F., et al. 1982. *Nonvascular Plants: An Evolutionary Survey.* Belmont, California: Wadsworth.

Sheetz, M., et al. 1976. *The Journal of Cell Biology* 70:193.

Shih, C. Y., and R. G. Kessel. 1982. *Living Images.* Boston: Science Books International/Jones and Bartlett Publishers.

Starr, C. 1991. *Biology: Concepts and Applications.* Belmont, California: Wadsworth.

Starr, C. 2000. *Biology: Concepts and Applications.* Fourth Edition. Belmont, California: Wadsworth.

Starr, C., C. Evers, and L. Starr. 2010. *Biology Today and Tomorrow.* Third Edition. Belmont, California: Brooks/Cole.

Starr, C., C. Evers, and L. Starr. 2011. *Biology: Concepts and Applications.* Eighth Edition. Belmont, California: Brooks/Cole.

Starr, C. and B. McMillan. 2012. *Human Biology.* Ninth Edition. Belmont, California: Brooks/Cole.

Starr, C., and R. Taggart. 1984. *Biology.* Third Edition. Belmont, California: Wadsworth.

Starr, C., and R. Taggart. 1989. *Biology.* Fifth Edition. Belmont, California: Wadsworth.

Starr, C., and R. Taggart. 2001. *Biology: The Unity and Diversity of Life.* Ninth Edition. Belmont, California: Wadsworth.

Starr, C., and R. Taggart. 2006. *Biology: The Unity and Diversity of Life.* Eleventh Edition. Belmont, California: Brooks/Cole.

Starr, C., R. Taggart, C. Evers, and L. Starr. 2009. *Biology: The Unity and Diversity of Life.* Twelfth Edition. Belmont, California: Brooks/Cole.

Villee, C. A., et al. 1973. *General Zoology.* Fourth Edition. Philadelphia: W. B. Saunders.

Weller, H., and R. Wiley. 1985. *Basic Human Physiology.* Boston: PWS Publishers.

Wischnitzer, S. 1979. *Atlas and Dissection Guide for Comparative Anatomy.* Third Edition. New York: W. H. Freeman.